# Natural Resource Management and Policy

Volume 52

**Series Editors**

David Zilberman, California, CA, USA
Renan Goetz, Girona, Spain
Alberto Garrido, Madrid, Spain

There is a growing awareness to the role that natural resources, such as water, land, forests and environmental amenities, play in our lives. There are many competing uses for natural resources, and society is challenged to manage them for improving social well-being. Furthermore, there may be dire consequences to natural resources mismanagement. Renewable resources, such as water, land and the environment are linked, and decisions made with regard to one may affect the others. Policy and management of natural resources now require interdisciplinary approaches including natural and social sciences to correctly address our society preferences.

This series provides a collection of works containing most recent findings on economics, management and policy of renewable biological resources, such as water, land, crop protection, sustainable agriculture, technology, and environmental health. It incorporates modern thinking and techniques of economics and management. Books in this series will incorporate knowledge and models of natural phenomena with economics and managerial decision frameworks to assess alternative options for managing natural resources and environment.

More information about this series at http://www.springer.com/series/6360

Leslie Lipper • Nancy McCarthy
David Zilberman • Solomon Asfaw
Giacomo Branca
Editors

# Climate Smart Agriculture

Building Resilience to Climate Change

Food and Agriculture
Organization of the
United Nations

Springer

OPEN

*Editors*
Leslie Lipper
ISPC-CGIAR
Roma, Italy

Nancy McCarthy
Lead Analytics Inc.
Washington, DC, USA

David Zilberman
Department of Agriculture and Resource
  Economics
University of California Berkeley
Berkeley, CA, USA

Solomon Asfaw
FAO of the UN
Roma, Italy

Giacomo Branca
Department of Economics
University of Tuscia
Viterbo, Italy

ISSN 0929-127X             ISSN 2511-8560  (electronic)
Natural Resource Management and Policy
ISBN 978-3-319-61193-8      ISBN 978-3-319-61194-5  (eBook)
ISBN 978-92-5-109966-7 (FAO)
DOI 10.1007/978-3-319-61194-5

Library of Congress Control Number: 2017953417

© FAO 2018
**Open Access** This book is distributed under the terms of the Creative Commons Attribution-NonCommercial-ShareAlike 3.0 IGO license (https://creativecommons.org/licenses/by-nc-sa/3.0/igo/), which permits any noncommercial use, duplication, adaptation, distribution, and reproduction in any medium or format, as long as you give appropriate credit to the Food and Agriculture Organization of the United Nations (FAO), provide a link to the Creative Commons license and indicate if changes were made. If you remix, transform, or build upon this book or a part thereof, you must distribute your contributions under the same license as the original. Any dispute related to the use of the works of FAO that cannot be settled amicably shall be submitted to arbitration pursuant to the UNCITRAL rules.
The designations employed and the presentation of material in this information product do not imply the expression of any opinion whatsoever on the part of the Food and Agriculture Organization of the United Nations (FAO) concerning the legal or development status of any country, territory, city or area or of its authorities, or concerning the delimitation of its frontiers or boundaries. The mention of specific companies or products of manufacturers, whether or not these have been patented, does not imply that these have been endorsed or recommended by FAO in preference to others of a similar nature that are not mentioned. The views expressed in this information product are those of the author(s) and do not necessarily reflect the views or policies of FAO.
In any use of this work, there should be no suggestion that FAO endorses any specific organization, products or services. The use of FAO's name for any purpose other than for attribution, and the use of FAO's logo, shall be subject to a separate written license agreement between FAO and the user and is not authorized as part of this CC-IGO license. Note that the link provided above includes additional terms and conditions of the license. The images or other third party material in this book are included in the work's Creative Commons license, unless indicated otherwise in the credit line; if such material is not included in the work's Creative Commons license and the respective action is not permitted by statutory regulation, users will need to obtain permission from the license holder to duplicate, adapt or reproduce the material.
The use of general descriptive names, registered names, trademarks, service marks, etc. in this publication does not imply, even in the absence of a specific statement, that such names are exempt from the relevant protective laws and regulations and therefore free for general use. The publisher, the authors and the editors are safe to assume that the advice and information in this book are believed to be true and accurate at the date of publication. Neither the publisher nor the authors or the editors give a warranty, express or implied, with respect to the material contained herein or for any errors or omissions that may have been made. The publisher remains neutral with regard to jurisdictional claims in published maps and institutional affiliations.

Printed on acid-free paper

This Springer imprint is published by Springer Nature
The registered company is Springer International Publishing AG
The registered company address is: Gewerbestrasse 11, 6330 Cham, Switzerland

# Foreword

Eradicating poverty, ending hunger, and taking urgent action to combat climate change and its impacts are three objectives the global community has committed to achieving by 2030 by adopting the sustainable development goals. Agriculture, and the way we manage it in the years leading up to 2030, will be a key determinant of whether or not these objectives are met. Agriculture has been, and can be further, used as an important instrument in eradicating hunger, poverty, and all forms of malnutrition. Climate change however is expected to act as an effective barrier to agricultural growth in many regions, especially in developing country contexts heavily dependent on rain-fed agriculture.

Climate change impacts agriculture through a number of pathways. According to the 2013 IPCC report, all four dimensions of food security are potentially affected by climate change through their effects on agricultural production and the incomes of rural households, food prices and markets, and in many other parts of the food system (e.g., storage, food quality, and safety) (IPCC WGII AR5 Ch 7). Reducing the vulnerability of agricultural systems to climate change – including the increased incidence of extreme weather events – and strengthening its adaptive capacity are therefore important priorities to protect and improve the livelihoods of the poor and allow agriculture to fully play its role in ensuring food security. Reducing emissions that contribute to global warming is crucial to securing global wellbeing, and the agricultural sector has considerable potential for emissions reductions while at the same time playing its important role in poverty reduction and food security. In short, agriculture lies at the nexus of resolving urgent global priorities.

FAO is actively working to support countries in grappling with the challenge of managing agriculture to reduce hunger and poverty in an increasingly climate-constrained world. FAO launched the concept of climate smart agriculture (CSA) in 2009 to draw attention to linkages between achieving food security and combating climate change through agricultural development, and the opportunities for attaining large synergies in doing so. In practice, the CSA approach involves integrating the need for adaptation and the potential for mitigation into the planning and implementation of agricultural policies, planning, and investments. The point of departure for the CSA approach is the emphasis on food security and poverty reduction as the priority in developing countries through enhanced capacity of their agri-food sectors and institutional and technological innovations. This capacity cannot be attained without adaptation to changing conditions. At the same time, reducing the emissions associated with conventional agricultural growth models is one of the largest and most cost-effective means of reducing GHG emissions, and thus the CSA approach integrates the potential for obtaining mitigation co-benefits from agricultural growth strategies.

The CSA concept has gained considerable traction at the international and national levels; however, there is still a fair amount of confusion regarding the concept and its theoretical underpinning. In addition, the empirical evidence base to support country implementation strategies is lacking. In particular, there is a need for defining and operationalizing the concept of resilience and adaptive capacity in the context of agricultural growth for food security. For these reasons, the Economic and Social Development Department of FAO has supported the development of this book, which represents a significant step forward in shedding light to the issues raised above. This volume brings together research, analysis, and opinions of leading agricultural and resource economists and policy experts to develop the conceptual, empirical, and policy basis for a better understanding of CSA and enhanced potential for achieving it on the ground.

The first section of this book provides conceptual frameworks as well as methodological approaches for operationalizing CSA at the country level. Its main focus is comparing and contrasting the conceptual approaches to risk management and resilience used in the agricultural development context with that used in the context of climate change and proposing a consistent approach. It also provides an overview of the development of the CSA concept, the controversies it has sparked, and how they relate to the broader debate of sustainable development.

The second section consists of 19 case study chapters focusing on issues of vulnerability measurement and assessment, as well as ways of improving the adaptive capacity at farm and system level and what could be some of the policy responses to achieve them. These empirical studies showcase a wide range of options (policy instruments) that contribute to building resilience to climate risk. They include policy instruments aimed at changing agricultural practices but also policy instruments in other sectors. Examples include social protection, micro-finance, input subsidies, micro-insurance, and agricultural knowledge and information systems. The case studies cover a wide geographic range and scale, from Asia to Africa and the USA and from households to markets and institutions and the national and global economy. They draw upon the CSA project work of FAO, as well as that of other agencies applying the CSA approach. The breadth of the case studies provides a basis for lessons learned in which contribute to a more comprehensive understanding of policy options to improve the resilience of livelihoods of the rural poor to climate change. They indicate that we do have considerable tools available to measure, reduce, and effectively react to climate change–related vulnerability in the agricultural sector, and that it is essential to utilize these instruments in seeking to improve the agriculture sector's capacity to support hunger, poverty eradication, and sustainable development.

The third and final section of this book presents the results of a consultation with a panel of leading thinkers and practitioners on agricultural and climate change policy. This section is comprised of the responses of these experts to a set of questions based on the main findings, conclusions, insights, and questions that emerged from the set of case studies and conceptual papers. Their varied responses to the issues provide considerable insights into the different approaches and policy priorities for CSA across varying contexts, as well as practical ideas on how to operationalize them.

The FAO is committed to providing support to agricultural and climate change policy-makers and the agricultural producers they serve in their ongoing efforts to end hunger and poverty and effectively combat climate change effects now and in the future. This book offers tools and insights for a range of stakeholders to help meet these challenges in the many forms they are manifested.

Rome, Italy                                                              Kostas Stamoulis

# Acknowledgments

This book is the outcome of a cooperation between Economic and Policy Innovation of Climate-Smart Agriculture (EPIC) team of FAO, Department of Agricultural and Resource Economics of University of California (Berkeley) and the Department of Economics and Business (DEIM) of Tuscia University (Viterbo, Italy). We express sincere gratitude to Professors Alessandro Mechelli and Alessandro Sorrentino (Departmental Faculty) for their continuous support. This publication would not have been possible without the administrative and organizational help of Laura Gori, Cristina Mastrogregori, and Giuseppe Rapiti (Departmental Staff). We would also like to thank the Italian Institute for International Political Studies (ISPI) which hosted the Book Authors' Workshop "Climate Smart Agriculture: Building Resilience to Climate Change" held in Palazzo Clerici, Milan (Italy) on August 6, 2015.

We would also like to sincerely thank FAO-HQ staff particularly Jessica Mathewson, Liliana Maldonado, Paola DiSanto, and Alessandro Spairani for their administrative and organizational support throughout the whole publication process. We finally would like to acknowledge the financial support of FAO.

# Contents

**Part I  Overview and Conceptual Framework**

**Introduction and Overview** ................................................................. 3
Solomon Asfaw and Giacomo Branca

**A Short History of the Evolution of the Climate Smart Agriculture Approach and Its Links to Climate Change and Sustainable Agriculture Debates** ............................. 13
Leslie Lipper and David Zilberman

**Economics of Climate Smart Agriculture: An Overview** ........................ 31
Nancy McCarthy, Leslie Lipper, and David Zilberman

**Innovation in Response to Climate Change** .......................................... 49
David Zilberman, Leslie Lipper, Nancy McCarthy, and Ben Gordon

**Part II  Case Studies: Vulnerability Measurements and Assessment**

**Use of Satellite Information on Wetness and Temperature for Crop Yield Prediction and River Resource Planning** ....................... 77
Alan Basist, Ariel Dinar, Brian Blankespoor, David Bachiochi, and Harold Houba

**Early Warning Techniques for Local Climate Resilience: Smallholder Rice in Lao PDR** ................................................................. 105
Drew Behnke, Sam Heft-Neal, and David Roland-Holst

**Farmers' Perceptions of and Adaptations to Climate Change in Southeast Asia: The Case Study from Thailand and Vietnam** ............ 137
Hermann Waibel, Thi Hoa Pahlisch, and Marc Völker

**U.S. Maize Yield Growth and Countervailing Climate Change Impacts** ................................................................................. 161
Ariel Ortiz-Bobea

**Understanding Tradeoffs in the Context of Farm-Scale Impacts:
An Application of Decision-Support Tools for Assessing
Climate Smart Agriculture**..........................................................................  173
Susan M. Capalbo, Clark Seavert, John M. Antle, Jenna Way,
and Laurie Houston

**Part III   Case Studies: Policy Response to Improving Adaptation
            and Adaptive Capacity**

**Can Insurance Help Manage Climate Risk and Food Insecurity?
Evidence from the Pastoral Regions of East Africa**...................................  201
Michael R. Carter, Sarah A. Janzen, and Quentin Stoeffler

**Can Cash Transfer Programmes Promote Household Resilience?
Cross-Country Evidence from Sub-Saharan Africa**...................................  227
Solomon Asfaw and Benjamin Davis

**Input Subsidy Programs and Climate Smart Agriculture:
Current Realities and Future Potential**........................................................  251
Tom S. Jayne, Nicholas J. Sitko, Nicole M. Mason, and David Skole

**Part IV   Case Studies: System Level Response
            to Improving Adaptation and Adaptive Capacity**

**Robust Decision Making for a Climate-Resilient Development
of the Agricultural Sector in Nigeria**...........................................................  277
Valentina Mereu, Monia Santini, Raffaello Cervigni,
Benedicte Augeard, Francesco Bosello, E. Scoccimarro,
Donatella Spano, and Riccardo Valentini

**Using AgMIP Regional Integrated Assessment Methods
to Evaluate Vulnerability, Resilience and Adaptive Capacity
for Climate Smart Agricultural Systems** ....................................................  307
John M. Antle, Sabine Homann-KeeTui, Katrien Descheemaeker,
Patricia Masikati, and Roberto O. Valdivia

**Climate Smart Food Supply Chains in Developing Countries
in an Era of Rapid Dual Change in Agrifood Systems
and the Climate** ............................................................................................  335
Thomas Reardon and David Zilberman

**The Adoption of Climate Smart Agriculture:
The Role of Information and Insurance Under Climate Change**..............  353
Jamie Mullins, Joshua Graff Zivin, Andrea Cattaneo, Adriana
Paolantonio, and Romina Cavatassi

**A Qualitative Evaluation of CSA Options in Mixed
Crop-Livestock Systems in Developing Countries** ................................. 385
Philip K. Thornton, Todd Rosenstock, Wiebke Förch, Christine Lamanna,
Patrick Bell, Ben Henderson, and Mario Herrero

**Identifying Strategies to Enhance the Resilience
of Smallholder Farming Systems: Evidence from Zambia** ..................... 425
Oscar Cacho, Adriana Paolantonio, Giacomo Branca,
Romina Cavatassi, Aslihan Arslan, and Leslie Lipper

**Part V   Case Studies: Farm Level Response to Improving Adaptation
and Adaptive Capacity**

**Climate Risk Management through Sustainable Land
and Water Management in Sub-Saharan Africa** ...................................... 445
Ephraim Nkonya, Jawoo Koo, Edward Kato, and Timothy Johnson

**Improving the Resilience of Central Asian Agriculture
to Weather Variability and Climate Change** .............................................. 477
Alisher Mirzabaev

**Managing Environmental Risk in Presence of Climate Change:
The Role of Adaptation in the Nile Basin of Ethiopia** .............................. 497
Salvatore Di Falco and Marcella Veronesi

**Diversification as Part of a CSA Strategy: The Cases
of Zambia and Malawi** ..................................................................................... 527
Aslihan Arslan, Solomon Asfaw, Romina Cavatassi, Leslie Lipper, Nancy
McCarthy, Misael Kokwe, and George Phiri

**Economic Analysis of Improved Smallholder Paddy
and Maize Production in Northern Viet Nam
and Implications for Climate-Smart Agriculture** ...................................... 563
Giacomo Branca, Aslihan Arslan, Adriana Paolantonio,
Romina Cavatassi, Nancy McCarthy, N. VanLinh, and Leslie Lipper

**Part VI   Policy Synthesis and Conclusion**

**Devising Effective Strategies and Policies for CSA:
Insights from a Panel of Global Policy Experts** ........................................ 599
Patrick Caron, Mahendra Dev, Willis Oluoch-Kosura, Cao Duc Phat,
Uma Lele, Pedro Sanchez, and Lindiwe Majele Sibanda

**Conclusion and Policy Implications to "Climate Smart Agriculture:
Building Resilience to Climate Change"** .................................................... 621
David Zilberman

**Index** ................................................................................................................. 627

# Contributors

**John M. Antle** College of Agricultural Sciences, Oregon State University, Corvallis, OR, USA

**Aslihan Arslan** International Fund for Agriculture Development (IFAD), Rome, Italy

**Solomon Asfaw** FAO of the UN, Rome, Italy

**Benedicte Augeard** The French National Agency for Water and Aquatic Environments, Vincennes, France

**Alan Basist** EyesOnEarth, Asheville, NC, USA

**Drew Behnke** Department of Economics, University of California Santa Barbara, Santa Barbara, CA, USA

**Patrick Bell** Ohio State University, Columbus, OH, USA

**Brian Blankespoor** World Bank, Washington, DC, USA

**Francesco Bosello** Euro-Mediterranean Center on Climate Change, Lecce, Italy

**Giacomo Branca** Department of Economics, University of Tuscia, Viterbo, Italy

**Oscar Cacho** University of New England Business School, Armidale, Australia

**Susan M. Capalbo** College of Agricultural Sciences, Oregon State University, Corvallis, OR, USA

**Michael R. Carter** Department of Agricultural and Resource Economics, University of California Davis, USA, NBER and the Giannini Foundation, Davis, CA, USA

**Andrea Cattaneo** FAO of the UN, Rome, Italy

**Romina Cavatassi** International Fund for Agriculture Development (IFAD), Rome, Italy

**Raffaello Cervigni** Environment and Natural Resources Global Practice, Africa Region, The World Bank, Washington, DC, USA

**Benjamin Davis** Food and Agricultural Organization (FAO) of the United Nations, Rome, Italy

**Katrien Descheemaeker** Wageningen University, Wageningen, Netherlands

**Salvatore Di Falco** Department of Economics, University of Geneva, Geneva, Switzerland

**Ariel Dinar** School of Public Policy, University of California Riverside, Riverside, CA, USA

**Wiebke Förch** Deutsche Gesellschaft für Internationale Zusammenarbeit (GIZ) GmbH, Gaboron, Germany

**Ben Gordon** Department of Agriculture and Resource Economics, University of California Berkeley, Berkeley, CA, USA

**Joshua Graff Zivin** School of Global Policy and Strategy, University of California San Diego, San Diego, CA, USA

**Sam Heft-Neal** Department of Agricultural and Resource Economics, University of California Berkeley, Berkeley, CA, USA

**Ben Henderson** Commonwealth Scientific and Industrial Research Organization (CSIRO), Australia

**Mario Herrero** Commonwealth Scientific and Industrial Research Organization (CSIRO), Australia

**Sabine Homann-KeeTui** International Crops Research Institute for the Semi-Arid Tropics, Zimbabwe

**Harold Houba** Free University of Amsterdam, Amsterdam, Netherlands

**Laurie Houston** College of Agricultural Sciences, Oregon State University, Corvallis, OR, USA

**Sarah A. Janzen** Department of Economics, Montana State University, Bozeman, MT, USA

**Tom S. Jayne** Department of Agricultural, Food and Resource Economics, Michigan State University, East Lansing, MI, USA

**Timothy Johnson** Environment and Production Technology, IFPRI, Washington, DC, USA

**Edward Kato** Environment and Production Technology, IFPRI, Washington, DC, USA

**Misael Kokwe** FAO of the UN, Lusaka, Zambia

**Jawoo Koo** Environment and Production Technology, IFPRI, Washington, DC, USA

**Christine Lamanna** World Agroforestry Centre, Nairobi, Kenya

**Leslie Lipper** ISPC-CGIAR, Rome, Italy

**Patricia Masikati** World Agroforestry Centre, Lusaka, Zambia

**Nicole M. Mason** Department of Agricultural, Food and Resource Economics, Michigan State University, East Lansing, MI, USA

**Nancy McCarthy** Lead Analytics Inc., Washington, DC, USA

**Valentina Mereu** Euro-Mediterranean Center on Climate Change, Change, Italy

**Alisher Mirzabaev** University of Bonn, Bonn, Germany

**Jamie Mullins** Department of Resource Economics, University of Massachusetts Amherst, Amherst, MA, USA

**Ephraim Nkonya** Environment and Production Technology, IFPRI, Washington, DC, USA

**Ariel Ortiz-Bobea** Cornell University, Ithaca, NY, USA

**Thi Hoa Pahlisch** Institute of Development and Agricultural Economics, Leibniz University Hannover, Germany

**Adriana Paolantonio** International Fund for Agriculture Development (IFAD), Rome, Italy

**George Phiri** FAO of the UN, Lilongwe, Malawi

**Thomas Reardon** Department of Agricultural, Food and Resource Economics, Michigan State University, East Lansing, MI, USA

**David Roland-Holst** Department of Agriculture and Resource Economics, University of California Berkeley, Berkeley, CA, USA

**Todd Rosenstock** World Agroforestry Centre, Nairobi, Kenya

**Monia Santini** Euro-Mediterranean Center on Climate Change, Lecce, Italy

**E. Scoccimarro** Euro-Mediterranean Center on Climate Change, Lecce, Italy

**Clark Seavert** College of Agricultural Sciences, Oregon State University, Corvallis, OR, USA

**David Skole** Department of Forestry, Michigan State University, East Lansing, MI, USA

**Nicholas J. Sitko** Department of Agricultural, Food and Resource Economics, Michigan State University, East Lansing, MI, USA

**Donatella Spano** Euro-Mediterranean Center on Climate Change, Italy

**Kostas Stamoulis** FAO, Economic and Social Development Department, Rome, Italy

**Quentin Stoeffler** Department of Economics, Istanbul Technical University, Istanbul, Turkey

**Philip K. Thornton** CGIAR Research Program on Climate Change, Agriculture and Food Security (CCAFS), ILRI, Nairobi, Kenya

**Roberto O. Valdivia** Department of Applied Economics, Corvallis, OR, USA

**Riccardo Valentini** Euro-Mediterranean Center on Climate Change, Lecce, Italy

**N. VanLinh** Food and Agriculture Organization of the United Nations, Viet Nam, Rome, Italy

**Marcella Veronesi** Department of Economics, University of Verona, Verona, Italy

**Marc Völker** Institute for Population and Social Research, Mahidol University, Salaya, Thailand

**Hermann Waibel** Institute of Development and Agricultural Economics, Leibniz Universität Hannover, Hanover, Germany

**Jenna Way** College of Agricultural Sciences, Oregon State University, Corvallis, OR, USA

**David Zilberman** Department of Agriculture and Resource Economics, University of California Berkeley, Berkeley, CA, USA

# Part I
# Overview and Conceptual Framework

# Introduction and Overview

**Solomon Asfaw and Giacomo Branca**

**Abstract** The climate-smart agriculture (CSA) concept is gaining considerable traction at international and national levels to meet the challenges of addressing agricultural planning under climate change. CSA is a concept that calls for integration of the need for adaptation and the possibility of mitigation in agricultural growth strategies to support food security. Several countries around the world have expressed intent to adopt CSA approach to managing their agricultural sectors. However there is considerable confusion about what the CSA concept and approach actually involve, and wide variation in how the term is used. It is critical to build a more formal basis for the CSA concept and methodology and at the same time providing illustrations of how the concept can be applied across a range of conditions. This book expand and formalize the conceptual foundations of CSA drawing upon theory and concepts from agricultural development, institutional and resource economics. The book is also devoted to a set of country level case studies illustrating the economic basis of CSA in terms of reducing vulnerability, increasing adaptive capacity and ex-post risk coping. It also addresses policy issues related to climate change focusing on the implications of the empirical findings for devising effective strategies and policies to support resilience and the implications for agriculture and climate change policy at national, regional and international levels. The book provide development agencies and practitioners, policymakers, civil society, research and academia as well as private sector with tested good practices and innovative approaches of promoting CSA system at country level.

---

S. Asfaw (✉)
FAO of the UN, Rome, Italy
e-mail: Solomon.Asfaw@fao.org

G. Branca
Department of Economics, University of Tuscia, Viterbo, Italy
e-mail: branca@unitus.it

Climate change poses a major and growing threat to global food security. Population growth and rising incomes in much of the developing world have pushed demand for food and other agricultural products to unprecedented levels. FAO has estimated that, in order to meet food demand in 2050, annual world production of crops and livestock will need to be 60% higher than it was in 2006. In developing countries, about 80% of the required increase will need to come from higher yields and increased cropping intensity and only 20% from expansion of arable land[1].

Meeting food demand for a growing population is already a formidable challenge for the agriculture sector, but it will be further exacerbated by climate change. The expected effects of climate change – higher temperatures, extreme weather events, water shortages, rising sea levels, the disruption of ecosystems and the loss of biodiversity – will generate significant effects on the different dimensions and determinants of food security by affecting the productivity of rainfed crops and forage, reducing water availability and changing the severity and distribution of crop and livestock diseases. The fifth assessment report of the IPCC released in 2014 found that climate change effects are already being felt on agriculture and food security, and the negative impacts are most likely in tropical zones where most of the world's poor agricultural dependent populations are located. Through its impacts on agriculture, climate change will make it more difficult to meet the key Sustainable Development Goal of ending hunger, achieving year-round food security, and ensuring sustainable food production systems by 2030.

The magnitude and speed of climate change, and the effectiveness of adaptation and mitigation efforts in agriculture, will be critical to the future of large segments of the world's population. Integrating the effects of climate change into agricultural development planning is a major challenge. This requires technology and policy measures to reduce vulnerability and increase the capacity of producers, particularly smallholders, to effectively adapt. At the same time, given agriculture's role as a major source of greenhouse gas emissions and the high rate of emissions growth experienced with recent conventional intensification strategies, there is a need to look for low emissions growth opportunities and adequate policies. Policymakers are thus challenged to ensure that agriculture contributes to addressing food security, development and climate change.

In this frame, Climate Smart Agriculture (CSA) is an approach that calls for integration of the need for adaptation and the possibility of mitigation in agricultural growth strategies to support food security. The concept was launched by FAO in 2010[2], gaining rapid and widespread interest and attention. CSA goes beyond agricultural practices and technologies to include enabling policies and institutions as well as identification of financing mechanisms. There are significant intellectual and policy gaps to be filled in CSA literature. An economic decision-making framework will also assist in identifying challenges for CSA application.

---

[1] See http://www.fao.org/fileadmin/templates/wsfs/docs/expert_paper/How_to_Feed_the_World_in_2050.pdf.

[2] See http://www.fao.org/docrep/013/i1881e/i1881e00.pdf.

Introduction and Overview

# 1  Overview of the Book

This book expands and formalizes the conceptual foundations of CSA drawing upon theory and concepts from agricultural development, institutional and resource economics. The book focuses particularly on the adaptation/resilience dimension of CSA, since this is the least well developed in the economics literature. A mixture of conceptual analyses, including theory, empirical and policy analysis, and case studies look at: (1) ex-ante reduction of vulnerability, (2) increasing adaptive capacity through policy response, (3) increasing adaptive capacity through system level response and (4) increasing adaptive capacity through farm level response.

The book provides a wide array of case studies to illustrate that these concepts have strong real-world applicability. The case study approach will provide concrete illustrations of the conceptual and theoretical framework, taking into account the high level of diversity in agro-ecological and socioeconomic situations faced by agricultural planners and policy-makers today. Some case studies assess issues of measurement of vulnerability to climate change and damage caused by it. Others address issues of improving adaptive capacity, and the ex-post impact of different policy measures.

In the book, economists and policy-makers will find an interpretation and operationalizing of the concepts of resilience and adaptive capacity in the context of agricultural growth for food security. The combination of methodological analysis of CSA and an empirical analysis based on a set of case studies from Asia and Africa is unique. We are not aware of other books that contain all of this integrated knowledge in one place and provide a perspective on its lessons.

The book is structured as follows. Part I illustrates the conceptual framework, giving an overview of CSA concept, approach, and its main components. This part relates the main features of the CSA paradigm to core economic principles and seeks to clarify how the concepts of resilience, adaptive capacity, innovation, technology adoption and institutions relate to each other and the economic principles of CSA. Part II reports a set of case studies from leading agricultural development economists aimed at illustrating the economic basis of CSA in terms of reducing vulnerability and increasing adaptive capacity. It makes a clear distinction between responses to building adaptive capacity at policy, system and farm levels. Last, part III addresses policy issues related to climate change and provides a synthesis of the key messages of the book. A detailed overview of each part is presented next.

## 1.1  Part I. Conceptual Chapters

Chapter 2 presents an overview of the evolution of CSA concept, introduces its major components, and summarizes the key issues associated within the context of climate change and agricultural policy debates. The main message of this chapter is that CSA concept has been reshaped through inputs and interactions of multiple stakeholders involved in developing and implementing it. The first section provides an overview of international climate change policy followed by an introduction and analysis of CSA and its history. This is then followed by a discussion of three broad controversies related to CSA, namely the role of mitigation, the relationship of CSA to sustainable agriculture, and how biotechnology is treated in the CSA approach. CSA provides a tool to identify locally appropriate solutions to managing agriculture for sustainable development and food security under climate change.

Chapter 3 tackles the economic considerations of CSA in addressing sustainable agricultural growth for food security under climate change. It addresses the lack of coherence of the CSA approach by building a conceptual framework to rooted in agricultural development economic theories and concepts. The chapter begins by highlighting the key features of climate change that require a shift in emphasis in research, and for innovations in technologies, institutions, and government policies and programs to consider heterogeneity of impacts and implications of decision-making under uncertainty. The chapter does this by posing a dynamic constrained optimization problem wherein a social planner seeks to maximize expected discounted welfare associated with agriculture of the population they serve, both now and in the future. The objectives are the four pillars of food security, food availability, accessibility, utilization, and stability, as well as reducing emissions growth. The problem is also characterized by current constraints that bound the feasible outcomes, including bio-physical, behavioral, political, institutional and distributional constraints. The chapter stresses that the nature of the optimization, and thus adaptation strategies, are context specific and highlight that the solution to the social planner's problem for climate change must balance adaptation and responsiveness to uncertain climate change with the needed growth and food security objectives of the agricultural sector.

Chapter 4 provides more detailed guidance on the key role of innovation to address the negative impact of climate change. Innovation in agriculture is clearly an important response for effective and equitable adaptation and mitigation – and the chapter highlights the need for managerial and institutional changes that promote innovation to address the heterogeneity and uncertainty of climate change impacts. The chapter discusses the main features and the nature of innovation needed to align these actions with a CSA strategy, suggesting several principles to guide the introduction of innovation and develop capacity and policies to address climate change.

## 1.2 Part II. Country Case Studies

### 1.2.1 Vulnerability Measurement and Assessment

Chapter 5 shows that near real-time satellite observations can be used to mitigate impacts of extreme events and promote climate resilience. First, the early detection of growing conditions and predicting the availability of food directly improves climate resilience and food security. Second, insurance (risk management) programs can use the indexes in triggers for a quick release of catastrophic bonds to farmers to mitigate impacts of crop failure. Third, these tools provide information useful for farmers in assessing yield potential from various crops under current and changing climatic conditions. Fourth, an early warning system distributed across the globe can help identify and expedite the exportation of food supplies from areas where they are in excess into areas where a deficiency is likely to occur. The chapter also discusses ways of integrating these products with various datasets, such as in situ surface temperature, the greenness index, and soil moisture data, in order to expand their complementary value and utility.

Chapter 6 presents key findings from advanced econometric models of long-term impacts of climate change on rice production in Lao PDR. Results are consistent with previous work in the region, where there is weak evidence that elevated minimum night-time temperatures are highly damaging to rice yields. Conversely, it is found that elevated maximum daytime temperatures increase yields. Overall, the size of the impact and statistical significance is larger for increased maximum temperatures, suggesting that elevated temperatures might have a net positive impact on rice yields in Lao PDR. The chapter also discusses some major caveats to these findings in particular the limitation with the quality data used for the analysis.

The perception of climate change and adaptation choices made by farmers are important considerations in the design of adaptation strategies. Chapter 7 uses a comprehensive dataset of farm households from Thailand and Vietnam to show that farmers do perceive climate change, but describe it in quite distinct ways. Further, adaptation measures are informed by perception and, at least in the case of Vietnam, perceptions are shaped by the respondent's characteristics, location variables and recent climate related shocks.

Chapter 8 illustrates how to assess the yield growth rate requirements needed to compensate yield losses due to climate change. The crop statistical model employed allows for nonlinear effects of temperature on yields. In line with the literature, it suggests that exposure to temperature exceeding 30 °C is detrimental to maize yields in the US Midwest. The chapter reports that a historical rate in maize yield growth in the US Midwest of 17.4%/decade exceeds the rate (6.56%/decade) needed to compensate a plausible warming of 3 °C within the next 3 decades. However, the net yield trend would be substantially diminished under this scenario due to the countervailing effect of a warming climate. The chapter also discusses the possibilities of extending the analysis with a cost-benefit analysis of alternative mean-increasing or variance-reducing technological change.

Chapter 9 shows that a fine-tuned integrative decision support tool can better inform growers and landowners of how changes in climate will impact their operations and their environmental outcomes. The use of a decision support tools such as *AgBiz Logic* can provide farmers better information on the relative impacts of adapting to a change as reflected in changes in future climate conditions, changes in future policies, prices, and costs or changes in terms of lease arrangements. By incorporating both climate change and environmental outcomes, these decision tools can be used to evaluate climate smart options at the farm-scale. The authors discuss the use of different tools such as *AgBizClimate, AgBizProfit, AgBizFinance, AgBizLeasee and AgBizEnvironment* to measure the impacts of climate change to wheat production, the role of adaptation strategies to an annual cropping system, the feasibility of purchasing additional equipment to farm the annual cropping system and also estimate the trade-offs of economic returns to environmental impacts.

### 1.2.2 Policy Response to Improving Adaptation and Adaptive Capacity

Chapter 10 uses empirical evidence from the Index-based Livestock Insurance (IBLI) project in the pastoral regions in East Africa to answer if insurance can cost-effectively mitigate the increasingly deleterious impacts of climate risk on poverty and food insecurity. The theory reviewed in this chapter suggests an affirmative answer if well-designed insurance contracts can be implemented and priced at a reasonable level despite the uncertainties that attend climate change. At the same time, much remains to be done if quality index insurance contracts are to be scaled up and sustained. Demand has often been tepid and unstable. Outreach and administration costs have been high. Pricing by a private insurance industry made nervous by climate change has pushed costs up. Finally, the effective quality of the IBLI contact has been scrutinized and found wanting. The chapter concludes that insurance is not an easy, off-the-shelf solution to the problem of climate risk and food insecurity. Creativity in the technical and institutional design of contracts is still required.

Chapter 11 synthesizes the key findings of From Protection to Production Project (PtoP) of FAO to show the potential role of cash transfer programmes as a tool to support risk management and build resilience in sub-Saharan Africa. Such programs address household resilience by building human capital and improving food security and potentially strengthening households' ability to respond to and cope with exogenous shocks. This may allow households to mitigate future fluctuations in consumption. Many of the programmes studied increased investment in agricultural inputs and assets, including farm implements and livestock, and improved food security indicators, though results differed across countries. This too was met by increases in consumption and dietary diversity. Although the impacts on risk management are less uniform, the cash transfer programmes seem to strengthen community ties, allow households to save and pay off debts, and decrease the need to rely on adverse risk coping mechanisms. Finally, using the case study of Zambia the authors demonstrates the potential for cash transfers to help poor households manage climate risk.

Chapter 12 shows that Input Subsidy Programs (ISPs) may provide a potentially useful means to encourage system-wide and farm-level changes to achieve CSA objectives in Africa. While many ISPs have not contributed significantly to *ex-ante* risk management at the household level, recent innovations in ISPs may enable them to be more climate smart. In particular, moves toward open voucher systems that induce greater private sector participation hold potential to support the development of profitable and more sustainable input distribution systems providing more heat-, drought- and saline-tolerant seed types. Moreover, moving from a limited range of options to a system that provides farmers with a wide range of input choices has the potential to promote greater livelihood diversification and resilience. Programs that make farmer participation in ISPs conditional on the adoption of certain climate smart practices also have some potential but would require more robust monitoring and setting of targets. These two requirements currently limit the potential of ISPs to achieve widespread CSA benefits. Moreover, using ISPs to contribute to CSA objectives would need to be evaluated against the potential benefits of using comparable resources for investments in irrigation, physical infrastructure, and public agricultural research and extension programs, which may generate higher comprehensive social benefits.

### 1.2.3 System Level Response to Improving Adaptation and Adaptive Capacity

The expansion of irrigation is often considered as a complementary strategy to enhance the resilience of agriculture to climate. However, irrigation entails large capital expenditures and an adequate sizing of any given irrigation scheme cannot neglect the expected changes in climate trends and variability. Chapter 13 explores these issues using historical climate records as a basis for determining what investment is adequate in water storage or in area equipped for irrigation is likely to result in "regrets," because the investment will be undersized/oversized, if the climate turns out to be drier/wetter than expected. An investment strategy that minimizes the risk of misjudgements across multiple climate outcomes reduces regrets and allows for greater flexibility of the system: cropping patterns, water use, or other parameters can be adapted for wet or dry years to increase the return on irrigation investment.

Chapter 14 shows how the use of the new simulation-based technology impact assessment methods, developed by the Agricultural Model Inter-comparison and Improvement project (AgMIP), can evaluate the potential for currently available or prospective agricultural systems to achieve the goals of CSA. The approach combines available data (observational and farm performance indicators), with bio-physical and economic models and future climate and socio-economic scenarios. A case study of crop-livestock systems in Zimbabwe illustrates the potential for these methods to test the usefulness of specific modifications to raise incomes, reduce vulnerability to climate change and to enhance resilience. It is important to note that the framework presented can also incorporate greenhouse gas emissions as part of a technology assessment. The authors point out the need to incorporate livestock herd dynamics and interaction of crop and livestock systems into the methodology.

Chapter 15 tackles four major issues with respect to food supply chain in the context of climate change. First, the importance of analysing climate short-term shocks and long-term change on the full food supply chain (inputs, farms, processing, and distribution). Second, the authors show the importance of viewing a given supply chain as an interdependent set of segments and sub-segments.

Climate shocks upstream in the supply chain can disrupt a wide complex of midstream and downstream activities. Third, supply chain analysis is greatly benefited by using "hot spots" of vulnerability to understand climate impacts, both before and after the farm gate. Fourth, climate shocks, and strategies to mitigate them, can be viewed from as (i) strategic supply chain design choices by actors along the supply chain, of sourcing and marketing systems, geography, institutions, and organization; and (ii) threshold investments by actors (firms and farms) along all supply chains.

Chapter 16 uses a conceptual model and empirically-based simulations to investigate the effectiveness of extension-driven informational programs, rain-indexed crop insurance, and the interaction of the two programs in driving adaptation and providing a safety net for farmers. Based on options between diversification strategies and land management practices, different potential welfare outcomes for agricultural households are investigated. The findings show that CSA techniques, including advanced information, about changing conditions in Malawi can mitigate expected losses. The value of this information is greater for farmers with less-binding subsistence constraints and under scenarios for which the effects of climate change are larger. Rain-indexed insurance appears to drive farmers to increase their usage of cash crops and higher yield/higher variability hybrid crop options. Such information is even more important in addressing larger expected losses among farmers with greater flexibility.

The mixed crop-livestock systems of the developing world will become increasingly important for meeting food security challenges of the coming decades. Chapter 17 addresses the gap in understanding of the synergies and trade-offs between food security, adaptation, and mitigation objectives based on a systematic review protocol coupled with a survey of experts. The chapter also discusses constraints to the uptake of different interventions and the potential for their adoption, and highlights some of the technical and policy implications of current knowledge and knowledge gaps.

The effectiveness of a policy depends on specific climate, demographic, environmental, economic and institutional factors. Chapter 18 introduces temporal aspects of household vulnerability to a conceptual model building on available econometric results. The method is based on a factorial design with two vulnerability levels and two production methods. Farms are classified into groups based on cluster analysis of survey data from Zambia. The chapter shows that small, vulnerable farms are more likely to face labor and cash constraints, which may prevent them from adopting technologies that have the potential to sustainably improve food security and enhance their adaptive capacity, i.e. be climate-smart. Widespread adoption, however, will require policies that address the barriers identified here to provide: (i) improved techniques that are less labor intensive, (ii) improved availability of fertilizers, and (iii) credit to cover the up-front costs of investing in soil health that takes several years to bear fruit.

Introduction and Overview                                                                                      11

## 1.2.4 Farm Level Response to Improving Adaptation and Adaptive Capacity

Chapter 19 uses Mali and Nigeria as case study countries to show that sustainable land and water management (SLWM) could more than offset the effect of climate change on yield under the current management practices. Despite the benefits, adoption rates of SLWM remain low. The authors discuss policies and strategies for increasing their adoption including improvement of market access, enhancing the capacity of agricultural extension service providers to provide advisory services on SLWM, and building an effective carbon market that involves both domestic and international buyers.

Chapter 20 identifies the key barriers, opportunities and impacts for a wider adoption of climate smart technologies by differentiated groups of agricultural producers, with a focus on the poor in Central Asia. It is found that access to markets and extension, and higher commercialization of household agricultural output, may serve as major factors facilitating the adoption of CSA technologies. The adoption of CSA technologies has a positive impact on the farming profits of both poorer and richer households, although these positive impacts may likely to be higher for the richer households. Even still, adoption rates among the poorer households are lower than among the richer households.

Chapter 21 shows the implications of farm households' past decision to adapt to climate change on current downside risk exposure in the Nile Basin of Ethiopia. Using moment-based specification to capture the third moment of a stochastic production function as measure of downside yield uncertainty, it finds that past adaptation to climate change (i) reduces current downside risk exposure, and so the risk of crop failure; (ii) would have been more beneficial to the non-adopters if they had adopted, **in terms of reduction in downside risk exposure**; and (iii) is a successful risk management strategy for adopters.

Chapter 22 uses case studies from Zambia and Malawi to discuss the drivers of diversification and its impacts on selected welfare outcomes with a specific attention to climatic variables and institutions. Geo-referenced farm-household-level data merged with data on historical rainfall and temperature as well as with administrative data on relevant institutions are used to demonstrate that diversification is an adaptation response, as long term trends in climatic shocks have a significant effect on livelihood diversification, albeit with different implications. Access to extension agents positively and significantly correlates with diversification in both countries. The results also demonstrate that the risk-return trade-offs are not as pronounced as might be expected.

Chapter 23 presents a case study on potential impacts and implications for adoption of CSA solutions in the Northern Mountainous Region (NMR) of Viet Nam. The authors use primary data collected through *ad hoc* household and community surveys in the study area, on the costs and benefits of agricultural practices, as well as on socio-economic information relevant for households' adoption decisions. A profitability estimate and technology adoption analysis indicate that the potential of some sustainable farming practices to increase productivity and incomes and provide adaptation benefits under the specific climate patterns being experienced in NMR of Viet Nam, particularly in "critical growing periods" of crops. However, such practices often have higher capital and labour requirements, which are likely to prevent or impede adoption. The findings suggest the importance of local climate and socio-economic contexts in determining which practices will actually be climate-smart. Results highlight the importance of using climate information for targeting the promotion of improved practices, and building adaptive capacity amongst farmers.

## 1.3  Part III. Policy Synthesis and Conclusion

Chapter 24 focuses on the implications of the empirical findings for devising effective strategies and policies to support resilience and the implications for agriculture and climate change policy at national, regional and international levels. This section is built upon the analysis provided in the case studies as well as short "think" pieces on specific aspects of the policy relevance issues from policy makers as well as leading experts in agricultural development and climate change. Lastly, Chapter 25 is a synthesis to identify and reconcile the common themes across all the chapters and draws some major economic conclusions and policy recommendations.

**Open Access** This chapter is distributed under the terms of the Creative Commons Attribution-NonCommercial-ShareAlike 3.0 IGO license (https://creativecommons.org/licenses/by-nc-sa/3.0/igo/), which permits any noncommercial use, duplication, adaptation, distribution, and reproduction in any medium or format, as long as you give appropriate credit to the Food and Agriculture Organization of the United Nations (FAO), provide a link to the Creative Commons license and indicate if changes were made. If you remix, transform, or build upon this book or a part thereof, you must distribute your contributions under the same license as the original. Any dispute related to the use of the works of the FAO that cannot be settled amicably shall be submitted to arbitration pursuant to the UNCITRAL rules. The use of the FAO's name for any purpose other than for attribution, and the use of the FAO's logo, shall be subject to a separate written license agreement between the FAO and the user and is not authorized as part of this CC-IGO license. Note that the link provided above includes additional terms and conditions of the license.

The images or other third party material in this chapter are included in the chapter's Creative Commons license, unless indicated otherwise in a credit line to the material. If material is not included in the chapter's Creative Commons license and your intended use is not permitted by statutory regulation or exceeds the permitted use, you will need to obtain permission directly from the copyright holder.

# A Short History of the Evolution of the Climate Smart Agriculture Approach and Its Links to Climate Change and Sustainable Agriculture Debates

**Leslie Lipper and David Zilberman**

**Abstract** Climate Smart Agriculture (CSA) is an approach to guide the management of agriculture in the era of climate change. The concept was first launched in 2009, and since then has been reshaped through inputs and interactions of multiple stakeholders involved in developing and implementing the concept. CSA aims to provide globally applicable principles on managing agriculture for food security under climate change that could provide a basis for policy support and recommendations by multilateral organizations, such as UN's FAO. The major features of the CSA approach were developed in response to limitations in the international climate policy arena in the understanding of agriculture's role in food security and its potential for capturing synergies between adaptation and mitigation. Recent controversies which have arisen over CSA are rooted in longstanding debates in both the climate and sustainable agricultural development policy spheres. These include the role of developing countries, and specifically their agricultural sectors, in reducing global GHG emissions, as well as the choice of technologies which may best promote sustainable forms of agriculture. Since the term 'CSA' was widely adopted before the development of a formal conceptual frame and tools to implement the approach, there has been considerable variation in meanings applied to the term, which also contributed to controversies. As the body of work on the concept, methods, tools and applications of the CSA approach expands, it is becoming clearer what it can offer. Ultimately, CSA's utility will be judeged by its effectiveness in integrating climate change response into sustainable agricultural development strategies on the ground.

L. Lipper (✉)
ISPC-CGIAR, Rome, Italy
e-mail: leslie.lipper@fao.org

D. Zilberman
Department of Agriculture and Resource Economics, University of California Berkeley, Berkeley, CA, USA
e-mail: zilber11@berkeley.edu

# 1 Introduction

Climate Smart Agriculture (CSA) is an approach to guide the management of agriculture in the era of climate change. The concept was first launched in 2009, and since then has been reshaped through inputs and interactions of multiple stakeholders involved in developing and implementing the concept. CSA aims to provide globally applicable principles on managing agriculture for food security under climate change that could provide a basis for policy support and recommendations by multilateral organizations, such as UN's FAO. The major features of the CSA approach were developed in response to debates and controversies in climate change and agricultural policy for sustainable development.

The purpose of this paper is to give an overview of the evolution of CSA, introduce its major components, and summarize the key debates associated with it within the context of climate change and agricultural policy debates The first section provides an overview of international climate change policy followed by an introduction and analysis of CSA and its history. This is then followed by a discussion of three broad controversies related to CSA, namely the role of mitigation, the relationship of CSA to sustainable agriculture, and way biotechnology is treated in the CSA approach.

## 1.1 The Evolution of Climate Change Policy

To put CSA and its controversies in context, it is necessary to understand the evolution of global climate change policies over recent years. We use the framing of Gupta (2010), who traces the history of international climate change policy, from 1979 to 2010. He distinguishes between five phases of evolution. He refers to the pre-1990 phase as the period of framing the problem, beginning with the World Climate Conference in 1979 and including the establishment of the International Panel on Climate Change (IPCC) in 1988. The main focus of global climate change policy during this period was the need for global action to stabilize greenhouse gas (GHG) emissions, to be supported and guided by a globally cooperative framework for undertaking scientific research in the form of the IPCC, and with the understanding that developed and developing countries would bear different responsibilities to mitigate climate change. Because of the high uncertainty associated with climate change, a precautionary approach to climate change policy was adopted. This implies the need to take preventive action even before full certainty about human-induced climate change was obtained, and secondly, to emphasize no-regrets actions that would be valuable even in the absence of climate change. The publication of the Bruntland Commission Report on Sustainable Development in 1987 (WCED 1987) also led to the realization of the links between climate change and sustainable development and the benefits of considering them in an integrated fashion.

During the second period of international climate policy between 1991 and 1996, the initial articulation of a global policy framework was introduced, signified by the Rio Convention in 1992 and the adoption of Agenda 21. An important outcome of the Rio Conventions was the establishment of the UN Framework Convention on Climate Change (UNFCCC) which entered into force on 21 March 1994. The ultimate aim of the convention is preventing "dangerous" human interference with the climate system. Article 2 of the convention says this objective should achieved while ensuring that "food production is not threatened". There was much debate on equity and the principle of common but differentiated responsibilities.[1]

Developed countries were assumed to bear much of the responsibility for both causing and reducing GHG emissions. However their response could also include helping developing countries pay for mitigation actions in the developing world. As the policy formation process moved forward, countries began to form coalitions around common interests. For example, small island nations formed one coalition, as did the G77, representing a block of 130 developing countries. Among the developed nations there was clear difference between the EU and the US and furthermore, the division grew between the EU and non-EU nations. Civil society organizations became a major player in the climate change debate with a major division between the northern organizations pursuing environmental and the southern organizations emphasizing development objectives.

The period between 1997 and 2001 saw the emergence of the first global agreement: the Kyoto Protocol. The Protocol emphasized comprehensive targets for GHG reduction in terms of $CO_2$ equivalence rather than individual GHGs. Developed countries were assigned different GHG reduction targets and there was emphasis on flexibility in achieving these via mechanisms including emission trading, joint fulfillment and implementation (countries could form a bloc to share responsibilities to meet their joint targets). There was also recognition of the importance of financial mechanisms to promote the implementation of the agreements. The clean development mechanisms (CDM) was established, which allowed developed countries to use financial incentives to finance GHG emission reductions in developing countries and then use the credits to meet their own targets.

The establishment of the CDM provided a basis for expanding the use of payment for ecosystem services to meet GHG reduction targets. One important category of actions for emissions reductions highly relevant to agricultural development is that of sequestering carbon in soils and forestry. Many opportunities for agricultural related carbon sequestration were identified through improved soil manage-

---

[1] The Rio Declaration states: "In view of the different contributions to global environmental degradation, States have common but differentiated responsibilities. The developed countries acknowledge the responsibility that they bear in the international pursuit of sustainable development in view of the pressures their societies place on the global environment and of the technologies and financial resources they command."

Similar language exists in the Framework Convention on Climate Change; parties should act to protect the climate system "on the basis of equality and in accordance with their common but differentiated responsibilities and respective capabilities." http://cisdl.org/public/docs/news/brief_common.pdf.

ment and forestry (McCarl and Schneider 2001). One of the challenges of implementing the Kyoto Protocol (KP) was the need for reliable and cost-effective mechanisms for carbon accounting, monitoring and validation which proved particularly difficult in the case of carbon sequestration. The issue of soil carbon inclusion was hotly debated in the discussions on establishing the CDM (Post et al. 2001; Ringius 2002).

The US, Canada, Brazil, and other countries advocated for the inclusion of soil carbon sequestration as part of the Protocol and developed mechanisms to improve its accounting (Paustian et al. 2004). Lal (2004) argued that payment for carbon sequestration could provide farmers, especially in developing countries, with significant supplementary income. However the EU and others were against its inclusion and ultimately the decision was taken to exclude this category from the international carbon offset markets.

Even more importantly, the global significance of the Kyoto Protocol suffered with the US withdrawl from it in 2001, since the two biggest carbon emitters (US and China) were not a part of it. Nevertheless, the Protocol provided a foundation for international collaboration and established many principles for future policy implementation.

The period between 2002 and 2007 saw a retreat from a global agreement to many bi- and multi-lateral agreements, many of which were initiated by the U.S. The period was characterized by competition for leadership among countries regarding climate change policy strategies. While the EU continued to push for extension and expansion of the Kyoto Protocol, the U.S. emphasized multi-lateral agreements. In particular, the Asia-Pacific Partnership on Clean Development and Climate, signed in 2005 (and concluded, with many of its projects canceled, in 2011) emphasized the desire to introduce technological solutions to reduce greenhouse gases (GHG) through, for example, collaboration on R&D aiming towards 'clean coal' (Tan 2010).

The growing emphasis on government support to pursue alternative energy sources also had significant impact on agriculture, especially with the introduction of biofuel policies in much of the world (U.S., Brazil, EU and many other countries). While GHG reduction was one justification for the subsidization of biofuels, perhaps more important was the need to combat rising energy prices, to improve the balance of trade, and to increase the income of the agricultural sector (Zilberman et al. 2014). The increase in the price of food in 2008 as well as the concern about indirect land use led to the curtailment of biofuel policies, but some studies (Huang et al. 2012) found that biofuels can be beneficial for the poor, as long as mechanisms exist to protect vulnerable populations against extreme price shocks. Since national governments were not able to initiate potent global climate change actions during the period, subnational entities like U.S. states and Canadian provinces have established their own climate change programs. Both national and provincial plans have significantly impacted agriculture by introducing demand for biofuel and biomass as well as subsidizing carbon sequestration activities.

The final period of climate policy evolution considered by Gupta (2010) is the financial crisis period (from 2008 and on). In this time period the UNFCCC has moved away from a system where mitigation actions were solely the responsibility of rich countries, to one where mitigation actions in developing countries are now being articulated as part of national policy processes to meet the nation's own mitigation aspirations. The policy and financing issues are significantly different in this context, compared with the situation when developing countries were only participating in greenhouse gas reductions on behalf of rich countries, in the form of a carbon offset.

The main issue on the international climate policy agenda for the UNFCCC COP 15 negotiation held in Copenhagen in 2009 was agreement on a global climate treaty which would lay out responsibilities for reducing emissions. Although COP 15 failed to achieve a global climate agreement, it did produce the "Copenhagen Accord" which called for developing countries to develop mitigation targets to 2020 and included financing commitments of $100 billion/year by 2020 as well as $30 billion for urgent actions up to 2012. In the following year at COP 16, the Green Climate Fund was established as an operating entity of the Financial Mechanism of the UNFCCC to support projects, programmes, policies and other activities in developing countries. Developing countries – including both emerging and least developed countries – have articulated mitigation actions through Nationally Appropriate Mitigation Actions (NAMAs) (result of COP 18 2011), as well as more recently through their Intended Nationally Determined Contributions (INDCs).

It is also important to note that during this period, CDM operations had expanded considerably, with new methodologies and accounting procedures accompanying the expansion. At the same time the volume and value in the voluntary (e.g. non-compliance) carbon offset markets, which generally does allow for the inclusion of agricultural soil carbon, also expanded rapidly, although still only representing a small percentage of the value of the trading in compliance markets (Hamrick and Goldstein 2016) Opposition to soil carbon credits in the context of developing country agriculture was raised by civil society actors. This opposition was based on the argument that soil carbon offsets were a means of putting the mitigation burden on low income developing country farmers and that farmers were unlikely to see any benefit from participating in such markets, but rather could be exposed to losing rights to their land (Action Aid 2011).

In the most recent period of climate policy development, there is a growing realization that significant impacts of climate change are already being felt, and are likely to continue and deepen. The Paris Agreement reached at the 21st Conference of Parties of the UNFCCC in 2015 signifies an increased global commitment to address climate change, as countries agreed to establish legally binding constraints on GHG emissions that aim to contain average global temperature rise by the use of a mixed market approach that induces both introduction of clean energy and conservation (Cooper 2016). All parties recognize the urgency of establishing adaptation strategies, especially to protect the poor and the vulnerable. As of 31 March 2016, 188 countries had submitted "Intended Nationally Determined Contributions" (INDCs) to the UNFCCC which includes statements of intended actions for mitigation as well as adaptation. More than 90% of the countries explicitly include agriculture in their mitigation and adaptation plans, with a particularly strong focus amongst least developed countries (LDCs) (FAO 2016). Adaptation in the agriculture sector is given high priority, and mitigation from agriculture, including sequestration is also quite prominent in the submissions. Thus the importance of considering adaptation and mitigation together and capturing the potential synergies between them is more important than ever. The potential of the CSA approach for supporting this is also increasingly recognized; 31 of the INDCs explicitly mention CSA in the context of seeking joint poverty reduction and environmental benefits (FAO 2016).

## 2 Overview of CSA

The CSA concept emerged at a moment in time of considerable controversy around the concept and approaches to sustainable agricultural development, and when the specificities of agriculture and its role in food security were not well articulated in the climate change policy process. The former was clearly reflected in the debates and controversies of the development of the International Assessment of Knowledge, Science 2009) Technology for Development (IAASTD) which ran from 2003 to 2008 (Scoones 2009). The main arguments in this fora centered around the role of top-down expert assessments versus local participatory approaches to knowledge generation, as well as the role of biotechnology and specifically transgenic crops in sustainable development. In the global climate change policy arena, agriculture's key role in food security was not clearly articulated and the consideration of adaptation and mitigation in two separate negotiation streams limited capacity to build synergies between them.

The first articulation of the CSA concept was presented in the 2009 FAO report entitled "Food Security and Agricultural Mitigation in Developing Countries: Options for Capturing Synergies, which was launched at the Barcelona Climate Change workshop held in November of that year. In 2010, the FAO paper entitled "Climate-Smart" Agriculture, Policies, Practices and Financing for Food Security, Adaptation and Mitigation" was released as a background paper for the Hague Conference on Agriculture, Food Security and Climate Change held in October of that year (FAO 2010). The conference was organized as a follow up to the Shared Vision Statement agreed at the Seventeenth Session of the Commission on Sustainable Development (CSD-17) in May 2009 and to further develop the agriculture, food security and climate change agenda.

These first expressions of the climate smart agriculture concept argue that the agricultural sector is key to climate change response, not only because of its high vulnerability to climate change effects, but also because it is a main contributor to the problem. It also argued that sustainable transformation of the agricultural sector is key to achieving food security, and thus it is essential to frame climate change responses within this priority. Analysis of the state of knowledge on the adaptation, mitigation and food security benefits of a range of agricultural practices, as well as their potential tradeoffs was given as well (e.g. see table 2.2 of the 2009 report as well as FAO 2010). Finally these reports focussed on one of the key issues that arose in CSD-17 discussions – how to finance the transformative changes needed. The CSA work focused on the potential for linking the emerging and potentially huge new sources of climate finance – including but not limited to carbon markets – to support the transition to sustainable agriculture. However, important barriers such as high transactions costs for smallholder agricultural producers to access and benefit from climate finance were clearly identified as major issues (FAO 2011).

The CSA concept sparked considerable attention and debate in international and national agricultural and climate change policy arenas, and it was quickly taken up as a rallying point for mobilizing actions on climate change and agriculture. In the wake of the Hague conference, two parallel global processes related to policy and science of CSA were established. The policy process involved follow up conferences in 2012 in Hanoi Vietnam and 2014 in Johannesburg South Africa. The global CSA science process was initiated with a global CSA science conference at Wageningen in 2011, with subsequent CSA science conferences held at University of California at Davis in 2013 and at CIRAD Montpelier in 2015. One of the main outcomes of these processes was the proposal to establish a global alliance on climate smart agriculture (GACSA) which would bridge the policy and science aspects by focussing on three key action areas: (1) knowledge; (2) enabling environment and (3) investments.

After considerable debate, the GACSA was launched in September 2014 at the UN Climate Summit. Memberships in GACSA may include governments, civil society member/non-government organizations, farmers, fishers and forester organizations, intergovernmental organization (including UN entities), research/extension/education organizations, financing institutions and private sector organizations. As of January 2016 the GACSA has 122 members, including 22 countries.

CSA developments were not only at international level however, with CSA projects initiated at country and regional levels, generally in partnership with international organizations such as FAO, World Bank, local and international NGOs and the Climate Change and Food Security program of the CGIAR.

The rapid and widespread uptake of the CSA concept took place in advance of a clearly defined methodology and definition of CSA, and thus differences in meanings and application of the concept have arisen, and given rise to controversies, which further clarification and development of the CSA concept could ostensibly resolve. However much of the controversy around the CSA concept is related to more fundamental disagreements in global policy debates on climate change and sustainable agriculture.

## 3  Key Features and Evolution of the CSA Concept

One of the main features of the CSA concept is that it calls for meeting three objectives: sustainably increasing food security through increases in productivity and incomes, building resilience and adapting to climate change, and reducing greenhouse gas emissions compared to a business as usual or baseline scenario.

From its inception, recognition of possible trade-offs between the three objectives, and the potential to increase synergies amongst them through policies, institutions and financing was a key feature of the CSA concept (FAO 2009). The need for locally specific solutions was also an important component. A general framework for assessing trade-offs and synergies was provided in FAO (2009, p. 25), along with several examples of sustainable land management practices and "modern" inputs. However, no specific guidance was provided on how to define a CSA practice, or prioritize amongst objectives, to develop the site specific solutions. A clear conceptual framing of the link between sustainable agriculture and CSA was also missing, hindered by the complexity of tying together the three main objectives. The lack of a clear methodology together with a rapid uptake of the concept resulted in considerably variability in the use of the term and confusion, which in turn has been a major source of controversy around the concept.

By the second global CSA policy conference held in Hanoi in 2012, the beginnings of a CSA methodology and principles were emerging. A CSA methodology presented in one of the background papers to the conference consisted of three major elements included: (1) building a relevant evidence base for assessing trade-offs and synergies amongst the three main objectives, (2) creating an enabling policy environment that required coordination of climate change and agricultural policies and (3) guiding investments and linking to climate finance. The methodology was based on lessons learned from a CSA project funded by the EC in 2010 and jointly implemented by FAO and three partner countries. As such, it focussed on national level actions; e.g. building evidence on climate impacts and vulnerabilities for the agricultural sector at country level; analysing the effectiveness of varying actions on productivity and incomes and their resilience to site specific climate shocks, and their effects on reducing emissions compared to a business as usual agricultural growth path for the country. Enhanced coordination between national climate change and agricultural policies and strategies is key to creating an enabling policy environment, while analysis of the marginal abatement costs of nationally appropriate mitigation actions gives a clear indication of where potential synergies between the three CSA objectives can best be obtained, and the potential of using mitigation finance to support them.

The Climate Smart Agriculture sourcebook, which was a joint effort of several international organizations, came out in 2013 and provided principles for defining CSA practices as well as conceptual links to sustainable agriculture processes and a wide range of examples from livestock, cropping, fishery and forestry sectors (FAO 2013). The first chapter of the sourcebook lays out two major principles defining CSA practices: (1) increasing resource use efficiency in agricultural systems and (2) enhancing the resilience of agricultural systems and the people who depend upon them. Resource use efficiency is a key component of sustainable agricultural intensification strategies. By using resources such as nitrogen fertilizer, feed for livestock, land and water more efficiently, the net return to farmers and thus incomes increase, while pressure on scarce resources and emissions per unit produced are reduced. Increasing resilience involves reducing vulnerability as well as enhancing adaptive capacity. CSA strategies require that resilience and resource use efficiency are pursued together, although specific technologies and institutional arrangements may affect only one or the other. Rather, efficiency and resilience need to be considered in an overall systems perspective that considers different spatial and temporal scales. The importance of ecosystem services provided through for example, improved soil management, agro-biodiversity and landscape management, in achieving resource use efficiency and resilience is also a major tenet of CSA approaches outlined in the sourcebook.

The CSA methodology and principles were further defined through a consultative process involving representatives from a broad spectrum, including international organizations such as FAO, CCAFS and World Bank, national agricultural and climate change policy-makers, academics, and civil society. This consultative process resulted in the publication of a perspectives piece in Nature Climate Change in 2014 that reaffirmed the key components of a CSA methodology, but also addressed some of the emerging controversies associated with the concept (Lipper et al. 2014). One of these was a response to the heavy emphasis on exante identification of farm level practices that could meet all three CSA objectives. The paper argued that CSA did not imply that every practice in every field would have to contribute to food security, adaptation and mitigation, but that meeting these objectives should be considered at broader spatial and temporal scales. It also highlighted the controversy around mitigation in developing countries.

More recently, the World Bank and the CCAFS program have launched a set of "country CSA profiles".[2] These provide critical stocktaking of ongoing and promising practices for the future, and of institutional and financial enablers for CSA adoption. The profiles provide information on CSA terminology and how to contextualize it under different country conditions. The knowledge product is also a methodology for assessing a baseline on climate smart agriculture at the country level (both national and sub-national) that can guide climate smart development.

The CSA concept and methods were developed by international technical agencies, including FAO, the World Bank, the Climate Change and Food Security Programme of the CGIAR. As such, the concept was built to provide a framework for formulating and taking actions to respond to climate change in agriculture that was broad enough to encompass a wide spectrum of political and economic approaches to managing agriculture. In this way, the concept could be relevant to the wide range of clients served by international agencies and adapted to their specific needs and circumstances. At the same time however, the generality of the

---

[2] http://sdwebx.worldbank.org/climateportal/index.cfm?page=climate_agriculture_profiles.

concept has led to multiple interpretations of its core meaning and thus some confusion and controversy. In the next section we look more closely at the most prominent of these.

## 4 CSA Controversies in the Broader Policy Context

### *4.1 The Role of Mitigation and Carbon Finance in CSA*

One of the main criticisms of the CSA approach has been that it prioritizes mitigation over food security and adaptation, and it mandates a link to carbon offset markets (Action Aid 2011, Neufeldt et al. 2013). By explicitly calling attention to the potential of agricultural transformation to generate mitigation benefits, and actively pursuing links to mitigation finance, the CSA approach raised suspicions that it was a means of pushing the mitigation burden on the world's poorest people (Action Aid 2010). The argument was made that CSA advocated pushing carbon offsets for soil carbon sequestration on poor farmers, and this would shift the burden of reducing greenhouse gas emissions from rich, industrialized countries who had actually created the problem, to poor developing countries that already are facing the biggest burden in adapting to climate change. This argument is rooted in controversies over soil carbon sequestration and the role of developing countries in mitigation in the global climate policy debate (see previous section) as well as misconceptions of the framing of climate finance in CSA.

Before discussing misconceptions and policy debates, it is useful to understand the impetus for connecting mitigation finance to agricultural development. In 2008 the fourth assessment report of the IPCC was released. The report included a detailed analysis of the state of knowledge at the time on the technical and economic potential of mitigation from agriculture (Smith et al. 2008). They found an estimated global economic mitigation potential for 2030 from agriculture of 1500–1600, 2500–2700, and 4000–4300 MtCO2-eq/year at carbon prices of up to 20, 50 and 100 US\$/tCO2-eq. The activities with highest economic potential were restoring cultivated organic soils, cropland management, grazing land management, restoration of degraded lands, rice management and livestock. Sequestration of carbon in agricultural soils is a key feature of most of these practices. Within each of these categories the actions analysed had high correspondence with actions promoted for sustainable agriculture, e.g. crop rotation, minimum tillage, nutrient use efficiency, feed efficiency. This analysis from the leading science body on climate change indicated the potential to capture huge synergies between mitigation and sustainable agricultural development.

At the same time, the rapid growth in the development of international carbon offset markets represented a major new and potentially huge source of finance to support sustainable agricultural activities with mitigation co-benefits. At the time of the launching of the CSA concept, the valuation of global carbon markets was $141 billion, composed principally of the clean development mechanism of the Kyoto Protocol and the European ETS system (World Bank 2011). However, as noted in the section on climate policy above, neither of these major financing mechanisms allowed soil carbon sequestration from agricultural practice change as a source of mitigation.

Outside of the formal carbon markets, an alternative voluntary market for carbon offsets was springing up, including projects sponsored by the World Bank Biocarbon Fund, NGOs in developed and developing countries, as well as some regional exchanges. The Chicago Climate Exchange which developed a protocol for soil carbon offsets from reduced tillage and improved pasture management (FAO 2012). However the financing flows through these voluntary markets was miniscule compared with those of the formal carbon markets (FAO 2012).

Essentially, there was very little demand for carbon offsets from soil carbon sequestration from developing country farmers due to their exclusion from the major carbon financing mechanisms. However the question of whether or not they should be allowed in order to open the doors to new financing that could generate both mitigation and development outcomes was an important thrust of early CSA work. If the barrier to accessing a significant new source of financing was simply a lack of good research on how much soil could be sequestered from changes in developing country farming systems, then surely the response should be developing a research agenda to provide the needed science. However as research into the potential of carbon offsets as a source of finance for developing country farmers proceeded, it became clear that issues of weak institutional capacity in developing countries was a more serious barrier. In particular, the rights of people with unclear and informal systems of land tenure to reap carbon benefits was very problematic Leach & Scoones 2015). Experience with payment for environmental service programs, and particularly the REDD+ process had indicated this was a particularly difficult issue to address, but very commonly found. The REDD+ experience indicated that there was indeed potential for poor farmers and land managers with insecure title to land to be dispossesed through the implementation of a REDD+ program, but that there was also potential for stimulating improvements in tenure systems through the impetus of such programs (Larson et al. 2013). Ultimately, it was well recognized that weak and inequitable institutions were a key barrier to making carbon finance work for small and poor farmers, and thus greater attention should be given to linking international public sources of finance such as the Global Environment Fund to support climate smart agriculture (FAO 2013). At the same time, major shifts in the international climate policy negotiations reduced the importance of international carbon offset markets as the main source of climate finance. The newly reconfigured international climate policy regime with its emphasis on nationally determined contributions to mitigation and adaptation and the prominence of agriculture in the contributions from developing countries has created interest in the capacity of agricultural mitigation sources to contribute to developing country's own nationally determined contributions. It also implies a greater need for an approach that can identify how mitigation can be integrated into agricultural transformation strategies without compromising food security, which is of course a major focus of CSA.

To summarize, a major thrust of CSA is building the enabling conditions for a major transformation in agriculture, and developing adequate financing streams adapted to the specific conditions of agriculture is important in this regard. At the time of the launching of the CSA concept, the international carbon offset markets were the largest source of climate finance and thus much attention initially was given to its potential for supporting agricultural transformation in developing countries. Due to the problems with linking carbon finance to smallholder agriculture countries, together with the emergence of new funds for supporting mitigation actions on the part of developing countries in recent years, the emphasis of CSA has shifted away from carbon markets to international public climate finance such as the Green Climate Fund and the Global Environmental Facility. Given the high importance of agriculture in the national expressions of mitigation actions on the part of developing countries, the importance of identifying mitigation actions that are synergistic with food security and adaptation and building financing mechanisms to support them is of greater importance than ever.

## 5 CSA and Sustainable Agriculture

Another major criticism of CSA has been the lack of clear principles by which to define a CSA practice, and thus concerns that the concept and branding could to be used to advance non-sustainable and non-desirable forms of agricultural development. This debate was fuelled by the mistaken notion that CSA was essentially a proposal for a new type of agricultural practice, giving rise to concerns directly related to ongoing and fierce debates about technologies for sustainable agriculture.

CSA is not intended to provide a new set of sustainability principles, but rather a means of integrating the specificities of adaptation and mitigation into sustainable agricultural development policies, programs and investments. CSA strategies and practices then should adhere to the principles that underpin sustainable agriculture and food systems. Recently FAO published a new set of guidelines and approach to achieving sustainable agriculture and food systems (SFA) as ones which meet the following criteria: (1) improving the efficiency of resource use, (2) conserving, protecting and enhancing natural resources, (3) protecting and improving rural livelihoods, (4) enhancing resilience of people, ecosystems and communities and (5) responsible and effective governance mechanisms.

Of course, these principles are very broad and do not mandate any specific balance or weighting between them in terms of defining a sustainable technology. Nonetheless, the links between the sustainability principles and CSA can be seen. Increasing resilience, conservation and protection of natural resources and increasing resource use efficiency are key components of adaption and mitigation. Protecting and improving rural livelihoods is closely related to the CSA objective of sustainably increasing productivity and incomes. A major thrust of CSA is improvement of climate change and agricultural governance through better coordination and institutional strengthening.

With its emphasis on assessing trade-offs and synergies between its three main objectives, as well as the barriers to adoption, CSA actually addresses one of the most essential issues in sustainable agriculture: what will it take to actually achieve a large scale transformation? The emphasis on explicitly identifying trade-offs in the CSA approach is a reaction to the lack of such consideration in many of the sustainable agricultural approaches which focus only on the benefits obtainable, ignoring costs and barriers. The result has been disappointingly low adoption of sustainable agricultural techniques, despite decades of efforts and funds to support them. In the end it is the farmers, fishers, livestock keepers and forest managers that are assigning weights to environmental, social and economic criteria through the decisions they make on how to manage their production systems. However the tradeoffs they face between the objectives are determined by the institutional environment they operate under. For example, sustainable land management techniques such as land restoration or agroforestry can take some years to generate benefits, and they require up-front investments and can involve reductions in income during the initial phase. While over a 20 year time frame such actions can result in higher economic, environmental and social benefits, in the initial phases there are significant tradeoffs between them. This is essential to understanding how to effectively induce transformative change – and it has all too often been ignored in the literature on sustainable agricultural development.

A key issue in the debate on technologies for sustainable agricultural growth focuses on the relationship between natural capital inputs (e.g. ecosystem services such as soil quality or genetic diversity) and manufactured capital inputs (inorganic fertilizer, machinery, improved seed) in an agricultural production system. This debate is rooted in a reaction to the great push in capital inputs (improved seed and inorganic fertilizers) which began in the 1960s, which to a large extent built upon a model of substituting manufactured capital inputs for natural capital; e.g. inorganic fertilizer use could substitute for soil quality, or pesticides for genetic diversity (Tilman et al 2002; IAASTD 2009). Particularly in initial phases, increasing manufactured capital inputs to agricultural production systems was the main thrust of this model of development, although in later phases, the focus has shifted in most cases to increasing the efficiency of manufactured capital inputs (FAO 2012). While the results in terms of production increases have been dramatic, these positive results have been accompanied by high rates of natural resource depletion and degradation, as well as negative environmental impacts on land, air and water (Tilman et al. 2002,

IAASTD 2009). The social impacts have been the subject of much debate. On the one hand the expansion of food production and lowering of food prices a major benefit to the consumers, particularly the poor (Pingali 2012). On the other hand, the model of a top down technology delivery focussed primarily on favorable production areas, excluded many of the poorest from its benefits.

Sustainable agriculture is part of the larger concept of sustainable development that according to the Brundtland Commission is a development strategy that aims to ensure that future generations would not be worse off compared to the present generation. Sustainable development contains economic, social, and environmental elements, but in principle has limited restrictions on technology, per se, and the use of technologies are judged based on their impacts. Zilberman (2014) argues that one of the major features of sustainable development is the emphasis on conservation technologies that enhance input use efficiency and reduce pollution, introduction of strategies that include resilience and ability to withstand environmental risk, adoption of recycling technologies, and transition from non-renewable to renewable technologies. Renewable technologies include both energy production using solar and wind as well as extension of the bioeconomy, which relies on biological processes to produce food, fuel, and fine chemicals. This approach to sustainable development that allows some substitution among resources and encourages production systems that enhance human welfare subject to constraints should have bearing on the definition of CSA.

The CSA approach is criticized by some advocates of alternative development models, because it does not explicitly exclude the use of manufactured capital inputs and while incorporating participatory and bottom up approaches, it also allows for integration of science-based technology transfers. The CSA literature does however explicitly call for enhancing the complementarity between ecosystem services and manufactured capital, such as improving soil quality to enhance the productivity gains from inorganic fertilizer use, improving livestock breeds to enhance their feed conversion efficiency, or planting trees in agricultural landscapes to reduce flood risks.

The issue of biotechnology use in agriculture is perhaps the most highly contested, with most of the focus on genetically modified organisms (GMOs). The use of GMOs has been limited to few crops, used mostly for fiber (cotton) and feed and oil (maize, soybean, canola) with limited use for direct human consumption (papaya, maize, canola). Furthermore, while adoption of GMOs on farm has been quite broad in the U.S., Canada, Brazil, Argentina, and South Africa, and in cotton in other major countries (India, China), its use in Europe and most of Africa has been limited or even practically banned. Most major national academies of science and international organizations have argued that it poses no new health risks compared to other sources of food, and there is evidence that GMOs have reduced the price of major agricultural commodities as well as the extent of GHG emissions (Barrows et al. 2014). There is also significant evidence that it has improved the well-being of poor farmers, especially in cotton production (Klümper and Qaim 2014; Qaim 2015).

Nonetheless, significant concern about environmental and social effects of GMOs persists and there is ongoing debate on the application of the precautionary principle by opponents of the technology. Another source of concern is the large role of the private sector in the development of the technology and its control of intellectual property rights. But the heavy regulatory requirements associated with the development of GMOs has led to the concentration of the industry in the hands of a few major companies (Bennett et al. 2013). More recently however, the reduction of the cost of genome mapping and the introduction of new technologies like gene editing increase the capacity of a broader range of stakeholders to utilize and control modern biotechnology to provide effective and quick solutions to address the challenges of climate change.

The issue of which technologies to consider, and specifically whether biotechnologies should be included has been addressed in different ways under current applications of the CSA approach. To a large extent, the technologies and practices considered under CSA approaches are ones that governments have already included in their national agricultural plans, which often do not include biotechnology at present. Under the EC funded FAO CSA project, consultations with national policymakers and stakeholders including representatives from farmer's associations and other civil society groups have been held to identify a set of possible options for further detailed analysis. The World Bank/CCAFS profiles analyse a range of technologies and practices that are currently being practiced in the country or that are likely to be beneficial under projected climate change conditions, including from traditional as well as science based sources. They also provide a set of country specific criteria for identifying climate smartness of the technologies which also give information on the economic, environmental and social impacts of the technologies in that country. Ultimately, CSA neither mandates nor excludes the use of biotechnology or GMOs for any specific user of the approach, but it can provide a basis for helping potential users identify the risks and benefits of its use in addressing the challenges of achieving food security under climate change.

## 6 Conclusion

Climate smart agriculture is a relatively new concept which was launched in 2009 advocating for better integration of adaptation and mitigation actions in agriculture to capture synergies between them and to support sustainable agricultural development for food security under climate change. The rapid uptake of the concept after its launch indicates the tremendous demand for a framework to guide policy and technical interventions in agriculture that integrates the effects of change, the challenges of achieving sustainable agricultural development and the critical role of agriculture in attaining food security. At the same time, the widespread adoption of the CSA term prior to the development of a formal conceptual framing and methodology has lead to considerable variation in meanings applied to the term, as well as confusion and controversy.

The CSA concept has been reshaped through inputs and interactions of multiple stakeholders involved in developing and implementing the concept. At this point there is greater clarification on the definition of the concept and methodology for its application. However controversies over CSA remain. Most of these are related to the controversies in climate change and sustainable agricultural policies. In particular, the role of agricultural mitigation and its financing in developing countries, as well as the development and deployment of technologies for agricultural development are two key areas of continuing controversy in the respective policy circles. CSA does not attempt to provide a prescription to any user of the approach for resolving the controversies, but rather a tool to identify locally appropriate solutions to managing agriculture for sustainable development and food security under climate change. Ultimately the utility of the concept and its implementation will be judged by its effectiveness in integrating climate change responses into sustainable agricultural development actions on the ground.

# References

Action Aid 2011 'Fiddling with carbon markets while Africa burns' Action Aid Johannesburg http://www.actionaid.org/publications/fiddling-soil-carbonmarkets-while-africa-burns

Barrows, Geoffrey, Steven Sexton, and David Zilberman. "Agricultural biotechnology: the promise and prospects of genetically modified crops." *The Journal of Economic Perspectives* 28, no. 1 (2014): 99–119.

Bennett, Alan B., Cecilia Chi-Ham, Geoffrey Barrows, Steven Sexton, and David Zilberman. "Agricultural biotechnology: economics, environment, ethics, and the future." *Annual Review of Environment and Resources* 38 (2013): 249–279.

Cooper, Mark. "The Economic and Institutional Foundations of the Paris Agreement on Climate Change: The Political Economy of Roadmaps to a Sustainable Electricity Future" Available at SSRN 2722880 (2016).

David Tilman, Kenneth G. Cassman, Pamela A. Matson, Rosamond Naylor, Stephen Polasky, (2002) Agricultural sustainability and intensive production practices. Nature 418 (6898):671-677

FAO 2009 Food Security and Agricultural Mitigation in Developing Countries: Options for Capturing Synergies FAO Rome http://www.fao.org/docrep/012/i1318e/i1318e00.pdf

FAO 2010. Climate-Smart Agriculture: Policies, Practices and Financing for Food Security, Adaptation and Mitigation. Rome, FAO.

FAO 2011 Climate Change Mitigation Finance for Smallholder Agriculture: A guide book to harvesting soil carbon sequestration benefits FAO Rome http://www.fao.org/docrep/015/i2485e/i2485e00.pdf

FAO. (2013). Climate smart agriculture sourcebook. FAO Rome.

FAO 2016 State of Food and Agriculture Report "Climate change Agriculture and Food Security" FAO Rome http://www.fao.org/3/a-i6030e.pdf

Gupta, Joyeeta 2010 A history of international climate change policy Wiley Interdisciplinary Reviews: Climate Change Vol. 1 Issue 5 pp 621–763

Hamrick K. and A. Goldstein 2016 "Raising Ambition: State of the Voluntary Carbon Markets 2016" Ecosystem Marketplace Washington DC http://www.forest-trends.org/documents/files/doc_5242.pdf

Henry Neufeldt, Molly Jahn, Bruce M Campbell, John R Beddington, Fabrice DeClerck, Alessandro De Pinto, Jay Gulledge, Jonathan Hellin, Mario Herrero, Andy Jarvis, David LeZaks, Holger Meinke, Todd Rosenstock, Mary Scholes, Robert Scholes, Sonja Vermeulen, Eva Wollenberg, Robert Zougmoré, (2013) Beyond climate-smart agriculture: toward safe operating spaces for global food systems. Agriculture & Food Security 2 (1):12

Huang, Jikun, Jun Yang, Siwa Msangi, Scott Rozelle, and Alfons Weersink. "Biofuels and the poor: Global impact pathways of biofuels on agricultural markets." Food Policy 37, no. 4 (2012): 439–451.

International Assessment of Agricultural Science, Knowledge and Technology for Development (IAASTD) 2009. Agriculture at a Crossroads Synthesis Report Island Press Washington DC

Klümper, Wilhelm, and Matin Qaim. "A meta-analysis of the impacts of genetically modified crops." *PLoS One* 9, no. 11 (2014): e111629

Lal, Rattan. "Soil carbon sequestration to mitigate climate change." *Geoderma* 123, no. 1 (2004): 1–22.

Larson, A.M., Brockhaus, M., Sunderlin, W.D., Duchelle, A.E., Babon, A., Dokken, T., Pham, T.T., Resosudarmo, I. A. P., Selaya, G., Awono, A., Huynh T-B. "Land tenure and REDD+: the good, the bad and the ugly." *Global Environmental Change* 23, no. 3 (2013): 678–689.

Leach, M. and I. Scoones 2015 Political Ecologies of Carbon in Africa in Carbon Conflicts and Forest Landscapes in Africa in June 2015 https://www.routledge.com/products/9781138824836

Leslie Lipper, Philip Thornton, Bruce M. Campbell, Tobias Baedeker, Ademola Braimoh, Martin Bwalya, Patrick Caron, Andrea Cattaneo, Dennis Garrity, Kevin Henry, Ryan Hottle, Louise Jackson, Andrew Jarvis, Fred Kossam, Wendy Mann, Nancy McCarthy, Alexandre Meybeck, Henry Neufeldt, Tom Remington, Pham Thi Sen, Reuben Sessa, Reynolds Shula, Austin Tibu, Emmanuel F. Torquebiau, (2014) Climate-smart agriculture for food security. Nature Climate Change 4 (12):1068–1072

McCarl, Bruce A., and Uwe A. Schneider. "The cost of greenhouse gas mitigation in US agriculture and forestry." *Science* 294, no. 21 (2001): 2481–82.

Paustian, Keith, Bruce Babcock, J. Hatfield, Rattan Lal, B. A. McCarl, S. McLaughlin, A. Mosier et al. "Agricultural mitigation of greenhouse gases: science and policy options." *CAST (Council on Agricultural Science and Technology) Report* 141 (2004): 2004.

P. L. Pingali, (2012) Green Revolution: Impacts, limits, and the path ahead. Proceedings of the National Academy of Sciences 109 (31):12302-12308

Post, Wilfred M., R. Cesar Izaurralde, Linda K. Mann, and Norman Bliss."Monitoring and verifying changes of organic carbon in soil." In *Storing Carbon in Agricultural Soils: A Multi-Purpose Environmental Strategy*, pp. 73–99. Springer Netherlands, 2001.

Qaim, Matin. *Genetically Modified Crops and Agricultural Development*. Palgrave Macmillan, Basingstoke 2015.

Ringius, Lasse. "Soil carbon sequestration and the CDM: opportunities and challenges for Africa." *Climatic change* 54, no. 4 (2002): 471–495.

Scoones, Ian 'The politics of global assessments: the case of the International Assessment of Agricultural Knowledge, Science and Technology for Development (IAASTD)', Journal of Peasant Studies, (2009) 36: 3, 547–571.

P. Smith, D. Martino, Z. Cai, D. Gwary, H. Janzen, P. Kumar, B. McCarl, S. Ogle, F. O'Mara, C. Rice, B. Scholes, O. Sirotenko, M. Howden, T. McAllister, G. Pan, V. Romanenkov, U. Schneider, S. Towprayoon, M. Wattenbach, J. Smith, (2008) Greenhouse gas mitigation in agriculture. Philosophical Transactions of the Royal Society B: Biological Sciences 363 (1492):789–813

Tan, Xiaomei. "Clean technology R&D and innovation in emerging countries—experience from China." Energy Policy 38, no. 6 (2010): 2916–2926.
WCED, UN. "Our common future." *World Commission on Environment and Development Oxford University Press* (1987).
World Bank 2011 State and Trends of the Carbon Market 2011. World Bank Washington DC https://siteresources.worldbank.org/INTCARBONFINANCE/Resources/StateAndTrend_LowRes.pdf
Zilberman, David. "The economics of sustainable development." *American Journal of Agricultural Economics* 96, no. 2 (2014): 385–396.
Zilberman, David, Scott Kaplan, Gal Hochman, and Deepak Rajagopal. "Political Economy of Biofuels." In The Impacts of Biofuels on the Economy, Environment, and Poverty, pp. 131–144. Springer New York, 2014.

**Open Access** This chapter is distributed under the terms of the Creative Commons Attribution-NonCommercial-ShareAlike 3.0 IGO license (https://creativecommons.org/licenses/by-nc-sa/3.0/igo/), which permits any noncommercial use, duplication, adaptation, distribution, and reproduction in any medium or format, as long as you give appropriate credit to the Food and Agriculture Organization of the United Nations (FAO), provide a link to the Creative Commons license and indicate if changes were made. If you remix, transform, or build upon this book or a part thereof, you must distribute your contributions under the same license as the original. Any dispute related to the use of the works of the FAO that cannot be settled amicably shall be submitted to arbitration pursuant to the UNCITRAL rules. The use of the FAO's name for any purpose other than for attribution, and the use of the FAO's logo, shall be subject to a separate written license agreement between the FAO and the user and is not authorized as part of this CC-IGO license. Note that the link provided above includes additional terms and conditions of the license.

The images or other third party material in this chapter are included in the chapter's Creative Commons license, unless indicated otherwise in a credit line to the material. If material is not included in the chapter's Creative Commons license and your intended use is not permitted by statutory regulation or exceeds the permitted use, you will need to obtain permission directly from the copyright holder.

# Economics of Climate Smart Agriculture: An Overview

**Nancy McCarthy, Leslie Lipper, and David Zilberman**

**Abstract** Climate change, especially through greater frequency and intensity of climate extremes, is expected to negatively impact agriculture and food security, particularly in developing countries highly dependent on rain-fed agriculture. Promoting growth and food security must draw on the rich literature of the past 50–60 years while also addressing potential structural shifts in the factors that promote growth. This paper summarizes the economic considerations of Climate Smart Agriculture, a concept developed by the FAO to address the complex issue of how to achieve sustainable agricultural growth for food security under climate change. It addresses the lack of coherence on the CSA approach by building a formal basis of the CSA concept and methodology. We do this by posing a dynamic optimization problem wherein a social planner seeks to maximize expected discounted welfare associated with agriculture of the population they serve, both now and in the future. We analyze constraints, choices, and features of design of CSA to illustrate on the concept can be applied across a range of locations and conditions. This has implications for research, innovation, and policy design.

## 1 Introduction

Climate change is expected to have negative impacts on agriculture and food security in many regions, particularly in developing countries highly dependent on rain-fed agriculture. The fifth assessment report of the IPCC released in 2014 found that climate change effects are already being felt on agriculture and food security, and

N. McCarthy (✉)
Lead Analytics Inc., Washington, DC, USA
e-mail: nmccarthy@leadanalyticsinc.com

L. Lipper
ISPC-CGIAR, Rome, Italy
e-mail: leslie.lipper@fao.org

D. Zilberman
Department of Agriculture and Resource Economics, University of California Berkeley, Berkeley, CA, USA
e-mail: zilber11@berkeley.edu

the negative impacts are most pronounced in tropical zones where most of the world's poor and agricultural-dependent populations are located (IPCC 2012). And yet in the next 20 years, increasing the rate of agricultural growth in these regions is essential to reach the goals of eradicating poverty and meeting growing food demand associated with population growth and dietary transitions.

Over the last 50–60 years, a rich and extensive body of work on agricultural development economics has been developed, aimed at supporting agricultural growth and food security. Over time this work has been augmented with insights and techniques from natural resource and environmental economics, as well as behavioral and institutional economics. The evidence base has also expanded dramatically due to advancements in empirical research design, econometric techniques, data availability and computing power. At the same time, the public sector has invested in agricultural and rural development, accumulating practical experience and knowledge.

Climate change, with its potentially transformative impacts on agricultural systems, means that we need to revisit the key tenets of this accumulated body of knowledge and experience in order to identify its applicability to current and changing circumstances. Does climate change actually require a change in how we go about planning and investing in agricultural growth for food security and poverty reduction? The answer is not obvious – much research and policy design in agricultural development has been concerned not only with enhancing productivity, but also with reducing negative environmental impacts and providing public goods, as well as managing trade-offs between risk and returns and reducing vulnerability of farm households to a wide array of shocks. These are also some of the major concerns raised, perhaps to a more urgent level, with respect to addressing climate change in agriculture. However we need to consider whether the potential magnitude and scale of climate change will result in a structural shift in the factors that will promote growth – and thus how we go about promoting growth and food security.

The increased frequency and intensity of extreme events is clearly one of the most important game-changing effects of climate change. Recent work by Fischer and Knutti (2015) on the link between climate change and extreme events estimated that 75% of extreme hot days and 18% of days with heavy rainfall worldwide can be explained by the warming we've seen over the industrial period. The same study also finds that the probability of extreme events increases nonlinearly with increasing global warming. For instance, the probability of an extreme hot day under a scenario of 2 °C increase over pre-industrial levels is almost double the probability at a 1.5 °C increase, and is more than five times higher than with today's climate. Essentially, the vulnerability of the agricultural sector to adverse events is increasing at a rapid, steep and broad scale, which implies a need for innovative measures to reduce the exposure and sensitivity of the agricultural sector, and also to increase adaptive capacity.

Greater frequency and intensity of climate extremes has implications for research, innovation, and policy design. With respect to research, though the empirical evidence on households' responses to weather shocks is fairly large, most of the data collected has been undertaken under relatively normal weather conditions, with spatially limited idiosyncratic weather shocks. Thus, little is known about the impacts of generalized climate shocks on households' wellbeing, and even less is known about which mechanisms are most effective at minimizing those impacts. Additionally, evidence is lacking on which measures are most effective at increasing the resilience of the agricultural sector as a whole. Part of the problem is the lack of capacity to mobilize resources needed to collect relevant data in the immediate wake of disasters that occur at significant scale, as well as logistical, and potentially ethical, issues involved with collecting data under such circumstances. Valuable information could be obtained by those involved in disaster relief activities, but such information is generally not collected in a systematic manner nor widely shared. As noted by Scott et al. (2016), though everyone agrees that monitoring and evaluation (M&E) should be a critical element in disaster relief, most M&E systems remain weak and data collected remains little shared.

With respect to innovation and policy design, increased frequency and intensity of climate extremes dramatically increases the value of innovations and policies that increase the range of cost-effective options that allow rapid adjustments in the face of climate extremes. This implies a need for a strong shift towards investing in technological and institutional innovations that create options and increase flexibility. This also implies a need for designing policies and regulations that enable different actors – including government agencies as well as the private sector – to exercise various options in response to climate extremes.

The second potential game-changer arises from the possibility of major regional shifts in weather patterns, or "migration" of climate. This effect may be due to spatially and seasonally heterogeneous increases in average temperature and altered rainfall patterns. Such changes may have major consequences in terms of movement of pests and diseases, as well as loss of coastal and certain inland agricultural lands. We can expect that migration of climate will disproportionately affect resource-poor and marginalized farmers who have less adaptive capacity but depend primarily on agriculture for their livelihoods (Hitz and Smith 2004; Thornton et al. 2011). Experience has indicated that intensifying labor migration is a common response to prolonged and chronic environmental degradation, with permanent resettlement less common and generally considered less desirable. However this option is increasingly considered as an adaptation strategy in response to major shifts, such as sea level rise. Current empirical evidence indicates that the poor and most vulnerable to climate risks are again the least capable to undertake effective migration, since they lack the assets and social networks required (Adger et al. 2014; Taylor and Martin 2001).

Successfully adapting to emerging major shifts in weather means that research needs to focus on which factors facilitate the transition to new climate patterns while maintaining growth rates and reducing poverty. Research is needed to evaluate both adaptive, marginal changes within the system to confront such shifts, as well as far-reaching transformational changes. Research is also needed to generate sufficient evidence to compare the relative merits of pursuing incremental adaptation strategies versus transformational strategies. For instance, access to new crop varieties, more suitable livestock, irrigation systems, and pest management strategies can enable farmers to successfully adapt to new climate patterns. At the same time, enabling farm households to relocate may well be a better strategy, especially under more extreme shifts in climate patterns. While there is a fair amount of household-level research on internal and international migration and its impacts on migrant households, much less is known about which institutional structures and mechanisms best support peaceful relocations. While processes of movement in and out of agriculture are ongoing (Taylor and Martin 2001), future research should aim to understand the institutional challenges and planning requirements to address climate related migration within ongoing population transition processes.

More broadly, the interaction between climate change induced changes in agricultural production patterns and structural transformation in the larger food system and rural non-farm sectors need to be better understood (c.f. Haggblade et al. 2007; Reardon and Timmer 2007; Gollin et al. 2002). Given the systems-level focus of such research, this calls for greater integration of sub-discipline research, e.g. linking agro-ecosystem or agri-food sector-wide models with evidence from household surveys. To date, however, such models capture institutional structures and mechanisms in a fairly rudimentary way. While institutions are important for understanding marginal changes, they are particularly important for understanding and promoting transformational changes.[1] Large-scale household surveys and randomized experiments will be of limited value in answering many key questions about systems-level outcomes and optimal institutional structures and mechanisms. Instead improved methodologies for analyzing limited data, e.g. using case studies across disciplines will be required, echoing recommendations of Reardon and Timmer (2007) with respect to agrifood systems.

A third major transformation climate change imposes on agricultural development planning is the need to decouple agricultural growth from emissions growth, given the high share of agriculture in contributing to global emissions. World Resource Institute (WRI) estimated that emissions from agriculture could grow from approximately 6.5 GT in 2010 to 9.5GT per year in 2050 under a conventional agricultural growth strategy. At the same time, the development of the nationally appropriate mitigation actions (NAMAs) and Intended Nationally Determined Contributions (INDCs), has shown that developing countries are interested in pursuing low-emissions agricultural growth strategies, if financing to support such actions can be made available. Reducing emissions from the agricultural sector requires technologies and practices to increase efficiency and reduce leakage from agricultural production systems, and also enhance the sequestration capacity of the sector by increasing trees and shrubs. Improved soil management, sustainable rice intensi-

---

[1] Certain institutional mechanisms are relatively well-studied, such as various aspects of property rights. The impacts of increased access to institutions has also been well-studied but mostly in a rudimentary way, e.g. dummy variables capturing access to a health care center, credit, extension, etc. But, specific delivery mechanisms, the range of services offered, service quality, contract clauses etc. are much less well-studied. Such information is crucial to policy design. New research tools and methods are needed to help build this evidence base.

fication, precision farming, and restoration of degraded lands can all contribute to reduced GHG emissions and/or soil carbon sequestration under certain conditions (Burney et al. 2010; Lal 2004; Paustian et al. 2004; Antle and Diagana 2003). But, as many researchers have documented, there has been limited adoption of sustainable land management (SLM) practices that could also contribute to a low-emissions agricultural growth path, particularly in sub-Saharan Africa and parts of Southeast Asia (Barbier 2010; Pender et al. 2006; Barrett et al. 2002).

In terms of research, there is a great deal of evidence on the benefits to adopting SLM, but much less evidence on the costs and barriers that farmers face in adopting such practices (McCarthy et al. 2012; Pender et al. 2006; Nkonya et al. 2004). Given these costs and barriers, there is a need for the public sector to develop innovative policies and mechanisms that alter incentives for actors in the agricultural sector to pursue such strategies. One mechanism that has received a great deal of attention is a carbon-sequestration based payment (Seeberg-Elverfeldt et al. 2009). However, such programs often fail because of the difficulty in monitoring and verifying compliance, and with making and enforcing contracts with, and delivering payments to, many smallholders (Lockie 2013; Alix-Garcia et al. 2012; Cacho et al. 2005). Research needs to shift towards generating better evidence on a wider range of specific institutional structures and mechanisms that link smallholders to financing opportunities, including expanding the innovative use of information and communication technologies (ICTs) and geo-spatial information. This type of evidence is critical if poor smallholders are to benefit from international mitigation financing. At the country level, many governments are still leery of promises of mitigation financing – and the bureaucracy and conditionalities it brings – and there is a clear need to refine the international institutional mechanisms associated with such financing.

To summarize, the need to address an unprecedented level and magnitude of uncertain change poses a challenge to economic analyses aiming to support agricultural growth and food security, particularly as these changes will clearly differ across regions. Research that will identify methods to improve agricultural resource allocation and management strategies to address emerging climate change patterns, as well as empirical research that will identify the effectiveness of existing management tools in addressing some of the early manifestations of climate change, will be of high value. This research needs to be part of multidisciplinary efforts needed to expand the feasible set of technologies and agronomic management practices, explicitly accounting for decision-making under uncertainty. In addition to technologies and management practices aimed at the farm level, research will also be needed to assess the net benefits from investments in public infrastructure and services, and to evaluate the potential benefits from creating or reforming laws and regulations critical to the agricultural sector, such as those related to public and private land use, as well as the finance, communications and insurance sectors. Research is also needed to understand the role of key institutions in meeting growth objectives while minimizing negative impacts of climate change and securing GHG reductions where possible, and what new institutional forms may be required. Land tenure and property rights, water rights, extension and weather information dissemination services, cooperatives and farmers' unions, and credit and insurance markets are but a few such key institutions. Finally, we emphasize that the responses to climate change may consist both of incremental adaptation, primarily based on scaling up existing technologies and modifying institutions, laws and regulations, and transformative adaptation, including new institutions and major reallocation of resources over space and time. These responses vary in their time dimension and are interdependent (Nelson et al. 2007).

Since policy planning addresses multiple objectives, such as higher incomes, more stable incomes, and lower emissions, one of the key areas of focus is highlighting potential trade-offs in meeting multiple objectives. The goal is to be able to evaluate which policy actions can ameliorate trade-offs and harness synergies amongst the multiple objectives. The latter is particularly important since meeting increasing global food demand and local food security objectives requires continued growth in the agricultural sector. There are a number of potential trade-offs that can arise due to impacts from climate change. For instance, increased frequency of extreme weather events increases the value of policy actions that reduce household vulnerability to such events, but may also compromise strategies to enhance average growth levels of agricultural productivity and farmer incomes. Similarly, policies and public investments to address uncertain longer-term shifts in weather patterns can shift resources away from addressing current poverty alleviation goals. Pursuing low-emissions growth strategies can also involve trade-offs with near-medium term growth objectives, which need to be clearly understood – and externally financed – in order to avoid placing additional burdens on smallholders in developing countries.

Understanding the potential impacts of climate extremes and shifting climate patterns and evaluating how different options and strategies can best address these is a complicated process. As a beginning step, the Climate Smart Agriculture (CSA) concept was developed in order to address the complex issue of how to achieve sustainable agricultural growth for food security under climate change (FAO 2009, 2010; Lipper et al. 2014). The concept calls for integration of the need for adaptation and the possibility of GHG mitigation in agricultural growth and poverty reduction strategies. However there is considerable confusion about what the CSA concept and approach actually involve, and wide variation in how the term is used. At this time, it is critical to build a more formal basis for the CSA concept and methodology and at the same time provide illustrations of how the concept can be applied across a range of conditions. This is the primary focus of this book.

## 2 CSA: The Objectives of the Social Planner

The design of CSA can be analyzed as an economic decision-making problem from the perspective of a social planner. We will not solve the problem formally, but will identify its main features and some of the characteristics of potential solutions. The social planner is concerned with optimizing the welfare of the population they serve, both now and in the future. CSA then is a way of laying out this dynamic optimization problem and its constraints that explicitly incorporates effects of climate change. A plausible objective is maximization of expected discounted welfare associated with agriculture, from a basket of "goods" provided by agriculture. Of course, the agricultural sector is but one sector in the economy, and as noted above, the best option may be to help people transition out of agriculture. Thus, while we emphasize the agricultural sector, other sectors are clearly important. Welfare is comprised of several components. Here we focus on the four pillars of food security: food availability, access, utilization (e.g. food safety), and stability of food supplies. Stability of food supplies is related both to household-level vulnerability as well as resilience of the agricultural system.[2] Finally, we can include environmental objectives, including the global objective to reduce GHG emissions growth as well as local objectives related to improved land quality and water resource management.

The dynamic nature of the optimization problem captures potential trade-offs between choices to improve welfare now versus choices made now to improve welfare under uncertain future outcomes. It also highlights the impacts of uncertainty on decisions made now, and thus the value of additional information and/or the value of choices that increase the flexibility to adapt as more information becomes available. A dynamic framework also enables us to evaluate costs and benefits associated with alternative "weather-migration" scenarios and lower emissions growth strategies.

## 3 The Constraints Facing the Social Planner

When deciding on the extent and means of pursuing avenues for improving welfare outcomes, the social planner must take into consideration constraints in the form of biophysical relationships and behavioral, institutional and political constraints. The biophysical relationships consist of several elements. First is the production function, which links outputs to ecological inputs and weather. One of the key challenges in designing agricultural policies is in understanding the heterogeneous impacts of climate change on productivity. Furthermore, modeling of the production function needs to consider both continuous as well as discrete variables. This approach allows us to investigate technology adoption in response to climate change (Mendelsohn and Dinar 1999; Antle and Capalbo 2010; Arslan et al. 2015). Understanding the stochastic nature of the production function, particularly due to weather realizations, will also be important in designing programs, such as insurance and inventory, to address the challenges of climate change. The second biophysical element is the externality function, which expresses the relationships between economic activities and the various externalities generated by them

---

[2] We basically adopt the IPCC WGII AR5 definitions of vulnerability and resilience, as provided in Appendix 1. However, for conceptual convenience, we are defining vulnerability as a household-level characteristic, and resilience as a system-level characteristic.

(Zilberman 2014). In the context of CSA, the greenhouse gas emissions are the main, but not sole, externality considered. Various agricultural practices and investments also generate both positive and negative local externalities. Overuse of inorganic fertilizer generates greenhouse gas emissions and can also pollute local water sources (Norse 2012). Investment in soil and water conservation structures at the farm and ecosystem levels can generate positive spillover benefits to neighboring farmland productivity (Mirzabaev et al. 2015; McCarthy et al. 2012). Without effective coordination and collective action, too few positive spillovers, and too many negative spillovers, will be generated.

In analyzing both the production and externality functions, we recognize that agriculture is very diverse, and different sectors of agriculture (e.g. irrigated agriculture, rain-fed agriculture, etc.) will experience climate change differently. Livestock husbandry and fisheries will have unique challenges as well, and our analysis should strive to provide appropriate solutions that recognize specific contexts.

The behavioral constraints include market choices made by risk-averse individual agents (both inputs and outputs) operating in contexts where insurance markets are very thin or entirely absent. Our analysis will emphasize the importance of climate conditions on the supply and demand of various goods. The choices will be dependent on risk preferences and market conditions, as well as government policies. An important category of behavioral choices relates to decisions regarding technology adoption, including irrigation, seed varieties and production practices. Almost all empirical evidence suggests that uninsured risk and uncertainty leads to low levels of adoption of new technologies, and this behavioral constraint must be addressed if hoped-for wide-scale adoption is to be realized (Antle and Crissman 1990; Dercon and Christiansen 2011). Furthermore, adopting any new technology is often itself seen to be risky by the farmer who faces uncertainty about its performance (Foster and Rosenzweig 2010). Zilberman et al. (2012) note that, in addition to risk preferences, the diffusion of technology adoption as an adaptation to climate change will also be a function of heterogeneity in farmers' access to capital, the underlying agro-ecology, and prevailing institutions that can foster or hinder adoption.

Technology adoption and institutional innovations are also a function of political constraints. As Hayami and Ruttan (1971) emphasize, innovations of new technologies are outcomes of economic choices that are responsive to incentives and policies. Thus, the literature on innovation also emphasizes the role of learning in innovation and the evolution of new technologies, which in turn affect adoption. Political economic modeling suggests that government policy is affected by economic conditions as well as environmental and political considerations (Buchanan and Tollison 1984; Shepsle 1992; Rausser et al. 2011). These suggest that individual government policy choice problems are derived from their own political economy constraints so that the decision to implement policies that favor certain technologies over others will be a function of this political calculus. Where political weighting favors high economic growth, for instance, the technologies promoted may conflict both with resilience and low-emissions growth goals, for instance.

In addition to political economy considerations, additional political constraints will bound the range of feasible policy and legal actions to address climate change. Some policy solutions to climate change may not be politically feasible, and realistic policy design must consider feasibility of solutions within various local and global contexts. For example, it will be politically easier and it makes common sense to enact policies that improve human well-being and welfare regardless of climate change. A no-regret constraint may bind the set of policies that would be valuable under certain future conditions to those that also address pressing issues of food security or sustainable land use, thereby satisfying distributional and environmental objectives.

The institutional constraints include input, output and labor markets, property rights and tenure security, information dissemination systems such as agriculture extension and weather forecasting, credit and insurance markets and their regulatory framework, social safety net programs, environmental regulations, and the international trading system and local import, export, and foreign direct investment regulations. The institutional environment has a significant impact on farmers' incentives and ability to invest in agriculture practices with CSA characteristics and to adapt to climate change. Thin value supply chains limit farmers' ability to access inputs in timely fashion, and sell their output at a profit. Integrated supply chains can significantly reduce market price swings in response to extreme weather events, thereby reducing vulnerability of rural households to poor crop output and high food prices (Reardon and Timmer 2007). As discussed above, thin or absent credit markets, often combined with very limited insurance mechanisms, dampen incentives to make any types of investment on-farm, and limits the choices available to risk-averse farmers to adapt. Similarly, property rights systems that result in tenure insecurity also limit incentives to invest in land (Mirzabaev et al. 2015; Holden et al. 2009).

The ability to adapt to climate change will also be affected by the information dissemination system and farmers' ability to access weather forecasts and longer-term climate predictions and to incorporate that information into adaptation and coping strategies. Additionally, improving the resilience of the agricultural system as a whole will necessitate making investments and coordinating changing practices at scales higher than the household level. The ability to invest in larger-scale infrastructure to improve the resilience of a watershed (Bassist et al. forthcoming), or coordinating investments in tree planting or check dams across many small communities will depend on local property rights, land use regulations and powers of eminent domain, as well as environmental regulations. The ability to coordinate actions across communities will also be affected by collective active institutions and local-level governance structures (Meinzen-Dick et al. 2004; Pender et al. 2006). The ability to relax institutional constraints will be key in reducing household vulnerability and increasing system resilience in many contexts.

The optimization problem has several dynamic constraints as well. The first constraint is the dynamics of climate change. Because of the nature of agriculture, it is important to have an adequate assessment of climatic variation over space and time in order to make predictions of yields and outputs. There is much uncertainty in climate modeling and it must be incorporated into policy design. Thus, it is not sufficient to get average predictions of climatic patterns over time, but also some indication of variability and reliability thereof. Uncertainty of weather patterns is important because as Dixit and Pindyck (2001) suggested, the pattern and levels of uncertainty delay the optimal timing of investment. With uncertainty, decision-makers value additional information and are willing to wait some time for more information, which can lead to significant delays in investments. This compounds risk-averse farmers' disincentives to invest in land or adopt new technologies.

A second dynamic element is population growth, which affects demand for food as well as urbanization patterns, both of which are important determinants of optimal agricultural growth pathways. Human population growth is also behavioral to some extent and thus population dynamics must take account of behavioral parameters. Furthermore, population dynamics are subject to uncertainty so we must consider outcomes under several scenarios in assessing and designing climate change policies.

The third dynamic element is the ongoing transition in agriculture associated with globalization and the spread of information and technological advances. Global supply chains are spread everywhere, and the expanded use of the internet, cell phones, and improved transportation mechanisms are likely to continue. Technological change is especially important given the role of innovation and adoption in adaptation to climate change, but its diffusion will be a function of both political constraints as well as the need to adapt technologies to site-specific characteristics. One also needs to understand the workings of the supply chain innovations in different regions and how they can be utilized to introduce new technologies in response to climate change. While further integration and connectivity can increase agricultural system resilience by reducing, pooling and transferring risks, positive results will nonetheless be a function of the international and national level regulatory frameworks. To achieve food security objectives, such frameworks need to incorporate regulations that limit monopolistic/oligopolistic power and instead harness the risk-reducing benefits for everyone in the agricultural system, as well as effective enforcement mechanisms.

## 4 The Social Planner's Choice Set

Returning to the social planner's problem - to maximize constrained expected welfare - the social planner can take actions at the system level, or actions that alter incentives for farmers and other actors in the agricultural sector to adopt technologies and practices that improve welfare outcomes. With respect to system-level actions, the social planner can invest in providing a wide range of public goods that improve welfare and increase system resilience in the face of climate change, including: investing in CSA research and development; investing in large-scale infrastructure projects to increase system resilience to climate extremes and longer-term changes in weather patterns such as irrigation systems and flood control structures; investing in weather information systems; investing in disaster risk management systems, including restructuring social safety-net programs to explicitly incorporate payouts related to climate disasters; and, creating or amending laws and regulations regarding property rights, land use and zoning, contract farming, and insurance markets. At the system-level, improved risk coping measures include the design and implementation of disaster risk management plans at various government scales, rapid repair of damaged infrastructure, and, development of insurance instruments targeted for national and municipal governments.

Reducing household vulnerability and increasing system resilience can be accomplished through expansion and promotion of *ex ante* risk management strategies and/or *ex post* coping strategies. At the household level, *ex ante* risk management strategies include adopting SLM techniques; irrigation; drought, heat and/or flood resistant crop varieties and livestock breeds; and, diversifying land and labor activities. Measures that can be undertaken to improve the capacity of farm households to cope with shocks when they do occur include access to social safety net programs, access to attractive insurance instruments, and access to information and infrastructure to re-allocate labor to less affected areas. With respect to actions that affect farmers' incentives, potential actions include payment for environmental services programs; direct subsidies for adoption of certain investments and/or practices such as irrigation or SLM practices; and subsidies for inputs or participation in insurance schemes.

The social planner can also undertake actions to increase adaptive capacity and to pursue least-cost strategies of adaptation under an uncertain future climate, including the possibility of "weather migration". Adaptive capacity is a function of available risk management and risk coping mechanisms, but also includes broader measures to improve decision-making under uncertainty. Uncertainty increases the value of putting in place sophisticated monitoring and evaluation systems and continual learning (IPCC 2012) Greater adaptive capacity is associated with increasing the range of options to manage climate extremes and potentially changed climate patterns, and increasing the ability to exercise those options when needed. It should be stressed that the ability to exercise options when needed is often as critical as having options to begin with. For instance, many researchers find that it is precisely wealthier farmers who are more able to diversify their income sources, reconfirming longstanding findings in most sub-Saharan African countries (Davis et al. 2014; Arslan et al. 2015). So, allocating labor off-farm in response to a weather shock means not only that there are labor opportunities somewhere in the country, but also that farmers know where those opportunities are, can afford transportation, and have sufficient skills to be hired.

Resilience and adaptive capacity are complementary traits. Greater adaptive capacity can increase a system's capacity to recover from swings in climatic and biophysical conditions. But when the pressures exceed some threshold, adaptive capacity can also enable systems to change completely, to adapt through structural transformation, thereby enabling the people to survive and even flourish. Similarly, greater adaptive capacity can enable farm households to reduce vulnerability, but at some point, the best option may be for at least some family members to leave the agricultural sector or diversify their livelihood in order to best adapt to changing climate conditions. At the system-level, adaptive capacity will also be required to address potential mass migration from areas no longer suitable for agricultural production.

The above discussion on adaptive capacity and adaptation captures a major potential trade-off between pursuing strategies that enable farmers to improve their well-being in the face of climate change within the current agricultural system versus strategies that allow for the system itself to change in response to climate change e.g. the difference between incremental and transformative adaptation strategies (Adger et al. 2014). Insurance and safety net payments are classic examples of policies that enable people to better withstand extreme events within the current system. Access to irrigation, improved tenure security, and investments in flood control infrastructure all have similar impacts. In certain circumstances, particularly changes in weather patterns that make current production systems impossible or unprofitable, the social planner will have to determine whether to continue pursuing incremental strategies, or whether to accommodate and manage migration or promote a structural transformation in the production system.

Finally, the social planner can assess opportunities for pursuing low-emissions growth strategies. Certain practices, such as most sustainable land investments and practices, can generate both greater food security and lower emissions, though as noted above, current incentives are too low to foster wide-spread adoption in many countries. Low-emissions growth strategies that pose greater trade-offs with both immediate and long-term food security objectives require international financing, particularly given that most developing countries have contributed very little to cumulative GHG emissions. Where suitable and/or external financing is available, adaptive capacity will need to be built to foster a switch to low-emissions agricultural growth strategies.

## 5 Towards a Socially Optimal Solution: Expected Features of Model Outcomes

Optimizing welfare over multiple objectives that include all four elements of food security and potentially reduced GHG emissions first implies that the impacts of any potential policy action be evaluated for each objective, with the aim of identifying synergies and trade-offs. And, by inserting alternative solutions to this constrained optimization problem, we are able to evaluate their relative merits by comparing the balance of outcomes across a range of objectives from each of these proposed solutions, under a wide range of climate change scenarios. Evaluating outcomes across the multiple objectives will highlight the role of weighting these objectives in arriving at a solution, particularly where there are trade-offs. Assigning weights is a necessary step toward defining a socially optimal solution. The modeling exercise provides a framework for highlighting these weighting choices and can thus feed into climate change policy debates at national and international levels.

A second important outcome of this model is the implication that shadow prices of various constraints will allow us to consider alternative policies by changing the constraints and parameters of the system. The most valuable reforms are implied by the solution to the constrained optimization problem and resulting shadow prices. Business-as-usual scenarios can then be contrasted with scenarios under various types of policy reform that relax various constraints, which may induce either incremental or transformative changes.

This formulation provides us a starting point for our analysis and the type of solutions and research needed to inform it. Because of the increased importance of uncertainty, the solution strategy to this problem will involve adaptive learning. The decision makers have the capacity to learn from the past—and improve their estimation of key parameters over time as knowledge is accumulated—so data accumulation and learning will be part of the policy making process, and decision-makers may experiment with various policies to learn more about the system and its constraints. The random pressures on the system give rise to incentives to invest in adaptive capacity—solutions that will allow decision making to respond effectively to a wide range of potential outcomes. Adaptive capacity may include the ability to learn, analyze, and respond effectively. In many situations, it may be through increasing flexibility and adaptability of institutions, capital goods, and the population through enhancing human capital and reducing transactions costs associated with re-allocating resources (e.g. labor, money, goods), including effective information systems that reach all actors in the system.

## 6 Concluding Comments

In this chapter, we have attempted to lay out a conceptual framework to underpin the CSA concept rooted in agricultural development economic theories and concepts. We began by highlighting the key features of climate change that require a shift in emphasis in research, and for innovations in technologies, institutions, and government policies and programs. These changes include: (1) increased frequency and intensity of climate extreme events, with potentially disastrous impacts on already vulnerable smallholders dependent on rainfed agriculture, (2) permanent changes in weather patterns making certain areas unsuitable for agricultural production under existing conditions, and (3) the need to reduce emissions from the agricultural sector as a whole, while ensuring growth in the sector. These changes strongly highlight the need to consider the heterogeneity of impacts and to understand the implications of decision-making under uncertainty. They also point to the increased value of an expanded set of technological and institutional options to deal with both heterogeneity and uncertainty, and particularly to the increased value of flexibility broadly understood.

To set the framework, we began by viewing CSA as a welfare optimization problem. The problem has multiple objectives, namely the four pillars of food security, food availability, accessibility, utilization, and stability, as well as reducing emissions growth in the sector as a whole. The problem is also characterized by current constraints that bound the feasible outcomes, including bio-physical, behavioral, political, institutional and distributional constraints. Achieving better outcomes can occur by directly increasing food security, for instance by introducing technologies that increase yields and reduce yield losses in extreme years. Or, better outcomes can be achieved by relaxing key constraints. We also stress that the nature of the optimization, and thus adaptation strategies, are context specific.

Adaptation to climate change may take several forms: innovation and adoption of new technologies, adoption of existing technologies, temporary or permanent migration, changes of agricultural activities and trade patterns, and increased range of attractive and viable insurance products. Adaptation in most cases will also include addressing institutional failures and constraints such as reducing tenure insecurity, increasing access to relevant information, and improving the ability to coordinate actions across a watershed or ecosystem. And, some adaptation strategies will imply a discrete system-level change realized through broad-based structural transformation. While the solution cannot provide the exact changes in technologies or institutions that would result in the best outcomes, it can help to define the characteristics, or principles, associated with improved technologies or highly effective institutional structures and mechanisms.

Finally, we highlight that the solution to the social planner's problem for climate change must balance adaptation and responsiveness to uncertain climate change with the needed growth and food security objectives of the agricultural sector. Weighting the multiple objectives is essentially a political process.

## References

Adger, W.N., J.M. Pulhin, J. Barnett, G.D. Dabelko, G.K. Hovelsrud, M. Levy, Ú. Oswald Spring, and C.H. Vogel. 2014. Human security. In: *Climate Change 2014: Impacts, Adaptation, and Vulnerability. Part A: Global and Sectoral Aspects*. Contribution of Working Group II to the Fifth Assessment Report of the Intergovernmental Panel on Climate Change [Field, C.B., V.R. Barros, D.J. Dokken, K.J. Mach, M.D. Mastrandrea, T.E. Bilir, M. Chatterjee, K.L. Ebi, Y.O. Estrada, R.C. Genova, B. Girma, E.S. Kissel, A.N. Levy, S. MacCracken, P.R. Mastrandrea, and L.L. White (eds.)]. Cambridge University Press, Cambridge, United Kingdom and New York, NY, USA, pp. 755–791.

Alix-Garcia, J.M., Shapiro, E.N. and Sims, K.R., 2012. Forest conservation and slippage: Evidence from Mexico's national payments for ecosystem services program. *Land Economics*, 88(4), pp.613–638.

Antle, J.M. and S.M. Capalbo. 2010. Adaptation of agricultural and food systems to climate change: an economic and policy perspective. *Applied Economic Perspectives and Policy*, doi: 10.1093/aepp/ppq015.

Antle, J.M. and B. Diagana. 2003. Creating incentives for the adoption of sustainable agricultural practices in developing countries: the role of soil carbon sequestration. *American Journal of Agricultural Economics*, 85(5): 1178–1184.
Antle, J.M. and C.C. Crissman. 1990. Risk, efficiency, and the adoption of modern crop varieties: Evidence from the Philippines. *Economic Development and Cultural Change*, 38(3):517–537.
Arslan, A., McCarthy, N., Lipper, L., Asfaw, S., Cattaneo, A. and Kokwe, M. (2015): Climate Smart Agriculture: Assessing the Productivity and Adaptation Implications in Zambia. *Journal of Agriculture Economics*, 66(3): 753–780.
Barbier, E.B. 2010. Poverty, development, and environment. *Environment and Development Economics*, 15(06): 635–660.
Barrett, C.B., F. Place, and A. Aboud. 2002. The challenges of stimulating adoption of improved natural resource management practices in African agriculture. In: C. Barrett, F. Place, and A. Aboud (eds), *Natural Resources Management in African Agriculture*. Nairobi, Kenya: ICRAF and CABI.
Bassist, A., B. Blankespoor, A. Dinar, and S. Dinar. 2017. Assessing Technical, Economic and Policy Aspects of Water Scarcity Using Surface Wetness with Application to the Zambezi, Mekong and Red River Basins. In: L. Lipper, N. McCarthy, D. Zilberman, S. Asfaw, and G. Branca: *Climate Smart Agriculture - Building Resilience to Climate Change*. New York: Springer.
Buchanan, J.M., and R.D. Tollison. 1984. *The Theory of public choice--II*. University of Michigan Press.
Burney, J.A., S.J. Davis, and D.B. Lobell. 2010. Greenhouse gas mitigation by agricultural intensification. *Proceedings of the national Academy of Sciences*, 107(26): 12052–12057.
Cacho, O.J., Marshall, G.R. and Milne, M. 2005. Transaction and abatement costs of carbon-sink projects in developing countries. *Environment and Development Economics*, 10(05): 597–614.
Davis, B., S. Di Giuseppe, and A. Zezza. 2014. Income Diversification Patterns in Rural Sub-Saharan Africa: Reassessing the Evidence. World Bank Policy Research Working Paper No. 7108. Available at SSRN.: http://ssrn.com/abstract=2524162
Dercon, S., and L. Christiaensen. 2011. Consumption risk, technology adoption and poverty traps: Evidence from Ethiopia. *Journal of development economics*, 96(2): 159–173.
Dixit, A.K., and R.S. Pindyck. 2001. The options approach to capital investment. *Real options and investment under uncertainty: Classical readings and recent contributions*, pp. 61–78.
FAO. 2009. Food Security in Agricultural Mitigation in Developing Countries: Options for Capturing Synergies. October, 2009. Rome: FAO.
FAO. 2010. Climate-Smart Agriculture: Policies, Practices and Financing for Food Security, Adaptation and Mitigation. Rome, FAO.
Fischer, E. M., and R. Knutti. 2015. Anthropogenic contribution to global occurrence of heavy-precipitation and high-temperature extremes. *Nature Climate Change* 5: 560–564, doi:10.1038/nclimate.
Foster, A.D. and M. R. Rosenzweig. Microeconomics of technology adoption. *Annual Review of Economics* 2: 2010, doi: 10.1146/annurev.economics.102308.124433.
Gollin, D., Parente, S. and Rogerson, R. 2002. The role of agriculture in development. *The American Economic Review*, 92(2): 160–164.
Haggblade, S., Hazell, P.B. and Reardon, T. eds. 2007. *Transforming the rural nonfarm economy: Opportunities and threats in the developing world*. Washington, DC: IFPRI.
Hayami, Y., and V.W. Ruttan. 1971. Agricultural development: an international perspective. *Agricultural development: an international perspective*. Baltimore, MD/London: The Johns Hopkins Press.
Hitz S., and J. Smith. 2004. Estimating global impacts from climate change. *Global Environmental Change* (Part A) 14(3): 201–218.
Holden, S.T., K. Deininger, and H. Ghebru. 2009. Impacts of low-cost land certification on investment and productivity. *American Journal of Agricultural Economics* 91(2): 359–373.

IPCC. 2012. Managing the Risks of Extreme Events and Disasters to Advance Climate Change Adaptation. A Special Report of Working Groups I and II of the Intergovernmental Panel on Climate Change. Field, C.B., V. Barros, T.F. Stocker, D. Qin, D.J. Dokken, K.L. Ebi, M.D. Mastrandrea, K.J. Mach, G.-K. Plattner, S.K. Allen, M. Tignor, and P.M. Midgley (eds.). Cambridge, UK, and New York, NY: Cambridge University Press.

Lal, R. 2004. Soil carbon sequestration impacts on global climate change and food security. Science 204: 1623–1627.

Lipper, L., P. Thornton, B.M. Campbell, T. Baedeker, A. Braimoh, M. Bwalya, P. Caron, A. Cattaneo, D. Garrity, K. Henry, R. Hottle, L. Jackson, A. Jarvis, F. Kossam, W. Mann, N. McCarthy, A. Meybeck, H. Neufeldt, T. Remington, P. Thi Sen, R. Sessa, R. Shula, A. Tibu, and E. Torquebiau. 2014. Climate-smart agriculture for food security. *Nature Climate Change*, 4(12): 1068–1072.

Lockie, S. 2013. Market instruments, ecosystem services, and property rights: assumptions and conditions for sustained social and ecological benefits. *Land Use Policy*, 31: 90–98.

McCarthy, N., L. Lipper, and G. Branca. 2012. Climate-smart agriculture: smallholder adoption and implications for climate change adaptation and mitigation. *Mitigation of Climate Change in Agriculture Working Paper* 3.

Meinzen-Dick, R., M. DiGregorio, and N. McCarthy. 2004. Methods for studying collective action in rural development. *Agricultural systems* 82(3): 197–214.

Mendelsohn, R., and A. Dinar. 1999. Climate change, agriculture, and developing countries: does adaptation matter? *The World Bank Research Observer* 14(2): 277–293

Nelson D.R., W.N. Adger and K. Brown. 2007. Adaptation to environmental change: contributions of a resilience framework. *Annual Review Environment and. Resources*, 32:395–419.

Norse, D. 2012. Low Carbon Agriculture: Objectives and Policy Pathways. *Environmental Development*, 1(1): 25–39.

Nkonya, E., J. Pender, P. Jagger, D. Sserunkuuma, C. Kaizzi, and H. Ssali. 2004. *Strategies for sustainable land management and poverty reduction in Uganda*. Research Report 133. Washington, DC: IFPRI.

Paustian, L., B. Babcock, J.L. Hatfield, R. Lal, B.A. McCarl, S. McLaughlin, A. Mosier, C. Rice, G.P. Roberton, N. Rosenberg, and C. Rosenzweig. 2004. Agricultural mitigation of greenhouse gases: science and policy options. In: *2001 Conference Proceedings, First National Conference on Carbon Sequestration*. Washington, DC: Conference on Carbon Sequestration.

Pender, John, Frank Place, and Simeon Ehui, eds. 2006. *Strategies for sustainable land management in the East African highlands*. Washington, DC: IFPRI.

Rausser, G.C., J. Swinnen, and P. Zusman. 2011. Political power and economic policy: theory, analysis, and empirical applications. Cambridge, UK and New York, NY: Cambridge University Press.

Reardon, T. and C.P. Timmer. Transformation of markets for agricultural output in developing countries since 1950: How has thinking changed? *Handbook of agricultural economics*, 3(2007): 2807–2855.

Scott, Z., Wooster, K., R. Few, A. Thomson, and M. Tarazona. 2016. Monitoring and evaluating disaster risk management capacity. *Disaster Prevention and Management* 25 (3): 412–422.

Seeberg-Elverfeldt, C., S. Schwarze, and M. Zeller. 2009. Payments for environmental services–Carbon finance options for smallholders' agroforestry in Indonesia. *International Journal of the Commons*, 3(1).

Shepsle, K.A. 1992. Congress is a "they," not an "it": Legislative intent as oxymoron. *International Review of Law and Economics* 12(2): 239–256.

Taylor, J.E. and P.L. Martin. 2001. Human capital: Migration and rural population change. *Handbook of agricultural economics*, 1(2001): 457–511.

Thornton, P.K., P.G. Jones, P.J. Ericksen, and A.J. Challinor. 2011. Agriculture and food systems in sub-Saharan Africa in a 4°C+ world. *Philosophical Transactions of the Royal Society A: Mathematical, Physical and Engineering Sciences* 369(1934):117–136.

Zilberman, D., Jinhua Zhao, and Heiman, A. 2012. Adoption versus adaptation, with emphasis on climate change. Annual Review of Resource Economics, 4(1): 27–53.

Zilberman, D. 2014. The economics of sustainable development. *American Journal of Agricultural Economics*, 96 (2): 385–396.

**Open Access** This chapter is distributed under the terms of the Creative Commons Attribution-NonCommercial-ShareAlike 3.0 IGO license (https://creativecommons.org/licenses/by-nc-sa/3.0/igo/), which permits any noncommercial use, duplication, adaptation, distribution, and reproduction in any medium or format, as long as you give appropriate credit to the Food and Agriculture Organization of the United Nations (FAO), provide a link to the Creative Commons license and indicate if changes were made. If you remix, transform, or build upon this book or a part thereof, you must distribute your contributions under the same license as the original. Any dispute related to the use of the works of the FAO that cannot be settled amicably shall be submitted to arbitration pursuant to the UNCITRAL rules. The use of the FAO's name for any purpose other than for attribution, and the use of the FAO's logo, shall be subject to a separate written license agreement between the FAO and the user and is not authorized as part of this CC-IGO license. Note that the link provided above includes additional terms and conditions of the license.

The images or other third party material in this chapter are included in the chapter's Creative Commons license, unless indicated otherwise in a credit line to the material. If material is not included in the chapter's Creative Commons license and your intended use is not permitted by statutory regulation or exceeds the permitted use, you will need to obtain permission directly from the copyright holder.

# Innovation in Response to Climate Change

### David Zilberman, Leslie Lipper, Nancy McCarthy, and Ben Gordon

**Abstract** Climate change impacts on agriculture are varied over space and time. The effects are heterogeneous and highly uncertain. Innovation in agriculture is clearly an important response for effective and equitable adaptation and mitigation – and we need to rethink how to promote innovation to address the heterogeneity and uncertainty of climate change impacts. In moving towards climate smart agricultural (CSA) systems in developing and developed countries, innovation will be key. For CSA we will need greater resilience in agricultural systems and also greater efficiency of resource use for both adaptation and mitigation. Technological innovation will need to play a key role – but its not enough. Managerial and institutional innovations are likely to be even more important in dealing with the heterogeneous and uncertain impacts of climate change. Innovation can complement other forms of adaptation to climate change to form CSA practices. In particular innovation can enhance technology adoption, may prevent or facilitate migration of production/population, enhance trade & aid, and increase efficiency of insurance & feasibility of inventories. We discuss their main features and the nature of innovation needed to align these actions with a CSA strategy.

## 1 Introduction

The evolution of agriculture in the future will be shaped by its response to climate change. Farmers need to adapt their practices to accommodate climatic conditions, and agricultural activities will need to be modified to reduce greenhouse-gas (GHG)

---

D. Zilberman (✉) • B. Gordon
Department of Agriculture and Resource Economics, University of California Berkeley, Berkeley, CA, USA
e-mail: zilber11@berkeley.edu; benjamingordon@berkeley.edu

L. Lipper
ISPC-CGIAR, Rome, Italy
e-mail: leslie.lipper@fao.org

N. McCarthy
Lead Analytics Inc., Washington, DC, USA
e-mail: nmccarthy@leadanalyticsinc.com

emissions. But climate change is only one of the major forces that will change the future of agriculture. Others include population growth and increases in income as well as changes in human capital, knowledge, and infrastructure. Much of the change in agriculture will stem from new innovations, both in terms of technologies and institutions.

This paper aims to provide the background and analyze some of the challenges associated with the development and introduction of new innovations in agriculture and food systems in response to climate change. The analysis will emphasize the role of innovations in CSA. The first section will provide an overview of the impact of climate change and possible mechanisms in response to it. The next section will identify the major categories of innovation associated with CSA. We distinguish between technological, managerial, and institutional innovations and between micro (farm level) vs. macro (farm-system) innovations. This will be followed by a discussion of the barriers to introduction faced by these innovations, and a conclusion.

## 2 The Impact of Climate Change on Agriculture and the Implications

The research on climate change has identified several avenues that will affect agriculture. They include (1) rising temperatures around the world that lead to migration of climate from regions closer to the tropics to regions closer to the poles, (2) rising sea levels, (3) increased snowmelt and change in the volume and timing of water use for irrigation, and (4) increased probability of extreme events. We will next analyze the implications of each of these events and what they imply for the evolution of agricultural systems focusing on innovations, which are a crucial component for adaptation to climate change (Stern 2006).

### 2.1 Rising Temperatures and Migrating Weather

Depending on the range of mitigation actions taken in the next decades, we can expect that climate change will lead to increased temperatures throughout the world by 1–3 °C, which is equivalent to a shift of 300–500 km of weather patterns away from the equator and towards the poles. Similarly, temperature variability in regions at higher altitudes will also increase (Ohmura 2012). While climate change may have negative overall impact on agricultural production, the distributional impacts are much more substantial than the aggregate affect. Thus, for instance, some warm agricultural areas in Texas, Oklahoma, Mexico, and Western Africa will become unviable for crop production. While at the same time, regions in Russia, Canada, and even the Arctic will become suitable for agricultural production. Innovations to respond to changes in temperature may involve adopting new crops and varieties in some areas, to migration away from regions unviable for agricultural production in others, or investment in infrastructure and other activities in new regions. The effect of weather migration will not be limited to plants, but rather felt across multiple species. For example, temperature serves as an important barrier to prevent pest infestations and while insects and other pests can move in response to changing conditions, trees are stationary. Pest migration can endanger viable tree-based economies and will require monitoring and interventions (Porter et al. 1991). The people displaced because of these trends may not be the ones that are able to take advantage of new opportunities presented by climate change. Development of new technologies and other economic activities to facilitate adaptation to climatic changes and amelioration of painful displacement will be valuable. Innovations to adapt to migration of weather will vary across location reflecting spatial heterogeneity. In some areas, new solutions will be required to address movement of pests as well as to modify crop varieties to adjust to changing weather conditions. In other areas, entirely new crops may need to be introduced. Finally, in some regions mechanisms may need to be introduced to facilitate out migration of people. The design and implementation of these solutions is challenged due to uncertainty about magnitude and timing of change.

## 2.2 Rising Sea Levels

Sea level rise (SLR) may lead to loss of high value agricultural land as well as important infrastructure that is crucial for exporting and importing food in many regions throughout the world. An estimated 10% of the world's population lives in coastal zones (i.e. at less than 10 m altitude), with wide variation in share of population by country, representing 14% of global GDP (McGranahan et al. 2007). Most notably, close to half of Vietnam, Bangladesh, and Egypt's populations live in these zones, while China and India, with a far smaller portion of overall population, contain over 200 million people living in these zones. The population impacted by SLR will vary significantly by actual rise in sea level – from 56 million people (1.28% of world population) with a 1-m rise to 245 million (5.57%) with a 5-m rise (Dasgupta et al. 2009). Also, large tracts of prime agricultural land will be threatened by rising sea levels especially in tropical regions (Kurukulasuriya and Rosenthal 2013). Given heterogeneity across location, it is important to develop location specific solutions. In areas especially vulnerable to SLR, transformational innovation may be required rather than incremental approaches in order to spur adaptation and protect vulnerable populations (Kates et al. 2012). In few areas, vulnerable coastal regions may be saved by investment in protective infrastructure (e.g. dikes, dams), but in many cases vulnerable areas will need to be abandoned causing problems of displacement. In some areas, there may be opportunities to adopt different types of agricultural production, but these will require innovation.

## 2.3 Increased Snowmelt and Timing of Irrigation

In addition to changes in precipitation patterns, increased temperatures will increase snowmelt, decreasing the possibility of using water stored in snow accumulated during the wet season to be available for irrigation during the dry season. Furthermore, the likelihood of flooding may increase. Given the relative importance of irrigated agriculture during dry seasons in many parts of the world, this change may have significant impact on food supply, unless some remedial measures are taken. These solutions are dependent on the conditions at each location. Solutions may include investment in new forms of water inventories and storage, for example dams for flood control and storage as well as diversion of water to underground reservoirs. These changes may also prompt changes in crop timing and selection to adjust to water availability. Furthermore, changes in water availability may also affect availability of hydroelectric power for irrigation, which will also affect agricultural supply (Xie et al. 2015). Thus climate change will prompt re-arrangement and new management of agricultural water supplies (Grafton et al. 2013; Chartzoulakis and Bertaki 2015; Basist er al. forthcoming). The substitution of snow as water storage will require significant investment under conditions of uncertainty and require innovative approaches to financial, institutional, and physical structures applying and extending the option-value approach of Dixit and Pindyck (1994).

## 2.4 Increased Probability of Extreme Events

In addition to the changes in average temperature as well as water availability, climate change is likely to shift the climatic distribution that will increase the probability of extreme events, such as heatwaves, heavy rainfall, storms and coastal flooding. Furthermore, climate change is a gradual process. While average conditions may be changing gradually, there may be increased variability of climatic conditions (Fischer and Schär 2009). There is already evidence of such changes and they require a higher degree of resilience of farmers to fast changing conditions. This requires both innovative efforts in terms of new technologies and management practices, as well as capacity to adopt these technologies and thus enhance resilience.

Furthermore, there is a risk of climate change triggering a tipping point that will lead to abrupt and irreversible changes that increase in severity with rising temperature (IPCC 2014; Barnosky et al. 2012). Such very low probability catastrophic events may include, for example, drastic rise in temperature (of 6 °C and beyond) because of sudden release of methane gas resulting from the loss of permafrost (Lenton et al. 2008). Such extreme events may devastate agriculture throughout much of the world. Nevertheless there is a need for continued research to develop agricultural production and storage systems suitable for more extreme climate conditions as well as institutions for emergency responses that include movement of people and other living creatures and relocation of resources.

## 2.5 Discussion

As emphasized above, the nature of innovative responses to climate change impacts need to adapt to two characteristics of these impacts. The first is *heterogeneity*. Different regions are affected differentially by climate change: for some desert or low-lying coastal region climate change may be devastating, while for other cold region, climate change may be perceived as "climate improvement". These differences in impacts, as well as differences in gains and losses from engagement in mitigation activities, may contribute to the diverse responses and willingness to participate and contribute to coordinated efforts to avert or slow climate change. Weitzman (2009) studies the economic significance of catastrophic climate change and argues that regardless of the differential impacts of likely climate change scenarios on various regions, humanity as a whole needs to take action to prevent some low probability catastrophic outcomes.

The second factor that affects engaging in action addressing the climate change challenges is *uncertainty*. The timing, magnitudes and locations of different impacts of climate change are not known with certainty. At the same time, there is a wide body of literature that suggests that farmers and other agricultural actors behave in a manner consistent with risk aversion. Sandmo (1971) suggests, in a static framework, that risk aversion reduces the magnitude of actions taken by risk averse enterprises as the risks they face increase. The real option approach of Dixit and Pindyck (1994) argues, within a dynamic setup, that higher uncertainty about future outcomes will lead to a delay of actions. Thus, the uncertainty surrounding the impacts of climate change tend to delay and reduce the magnitude of activities aimed to adapt to and mitigate it. Uncertainty about possible impacts of climate change also increases the need for further research (Dixit and Pindyck 1994) to reduce the uncertainties surrounding climate change.

Heterogeneity and uncertainty will thus increase the difficulty of identifying the full range of responses to climate change from observable data, especially at the present when some of the impacts of climate change (e.g., migration of warm weather toward the pole and a significant rising sea level, triggering of tipping points leading to irreversible changes) are more likely to occur in the longer run—2050 and beyond. Others, for example, that increase the likelihood of extreme events, like flood and droughts, might have already started to occur and are more likely in the near future.

The investment in innovative activities to address the challenges of climate change will evolve over time as knowledge accumulates. The innovative approach must consider new technological and institutional options but also the changes in behavioral responses to climate change and related solutions over time.

We can learn from the responses thus far on some activities, the capacity to adapt to climate change in the future, and the factors that affect responses. The empirical case studies in these chapters cover lessons that have analyzed responses to climate change thus far and their implications for innovation, including technology adoption and adaptation, insurance schemes, and diversification of land and labor, and to a lesser extent internal migration. While these case studies cover a subset of adaptation options for which there is solid empirical evidence in developing country contexts, there is a broader range of adaptation activities that we will also cover, including external migration, use of trade and aid policies, and physical inventories.

## 3 Innovations for Climate Smart Agriculture

There are many ways to categorize innovations (Sunding and Zilberman 2001). Economic growth theory distinguishes among technologies depending on their impact on inputs and outputs. For example, distinctions can be made between capital saving, labor saving, quality improving, and risk reducing innovations. Another way of distinguishing innovations is according to their form, e.g. technological, managerial, and institutional innovations. Technological innovations are embodied in new machinery, and can be further divided into mechanical (e.g. tractors), biological (e.g. seeds), and chemical (e.g. fertilizers) innovations. Managerial innovations are not embodied in physical capital, but rather are described by better practices such as Integrated Pest Management, improved pruning techniques, and crop rotation. Institutional innovations may include new organizational forms (e.g. cooperatives) and arrangements for trading (e.g. future markets and contract farming). Because of the heterogeneity and randomness of climate change impacts, there are several types of innovation that will be especially valuable, and the following section outlines many of these innovations. Below we present and analyze the innovations that are likely to be required to adapt to climate change. We classify them in three categories: technological innovations, managerial innovations and institutional innovations. The technological and managerial innovations are divided into micro–farm level innovations and macro-farm system innovations. All the institutional innovations we consider are at the macro level.

### *3.1 Technological Innovations*

#### 3.1.1 Micro, Farm-Level Approaches

*Resilient crops and livestock* Because of rising temperatures and increased variability, development of new crop varieties and livestock breeds that can tolerate these changes will be very important. Due to the frequency of change, it will be important to detect change and develop genetic material that can adapt to this change relatively fast.

*Pest control* The migration of pests may prompt the need to develop new pest management techniques, which are both environmentally friendly, cost-effective, easy to use, and efficacious. A diverse approach utilizing biological, mechanical, and chemical control, in concert with genetic approaches, will be needed. An on going effort to identify emerging pest problem will need to guide the development these pest control innovations.

*Input use efficiency enhancing technologies* Frequently, there is a significant gap between the level of applied inputs and the amount utilized by the crop. For example, with flood irrigation, input use efficiency may be 50%, but with technologies like drip irrigation, efficiency may increase to 90%. Frequently the residue (i.e. the input not taken up by the crop) is a source of externalities. Khanna and Zilberman (1997) suggest that adoption of input use efficiency enhancing technologies tend to increase yield, save input, and reduce pollution. Better application technologies may reduce water, fertilizer, and chemicals while reducing the side effect associated with their use. The notion of input use efficiency enhancing technologies applies to crops and even livestock. Some crop varieties may increase output while the change in feeding regimes for livestock may decrease greenhouse gas emissions.

*On-farm storage* Parfitt et al. (2010) suggest that there is significant post-harvest loss on the farm and much of it occurs among subsistence farmers in developing countries that lack basic storage capacity. Innovative on-farm storage infrastructure can help address yield losses brought on by increased temperature as well as increased frequency of shocks. The challenge is to design systems that are affordable, easy to install and operate, and reliable. The design of the system must address heterogeneity in bioclimatic conditions.

*Higher yield and longer shelf life* Crop varieties, as well as livestock, that increase yield per area tend to reduce agricultural footprint and the effort required to compensate for production loss due to climate change. Longer shelf life would decrease transportation costs, storage costs, and, especially, waste associated with agricultural distribution. Shelf life enhancement is important in the context of climate change because increased temperatures increase the likelihood of spoilage.

*Sustainable Land Management (SLM)* Frequently, agricultural practices in developing countries lead to reduced soil quality. Extreme weather associated with climate change may worsen this problem unless improved agronomic practices are introduced. SLM practices aim to increase yield without degrading soil and water resources. In addition, they aim to sequester carbon. There are already several SLM practices such as organic fertilization, minimum soil disturbance, and incorporation of residues, terraces, water harvesting and conservation, and agroforestry (Branca et al. 2013), but there are many opportunities for developing new SLM practices and refining existing ones to accommodate spatial and climatic variability.

### 3.1.2 Farm System Approaches

*Low-cost flood protection and water storage facilities* Because of the concern of rising water level, and the resulting instability due to floods, innovation that reduces the cost of protection against rising water levels and floods will be a priority. In assessing such investments, it is important to consider the benefit of avoided conflict due to reduced climate migration.

*Weather information distribution technologies* There is significant evidence that availability of weather information, including its implications on irrigation (evapotranspiration losses), enable farmers to modify their irrigation and pest control strategies which lead to significant increases in yield and saving of water and other inputs (Parker and Zilberman 1996). Reliable weather information will be especially important during periods of heightened climate change during which farmers face greater uncertainty of weather patterns. But information about weather systems requires both weather stations as well as delivery systems that provide useful and reliable information across many users. This system must be affordable and fit the needs and capacity of poor farmers.

*Improved mitigation* Reducing GHGs is a key to effective adaptation to climate change in the long run, and an important CSA goal and thus it includes innovation and adoption of cultural practices, crop varieties, management practices, and institutions that will accelerate mitigation. Already, the transition to no- or low-tillage practices has been considered a major source of carbon sequestration, and adoption of higher yield varieties and conservation technologies that reduce the land, atmospheric, and fossil fuel footprint of agriculture is another important mitigation strategy (Lal 2011; McCarthy et al. 2012).

## 3.2 Managerial Innovations

### 3.2.1 Micro, Farm-Level Approaches

The differences between technological and managerial innovations are not clear cut. New machinery or input require innovative management practice to be effective and adopted. Here we will emphasize innovation that mostly emphasize improve management – but may also involve use of new technologies.

*Input use efficiency management techniques* The efficiency of water use or chemical input can be significantly increased through the adoption of information intensive management practices that optimize the timing and quantities of application of inputs. Precision technologies vary variable input application over space and time based improved monitoring of field and weather conditions. Dobermann et al. (2004) suggest that precision farming may save input and/or increase yield and that both mechanisms for monitoring spatial or other sources of variability and methods to utilize this information have a large potential for further improvement. Development of precision techniques for resource poor developing countries is a special challenge as they may be the major beneficiary from these techniques.

*Integrated Pest Management (IPM)* The likely increases in pest pressure because of climate change may require new technical solutions but also increase effectiveness of pest management in terms of detection and coordination of pest control activities. IPM emphasizes measurement of pest pressure and integration of alternative approaches (cultural practices, chemical, genetic modification and biological) to optimize the net benefits of treatment, taking into account pest dynamic and environmental side effects. The adoption of IPM is constrained by the cost of monitoring pests and difficulty of tailor-made IPM approaches specific to bioclimatic conditions (Waterfield and Zilberman 2012). The effectiveness of responses to climate change will benefit from the development of affordable and easy to implement IPM strategies.

*Land use and on-farm management practices* Changes in both the mean and variability of climatic conditions accompanied by changes in technologies and economic conditions will require improved management tools used to facilitate the selection of crop types and crop varieties, allocation of land among crops, and selection and implementation of production practices. The improvement of quality of data, computation capabilities and communication will provide opportunities for introducing new management tools that are affordable and accessible even to small farmers in developing countries.

### 3.2.2 Farm System Approaches

*Local collective action for improved input use and management* Management practices like IPM, SLM and improved input use efficiency require a knowledge base that is shared by many farmers. For example, both IPM and improved water use efficiency rely on weather information that may be collected by regional weather stations. Developing strategies to address crop diseases as well as controlling build-up of resistance to pest control will require collective action. Effective land use management should take into account externalities among crops and other production activities within a region. Therefore, development of regional institutions for collaboration that will allow for the provision of public goods and capturing economies of scale among small producers will be of high value. Poteete, Janssen, and Ostrom (2010) provide multiple forms of institutions to address various collective action challenges in the development context, but different situations may require different solutions and there are many opportunities for innovative institutional designs to address emerging climate change challenges.

*Insurance Products* The decreased stability of weather due to climate change raises the value of risk management strategies. For example, Mendelsohn (2006) suggests that crop insurance can be a good strategy to cope with increased risk. Golden et al. (2007) suggest that using weather derivatives and similar financial instruments can be an effective mechanism to address climate change related risk. The story of Joseph in the Bible illustrates the role of inventory as mitigating weather variability; similarly, there is a large literature on the economics of storage management in agriculture (Williams and Wright 2005) that applies to increased weather instability.

The implementation of insurance as an adaptation mechanism is quite challenging. First, risks associated with climate change are difficult to quantity – risks are dynamic, rather than static, and the parameters of key variables change over time and cannot be predicted reliably (Patt et al. 2009). Furthermore, Millner et al. (2010) suggest that some impacts of climate change cannot be captured well by a standard probability distribution, which makes actuarial computation even more challenging. Second, insurance may affect other adaptation strategies. It may lead to a moral hazard by reducing precautionary activities, while other adaptation strategies may reduce the need for insurance. Thus risk and adaptation strategies must be designed simultaneously (Tol 2009). Third, implementation of insurance may require good monitoring of behavior to overcome adverse selection. The design of mechanisms to adverse selection is especially challenging when distributions of risks are evolving or partially unknown. Finally, agricultural insurance programs have served as rent seeking mechanisms (transferring income) indicating that their efficiency has been questionable (Schmitz 2010; Krueger 1990). Thus, the development of insurance strategies to address climate change must proceed with caution.

*Resilient supply chain management* Design of appropriate supply chains is essential to enhance effective adoption (Lu et al. 2015). Agriculture in developing countries is going through a food system revolution characterized by the introduction of new rationalized supply chains that enable better storage and allow for product differentiation and link farmers in developing countries with super markets (Reardon and Timmer 2012). This modern supply chain led to the adoption of many innovative practices and a substantial effort must exist to enhance supply chains further to allow for coping with the effects of climate change.

## 3.3 *Institutional Innovations*

Institutional innovations occur at the macro, *farm system* level. We can distinguish between two types of institutional innovations: (1) Institutions that will enable innovation processes. Some of these institutions that are part of CSA innovations themselves are discussed in this section. Institutional innovations that address the limitations of the existing systems are discussed in next section on 'Overcoming Barriers to Innovation in the Era of Climate Change'. (2) Institutions that will allow implementing other elements of adaptation strategies besides innovation and adoption.

### 3.3.1 Innovations as Part of CSA Programs

*"Climate Smart" extension programs* Innovations are mostly concepts that present new ways of doing things within a context. To be implemented, innovations must be developed, upscaled, and then tested at the implementation level. A program of

marketing and education is then needed to bring an innovation to practitioners. Different countries have their own innovation systems, which are adapted to different types of innovations and contexts (Nelson 1993). The implementation of CSA may require innovative design of networks that will extend the technology from the scientists to the practitioners and this extension effort should include not only the public extension service, but also private firms, cooperatives, and NGOs.

*Integrated Pest Management at relevant ecosystem scale* Pest control activities generate externalities, especially given the small scale of farms and the movement of pests. These externalities may be positive, for instance through pollination, or negative, for instance through the build-up of resistance. There are some activities that require the full spatial coordination among farmers, such as pest eradication plans (Waterfield and Zilberman 2012). The introduction of CSA pest management programs may require innovative efforts to identify and monitor their possible externalities and develop mechanisms to control them.

*Land use regulations and management at ecosystem scale* Agricultural production have significant environmental externalities, including chemical contamination of bodies of water and soil erosion, as well as damage to ecosystems and wildlife. The introduction of CSA activities without considering and addressing their potential side effects may lead to counter-productive outcomes. Therefore, innovative efforts are required to design systems of education and regulation to design and implement systems of regulation and implementation that will monitor the externalities of CSA and control them.

### 3.3.2 Institutions for Enhancing Various Adaptation Strategies

*Trade regulations* International trade results from differences in relative advantage between regions and is a risk sharing mechanism. Climatic changes and shifts in weather patterns, may result in crop production patterns that will lead to changes in trade. For example, Aker (2012) finds that increases in trade ameliorate the impact of drought in West Africa. A region with a warming climate may switch from growing wheat to corn, export the corn, and import wheat. Changes in trade patterns resulting from climate change may have significant distributional implications. Innovative frameworks that are able to identify new trade opportunities, their implications, and barriers to its implementation will be of importance. The capacity to utilize trade in response to climate change depends on infrastructure (e.g. availability of transportation and processing facilities) as well as international trade policies and institutions (Zilberman et al. 2012). New innovative frameworks can identify, for example, new infrastructure requirements and how to implement them and institutional arrangements that will provide an enabling environment for new trade opportunities.

*Aid distribution mechanisms* While trade is an exchange between two parties, aid is a transfer from one party to another. Even still, aid can play an important role as a mechanism to address risk associated with climate change. Like trade, the capacity of aid to address climate change depends on the availability of efficient transportation as well as accurate detection and response systems (Donaldson 2010). Both aid and trade could serve as substitutes to migration as a response to climate change. Research and development may lead to innovations that enable trade or to mechanisms that facilitate provision of aid in times of crisis while maintaining overall social welfare. Innovative approaches that reduce the cost of implementation and increase the effectiveness of aid mechanisms is especially important given financial constraints on such efforts.

*Movement of water resources (management and conflict resolution)* Climate change may drastically change precipitation patterns, as well as lead to significant melting of snow packs, and thus lead to changes in water availability over space and time, water movement and storage patterns. These changes will occur both within and between countries. It will raise issues of property rights that have to be sought and solved before they lead to conflicts. Furthermore, the institutions that currently own and distribute water will lose capacity, and some of them will get into severe financial troubles, as they would not be able to meet their obligations. At the same time, there will be a need to design and develop new water facilities and water distribution organizations that will be able to address the new reality.

Addressing these challenges require significant institutional innovations. There will be a need to develop insurance mechanisms for water districts and other water suppliers against the hydrological risks faced, as well as the resulting financial losses. As the knowledge about the changes in water supply and storage patterns emerge, there will be a need to rethink water infrastructure and supply. Designing water systems is a lengthy process and an early start may provide significant edge. The work of Xie and Zilberman (2016) shows that the investment in water project capacity is affected both by changes in water availability as well as the investment in water technology and thus regional planning of water systems is needed prior to the investment in water system modification.

One of the most challenging aspects of water resource management is the assignment of water rights. Traditional water rights systems, established during periods of water abundance and under colonial arrangements, can be an obstacle to efficient development of water resources (Schoengold and Zilberman 2007), and water right reform is essential for improvement to allocation. Legal and policy research that lead to innovative water right reform will be an important step in designing and implementing strategies to address water supply implications of climate change.

*Insurance regulations* Risk and uncertainty are the most challenging aspects of climate change. New designs of institutions to address these two facets are a major challenge. It is especially important to develop mechanisms that ensure farmers have insurance against extreme events. Much of the literature on crop insurance argues that it serves frequently as a subsidy rather than insurance *per se*, and farmers tend to undersubscribe to insurance schemes that are self-supporting. Furthermore, subsidized insurance may lead to engaging in risky and environmentally damaging behavior (see survey by Smith and Goodwin 2013). There are new forms of index-based agricultural insurance, but thus far, the quality of their performance has been questionable and there remains a significant need to redesign them (Binswanger-Mkhize 2012). With new sources of information and improved communication technologies, the continued redesign of various forms of insurance is a major challenge for interdisciplinary research and practitioners alike.

*Social safety nets* A higher frequency of extreme events and loss of livelihood due to changing weather may cause farmers to loss their main sources of income, and in many cases food for subsistence. Society will need to design innovative approaches to sustain individuals and communities that experience significant loss as a result of climate change. These approaches must enable them to survive through tough transitional periods while also providing the foundation for re-engaging in the economy. The design of safety net mechanisms may consist of emergency intervention, relocation, insurance arrangements, credit and financial products, and job training. These mechanisms need to be able to adjust to varying conditions and to recognize the limited capacity of the poor to utilize such assistance and insurance while also having rapid response times in order to be effective (Dercon 2002).

*Incentives for farmer-level adoption* The most important factor that affects adoption of new technologies is incentives. There is growing research to introduce innovative policies that will provide farmers the incentives to utilize new technologies, engage in preventive practices to reduce the risks of climate change, and adopt resilient new varieties and activities most appropriate for the challenges posed by climate change.

Adoption of existing and new technologies is a crucial element of mitigation of and adaptation to climate change. There is evidence that many barriers to adoption of new valuable technology exist, which are discussed in the literature (Zilberman et al. 2004). New information and communication technologies provide new opportunities to improve the ways that new technologies are introduced and marketed to enhance adoption. These technologies can be used to improve the information that farmers have of new technologies, accelerate the learning curve of using technologies efficiently and effectively, and reduce the fit and reliability risk associated with these technologies. Innovative approaches may be applied by cooperative extension as well as the private sector.

*Migration* Since climate change will result in relocation of people, design of mechanisms and institutions to facilitate peaceful migration and relocation will become important. As the 2015 migration crisis[1], resulting from the Syrian war and other problems, in Europe suggests, accommodating immigrants is a major policy challenge. Mechanisms to address the increase in migration due to climate change will be a priority of climate smart policy. According to Docherty and Giannini (2009), there is an urgent need to develop innovative approaches to address the climate change refugee problem. They call for a new legal instrument that will establish the

---

[1] See for example: "How Climate Change is Behind the Surge of Migrants to Europe" Time Magazine, September 7, 2015.

human rights of climate refugees, mechanisms for humanitarian aid, and develop criteria to share the burden of relocating climate refugees, as well as financing the relocation efforts. Because climate change will also create new agricultural opportunities, it will be ideal to develop an institutional framework that will enable farmers, especially within regions, to relocate from locations that suffer from climate change to ones that offer new opportunities. The development of institutions to address migration and relocation requires multi-disciplinary efforts and international collaboration and it is a major and urgent challenge.

## 4 Overcoming Barriers to Innovation in the Era of Climate Change

Practitioners have been a major source of innovations throughout history. For example, the wheel, crops for cultivation, and initial farming practices were identified and improved by practitioners. However, science and research are becoming major sources for new innovations in the modern era (Harari 2014). Still further, in the case of climate change, it is important to accelerate the innovative process so that new solutions will be available when and where climatic changes materialize. Scientific research has contributed to the development of new forms of engines, electric appliances, and new medicines, as well as fertilizers and new crop varieties. The innovation process goes through multiple stages. In the case of technological innovation, the process begins with research activities that lead to discoveries of ideas, which are at the core of new innovations. Then through the development process, ideas are refined, tested, and scaled up through further experimentation. For many biological and chemical innovations, the development process also includes government approval for use before commercialization. Upon product feasibility and approval, it is commercialized through activities of production and marketing. Consumers begin to adopt the product, both using and evaluating it, and their feedback leads to product refinement and further innovations. This mostly linear characterization ignores feedbacks and interactions (Etzkowitz 2010) but provides a useful framework to consider some of the major challenges faced by new innovations. In the case of managerial and institutional innovation, the innovation process may also start with research activities that identify alternative options to solve a problem, for example, through economic research or decision theory. Once solutions are identified, there will be a process of experimentation. Managerial and institutional innovations are frequently introduced gradually, for example the reforms in China were first introduced in one location and then spread gradually (Rozelle 1996). The recent increasing use of randomized controlled trials is another mechanism that exist for the introduction and diffusion of new managerial and institutional innovation.

A viable and effective research infrastructure contributes significantly to the introduction of new innovations. The theory of induced innovation suggests that the selection of research priorities is affected by the potential economic gains from innovation and the relative effort required to attain the desired outcome. But obtaining basic research results is not sufficient to achieve practical innovations. The stage of development in scaling up results often requires more funding than the basic research. It requires organization that has the resources necessary to carry out this process. In the developed world, the public sector is more dominant in the research stage while the private sector (start-ups and multinationals) is more dominant in product development and commercialization. Because of the significant investment associated with development, companies would not otherwise engage in it absent some assurance of economic benefit from its outcome, such as intellectual property rights. This assurance is a major reason behind technology transfer from universities and research institutions, through offices of technology transfer, to the private sector (Graff et al. 2002).

The commercialization effort and investment in establishing a supply chain, which includes manufacturing, distribution, and retail outlets, for new product distribution may be more significant than the development of the product itself (Reardon and Timmer 2012). The development of the supply chain, and its subsequent patterns of production and marketing, may vary across products and locations. The private sector will not engage in development of such supply chains without the expectation that investment will result in a positive net return of capital. The private sector is more likely to invest in innovations that are directed to the needs and wants of the developed world than the developing world. For example, the higher willingness to pay by consumers in developed countries for high quality agricultural products may lead the private sector to invest more in innovations that are targeted towards these markets. Research may lead to innovation that will reduce the cost of establishing new supply chains that facilitate a faster adaptation to climate change as part of CSA.

The above analysis suggests that several barriers exist to selecting and implementing climate smart agriculture innovations that will meet the need for growth in agriculture to meet food demand and contribute to poverty reduction in developing countries. The following section presents specific barriers organized by (i) research, (ii) refinement, and (iii) commercialization, approximating the rough order of progression of an innovation.

## 4.1 Research and Refinement

*Knowledge and technology* The development of production practices as well as new crop varieties that may enable adaptation to climate change require knowledge that combines understanding of crop systems, current and alternative practices, and biophysical constraints for a given location. Thus, it is important to invest both in basic research as well as applied development efforts especially because the private sector is less likely to tend to the problems of developing countries. The Consultative Group on International Agricultural Research (CGIAR) centers emphasize research on the challenges of the developing world, and national agriculture research centers are supposed to focus on the application of innovations to local needs. However, while this bifurcated system had significant achievements during the Green Revolution, it is unclear to what extent it can meet the challenges posed by climate change. The system was not designed to withstand larger shocks and the increased degree of uncertainty and variability that are associated with climate change. It has not emphasized climate science and building large capacity to adapt to varying conditions. While this system provides a good foundation to local research and innovation, the extra benefit from extra knowledge because the growing risk of climate change suggests that this system should be reevaluated and strengthened (Sanchez 2000).

Many of the technologies required to adapt to and mitigate climate change are developed at universities in the developed world. Developing of mechanisms to accelerate the transfer of knowledge to action in developing countries coping with climate change problems is a major challenge. But to be effective, technology transfer should include local adaptation and adjustments. Furthermore, a key challenge is to develop systems that will incorporate local and traditional knowledge in agricultural production systems. Thus, new systems will incorporate modern methods with traditional models adjusted to local conditions (Nyong et al. 2007). It requires enhancing human capital and research capacity at universities in developing countries, engaging developing mechanisms to identify local knowledge to innovation systems and providing ongoing support for collaborative research between universities.

*Intellectual property rights* One of the main challenges associated with transfer of information is that much of it is proprietary and thus protected by intellectual property rights. However, several mechanisms exist to address this situation. First, much of the innovation, especially in the area of biotechnology, was generated at universities that sold some of these rights to the private sector (Graff et al. 2003). However, the licensing frequently does not cover application to crops for use in developing countries. And thus, establishment of a clearinghouse would serve to facilitate the transfer of public control intellectual protection rights for use in developing countries can go a long way to solve the IPR challenge (Graff and Zilberman 2001). Indeed, some facilitating organizations for technology transfer exist, including Public-Sector Intellectual Property Resource for Agriculture (PIPRA) and African Agricultural Technology Foundation (AATF). Here should also raise the international treaty for plant genetic resources.

*Fit* One of the major barriers of technology is that technologies may not fit the specific needs, preferences, or capacities of the intended adopters. Much of the effort of marketing is to reduce fit risk (i.e. probability that the technology is not adopted) through demonstrations, return policies, education & training, etc. (Zhao et al. 2012). However, lack of fit may arise from inappropriate design that does not take account of the needs and desires of the particular population. Therefore, there exists a place for participatory research and wide engagement of community in product design and introduction. This approach builds a bridge between the innovation and extension of the technology. One of the major factors of success of drip irrigation in some regions is that cooperative extension worked with practitioners to redesign complementary aspects of the production system so that the new irrigation system would fit with other components of the extant system. Venot et al. (2014) argue that for a technology to be successfully adopted, the production system and technology must be re-designed to incorporate the multiple contexts and practices of the specific location.

*Financing* The innovation process serves as an investment to produce new procedures and institutions that can help address climate change. Each stage of the innovation process requires finance, often in unique ways for research, development, production, and adoption. Because mitigation and adaptation to climate change have properties of public goods (as we argued, climate change may result in damage to public infrastructure and human life throughout the world), the finance should rely on public sources in addition to private ones. The role of public finance may be more essential in some aspects of the innovation process (e.g. basic research). But since much of the technological innovations associated with climate smart agriculture will be introduced in developing countries, development of targeted funds to facilitate adoption will be a major priority. For example, this can be accomplished through financial mechanisms[2] that support innovations and adaptations to climate change in the developing world.

## 4.2 Commercialization/Adoption

*Knowledge dissemination systems* Dissemination of new technologies in developed countries is done jointly by the public and private sector (Wolf et al. 2001). Farmers receive information about new technologies from agricultural media, commercial vendors, cooperative extension, and commodity associations. Frequently media processes information obtained from cooperative extension. Different sources of information have varying degrees of reliability while also highlighting different aspects of some technology (Just et al. 2002). In many developing countries especially vulnerable to climate change, the knowledge dissemination system may be lacking. For example, the private sector may not invest in distribution networks, extension services may be understaffed and underfunded, and access to information from media may be limited. Frequently, the introduction of new technologies will require the development of a dissemination system. Dissemination will improve with investment in extension services and a communication network.

*Limited incentives for farmers to adopt innovations* Many of the innovations that are associated with CSA address problems of externalities and public goods. For example, innovations that lead to a reduction of GHG emissions provide a public good. When externalities or public goods exist, there are likely to be problems of market failure. In particular, adopters will not capture the social benefit associated

---

[2] a la the Clean Development Mechanism of the Kyoto Protocol that is well-designed.

with reduction of externalities or provision of public goods. Thus, policy interventions are needed to incentivize and enhance adoption. Mechanisms suggested by environmental economists (e.g. financial incentives, direct control, subsidies, voluntary agreements) require design of policies that take into account financial and institutional arrangements (Hanley et al. 2007). The new knowledge of behavioral economics suggests the value of nudges (positive reinforcement and indirect suggestion) as a mechanism to enhance adoption and utilization of new innovations (Thaler and Sunstein 2008).

*Limited incentives for governments to adopt progressive regulatory regimes* Because climate change may require introduction of new varieties and new crop production systems at various locations, and changes may occur frequently over time, capacity to innovate and adopt in a timely matter will be important. One of the major barriers to introduction of new varieties is a regulatory that hinders dynamic growth. Regulations are of prime importance because much of agricultural technology may pose unforeseen risks. However, the regulatory process may be too lengthy and costly and hinder the creation of institutions that accelerate innovation, such as CSA practices. Efficient regulation should balance risks and benefits, taking account of precautionary measures,[3] but also take into account the cost of not implementing a new technology.[4] A regulatory system should be designed to avoid bureaucratic redundancy and to be transparent. One of the challenges of introducing a portfolio of technologies within CSA is to design and build human capital and procedures to ensure effective implementation with appropriate safety mechanisms (Rennings 2000).

The challenge of regulatory systems is in adjustment of regulation and policy to account for variability of conditions within agriculture and the heterogeneity of impact as well as the uncertainty not only with technology vis-à-vis climate change but also the need for technology to be able to adjust to diverse conditions and respond to unexpected random shocks. A flexible system of regulation would include insurance, credit, land use and property right regimes similar to those described in this chapter, thus acknowledging the challenges of implementing innovations that adequately address the impacts of climate change.

*Finance* The literature on adoption recognizes credit constraint as a major obstacle to adoption of new agricultural technologies, especially for the poor in developing countries who are further among the most vulnerable to the effects of climate change (Zilberman et al. 2012). Availability of credit depends on an individual's capacity to repay loans with income generated by the technology financed. When CSA does not increase significantly the expected profitability or earned income, but mostly serves to decrease risk or reduce externalities, financial constraints will be even more binding. This constraint can be relaxed through policies that provide increased availability

---

[3] For example, using a risk threshold that may occur at 1%, or even lower, for risk analysis (Lichtenberg and Zilberman 1988).

[4] The regulatory delay on the introduction of golden rice is an example of the cost of excessive regulation of a new technology that has the potential to benefit the poor.

of credit directly or by paying for environmental services associated with adoption of the technology.

*Certification* Innovation or adoption of strategies that will enable mitigation of or adaptation to climate change is likely to be greater if the innovators or adopters are rewarded. Economists prefer to use financial incentives to encourage environmental stewardship. But, when mandatory environmental policies are not feasible, voluntary policies may be attractive. For example, innovative environmental certification has enhanced environmental practices and tourism in Costa Rica (Rivera 2002). In the case of climate change, economists have advocated for introduction of a carbon tax because it provides incentives to reduce emissions of GHGs and enhance mitigation. However, carbon tax mechanisms in agriculture do not yet exist. An alternative mechanism to encourage adoption of climate change reducing strategies is to develop a voluntary mechanism such as certification that increases the value of products produced with practices deemed to effectively address climate change challenges.

A key component of CSA may be to identify practices that are desirable within this context and to develop a mechanism for certification that will reward policy makers that pursue such practices. While this approach has much merit, its implementation is challenging due to issues of fraud and the cost of monitoring (Hamilton and Zilberman 2006). For example, de Janvry and Sadoulet (2015) show how the implementation of a certification program, in this case Fair Trade, may not lead to the desired outcomes. Furthermore, in the case of CSA, the program may backfire if it does not correctly identify activities that contribute to effective management of climate change challenges. Therefore, the design of any certification program must be done in consultation with the latest scientific information available and the performance of the program must be reassessed periodically to ensure it takes into account new knowledge.

*Unintended consequences of conservatism* While environmental groups are among the most concerned about climate change, and were on the forefront of developing mechanisms to finance mitigation, sometimes they may oppose many innovative technologies and institutions that may be part of the solution to the challenges of climate change. This cautious response is not surprising because the traditional instinct of such groups is to protect and conserve (Douglas and Wildavsky 1983). Yet scientific progress may lead to new outcomes that may change reality and have uncertain outcomes. It is prudent to develop regulatory systems to pre-test new technologies, monitor and reevaluate their performance and then design regulations. But over regulation may lead to underinvestment in research that may stymie the development and implementation of new innovations. The risk of implementing new innovative concepts should be compared with the cost of not utilizing them. There are some special examples where strong objection to new innovations on environmental grounds may be especially counter productive. Changes in weather may lead to initiatives to change land use and in some cases conversion of wilderness areas to agricultural production. These initiatives should be considered and adopted if their expected benefits significantly exceed their costs. New technologies that take advantage of modern molecular biology, including genetic modification, should be considered as part of the solution to climate change (Zilberman 2015) These new technologies have significant potential for fast adaptation and reduced human footprint, and the resistance to such technologies can be counterproductive.[5]

The notion of sustainable development recognizes that dynamic processes are occurring and realities are changing. It aims to enhance human development and growth while protecting human well-being and environmental quality (Zilberman 2014). A defensive environmental strategy justifies mitigation and mechanisms to address it, such as carbon tax, but may provide obstacles to adaptation. For example, with climate change, some areas that are considered wilderness will have to be converted to agricultural use. Thus, zoning will need to be flexible to accommodate changing conditions.

## 4.3 Discussion

Barriers to innovation may vary across different categories of innovation, as well as over space and time. Scientific knowledge in the biophysical fields may be a significant barrier to cutting edge technological innovation and thus require significant investment in research. Furthermore, the knowledge gap varies across fields and different types of innovation. The knowledge gap in social sciences on understanding human behavior may hinder the development of management innovations. It can be addressed by both advanced conceptual understanding as well as experimentation with various types of management schemes under different conditions. Lack of information on behavior of both socioeconomic and biophysical systems under different conditions is another constraint on further development of innovations and especially refining it to address the specific needs of the end users. Thus improved data collection and methods can reduce these constraints. Financial constraints may be especially limiting for the development of capital intensive technological innovation but also may limit the development of managerial or institutional innovations that require investment in infrastructure. For example, the introduction of a carbon tax or incentive for carbon reduction that would lead to carbon saving practices, might require investment in monitoring to implement the policy.

Policies to reduce barriers to innovation require significant amounts of research on the institutional framework, technology transfer and adoption. This research should investigate the design of institutions that allocate research funding to

---

[5] The case of genetically modified (GM) organisms is one example. As Bennett et al. (2013) have shown, GM technologies increase yield and reduce agricultural footprint as well as having a big potential to have environmental protection and adaptation to climate change. Their further use is slowed down by objections from environmental groups. Some of the objections to adoption of GMOs are based on the fact that much of the technology was developed by private sector. Yet there are mechanisms that allow access to the technology to develop new varieties for farmers in developing countries (Graff et al. 2003).

innovative activities in a fair, efficient manner that take into account both costs and benefits as well as various levels of assessed risk. The allocation of resources must have a strong spatial element capable of addressing the needs of remote areas, local communities, and have a cultural understanding to get buy-in for new solutions. Furthermore, a key element in developing policy is alliance between the private and public sector that will allow smooth technology transfer and efficient commercialization of new innovations.

# 5 Conclusion

Climate change is a dynamic process and its evolution and impacts depend on human actions. Without mitigation and with continuing build up of GHGs in the atmosphere, the severity of climate change impacts increase over time. At the early stages of climate change, adaptation may be *incremental*. It mostly consists of responses to changes in variability, increased mitigation efforts, better learning and understanding of climate change, development of new technologies and design of infrastructure and more transformative adaptation in anticipation of more drastic changes (Sea level rise, significant migration of weather). During these periods the challenge is in the response to crisis, mitigation, and development of capacity that may allow for adaptation to more drastic changes.

At future dates for many parts of the world, the new capacity and preparation in terms of technology and institutions in the near future will allow regional transformations of agriculture, peaceful migration and resettlement, and new reallocation and better management of water and other resources in response to more drastic changes. However, the timing for *transformational* adaptation varies by location. For instance, in low-lying coastal areas, such as Bangladesh, this form of adaptation may be required in the near future (Kates et al. 2012).

Adaptation to climate change does not occur in isolation, but rather in parallel with other dynamic processes. The impact of climate change, and the design of adaptation strategies, depends on these processes. Three processes are of particular mention: technological change, population growth, and consumption per capita. If technological change in agriculture is moving relatively fast and productive capacity outpaces growth in demand for agricultural products (resulting from population growth and growth in per capita demand), then adaptation to climate change will be less painful in terms of its impact on social welfare. If overall demand for agricultural production outpaces the rate of technological change in agriculture, then the attempts to adapt to climate change will be more painful and the challenges of climate smart agriculture will be exacerbated. If and where migration from rural to urban areas continues in many parts of the world and average farm size increases over time,[6] then climate smart agricultural strategies may be more affordable and the impact of climate change may be less harmful than when the landholding of

---

[6] As the next generation of people that grew up on farms leave them for the cities.

individual farmers declines. The overall geopolitical situation will be crucial to the ability of technology transfer and peaceful relocation programs in response to climate change. Thus a more peaceful, collaborative world is a necessary condition for the implementation of climate smart agriculture.

While climate change affects average conditions and variability at each location, the impacts of climate change are heterogeneous and uncertain. The heterogeneity suggests that some regions gain, others lose and the magnitude of the impacts vary as well. Furthermore, adaptation and the innovations that are associated with it vary by location.

Climate change will increase the value of good management and flexibility, especially in agriculture. Adaptation, including mitigation, to climate change will require a high degree of technological innovation, both in terms of physical technologies as well as institutions and policies. Thus, a key element to develop policies to adapt to climate change is investment in R&D as well as international collaboration. As CSA requires investments, namely some sacrifice in the present for future benefit, it requires buy-in, education, and building awareness about climate change and the gain from adaptation.

The analysis here suggests several principles to guide the introduction of innovation and develop capacity and policies to address climate change. First, pick up the low-lying fruit. Namely, identify no-regret strategies of R&D and innovation that will address climate change and other pressing needs as well as emphasize cost-effective strategies to mitigate and delay the effects of climate change. Second, invest in R&D focused on the development of resource-conserving technologies and monitoring technologies. Third, emphasize innovations (technological, managerial and institutional) that increase the resilience of agriculture and allow it to withstand severe weather events. Fourth, take advantage of the frontier of knowledge of all types and utilize technologies that enhance human welfare and improve capacity to mitigate and adapt to climate change. Restricting the set of allowable solutions will reduce the capacity to sustain the effects of climate change. Fifth, emphasize the use of efficient mechanisms to incentivize farmers and other contributors to the agricultural sector to adopt smart agricultural practices. Sixth, emphasize adaptive management, which includes continuous monitoring, learning through experience, and adaptation of policies as you go. Seventh, distinguish between short-term emphasis on improved resilience in response to increased variability and long-term changes in spatial patterns that may include relocation of activities and people. Finally, harmonize agricultural and climate change policies that aim towards consistent outcomes.

# References

Aker, Jenny C. "Rainfall shocks, markets and food crises: the effect of drought on grain markets in Niger." *Center for Global Development, working paper* (2012).

Barnosky, Anthony D., Elizabeth A. Hadly, Jordi Bascompte, Eric L. Berlow, James H. Brown, Mikael Fortelius, Wayne M. Getz et al. "Approaching a state shift in Earth/'s biosphere." Nature 486, no. 7401 (2012): 52–58.

Basist, Alan, Ariel Dinar, Brian Blankenspoor, and Harold Houba. "Global Land Surface Wetness and Temperature from Space, using Passive Microwave Emission: the Value of Satellite Information in Crop Yield Prediction and River Discharge Models." FAO. Forthcoming.

Bennett, Alan B., Cecilia Chi-Ham, Geoffrey Barrows, Steven Sexton, and David Zilberman. "Agricultural biotechnology: economics, environment, ethics, and the future." Annual Review of Environment and Resources 38 (2013): 249–279.

Binswanger-Mkhize, Hans P. "Is there too much hype about index-based agricultural insurance?" *Journal of Development Studies* 48, no. 2 (2012): 187–200.

Branca, Giacomo, Leslie Lipper, Nancy McCarthy, and Maria Christina Jolejole. "Food security, climate change, and sustainable land management. A review." *Agronomy for sustainable development* 33, no. 4 (2013): 635–650.

Chartzoulakis, Konstantinos, and Maria Bertaki. "Sustainable Water Management in Agriculture under Climate Change." Agriculture and Agricultural Science Procedia 4 (2015): 88–98.

De Janvry, Alain, Craig McIntosh, and Elisabeth Sadoulet. "Fair trade and free entry: can a disequilibrium market serve as a development tool?." Review of Economics and Statistics 97, no. 3 (2015): 567–573

Dasgupta, Susmita, Benoit Laplante, Craig Meisner, David Wheeler, and Jianping Yan. "The impact of sea level rise on developing countries: a comparative analysis." Climatic change 93, no. 3–4 (2009): 379–388.

Dercon, Stefan. "Income risk, coping strategies, and safety nets." *The World Bank Research Observer* 17, no. 2 (2002): 141–166.

Dixit, Avinash K., and Robert S. Pindyck. Investment under uncertainty. Princeton university press, 1994.Princeton, New Jersey

Dobermann, Achim, Simon Blackmore, Simon E. Cook, and Viacheslav I. Adamchuk. "Precision farming: challenges and future directions." In *Proceedings of the 4th International Crop Science Congress*, vol. 26. 2004.

Docherty, Bonnie, and Tyler Giannini. "Confronting a rising tide: a proposal for a convention on climate change refugees." Harv. Envtl. L. Rev. 33 (2009): 349.

Donaldson, Dave. *Railroads of the Raj: Estimating the impact of transportation infrastructure*. No. w16487 National Bureau of Economic Research, 2010.

Douglas, Mary, and Aaron Wildavsky. Risk and culture: An essay on the selection of technological and environmental dangers. Univ of California Press, 1983.

Etzkowitz, Henry. *The triple helix: university-industry-government innovation in action*. Routledge, 2010.

Fischer, Erich M., and Christoph Schär. "Future changes in daily summer temperature variability: driving processes and role for temperature extremes." Climate Dynamics 33, no. 7–8 (2009): 917–935.

Golden, Linda L., Mulong Wang, and Chuanhou Yang. "Handling weather related risks through the financial markets: Considerations of credit risk, basis risk, and hedging." *Journal of Risk and Insurance* 74, no. 2 (2007): 319–346.

Graff, Gregory, and David Zilberman. "An intellectual property clearinghouse for agricultural biotechnology." *Nature Biotechnology* 19, no. 12 (2001): 1179–1180.

Graff, Gregory, Amir Heiman, David Zilberman, Federico Castillo, and Douglas Parker. "Universities, technology transfer and industrial R&D." In *Economic and social issues in agricultural biotechnology* Wallingford: CABI Publishing Wallingford (2002): 93–117.

Graff, Gregory D., Susan E. Cullen, Kent J. Bradford, David Zilberman, and Alan B. Bennett. "The public–private structure of intellectual property ownership in agricultural biotechnology." *Nature biotechnology* 21, no. 9 (2003): 989–995.

Grafton, R. Quentin, Jamie Pittock, Richard Davis, John Williams, Guobin Fu, Michele Warburton, Bradley Udall, et al. "Global insights into water resources, climate change and governance." Nature Climate Change 3, no. 4 (2013): 315–321.

Hamilton, Stephen F., and David Zilberman. "Green markets, eco-certification, and equilibrium fraud." *Journal of Environmental Economics and Management* 52, no. 3 (2006): 627–644.

Hanley, Nick, Jason F. Shogren, and Ben White. Environmental economics: in theory and practice. New York: Palgrave macmillan, 2007.

Harari, Yuval Noah.: *Sapiens: A brief history of Humankind* Random House, New York 2014.

IPCC "Climate Change 2014: Impacts, Adaptation and Vulnerability, summary for Policymakers" WGII Contriubtion ot the Fifith assessment report of the IPCC, WHO, UNEP.

Just, David R., Steven A. Wolf, Steve Wu, and David Zilberman. "Consumption of economic information in agriculture." *American Journal of Agricultural Economics* 84, no. 1 (2002): 39–52.

Kates, Robert W., William R. Travis, and Thomas J. Wilbanks. "Transformational adaptation when incremental adaptations to climate change are insufficient." Proceedings of the National Academy of Sciences 109, no. 19 (2012): 7156–7161.

Khanna, Madhu, and David Zilberman. "Incentives, precision technology and environmental protection." Ecological Economics 23, no. 1 (1997): 25–43.

Krueger, Anne O. Government failures in development. No. w3340. National Bureau of Economic Research, 1990.

Kurukulasuriya, Pradeep, and Shane Rosenthal. "Climate change and agriculture: A review of impacts and adaptations." (2013).

Lal, Rattan. "Sequestering carbon in soils of agro-ecosystems." Food Policy 36 (2011): S33-S39.

Lenton, Timothy M., Hermann Held, Elmar Kriegler, Jim W. Hall, Wolfgang Lucht, Stefan Rahmstorf, and Hans Joachim Schellnhuber. "Tipping elements in the Earth's climate system." *Proceedings of the National Academy of Sciences* 105, no. 6 (2008): 1786–1793.

Lichtenberg, Erik, and David Zilberman. "Efficient regulation of environmental health risks." The Quarterly Journal of Economics (1988): 167–178.

Lu, Liang, Thomas Reardon, and David Zilberman. "Supply Chain Design and Adoption of Indivisible Technology." Presentation at Allied Social Sciences Association Annual Meeting (2015).

McCarthy, N., L. Lipper, W. Mann, G. Branca, and J. Capaldo. "Evaluating synergies and trade-offs among food security, development and climate change." Climate Change Mitigation and Agriculture (2012): 39–49.

McGranahan, Gordon, Deborah Balk, and Bridget Anderson. "The rising tide: assessing the risks of climate change and human settlements in low elevation coastal zones." Environment and urbanization 19, no. 1 (2007): 17–37.

Mendelsohn, Robert. "The role of markets and governments in helping society adapt to a changing climate." *Climatic change* 78, no. 1 (2006): 203–215.

Millner, Antony, Simon Dietz, and Geoffrey Heal. Ambiguity and climate policy. No. w16050. National Bureau of Economic Research, 2010.

Moschini, Giancarlo, and David A. Hennessy. "Uncertainty, risk aversion, and risk management for agricultural producers." *Handbook of agricultural economics* 1 (2001): 88–153.

Nelson, Richard R., ed. *National innovation systems: a comparative analysis*. Oxford university press, 1993.

Nyong, Anthony, Francis Adesina, and B. Osman Elasha. "The value of indigenous knowledge in climate change mitigation and adaptation strategies in the African Sahel." Mitigation and Adaptation Strategies for Global Change 12, no. 5 (2007): 787–797.

Ohmura, Atsumu. "Enhanced temperature variability in high-altitude climate change." Theoretical and Applied Climatology 110, no. 4 (2012): 499–508.

Parfitt, Julian, Mark Barthel, and Sarah Macnaughton. "Food waste within food supply chains: quantification and potential for change to 2050." *Philosophical Transactions of the Royal Society of London B: Biological Sciences* 365, no. 1554 (2010): 3065–3081.

Parker, Douglas D., and David Zilberman. "The use of information services: The case of CIMIS." *Agribusiness* 12, no. 3 (1996): 209–218.

Patt, Anthony, Nicole Peterson, Michael Carter, Maria Velez, Ulrich Hess, and Pablo Suarez. "Making index insurance attractive to farmers." Mitigation and Adaptation Strategies for Global Change 14, no. 8 (2009): 737–753.

Porter, J. H., M. L. Parry, and T. R. Carter. "The potential effects of climatic change on agricultural insect pests." *Agricultural and Forest Meteorology* 57, no. 1 (1991): 221–240.

Poteete, Amy R., Marco A. Janssen, and Elinor Ostrom. *Working together: collective action, the commons, and multiple methods in practice*. Princeton University Press, Princeton, New Jersey 2010.

Reardon, Thomas, and C. Peter Timmer. "The economics of the food system revolution." *Annu. Rev. Resour. Econ.* 4, no. 1 (2012): 225–264.

Rennings, Klaus. "Redefining innovation—eco-innovation research and the contribution from ecological economics." Ecological economics 32, no. 2 (2000): 319–332.

Rivera, Jorge. "Assessing a voluntary environmental initiative in the developing world: The Costa Rican Certification for Sustainable Tourism." *Policy Sciences* 35, no. 4 (2002): 333–360.

Rozelle, Scott. "Gradual reform and institutional development: The keys to success of China's agricultural reforms." Reforming Asian Socialism. The Growth of Market Institutions (1996): 197–220.

Sandmo, Agnar. "On the theory of the competitive firm under price uncertainty." The American Economic Review 61, no. 1 (1971): 65–73.

Sanchez, Pedro A. "Linking climate change research with food security and poverty reduction in the tropics." Agriculture, Ecosystems & Environment 82, no. 1 (2000): 371–383.

Schmitz, Andrew. *Agricultural policy, agribusiness, and rent-seeking behavior*. University of Toronto Press, Toronto 2010.

Schoengold, Karina, and David Zilberman. "The economics of water, irrigation, and development." *Handbook of agricultural economics* 3 (2007): 2933–2977.

Smith, Vincent H. and Barry K. Goodwin. "The Environmental Consequences of Subsidized Risk Management and Disaster Assistance Programs." *Annual Review of Resource Economics vol. 5* (2013): 35.60.

Stern, Nicholas Herbert. *The economics of climate change: the* Stern Review. Vol. 30. London: HM treasury, 2006.

Sunding, David, and David Zilberman. "The agricultural innovation process: research and technology adoption in a changing agricultural sector." Handbook of agricultural economics 1 (2001): 207–261.

Thaler, Richard H. and Sunstein, CR "Nudge: Improving decisions about health, wealth, and happiness." Yale University Press Connecticut, New Haven (2008).

Tol, Richard SJ. "The economic effects of climate change." *The Journal of Economic Perspectives* (2009): 29–51.

Venot, Jean-Philippe, Margreet Zwarteveen, Marcel Kuper, Harm Boesveld, Lisa Bossenbroek, Saskia Van Der Kooij, Jonas Wanvoeke et al. "Beyond the promises of technology: A review of the discourses and actors who make drip irrigation." Irrigation and drainage 63, no. 2 (2014): 186–194.

Waterfield, Gina, and David Zilberman. "Pest management in food systems: An economic perspective." *Annual Review of Environment and Resources* 37 (2012): 223–245.

Weitzman, Martin L. "On modeling and interpreting the economics of catastrophic climate change." *The Review of Economics and Statistics* 91, no. 1 (2009): 1–19.

Wessler, Justus, and David Zilberman. "The economic power of the Golden Rice opposition." Environment and Development Economics 19, no. 06 (2014): 724–742.

Williams, Jeffrey C., and Brian D. Wright. *Storage and commodity markets*. Cambridge university press, Cambridge 2005.

Wolf, Steven, David Just, and David Zilberman. "Between data and decisions: the organization of agricultural economic information systems." *Research policy* 30, no. 1 (2001): 121–141.

Xie, Yang, David Zilberman, and David Roland-Holst. "Implications of Climate Change for Adaptations through Water Infrastructure and Conservation" (2015).

Xie, Yang, and David Zilberman. "Theoretical implications of institutional, environmental, and technological changes for capacity choices of water projects." *Water Resources and Economics* 13, no. 4 (2016): 19–29.

Zilberman, David. "IPCC AR5 overlooked the potential of unleashing agricultural biotechnology to combat climate change and poverty." *Global change biology* 21, no. 2 (2015): 501–503.

Zilberman, David, Xuemei Liu, David Roland-Holst, and David Sunding. "The economics of climate change in agriculture." *Mitigation and Adaptation Strategies for Global Change* 9, no. 4 (2004): 365–382.

Zilberman, David, Jinhua Zhao, and Amir Heiman. "Adoption versus adaptation, with emphasis on climate change." *Annu. Rev. Resour. Econ.* 4, no. 1 (2012): 27–53.

Zilberman, David. "The economics of sustainable development." American Journal of Agricultural Economics 96, no. 2 (2014): 385–396.

**Open Access** This chapter is distributed under the terms of the Creative Commons Attribution-NonCommercial-ShareAlike 3.0 IGO license (https://creativecommons.org/licenses/by-nc-sa/3.0/igo/), which permits any noncommercial use, duplication, adaptation, distribution, and reproduction in any medium or format, as long as you give appropriate credit to the Food and Agriculture Organization of the United Nations (FAO), provide a link to the Creative Commons license and indicate if changes were made. If you remix, transform, or build upon this book or a part thereof, you must distribute your contributions under the same license as the original. Any dispute related to the use of the works of the FAO that cannot be settled amicably shall be submitted to arbitration pursuant to the UNCITRAL rules. The use of the FAO's name for any purpose other than for attribution, and the use of the FAO's logo, shall be subject to a separate written license agreement between the FAO and the user and is not authorized as part of this CC-IGO license. Note that the link provided above includes additional terms and conditions of the license.

The images or other third party material in this chapter are included in the chapter's Creative Commons license, unless indicated otherwise in a credit line to the material. If material is not included in the chapter's Creative Commons license and your intended use is not permitted by statutory regulation or exceeds the permitted use, you will need to obtain permission directly from the copyright holder.

# Part II
# Case Studies: Vulnerability Measurements and Assessment

# Use of Satellite Information on Wetness and Temperature for Crop Yield Prediction and River Resource Planning

**Alan Basist, Ariel Dinar, Brian Blankespoor, David Bachiochi, and Harold Houba**

**Abstract** Satellite derived measurements are essential inputs to monitor water management and agricultural production for improving regional food security. Near real-time satellites observations can be used to mitigate the adverse impacts of extreme events and promote climate resilience. Population growth and demand of resources in developing countries will increase vulnerability in agriculture production and are likely to be exacerbated by the effects of climate change. This paper introduces wetness and temperature products as important factors in decision and policy making, especially in regions with sparse surface observations. These objective satellite data serve as: (1) an early detector of growing conditions and thus food supply; (2) an index for insurance programs (i.e. risk management) that can more quickly trigger release of catastrophic bonds to farmers to mitigate crop failure impact; (3) an important educational and informational tool in crop selection, resource management, and other adaptation or mitigation strategies; (4) an important tool in food aid and transport; (5) and management of water resource allocation. The two new indices (surface wetness and temperature) are meant to complement currently available datasets, such as the greenness index, soil moisture measurements, and river guages.

---

A. Basist (✉) • D. Bachiochi
EyesOnEarth, Asheville, NC, USA
e-mail: alan@eyesonearth.org; dave.bachiochi@weatherpredict.com

A. Dinar
School of Public Policy, University of California Riverside, Riverside, CA, USA
e-mail: ariel.dinar@ucr.edu

B. Blankespoor
World Bank WeatherPredict Consulting, Washington, DC, USA
e-mail: bblankespoor@worldbank.org

H. Houba
Free University of Amsterdam, Amsterdam, Netherlands
e-mail: harold.houba@vu.nl

# 1 Introduction

As world population grows and income increases in developing countries, food consumption habits change, requiring more feedstock for animal production. Furthermore, climate change will have a direct impact on primary and secondary food production, caused by extreme temperatures, precipitation and river flow. This variability will have a direct impact on regional and global food and water supplies. To help vulnerable regions of the world cope with such challenges the concept of climate smart agriculture (CSA) directly addresses the need for adaptation in order to mitigate exposure to the hazards associated with interannual variability and climate change.

The information contained in this chapter demonstrates the value of satellite data (the wetness and temperature products) for monitoring crop production, food security, river flow, and river basin planning in many regions of the world. These products can serve as valuable climate smart decision-making tools in CSA. Specifically, there are several benefits to monitoring growing conditions from objective satellite derived observations:

1. They provide early warning to the available food supply, which mitigates the impact of reduced yields;
2. The wetness and temperature anomalies can be used as indexes in insurance programs as triggers in catastrophic bonds used to compensate the farmers for their losses in near real time;
3. The historic record of growing conditions can be used to identify the return period for various levels of crop failure, which can be used to define vulnerability and return periods for various levels of crop failure, which is essential information for risk management and premium calculation in the insurance industry;
4. Use of the climatology identifies the viability of alternative crop production, beyond the crops traditionally grown in the region. The production of multiple crops is a valuable hedge against catastrophic crop failure. Benefits may be complementary to mitigation activities, agricultural productivity, climate resiliency and natural resource management (Larson et al. 2015).

Since clouds at any one time covers over half of the world, clouds impact most of the surface signal of remotely sensed data across the world (Jackson 2005). Therefore, this study uses satellite derived microwave signals, since they penetrate through most cloud types. Consequently, they are effective in monitoring the surface through most sky conditions. In contrast, before infrared and visible signals can be used, they must be processed by sophisticated and complex cloud clearing algorithms, and can only effectively detect the surface under clear skies (Tucker et al. 2005). Moreover, the most interesting weather usually occurs under partly cloudy to overcast conditions. The microwave signal allows us to observe these events.

In an effort to derive surface temperature from microwave observations, it is necessary to overcome the primary source of noise in the satellite signal: water near the surface. Therefore we developed a technique to identify the magnitude of the water and filter its influence (liquid water reduces emissivity in the microwave spectrum). Specifically, in order to detect land surface temperatures, this low temperature bias must be removed. In the process of accurately identifying the emissivity reduction associated with liquid water and removing its effect on reduction in temperature observations, we were able to accurately identify the magnitude of liquid water near the surface. This byproduct may be more relevant and useful than the surface temperature product we were attempting to observe. Therefore, this chapter will primarily focus on the utility of the surface wetness product and its applications. The wetness product detects: (1) Upper-level soil moisture; (2) Water accumulating into the drainage basins (rivers) of the world; (3) Melting snow packs; (4) Lakes and bogs; (5) Water in the canopy. Upper level soil moisture is effectively used to monitor agricultural yields and river discharge. Consequently, these measurements are essential to water resources management and food production.

There is a need for improvements in crop prediction models, both at high (field level) (Becker-Reshef et al. 2010) and moderate (district level) resolution (Deryng et al. 2011). The satellite-derived wetness index provides data at a moderate spatial resolution. It has been applied in the insurance industry for monitoring likelihood of crop failure throughout the world, and by various governmental and international organizations (e.g. United States, Canada, China, World Bank and UNDP) for assessing yield and food security around the globe, as well as to monitor flow discharge in rivers (e.g. Blankespoor et al. 2012). The goal is to expand the application to a larger client base and provide accurate yield predictions during the growing season. The product can also provide valuable information about adversity thresholds for various levels of crop failure, which is essential for determination of rates for crop insurance underwriting. Moreover, accurate near real monitoring program has several important benefits for CSA: (1) The prediction of yield directly impacts food security and activates infrastructure to move food from where it is in surplus to areas in need; (2) Knowing the wetness and temperature and how they impacts development of the various crops, can be used to optimize the crop types to field conditions, the information can be spread by agricultural extension agents; (3) Planting is one of the most important periods in crop production, it has been shown that the wetness and temperature can be used to optimize planting decisions.

Weather, climate, topography, and vegetation cover have the greatest impacts on the hydrology of a river basin and the variability of natural flow. However, human diversions on river discharge and the effects of climate change confound the predictability of water in the future (Jury and Vaux 2005; Miller and Yates 2006). Since changes in flow affect populations and society in profound social and economic ways, our lack of confidence in future water resources requires mitigations strategies to address the uncertainty (Palmer et al. 2008). Specifically, hydrologic variability creates a significant challenge to countries, since high or low flow events may lead to flooding damage, severe drought, destruction of infrastructure, and/or fatalities. These events promote economic shocks and even generate intra-state violent conflict (Drury and Olson 1998; Nel and Righarts 2008; Hendrix and Salehyan 2012). Moreover, water variability affects international political tensions (Adger et al. 2005; Intelligence Community Assessment 2012). This may even occur in basins where mitigating institutions (like water treaties) have been negotiated (Drieschova et al. 2008). In other words, uncertainty and lack of predictability in flow increases tensions between sectors within a society, as well as between riparian states (Ambec et al. 2013), and the availability of water resources is central to CSA in many areas of the world.

The importance of having a good estimate of the water supply is the foundation of allocation and distribution of irrigation supplies. Since the wetness index is highly sensitive to liquid water near the surface, it effectively quantifies the melting snowpack, and this water feeds many irrigation supplies around the world. Since the origin of the water is monitored, there is a valuable lead-time to communicate with decision makers and allocate the water based on CSA principals and guidelines.

Lakes and bogs are generally permanent features observed by the wetness index, although they may slowly change in size. Since they are a significant component of the surface wetness signal, it is useful to remove these permanent features from the variable signal observed by the index:. specifically, water on the upper section of the soil and held in the canopy. Since water in the canopy has an association with leaf area, part of the signal represents the health of the crop. Our goal is to filter the permanent features, the climatology, and the annual cycles, and focus on the inter-annual variability in wetness, which is driven by the weather. Anomalies are the best tool to achieve this goal. Therefore, the crop models are based on anomalies.

The wetness product is hereafter noted as the Basist Wetness Index (BWI), which detects water near the surface from multiple sources (as mentioned above). In order to simplify the interpretation of the BWI, it is calculated as the percentage of the radiating surface that is liquid water. A reasonable spectrum of this value would be zero percent in desert regions, while agricultural areas have values ranging between 2 and 10% of the surface that is liquid water. Values above 10 usually indicate a very wet surface, such as recently melted snow cover or recent rain.

The following section presents the methodology used to define the BWI, and as well as how it can be used to estimate present and future water supplies under situations where traditional (surface based) observations of surface water are not available, as is the case in many countries. Section 3 illustrates the use of these satellite drived monitoring tools in three different applications (predicting yield of agricultural crops, estimating river flow, and planning in a river basin). The chapter discusses several other applications without demonstrating them, for space consideration.

## 2 Methodology

The BWI index is derived from a linear relationship between channel measurements (Eq. 1), where a channel measurement is the value observed at a particular frequency and polarization, i.e. the Special Sensor Microwave Imager (SSM/I) observes seven channels (Basist et al. 1998).

$$BWI = \Delta\varepsilon \cdot T_s = \beta_0 \left[ T_b(v_2) - T_b(v_1) \right] + \beta_1 \left[ T_b(v_3) - T_b(v_2) \right] \quad (1)$$

where the BWI is the percentage of the surface that is liquid water (Basist et al. 2001), $\Delta\varepsilon$, is empirically determined from global SSM/I measurements, $T_s$ is surface temperature from station measurements, $T_b$ is the satellite brightness temperature at a particular frequency (GHz), $\vartheta_n$ ($n = 1, 2, 3$) is a frequency observed by the SSM/I instrument, $\beta_0$ and $\beta_1$ are estimated coefficients that correlate the relationship of the various channel measurements with observed in situ surface temperature at the time of the satellite overpass. Specifically, as wetness values increase, the differences between the observed surface temperature and the observed channel measurements also increase (Williams et al. 2000).

Weekly and monthly average BWI values are very good indicators of the magnitude of water near the surface, which has a relationship to water at greater depths. These observations have proven valuable in agricultural monitoring during the previous 25 years of analytical work. The wetness anomalies have proven valuable in predicting agricultural yields in many areas of the world (Curt Reynold USDA, personal correspondence). Research indicates the wetness product has a gamma distribution, much like precipitation (Gutman 1999); therefore a gamma distribution is used to derive the variation of wetness from the expected value.

Since most regions of the world have annual cycles associated with their liquid water near the surface, it is best to calculate anomalies for each pixel, location and time of year. The resolution of the pixel is 33 km by 33 km, and anomalies are calculated on a monthly and weekly basis. A value of 0.01 means that only 1 year in a 100 would realize a value so low (extremely dry) at the location for a particular time of year. Conversely, a value of 0.99 corresponds with an excessively wet event that only occurs one out of a 100 years. In summary, values progressively less than 0.5 indicate increasingly drier conditions and values progressively greater than 0.5 indicate increasingly wetter conditions than the expected value (Fig. 1).

The period of record for these wetness and temperature products begins in 1988 and they have been maintained in near real time for decades.[1] There is a period of 2 years, 1990 and 1991, when the stability of the microwave satellite instrument was deemed unreliable. Therefore, these 2 years are removed from the analysis. The climatology we use is based on the 23 years of data from 1988 to 2010. A series of operational satellite instruments flown by the United States Meteorological Satellite Service comprise the period of observations. Great effort has been made to seam the observations between the various satellite instruments into one contiguous record. A daily set of observations is composed of 14 orbits across the globe. These observations are sun synchronous over the equator, at an overpass time around 6 a.m. and 6 p.m. every day. The morning and afternoon overpasses are processed independently and then combined together into one set of observations across the globe. Each set of observations is added to this record in near real-time, as both weekly and monthly fields of temperature and wetness values.

---

[1] SSMI based temperature and wetness data and algorithms discussed in this chapter are a proprietary technology owned by WeatherPredict Consulting, Inc.

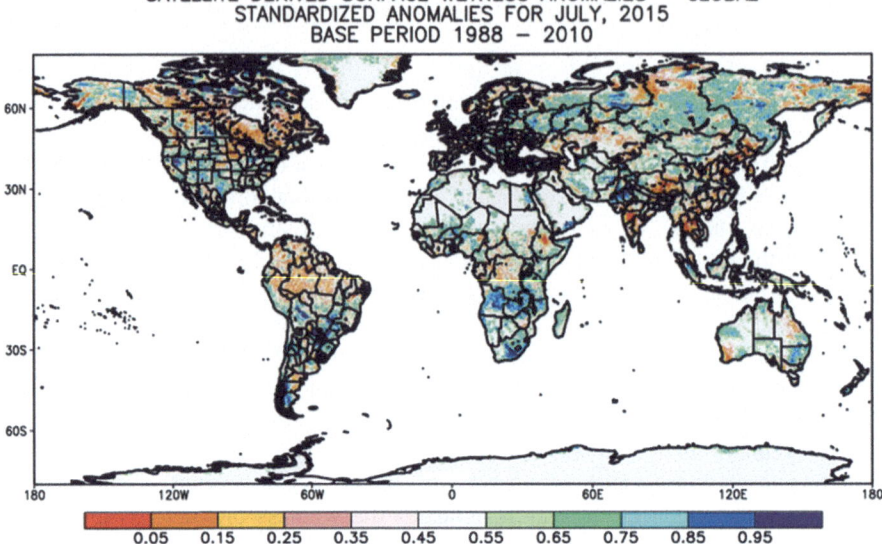

**Fig. 1** Global surface wetness anomalies for July 2015. Note: The *grey* shade of the legend corresponds with the expected value, while values to the *left* (*right*) of the *grey* shade correspond with increasingly drier (wetter) than average conditions. For example, the value of 0.05 means that only 5% if the time is it that dry at a location and time of year. Inversely, a value of 0.95 mean that only 5% of the time is it that wet at a location and time of the year

The actual wetness observations (not the anomaly) are valuable for measuring river discharge. These values identify the percentage of the radiating surface that is liquid water. Moreover, in many river basins there is 1–2 months lag in the time it takes for water in the upper section of the watershed to pass a monitoring gauge in the lower section of a river basin (where most people live and economic activity takes place). This lag, which averages prior month(s) BWI with the concurrent month (hereafter noted as the cumulative lag) improves the skill of the model to predict the flow passing through a river gauge. It also provides valuable lead-time to predict and mitigate the magnitude of drought or flood heading into the lower basin, where the impacts are generally most severe. Therefore, the early warning can be used to mitigate the impact of extreme events on society. An added advantage of applying a quantitative flow model, which can predict flow downstream, is that a consortium of riparian states can use the information to determine how the water resources will be distribution under various flow regimes. Therefore, treaties have the capacity to allocate water as a function of an independent and quantitative measure of flow, providing a simple and accurate predictive model for a fair and transparent distribution of water under times of scarcity.

The observations of the BWI spanning national borders allows for an objective (independent of national influence) calculation of water resources under almost all sky conditions. Since the wetness index is an independent tool that integrates the accumulation of water across large areas, it has the potential to be used as an index and/or trigger for: (1) implementation or call to action in mitigation strategies; (2) insurance compensation; (3) allocation of water between sectors of society; (4) distribution of water between riparian states. These are important applications that warrant further research.

The following section demonstrates the use of the BWI tool for: monitoring crop yield, monitoring river flow, and river basin management. The Mekong River is used as an example. While these applications are site specific, the extrapolation from one site to another is easily done and can be accomplished with minimal cost to the agency.

## 3 Application

Currently, the wetness and temperature anomalies have proven valuable for monitoring crop development and assessing potential yields during the growing season, and have been effectively applied in crop yield prediction models. These models are statistically-based, using linear relationships between the wetness and temperature anomalies and yield, which serves as the calibration. The statistically-derived model parameters are used to predict yield during real time growing conditions and have been applied by many organizations around the world to assess future yields, as well as support planning policies related to the regional, national and global food security (Fig. 2).

There are several limitations in applying the wetness and temperature anomalies across various regions of the world. The first is the large footprint (33 km × 33 km), which is about 1000 km$^2$. This limits the application into a mesoscale analysis and has limited value for high-resolution assessments. Another limitation is coastal boundaries. Specifically, locations within 30 km of a coastline (ocean or large inland water bodies) will unduly influence the temperature and wetness products, since the presence of more than 50% water destabilizes the model, requiring that those signals be recognized and removed from the data sets. Exposed soils or rocks (dry areas) where minerals are exposed on the surface, introduces noise in the signal. This is particularly true when limestone is exposed on the surface. In these instances the product should be used with caution.

### 3.1 Monitoring Crop Yield

The yield prediction models are uniquely calibrated for each crop and particular locations. Specifically, yield prediction models are calibrated on historical values, using the linear variations of temperature and wetness anomalies as predictors. In addition, the quadratic of the wetness and temperature interaction is a predictor in the model. The models are run as the crop enters the reproductive stage, and

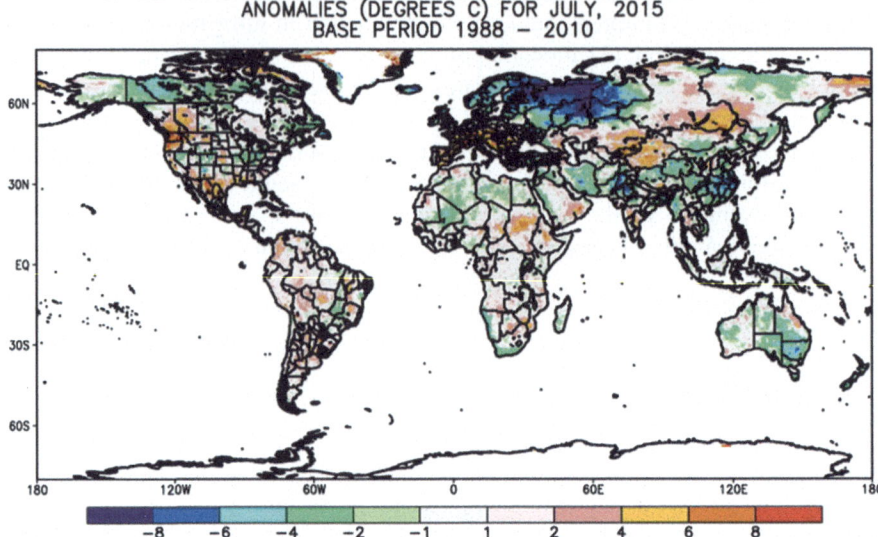

**Fig. 2** Global surface temperature anomalies for July 2015. Note: The *grey* shade in the legend corresponds with the expected value, while values to the *left* (*right*) of the *grey* shade correspond with increasing colder (warmer) than average values. For example the value of −8 means that temperatures were −8°C colder than average at the location and time of year. Inversely, a value of 8 means that it was 8°C warmer than average at a location and time of the year

continues to be updated on a monthly basis through the maturation stage of the crop. The most important month of the growing season is usually reproduction, and therefore the influence of this period has a strong relationship to yield. The benefit of the interactive term is multifold. Specifically, linear statistical models tend to be meancentric, which means they are challenged to capture extreme events. The quadratic component of their interaction generally captures these extreme events in the model.

The models are generally run at the district level. Moreover, each country is unique in the way that it reports yield data. The spatial resolution of the yield data provided by a country serves as the basis of calibration in the model. Both deviation from expected yield and actual yield prediction are presented in the findings of the report. The expected yield has been trended to account for linear improvement of seed stock and improved agricultural practices. These trends are removed, since they are independent of the weather. An example report or the corn belt of the USA during the 2015 growing season is presented below.

Figure 3a shows the predicted deviation from trended (expected) corn yields for the center of the corn-belt in the United States at the end of August 2015. The reasons this region is chosen are twofold; it produces one of the highest yields and is one of the most important growing areas for corn in the world and the sophisticated procedure for calculating yield by the United States Department of Agriculture (USDA) provides one of the best data sets for calibrating the yield prediction

Corn: USA
ASDS Based Crop Districts

SSMI Data Collection Date
8/26/2015

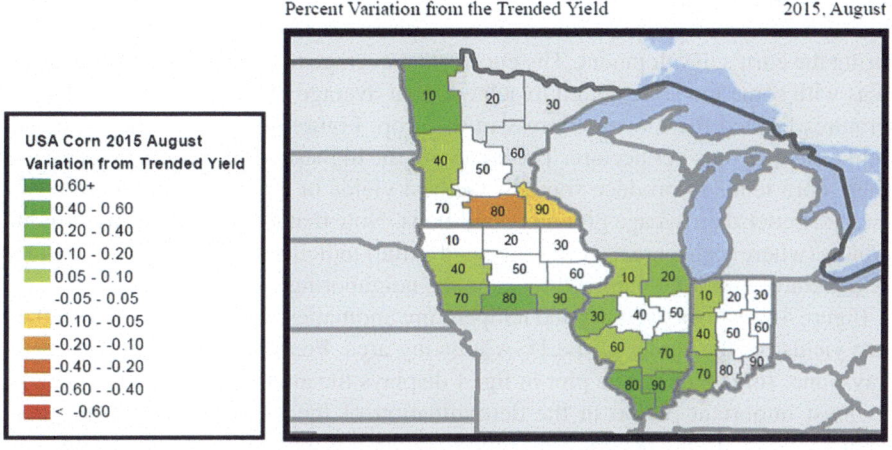

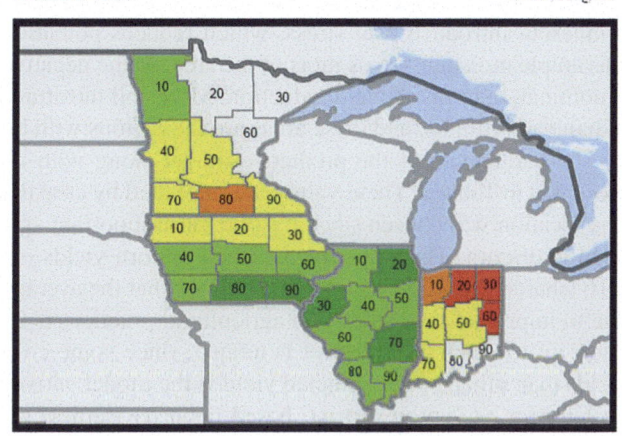

**Fig. 3** (**a**) The percentage departure from the expected (trended) yield. (**b**) The predicted yield in Mt/ha. Note: Zero departures are *white*, and the departures are more amplified the color gets darker towards *red* (*below*) expected, or *green* (*above*) expected yields. They are displayed percentages from the expected value

models. August was chosen, as it provides an early warning to projected yield, as the crop has already entered seed-pod filling.

Generally, the predictions in this report range from average to above average yields for the primary growing regions in the United States. The exceptions are in southeastern Minnesota, where predictions are generally below the expected value. Yields, which have the greatest deviation above the expected values, include much of Illinois and southern Iowa. These areas had near average wetness and slightly below average temperatures, thereby promoting healthy growing conditions during the corn's development. The cooler than average temperatures allowed many areas with some moisture deficit to achieve near average yields, since the cool temperatures limited the moisture stress in the crop. Figure 3b displays the predicted yield as metric tons per hectare. The area with the highest yields occurs in locations where corn tends to produce some of the best yields in the world, and these areas also had better than aveage growing conditions. Note that the low yields in northern Indiana (where yields are near the expected value) indictate that growing conditions are generally inferior, compared to some the neighboring crop districts.

Figure 4 shows the wetness and temperature anomalies, which are used to predict corn yields for the center of the USA growing area. Predictions include data from May, June, July, August, the plot in fig. 4 displays the anomalies for July, which is the most important period in the determination of the yield. August is the time when seed pod filling occurs, after reproduction, it is the most critical period in the development of corn yield.

The above-average temperatures in July across areas of Iowa and most of Minnesota introduce heat stress, which reduces potential yield. Fortunately, there was ample moisture across most of the area, so the negative impact of excessive heat is nominal, in terms of yield reduction. More soil mositure is available in portions of Indiana and Illinois, and these areas are the regions with better than expected yields.

The parameters of the predictive model along with its calculation of yield are presented in Table 1. These values are presented by crop district for the state of Iowa. The location was chosen since it is the most important agricultural state for the production of corn. The slope for the trend of corn yields over the period of record is 0.16 (shared across the state), which means that the average annual increase in yield, due to improved seed stock and agricultural practices is 0.16 metric tons/ha/yr. The intercept for each crop district is unique, since some crop districts produce higher yields than others. The predicted yield is the model derived yield, in metric tons per hectare, for each crop district, based upon its wetness and temperature anomalies throughout the growing season to August 2015. The trended (expected) yield value is based on the 2015 crop season. The last column on the right is the percent variation from the expected yield, the parentheses means the value is negative.

Figure 5 illustrates that some crop districts are slightly below the expected value in terms of yield. However, the majority of the crop districts had higher than expected yield. Therefore, at the end of August the state of Iowa as a whole is predicted to have higher than expected yield. At this time of the growing season the seedpods are approaching maturity, and they provide a reliable measurement of the final yield.

The regression equation and statistical significance of each predictor variable in the model are presented in Table 2. The adjusted $R^2$ for the model is 0.60 with an F-statistic of 28.46. The model has 211 degrees of freedom. The predictive variables

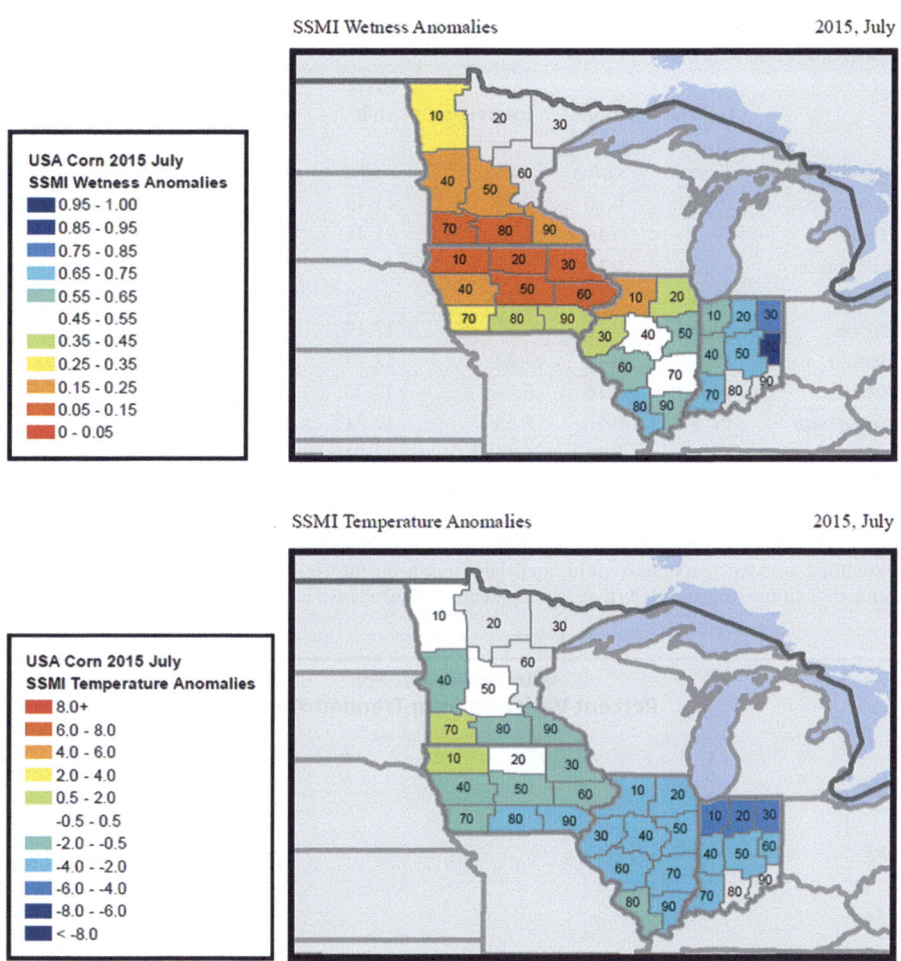

**Fig. 4** July values are presented by crop districts: (**a**) Surface wetness anomalies are displayed by color, where shades towards *blue* (*red*) are increasingly *above* (*below*) the expected surface wetness value (see text for more details). (**b**) Surface temperature anomalies are displayed by color, where shades towards *blue* (*red*) are increasingly *below* (*above*) the expected surface temperature

are temperature and wetness anomalies from May, June, July and August. Also, the interaction of temperature and wetness is included as an independent variable in the model. The negative coefficients are portrayed in red and are inside parentheses. Predictive variables that are significant at the 0.90 confidence level are checked in the right-hand column. The most important variables in the model are the interaction of temperature and wetness in June and July, and the temperature in August. These three variables are all significant above the 99 percent confident interval.[2]

---

[2] The interactions of temperature and wetness for June and July are two of the strongest predictor variables in the model.

**Table 1** Regression-model derived parameters for Iowa

| Corn | | | | | | |
|---|---|---|---|---|---|---|
| United States, Iowa | | | | | | |
| Percent variation from trended yield | | | | | | |
| Crop districts, ASDS based | | | | | | |
| SSMI collection data date 8/26/2015 | | | | | | |
| Admin region | GeoID Crop district | Slope mt/ha | Intercept mt/ha | Pred yield mt/ha | Trend yield mt/ha | Percent variation from trended |
| Buena Vista | 19_10 | 0.16 | 7.53 | 11.45 | 12.05 | (0.05) |
| Butler | 19_20 | 0.16 | 7.46 | 11.48 | 11.98 | (0.04) |
| Allamakee | 19_30 | 0.16 | 7.26 | 11.53 | 11.78 | (0.02) |
| Audubon | 19_40 | 0.16 | 7.10 | 12.27 | 11.62 | 0.06 |
| Boone | 19_50 | 0.16 | 7.51 | 12.19 | 12.03 | 0.01 |
| Benton | 19_60 | 0.16 | 7.22 | 12.29 | 11.75 | 0.05 |
| Adair | 19_70 | 0.16 | 6.54 | 12.28 | 11.06 | 0.11 |
| Appanoose | 19_80 | 0.16 | 5.69 | 12.81 | 10.21 | 0.25 |
| Davis | 19_90 | 0.16 | 6.45 | 12.74 | 10.97 | 0.16 |

Identifies the slope and intercept for the linear trend in yield derived by the USDA yield values from 1988 to 2014

Note: The three columns to the right are predicted yield derived from the wetness and temperature anomalies, trended (expected) yield, and the column on the right is the ratio of the predicted/trended yield for August 2015 (parenthesis means the values are negative).

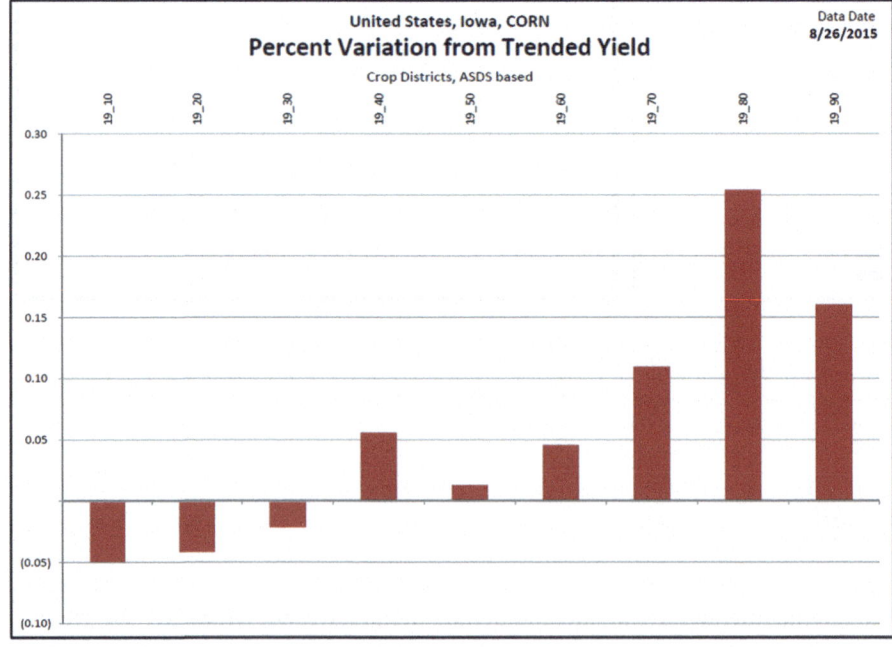

**Fig. 5** Graphical representation of the variation from trended yield, in Iowa plot is conveyed by crop district in the state

**Table 2** Model coefficients and significance values

| Corn | | | |
|---|---|---|---|
| United States, Iowa | | | |
| Statistical model output | | | |
| Crop districts, ASDS based | | | |
| Data date 8/26/2015 | | | |
| # observations | 225 | R-squared | 0.62 |
| # variables | 13 | Adjusted R-squared | 0.60 |
| Degrees of freedom | 211 | F-Statistic | 28.46 |
| **Variables** | **Coefficients() negative values** | **Significance (in percent probability)** | **Significance @ 90% confidence** |
| Constant | 13.28 | 0.00 | × |
| Temp May | 0.05 | 0.01 | × |
| Temp Jun | 0.01 | 0.69 | |
| Temp Jul | (0.05) | 0.03 | × |
| Temp Aug | (0.17) | 0.00 | × |
| Wet May | (0.19) | 0.58 | |
| Wet Jun | (1.06) | 0.00 | × |
| Wet Jul | (0.57) | 0.24 | |
| Wet Aug | 0.11 | 0.78 | |
| Interact May | (0.00) | 0.10 | |
| Interact Jun | (0.02) | 0.00 | × |
| Interact Jul | (0.02) | 0.00 | × |
| Interact Aug | (0.01) | 0.10 | |

The degrees of freedom in the model, along with its predictive skill, regression coefficients, their significance level for each predictor variable Negative coefficients are in parenthesis

Finally, a scatterplot of the wetness and temperature anomalies for the months of July and August at the crop district level is presented (Fig. 6). Note that in the month of July the majority of Iowa had slightly below normal temperatures, while wetness values were drier than normal during the month. The lack of heat stress during reproduction was for yields. August continued to bring drier than average conditions to the majority of the state, while near average temperatures helped minimize soil moisture stress. Therefore yields predictions were near-normal. The forecast generally remained the same between the end of July and the end of August, since July is the most important month for yield prediction. Although there were changes in field conditions across a few crops districts during the August, the addional information in August improves the model skill as the crop reached maturity.

## 3.2 Monitoring River Flow

Quantitative and indepenedent measurements of river flow levels are essential for water rights and planned allocations. Moreover, reliable and independent measurements of available water resources are required for mitigation strategies and

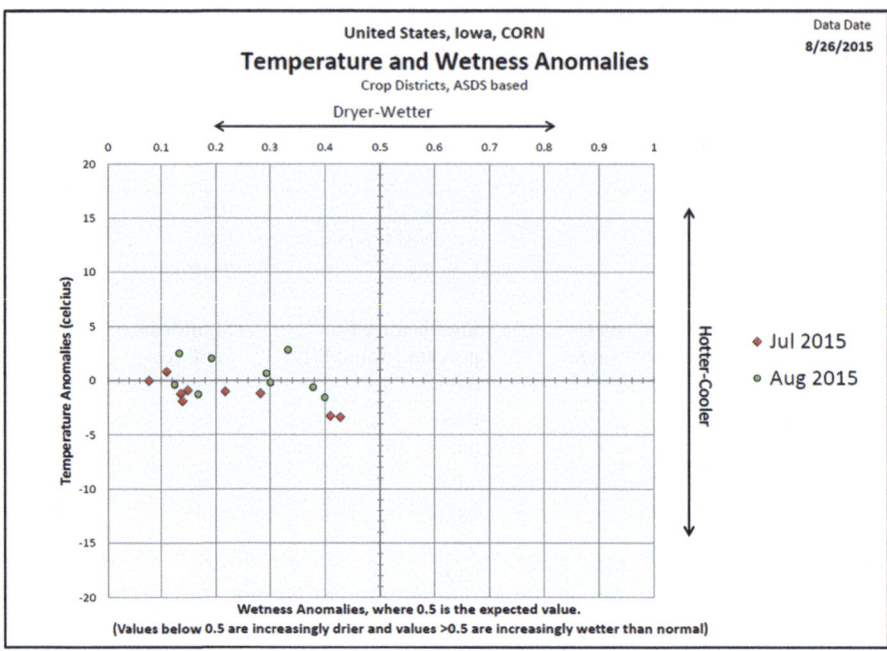

**Fig. 6** Scatter plot of wetness and temperature anomalies by crop district for the months of July and August. Note: *Top left* quadrant is above temperature and below wetness, *bottom left* is below both temperature and wetness, *top right* is above both temperature and wetness, and *bottom right* is below temperature and above wetness

insurance compensation, which are a fundamental component of an effective treaty (Dinar et al. 2010) that allows proper planning and allocation of the basin water to various water consuming activities. Also, independent monitoring of flow measurements is required to implement an effective treaty, which is based on triggers, response and compensation, or to operate reservoirs used for irrigation projects. Therefore, high quality flow data are a necessary component of effective treaty stipulations and institutional mechanisms (Dinar et al. 2015), as well as infrastructure for reservoirs that can deal with future challenges. Real time data can also provide policy makers and researchers with the ability to predict extreme weather events, and cooperatively address economic impacts on existing projects. In addition, models can increase institutional capacity by providing timely (near real time) flow information to build climate resilience and effective sharing and allocation of limited water resources.

Considering the challenges to estimate flow where standard measurements are not available, we demonstrate a simple, yet robust model to predict both present and future flow measurements, using the wetness product in two basins: Zambezi and Mekong. The period of record for calibration of the models is from historic river gauge values, and these flow values are regressed on the BWI values (the predictor of flow). In order to keep the equation as simple as possible, yet robust, the regression is based on one variable and tested in two basins of very different climatology's, topography's, land use patterns and annual water supply cycles. An important consideration between the gauge and BWI values is a lagged relationship between water accumulating near the surface and detected downstream at the gauge. The lag between the water input upstream and the detection of changes in flow downstream is based on numerous empirical observations and theory that flow models are more accurate when they include the prior month(s) due to the time lapse for the water accumulate into the major stem of the river (Demirel et al. 2013). The number of prior months used in the predictions of flows is directly related to the size of the basin, the influence of snow melt and its topography. Therefore, a lagged term is included in Equation 2, where $Q_{m(BWI)}$ is the discharge at a station for month $m$ While $n$ is the number of previous month(s) averaged together with the concurrent month BWI value.

$$Q_{m(BWI)} = g(d) \qquad (2)$$

where $d = \dfrac{\sum_{i=0}^{n} BWI_{m-n}}{n}$.

Table 3 lists model statistics and parameters for the two river basins. The number of month(s) lagged prior to the gauge observations is included, along with the parameters of the regression model. Our goal is to define a simple and robust prediction from one variable and explore the utility of the predictor in areas of society that could benefit from the models.

The Zambezi model flow signature is clearly curved (Fig. 7a); it has a quadratic structure of high wetness values and extremely high flow. High values display considerable heteroscedasticity (from the studentized Breusch-Pagan test), which implies that numerous factors impact the high rate of flow past the gauge. In contrast, low BWI values (less than 1) contain a high confidence that the flow will be near the base flow. These results compared favorably to model prediction for the Zambezi presented by Winsemius et al. (2006), whose predictions were based on a more complex model. As a result, the BWI can be a quantitative indicator for periods and frequencies of flow associated with limited water – of particular relevance to obligations and commitments agreed upon in international water treaties.

**Table 3** Parameters from Zambezi, Mekong predictive river flow models

| Model | Zambezi (BWI) | Zambezi (precip) | Mekong (BWI) | Mekong (precip) |
|---|---|---|---|---|
| Linear term | −420.2 | 71.9 | 303.8 | 75.9 |
| Quadratic term | 748.6 | 0.78 | 886.6 | 0.297 |
| Months lagged | 2 | 2 | 2 | 2 |
| month observation | 148 | 198 | 44 | 44 |
| Predictive skill (R2) | 0.89 | 0.52 | 0.95 | 0.97 |
| Residuals | 485 | 1020 | 645 | 523 |

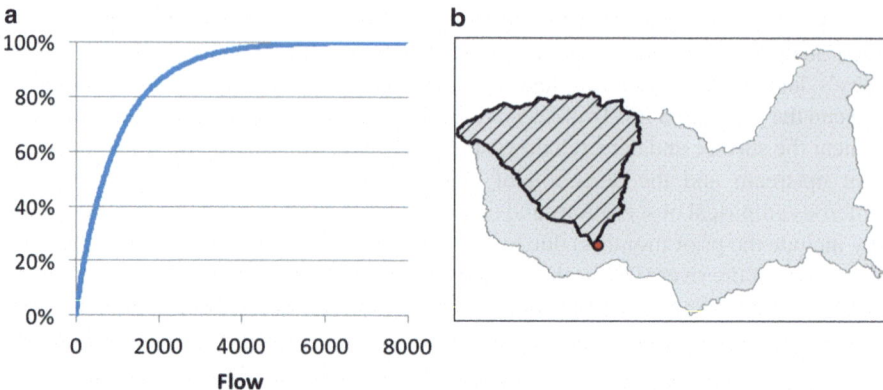

**Fig. 7** (a) Cumulative distribution of flow using a gamma distribution (percent. y-axis) and flow (m$^3$/s per month. x-axis) of the Zambezi river basin sample area; (b) Map of Zambezi basin (*grey*) with the selected gauge data (point), international border (line) and respective catchment (hatched) used in the model

The lower bound of predicted flow is 288 m$^3$/s (BWI = 1.0) occurs approximately 28% of the time. Therefore, for the Zambezi River at the Katima Mulilo station, approximately 28% of the time the flow is less than 288 m$^3$/s averaged over the 3 months. The area feeding water to the gauge is defined in Fig. 7b.

Since the SSM/I instrument is currently operational, it is possible to use the fitted model to predict recent runoff from monthly wetness values, based on the calibration period. Due to the accuracy and significance of the models, we chose to explore the ability of the BWI to predict seasonality, low flow (e.g. droughts), and high flow events (e.g. floods). This analysis was used to explore the utility of the model in *serving as an early warning indicator.*

With regards to the Zambezi, the BWI model identified and predicted a flood in 2010, which according to the model is higher than any previous flood over the period of the SSMI record (Fig. 8). In April 2010, there is a pattern of large positive surface wetness anomalies in Western Zambia (Fig. 9). This broad pattern of purple indicates that the area was extremely wet conditions. This extreme event occurred across a large section of the basin. In rare instances, when there is an extreme flood on the Zambezi, due to heavy rainfall on the highlands in Angola and Zambia, the flow can actually accumulate at the Mambova fault. During this instance, the river expands over the flat floodplain behind the fault until the waters meet the channel cut by the Chobe River in the south. During this extreme flood, the accumulation of water from the Zambezi River overcomes the Chobe River, and water begins to flow upstream on the Chobe, flowing into Lake Liambezi. At the height of the flood, water flowed directly into Lake Liambezi from the Zambezi River through the Bukalo Channel on May 8, 2010 (NASA 2010), which is the same time the BWI predicted the highest flow over the period of record.

Next is discussed the Mekong model, which is presented in Table 3. The section of the river basin that feeds the Mekong gauge station is presented in Fig. 10b. The best explanatory model has a non-linear relation. The Mekong models also used a

Use of Satellite Information on Wetness and Temperature for Crop Yield Prediction... 93

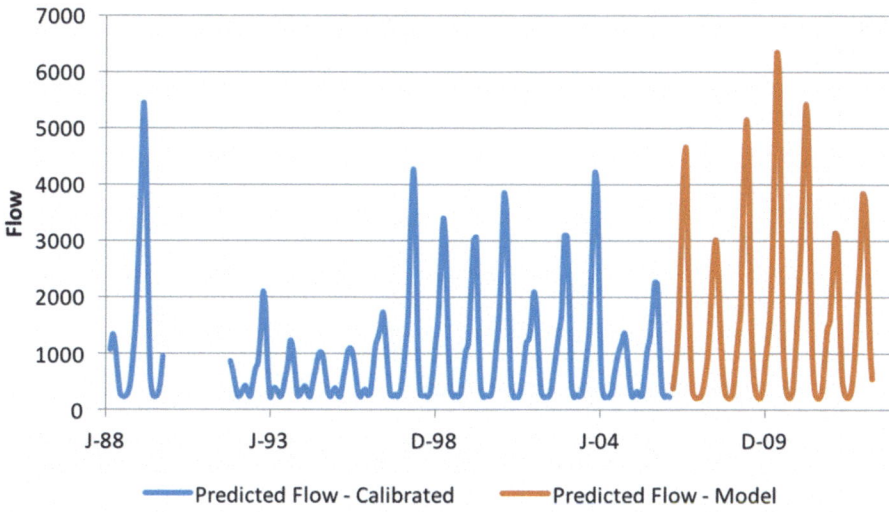

**Fig. 8** The Zambezi values of runoff (m³/s per month, y-axis) and time ( x-axis, January 1988 through July 2013). The time series displays seasonality and interanual variability over the predicted (calibration) period in red (*blue*). The highest flow occurred in April/May 2010. Missing values are due to the lack of reliable SSM/I data

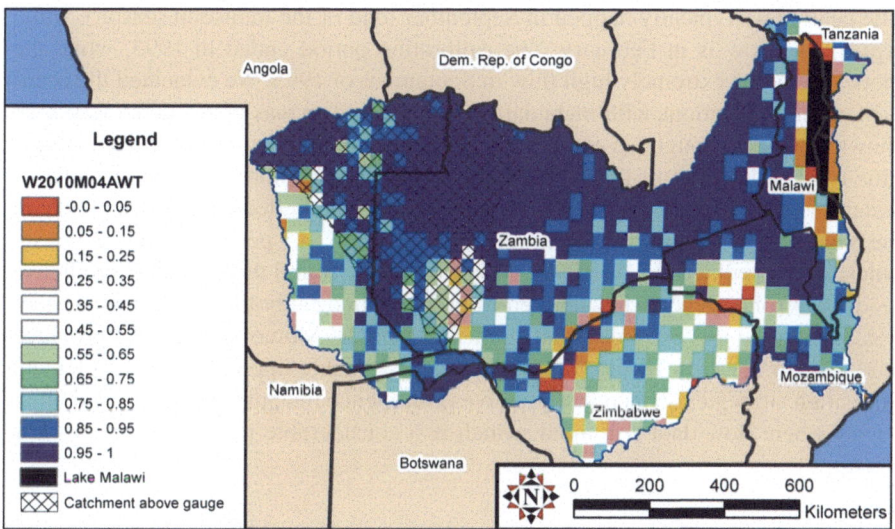

**Fig. 9** Surface wetness Values for a section of the Zambezi River: April 2010, where 0.00–0.05 (*red*) means that less than 5% of the time is it this dry, 0.45–0.55 (*white*) is the expected normal soil moisture, and 0.95–1.0 (*purple*) means less than 5% of the time is it this wet

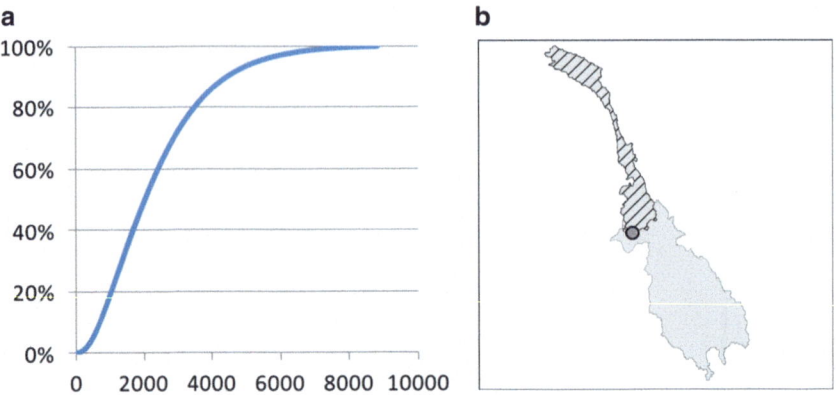

**Fig. 10** (a) Cumulative distribution of flow using a gamma distribution (percent. y-axis) and flow (m³/s per month. x-axis) of the Mekong river basin sample area. (b) Map of Mekong basin (*grey*) with the selected gauge data (point) and respective catchment (hatched)

quadratic form. It also implies that predicted flow below 1215 m³/s (BWI = 1.0) occurs less than 25% of the time. There is a limited period of calibration data, and some concern about the accuracy of the model. Therefore, an evaluation of the skill during the predictive preiod will demonstarte the robustness of this approach to monitor flow from the BWI data.

The Mekong river model captures the seasonal hydrologic variation (Fig. 11). The peak flows typically happen in September (end of the monsoon season), while typical low flow is in February. The calibration period ended in 1993, while the model predicted extremely high flow in September of 1995. We evlauated the accuracy of this predictions with meta data, since guage data was unavailable. Research shows that 1995 brought an extreme flood, which was predicted by the BWI. At this time over 100,000 ha of the Vientiane Plain was under more than a half-meter of water for up to 8 weeks. In human terms, the 1995 flood affected 153,398 people in the Vientiane Plain (out of a total population of 653,013 persons), 26,603 households, or 427 villages (FAO 1999). Importantly, we found that the BWI predictive model was robust, even when derived from the limited calibration period. None-the-less, it captured this extreme event and its magnitude. Moreover, the BWI provided lead-time to the crest of the event, allowing a valuable opportunity to implement mitigation strategies. This result promotes confidence in applying the BWI to other basins where flow data is limited, which is a considerable number of the world's river.

## 3.3 River Basin Management: The Case of the Mekong

In locations where irrigation is a major component of agricultural production, economic planning around limited water resources is critical to the success of Climate Smart Agriculture. Specifically, it applies to allocation of river water to promote

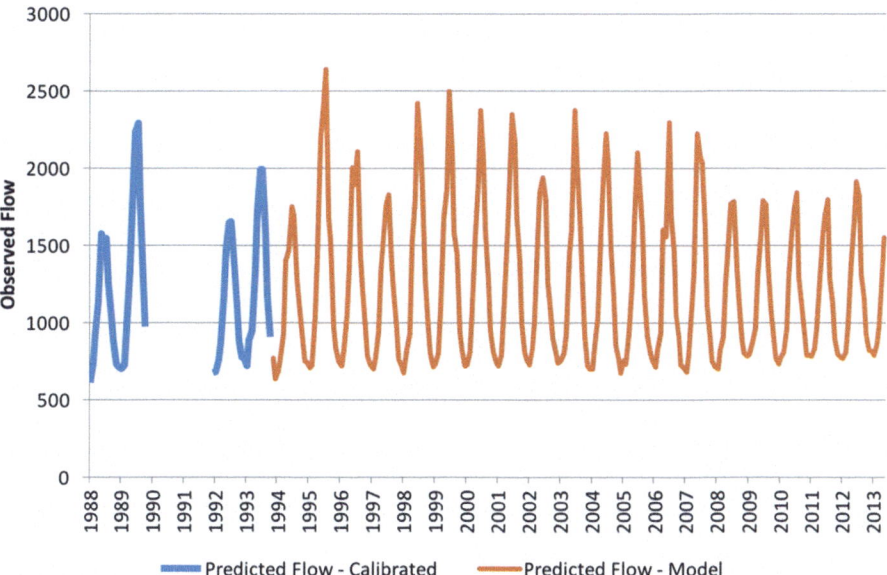

**Fig. 11** The Mekong values of runoff (m$^3$/s per month, y-axis) and time (January 1988 through July 2013) display seasonality and the interannual variability over the calibration (predicted) in blue (red) period of the time series. Missing values are due to the lack of reliable SSM/I data

resilience to climate variability and optimize water allocation for economic growth. We provide a modified version of the empirical model used in Houba et al. (2013). The range of flow probabilities as measured by the BWI and at the gauging station Chiang Saen in Thailand are presented. These probabilities are used to calculate the expected value of basin benefits under various climatic scenarios. While the application of the BWI is demonstrated with the Mekong River Basin, we argue that it is a very simple process to apply the BWI to assist policy guidance in any of the river basins around the world, due to the fact that the main information needed for the analysis comes from satellite-based data, which is readily available. This application can benefit river basin planning, economic opportunities, resource management, and agricultural resilience.

### 3.3.1 Description of the Model

The model is based on a simplified hydrological structure of the basin, where water flows from China, hereafter noted as the Upper Mekong Basin (UMB) to the Lower Mekong Basin (LMB) and its tributaries, which originate in Thailand, Laos, Cambodia, and Vietnam, before the river enters the Delta (estuary), as seen in Fig. 12.

Basin-wide water availability is determined by water arriving from the UMB, and precipitation received in tributaries of the LMB. Water uses are aggregated in each sub region of the model into (1) industry and households, (2) hydropower

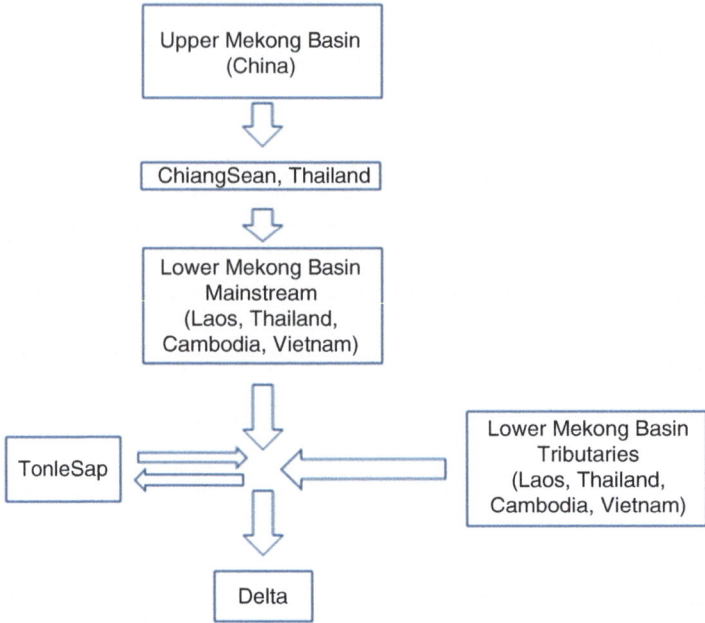

**Fig. 12** Simple representation of the Mekong river basin used in our model (Modified from Houba et al. 2013). Note: We exclude Burma (Myanmar) from the analysis because it has a negligible share of water and land in the basin

generation, (3) irrigated agriculture, and (4) fisheries (Table 4). Water quality is measured in terms of salinity in Houba et al. (2013). In this paper we assume that salinity impacts fishery and irrigated agriculture. Hydropower generation is considered to be an in-flow user, while providing economic opportunities and growth. Moreover, water entering the first reservoir of a cascade can be reused and stored, over time, in all downstream reservoirs, which expanding capacity for economic growth along the river.

The model is calibrated on flow data from 2010 and it is static with an annual setup, represented by two seasons' dynamics (wet and dry) across the entire basin. All modifications introduced in this paper comply with the original calibration. The water inflow for the mainstream of the LMB consists solely of the outflow received from China. Reservoirs/dams are filled in the wet season and the water is used during the dry season mainly for irrigation. During the wet season the Mekong water in UMB (China) can be used for industrial and household activities, fish production, storage for use in the dry season, and non-consumptive hydropower generation. Moreover, the wet season water supplies dry season irrigation for Climate Smart Agriculture. Moreover, effectively monitored outflow from mainstream UMB and tributary dams can promote inundations of wetlands in the delta. This nurtures fisheries production and flushes salinity from the estuary (Delta), which improves water quality and irrigation supplies.

Use of Satellite Information on Wetness and Temperature for Crop Yield Prediction... 97

**Table 4** Water balances and use by sectors (km³/year) for mean flows at UMB and LMB tributaries

| Variable | UMB wet season | UMB dry season | LMB wet season | LMB dry season |
| --- | --- | --- | --- | --- |
| Inflow water | 66.737 | 9.534 | 375.920 | 53.703 |
| River flow from upstream | | | 60.522 | 7.151 |
| Water availability | 66.737 | 9.534 | 436.442 | 60.854 |
| Stored water total[a] | 5.474 | | 12.888 | |
| H&I water use | 0.741 | 0.529 | 1.895 | 1.352 |
| Outflow water from dams | 60.522 | 13.565 | 421.659 | 69.735 |
| Irrigation | | 6.414 | | 6.579 |
| River flow to Tonle Sap | | | 86.950 | −86.950 |
| River flow to downstream/estuaries | 60.522 | 7.151 | 334.709 | 150.107 |
| Hydropower water total[b] | 69.226 | 74.912 | 60.003 | 42.860 |

Source: Houba et al. (2013)
[a]Water is stored on main river in UMB and on tributaries in LMB
[b]Hydropower is produced on main river in UMB and on tributaries in LMB

Following Houba et al. (2013) the benefit, cost and loss functions in the model are quadratic, with the benefit function being concave (same as the flow parameters in the BWI model) and the cost and loss functions being convex to the origin. The volume of water that enters the Tonle Sap and then flows out into the Delta wetlands is a linear function of the river flow. Benefit functions were used for industry and households, hydropower generation, irrigated agriculture, and fisheries. The value function of the Tonle Sap and Delta/Wetlands assumes that all fishery production concentrates in that lake and surrounding wetlands. Salinity losses are modeled only in the LMB agricultural sector.

### 3.3.2 Applying the BWI to the Mekong Economic Model

A regression equation calibrates the BWI on gauge data from the UMB at Chiang Saen. The upper and lower basins have appreciably different geographies, sizes, and rainfall. Nonetheless, we applied the upstream hydrological model to the lower basin. Our assumption in doing so is that the BWI signal is designed to detect liquid water from all sources, and is defined as the percentage of the surface that is liquid water near the surface. Therefore, we explore the robustness of the model to detect that amount of water moving through the lower basin. Our hypothesis is that BWI values are a robust signal and the model parameters could effectively transcend different geographies.

There was the possibility of shifting the intercept, since the lower basin is appreciably larger, and therefore its base flow should be higher. However, we wanted to minimize any tuning, in order to test the robustness of the model. The only change is the lag was reduced from 2 to 1 month, to allow for better integration (time to flow) from the upper basin into the lower basin. This, in turn, would allow us to model the flow as one kinematic wave based on the speed of flow.

In order to calculate the magnitude of water moving through the entire basin, the upper and lower basins were weighed in terms of their area (the large lower basin is a much larger area, and therefore has higher weights). This allowed us to integrate the upper and lower basins into one combined flow. Since the upper basin has a two-month lag, the first 2 months of 1988 and 1992 were set to be missing. A simple interpolation technique could easily and effectively be applied, since the beginning of the year is not a critical period of flow, however we did not apply it in order to minimize assumptions.

The average flow was derived from the BWI values and the model parameters over the period of record, in terms of cubic meters/second. To keep our economic optimization comparable with previous work Houba et al. 2013, we express water in cubic kilometers per year rather than in cubic meters per second (1 m$^3$/s = 0.031556926 km$^3$/year). The mean annual flow over the period of record derived by the BWI for the UMB and LMB is 424 km$^3$, which is reasonably close to the independent assessments of annual mean flow on the Mekong, which range from 410 (Houba et al. 2013) to 475 (Mekong Water Commission 2009).

We were very encouraged by the fact that the flow numbers derived through the BWI wetness values were congruent with the expected flow values. Equally important, the monitored variation of flow from month to month, and year to year was accurately captured by the BWI values. For example, the major flood of of 1995 and smaller flood of 2000 was also predicted by the BWI, providing a one-month lead-time to the magnitude of the flood, allowing time to mitigate its consequences.

We performed a similar analysis using precipitation inputs to predict mean annual flow for the Mekong. Specifically, we used the flow model parameters derived from the upper basin and applied them to the LMB, in order to determine integrated flow for the River as a whole. The calculated flow based on rainfall is 359, while the BWI provided a value of 424 km$^3$/year (i.e. the BWI value is much closer to the consensus of the mean annual flow). This result was surprising; since the precipitation model had a slightly better explanatory power of flow in the upper basin, see Blankespoor et al. 2012. We interpreted this finding as demonstrating the robustness of the wetness index, and the ability to apply the model in areas outside of the region where they are calibrated. Consequently, we use the BWI flow predictions to enhance CSA, climate resilience, and calculate return periods of extreme events (Table 5).

### 3.3.3 Results of the Economic Model

We ran four scenarios, following the pairs ($a_i$; $b_i$, $i$= 1,…,4) of flow values from Table 5, which correspond to distribution of the flow in both the UMB and the LMB tributaries. As can be seen from Table 5, the distribution of the LMB tributaries flow is much more skewed towards lower values (drought) than the flow of the UMB. Table 6 presents the net welfare in each region for various distributions of the flow as obtained from the basin optimization model we run.

**Table 5** Flow data in the UMB and LMB as calculated by the BWI

| Description | km³/year | m³/sec | Cumulative probability | Probability |
|---|---|---|---|---|
| **a. Flow at Chiang Saen (UMB coming from China)** | | | | |
| a1: Mean − 1 SD | 27.863 | 882 | 0.117 | 0.117 |
| a2: Mean | 76.271 | 2416 | 0.588 | 0.471 |
| a3: Mean + 1 SD | 124.679 | 3950 | 0.862 | 0.274 |
| a4: Mean + 2 SDs | 173.087 | 5484 | 0.961 | 0.099 |
| **b. Flow of LMB tributaries** | | | | |
| b1: Mean − 1 SD | 345.536 | 10,949 | 0.414 | 0.414 |
| b2: Mean | 429.623 | 13,614 | 0.576 | 0.162 |
| b3: Mean + 1 SD | 513.710 | 16,278 | 0.710 | 0.134 |
| b4: Mean + 2 SD | 597.797 | 18,943 | 0.809 | 0.099 |

**Table 6** Net benefit calculations for various flow values in the Mekong basin (billion $)

| | Mean flow − 1 SD | | Mean flow | | Mean flow +1 SD | | Mean flow +2 SD | |
|---|---|---|---|---|---|---|---|---|
| | UMB | LMB | UMB | LMB | UMB | LMB | UMB | LMB |
| km³/year | 27.863 | 345.536 | 76.271 | 429.623 | 124.679 | 513.710 | 173.087 | 597.797 |
| Net welfare created | 2.376 | 3.222 | 2.656 | 6.663 | 2.544 | 6.445 | 2.313 | 6.336 |
| Aggregated economic value | 2.376 | 6.355 | 2.656 | 6.663 | 2.544 | 6.445 | 2.313 | 6.336 |
| Econ value households and industry | 0.408 | 1.957 | 0.408 | 1.957 | 0.408 | 1.957 | 0.408 | 1.957 |
| Econ value fishery | 0.128 | 2.772 | 0.241 | 2.728 | 0.167 | 2.077 | 0.082 | 1.109 |
| Econ value irrigation | 1.193 | 1.421 | 1.193 | 1.772 | 1.193 | 2.206 | 1.193 | 3.065 |
| Econ value of hydro in main | 0.647 | | 0.815 | | 0.776 | | 0.629 | |
| Econ value of hydro in tributaries | | 0.205 | | 0.206 | | 0.206 | | 0.206 |
| Aggregated economic costs | 3.133 | | 0.000 | | | | | |
| Costs saltwater intrusion | 3.133 | | 0.000 | | | | | |

Source: Authors' calculations
Note: *SD* standard deviation, *UMB* upper Mekong basin, *LMB* lower Mekong basin

As is apparent from Table 6, the net welfare generated in the UMB is $2.656 billion and that of the LMB is $6.663 billion, annually. Of the net welfare produced annually in the UMB, hydropower comprises 31%, irrigation 45%, fisheries 9% and households and industry 15%. For the LMB the values are 3%, 27%, 41%, and 30%, respectively. Table 6 also suggests that the damage from salinity due to seawater intrusion in the LMB is 0 for mean flow or above mean flow runs. However, losses of $3.133 billion are encountered in the LMB in the case of the below mean flow run. It appears that the LMB is much more sensitive to flow fluctuations than the UMB. This is also apparent from Fig. 13, which summarizes the results in aggregate terms for different flow distributions by the Mekong regions. Both high and low levels of flow have a negative impact on net welfare of the basin.

Using the probabilities in Table 5 and the net benefits in Fig. 13 the expected total basin net benefit value at $6.359 billion at one standard deviation below mean flow. This figure represents only 68% of the basin-wide net benefits ($9.313 billion) that was estimated under the mean flow. Having the flow distribution information (as provided by the BWI) allows the basin riparians to reconsider arrangements that will secure their economies rather than face significant losses under extreme flow situations. Having probabilities assigned to the various flow values allows a cost-benefit analysis by policy makers who consider their interventions. The information can be used directly in Climate Smart Agriculture to promote cooperation for efficient and equable water use in agriculture, as well as serve as a quantitative measure to implement early warning strategies to mitigate the losses from limited water supplies.

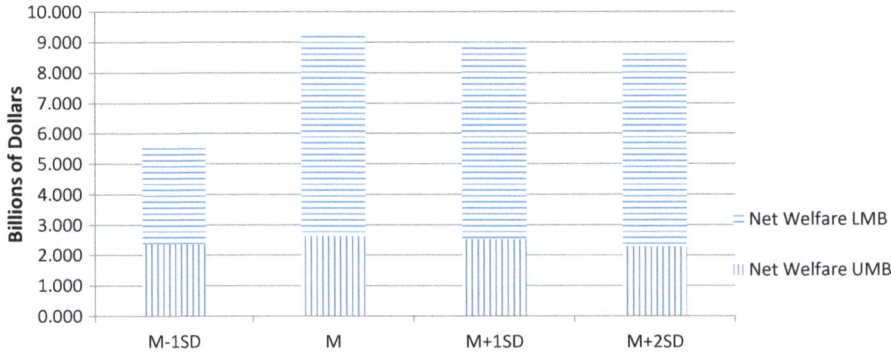

**Fig. 13** Net benefits in the Mekong basin as a function of flow distribution. *M* mean, *SD* standard deviation

## 4 Concluding Discussion

This chapter demonstrates several applications of the satellite derived surface wetness and temperature data to promote CSA. First, the early detection of growing conditions and predicting the availability of food directly improves climate resilience and food security. Second, insurance (risk management) programs can use the indexes in triggers for a quick release of catastrophic bonds to farmers adversely impacted by the weather in order to mitigate the impact of crop failure. Third, these tools provide information to educate farmers about the viable yields from various crops under current and changing climatic conditions. Fourth, an early warning system distributed across the globe can help identify and expedite the exportation of food supplies from areas where they are in excess into areas where a deficiency is likely to occur.

The BWI has skill to predict river flows in several geographies and locations around the world, where it captured the integration of rainfall, melting snow cover, the change in wetland areas in a quantitative measure of river flow. It also provides a quantitative measurement that is independent of local governmental reports. We realize that more sophisticated models can generate more accurate calculations of flow. However these models require detailed parameterizations and assumptions, which means they are difficult to run and maintain, and they must be trained for each basin. Whereas the approach taken in this study is a simple, yet robust variable that has expanded application and portability to other basins and periods of time beyond the calibration time and location. This expands the accuracy and utility of the product for CSA.

In terms of adding new variables to interact with the wetness and temperature products, the Normalized Difference Vegetative index (NDVI) is a natural complement, since it is a direct measurement of canopy greenness. The three products together can be used as a superior signal of crop conditions and potential yields. The CSA will benefit directly by improving near real time monitoring capacity. In this situation the synergy between the three observations can create a superior tool for crop yield predictions, insurance triggers, trends and return period of extreme events, all of which improve climate resilience.

In order to maximize the skill of crop prediction models, it is essential to calibrate the models with reliable yield data from at least 10 years and preferably 20–25 years. Most countries collect field data and calculate yields, however the spatial resolutions of the values can range from county (districts) to province (states, oblast), all the way to country-wide estimates. Since these yield values are always best guesses, CSA needs independent, objective and transparent tools to assess the food production at the regional level in across the globe in near real time. This is a particularly important requirement, since many countries do not release their best estimates; instead the data they do release is manipulated data for national security, political and economic reasons. Consequently, models based on these yield data lack both skill and confidence in their predictions. One approach is to use analogues from areas that grow the same crop and share similar climate, soils, and irrigation practices. In this case, the models developed in the analogue region can be applied to the target area.

Another application to the CSA is using the indexes and predictions as triggers to release catastrophic bonds to farmers having substantial crop failure. There are several advantages to index-based insurance that support CSA.

1. The cost of the premium is substantially lower than the traditional indemnification insurance programs, since no adjuster or field survey are required.
2. The funds are released in near real time, mitigate the impact of the financial losses of the harvest.
3. It is an objective program that can be readily underwritten by numerous sources, thereby the distribution of the losses through various government and financial institutions, reducing exposure to a particular organization. Insurance based on a composite of indexes (used as triggers) has been tried with some success. However, one of the major obstacles is confidence in the triggers by both the insurance companies and the farmers. One intention of the study is to support the CSA's ability to identify reliable and easy to apply triggers in the crop insurance industry.

The value of the wetness index for monitoring and predicting river flow is multifold.

1. Improved knowleddge on the distribution of water resources and the probability of various levels of water for agriculture, commercial, industrial and human consumption is critical to sustainability and development strategies.
2. Mitigate the impact of flood and drought with a reliable early warning system, which provides valuable lead-time about upcoming extreme events.
3. Provide a reliable and objective source of information about the available water resources, in planning and promoting water sharing between riparian states .
4. Use objective measurements to establish an insurance program that protects sectors of society against extreme events, and provides financial compensations for mitigating impacts on infrastructure and society's welfare.

We introduced a model to demonstrate how to qunatify the value on water resources in various sectors of society. The model broke the impacts across the agriculture, fishing, commercial and human consumption. Ther are many benefits to use the BWI to quantify these relationships, in terms of social and economic costs/benefits related to water resource management and mitigation strategies against extreme events. This chapter demonstrates the application of both the wetness and temperature data for monitoring growing conditions and predicting yields, which directly support CSA around the world. We plan to integrate these products with various datasets, such as in situ surface temperature, the greenness index, and soil moisture data, in order to expand their complementary value and utility. We are excited about collaborating with organizations that would like to apply these products in various sectors. Since the data is global and has more than 25 years of observations, we believe that the potential for application is vast and look forward to developing that potential in many areas. The goal is to assist the CSA by applying these products to support resource management, food security, climate resilience, as well as mitigate the adverse impacts of extreme events.

# References

Adger, N., T. Hughes, C. Folke, S. Carpenter, and J. Rockström (2005), Social ecological resilience to coastal disasters. *Science*, 309, 5737,1036–1039.
Ambec, S., A. Dinar, and D. McKinney (2013), Water sharing agreements sustainable to reduced flows, *Journal of Environmental Economics and Management*, 66(3), 639–655.
Basist, A., Grody, N. C., Peterson, T. C., and Williams, C. N. (1998), "Using the Special Sensor Microwave / Imager to Monitor Land Surface Temperatures, Wetness, and Snow Cover," Journal of Applied Meteorology, 37(September): 888–911.
Basist, A., C. Williams Jr, T. F. Ross, M. J. Menne, N. Grody, R. Ferraro, S. Shen, and A. T. C. Chang (2001), Using the Special Sensor Microwave Imager to monitor surface wetness, *Journal of Hydrometeorology*, 2(3), 297–308.
Becker-Reshef I, Vermote E, Lindeman M, Justice C (2010) A generalized regression-based model for forecasting winter wheat yields in Kansas and Ukraine using MODIS data. Remote Sensing of Environment 114: 1312–1323
Blankespoor, B., A. Basist, A. Dinar and S. Dinar (2012), Assessing Economic and Political Impacts of Hydrological Variability on Treaties: Case Studies of the Zambezi and Mekong Basins Policy Research Working Paper No. 5996, 1–56 pp, World Bank, Washington, DC.
Demirel Mehmet C., Martijn J. Booij and Arjen Y. Hoekstra (2013) Identification of appropriate lags and temporal resolutions for low flow indicators in the River Rhine to forecast low flows with different lead times Hydrological Processes. 27(19): 2742–2758,
Deryng, D., W. J. Sacks, C. C. Barford, and N. Ramankutty, 2011: Simulating the effects of climate and agricultural management practices on global crop yield. GLOBAL BIOGEOCHEMICAL CYCLES, VOL. 25, GB2006, 1-18.
Dinar, A., B. Blankespoor, S. Dinar, and P. Kurukulasuriya (2010), Does precipitation and run-off variability affect treaty cooperation between states sharing international bilateral rivers?, *Ecological Economics*, 69(12), 2568–2581.
Dinar, S., D. Katz, L. De Stefano, and B. Blankespoor (2015), Climate Change, Conflict, and Cooperation: Global Analysis of the Effectiveness of International River Treaties in Addressing Water Variability. *Political Geography*.
Drieschova, A., M. Giordano and I. Fischhendler (2008), Governance mechanisms to address flow variability in water treaties. *Global Environmental Change*, 18, 285–295.
Drury, A. C., and R. S. Olson (1998), Disasters and Political Unrest: An Empirical Investigation, *Journal of Contingencies & Crisis Management*, 6(3), 153.
(FAO) Food and Agricultural Organization of the United Nations, Mekong River Commission Secretariat and Department of Irrigation, Ministry of Agriculture and Forestry of LAO P.D.R. (1999), Flood Management and Mitigation in the Mekong River Basin, 40pp, FAO, Bangkok. Accessed 2014–10 at: http://www.fao.org/3/a-ac146e/AC146E01.htm
Gutman, Nsthaniel B. 1999: Accepting the standardized precipitation index: A Calculation algorithm, Journal of the American water resources association. Vol. 35, No.2, 311–322.
Hendrix, C. S., and I. Salehyan (2012), Climate change, rainfall, and social conflict in Africa, *Journal of Peace Research*, 49(1), 35–50.
Houba, H., Kim Hang Pham Do, and X. Zhu (2013), Saving a river: a joint management approach to the Mekong River Basin, *Environment and Development Economics*, 18:93–109.
Intelligence Community Assessment (2012), Global water security. Office of the Director of National Intelligence, February 2.
Jackson, T. Passive microwave remote sensing of soil moisture and regional drought monitoring, (2005). V:89–104. in Boken, V. (ed.) Monitoring and Predicting Agricultural Drought. Oxford Univ. Press
Jury, W. A., and H. Vaux (2005), The role of science in solving the world's emerging water problems, *Proceedings of the National Academy of Sciences of the United States of America*, 102(44), 15715–15720.

Larson, D. F., A. Dinar, and B. Blankespoor (2015), Aligning Climate Change Mitigation and Agricultural Policies in ECA, in Asia and the World Economy, edited by J. Whalley, pp. 69–151, World Scientific, Singapore.

Mekong Water Commission (2009), Annual Report. http://mwcmekong.org.

Miller, K., and D. Yates (2006), *Climate change and water resources: a primer for municipal water providers*, 83 pp., American Water Works Research Foundation and UCAR, Denver, CO.

NASA (2010), Flooding on the Zambezi River: Natural Hazards, edited, NASA, http://earthobservatory.nasa.gov/IOTD/view.php?id=44132.

Nel, P., and M. Righarts (2008), Natural Disasters and the Risk of Violent Civil Conflict, *International Studies Quarterly*, 52(1), 159–185.

Palmer, M. A., C. A. Reidy Liermann, C. Nilsson, M. Flörke, J. Alcamo, P. S. Lake, and N. Bond (2008), Climate change and the world's river basins: anticipating management options, *Frontiers in Ecology and the Environment*, 6(2), 81–89.ds

Tucker, C.J., M. E. Brown, J. E. Pinzon, D. A. Slayback, R. Mahoney, N. E. Saleous, and E. F. Vermote: 2005, "An extended AVHRR 8-km NDVI dataset comparable with MODIS and SPOT Vegetation NDVI data," *Int. J. Remote Sens.*26:4485–4498.

Williams, C., A. Basist, T. C. Peterson, and N. Grody 2000: Calibration and Verification of Land Surface Temperature Anomalies Derived from the SSMI, Bull. Of the Amer. Meteor. Soc. 2141–2156.

Winsemius, H. C., H. H. G. Savenije, A. M. J. Gerrits, E. A. Zapreeva, and R. Klees (2006), Comparison of two model approaches in the Zambezi river basin with regard to model reliability and identifiability, *Hydrol. Earth Syst. Sci.*, 10, 339–352.

**Open Access** This chapter is distributed under the terms of the Creative Commons Attribution-NonCommercial-ShareAlike 3.0 IGO license (https://creativecommons.org/licenses/by-nc-sa/3.0/igo/), which permits any noncommercial use, duplication, adaptation, distribution, and reproduction in any medium or format, as long as you give appropriate credit to the Food and Agriculture Organization of the United Nations (FAO), provide a link to the Creative Commons license and indicate if changes were made. If you remix, transform, or build upon this book or a part thereof, you must distribute your contributions under the same license as the original. Any dispute related to the use of the works of the FAO that cannot be settled amicably shall be submitted to arbitration pursuant to the UNCITRAL rules. The use of the FAO's name for any purpose other than for attribution, and the use of the FAO's logo, shall be subject to a separate written license agreement between the FAO and the user and is not authorized as part of this CC-IGO license. Note that the link provided above includes additional terms and conditions of the license.

The images or other third party material in this chapter are included in the chapter's Creative Commons license, unless indicated otherwise in a credit line to the material. If material is not included in the chapter's Creative Commons license and your intended use is not permitted by statutory regulation or exceeds the permitted use, you will need to obtain permission directly from the copyright holder.

# Early Warning Techniques for Local Climate Resilience: Smallholder Rice in Lao PDR

**Drew Behnke, Sam Heft-Neal, and David Roland-Holst**

**Abstract** As part of the Regional Rice Initiative Pilot Project, UNFAO has committed resources to support policy dialog and decision capacity related to climate change adaptation and mitigation in agriculture, with particular attention to food security and the rice sector in Asia and the Pacific. This initiative includes sponsorship of research to deliver information and knowledge products for policy makers to better manage climate risks to the rice sector and identify adaptation needs for the rice sector in Lao PDR. In the following pages, we report on progress of one component of this activity, econometric estimation of long term impacts that climate change can be expected to have on rice yields. The work reported here is preliminary and should not in its current form be used as a basis for policy.

## 1 Introduction

The report presents a new approach to estimating how climate conditions affect rice production in Lao PDR and modeling the associated potential future impacts of climate change in the rice sector. To conduct our analysis, we use advanced econometric models to estimate the historical relationship between observed rice yields and weather inputs. We then downscale projections from leading climate models to evaluate potential future climate conditions in Lao PDR and implement the econometric models to estimate rice yields under these climate scenarios.

The organization of this report is as follows. First, we provide background and review weather and rice production conditions in Lao PDR as well as summarize the role of weather inputs in rice yields. In addition to average weather conditions,

---

Originally published by UNFAO as RR Nr. 10-13-1; November 2013

D. Behnke (✉)
Department of Economics, University of California Santa Barbara, Santa Barbara, CA, USA
e-mail: dbehnke@umail.ucsb.edu

S. Heft-Neal • D. Roland-Holst
Department of Agricultural and Resource Economics, University of California Berkeley,
207 Giannini Hall, Berkeley, CA 94720-3310, USA
e-mail: drwh@berkeley.edu

special attention is devoted to extreme events such as floods and droughts that can play disruptive roles in rice production. Next we review methodologies used in the literature and discuss the statistical approach employed here in order to estimate the relationship between weather and observed rice yields. Again, we include both average weather and measures of natural disasters in our analysis. Finally, we provide an overview of climate models and apply climate projections to our statistical models of rice yields in order to evaluate potential impacts of climate change on rice yields in Lao PDR.

## 2 Background

The following section provides an overview of rice growing conditions in Lao PDR. Weather inputs, the occurrence of extreme events, and rice production systems are all discussed in order to provide context for the subsequent analysis.

### 2.1 Overview of Climate Conditions

Total rainfall during the rice-growing season in Lao PDR ranges from about 100–170 cm. However, year-to-year rainfall is highly variable. Moreover, even years with identical levels of total rainfall can have very different growing conditions depending on the pattern of rainfall arrival. Monthly rainfall generally rises each month from the beginning of the growing season until it peaks in August and then decreases thereafter as illustrated in Fig. 1 (both panels).

There is also significant variation in growing season temperatures across Lao PDR. Figure 1 shows the geographical distribution of growing season conditions across space and time. Average minimum (nighttime) temperatures during the growing season range from approximately 20–24 °C, while average maximum (daytime) temperatures range from 28–32 °C. It should be noted however, that these averages mask much of the underlying variability in temperature. For example, average temperature varies across the growing season, where the beginning of the season is typically several degrees hotter than the end of the growing season. Moreover, daily maximum temperatures can exceed 40 °C. Extreme heat, particularly if sustained over several days, puts additional stress on rice growth and may cause large damages (Wassmann et al. 2009b).

### 2.2 Extreme Events

While average climate conditions play an important role in average rice yields, extreme events can cause large impacts that may not be captured by seasonal averages. For example, a year with early season drought and late season floods may

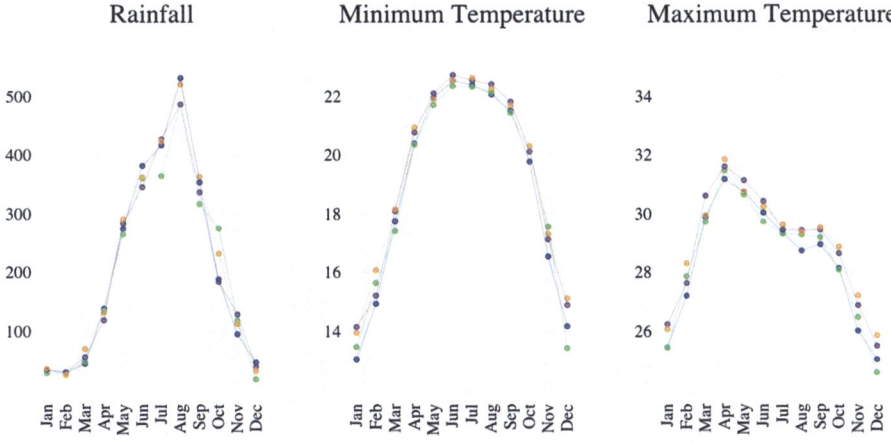

blue = 1970s, green = 1980s, purple = 1990s, orange = 2000s

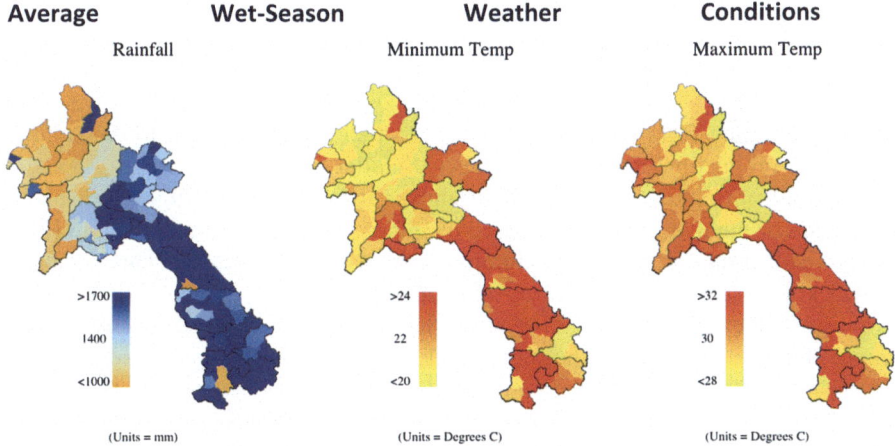

*Source: CRU (see section 2.2)*

**Fig. 1** Decadal changes in seasonal weather conditions (two panels)

record normal growing season rainfall totals while resulting in significant crop damage. Furthermore, rather than contributing to lower annual yields, extreme events may cause the rice planted area to be damaged, resulting in significant loss of the planted crop, which can be devastating to farmer livelihoods. In order to address this important facet of the climate-rice production relationship, we incorporate effects of both average climate and extreme extreme weather events on rice yields.

The majority of rice production in Lao PDR is rain-fed and consequently droughts pose a serious threat. In addition to water shortage, flooding is also a common danger to Lao and other Southeast Asian rice production. In fact, regular

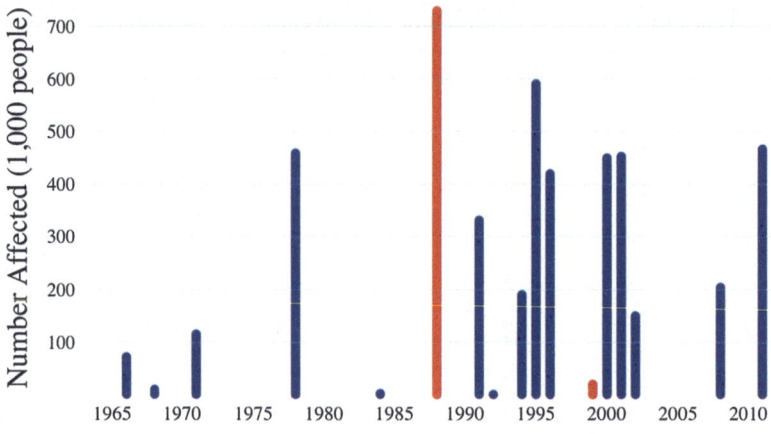

Source: EMDAT database (www.emdat.be)

**Fig. 2** Population affected by major flood or drought events in Lao PDR. *Blue* represents floods and *red* represents droughts. Note that regional floods and droughts are not included in the figure. Consequently, the figure represents only the largest scale events that have been recorded in this international database of natural disasters

seasonal flooding from the Mekong River is often a greater threat to the central region rice production than water shortages (Schiller et al. 2001).

The toll from extreme flooding and droughts can be significant. Figure 2 displays the estimated number of people affected by major floods and droughts in Lao PDR as recorded in the international natural disaster database EMDAT.[1] This database provides statistics for the number of people affected by particular large-scale extreme weather events. It should be noted that smaller regional scale events are not recorded in the database and thus not included in the figure. It should also be noted that many of the people affected by these disasters may not be farmers. That being said, farmers are particularly vulnerable to droughts and floods because their livelihoods can be negatively affected. Nonetheless, the EMDAT database provides insight into the potential magnitude of these effects. According to the database, there have been six floods in the last 20 years that affected at least 300,000 people in Lao PDR. Major droughts, although less common than floods, can also exact large damages. In fact, the biggest event in the database is a late 1980s drought that affected more than 700,000 people in Lao PDR.

To address the shortcomings of the EMDAT data we consider the direct impact of flooding and droughts on rice yields in subsequent sections. The data that we use in our analysis, which comes from the Department of Agriculture and is described further in Section 4, is more precise and includes annual damaged rice area for each district that resulted from drought, floods, or pests (Fig. 3).

---

[1] Available online at www.emdat.be.

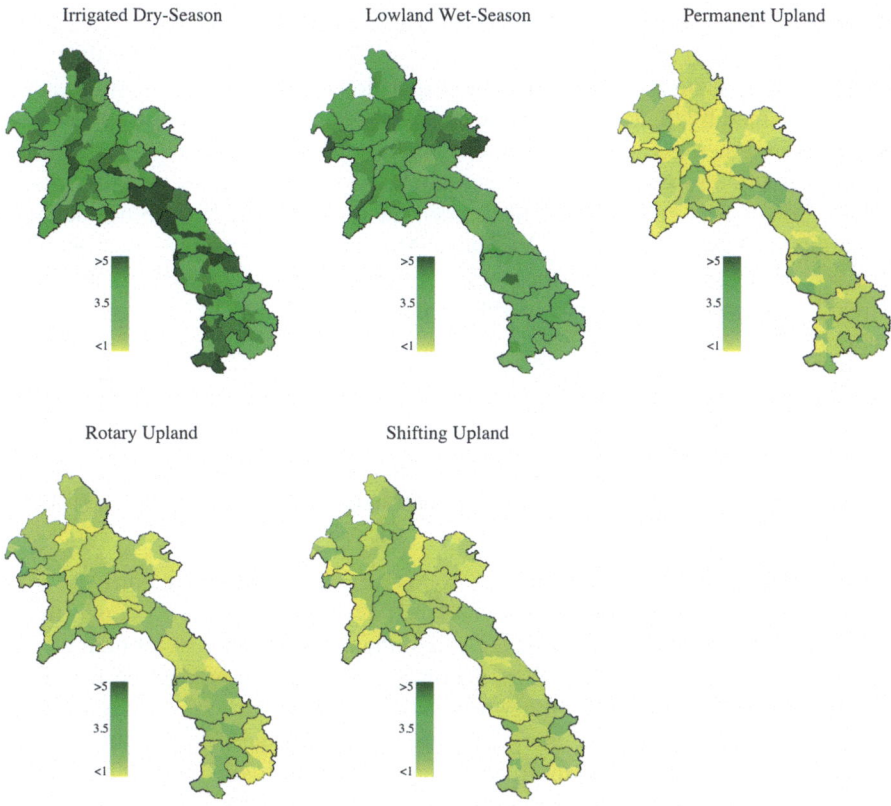

Source: Crop Statistics Yearbook (DOA, Lao PDR)

**Fig. 3** Average rice yields. Maps show average rice yields by rice production system. Data cover the period 2006–2012

## 2.3 Rice Production

As a culturally significant, staple food crop, rice has an important role in the economy of Lao PDR. Because of this, the rice production sector has been the focus of various political policies in order to increase production and maintain food security. As a result, Lao PDR has undergone significant transitions in the sector over the past several decades, moving from a net rice importer in the 1970s and 1980s, to a stable and increasing surplus over the last decade.

The introduction of improved seed varieties in the 1970s as well as loosening of price controls in the early 1980s led to some production increases, but the majority of growth occurred in the 1990s. Over the last 20 years, rice production has more than doubled to reach nearly 3.5 million tones of paddy in 2012 (DOA 2012). This represents an average of 5.1% annual growth, which is one of the highest in the region over this time period. This high growth can be attributed both to the yield improvements (from new, improved seed varieties and increased use of fertilizer) as well as land expansion. Growth from land expansion over the previous two decades can be explained by the steady increase in lowland, rain fed production systems as well as a rapid increase in dry season irrigated production. Concurrently, the lower yield, upland rice production system saw total area steadily fall. Regionally, much of this growth was concentrated in the central plain provinces of Savannakhet, Khammuane, Vientiane, and the Vientiane Municipality as well as the southern province of Saravan. In total, these five provinces comprised 70% of the total increase in rice production between 1995 and 2010 (MAF 2012).

### 2.3.1 Production Systems

Rice production systems can be categorized into one of five different categories: lowland wet-season, lowland irrigated dry-season, upland permanent, upland rotary, and upland shifting.

**Lowland Wet-Season** Lowland wet-season is responsible for the majority of production, representing 79% of the total yield in 2012. This production system is most common in the central and southern regions of the country with 83% of total yields coming from these areas (DOA 2012). Lowland wet-season production has relatively high yields compared to other production systems with an average of 3.91 tons per ha in 2012. Given the comparatively high yields, and ubiquity of production along the populated Mekong River Valley, lowland wet-season will remain the most important ecosystem for rice cultivation in the foreseeable future.

That being said, lowland wet-season production faces a variety of production constraints. First and foremost, is the constraint from climatic variability, as the production system is reliant on weather inputs for the production process. Rainfall is identified as a particular concern among farmers, as the rainfall pattern can vary from year-to-year, resulting in large fluctuations in production. Furthermore, the permeable nature of the sandy soils that prevail in much of the Mekong River Valley means drought is common occurrence. Temperature is of course an issue as well, as extreme temperature events are known to be harmful to rice production and the random nature of such events means farmers and unable to anticipate temperature shocks (Schiller et al. 2001).

Related to climatic variability, is the problem of insect pests that are rated by farmers as being among the top three production constraints. The relationship between pests, climatic variability, and production is not clearly understood, although it is understood that pests are believed to significantly impact yields and climate plausibly affects the prevalence of pests (Schiller et al. 2001).

**Irrigated Dry-Season** Dry-season production occurs under irrigated conditions only. During the 1990s, the irrigated dry-season production system saw a rapid increase in production as part of the official national policy to support the continued development of small-scale irrigation schemes. The expansion of the irrigated system was promoted in order to increase national rice production, while at the same time reducing the year-to-year variability associated with wet-season production. Over the 2011–2012 dry-season growing season planted area totaled 108,000 ha representing approximately 11% of the national crop. Although this is a large increase from the 13,000 ha planted in 1992–1993, it represents only a modest increase from the 87,000 ha planted in 1998–1999. Furthermore, there is a large disparity from the MAF's projected goal of 180,000 ha of production by 2005 (DOA 2012; Schiller et al. 2001).

Due to the intensive nature of irrigated production, the majority of production is concentrated in a few provinces that can support this system. The central region is home to nearly 68% of the total irrigated dry-season planted area, with production being highly concentrated in the Vientiane Capital and Savannakhet (19% and 29% of total area planted respectively). Yields are the highest in this production system with 4.72 tons per ha on average over the 2011–2012 season (DOA 2012). This is unsurprising as the adoption of improved rice production technology is highest in the irrigated areas both as a combination of better extension services and higher farm incomes.

In regards to production constraints, temperature likely plays a larger role for the irrigated production system, as dry-season temperatures are initially cool before dramatically increasing toward the end of the season. Especially of concern are low temperatures in the north where temperatures can fall below 5 °C. In southern and central Lao PDR, the high temperatures during March and April that can coincide with flowering and grain filling are of primary concern (Schiller et al. 2001).

**Upland** Upland rice cultivation in Lao PDR is split between three production categories; permanent, rotary, and shifting. Estimates vary about the size of these systems, as they are predominantly located in the remote, mountainous northern and eastern regions of the country. Furthermore, due to remote nature of these systems accurate yield measurements are next to impossible. Often upland rice plots are not clearly marked and typically grow in combination with forest trees and other crops. Furthermore, much of the production is in remote areas with limited to no road access and inadequate resources and staff to accurately record yields.

That being said, some estimates for upland production do exist. In the early 1990s it was estimated that 2.1 million ha (or 8.8% of the national territory) was being used for slash-and-burn cultivation (Schiller et al. 2001). By 2000, it was estimated that about one third of the population still relied on shifting cultivation systems, covering about 13% of the of the total land area of the country (ADB 2006). In regards to rice production only, official data reports there was 119,000 ha of upland rice planted in 2012 representing approximately 12% of the total planted area of rice. Of this, approximately 47% was classified as a permanent upland system (DOA 2012). Furthermore, the DOA reports data on two types of slash-and-burn systems referring to them as either "rotary" or "shifting," but has no explicit information on the differences between these systems.

Much like other production systems, there is a strong regional trend in the upland production system. The northern provinces accounted for over 73% of the total area planted, with Luangprabang responsible for 18% of the total area alone. Yields are low in the upland system and relatively much lower than the other production types with an average yield of 1.8 ton per ha (DOA 2012).

In regards to production constraints, the upland production system has both similar and unique limitations to production. Climatic variability is again a major concern, as farmers must rely on the weather for inputs into production. However, biotic constraints are a much larger concern for the upland system than others. Weeds and rodents were highlighted as the two largest limitations to production for upland farmers (Schiller et al. 2001). Additionally land pressure and pressure for the government have limited production. Traditionally, farmers would clear the forest with fire and after growing rice for a year or two, land would be left to fallow for 10–20 years before returning. However, increased population pressure and land-use restrictions have led to a reduction in fallow periods to as short as 3 years (ADB 2006). Without the necessary time for the land to restore fertility, production is adversely impacted and furthermore such a system is unsustainable ecologically.

### 2.3.2 Irrigation

As previously discussed, irrigation in Lao PDR increased dramatically during the mid-1990s and early-2000s under the government's official policy to expand coverage. During this time, large investments were made to install high-capacity pumps along the Mekong River and its tributaries to expand small-scale irrigation opportunities for smallholders. As a result of the government's expansion efforts, irrigated area increased from about 12,000 ha in 1990 to 87,000 in 1999, representing a seven-fold increase (Pandey 2001). Growth was even more rapid in the early 2000s, eventually reaching peak coverage of over 500,000 ha in 2006 before declining slightly to the current 400,000 ha of coverage in 2012 (DOA 2012).

## 2.4 The Physiological Relationship Between Rice and Weather Inputs

### 2.4.1 The Role of Water

Rice production, more than most crops, is highly dependent on water availability, both in terms of quantity and timing of application. At some points during the growing season rainfall is highly beneficial, while at other times during the season it can be harmful. Too much or too little rainfall at any stage of rice growth can cause partial or total crop failure (Belder et al. 2004). Excessive water can lead to partial submergence of the rice plant, which reduces yields. In one experiment, Yoshida (1981) reports that 50% of plant submergence during any of the growth phases led to a 30–50% reduction in yields. However, while excessive water damages rice crops, drought is widely recognized as the primary constraint for rain-fed rice production (Bouman et al. 2005, 2007). Insufficient water causes plant mortality and a wide range of stresses that can lead to spikelet sterility, incomplete grain filling, stunting (Yoshida 1981), delayed heading (Homma et al. 2004), and other adverse yield effects.

Prior to planting, water is also important for rice production as an input to field preparation. In rain-fed production systems, insufficient early rainfall can force farmers to delay planting. Although data in Lao PDR are not available, Sawano et al. (2008) studied the relationship between rainfall and planting dates in rain-fed areas of northeast Thailand, an area that is geographically similar to the central plains of Lao PDR. The authors concluded that, depending on field-level water availability from rainfall, planting dates were locally distributed over an approximately two-month period, while local harvesting took place around the same time everywhere. The implication is obvious – delayed planting from insufficient early season water resources can significantly shorten the growing season and thus reduce output. It remains unclear why farmers who delayed planting did not delay harvest. While the authors did not offer any conclusive answers for this question, they suggested that farmers may not want to delay harvesting in order to prevent interference with subsequent growing seasons, marketing considerations, and other farm and nonfarm activities.

### 2.4.2 The Role of Temperature

Sunlight is another essential input into rice production – rice plants require solar radiation for photosynthesis and heat to promote tissue growth. There are a number of ways to measure energy requirements, the simplest being average temperature. Other related measures include other temperature boundaries (e.g., daily min T, daily max T), agronomic measures such as Growing Degree Days (GDD), and radiation measures.

Generally, extreme highs and lows are of concern to crop growth. However, at the range of temperatures experienced by rice growers in Lao PDR, extreme lows are unlikely to harm rice growth, but extreme highs are a greater threat.[2] Extreme high temperatures hurt plant growth because it causes heat stress, which delays the growth process (Yoshida 1981; Wassmann et al. 2009b). Furthermore, researchers have highlighted the difference between extreme high nighttime (minimum) temperatures and extreme high daytime (maximum) temperatures. The respiration process appears to make rice plants particularly sensitive to nighttime temperature (Yin et al. 1996). Several studies have highlighted nighttime temperatures as a driving factor of rice growth, where elevated minimum nighttime temperatures greatly reduce rice yields (Yin et al. 1996; Peng et al. 2004; Welch et al. 2010). Using a laboratory experiment to artificially manipulate temperatures, Yin et al. (1996) demonstrate that a one-degree increase in nighttime temperature has a large negative

---

[2] Both daytime (daily maximum) and night-time (daily minimum) extreme highs are potentially harmful to rice yields.

effect on rice yields whereas a one-degree increase in daytime temperature has a slightly positive effect. In fact, across most observed ranges of maximum temperatures, higher daytime temperatures have generally been found to positively affect rice growth (Peng et al. 2004; Welch et al. 2010), however, as temperatures continue to rise, they eventually become harmful. The threshold where maximum daytime temperatures become detrimental to rice growth depends largely on genotype and local growing conditions (including e.g., soils and water availability). For example, depending on genotype and field conditions, Wassmann et al. (2009a) estimated an average cutoff for maximum temperature of 31 °C, beyond which "growth and productivity (yield) rapidly decrease". However, these estimates come from experimental rather than field results, which may not be representative of adaptive, farmer-managed fields where some precautions may be taken when temperatures become potentially harmful. Consequently, if we believe that farmers can effectively ameliorate the effects of extreme temperature through management practices, or through use of local varieties selected for heat resistance qualities, then we might expect observed field data to exhibit higher thresholds.

## 3 Analysis I: Estimating the Relationship Between Rice and Climate Change

This section constitutes the first part of our analysis, where we estimate the relationship between observed historical rice yields and weather conditions in Lao PDR. The following section will use the observed relationship to project yields under potential future climate scenarios. In this we first describe the data and methods used, then describe our primary results. Full model results are presented in tables in the appendix.

### 3.1 Methods

Climate change is a long run phenomenon and it is difficult to distinguish historical climate change from short to medium run weather cycles. In order to estimate potential climate change impacts on agriculture, researchers often estimate the short-term relationship between weather inputs and yields and then apply this relationship over the range of future conditions predicted by climate models. While this approach is imperfect[3], it allows us to provide an approximate estimate of future climate impacts.

In general, two approaches have been taken to characterize the relationship between weather inputs and rice yields. First, in agronomic studies, usually involv-

---

[3] One needs to be particularly careful about extrapolating current relationships to future unexperienced ranges of climate conditions.

ing laboratory or experimental fields, rice plants are placed under different types of environmental stresses and physiological responses are measured (e.g., Borrell et al. 1997; Homma et al. 2004; Yin et al. 1996). An extension of this approach is to use field data to calibrate crop models that simulate the physiological growth process. Perturbing the inputs in these models can in turn generate predictions of crop growth under potential future climate conditions.

The second approach, which we take here, applies statistical models using plausibly random variations in weather to estimate the effects of weather conditions on observed rice yields. We exploit the presumably random year-to-year variation in temperature and precipitation to estimate whether rice yields are higher or lower in years that are warmer and wetter. With the relationship firmly established, we then use climate projections to model how climate change will affect yields.

In a controlled lab experiment, scientists repeatedly carry out procedures that are identical except for one factor of interest, which is manually manipulated in order to measure the causal impact of said factor on the outcome. As with many social science settings, this type of experiment is not possible for the Lao PDR rice sector. Thus we rely on existing data to demonstrate the impact of historical weather realizations on yields and model the impacts of climate change once this relationship has been established. It should be noted that overall yields have increased over the study period due in large part to technological advances. Consequently, our estimates represent losses with respect to the counterfactual scenario of no climate change. Losses due to climate change do not imply that the yield trends are downward sloping, only that yields have been, and will continue to be, lower in the face of climate change than they would be otherwise. This distinction does not change the fact that climate change has potential to have strongly negative impacts on the rice sector in Lao PDR.

Typically, statistical studies use average growing season (or sub-season) conditions, to represent the weather inputs in the production function. The simplest approach estimates yields (calculated as *log(yield)*) as a function of mean temperature, mean precipitation, and their squares. However, several studies have emphasized the differential effects of minimum and maximum temperature (Yin et al. 1996; Peng et al. 2004; Welch et al. 2010), the importance of including radiation (Sheehy et al. 2006; Welch et al. 2010), and the differential effects across phases of the growing season (Welch et al. 2010). In addition, there has been extensive research on water requirements for rice production in irrigated (Bouman et al. 2005, 2007) or rain-fed settings (Xu and Mackill 1996; Sharma et al. 1994; Wade et al. 1999).

Our goal is to provide a localized analysis for Lao PDR. In order to do so, we seek to incorporate the main methods and findings from these disparate sources into statistical models that estimate the impact of climate on rice types grown particularly in Lao PDR. This analysis, in turn, will be used to inform policy prescriptions and identify the production systems and rice growing areas that are most vulnerable to adverse changes in growing conditions.

### 3.1.1 Average Weather Models

We begin with an approach of estimating the effects of climate on rice yields using a panel regression with a single growing season metric for each weather covariate (average min T, max T, and precipitation across the growing season). Using average seasonal conditions, we estimate a linear model for each rice production system. These are later used to predict yields under various climate scenarios.

Here, we present a variation of the panel fixed effects (FE) model. This model is an accepted and commonly applied model in the literature (see e.g. Lobell and Burke 2010). Panel data contains repeated observations of the same units over time. In this case we repeatedly observe district rice outcomes. Panel data allows the use of fixed effects, which control for a variety of observations that are unobserved. By conditioning on fixed effects, county specific deviations in weather from the county averages are used to identify the effect of weather on yields. Specifically we chose to control for district and year fixed effects. District fixed effects control for any unobservable characteristic that varies across district but is constant over time. This accounts for important differences across districts such as soil conditions or areas with a higher prevalence of intensive production systems. Year fixed effects control for any unobservable characteristic that varies across years but is constant across all districts. This includes national time trends such as improved technology (irrigation, fertilizer use, or the introduction of improved seed varieties for example).

Within this framework there are a number of choices/assumptions to be made. In each case, there is a tradeoff between controlling for unobserved factors and observing enough variation in the data to be able to make econometric estimations. In reality, we know that there are many factors that affect crop yields, including soil quality, technology, agrochemicals, endogenous behavior, etc. Here, we are only considering the impact of weather, while the other factors are unobserved by us. Thus we are trying to estimate the disaggregated yield impact of weather holding constant other explanatory variables. If district-level time-series data were available on other factors such as agricultural investment, fertilizer use, or pesticides, then we could include these explanatory variables in our model. However, to our knowledge these data do not exist at the required resolution. Fortunately, the fixed effects model attempts to control for these unobserved factors, so that we can still produce unbiased estimates of climate effects. In other words, we can control for a variety of unobserved characteristics but cannot estimate them in our model. We are not attempting to explain every factor that affects yields, but merely to identify the effect of temperature and rainfall. Given our interest is ultimately how yields will change in the face of new climate conditions this does not affect our analysis.

The following reduced form model is our primary empirical specification. In our ideal specification we would have a vector of controls for the other factors that affect yields that we have previously discussed. This would include characteristics such as fertilizer use, pesticide use, soil quality, etc. However, data of this quality does not exist in Lao PDR, which is why we rely on fixed effects.

Equation 1: Panel Model of Average Weather Effects

$$\log(Y_{dt}) = \gamma_d + \theta_t + \beta_1 \text{MinT}_{dt} + \beta_2 \text{MaxT}_{dt} + \beta_3 P + \varepsilon_{dt} \quad (1)$$

$Y_{dt}$ is yield for district d in year t. The model includes district fixed effects $\gamma_d$ and year fixed effects $\theta_t$. $\beta_{1-3}$ represent the coefficients on our weather variables

One of the fundamental assumptions we have to make is that individual specific time series variation is a valid source of variation for identifying causal effects. In other words, our model assumes that, for each district, weather variation from year-to-year is random. It is obviously not true that weather is random over space (i.e., we expect that some parts of the country to get more rain than other parts every year) but we argue that it is reasonable to assume that deviations from local averages in one year are unrelated to deviations from local averages in the next year.

The modeling approach in equation 1 makes the strong assumption that the effect of weather on yields is the same over different ranges. For example, the linear model assumes an increase in maximum temperature from 29 to 30 has the same effect as an increase from 33 to 34. This is a very strong assumption and other researchers (Schlenker and Roberts 2009) have found a nonlinear relationship between temperature and yields. Therefore, to add robustness to our analysis we also consider a non-linear model as seen in equation 2. This model adds square terms for the climate variables used in equation 1, which allows us to consider if there is a threshold at which the relationship between weather and yields changes. Ideally, we would like to estimate a piece-wise linear model that estimates different slopes over different ranges of covariates. However, given our limited number of observations, a piece-wise model is not advised as it will increase the number of covariates and reduce the necessary power for statistical inference.

Equation 2: Panel Model of Average Weather Effects

$$\log(Y_{dt}) = \gamma_d + \theta_t + \beta_1 \text{MinT}_{dt} + \beta_2 \text{MinT}^2_{dt} + \beta_3 \text{MaxT}_{dt} + \beta_4 \text{MaxT}^2_{dt} + \beta_5 P + \varepsilon_{dt} \quad (2)$$

$Y_{dt}$ is yield for district d in year t. The model includes district fixed effects $\gamma_d$ and year fixed effects $\theta_t$. $\beta_{1-5}$ represent the coefficients on our weather variables

### 3.1.2 Modeling Extreme Events

In addition to modeling the effects of average weather conditions on average rice yields, we can model the effects of drought and floods on rice losses with the same methodology. In equation 2, $L_{dt}$ represents rice losses[4] and $Dr_{dt}$ measures drought severity in district *d* and year *t*. Since our yield measures are annual, drought and

---

[4] Planted area that could not be harvested.

flood measures need to be aggregated annually. We will experiment with different aggregation methods.

Equation 3: Panel Model of Extreme Event Effects

$$\log(L_{dt}) = \gamma_d + \theta_t + \beta_1 Dr_{dt} + \beta_2 X_{dt} + \varepsilon_{dt} \qquad (3)$$

$Y_{pt}$ is yield for district d in year t. The model includes province fixed effects $\gamma_d$ and year fixed effects $\theta_t$. $\beta_1$ represents the coefficients on our drought measure. $X_{dt}$ are other controls.

## 3.2 Data

### 3.2.1 Rice Yields

Our rice yield data for Lao PDR come from the "Crop Statistics Year Book" published by the Department of Agriculture (DOA) within the Ministry of Agriculture and Forestry (MAF). These reports contain a wide variety of detailed crop production data at the district level and have been published annually since 2005. Unfortunately, rice production data before 2005 in Lao PDR is limited to province level aggregates that are of little use to our analysis, and district level rice production data is only available from 2005 through 2011. Although our panel is limited, it represents the most accurate and detailed rice production data in existence for this country. Rice production data is split between the five distinct production systems used in Lao PDR and these contain a variety of important statistics useful to our analysis. The variables in the data include planted area, harvested area, yield, and damaged area by source (drought, flood, etc).

### 3.2.2 Weather Conditions

It is inherently difficult to measure weather over space. Weather is observed at individual weather stations, and ideally want to have weather stations collecting data every few meters in order to capture variation in conditions over space. Of course, managing so many weather stations is impractical, and instead observed values are interpolated over locations in between weather stations. There are many different forms of weather data sets that have carried out this interpolation over different spatial and temporal resolutions. Each data set has its own advantages and drawbacks. Here we carry out our analysis with two separate weather data sets, known by the acronyms CRU and APHRODITE, described below. CRU data provide more weather variables (i.e., MIN, MAX) but at a lower temporal and spatial resolution. By including two completely different weather data sets we decrease the likelihood that our results will rely on the peculiarities of a particular data set.

The first weather data come from the Climatic Research Unit (CRU) at the University of East Anglia. The research group produces several global data products that include *monthly average* minimum (nighttime) temperature, maximum (daytime) temperature, mean temperature, and *monthly total* rainfall. We utilize the high-resolution gridded data sets[5] that have a resolution of 0.5 × 0.5 degrees globally. This translates to approximately 55 × 55 km at the equator. Each Lao PDR district is overlapped on the grid and area weighted averages are calculated in order to estimate monthly weather conditions for each district over the sample period.

The second data set, APHRODITE[6], is described by Yatagai et al. (2012). Researchers in Japan utilized a high density cluster of proprietary station data in order to create a high-resolution data set that includes daily average temperature and daily rainfall at a resolution of 0.05 × 0.05 degrees (~5 × 5 km). Although daily temperatures are useful, this data set does not contain minimum and maximum temperature information, and covers only Asia.

### 3.2.3 Extreme Events

Droughts

Although difficult to measure from seasonal rainfall and temperature data, researchers have begun to use remote sensing data from satellites to estimate drought severity. In the present analysis, we utilize a new measure developed by Mu et al. (2013) called the Drought Severity Index (DSI). Mu and colleagues produce global DSI measures from satellite data covering the globe averaged over eight day periods from 2000 through 2011 at a resolution of 0.05 × 0.05 degrees (~5.5 × 5.5 km). In theory, DSI values range from negative infinity to positive infinity, however, in practice most values are clustered around zero. Negative DSI values signify drier-than-normal conditions while positive values signify wetter than normal conditions. A zero value for DSI implies normal conditions. While it is an imperfect measure, DSI allows us to estimate district level drought severity across the rice-growing season and therefore estimate the effects of droughts on rice losses. Moreover, the drought patterns suggested by the DSI appear to be consistent with precipitation patterns observed in other data sets.

Floods

Like droughts, measuring flood extent is a practical difficulty that we address by using remotely sensed satellite data processed to estimate standing water extent. As far as the authors know, there are no available global remotely sensed flood measures. Consequently, as a second best option, we utilize DSI as a flood measure

---

[5] http://www.cru.uea.ac.uk/data.
[6] http://www.chikyu.ac.jp/precip/products/index.html.

where large positive values for DSI imply flooding. The developers of DSI note that flood measurement is a potential extension of DSI, but also caution that DSI has not been fully evaluated as a flood measure. Consequently, we proceed with caution using the best available flood measures to estimate the impact of flooding on rice production.

### 3.2.4 Data Limitations

There are significant constraints on data availability (and, inevitably, quality) for Lao PDR. First and foremost, detailed rice production statistics have only begun to be collected in recent years. Therefore, although we have a more than 40-year panel for weather, our analysis is limited given extreme constraints on availability of rice production statistics. For example, the small number of observations makes it difficult for us to detect non-linearities in the weather-rice relationship. That being said, the DOA has done an excellent job of identifying the data shortcomings, and there appears to be a serious effort underway to improve data availability across the country. Therefore, we believe that despite having a limited panel, this represents the single best quality data currently available.

We have also been unable to locate other data that would have improved our analysis. We hoped, for example, to obtain rice crop calendar information on the length of growing period for each district in the country, but no data like this currently exists. The closest data of use came from the National Agricultural and Forestry Research Institute (NAFRI), which had crop calendar information for just a single province, based on their own recent field study. Although this is of value, we do not incorporate into this analysis as we model yields for the entire country, which has diverse geographical regions and growing climates. Another potential area of further exploration we hoped to explore was the affect of changes on rice yields on different socio-economic variables. In order to examine this however, we would need access to the Lao Expenditure and Consumption Survey (LECS), which has been conducted every five years since 1997/98.

Given the serious data concerns over the quality of upland rice production data we chose to omit upland production from our analysis. Data collection in Lao PDR suffer from imperfect systems and data collection is often a highly political issue. Reliable data on yields at the district level require a dedicated support staff and systems in place to ensure accurate reporting. Furthermore, upland production faces a variety of constraints that severely limit the accuracy of data collection. Considering these issues, we instead focus our analysis on lowland systems where data quality is believed to be much higher.

## 3.3 Results

Consistent with previous statistical studies (e.g., Peng et al. 2004; Welch et al. 2010), the **preliminary** results of our linear fixed-effects regression model of average weather (equation 1) suggest that elevated minimum nighttime temperatures[7] are highly damaging to rice yields as seen in Table 1. With regards to different production systems we find these trends are largely similar, although varying in their severity and significance. For lowland rain-fed production we find that that a 1-degree rise in the nighttime temperature reduces rice yields by 4.6% holding all else constant. Although this result is not statistically significant at conventional levels it is consistent with results from previous studies that suggest an increase in average nighttime temperature leads to reduction in yields. Given the limited amount of data and associated low statistical power, non-significant effects are unsurprising. Looking at daytime temperatures, we find that a 1-degree rise in temperature increase yields by 11.8% holding all else constant, and these effects are significant at the 10% level. Based on this evidence, this might suggest that increasing temperatures could have an overall positive impact on rice yields for the most important and common rice production system in the country. Furthermore, we find statistically significant evidence that increases in precipitation increase yields, although the effect is very small. We show that increasing precipitation by 1 cm over the growing season increases yields by approximately 0.1% holding all else constant.

We find that changes in temperature appear to have no effect on yields for irrigated dry season production. This might be suggestive of the fact that irrigated

| Table 1 Impact of weather on log rice yields, district level, 2006–2011 | | (1) Dry season | (2) Wet season |
|---|---|---|---|
| | Min temperature | 0.045 | −0.046 |
| | | (0.028) | (0.038) |
| | Max temperature | −0.013 | 0.118* |
| | | (0.053) | (0.066) |
| | Precipitation | −0.001** | 0.001*** |
| | | (0.000) | (0.000) |
| | Mean log-yield | 1.530 | 1.277 |
| | No obs | 578 | 683 |
| | $R^2$ | 0.691 | 0.732 |

Standard errors in parentheses
Significance levels indicated by *0.1, **0.01, ***0.05

---

[7] For the purpose of this study, minimum nighttime temperature is defined as the lowest temperature recorded by weather stations at night. Some stations record several observations per night while other stations record a single nighttime observation.

**Table 2** Non-linear impact of weather on log rice yields, district level, 2006–2011

|  | (1) Dry season | (2) Wet season |
|---|---|---|
| Min temperature | −0.099 | 1.007* |
|  | (0.481) | (0.427) |
| Min temperature square | 0.003 | −0.024* |
|  | (0.010) | (0.010) |
| Max temperature | 0.249 | −0.490 |
|  | (0.692) | (0.358) |
| Max temperature square | −0.004 | 0.010** |
|  | (0.011) | (0.005) |
| Precipitation | −0.000*** | 0.000* |
|  | (0.000) | (0.000) |
| Mean log-yield | 1.530 | 1.277 |
| No obs | 578 | 683 |
| $R^2$ | 0.691 | 0.739 |

Significance levels indicated by *0.05, **0.1, ***0.01

production systems are typically market oriented, intensive systems, and thus farmers are better able to withstand extreme temperature events. However, we find there is a small effect that increased precipitation decreases yields in the dry season.

In regards to the non-linear approach modeled in eq. 2, we find some evidence that there is a non-linear relationship between temperature and yields as seen in Table 2. For lowland rain-fed production, we find that elevated nighttime temperatures improve yields up to approximately 21 °C, after which increased nighttime temperatures reduce yields. Given that the average minimum temperature across our sample is greater than 21 °C, we see the large negative effect in Table 1. For daytime temperatures we find weak evidence of the opposite effect. The results in Table 2 suggest that elevated daytime temperatures decrease yields until approximately 24.5 °C, after which they have a positive effect. Once again, average daytime temperatures are above 24.5 °C, which adds robustness to the effect we find in Table 1.

### 3.3.1 Evaluating the Model

While the results are broadly consistent with previous studies (i.e., negative coefficients on minimum temperature, positive coefficients on maximum temperature), limited data sources mean that our analysis may lack sufficient power to precisely identify these effects. Consequently, many of the coefficients are not statistically significant. The $R^2$ and adjusted $R^2$ are generally similar to studies carried out in other settings, if not slightly lower here.

As a robustness check, we also estimated Equation 1 for provincial level rice yields from 1990 through 2008 as seen in Table 4. These data represent all rice types across all growing seasons and comes from the IRRI World Rice Statistics database.

**Table 3** Rice area, production, and yield (2012)

| Region/province | Area (% of total) | Production (% of total) | Yield |
|---|---|---|---|
| **A. Northern** | **21.55** | **18.91** | **3.26** |
| Phongsaly | 1.98 | 1.50 | 2.81 |
| Luangnamtha | 1.78 | 1.78 | 3.7 |
| Oudomxay | 2.62 | 2.21 | 3.13 |
| Bokeo | 2.76 | 2.70 | 3.63 |
| Luangprabang | 3.91 | 2.63 | 2.51 |
| Huaphanh | 3.21 | 2.83 | 3.28 |
| Xayabury | 5.28 | 5.27 | 3.7 |
| **B. Central** | **52.63** | **54.18** | **3.85** |
| Vientiane Municipality | 8.14 | 9.82 | 4.49 |
| Xiengkhouang | 3.16 | 3.04 | 3.58 |
| Vientiane | 7.10 | 7.76 | 4.12 |
| Borikhamxay | 4.69 | 4.60 | 3.79 |
| Khammuane | 7.61 | 7.30 | 3.56 |
| Savannakhet | 21.92 | 21.66 | 3.67 |
| Xaysomboun | 25.82 | 26.92 | 3.91 |
| **C. Southern** | **9.32** | **8.74** | **3.51** |
| Saravan | 1.18 | 1.09 | 3.43 |
| Sekong | 12.74 | 15.07 | 4.45 |
| Chmpasack | 2.58 | 2.02 | 2.91 |
| Attapeu | 21.55 | 18.91 | 3.26 |

Source: DOA 2012

**Table 4** Impact of weather on log rice yields, province level, 1990–2008

|  | (1) |
|---|---|
| Min temperature | −0.074* |
|  | (0.032) |
| Max temperature | 0.052** |
|  | (0.025) |
| Precipitation | 0.000** |
|  | (0.000) |
| Mean log-yield | 7.89 |
| No obs | 337 |
| $R^2$ | 0.854 |
| Adjusted $R^2$ | 0.836 |

Significance levels indicated by *0.01, **0.05, ***0.1
Standard errors in parentheses

The results are displayed in the appendix. With the IRRI provincial data, all coefficients are found to be statistically significant and the $R^2$ values are significantly higher. This exercise suggests that a longer time series may provide more power to estimate these relationships relative to a larger cross-section.

## 4 Analysis II: Projecting Future Rice Production Under Climate Change

### 4.1 Climate Projections

The Intergovernmental Panel on Climate Change (IPCC 2007a) predicts that Southeast Asia will experience warmer temperatures, increased frequency of heavy precipitation, increased droughts, and lower annual levels of rainfall in the next century. Changes in the climate are most likely to affect Lao rice yields through harmful extreme temperatures, reduction in water availability from lower levels of rainfall, and a reduced growth period attributed to higher temperatures and radiation levels. Rice in Lao PDR is presently grown at the upper end of the optimal temperature range for rice production. This suggests that Lao rice production is likely to be harmed if future temperatures rise as expected (Wassmann et al. 2009b).

On a global scale, researchers estimate that minimum temperatures have risen faster than maximum temperatures over the last century. Easterling et al. (1997) dissects the trend of increasing diurnal temperatures and attributes it to increased $CO_2$ concentration in the atmosphere. However, in our data set we observe maximum temperatures *rising faster* than minimum temperatures in the last 30 years. For more detailed predictions of future conditions we turn to the Global Climate Models (GCM) published by the IPCC.

**Overview of Global Climate Models (GCMs)** GCM[8] are mathematical models used to simulate the dynamics of the climate system including the interactions of atmosphere, oceans, land surface, and ice. They take into account the physical components of weather systems and use these relationships to model future climate conditions. While there are high levels of uncertainty involved in GCMs, these models can help provide insights into future climate scenarios.

The IPCC serves as a central organization for research groups around the world to submit their models. Each research group must choose an approach to modeling physical climate interactions, spatial and time resolutions, and future economic conditions, among other things. Variation in model choice can result in a wide variety of predictions. Fortunately, the IPCC has attempted to standardize economic/emissions scenarios in order to increase comparability across models. However, while these scenarios limit the choices that modelers are faced with, there are still many assumptions to be made about how to model future climate. Differences in these choices result in a still wide variation in predictions across models, even within economic scenarios.

In order to improve comparison across GCMs from different research groups across the world, the IPCC publishes baseline greenhouse gas emissions scenarios, the most recent of which is called the Special Report on Emissions Scenarios (SRES), for all groups to utilize. Here we use three of the baseline scenarios established in the IPCC Fourth Assessment Report (AR4), published in 2007 (IPCC 2007b).

---

[8] Also referred to as Global Circulation Models with the same acronym.

The B1 scenario depicts increased emphasis on global solutions to economic, social, and environmental stability, but without additional climate initiatives. It assumes rapid global economic growth, but with changes toward a service and information economy with a population rising to 9 billion in 2050 and then declining thereafter. Clean and resource efficient technologies are introduced limiting future emissions. This scenario estimates an increase in global mean temperatures of 1.1–2.9 °C by 2100.

The A1B scenario also assumes global economic growth and a more homogenous future world but with less global emphasis on the information and service economy. Instead, it assumes a continuation of current economic activities, but with more efficient technologies and a balanced emphasis on all energy sources. It assumes similar population increase to 2050, followed by a decline in global birth rates. This scenario predicts, on average, a 2–6 °C warming of global temperatures by 2100.

The A2 scenario depicts a more heterogeneous world with uneven global economic develop and an emphasis on self-reliance and preservation of local identities. Fertility patterns across regions converge slowly, resulting in a continuous increase in global population. Economic development is regionally fragmented and there is less global cooperation. This scenario predicts a global increase in temperature of 2–5.4 °C by 2100.

### 4.1.1 Selecting GCM Models

It is unclear whether any one model is more 'valid' than others (Burke et al. 2015). However, some argue that models have different strengths and weaknesses and should thus be carefully selected for specific applications (e.g. Knutti et al. 2010). While many studies choose one (or a few) models, and make predictions based on those scenarios, it is unclear how one would select the 'best' model. To add to these difficulties, different models offer widely different future predictions of climate conditions. Consequently, predicted future yields will depend highly on which GCM is utilized to forecast future climate conditions. For the time being, we follow the recommendations made by Burke et al. (2015) and include as many models as possible with equal weights on the outcome predicted by each model. Our reasoning is that policy recommendations should be informed on the range of possibilities. However, by using many models the range of predicted outcomes can vary widely. Nonetheless, we argue that the alternative of counting on the predictions of one model underrepresents the uncertainty involved in predicting effects of future climate change, and that it would be unwise to make policy recommendations based on a single model. Instead, we incorporate predictions from the 14 models that offer predictions for our variables of interest (min temperature, max temperature, precipitation) under three economic scenarios (A1B, A2, B1). In total, we therefore have 42 future climate scenarios, one for each model-scenario pair, each of which can be evaluated for a range of time frames. Finally, we can calculate the yield outcomes under each of these scenarios and the median outcomes for each economic scenario represent our estimates for future yields assuming low, medium, or high emissions in the future.

### 4.1.2 Downscaling Methods

For each model-scenario combination we first calculate the model estimated monthly average weather conditions (min/mean/max temperature and precipitation) over the previous decade (2000–2010) for each district. We do this by matching each district to the four closest GCM grid cells and then weighting each GCM cell by the inverse distance of the center of the GCM cell to the center of the district where weights are forced to sum to 1. This provides us with a historical standard by which to measure future projections. Next, future period monthly averages are calculated for each decade up to 2050. Future average monthly conditions are then related back to the GCM estimated historical conditions for the 2000–2010 period to provide predicted climate change. Temperature changes are calculated as an absolute degree change in monthly averages while precipitation change is calculated as percentage change in average millimeters of rainfall per month.

Once we have estimated future changes in absolute (temperature), or percentage absolute (precipitation) terms, we add the predicted changes to the estimated historical data for each district, with changes separated by month. Once we have calculated historical conditions under climate change, we use our model to predict yields under the climate change weather conditions.

This process is repeated for all 42 model-scenario combinations (14 models, 3 scenarios) and the median outcomes are reported as the predicted yield changes under climate change for each decade. Although computationally tedious, incorporating 14 models provides a more representative range of possible future climate conditions, and of the high levels of uncertainty associated with predicting future climate. This issue is discussed in detail below.

### 4.1.3 Climate Projections for Lao PDR

Time-series of the climate projections for Lao PDR are displayed in Fig. 4. On average, growing season temperatures are predicted to increase approximately 1 °C by 2050 while growing season rainfall is expected to slightly decrease. However, some GCMs predict an increase in growing season rainfall over this period.

## 4.2 Yield Projections

### 4.2.1 Methods

In order to evaluate potential climate risk to rice production, we use our rice models to predict yields under future climate scenarios. Due to the resolution of our data, we are able to predict yields at the district level. We estimate future yields by using our estimated statistical model to predict yields at the values of weather variables

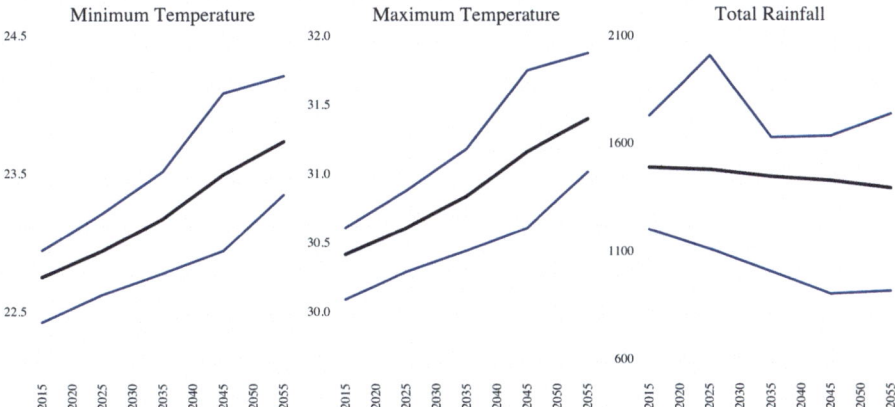

**Fig. 4** Forecast climate conditions across 14 GCMs. Average growing season climate conditions forecast up to 2055. The *black* line represents the median value across 14 GCMs. The *blue* lines represent the minimum and maximum values across GCMs

predicted by the climate model. In order to remain consistent, we use the same approach to estimate yields over the study period (i.e., the 2000s) and then calculate yield changes relative to this baseline.

**Quantifying Uncertainty with Yield Projections** There are two primary types of uncertainty associated with making yield-climate projections. First, there is uncertainty associated with our statistical models. Our models are linear approximations of the yield-weather relationship and thus are best suited to predict how yields respond to perturbations in weather variables only over the observed range of conditions. Fortunately, this type of uncertainty is quantifiable through standard errors and other measures such as Root Mean Squared-Error calculated by using our model to predict observed yields. The second type of uncertainty arises from unpredictability of future climate conditions. GCMs attempt to predict future conditions, however, the uncertainty associated with these predictions far exceeds the statistical uncertainties discussed above. In fact, simulations have shown that uncertainty arising from climate projections outweighs statistical uncertainty by several orders of magnitude (Burke et al. 2015). Quantifying model uncertainty is less straightforward. Here we follow the approach suggested in Burke et al. (2015) and use variation across yield projections utilizing different climate models to provide a measure of climate uncertainty.

### 4.2.2 Results

Figure 9 (see Appendix) displays the **preliminary** median yield projection across climate models using the statistical model described in equation 1 discussed above. Figure 9, panel 2 shows the time series of the yield changes. Yield changes are

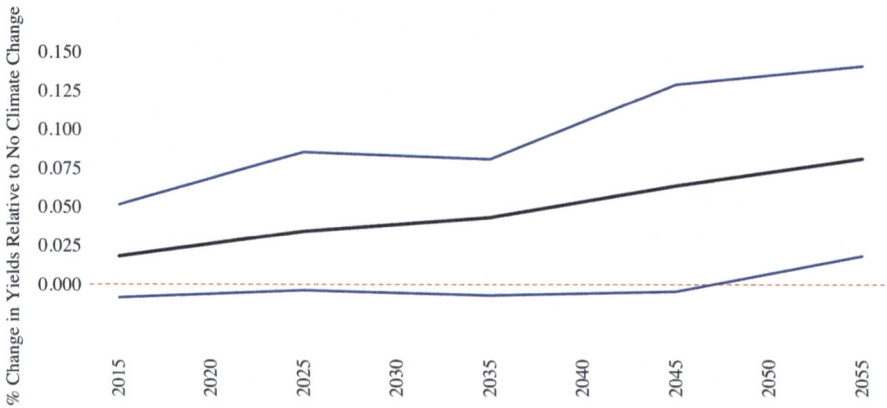

**Fig. 5** Time series of forecasted yield impacts (lowland wet rice). *Blue* lines represent minimum and maximum predicted yields across 14 climate models. *Black* line represents the median predicted yield change across models. Baseline scenario is that yield trends continue on their current path but temperatures and rainfall patterns continue to follow historical averages

measured relative to a baseline scenario where yields continue on their historical upward trends but where climate conditions continue to vary around their historical averages. The climate scenarios assume the same current yield trends but with changes in climate predicted by GCMs. Because maximum temperature is found to be strongly positively related to higher yields, future yields are predicted to be higher, on average, under climate change. This is likely a result of insufficient observations needed to estimate the historical relationship accurately. Here we find the benefits from rising maximum temperatures outweigh the negatives from rising minimum temperature. In other cases we have found the opposite to be true (Fig. 5).

## 5 Summary and Outlook

Given the extremely limited nature of data in Lao PDR we are hesitant to offer any precise policy recommendations. Our results come from a 6-year panel, which cannot be considered an entirely accurate representation of the historical relationship between climatic variables and yields. This is echoed in our results as we find only three significant effects across all specifications. Moreover, it should also be noted that our results rely on historical data and thus model accuracy is tied to (unobservable) data quality.

In regards to wet season production, we find that a 1-degree increase in daytime temperatures holding all else equal causes an 11.8% increase in yield. This would suggest that higher daytime temperatures as a result of climate change would in fact be beneficial for rice production in Lao PDR. Furthermore, given that Lao PDR has achieved self-sufficiency in rice production in recent years it appears that the impact of climate change on food security does not appear to be a major concern. Although the country appears to have met self-sufficiency at the national level, it is certainly clear that not all households are able to meet rice consumption requirements. According to some estimates, about 30% of the population has insufficient food for more than 6 months of the year. However, much of this deficiency is in the northern and eastern mountainous areas, while the Mekong River valley is an area of surplus (ADB 2006). Thus, based on our projections, yields in the Mekong River valley will increase as a result of climate change surpluses will be further extended. In regards to policy, marketing of the surplus will be the key policy challenge. According to the LECS only 8% of all rice produced is sold, and thus extending both domestic and international trade should be made a priority.

Of more concern are the individuals located in the mountainous regions of the country that rely on upland production systems. Our results suggest there is a high level of uncertainty between temperature and yields. For example, we find that an increase of 1 degree in average daytime temperature causes a 38% increase in yields, while an increase of 1 degree in average nighttime temperature causes a 30% reduction in yields. These large shocks can be incredibly damaging as individuals engaged in this production system are the most likely to be unable to reach self-sufficiency. Therefore, it appears that one clear policy option would be strategies to reduce variability. Crop diversification is one potential option, although our analysis does not consider other crops so we cannot comment wither there is less variability. Insurance mechanisms that protect against shocks are likely the best option. However, extending any type of insurance to individuals in such remote locations will likely be of extreme difficulty.

We also want to add the caveat that data from upland production systems are likely the most inaccurate. Due to the extremely remote nature of these systems the validity of the data should certainly be taken with a grain of salt. Furthermore, we would like to highlight the limited sample size and subsequent limited power of our results for the upland systems. Thus we offer these recommendations with reservations.

## 6 Conclusions and Extensions

This report adds support to the growing literature estimating the impacts of weather and climate change on rice production. We focus our analysis in Lao PDR, a country whose economy relies on the production of rice, but has had received little analysis on how climate change will impact the sector. This represents a crucial gap in the literature, as rice is instrumental to the Lao economy and will undoubtedly face challenges from climate change.

We use advanced econometric models to first estimate the historical relationship between observed rice yields and climatic variables. With this relationship established, we then downscale projections from the leading climate models to forecast the impact on rice yields under these climate scenarios. Our results are consistent with previous work in the region, as we find weak evidence that elevated minimum nighttime temperatures are highly damaging to rice yields. Conversely, we find support that elevated maximum daytime temperatures increase yields. Overall the size of the impact and statistical significance is larger for increased maximum temperatures, suggesting that elevated temperatures might have a net positive impact on rice yields in Lao PDR. Turning next to forecasting, our projections confirm this intuition, as future yields are predicted to be higher, on average, under climate change.

We offer some major caveats to these findings. First, our results are not significant at traditional levels although this not surprising given our limited panel. Our results come from a 6-year panel, which cannot be considered an entirely accurate representation of the historical relationship between climatic variables and yields. Second, there are major data quality issues surrounding rice yields. Although data quality is improving rapidly in Lao PRD, high-resolution rice yield data is only recently available, and is of unknown quality. Given our results rely on this historical data, our model accuracy is tied to the quality of the data. That being said, our results are in line with previous work in the region and serve as a useful preliminary first step to modeling how climate change will impact rice yields in Lao PDR. Over time as data quality improves, these results can be easily replicated to strengthen the analysis.

**Disclaimer and Contacts**  Regional Rice Initiative Research Reports have not been subject to independent peer review and constitute views of the authors only. For comments and/or additional information, please contact:
Sam Heft-Neal and David Roland-Holst
   Department of Agricultural and Resource Economics
   207 Giannini Hall
   University of California Berkeley
   CA 94720 - 3310 USA
   E-mail: dwrh@berkeley.edu
Drew Behnke
   Department of Economics
   University of California Santa Barbara
   CA, USA

# Appendix – Rice Yield Regression Model Results (Figs. 6, 7, 8, and 9)

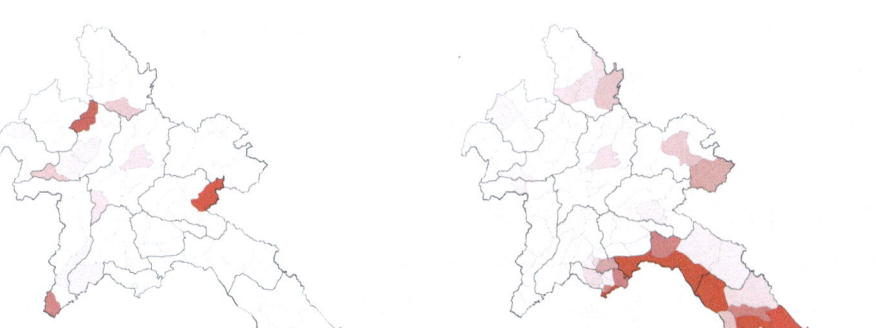

*Source: Crop Statistics Yearbook (DOA, Lao PDR)*

**Fig. 6** Largest rice area losses 2006–2012 by cause. Maps show the maximum wet-season lowland rice area lost from flood or drought in any year over the study period 2006–2012. The figure illustrates that over the seven-year study period a majority of districts experienced some losses from floods or droughts. Flood losses were more common and tended be to more severe with some districts reporting 100% losses in bad a flood year

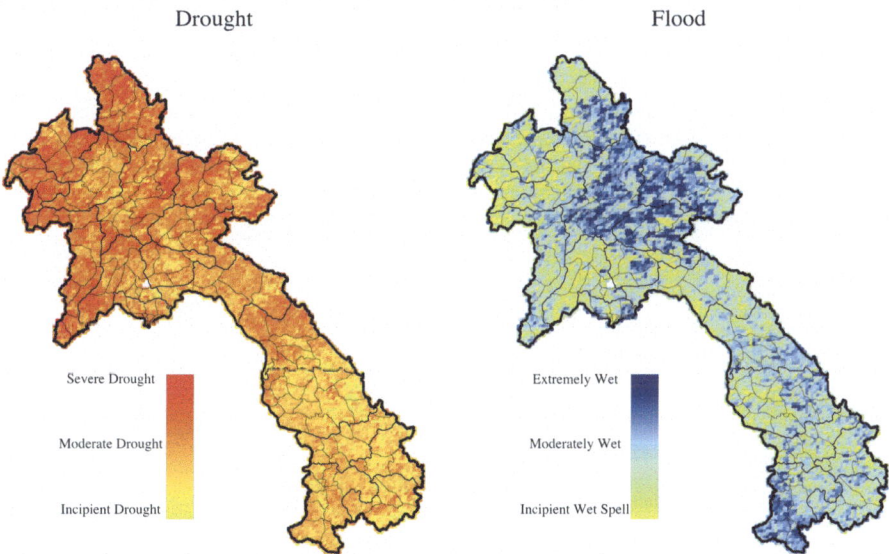

*Source: Drought Severity Index (DSI) described in Mu et al (2013)*

**Fig. 7** Most extreme growing-season weather conditions 2006–2012. Maps show the most extreme dry and wet conditions experienced during the rice-growing season over the study period. Categories correspond to the qualitative categories described in Mu et al.

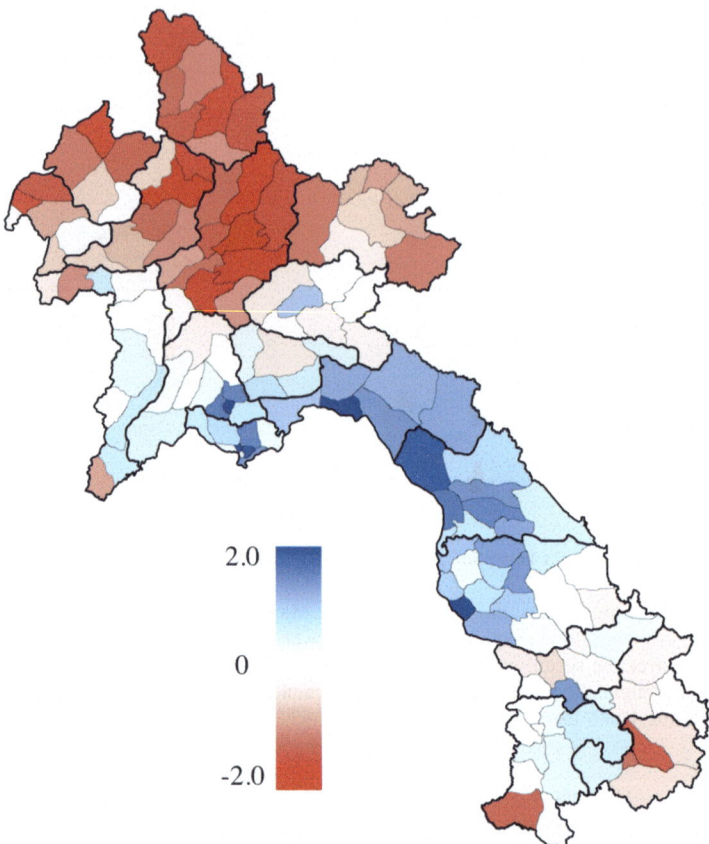

**Fig. 8** Average Drought Severity Index (DSI) for rainy season 2004. Average area-weighted DSI values for Lao PDR districts. Blue represents greater than normal and red represents less than normal water levels. This figure is meant to provide an illustration of the data source described in Mu et al. (2013). Data are averaged over rainy season in 2004. Note that the DSI map is roughly an inverse of the precipitation map in Fig. 1

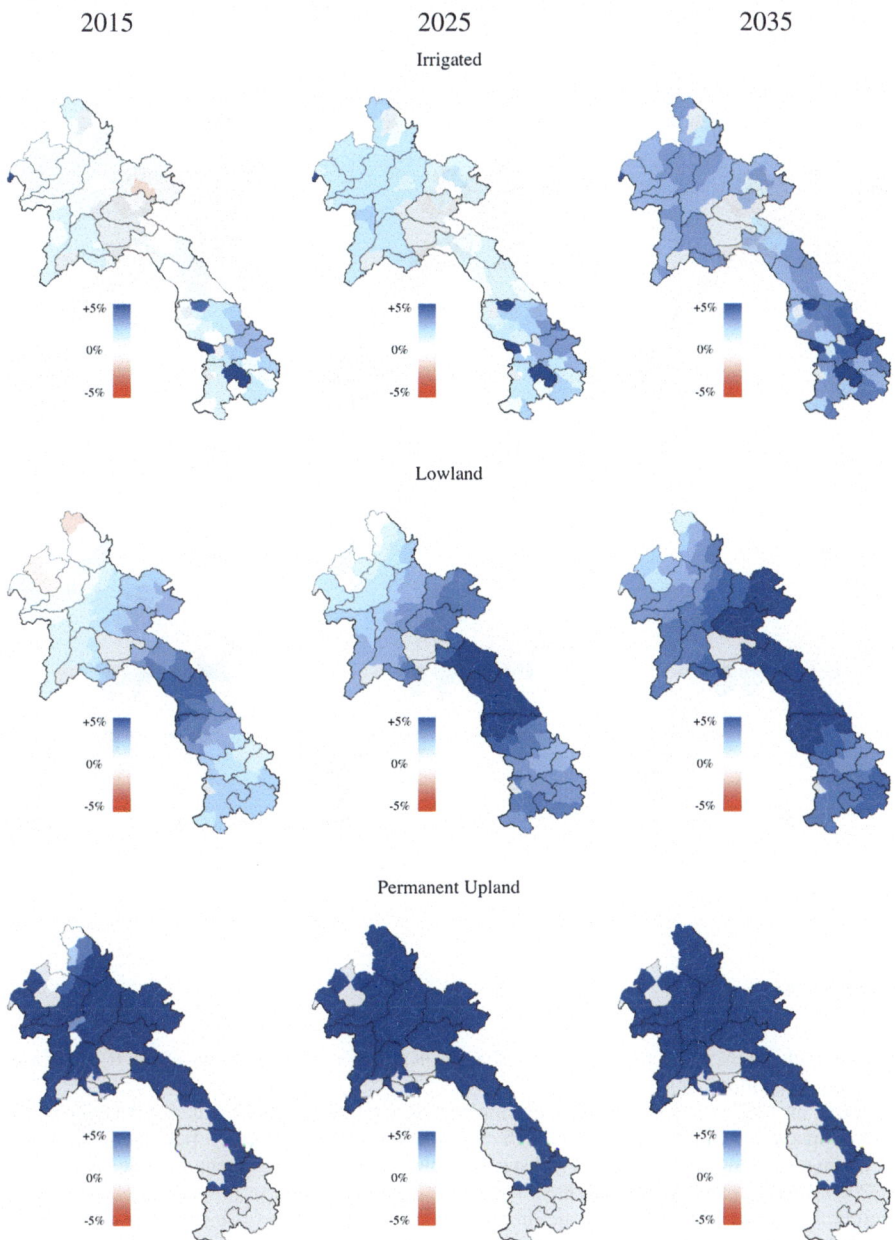

**Fig. 9** Preliminary projected yield changes 2015–2035

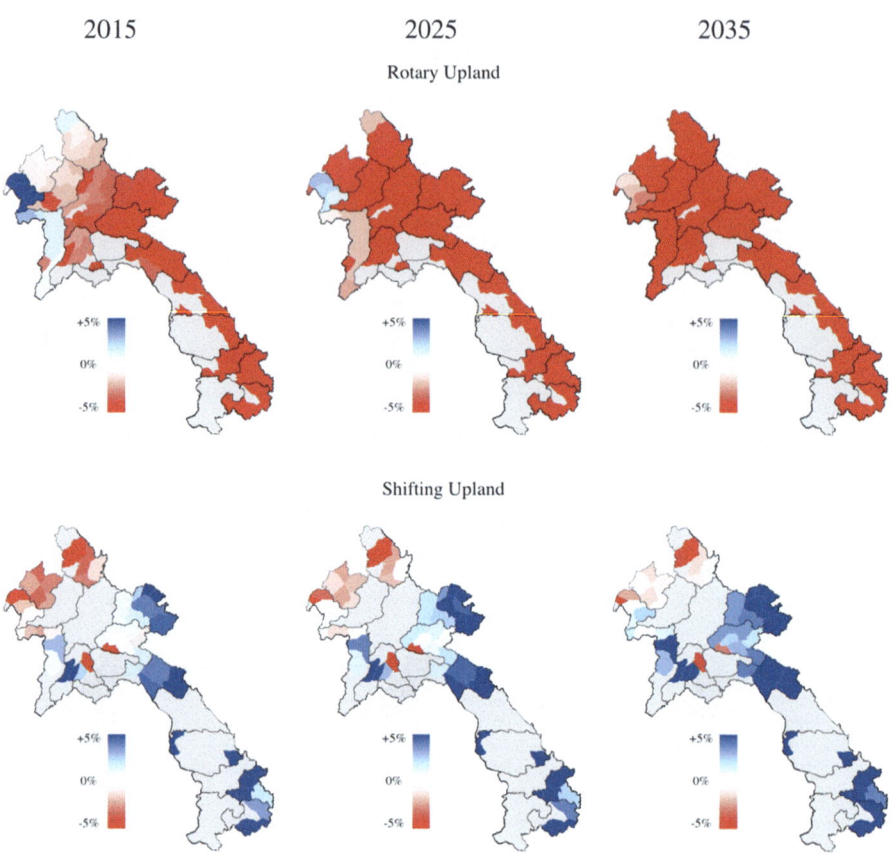

**Fig. 9** (continued)

# References

Asian Development Bank (ADB) (2006) "Lao PDR: An Evaluation Synthesis on Rice", Case Study, September 2006.

Belder et al. (2004) - P Belder, B.A.M Bouman, R Cabangon, Lu Guoan, E.J.P Quilang, Li Yuanhua, J.H.J Spiertz, T.P Tuong, Effect of water-saving irrigation on rice yield and water use in typical lowland conditions in Asia, Agricultural Water Management, Volume 65, Issue 3, 2004, Pages 193–210, ISSN 0378-3774.

Borrell, A., A. Garside, and S. Fukai, "Improving efficiency of water use for irrigated rice in a semi-arid tropical environment," *Field Crops Research*, 1997, 52 (3), 231–248.

Bouman, BAM, T.P. Tuong, E. Humphreys, TP Tuong, and R. Barker. (2007) "Rice and water," Advances in Agronomy, 92, 187–237.

Bouman, BAM, T.P. Tuong, S. Peng, AR Castaneda, and RM Visperas (2005) "Yield and water use of irrigated tropical aerobic rice systems," Agricultural Water Management, 74 (2), 87–105.

Burke, M., Dykema, J., Lobell, D. B., Miguel, E., & Satyanath, S. (2015). Incorporating climate uncertainty into estimates of climate change impacts. Review of Economics and Statistics, 97(2), 461–471.

Burke, M., J. Dykema, D. Lobell, E. Miguel, and S. Satyanath, "Incorporating Climate Uncertainty into Estimates of Climate Change Impacts, with Applications to US and African Agriculture," Review of Economics and Statistics, *Forthcoming*.

DOA (2012) – Department of Agriculture at Ministry of Agriculture and Forestry of the Lao People's Democratic Republic "Crop Statistics Year Book," 2006–2012.

Easterling, D. R., Horton, B., Jones, P. D., Peterson, T. C., Karl, T. R., Parker, D. E., ... & Folland, C. K. (1997). Maximum and minimum temperature trends for the globe. *Science*, 277(5324), 364–367.

Homma, K., T. Horie, T. Shiraiwa, S. Sripodok, and N. Supapoj, "Delay of heading date as an index of water stress in rainfed rice in mini-watersheds in Northeast Thailand," Field crops research, 2004, 88 (1), 11–19.

IPCC, "Fourth Assessment Report of the Intergovernmental Panel on Climate Change: The Impacts, Adaptation and Vulnerability (Working Group III)." Cambridge University Press, New York. 2007a.

IPCC. "Summary for Policymakers," in S. Solomon, D. Qin, M. Manning, Z. Chen, M. Marquis, K.B. Averyt, M.Tignor, and H.L. Miller, eds., Climate Change 2007: The Physical Science Basis. Contribution of Working Group I to the Fourth Assessment Report of the Intergovernmental Panel on Climate Change., Cambridge University Press, Cambridge, United Kingdom and New York, NY, USA, 2007b.

Knutti, R., Furrer, R., Tebaldi, C., Cermak, J., & Meehl, G. A. (2010). Challenges in combining projections from multiple climate models. Journal of Climate, 23(10), 2739–2758.

MAF – Ministry of Agriculture and Forestry of the Lao People's Democratic Republic "Lao People's Democratic Republic Rice Policy Study", Technical Report conducted by the International Rice Research Institute (IRRI), the Food and Agricultural Organization of the United Nations (FAO), and the World Bank, 2012.

Mu, Q., M. Zhao, J. S. Kimball, N. G. McDowell, and S. W. Running. "A remotely sensed global terrestrial drought severity index". Bulletin of the American Meteorological Society, 94(1):83–98, 2013.

Lobell, D. B., & Burke, M. B. (2010). On the use of statistical models to predict crop yield responses to climate change. Agricultural and Forest Meteorology, 150(11), 1443–1452.

Pandey, S., "Economics of Lowland Rice Production in Laos: Opportunities and Challenges" in Increased Lowland Rice Production in the Mekong Region edited by Shu Fukai and Jaya Bansnayake. ACIAR Proceedings 101, 2001.

Peng, S., J. Huang, J. Sheehy, R. Laza, R. Visperas, X. Zhong, G. Centeno, G. Khush, and K. Cassman. 2004. "Rice yields decline with higher night temperature from global warming." Proceedings of the National Academy of Sciences of the United States of America 101:9971.

Sawano, S., T. Hasegawa, S. Goto, P. Konghakote, A. Polthanee, Y. Ishigooka, T. Kuwagata, and H. Toritani, "Modeling the dependence of the crop calendar for rain-fed rice on precipitation in Northeast Thailand," Paddy and Water Environment, 2008, 6 (1), 83–90.

Schiller, J.M., B. Linquist, , K. Douangsila, P. Inthapanya, B. Douang Boupha, S. Inthavong, and P. Sengxua "Constraints to Rice Production Systems in Lao" in *Increased Lowland Rice Production in the Mekong Region* edited by Shu Fukai and Jaya Bansnayake. ACIAR Proceedings 101, 2001.

Schlenker, W. and M.J. Roberts, "Nonlinear temperature effects indicate severe damages to US crop yields under climate change," Proceedings of the National Academy of Sciences, 2009, 106 (37), 15594.

Sharma, P. K., Pantuwan, G., Ingram, K. T., & De Datta, S. K. (1994). Rainfed lowland rice roots: soil and hydrological effects. *Rice Roots Nutrient and Water Use. IRRI, Manila*, 55.

Sheehy, J., P. Mitchell, and A. Ferrer. 2006. "Decline in rice grain yields with temperature: Models and correlations can give different estimates." Field crops research 98:151–156.

Wade, L. J., Fukai, S., Samson, B. K., Ali, A., & Mazid, M. A. (1999). Rainfed lowland rice: physical environment and cultivar requirements. Field Crops Research, 64(1), 3–12.

Wassmann, R., S. Heuer, A. Ismail, E. Redona, R. Serraj, RK Singh, G. Howell, H. Pathak, and K. Sumfleth, "Climate change affecting rice production: the physiological and agronomic basis for possible adaptation strategies," Advances in Agronomy, 2009a, 101, 59–122.

Wassmann, R., SVK Jagadish, K. Sumfleth, H. Pathak, G. Howell, A. Ismail, R. Serraj, E. Redona, RK Singh, and S. Heuer, "Regional vulnerability of climate change impacts on Asian rice production and scope for adaptation," Advances in Agronomy, 2009b, 102, 91–133.

Welch, J., J. Vincent, M. Auffhammer, P. Moya, A. Dobermann, and D. Dawe. 2010. "Rice yields in tropical/subtropical Asia exhibit large but opposing sensitivities to minimum and maximum temperatures." Proceedings of the National Academy of Sciences 107:14562.

Xu, Kenong, and David J. Mackill. "A major locus for submergence tolerance mapped on rice chromosome 9." Molecular Breeding 2.3 (1996): 219–224.

Yatagai, Akiyo, Kenji Kamiguchi, Osamu Arakawa, Atsushi Hamada, Natsuko Yasutomi, and Akio Kitoh. "APHRODITE: Constructing a long-term daily gridded precipitation dataset for Asia based on a dense network of rain gauges." *Bulletin of the American Meteorological Society* 93, no. 9 (2012): 1401–1415.

Yin, X., M.J. Kropff, and Goudriaan J., "Differential effects of day and night temperature on development to flowering in rice," Annals of Botany, 1996, 77 (3), 203–213.

Yoshida, S., Fundamentals of rice crop science, Int. Rice Res. Inst., 1981.

**Open Access** This chapter is distributed under the terms of the Creative Commons Attribution-NonCommercial-ShareAlike 3.0 IGO license (https://creativecommons.org/licenses/by-nc-sa/3.0/igo/), which permits any noncommercial use, duplication, adaptation, distribution, and reproduction in any medium or format, as long as you give appropriate credit to the Food and Agriculture Organization of the United Nations (FAO), provide a link to the Creative Commons license and indicate if changes were made. If you remix, transform, or build upon this book or a part thereof, you must distribute your contributions under the same license as the original. Any dispute related to the use of the works of the FAO that cannot be settled amicably shall be submitted to arbitration pursuant to the UNCITRAL rules. The use of the FAO's name for any purpose other than for attribution, and the use of the FAO's logo, shall be subject to a separate written license agreement between the FAO and the user and is not authorized as part of this CC-IGO license. Note that the link provided above includes additional terms and conditions of the license.

The images or other third party material in this chapter are included in the chapter's Creative Commons license, unless indicated otherwise in a credit line to the material. If material is not included in the chapter's Creative Commons license and your intended use is not permitted by statutory regulation or exceeds the permitted use, you will need to obtain permission directly from the copyright holder.

# Farmers' Perceptions of and Adaptations to Climate Change in Southeast Asia: The Case Study from Thailand and Vietnam

**Hermann Waibel, Thi Hoa Pahlisch, and Marc Völker**

**Abstract** The perceptions of climate change and adaptation choices made by farmers are important considerations in the design of adaptation strategies by policy makers and agricultural extension services. This paper seeks to determine these perceptions and choices by farmers in already poor environmental regions of Thailand and Vietnam especially vulnerable to climate change. Overall findings were that farmers do perceive climate change, but describe it in quite distinct ways and that location influences how farmers recognize climate change. Our 2007 and 2013 surveys show that farmers are adapting, but it is difficult to determine if specific practices are "climate smart". Further, adaptation measures are informed by perception and, at least in the case of Vietnam, perceptions are shaped by the respondent's characteristics, location variables and recent climate related shocks. Finally, the three climate variables of rainfall, temperature, and wind are the most important factors in explaining specific adaptation measures chosen by farmers. Farmer participation is an essential part of public actions designed to allow adaptation to climate change. Our research can also contribute to understanding farmer constraints and tailoring good overall strategies to the local heterogeneity of vulnerable locations.

---

H. Waibel (✉) • T.H. Pahlisch
Institute of Development and Agricultural Economics, Leibniz Universität Hannover, Hannover, Germany
e-mail: waibel@ifgb.uni-hannover.de

M. Völker
Institute for Population and Social Research, Mahidol University, Salaya, Thailand
e-mail: marc.voe@mahidol.edu

## 1 Introduction

As established by the Intergovernmental Panel on Climate Change (IPCC 2014), climate change is affecting Southeast Asia through increasing average temperatures, sea level rise and changes in precipitation, although trends differ strongly across the region. Countries in Southeast Asia are especially vulnerable to the downside effects of global climate change because of (i) their long coastlines, (ii) high concentration of human and economic activities in coastal areas, (iii) large and growing populations, and (iv) the importance of agriculture as a source of employment and income (ADB 2009). Climate change can have especially negative consequences for agricultural productivity and food security (Iglesias et al. 2011). In Thailand, Boonpragob (2005) found that between 1991 and 2002 the country's agriculture experienced crop yield losses worth some 50 billion Thai Baht (approximately 1.3 billion EURO). In Vietnam, which ranks among the top five countries most affected by rising sea levels (Dasgupta et al. 2007), the impact of extreme weather has led to the damage of rice fields by frequent flooding, for example in the Red River Delta, Central Region, and the Mekong Delta. At the same time, rice areas affected by droughts doubled from some 77,000 ha in 1979–1983 to over 175,000 ha in 1994–1998 (Cuong 2008).

To reduce their vulnerability to the negative effects of climate change, farmers must adapt (Gbetibouo 2009). Adaptation measures should be both technically appropriate and economically feasible. In agriculture, adaptations to climate change will require new technologies and investments. Farmers may have to adopt new crop varieties and new livestock breeds, change their cropping systems and invest in new soil and water conservation methods.

In this paper, we explore climate change in Thailand and Vietnam from the perspective of households living in less favored rural areas who are especially vulnerable to the effects of climate change. We focus on three provinces in Northeast Thailand and three provinces in the Central Highlands and North Central Coast of Vietnam. The study makes use of a database of some 4000 households in these two countries collected as an ongoing research project since 2007 entitled "Impact of Shocks on the Vulnerability to Poverty: Consequences for Development of Emerging Southeast Asian Economies" (DFG FOR 756). We mainly use the 2013 survey as it contained a module on climate change. In addition, the survey included questions on household member characteristics, assets, income and consumption, past shock experience, expected risks and individual risk attitudes.

We aim to answer the following questions:

1. What climate-related shocks did farm households experience, what observations did they make about changes in climate over time and what indicators did they use to describe climate change?
2. What determines the farmers' perceptions of climate change and their decision to adjust agricultural production in response to the effects of perceived climate change?
3. What explains the choice of agricultural adaptation measures by farm households?

The answers to these questions are important for the design of policies and projects aimed to help farmers living in poor environments to adapt to climate change. The participation of farm households in public actions aiming to mitigate or adapt to the impacts of climate change depends on the willingness of these households to participate. Our research can also contribute to the interpretation of the results of climate change models that may have a good overall geographic perspective but may miss the heterogeneity that exists at local levels.

The paper proceeds as follows: Section 2 presents the theoretical background for the determinants of individual climate change perceptions and adaptation behavior. Section 3 describes data collection and Section 4 describes the methodology. Section 5 reports some descriptive results as background information. Section 6 discusses results of our models. Finally, in Section 7, summary and policy conclusions are submitted.

## 2  Theoretical Background

In principle farmers' adaptation to climate change can be modeled using the framework of technology adoption. Generally adoption of technologies depends on a number of factors such as financial incentives, access to extension services and markets but also perceptions and behavior. There is, however, a difference between conventional technology adoption and climate adaptation. While adoption of new technologies mostly aims at increasing profits, adjustments to climate change are often undertaken to reduce risks and to minimize future losses, both of which are directly affected by perceptions of current and future change. It is therefore necessary to incorporate farmers' perception of climate change in an adoption model (Maddison 2007).

Weber (2010) found that people's perception of climate change both in terms of its existence and extent are shaped by learning from personal experience and by making use of statistical information. The formation of perceptions depends on the trust that people attribute to climate scientists and their social amplifiers. Perceptions, however, are only meaningful when they can be linked to actual adaptation measures (Reilly and Schimmelpfennig 1999).

Theoretical insights about the relationship between risk perception and the adoption of risk management actions can be gained from the psychology and economics literature. The psychology literature (e.g. Fuster 2002) refers to the *perception-action cycle*, where people prepare themselves for perceived future outcomes, including the perceived seriousness of potential outcomes. From the economics literature, we can learn that it is necessary to distinguish between gain and loss domain (Kahneman et al. 1990). Tversky and Kahneman (1992) have shown people tend to weigh potential losses higher than potential gains.

Traditionally, adoption decisions have been analyzed in a utility maximization framework with profit as the primary motive (Greene 2003; Norris and Batie 1987). Accordingly, a technology is adopted when the perceived utility or net profit from adoption is significantly larger than not adopting it. The adoption decision is subject to a set of exogenous variables such as household characteristics, socioeconomic and physical factors (Feder et al. 1985). More recent models of climate change adaptations have been developed for African countries (Maddison 2007; Deressa et al. 2008; Gbetibouo 2009). These models incorporated climate change perceptions as explanatory variable. We follow this approach to model the factors that influence climate change perceptions and related adaptation measures as well as to explain specific climate change adaptation measures.

## 3 Study Regions and Data

We focus on the 2900 households from the DFG FOR 756 that are engaged in agricultural production because we are interested in the connection between climate change perception and consequences for agriculture. In Thailand, the provinces are Buri Ram, Nakhon Phanom and Ubon Ratchathani located in the Northeastern region of the country. In Vietnam, the provinces are Ha Tinh and Thua Thien Hue located in the North Central Coast region and Dak Lak situated in the Central Highlands. All six provinces are dominantly agricultural areas albeit with a large degree of heterogeneity in development potential. The provinces are bordering neighboring Laos and/or Cambodia. The choice of the provinces was motivated by the assumption that people in rural and geographically remote regions are more vulnerable than people in urban and central regions. Furthermore, these provinces belong to the poorer environments with less developed infrastructure in agriculture and a high potential for climate-related shocks and thus are more likely to be affected by climate change (Waibel et al. 2013).

The survey instruments comprise of a village head and a household questionnaires. The village head questionnaire contains information on the physical and social infrastructure of the village. The household questionnaire has a detailed shock section that included questions about past climate-related shock experience and details about shock severity in terms of income and asset loss (using a 4 point ordinal scale).[1] A special module on climate change was included where respondents were asked whether or not they had perceived a change in climate in the time that they had lived in their location. Respondents were also asked how they thought that changes in climate is affecting their agriculture (e.g. lower yield, more crop failure) and what measures they had taken to adapt to climate change (e.g. change crop varieties, invest more in irrigation, planting trees, etc.). Part of the household questionnaire was a simple risk item that measures respondents' general attitude towards risk on an 11 point Likert scale following Dohmen et al. (2011) and Hardeweg et al. (2013).

---

[1] 0 = no impact, 1 = low impact, 2 = medium impact, 3 = high impact.

## 4  Empirical Strategy

We address question 1 through a descriptive analysis of the household survey data, and question 2 by employing an econometric model (model 1) that allows us to establish a link between climate change perceptions and adaptation decisions. Question 3 is addressed through a second model (model 2).

The first model is a two-stage procedure. In the first stage, perception of climate change is specified as the outcome variable. In the second stage, adaptation is the outcome variable for respondents who reported awareness of climate change. Accordingly, households in the second stage are non-randomly selected from the entire sample.

To deal with potential selection bias, a Heckman's selection probit model was specified. We consider a random sample of $i$ observations. Equations for individual $i$ are:

$$Y_{1j} = X_{1j}\beta_1 + U_{1i} \tag{1a}$$

$$Y_{2i} = X_{2i}\beta_2 + U_{2i} \tag{1b}$$

where $X_{ji}$ is a $1 \times K_j$ vector of regressors, $\beta_j$ is a $K_j \times 1$ vector of parameters, and

$$E(U_{ij}) = 0, E(U_{ji}U_{j'j^*}) = \sigma_{jj'}, i = i'' \text{ and } E(U_{ji}U_{j'j^*}) = 0, i \neq i''$$

Suppose that $Y_{1i}$ is observed only if $Y_{2i} \geq 0$. In the case of independence between $U_{1i}$ and $U_{2i}$ or $E(U_{ji}U_{j'j^*}) = 0$ so that the data available on $Y_{1i}$ are missing randomly, the regression function for the selected subsample is the same as the population regression function. In the general case where $E(U_{ji}U_{j'j^*}) = \sigma_{jj'}$, least squares estimators yield biased results. Thus, the Heckman selection model as a solution in providing consistent, efficient estimates in the following way:

$$E(Y_{1i}|X_{1i}, Y_{2i} \geq 0) = X_{1i}\beta_1 + \frac{\sigma_{12}}{(\sigma_{22})^{1/2}} \lambda_i \tag{1c}$$

$$E(Y_{2i}|X_{2i}, Y_{2i} \geq 0) = X_{2i}\beta_2 + \frac{\sigma_{22}}{(\sigma_{22})^{1/2}} \lambda_i \tag{1d}$$

where $\lambda_i = \frac{\phi(Z_i)}{1-\Phi(Z_i)}$ and $Z_i = -\frac{X_{2i}\beta_2}{(\sigma_{22})^{1/2}}$ with $\phi$ and $\Phi$ are, respectively, the density and distribution function for a standard normal variable (Heckman 1979).

In our analysis, $Y_{1i}$ is a binary variable specifying whether or not household $i$ adapts their agricultural activities to climate change. $Y_{2i}$ is a binary variable taking on the value unity if respondent of household $i$ perceived climate change and zero otherwise. $X_{1i}$ is a vector of explanatory variables for the outcome Equation (1a).

$X_{2i}$ is a vector of explanatory variables for the selection Equation (1b). It is not absolutely necessary to have the exclusion restriction in the Heckman selection model (Wooldridge 2009) and in some cases the vectors of explanatory variables for selection equation and outcome equation are even identical (Puhani 2000). Thus, the justification for inclusion of variables for $X_{1i}$ and $X_{2i}$ is merely based on the expected effect of these variables on the dependent variables $Y_{1i}$ and $Y_{2i}$ respectively.

$X_{1i}$ includes household head characteristics (age, education, gender, membership of socio-political organization), household characteristic (agricultural member ratio, farm size, income, risk attitude and ethnicity in the model for Vietnam), and distance to district town and province dummies.

Based on the study of Gbetibouo (2009), there is no agreement in the adoption literature on the effect of age of household head. Age can be found to have negative influence on the adoption decision of new technologies because older farmers are more risk-averse than younger farmers and thus have a lesser likelihood of adopting. It is also possible however that older farmers have more farming experiences enabling them to better judge the merits of new technology.

Education is believed to increase the probability of accessing information (Norris and Batie 1987). Evidence from previous studies shows a positive influence of household head's education on the decision to adapt to climate change (Deressa et al. 2008; Maddison 2007). Therefore, we expect that education level of household head is positively related with adaptations to climate change.

We expect that male household heads are more likely to gain information on new technologies and are more likely to be risk takers (Asfaw and Admassie 2004). Therefore, the likelihood of male-headed households to adapt to climate change is believed to be higher than that of female-headed households.

Membership in a social-political organization is hypothesized to have a positive effect on the adaptation decision. It is considered as one kind of social capital of the farmers and as a member of such organization, household heads may have more opportunities to learn new agricultural practices than other members.

Household characteristics used in explaining the adaptation decision include agricultural member ratio, farm size, income and risk attitude. Agricultural member ratio is defined as the ratio between number of household members aged from 15 to 64 engaged in its own agricultural production and the total number of household members in that age range. This ratio is expected to positively influence the decision to adapt to climate change. This enables household to accomplish various agricultural tasks even at peak times. This hypothesis is based on the study of Croppenstedt et al. (2003) revealing that larger amount of labor increases the household's probability of adopting agricultural technology and using it more intensively.

The effect of farm size on the adaptation to climate change is ambiguous. Gbetibouo (2009) found a positive relationship between farm size and the adaptation to climate change. The author also argued that adoption of an innovation tends to take place earlier on larger farms than on smaller farms. On the contrary, farm size showed a negative effect on the adaptation decision in the study Deressa et al. (2008) which is perhaps due to plot level heterogeneity.

We hypothesize that households with higher income will be more likely to undertake adaptation measures. Similarly, if household has larger capital endowment, it has a better possibility to invest (e.g. Franzel 1999). We further hypothesize that in households where the respondent (household head) expresses a lower degree of risk aversion she is more likely to undertake adaption measures.

In the model for Vietnam, we included ethnicity as a binary variable taking on the value 1 if household is the majority Kinh and 0 if household belongs to any of the many ethnic minorities. We expect that ethnic minorities are less likely to invest in climate change related adaptation measures due to their living in the remote areas and villages less endowed with infrastructure (Hung et al. 2010).

To capture the effect of remoteness for all households we added the variable "Distance to district town" from the village head questionnaire. Here we expect a negative relationship with climate change adaptation. Finally, we added province dummy variables to capture other differences among the study regions.

In the selection Equation (1b), we use the respondent characteristics including age, education, gender and membership of socio-political organization as the independent variables. This is because the adaptation decision is made by the household head but the perception of climate change is given by the respondent of that household who in most cases is the household head. Age, a proxy of farming experience, is supposed to have a positive effect on the farmers' awareness. We expect that more experienced farmers are more likely to observe changes in climate over time. Likewise, better educated farmers are believed to have more access to information on climate change (Deressa et al. 2008). Household size is assumed to have a positive effect as the chance to obtain information increases with the number of household members and the same mechanism we assume for income (Deressa et al. 2008).

One important household characteristic included as an explanatory variable in the selection equation is the climate-related shock experience. This variable is computed by summing up the severity scores multiplied by the frequencies of all climatic events, namely drought, floods, storm and soil erosion experienced by a household in the reference period. We expect that more experience with negative climate-related shocks in the past increases the probability that a respondent is aware of climate change.

The inclusion of the ethnicity variable in the model for Vietnam is based on the same arguments as in Equation 1a. We expect that the Kinh majority is more likely to be aware of climate change. Likewise, we have added province dummy variables. In order to control for country heterogeneity we estimate models for Thailand and Vietnam separately.

In order to further explore the type of adaptation measures undertaken by farmers, we formulated a multinomial logit model (MNL) to assess the drives for four categories of adaptation measures, while not undertaking any adaptation was treated as the base category as follows:

$$\Pr(Y_i = j) = \frac{\exp(x\beta_j)}{\sum_{k=1}^{J} \exp(x\beta_k)} \qquad (2)$$

where the dependent variable $Y$ denotes adaptation categories taking on value $j = \{0,1,2,...J\}$ and $x$ is a vector of regressors (Greene 2003).

In our study, the adaptation categories include the following:

- 0 = No adaptation
- 1 = Crop diversification
- 2 = Chemical input management
- 3 = Water management
- 4 = Planting trees

The explanatory variables $x$ include different household head characteristics (i.e. age, education, gender, membership of socio-political organization), household characteristic (agricultural member ratio, farm size, income, risk attitude and ethnicity (only in model for Vietnam)), distance to district town and province dummies. The justification of these variables and their expected direction of influence are assumed to be identical with those in Equation 1a.

In addition, however, we include the respondent's perceptions of changes in climate-related parameters like rainfall, temperature and wind as these perceptions may influence the choice of adaptation measures in different ways. The multinomial logit model makes the assumption of *independence of irrelevant alternatives* (IIA) (Long and Freese 2006). We use the Hausman test to verify this assumption.

## 5 Descriptive Results

In the shock section of the survey, households were asked for the four most frequent types of climate-related shocks (i.e. droughts, floods, storms and soil erosion) experienced during the past 3 years (2010–2013). Table 1a reports these results for Thailand and Table 1b for Vietnam. As shown in Table 1a, drought was the major climate-related shock event reported with a considerable variation across the three provinces in Thailand. The province of Buri Ram was most affected. Flood was reported by over 10% of households in two provinces while storms and soil erosion was reported by only few households. Average frequency of climate events was little over one event

**Table 1a** Climate-related shocks experienced by farmers by province in Thailand

| Type of climate-related shocks | % of households reported | | | Average frequency | | | Average severity | | |
|---|---|---|---|---|---|---|---|---|---|
| | Buri Ram | Ubon Ratcha-thani | Nakhon Phanom | Buri Ram | Ubon Ratcha-thani | Nakhon Phanom | Buri Ram | Ubon Ratcha-thani | Nakhon Phanom |
| Drought | 58.57 | 21.27 | 16.84 | 1.00 | 1.00 | 1.08 | 2.49 | 2.43 | 2.39 |
| Flood | 6.96 | 11.21 | 13.68 | 1.02 | 1.00 | 1.05 | 2.37 | 2.51 | 2.63 |
| Storm | 4.41 | 1.21 | 3.16 | 1.00 | 1.00 | 1.00 | 2.54 | 2.00 | 1.78 |
| Soil erosion | 0.34 | 0.91 | 0.00 | 1.00 | 1.00 | – | 2.00 | 2.50 | – |

*Source*: DFG Household survey 2013

**Table 1b** Climate-related shocks experienced by farmers by province in Vietnam

| Type of climate-related shocks | % of households reported | | | Average frequency | | | Average severity | | |
| --- | --- | --- | --- | --- | --- | --- | --- | --- | --- |
| | Ha Tinh | Thua Thien Hue | Dak Lak | Ha Tinh | Thua Thien Hue | Dak Lak | Ha Tinh | Thua Thien Hue | Dak Lak |
| Drought | 13.23 | 14.37 | 47.48 | 1.00 | 1.00 | 1.04 | 2.37 | 2.58 | 2.65 |
| Flood | 36.38 | 13.97 | 3.47 | 1.03 | 1.03 | 1.00 | 2.55 | 2.60 | 2.59 |
| Storm | 8.56 | 8.58 | 0.79 | 1.00 | 1.00 | 1.00 | 2.43 | 2.51 | 1.80 |
| Soil erosion | 0.58 | 3.19 | 0.47 | 1.67 | 1.00 | 1.00 | 3.00 | 2.38 | 2.67 |

*Source*: DFG Household survey 2013

and quite consistent across the provinces. The same can be said for perceived severity which is mostly around 2.5 on average on scale from 0 to 3. This severity score implies that climatic extreme events affected farm households quite critically according to their subjective assessment. Overall, among the three provinces in Thailand, Buri Ram province located in the eastern part of the country and on the border with Cambodia had the highest degree of climate-related shocks reported.

From Table 1b it can be derived that results vary considerable across the three provinces in Vietnam. In the land locked province of Dak Lak where coffee is a major crop drought was reported by almost half of the households and storm was reported by just few households. On the other hand in Ha Tinh, the province located in the central coastal region with exposure to the sea, more households reported floods. Drought, flood and storm were reported with quite similar rates of households in Thua Thien Hue. This is also the province where soil erosion was most experienced. Frequency of events was similar to Thailand with the exception of soil erosion in Ha Tinh, which can be explained by the mountainous terrain where some of the sample households are located. This observation is also reflected in the perceived severity which is higher than for the other categories. Overall, severity is somewhat higher in the Vietnamese provinces compared to the provinces in Thailand. This seems reasonable as Vietnam is generally more severely affected by the climate change.

In the climate change module, we asked respondents whether or not they perceived changes in climate in general and changes in rainfall, temperature and wind in particular during the time they resided in the area. In Table 2, the different variants of climate change for the three climate categories are reported.

Overall, the vast majority of respondents in all six provinces in the two countries have recognized changes in climate and changes in rainfall and temperature were more frequently reported than changes in wind. Results do not differ much between the two countries although variation between provinces remains high.

Changes in rainfall patterns were described differently between provinces and countries. For example, in two provinces of Thailand respondents observed the length of the dry season to have increased while in Vietnam lower total rainfall was more noted. However, in Vietnam households perceived rainfall variability to increase. Differences among provinces in both countries may show the difference of their geographic conditions.

**Table 2** Climate change perceptions of farmers in Thailand and Vietnam by province, percentage of households reported

| Observations | Thailand | | | Vietnam | | |
|---|---|---|---|---|---|---|
| | Buri Ram | Ubon Ratchathani | Nakhon Phanom | Ha Tinh | Thua Thien Hue | Dak Lak |
| **Climate in general** | **94.57** | **90.61** | **74.74** | **81.52** | **82.04** | **90.69** |
| **Rainfall** | **94.51** | **88.79** | **68.98** | **78.30** | **80.40** | **89.19** |
| Less rain in the whole year | 40.08 | 24.26 | 11.63 | 25.95 | 42.44 | 46.09 |
| Less rain early in the season | 23.26 | 16.70 | 14.68 | 2.12 | 15.12 | 13.80 |
| Dry season becomes longer | 49.15 | 38.33 | 16.90 | 19.42 | 24.69 | 28.02 |
| Rain becomes more erratic | 16.43 | 33.18 | 9.97 | 30.35 | 19.91 | 37.13 |
| Fewer rainy days | 15.11 | 12.70 | 4.99 | 12.75 | 21.45 | 29.87 |
| **Temperature** | **94.41** | **90.27** | **72.85** | **76.93** | **77.16** | **86.77** |
| Getting hotter in summer | 86.86 | 87.64 | 55.68 | 55.08 | 61.57 | 63.02 |
| Cool season is shorter | 35.35 | 41.53 | 15.24 | 20.49 | 28.24 | 9.96 |
| More extreme temperature | 18.00 | 37.64 | 20.20 | 57.21 | 45.22 | 54.91 |
| More heat days | 59.53 | 62.36 | 17.45 | 23.07 | 52.47 | 56.19 |
| **Wind** | **80.81** | **67.39** | **54.85** | **34.14** | **27.93** | **37.84** |
| Wind speed higher | 71.62 | 60.18 | 46.54 | 21.4 | 19.60 | 32.43 |
| More frequent storms | 31.14 | 34.67 | 16.62 | 8.65 | 8.80 | 1.71 |
| Wind direction changes | 24.54 | 31.01 | 12.19 | 13.51 | 13.73 | 11.52 |

*Source*: DFG Household survey 2013

Temperature results generally follow those of rainfall. However, there is more agreement on the description of the type of temperature changes with most respondents observing higher summer temperatures. Both in Thailand and Vietnam over half the respondents in two provinces said that extreme temperatures have increased.

Changes in wind were less frequently mentioned especially in Vietnam while in the province of Buri Ram 80% of the respondents specified a higher wind speed as major change and 30% reported more frequent storms which was confirmed by respondents from the province of Ubon Ratchathani.

Comparing farmer observations with existing literatures supports the notion that their subjective perceptions match scientific data. This confirms findings from South Africa that farmers' perceptions of climate change are in line with the climatic data records (Gbetibouo 2009). Meteorological data from Thailand confirm that rainfall in Thailand decreased in the past three to five decades compared to the first half of

**Table 3** Effects of climate change on crop production and farmers' adaptation measures by province, percentage of households reported

|  | Thailand | | | Vietnam | | |
| --- | --- | --- | --- | --- | --- | --- |
|  | Buri Ram | Ubon Rathchathani | Nakhon Phanom | Ha Tinh | Thua Thien Hue | Dak Lak |
| **Effects on crop production** | **81.66** | **68.48** | **44.91** | **71.21** | **64.47** | **84.07** |
| Lower yields | 61.89 | 47.48 | 32.41 | 45.83 | 41.82 | 63.87 |
| More crop failures | 25.23 | 27.69 | 9.97 | 28.83 | 17.75 | 32.72 |
| More pests | 15.77 | 12.47 | 1.94 | 29.29 | 26.70 | 21.62 |
| More drought stress | 35.35 | 23.46 | 7.20 | 10.77 | 15.74 | 34.99 |
| **Adaptation measures** | **29.54** | **32.42** | **11.23** | **45.53** | **31.14** | **44.95** |
| Crop diversification | 19.69 | 21.82 | 6.67 | 13.62 | 11.38 | 20.82 |
| Chemical input management | 12.05 | 11.52 | 4.56 | 22.96 | 21.76 | 11.04 |
| Water management | 3.40 | 9.42 | 0.70 | 7.39 | 6.39 | 22.40 |
| Planting trees | 1.87 | 2.88 | 0.35 | 0.39 | 1.60 | 0.47 |
| Others | 0.00 | 0.30 | 0.00 | 11.09 | 1.80 | 2.05 |

*Source*: DFG Household Survey 2013

the last century. Also climate models predicted that precipitation will shift from the north to the south (Boonyawat and Chiwanno 2007). Based on climate data generated by a global circulation model temperature in Thailand projected to increase 2° C–4 °C by the end of the century (ADB 2009). Jesdapipat (2008) stated that storms in Thailand have become more intense which is consistent with the subjective perceptions of respondents in our sample.

In Vietnam it has been predicted that most regions will experience an increase in temperature of 2° C–4 °C by the end of the century (Cuong 2008). The same author also found that in most areas of Vietnam, overall rainfall intensity has increased considerably while monthly rainfall has decreased between the months of July and August, but has increased between September and November. It is also expected that the Southern part of Vietnam will become drier.

In Table 3, we illustrate the perceived impact of climate change by farmers on the performance of agriculture, in particular in crop production and their adaptation measures. It is striking that in all six provinces of the two countries a considerable share of households reports a decline in yields. The highest shares with over 60% of households reporting are in Buri Ram and Dak Lak, both provinces with a strong agricultural potential. In these two provinces the occurrence of drought stress was most frequent which is quite consistent with their observations on the change in climate generally and in rainfall reported in Table 2.

In spite of the high share of households who report an impact on crop production only between one fourth and two fifth undertake adaptation measures. This kind of discrepancy has also been observed in a study of farmers in Ethiopia

(Deressa et al. 2008). Adaptation measures include for example growing more (drought resistant) varieties, widening the crop portfolio, spraying more pesticides and applying more fertilizer. Although responses considerably vary by country and by province reflecting differences in agricultural systems, changes in crops and crops varieties and in the amount of chemical input used are the two dominant adaptation measures. In the province of Dak Lak, investment in irrigation was reported by over one fifth of households which is distinctively higher than in all other provinces. Here results are consistent with the perception of more droughts which however is not the case for the province of Buri Ram where 35.35% farmers reported drought stress but only 3.40% take a particular water management method.

In summary, what we can derive from the survey on subjective climate change perceptions is that there is a strong geographic effect of the perceived impacts of climate change. The fact that there is a fairly good congruence between the perceived effects of climate change and adaptations suggesting that farmers are well aware of climate change although the ratio of adaptations to perceptions is in the order of 1:3 only.

In Table 4, we have made use of the 2007 survey and compared farm management parameters related the use chemical inputs, irrigation practices and tools and tree plantation which can serve as proxy parameters for actual adjustment to climate change with the 2013 survey data. It shows that changes can be observed with more cases significant in Vietnam. While no causality to climate change perception can be established here and other factors can also play a role, results are consistent with respondents' climate change perceptions. For example, planting of trees has increased significantly in both countries.

Summarizing the results of the descriptive analysis suggests that farmers in poor and vulnerable environments in Thailand and Vietnam did experience climate-related shocks which on average are perceived as moderately severe. However, variation across locations exists. Furthermore, farmers are well aware of climate change and can describe the process by a range of indicators like "cool season getting shorter" or "rain become more erratic". These criteria differ from those used by scientists in climate models but they seem to correspond well with such findings.

**Table 4** Farm management practices in 2007 and 2013 across all provinces in Thailand and Vietnam

| Parameter | Thailand | | | Vietnam | | |
|---|---|---|---|---|---|---|
| | 2007 | 2013 | p-value | 2007 | 2013 | p-value |
| Chemical input (PPP$) | 35.41 | 55.45 | 0.02 | 118.36 | 93.83 | 0.02 |
| Irrigation tools (unit) | 1.89 | 1.73 | 0.63 | 0.90 | 2.29 | 0.00 |
| Newly-bought irrigation tools (unit) | 0 | 0.030 | 0.00 | 0 | 0.004 | 0.08 |
| Share of irrigated plots (%) | 13.98 | 7.71 | 0.00 | 50.64 | 71.31 | 0.00 |
| Share of tree areas (%) | 4.91 | 8.09 | 0.00 | 23.84 | 34.19 | 0.00 |
| Share of trees out of crop types (%) | 5.95 | 10.37 | 0.00 | 20.58 | 30.21 | 0.00 |

*Source*: DFG Household Survey 2007–2013

Also, farmers recognize that climate change has caused negative impacts on their agricultural production. Nevertheless, adaptation actions in response to the perceived downside effects are still few. This underlines the hypotheses established in Section 2 of the paper that perceptions are an important driver for adaptation decisions that aim at reducing risks and losses. In the next section the perception-adoption link will be explored further by means of econometric analysis.

## 6 Results of Econometric Analysis

With our first model we test the hypothesis that farmers' perception of climate change can be linked to the likelihood of farmer's respective adaption measures. Our two-step Heckman probit model shows a significant lambda for both Thailand and Vietnam dataset indicating the existence of sampling bias (Tables 5a and 5b). The perception model for Vietnam mostly shows the expected signs of the explanatory variables. Education and gender show positive and significant signs (Table 5a). In other words, better educated and male respondents are more likely to recognize climate change. Climate-related shock experience significantly increases the likelihood of respondents recognizing climate change suggesting that short term experience can shape perceptions for long term trends. Differences in province partly reflect the findings of the descriptive statistics. Relative to the base province of Ha Tinh, respondents in Dak Lak are significantly more likely to perceive climate change. This result is consistent with those presented in Tables 1b and 2 with increasing temperatures and an increase in droughts.

The outcome equation with the implementation of adaptation measures as the dependent variable also shows better statistical quality for Vietnam. Age of household head is negatively related to the likelihood of adaptation measures. It is plausible that older farmers are less likely to change their farming system in response to perceived climate change. Gender was significant suggesting that male household heads are more likely to implement adaptation measures which is consistent with the findings of Asfaw and Admassie (2004). As expected, membership in a sociopolitical organization has a positive influence on adaption measures. Likewise, the share of household members engaged in agriculture and ethnicity of household are positively correlated with likelihood of adaptation.

As shown in Table 5b, the perception model for Thailand overall performed poorly in terms of statistical tests. However, the climate-related shock variable was significant and the significant coefficients of the province dummy variables for Buri Ram (positive) and Nakhon Phanom (negative) were consistent with observations presented in Tables 1a and 2.

Similar to the selection equation, the adaptation model for Thailand showed poor explanatory power and the only significant variable (aside from a province dummy) was the respondent's individual attitude towards risk. The coefficient of risk attitude

**Table 5a** Perceptions of and adaptations to climate change by farm households in Vietnam, two-stage Heckman selection model

| Explanatory variables | Adaptation equation Coefficients | Selection equation Coefficients |
|---|---|---|
| **Household head characteristics** | | |
| Age (Years) | −0.004*** | |
|  | (−2.71) | |
| Education (Years of schooling) | −0.001 | |
|  | (−0.26) | |
| Gender (1 = Male, 0 = Female) | 0.058 | |
|  | (1.47) | |
| Member of socio-political organization (1 = Yes, 0 = No) | **0.090**** | |
|  | **(2.56)** | |
| **Respondent characteristics**[a] | | |
| Age (Years) | | 0.005 |
|  | | (1.45) |
| Education (Years of schooling) | | **0.027**** |
|  | | **(2.31)** |
| Gender (1 = Male, 0 = Female) | | **0.211**** |
|  | | **(2.56)** |
| Member of socio-political organization (1 = Yes, 0 = No) | | 0.035 |
|  | | (0.34) |
| **Household characteristics** | | |
| Agricultural member ratio | **0.227**** | |
|  | **(4.32)** | |
| Log of farm size (ha) | **0.029**** | |
|  | **(2.10)** | |
| Household size | | 0.022 |
|  | | (0.83) |
| Log of income (PPP$) | **0.029*** | 0.036 |
|  | **(1.72)** | (0.84) |
| Ethnicity (1 = Kinh, 0 = Minorities) | **0.095**** | −0.113 |
|  | **(2.25)** | (−0.97) |
| Climate-related shock experience (Ordinal score) | | **0.061**** |
|  | | **(2.44)** |
| Risk attitude (Likert scale) | −0.002 | |
|  | (−0.29) | |
| **Village characteristics** | | |
| Log of distance to district town (Km) | −0.016 | **0.089*** |
|  | (−0.80) | **(1.81)** |
| **Province dummies** | | |
| Thua Thien Hue | −**0.127**** | 0.087 |
|  | **(−2.96)** | (0.80) |
| Dak Lak | −**0.107**** | **0.405**** |

(continued)

Table 5a (continued)

| Explanatory variables | Adaptation equation Coefficients | Selection equation Coefficients |
|---|---|---|
|  | (−2.03) | (3.45) |
| Intercept | 0.408* | −0.219 |
|  | (1.80) | (−0.52) |
| Mills |  |  |
| Lambda | −0.487** |  |
|  | (−1.97) |  |
| rho | −0.87 |  |
| Total observations | 1529 |  |
| Wald chi2 | 77.86 |  |
| Prob > chi2 | 0.000 |  |

*Source:* Authors' own calculation
*Note*: $*p < 0.10$, $**p < 0.05$, $***p < 0.01$, z statistics in parentheses
[a]We tried to use the household head characteristics instead of respondent characteristics in the perception equation but the results are as not good as results in Tables 5a and 5b

shows that the higher the degree of risk-seeking, the higher the likelihood that a household adapts to climate change. While farmers in Buri Ram perceive a higher degree of climate change compared to the reference province of Ubon Ratchathani, fewer farmers undertake adaptation measures. Against this background the negative coefficient for the province dummy is surprising. However, this suggests that other factors such as poorer quality extension services or less attention given by other public institutions to the climate change phenomenon may cause this result.

To investigate the determinants for choosing different adaptation measures, we use a multinomial logit model for four groups of adaptations and "no adaptation" is the base category. The Hausman test for the validity of the *independence of the irrelevant alternatives* (IIA) was insignificant for both Thailand and Vietnam. This suggests that the multinomial logit model is an appropriate specification for modelling the choice of adaptation measures to climate change of farmers. The estimated coefficients along with the standard errors are presented in Table 6a for Vietnam and in Table 6b for Thailand.

In the model for Vietnam, the signs of the explanatory variables are largely consistent with the results of the outcome equation in the Heckman model (Table 5a). For all adaptation measures except for "planting trees" household head' age has a significant and negative signs which is consistent with expectations as older household heads are likely to stick to their traditional practices in spite of recognizing changes in climate conditions. On the other hand, changing water management practices is positively correlated with membership in a socio-political organization. This is plausible as water management in rural Vietnam is a collective action and usually requires good relationships with village authorities namely the people's

**Table 5b** Perceptions of and adaptations to climate change by farm households in Thailand, two-stage Heckman selection model

| Explanatory variables | Adaptation equation Coefficients | Selection equation Coefficients |
|---|---|---|
| **Household head characteristics** | | |
| Age (Years) | 0.001 | |
|  | (0.93) | |
| Education (Years of schooling) | 0.006 | |
|  | (1.01) | |
| Gender (1 = Male, 0 = Female) | 0.034 | |
|  | (0.96) | |
| Member of socio-political organization (1 = Yes, 0 = No) | −0.032 | |
|  | (−0.44) | |
| **Respondent characteristics** | | |
| Age (Years) | | −0.004 |
|  | | (−0.99) |
| Education (Years of schooling) | | 0.004 |
|  | | (0.28) |
| Gender (1 = Male, 0 = Female) | | 0.020 |
|  | | (0.20) |
| Member of socio-political organization (1 = Yes, 0 = No) | | −0.039 |
|  | | (−0.17) |
| **Household characteristics** | | |
| Agricultural member ratio | 0.030 | |
|  | (0.53) | |
| Log of farm size (ha) | −0.024 | |
|  | (−1.32) | |
| Household size | | 0.042 |
|  | | (1.39) |
| Log of income (PPP$) | 0.004 | 0.036 |
|  | (0.23) | (0.77) |
| Climate-related shock experience (Ordinal score) | | **0.090***  |
|  | | **(2.69)** |
| Risk attitude (Likert scale) | **0.013**** | |
|  | **(2.33)** | |
| **Village characteristics** | | |
| Log of distance to district town (Km) | 0.037 | −0.050 |
|  | (1.52) | (−0.72) |
| **Province dummies** | | |
| Buri Ram | **−0.085*** | **0.245**** |
|  | **(−1.88)** | **(2.01)** |
| Nakhon Phanom | −0.054 | **−0.643****  |
|  | (−0.54) | **(−5.51)** |
| Intercept | 0.149 | **1.057**** |

(continued)

**Table 5b** (continued)

| Explanatory variables | Adaptation equation Coefficients | Selection equation Coefficients |
|---|---|---|
| | (0.72) | **(2.10)** |
| Mills | | |
| Lambda | **−0.601*** | |
| | **(−1.65)** | |
| rho | −1.00 | |
| Total observations | 1361 | |
| Wald chi2 | 17.21 | |
| Prob > chi2 | 0.102 | |

*Source*: Authors' own calculation

*Note*: *$p < 0.10$, **$p < 0.05$, ***$p < 0.01$, z statistics in parentheses

committee. Among household characteristics it is shown in Table 6a that the higher the share of household members engaged in agriculture, the more likely the households undertake adaptation measures. The respective coefficient is positive and significant for all adaptation measures except for planting trees although the direction of influence is positive. This result is plausible as households whose major livelihood is in agriculture are more likely to actively meet the challenges of climate change. Indeed, the coefficients for all categories (i.e. changing crop diversity, chemical input management, and water management) are positive and highly significant for four categories. Income of households shows a significant and positive influence on adaptation measures "water management" and "planting more trees" which seems plausible as these measures are related to investments. The coefficients for the variables reflecting the perception of the respondent in the three indicators of climate change, i.e. rainfall, temperature and wind all show a positive sign although not all are significant. Consistent results are found for rainfall which is plausible as indeed rainfall is the major driving factor for productivity of agriculture and changing rainfall patterns may warrant adjustments in many agricultural practices. Temperature is significant for planting more trees and changes in crop diversification such as changing crops or crop varieties. The variable for farmer's perception in the change of wind conditions is significant for "crop diversification" and "planting trees" which seems plausible again. Overall, however, it can be argued that farmer's climate change perceptions prompt them to change their farming system. The significance of all climate related coefficients for planting more trees is a strong indicator that farmers recognize the need for climate change adaptation for a variety of reasons.

The ethnicity variable is only significant for water management which underlines again the importance of collective action which often relies on public support. This indicates that households belonging to the Kinh ethnic majority group may be more likely to undertake adaptation measures. Finally, the significant coefficient for the

Table 6a Results of multinomial logit model for the choice of adaptation measures, Vietnam

| Explanatory variables | Crop diversification coef/se | Chemical input management coef/se | Water management coef/se | Planting trees coef/se |
|---|---|---|---|---|
| **Household head characteristics** | | | | |
| Age (Years) | −0.015* | −0.017** | −0.016* | −0.045 |
| | (0.009) | (0.008) | (0.009) | (0.031) |
| Education (Years of schooling) | 0.014 | 0.010 | −0.010 | 0.070 |
| | (0.027) | (0.021) | (0.024) | (0.111) |
| Gender (1 = Male, 0 = Female) | 0.471 | 0.378 | 0.332 | −0.384 |
| | (0.298) | (0.233) | (0.257) | (0.759) |
| Member of socio-political organization (1 = Yes, 0 = No) | 0.178 | 0.329 | 0.568*** | 1.627 |
| | (0.222) | (0.219) | (0.198) | (0.990) |
| **Household characteristics** | | | | |
| Agricultural member ratio | 1.250*** | 0.986*** | 0.736** | 1.928 |
| | (0.364) | (0.299) | (0.324) | (1.357) |
| Log of farm size (ha) | 0.061 | 0.066 | 0.214** | 0.362 |
| | (0.098) | (0.073) | (0.084) | (0.220) |
| Log of income (PPP$) | 0.219** | 0.038 | 0.299*** | 0.678*** |
| | (0.105) | (0.087) | (0.094) | (0.262) |
| Rainfall perception (1 = Yes, 0 = No) | 1.607* | 17.775*** | 1.635** | 13.515*** |
| | (0.977) | (0.326) | (0.798) | (0.803) |
| Temperature perception (1 = Yes, 0 = No) | 0.973 | 0.631 | 0.953 | 15.283*** |
| | (0.756) | (0.393) | (0.650) | (0.581) |
| Wind perception (1 = Yes, 0 = No) | 0.736*** | 0.080 | 0.105 | 1.800*** |
| | (0.192) | (0.163) | (0.180) | (0.697) |
| Risk attitude (Likert scale) | 0.047 | −0.001 | 0.021 | −0.166 |
| | (0.043) | (0.029) | (0.035) | (0.105) |
| Ethinicity (1 = Kinh, 0 = others) | 0.102 | 0.291 | 0.374* | 0.714 |
| | (0.255) | (0.237) | (0.223) | (0.956) |
| **Village characteristics** | | | | |
| Log of distance to district town (Km) | −0.084 | 0.021 | −0.080 | 0.340 |

(continued)

Table 6a (continued)

| Explanatory variables | Crop diversification | Chemical input management | Water management | Planting trees |
|---|---|---|---|---|
| | coef/se | coef/se | coef/se | coef/se |
| | (0.118) | (0.094) | (0.099) | (0.284) |
| **Province dummies** | | | | |
| Thua Thien Hue | −0.220 | −0.137 | −0.083 | 1.687 |
| | (0.292) | (0.211) | (0.293) | (1.044) |
| Dak Lak | **0.556**** | **−1.070**** | **1.203**** | 0.398 |
| | **(0.262)** | **(0.260)** | **(0.240)** | (1.102) |
| Constant | **−7.200**** | **−20.009**** | **−7.371**** | **−40.993**** |
| | **(1.173)** | **(0.901)** | **(1.057)** | **(2.233)** |
| Base category | **No adaptation** | | | |
| Number of observations | 1529 | | | |
| Log likelihood | −1505.473 | | | |
| LR chi2 | **353.08**** | | | |
| Pseudo R2 | 0.136 | | | |

*Source*: Authors' own calculation
*Note*: *** $p < 0.01$, ** $p < 0.05$, * $p < 0.1$

province dummy of Dak Lak indicates the importance of irrigation and crop diversification is this land locked region compared to the coastal provinces of Thua Thien Hue and Ha Tinh.

In summary, the model for Vietnam shows satisfactory results. It largely confirms the finding of our first model (binary model 1a) and provides further information on the factors that drive specific adaption measures. The results can provide information for extension services to guide farmers in adopting more climate smart technologies.

The model for Thailand shows less explanatory power than the Vietnam model. Although the coefficients generally have the expected signs, much fewer of them are significant. Interestingly, however, individual attitude towards risk of the respondent pops up in two of the four categories of adaptation measures with a positive and significant coefficient. This is plausible as risk seeking behaviour may make farmers more likely to undertake climate change adaptation measures. This however was not observed in the Vietnam model. On the other hand, the coefficients for the three climate change indicators are quite consistent with the Vietnam model although wind speed seems to be a stronger factor in Thailand in explaining agricultural adjustments to climate change. The negative coefficient for the province dummy variable for Buri Ram is consistent with the binary model but does not match with the climate-related shock experience shown in the descriptive statistics. In summary, while the Thailand model is less satisfactory the main message that climate change perception is a major driver for specific adaption measures in agriculture can be confirmed.

**Table 6b** Results of multinomial logit model for the choice of adaptation measures, Thailand

| Explanatory variables | Crop diversification coef/se | Chemical input management coef/se | Water management coef/se | Planting trees coef/se |
|---|---|---|---|---|
| **Household head characteristics** | | | | |
| Age (Years) | 0.002 | 0.002 | **0.022**** | 0.017 |
| | (0.008) | (0.009) | **(0.010)** | (0.019) |
| Education (Years of schooling) | 0.038 | −0.005 | 0.010 | **0.176**** |
| | (0.032) | (0.037) | (0.047) | **(0.054)** |
| Gender (1 = Male, 0 = Female) | −0.075 | 0.273 | **0.670**** | 0.447 |
| | (0.200) | (0.252) | **(0.333)** | (0.467) |
| Member of socio-political organization (1 = Yes, 0 = No) | −0.845 | 0.358 | 0.212 | −0.924 |
| | (0.552) | (0.403) | (0.465) | (1.003) |
| **Household characteristics** | | | | |
| Agricultural member ratio | 0.182 | −0.185 | 0.451 | 0.348 |
| | (0.344) | (0.372) | (0.446) | (0.511) |
| Log of farm size (ha) | −0.113 | 0.050 | −0.150 | −0.279 |
| | (0.099) | (0.129) | (0.152) | (0.281) |
| Log of income (PPP$) | 0.088 | 0.036 | −0.000 | −0.071 |
| | (0.088) | (0.102) | (0.133) | (0.198) |
| Rainfall perception (1 = Yes, 0 = No) | 1.286 | **16.749**** | 0.944 | **14.083**** |
| | (1.115) | **(0.591)** | (1.025) | **(0.388)** |
| Temperature perception (1 = Yes, 0 = No) | 1.747 | −0.447 | **15.678**** | **12.952**** |
| | (1.558) | (0.719) | **(0.709)** | **(0.588)** |
| Wind perception (1 = Yes, 0 = No) | **0.453**** | **0.796**** | 0.476 | **2.443**** |
| | **(0.229)** | **(0.304)** | (0.328) | (1.042) |
| Risk attitude (Likert scale) | **0.085**** | 0.046 | **0.112**** | **−0.160*** |
| | **(0.033)** | (0.036) | **(0.045)** | **(0.094)** |
| **Village characteristics** | | | | |
| Log of distance to district town (Km) | 0.044 | **0.434**** | 0.044 | 0.034 |
| | (0.132) | **(0.152)** | (0.181) | (0.198) |

(continued)

**Table 6b** (continued)

| Explanatory variables | Crop diversification coef/se | Chemical input management coef/se | Water management coef/se | Planting trees coef/se |
|---|---|---|---|---|
| **Province dummies** | | | | |
| Buri Ram | −0.037 | 0.046 | **−1.259\*\*\*** | **−0.801\*** |
| | (0.191) | (0.217) | **(0.318)** | **(0.441)** |
| Nakhon Phanom | **−0.819\*\*\*** | −0.523 | **−2.441\*\*\*** | **−2.113\*\*** |
| | **(0.314)** | (0.365) | **(0.715)** | **(1.043)** |
| Constant | **−6.382\*\*\*** | **−20.852\*\*\*** | **−21.526\*\*\*** | **−33.101\*\*\*** |
| | **(1.298)** | **(1.214)** | **(1.441)** | **(2.448)** |
| Base category | **No adaptation** | | | |
| Number of observations | 1361 | | | |
| Log likelihood | −1174.558 | | | |
| LR chi2 | **176.10\*\*\*** | | | |
| Adjusted R2 | 0.089 | | | |

*Source*: Authors' own calculation
*Note*: \*\*\* $p < 0.01$, \*\* $p < 0.05$, \* $p < 0.1$

# 7 Summary and Conclusions

Using a comprehensive dataset of farm households in Thailand and Vietnam we have tried to answer three questions. Firstly, we wanted to explore what climate related shocks farm households experience in the more recent past and whether they perceive a change in the longer term climate conditions and what indicators they use to describe climate change. Secondly, what factors influence their climate change perceptions and can their perceptions be linked to their adaptation measures. Thirdly, we wanted to know to what extent the explanatory factors differ for specific climate change adaptation measures.

The answer to the first question is quite clear. The majority of farm households in both countries have experienced recent climate-related shocks and the vast majority does perceive that climate has changed. While the latter fact may not be very surprising our results however point out that farmers have their own way of describing the climate change related phenomenon. We can also see that quite consistent with differences in natural and economic conditions, the geographic location has an influence on how farmers recognize climate change. Furthermore, farmers reported adjustment measures which they are planning to undertake or have already undertaken in response to climate change. We have independently checked this claim by comparing some climate relevant agricultural practices from our 2007 survey with the most recent survey in 2013 and we found quite some differences that suggest that farmers are indeed climate-responsive although we cannot judge to what degree these changes fit the metaphor of "climate-smart".

To answer the second question we used a Heckman model that allows joint estimation of a selection and an outcome equation, separately for the two countries. Based on the results we can confirm that perceptions can be reasonably linked to farmers' decision to undertake adaptation measures. In the model for Vietnam we can show that perceptions are shaped by the respondent's characteristics, location variables and recent climate related shocks. Unfortunately, results for the Thailand model are less convincing. However, the climate-related shock variable is significant and consistent with the results in Vietnam. Similar results were found for the outcome equation where again the Vietnam model was more convincing. The difference could be attributed to the lower awareness among the Thai farmers as shown in the lower number of cases in spite of largely equal initial sample size between the two countries. From an objective point of view, Vietnam is indeed more exposed to climate change due to its geographic location along the South China Sea costal line.

Finally, the answer to the third question is that the factors that drive specific climate change related adaption measures differ among practices, provinces and countries. They are to be found in the characteristics of the respondent and the household head whenever there is a difference between the two. Perhaps the most important factor in explaining specific adaptation measures are the three specific climate variables namely rainfall, temperatures and wind, which are all significantly correlated with tree plantation. While for the other adaptation measures such as crop diversification, varietal change, etc. factors other than climate change may be more important, the clearest connection we find is with trees.

We believe our results can provide important information to policy makers and agricultural extension services who should improve their understanding of the farmers' interpretation of climate change and the constraints that have so far prevented them from undertaking more and better adaption measures. Further studies should take a more in-depth look at those constraints and provide a detailed assessment of the costs and benefits of farmer-based adaption measures.

## References

Asfaw A, Admassie A (2004) The role of education on the adoption of chemical fertilizer under different socioeconomic environments in Ethiopia. Agricultural Economics 30:215–228

Asian Development Bank (ADB) (2009) The Economics of Climate change in Southeast Asia: A Regional Review. Asian Development Bank, Manila

Boonpragob K (2005) Crisis or Opportunity: Climate Change Impacts and Thailand. Greenpeace Southeast Asia, Thailand

Boonyawat J, Chiwanno S (2007) Origin and One Decade of Global Change Study in Thailand. In: Boonyawat J (ed) Southeast Asia START Regional Center and a Decade of Global Change in Thailand. Southeast Asia Global Change System for Analysis, Research and Training.

Croppenstedt A, Demeke M, Meschi MM (2003) Technology adoption in the presence of constraints: The case of fertilizer demand of Ethiopia. Review of Development Economics 7:58–70

Cuong N (2008) Viet Nam Country Report—A Regional Review on the Economics of Climate Change in Southeast Asia. Report submitted for RETA 6427: A Regional Review of the Economics of Climate Change in Southeast Asia. Asian Development Bank, Manila

Dasgupta S, Laplante B, Meisner C et al (2007) The Impact of Sea Level Rise on Developing Countries: A Comparative Analysis. World Bank Policy Research Working Paper 4136,

Deressa T, Hassan RM, Alemu T et al (2008) Analyzing the Determinants of Farmer's Choice of Adaptation Methods and Perceptions of Climate Change in the Nile Basin of Ethiopia. International Food Policy Research Institute Discussion Paper 00798

Dohmen T, Falk A, Huffman D et al (2011) Individual risk attitudes: Measurement, determinants, and behavioral consequences. Journal of the European Economic Associations 9(3):522–550

Feder G, Just RE, Zilberman D (1985) Adoption of Agricultural Innovations in Developing Countries: A Survey. Economic Development and Cultural Change 33(2):255–298

Franzel S (1999) Socioeconomic factors affecting the adoption potential of improved tree fallows in Africa. Agroforestry Systems 47:305–321

Fuster J (2002) Physiology of executive functions: The perception-action cycle. In: Stuss DT, Knight R (eds) Principles of the frontal lobe. Oxford University Press, New York, p 96–108

Gbetibouo GA (2009) Understanding Farmers' Perceptions and Adaptations to Climate Change and Variability: The case of the Limpopo Basin, South Africa. International Food Policy Research Institute Discussion Paper 00849.

Greene, W. H. (2003). Econometric Analysis. New Jersey: Pearson Education.

Hardeweg B, Menkhoff L, Waibel H (2013) Experimentally Validated Survey Evidence on Individual Risk Attitudes in Rural Thailand. Economic Development and Cultural Change 61:859–888

Heckman J J (1979) Sample Selection as a Specification Error. Econometrica 47:153–161

Hung PT, Trung LD, Cuong N (2010) Poverty of the Ethnic Minorities in Vietnam: Situation and Chanlleges from the Poorest Communes. Munich Personal RePEc Archive.

Iglesias A, Quiroga S, Diz A (2011) Looking into the Future of Agriculture in a Changing Climate. European Review of Agricultural Economics 38(3):427–447

IPCC (2014) Climate Change 2014: Impacts, Adaptation and Vulnerability. Part B: Regional Aspects. Working Group II Contribution to the Fifth Assessment Report of the Intergovernmental Panel on Climate Change. Cambridge University Press, New York

Jesdapipat S (2008) Thailand Country Report—A Regional Review on the Economics of Climate Change in Southeast Asia. Report submitted for RETA 6427: A Regional Review of the Economics of Climate Change in Southeast Asia. Asian Development Bank, Manila

Kahneman D, Knetsch JL, Thaler RH (1990) Experimental Tests of the Endowment Effect and the Coase Theorem. The Journal of Political Economy 98(6): 1325–1348

Long JS, Freese J (2006) Regression Models for Categorical Dependent Variables Using Stata (2nd ed). Stata Press, Texas

Maddison D (2007) The Perception of and Adaptation to Climate Change in Africa. Policy Reseach Working Paper WPS4308, The World Bank

Norris E, Batie S (1987) Virginia farmers' solid conservation decisions: An application of Tobit analysis. Southern Journal of Agricultural Economics 19:89–97

Puhani, PA (2000) The Heckman Correction for sample selection and its critique. Journal of Economic Surveys 14(1):53–68

Reilly JM, Schimmelpfennig D (1999) Agricultural impact assessment, vulnerability and the scope for adaptation. Climate Change 43:745–788

Tversky A, Kahneman D (1992) Advances in Prospect Theory: Cumulative Representation of Uncertainty. Journal of Risk and Uncertainty 5(4):297–323

Waibel H, Tongruksawattana S, Voelker M (2013) Voices of the poor in climate change in Thailand and Vietnam. In: Ananta A, Bauer A, Thant M (eds) The Environments of the Poor in Southeast Asia, East Asia and the Pacific. Asian Development Bank, Singapore, p 170–186

Weber EU (2010) What shapes perceptions of climate change? Wires Climate Change 332–342

Wooldridge, JM (2009) Introductory Econometrics: A modern approach. South-Western Cengage Learning, p 562

**Open Access** This chapter is distributed under the terms of the Creative Commons Attribution-NonCommercial-ShareAlike 3.0 IGO license (https://creativecommons.org/licenses/by-nc-sa/3.0/igo/), which permits any noncommercial use, duplication, adaptation, distribution, and reproduction in any medium or format, as long as you give appropriate credit to the Food and Agriculture Organization of the United Nations (FAO), provide a link to the Creative Commons license and indicate if changes were made. If you remix, transform, or build upon this book or a part thereof, you must distribute your contributions under the same license as the original. Any dispute related to the use of the works of the FAO that cannot be settled amicably shall be submitted to arbitration pursuant to the UNCITRAL rules. The use of the FAO's name for any purpose other than for attribution, and the use of the FAO's logo, shall be subject to a separate written license agreement between the FAO and the user and is not authorized as part of this CC-IGO license. Note that the link provided above includes additional terms and conditions of the license.

The images or other third party material in this chapter are included in the chapter's Creative Commons license, unless indicated otherwise in a credit line to the material. If material is not included in the chapter's Creative Commons license and your intended use is not permitted by statutory regulation or exceeds the permitted use, you will need to obtain permission directly from the copyright holder.

# U.S. Maize Yield Growth and Countervailing Climate Change Impacts

**Ariel Ortiz-Bobea**

**Abstract** Over the past several decades, maize yields in the US Midwest have risen at about 17% per decade as a result of steady technological progress. Although the trend is expected to remain positive, climate change is expected to have an increasing countervailing effect. In this chapter, I compute the yield growth rates necessary to fully offset the potential negative effects of a warming climate. Relying on a statistical model allowing for nonlinear effects of temperature on yield, I find that maize yields would decrease by −4.2, −21.8 and −46.1% *around* the trend, under uniform warming scenarios of 1 °C, 3 °C and 5 °C, respectively. I find that an increase of 6.6%/decade in maize yields is required to fully offset the detrimental effects of a severe but still plausible 3 °C warming in the next three decades. This indicates that future maize yield trends could – all else equal – be substantially curtailed due to the climate change. This case study illustrates how agricultural policy analysts can assess the magnitude of potential climate change impacts relative to historical yield trends to help identify targets for agricultural research.

## 1 Introduction

Climate change is resulting in shifting rainfall patterns and rising temperatures that will increasingly challenge agricultural producers across the globe, including in temperate regions with high agricultural productivity such as the United States (US) Midwest region. Various statistical studies have found a strong longitudinal relationship between exposure to high temperature (>30 °C) and lower-than-average crop yields (Schlenker and Roberts 2009; Lobell et al. 2011). This historical evidence presages lower yields in the region under a warmer climate relative to a world without climate change.[1] At the same time, Midwest maize yields have risen at about 17% per decade in recent times as a result of steady technological progress. This chapter analyzes the extent to which these secular maize yield trends can help

---

[1] Evidence suggests that temperature affects yield by lowering the water supply in rainfed environments (see Lobell et al. 2013).

A. Ortiz-Bobea (✉)
Cornell University, Ithaca, NY, USA
e-mail: ao332@cornell.edu

offset the projected relative decline of maize yields resulting from a warming climate.[2] This case study illustrates how agricultural policy analysts can assess the magnitude of potential climate change impacts relative to historical yield trends to help identify targets for agricultural research and investments.

The case study is organized as follows. First, I estimate a statistical model of maize yields regressed on weather variables for the US Midwest. The model allows for nonlinear temperature effects on yield following the approach developed by Schlenker and Roberts (2009). This model accounts for distinct effects of temperature exposure to various temperature bins within *each day* of the growing season. The model is based on panel data and exploits the longitudinal covariance of maize yields and weather conditions at the county level. Second, I use the estimated climate sensitivity parameters to developed maize yield change projections under three uniform warming scenarios (1, 3 and 5 °C). Third, I use these projections to answer the following question. What yield growth rate would be necessary to fully offset the projected yield effects under warming scenario? Obviously, the answer depends on the time horizon of the warming, so I explore time frames ranging from one decade to a century. Finally, I discuss the magnitude of potential climate change impacts on maize yields in light of historical yield trends.

The chapter is organized as follows. First, I describe the data sources and provide summary statistics for key variables in the analysis. I also provide an overview of the warming scenarios. In the subsequent section I present the crop statistical model and describe how climate change impact projections are computed. I then present the model results and the associated impacts from a uniform warming and provide a discussion of the findings. I then conclude the chapter.

## 2 Data Sources and Summary Statistics

The empirical analysis in this chapter relies on agricultural and climate data. The agricultural data was obtained from *Quick Stats,* the US Department of Agriculture's (USDA) online database. This database provides data from historical surveys on county-level agricultural production variables such as acres planted and harvested as well as production. The dependent variable in the study, maize yield, is obtained by dividing total maize production by acres planted. For the 1929–2014 period, this information is complete for 644 counties in 13 Midwest states. This constitutes the set of counties in the study.

The climate data is obtained from the PRISM Climate Group, which provide USDA's official climatological data. The PRISM data is a detailed gridded dataset providing daily measurements of minimum, average and maximum temperature and total precipitation for each 4-by-4 km grid over the entire contiguous US since 1981. Because the data is gridded, it needs to be aggregated to the county level to

---

[2]Although crop yield does not directly reflect agricultural productivity, it provides a useful metric that is easily understood by a wide audience interested in agriculture and food security concerns.

**Table 1** Summary statistics for select variables

| Variable | Min | 25th pct. | 50th pct. | Mean | 75th pct. | Max |
|---|---|---|---|---|---|---|
| Corn yield (bu/acre) | 17.0 | 101.1 | 123.7 | 122.2 | 144.6 | 210.8 |
| Precipitation (mm) | 110 | 467 | 558 | 569 | 659 | 1254 |
| Temperature exposure (days) | | | | | | |
| <0 °C | 0.00 | 1.07 | 2.43 | 3.24 | 4.744 |
| 0–5 °C | 0.00 | 4.13 | 6.14 | 6.55 | 8.68 |
| 5–10 °C | 3.12 | 12.25 | 15.96 | 15.98 | 19.68 |
| 10–15 °C | 9.42 | 23.89 | 28.38 | 28.38 | 32.67 |
| 15–20 °C | 23.14 | 38.52 | 42.29 | 42.02 | 45.78 |
| 20–25 °C | 24.55 | 39.06 | 43.32 | 43.30 | 47.69 |
| 25–30 °C | 6.27 | 26.05 | 30.70 | 30.88 | 35.97 |
| >30 °C | 0.01 | 5.90 | 11.14 | 12.66 | 18.15 |

*Notes:* Summary statistics correspond to a balanced panel of 644 counties for the 1981–2014 period. Weather variables are aggregated between April and September of each year. For reference, 100 bu./acre of maize are roughly equivalent to 6.3 t/ha

match the agricultural observations. I perform this aggregation by weighting each PRISM grid by the amount of cropland it contains based on USDA's Cropland Data Layer (CDL). The CDL provides 30-m resolution land cover pixels corresponding to over 100 classes. The weights were based on cropland pixel counts falling within each PRISM data grid and the average of CDL cropland counts for years 2008–2014 were used. Note that temperature exposure to each temperature "bin" or interval is computed by fitting a double sine curve going through the minimum and maximum temperature of each consecutive day for each PRISM grid and subsequently counting the time spent within each degree bin over the growing season in each year. The temperature exposure was then aggregated to county using the aforementioned approach.

Key summary statistics are presented in Table 1 and correspond to a balanced panel of 644 counties over the 1981–2014 study period. This period is confined to years with complete climate data. The table shows maize yields vary considerably, ranging from 17.0 to 210.8 bu./acre. This variation obviously encompasses both cross-sectional (across counties) and longitudinal (within counties) dimensions. There is also a wide range of variation for precipitation over this time period with minimum and maximum levels of 110 and 1254 mm for the April–September period. Following conventional practice, these months correspond were chosen to approximate the maize growing season in the region.

Regarding air temperature, the present study relies on measurements of the temperature distribution across the entire growing season rather than average monthly temperature. In other words, the temperature variables correspond to the time spent within each temperature bin over the April–September period. This approach is arguably better suited to capture exposure to extreme temperatures than monthly average temperatures. Although the statistical analysis makes use of exposure data to each bin ranging from 0 to 36 °C, I only present summary statistics for aggregated

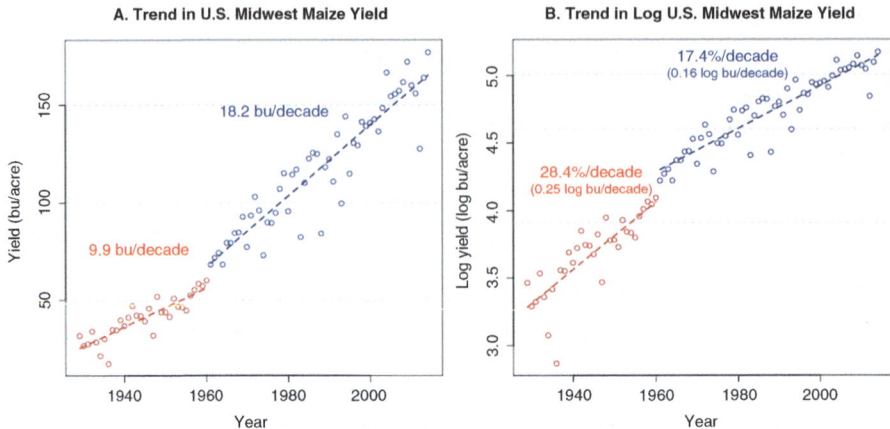

**Fig. 1** Corn yield trends in the US Midwest (*Notes:* Observations correspond to acreage-weighted maize yields for each year. The trend lines were fitted assuming a linear trend for years 1929–1960 and 1961–2014. The sample corresponds to a balanced panel of 644 counties across 13 states over the 1929–2014 period)

contiguous bins in Table 1. The table shows that the most frequent temperature range is between 15 and 25 °C, which corresponds to the bins with the highest mean exposures.

In this chapter I seek to compare the potential effects of a warming climate relative to historical yields trends. Figure 1 illustrates the rise in maize yields in the Midwest since 1929. Panel A shows the yield trend has roughly doubled in *absolute* terms between 1929–1960 and 1961–2014. However, this obscures the fact that the *rate* of this trend has slowed down by almost 40% during this period, as shown in panel B.[3] I will refer to these growth rates later on in the analysis. Also, it is worth noting that I do not detect a statistically significant trend in weather variables over the study period (1981–2014). This suggests that these yield trends are mostly a reflection of technological progress and not really of parallel climate trends.

Regarding climate change data, I adopt 3 uniform warming scenarios of 1, 3 and 5 °C with no precipitation change. The reason I focus on temperature rather than precipitation changes is that previous studies (e.g. Lobell et al. 2008; Schlenker and Roberts 2009) have found that temperature changes are the major explanatory factor explaining crop yield fluctuations in the US Midwest (and elsewhere). A possible reason is that high temperatures capture the effect of dry summer spells, which are crucial for maize production, but are not captured by the season-long precipitation variables. Figure 2 provides an overview of the temperature distribution for the baseline climate as well as under the warming scenarios (lower row). The maps illustrate the mean exposure above 30 °C in each county during the growing season. Under the baseline climate, very few counties have mean exposure exceeding 30 days over the April–September period (total of 183 days). However,

---

[3] The 1929–1960 period corresponds to the period of hybrid corn varieties adoption across the US.

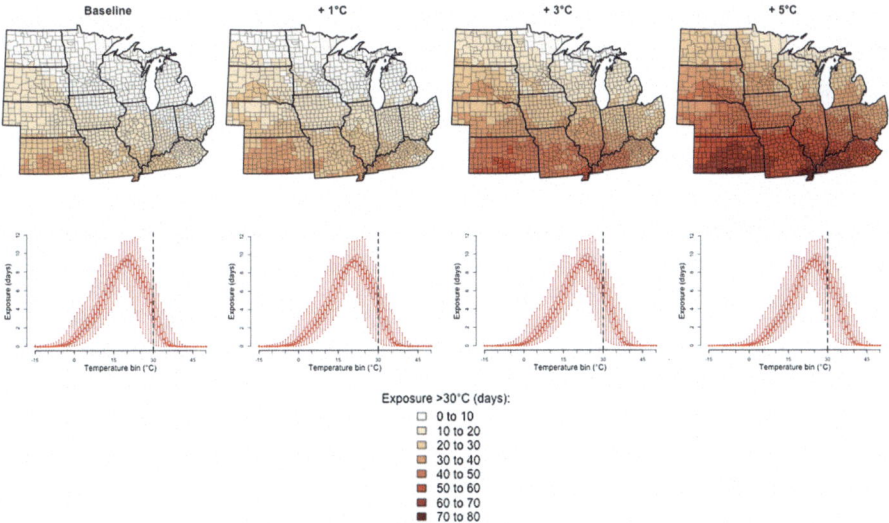

**Fig. 2** Exposure to extreme temperature under varying uniform warming scenarios (*Notes:* The upper row shows the yearly mean exposure (in days) to temperatures exceeding 30 °C during the April–September growing season for each county in 13 Midwest states for baseline and 3 uniform warming scenarios. The lower row presents the temperature distribution across the sample for each temperature bin. Each box represents the median and the first and third quantiles of the distribution. The whiskers extent to data extremes. The dotted vertical line indicates the 30 °C threshold for illustrative purposes)

exposure above this threshold substantially increases under the most severe warming scenario. This will have a major effect on the projected yield impacts as we will see shortly.

## 3 Crop Yield Model and Climate Change Impacts

Crop statistical models have re-emerged as an alternative approach to the traditional biophysical models for assessing the potential impacts of climate change on crop yields. A statistical crop yield model is basically a regression analysis of crop yields on weather variables. Early examples can be traced back to the early part of the last century (Wallace 1920; Hodges 1931). In this chapter, I adopt the approach developed more recently by Schlenker and Roberts (2009). These authors developed an innovative approach that separately estimates the effect of the cumulative exposure (over the growing season) to different temperature bins on crop yield.[4] Mathematically, the nonlinear effect of temperature on yield may be represented by

---

[4] This approach assumes that temperature effects on yield are cumulative and substitutable over time. This assumption may be relaxed.

a function of temperature $h$, denoted $g(h)$. Logged maize yield $y_{it}$ in county $i$ and year $t$ can thus be represented as:

$$y_{it} = \int_{\underline{h}}^{\overline{h}} g(h)\phi_{it}(h)d(h) + p_{it}\delta_1 + p_{it}^2\delta_2 + z_{it}\tau + c_i + \epsilon_{it} \qquad (1)$$

where $\phi_{it}(h)$ is the time distribution of temperature for April–September, $p_{it}$ is precipitation, $z_{it}$ is a quadratic time trend and the $c_i$ are county fixed-effects that capture time-invariant factors explaining yields level across counties (e.g. soil quality, etc). However, Eq. (1) cannot be estimated directly because of the integral. To make this model tractable one needs to approximate the integral with a summation over discrete temperature bins:

$$y_{it} = \sum_{h=0}^{36} g(h+0.5)\left[\Phi_{it}(h+1) - \Phi_{it}(h)\right] + p_{it}\delta_1 + p_{it}^2\delta_2 + z_{it}\tau + c_i + \epsilon_{it}$$

where $\Phi_{it}(h+1) - \Phi_{it}(h)$ represents the time spent over the $[h; h+1]$ interval, and $g(h+0.5)$ is a parameter to estimate. However, given the high number of temperature bins, collinearity between exposures to contiguous bins might create noisy estimates. As a result I assume that $g(h)$ is a smooth function over temperature bins which I can approximate with cubic B-spline with 8 degrees of freedom evaluated at each temperature bin. This can be written as:

$$y_{it} = \sum_{h=0}^{36}\sum_{j=1}^{8} \gamma_j B_j(h+0.5)\left[\Phi_{it}(h+1) - \Phi_{it}(h)\right] + p_{it}\delta_1 + p_{it}^2\delta_2 + z_{it}\tau + c_i + \epsilon_{it}$$

$$y_{it} = \underbrace{\sum_{h=0}^{36}\sum_{j=1}^{8} \gamma_j B_j(h+0.5)\left[\Phi_{it}(h+1) - \Phi_{it}(h)\right]}_{x_{it,j}} + p_{it}\delta_1 + p_{it}^2\delta_2 + z_{it}\tau + c_i + \epsilon_{it}$$

where $B_j$ is the $j$th column of the basis matrix of the natural cubic spline. The model effectively regresses yield on eight temperature variables, $x_{it,j}$. The model is estimated via Least Squares and errors are clustered by county and by year to account for heteroscedasticity and contemporaneous error dependence. Once parameters $\gamma_j$ are estimated, one can derive the marginal effects of temperature exposure by premultiplying estimated coefficients by the basis matrix. These marginal effects correspond to the marginal effects of each temperature bin on crop yield.

Obtaining climate change projections based on these marginal effects is straightforward and simply requires multiplying the marginal effects for each temperature bin by the change in exposure to each bin under a given warming scenario. The log yield changes can then transformed into percentage changes using well-known formulas.

## 4 Results and Discussion

### 4.1 Model Results and Warming Impacts

The main result of the model is the nonlinear effect of temperature on maize yields which is illustrated in Fig. 3. The effects of precipitation are not presented here because the scenarios do not alter the level of precipitation. Exposure to temperatures above 30 °C appear detrimental to maize yields. The response function reflects the fact that years with higher exposure to high temperature tend to be associated with lower than average maize yields in the study region. This is in line with previous findings in the literature.

The lower part of the Fig. 3 represents the baseline temperature distribution across temperature bins. This is somewhat similar to the distribution within bins illustrated in Fig. 2. Again, for the baseline climate, exposure beyond 30 °C is not very common. However, a uniform warming scenario shifts the temperature distribution to the right, which increases the frequency of high temperatures. The anticipated consequence is that maize yields would decrease as exposure to detrimental temperature levels rises.

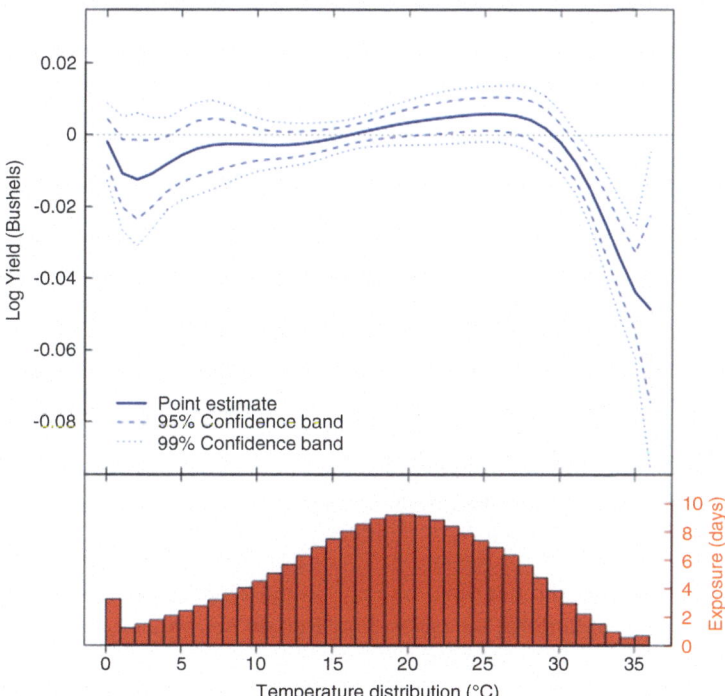

**Fig. 3** Nonlinear effects of temperature on maize yields

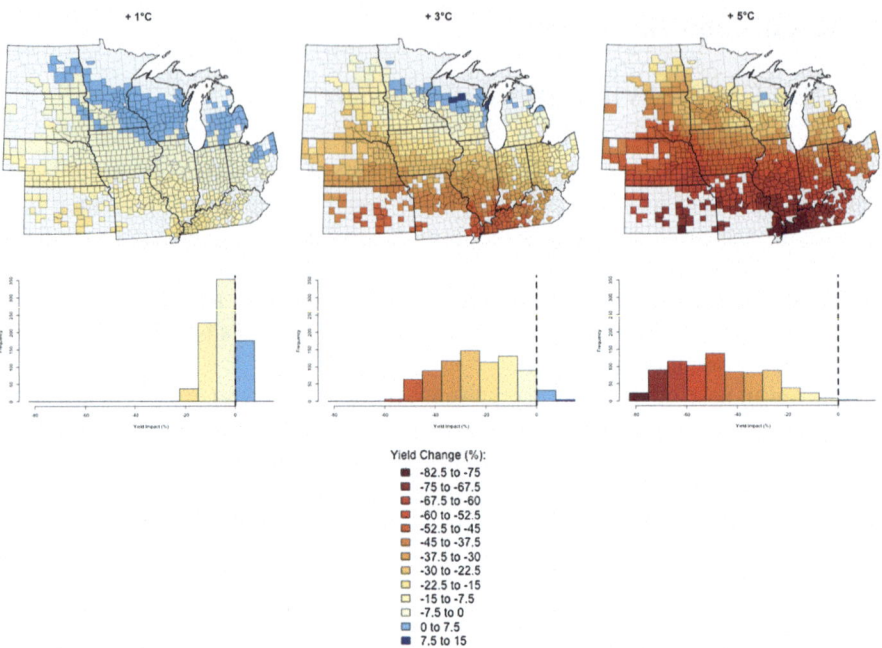

**Fig. 4** Maize yield impacts under alternative warming scenarios (*Notes:* The top row represents the projected effect of the corresponding warming scenario for each county in the sample. Grey counties are not in the sample. Some of these effects are not statistically significant when close to zero. The bottom row represents the distributions of these county-level effects)

Figure 4 illustrates the maize yields impacts for all counties in the sample (top row) as well as the distribution of impacts (bottom row) for each warming scenario. Because the statistical model regresses the log yield on weather variables conditional on a time trend, these impacts reflect percentage changes around the yield trend. A warming scenario of 1 °C has a relatively small effect with some northern counties experiencing small positive effects. However, more severe warming scenarios generate increasing crop yield losses. Interestingly, the model predicts rising heterogeneous effects across the sample as illustrated by the higher variance of projected impacts for the most severe warming scenario. The reason is that warming results in a disproportionately higher increase in the frequency of extreme temperatures in region that were warmer in the baseline climate.

The acreage-weighted maize yield impacts for the sample are −4.2%, −21.8% and −46.1% for the 1 °C, 3 °C, and 5 °C warming scenarios, respectively. Again, these impacts are around the trend so they do not represent net effects on yields. These impacts from uniform scenarios, however, do not provide information about their timing or the pace of warming.

## 4.2 Warming Impacts Against Technological Progress

To provide some context for the magnitude of these yield impacts, I compute the yield growth rate necessary to fully compensate these warming effects. This rate is computed as $r = 1/t\,((y_0 - y_t)/y_0 + 1)$ where $t$ is the time allowed for yield growth (in decades), and $(y_0 - y_t)/y_0$ is the share of acreage-weighted average yields loss in the projected climate relative to the baseline climate $(y_t < y_0)$.

I present these rates based on the historical yield sensitivity to temperature in panel A of Table 1. The table naturally shows that in order to compensate the impacts of a warming climate the growth rate in maize yields needs to be higher, the sooner this warming occurs. This explains the higher rates for lower time horizons (upper rows). Obviously, the rate needs to be even higher, to compensate larger damages from a more warming. This explains why higher rates are also found under more severe scenarios. Panel A shows that to compensate for a 3 °C warming within the next 3 decades (mid-century) the maize yield growth rate needs to be 6.56%/decade. This warming scenario approximately corresponds to climate change projections under higher emissions scenarios toward the middle of the century for the continental US. Recall that the recent historical yield trend shown in Fig. 1 is about 17.4%/decade. This is greater than the required growth rate to offset the warming impacts. However, these results show that climate change would have a sizable countervailing impact even if relatively high secular yield growth rates are maintained. More precisely, if the secular trend continues at this historical rate, the net yield growth might be reduced to about 17.4−6.6 = 10.8%/decade. This is a 38% reduction, which seems considerable.

The previous discussion assumed that only an increase in average yields is considered to counterbalance potential yield losses from a warming climate. However, breeding programs may be designed to reduce the vulnerability of maize yield to extreme conditions. This can be graphically represented as a reduction in the slope of the marginal effect of high temperature on crop yield in Fig. 3. I consider a case in which these marginal effects for temperatures exceeding 30 °C are reduced by half. Projected yield impacts will naturally be lower. Similarly, the required maize yield growth rates need to compensate a warming climate would also be lower. These rates are represented in panel B of Table 1. Indeed, with reduced extreme temperature sensitivity, the offsetting rates could be lower.

Panel C presents the difference between the compensating rates in the case based on historical heat sensitivity and with reduced heat sensitivity. These rates can be interpreted as the "secular yield growth rate equivalent" of an immediate reduction by half in extreme temperature sensitivity. In other words, the comparison of panels A and C provide insights into the tradeoff of combatting projected yield losses from warming by increasing average yield trends or by reducing the sensitivity of yields to extreme conditions. It is clear that the sooner and the more severe the warming is, the more appealing reducing the sensitivity to extreme becomes. Alternatively, if warming is mild or very distant, reducing yield sensitivity to high temperature present relatively small advantages (Table 2).

**Table 2** Maize yield growth rate required to fully compensate warming damages

| Time horizon | (A) Historical sensitivity | | | (B) Reduced sensitivity | | | (C) Difference | | |
|---|---|---|---|---|---|---|---|---|---|
| (in decades) | +1 °C | +3 °C | +5 °C | +1 °C | +3 °C | +5 °C | +1 °C | +3 °C | +5 °C |
| 1 | 4.14 | 19.69 | 37.90 | −0.24 | 5.88 | 18.66 | 4.38 | 13.81 | 19.24 |
| 2 | 2.07 | 9.84 | 18.95 | −0.12 | 2.94 | 9.33 | 2.19 | 6.90 | 9.62 |
| 3 | 1.38 | 6.56 | 12.63 | −0.08 | 1.96 | 6.22 | 1.46 | 4.60 | 6.41 |
| 4 | 1.03 | 4.92 | 9.47 | −0.06 | 1.47 | 4.66 | 1.09 | 3.45 | 4.81 |
| 5 | 0.83 | 3.94 | 7.58 | −0.05 | 1.18 | 3.73 | 0.88 | 2.76 | 3.85 |
| 6 | 0.69 | 3.28 | 6.32 | −0.04 | 0.98 | 3.11 | 0.73 | 2.30 | 3.21 |
| 7 | 0.59 | 2.81 | 5.41 | −0.03 | 0.84 | 2.67 | 0.62 | 1.97 | 2.74 |
| 8 | 0.52 | 2.46 | 4.74 | −0.03 | 0.74 | 2.33 | 0.55 | 1.72 | 2.41 |
| 9 | 0.46 | 2.19 | 4.21 | −0.03 | 0.65 | 2.07 | 0.49 | 1.54 | 2.14 |
| 10 | 0.41 | 1.97 | 3.79 | −0.02 | 0.59 | 1.87 | 0.43 | 1.38 | 1.92 |

*Notes:* The yield growth rate required to compensate damages is computed as $r = 1/t[(y_0 - y_t)/ y_0] + 1$ where t is the time allowed for yield growth (in decades), and $(y_0 - y_t)/ y_0$ is the share of acreage-weighted average yields loss in the projected climate relative to the baseline climate ($y_t < y_0$). The "Historical Heat Sensitivity" relies directly on the estimated parameters for computing climate change impacts. The "Reduced Heat Sensitivity" reduces by half the marginal effects of temperature exceeding 30 °C, i.e. the curve in Fig. 3 becomes less steep. "Difference" corresponds to the difference in rates between the "Historical Heat Sensitivity" rates and those for the "Reduced Sensitivity" rates

## 5 Conclusion

In this chapter I illustrate how to assess the yield growth rate requirements to fully compensate yield losses due to climate change based on statistical techniques. The crop statistical model employed allows for nonlinear effects of temperature on yields. In line with results in the literature, the statistical model suggests that exposure to temperature exceeding 30 °C is detrimental to maize yields in the US Midwest. A warming climate would therefore entail an increase in exposure to detrimental conditions and reduce yields. Indeed, I find sample-wide yield impacts around the yield trend of −4.2%, −21.8% and −46.1% for the 1 °C, 3 °C, and 5 °C uniform warming scenarios, respectively. The middle of the road-scenario is plausible by mid-century.

I find that a historical rate in maize yield growth in the US Midwest of 17.4%/decade exceeds the rate (6.56%/decade) needed to compensate a plausible warming of 3 °C within the next 3 decades. However, the net yield trend would be substantially diminished under this scenario due to the countervailing effect of a warming climate. In addition, I explore how the reduction in half of yield sensitivity to extreme temperature reduces the yield growth requirements to offset detrimental warming effects. I find that reducing sensitivity to extreme condition is a more attractive option when warming is imminent and severe. This case study highlights how agricultural policy analysis can assess the magnitude of potential yield losses due to climate change relative to historical yield trends.

The analysis could be extended with a cost-benefit analysis of alternative mean-increasing or variance-reducing technological change. The study also has important limitations including the fact that crop yield models cannot account for $CO_2$ fertilization or detailed management information that may be explicitly modeled with biophysical approaches. Other limitations include the assumptions about time separability of temperature effects as well as the omission of confounded effects of other inputs with weather conditions.

# References

Hodges, J. A., "The Effect of Rainfall and Temperature on Corn Yields in Kansas," Journal of Farm Economics, April 1931, 13 (2), 305–318.

Lobell, David B., Graeme L. Hammer, Greg McLean, Carlos Messina, Michael J. Roberts, and Wolfram Schlenker, "The critical role of extreme heat for maize production in the United States," Nature Climate Change, 2013, 3 (5), 497–501.

Lobell, David B., Marshall B. Burke, Claudia Tebaldi, Michael D. Mastrandrea, Walter P. Falcon, and Rosamond L. Naylor, "Prioritizing Climate Change Adaptation Needs for Food Security in 2030," Science, February 2008, 319 (5863), 607–610.

Lobell, David B., Wolfram Schlenker, and Justin Costa-Roberts, "Climate Trends and Global Crop Production Since 1980," Science, July 2011, 333 (6042), 616–620.

Schlenker, Wolfram and Michael J. Roberts, "Nonlinear temperature effects indicate severe damages to U.S. crop yields under climate change," Proceedings of the National Academy of Sciences, September 2009, 106 (37), 15594–15598.

Wallace, H. A., "Mathematical inquiry into the effect of weather on corn yield in the eight corn belt states," Monthly Weather Review, 1920, 48, 439.

**Open Access** This chapter is distributed under the terms of the Creative Commons Attribution-NonCommercial-ShareAlike 3.0 IGO license (https://creativecommons.org/licenses/by-nc-sa/3.0/igo/), which permits any noncommercial use, duplication, adaptation, distribution, and reproduction in any medium or format, as long as you give appropriate credit to the Food and Agriculture Organization of the United Nations (FAO), provide a link to the Creative Commons license and indicate if changes were made. If you remix, transform, or build upon this book or a part thereof, you must distribute your contributions under the same license as the original. Any dispute related to the use of the works of the FAO that cannot be settled amicably shall be submitted to arbitration pursuant to the UNCITRAL rules. The use of the FAO's name for any purpose other than for attribution, and the use of the FAO's logo, shall be subject to a separate written license agreement between the FAO and the user and is not authorized as part of this CC-IGO license. Note that the link provided above includes additional terms and conditions of the license.

The images or other third party material in this chapter are included in the chapter's Creative Commons license, unless indicated otherwise in a credit line to the material. If material is not included in the chapter's Creative Commons license and your intended use is not permitted by statutory regulation or exceeds the permitted use, you will need to obtain permission directly from the copyright holder.

# Understanding Tradeoffs in the Context of Farm-Scale Impacts: An Application of Decision-Support Tools for Assessing Climate Smart Agriculture

**Susan M. Capalbo, Clark Seavert, John M. Antle, Jenna Way, and Laurie Houston**

**Abstract** Climate change and enhanced climate variability will have differing impacts on agricultural producers worldwide. The increasing utilization of precision farming and mobile technologies, together with improvements in data management software, offer expanding opportunities for an integrated data platform that links farm-level management decisions and corresponding behavioral changes to site-specific biophysical data and analytical tools. The goals of this paper are to illustrate how decision support tools can be designed to address the farm-scale economic and environmental tradeoffs associated with changes in climatic conditions and how these farm-scale tools could be linked with regional based analyses to scale up to the information needed for better science-based policy.

We use the *AgBiz Logic*™ platform to evaluate farm-scale climate smart options for the dryland wheat producing area of the U.S. Pacific Northwest. A software tool like *AgBiz Logic* could also be utilized to provide higher quality, more timely data for landscape-scale and regional technology assessment. Decision support tools are at the very heart of the recommendations called for in the recent U.S. Government Accountability Office report 14–755 (U.S. GAO 2014), which speaks to USDA's ongoing efforts to better communicate information to growers in a timely down-scaled manner.

---

S.M. Capalbo (✉) • C. Seavert • J.M. Antle • J. Way • L. Houston
College of Agricultural Sciences, Oregon State University, Corvallis, OR, USA
e-mail: susan.capalbo@oregonstate.edu; john.antle@oregonstate.edu; jenna.way@oregonstate.edu

© FAO 2018
L. Lipper et al. (eds.), *Climate Smart Agriculture*, Natural Resource Management and Policy 52, DOI 10.1007/978-3-319-61194-5_9

# 1 Introduction

Climate change and enhanced climate variability will have differing impacts on agricultural sectors worldwide. Whether in the form of increased intra-seasonal variability, severe heat waves, long-term drought or warmer winters, farmers and growers need to be cognizant of the risks and opportunities that future weather patterns may bring to yields and profitability, as well as the possible environmental outcomes associated with changes in management regimes. Despite advances in applied research and analysis over the past half century, making informed management decisions based on integrating climate and environmental science findings at the farm scale remains a challenge. Critical information and data are often missing, and thus the consequences of changes in management practices across many dimensions are not easily identified.

Three key elements are required to improve the capability to make better management, and ultimately, policy decisions: (1) timely and accurate data on climate variability and its impact on yield and cost projections; (2) scientific understanding of the agro-ecological system at the farm scale; and (3) incorporation of those two elements into knowledge products that meet the needs of growers and policy decision makers. The increasing utilization of precision farming and mobile technologies, together with improvements in data management software, offer expanding opportunities for an integrated data platform that links farm-level management decisions and corresponding behavioral changes to site-specific biophysical data and analytical tools. Through the use of data technologies, farm-level information can be integrated with publically available data at the landscape scale for supporting science-based policy and sustainable management of agricultural landscapes.

The primary goal of this paper is to illustrate how decision support tools can be designed to address the farm-scale tradeoffs associated with changes in climatic conditions. We also explore how these farm-scale tools could be linked with regional based analyses to scale up to the information needed for better science-based policy. We illustrate how the three key elements noted above can be addressed within the *AgBiz Logic*™ platform and decision-support framework developed to aid growers in evaluating current and alternative management systems under future climate scenarios. By incorporating both climate change and environmental outcomes, these decision tools can be used to evaluate climate smart options. Our illustrative case study reflects the dry-land wheat producing area of the U.S. Pacific Northwest.

Decision tools and modules such as *AgBiz Logic*, provide essential analytical output for global and national efforts labeled climate-smart agriculture (CSA) which focus on making farms and farmers more resilient to a changing climate. These decision support tools are at the very heart of the recommendations called for in the recent U.S. Government Accountability Office report 14–755 (U.S. GAO 2014), which speaks to USDA's ongoing efforts to better communicate information to growers in a timely downscaled manner.

## 2 AgBiz Logic as a Decision Support Tool for Addressing CSA

*AgBiz Logic* is an integrated knowledge platform which collects and allocates grower data to enterprise budgets and saves the budgets. It also saves plans[1] and scenarios which can in turn be used in the economic, financial, climate and environmental modules. A simplified schematic of *AgBiz Logic* is provided in Fig. 1. Climate data from climate models and projections; environmental location-specific data on soil, slopes, rainfall etc.; and site-specific production data and other regional (public) data on prices, costs and transportation information are part of the information-base used and stored by *AgBiz Logic*. Outputs from each of the *AgBiz Logic* modules are inputted into another component of the software tool and/or used to generate metrics and other economic information. The economic and financial calculators are the means for farmers to better understand how climate change may impact their livelihood and their on-farm assets. The components are explained in greater detail in this paper.

*AgBiz Logic* (available online at http://www.agbizlogic.com/) consists of the following economic and financial calculators:

- *AgBizProfit*™ is a capital investment tool that evaluates an array of short-, medium-, and long-term investments. The module uses the economic concepts of net present value, annual equivalence, and internal rate of return to analyze the potential profitability of a given investment.
- *AgBizLease*™ is designed to help agricultural producers establish equitable short- and long-run crop, livestock and other capital investment leases. The module uses the economic concepts of net present value to analyze an equitable crop share or cash rent lease for a tenant and landowner.
- *AgBizFinance*™ is designed to help agricultural producers make investment decisions based on financial liquidity, solvency, profitability, and efficiency of the farm or ranch business. After an *AgBizFinance* analysis has been created, investments in technology, conservation practices, value-added processes, or changes to cropping systems or livestock enterprises can be added to or deleted from the current farm and ranch operation. Changes to a business' financial ratios and performance measures are also calculated.

Two recent additions to the *AgBiz Logic* decision support platform include the *AgBizClimate*™ and *AgBizEnvironment*™ modules:

- *AgBizClimate* delivers essential information about climate change to farmers and land managers that can be incorporated into projections about future net returns, via changes in expected yields. By using data unique to their specific farming operations, growers can develop management pathways that best fit their operations and increase net returns under alternative climate scenarios.

---

[1] Plans consist of a sequence of budgets that describe a particular management and or investment strategy. Plans can be compared to each other and saved as a scenario.

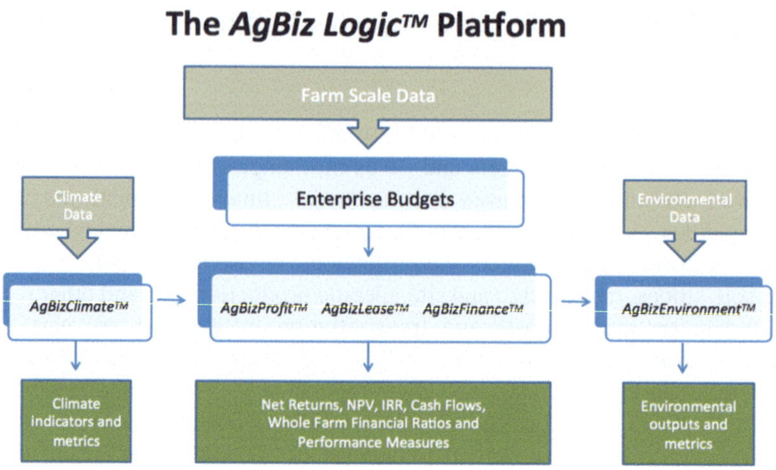

## The *AgBiz Logic*™ Platform

**Fig. 1** *AgBiz* logic platform

- *AgBizEnvironment* uses environmental models and other ecological accounting to quantify changes in environmental outcomes such as erosion, soil loss, soil carbon sequestration and GHG emissions resulting in the ability to incorporate on-farm and off-farm environmental outcomes into the decision support software and platform.

The *AgBiz Logic* platform provides both a farmer-level decision support tool and an assessment tool for researchers to realistically determine how climate change and climate change policies may influence and impact regional agricultural sectors. By incorporating regional downscaled climate change information, farm management and financial information, and on-and-off farm environmental impacts of land use changes and management decisions into an interconnected online program, actions of growers and data needs of researchers are linked. The downscaled climate change information influences projected yield and production inputs that change over time. These yield changes are the impetus for producer-generated adjustments in input use, management, and technology adoption that may lessen negative impacts or take advantage of positive opportunities.

## 3 Addressing the Farm-Scale Tradeoffs Associated with Changes in Climate

*AgBiz Logic* provides an internally consistent framework for evaluating climate change impacts and investment decisions at the farm scale. Farmers, growers, and land managers can use *AgBizClimate* to explore near-term projections for average weather conditions (e.g., growing degree days, chilling days) relevant to a commodity in their area. With knowledge of these projected changes, users have an opportunity to adjust their investments, yields and production inputs based on how such changes will affect their production and risk. *AgBizClimate* linked to *AgBiz Logic* allows users to step into the world of 20–30 years from present and consider how their current enterprises and operations would continue to serve them in the future, and whether there are any long-range planning decisions they may want to begin considering in order to maintain profitable operations.

What follows is an example of a case study in the mid-Columbia region of Umatilla County, Oregon using modules in the *AgBiz Logic* suite to observe the outcomes of climate change on current and alternative cropping systems (rotation) and on net returns (Seavert et al. 2012). We will first present an example of how *AgBizClimate* can be used to evaluate climate change impacts with changes in yields, tractor, combine and truck costs and production inputs, and we will also demonstrate how the *AgBizProfit* module can be used to evaluate investment decisions associated with changing a crop rotation.

## 3.1 Initial Setup and Baseline Scenario

The farm operation is a typical 3800-acre dryland wheat farm, in a region that receives between 12 and 18 inches of precipitation annually. In keeping with common practice, the producer uses a winter wheat and fallow crop rotation that includes direct seeding and chemical fallow to conserve soil moisture, increase wheat yields, reduce soil erosion, and reduce fuel usage. Weeds are controlled with glyphosate in the fallow years and other herbicides as needed during the crop years. Pesticides are applied as necessary. Fertilizer requirements are applied at planting using a direct-seed drill. The farm's average yield for winter wheat is 49.5 bushels per acre. One-half of the acres are leased and the farm operator owns the remaining acres. The leased land is based on the landowner receiving one-third of the crop and paying one-third of the weed control, fertilizer, and crop insurance costs (hail, fire and crop revenue coverage) and 100% of the property insurance and taxes. The yield levels are consistent with the yields from the 2007 USDA Agricultural Census for this area.

The data input needs and sequencing of steps are summarized in Appendix A. The producer selects previously generated crop and livestock enterprise budgets from *AgBiz Logic*; if these are not specific to this operation a grower can choose from a set that best reflects their returns and costs (Appendix A, Fig. 5). These previously generated/selected budgets serve as the baseline net returns scenario for comparison once weather variables are introduced. *AgBizClimate* is then used to select the weather station that is closest to the crop or livestock enterprises (Fig. 6). The result is downscaled, site-specific weather forecast information for the producer to use to best assess how climate change will impact the farm or enterprise.

After selecting the weather station in closest proximity to the farmed acres, the producer can select up to three weather variables that he/she believes will most impact wheat yields (Fig. 7). In this example, the number of nights below freezing, accumulated growing degree days and accumulated seasonal precipitation are chosen. Each weather variable has its own specific impacts, as shown in Appendix A, Figs. 8, 9, and 10. The modeled baseline weather condition (black line in Figs. 9 and 10) is an average for each weather variable chosen from 1970 to 1999. The modeled future climate variable is averaged over 2030–2059 for high and low emission scenarios. The solid red and yellow lines show the average, and the shading shows the 5-95th percentile range of resulting from 20 climate models (Figs. 10 and 11).

By the 2030s, the frequency of nights below freezing per year is expected to decrease by 29 nights for the low emissions future and by 34 nights for the high emissions future, as compared with the historical baseline (Fig. 8). From this information, predictions can be made regarding how wheat yields will be impacted from this specific weather variable, using either crop models or grower/expert estimates. In this example yields are increased 20% due to fewer nights below freezing; sensitivity analysis on fluctuations in yields can be incorporated into future analyses.

Figure 9 shows the results for changes in the number of growing degree days. By the 2030s, accumulated growing degree days from April 1 to October 31 are expected to increase by 525° hours for the low emissions future and by 620 degree hours for the high emissions future, as compared with the historical baseline. From this information, wheat yields are estimated to increase 15% due to a higher number of accumulated degree days above 50. Figure 10 shows the results for accumulated precipitation by month. Accumulated water year precipitation is expected to increase by 0.4 inches both for the low emissions and for high emissions future, as compared with the historical baseline. From this information, the producer estimates wheat yields will increase 25% due to an increase in precipitation combined with the time of year of the precipitation.

In Fig. 11, the producer can choose (observe from the available data) how likely his/her wheat yields will be impacted based on Crop Models, Grower Focus Groups, and from their own estimates of yields from Figs. 8, 9, and 10. The producer then enters a final yield estimate for each budget ("Your Changes"). This value will be leveraged to modify each budget used in the analysis. In the example shown, the user agrees with the Crop Models of an increase in wheat yields of 20.3%. However, the user also inserts an additional wheat budget and uses the Grower Focus Group value of 15.0% as a comparison. In *AgBizClimate* users can create new budgets by modifying selected inputs that are directly related to yields (Fig. 12). Examples of changing inputs related to yields include custom harvesting of hay or wheat crops, when paid by the ton.

## 3.2 Exploring Climate Change Impacts and Investments in Alternative Cropping Systems

Next, we evaluate the impact these changes in yields have on net returns. We also explore the profitability of changing the cropping system. For this region, research suggests that growers may benefit from climate change when they adapt to an annual cropping system of winter wheat and camelina. Camelina is a crop being studied for its potential use as a source of biodiesel fuel for aviation, particularly in regions where dryland cropping systems are predominant.

Using the *AgBizProfit* module we can run a scenario report (using the budgets that were modified using *AgBizClimate*). Each scenario consists of one to five individual plans that can be compared to each other simultaneously. In this case we compare four plans: (1) the current 2015 winter wheat fallow plan, (2) a winter wheat fallow plan with a 20% increase in wheat yields, (3) a winter wheat fallow plan with a 15% increase in wheat yields, and (4) a change from a winter wheat fallow system to a winter wheat and camelina rotation. On the latter cropping system wheat yields will decline from 50 to 39 bushels per acre (or about 13%) due to reduced soil moisture; however the revenues associated with the decline in wheat yields will be offset by the new revenues from the camelina crop. New crop budgets for these plans will be created for this scenario.

Table 1 reflects the yield changes under each scenario and shows how tractor, combine and machinery hours, truck miles driven, and expected years of life change as a result of the increased volumes of grain, annual acres harvested and the requirement of an additional combine when changing to an annual cropping system with camelina.[2] For the winter wheat and camelina rotation, an average camelina yield of 36 bushels (1800 lbs) per acre is used and the market price is $0.15/lb.; camelina is assumed to be grown in place of fallow. Even though the wheat yields are much less (38.71 bushels per acre, Table 1) and machinery costs higher (crop farming 3800 acres annually as compared to 1900 with the wheat and fallow rotation), the contributions to net returns from camelina compensate for the loss in wheat net returns.

Each of the winter wheat and fallow rotations in 2040 include the additional costs due to increased incidences of weeds, disease and insect infestations attributed to warmer temperatures and higher precipitation. Two additional applications (1 additional herbicide application and the addition of a pesticide application) with material costs are included as well as costs per acre for materials to control insects and diseases. These additional applications increase the tractor and sprayer hours in the wheat and fallow rotations in 2040. However, when camelina is included in an annual cropping system the applications and material costs for four herbicides are removed, which greatly reduces annual tractor and sprayer hours.

The *AgBizClimate* results for per acre returns, total variable cash costs, and net returns of the four cropping systems with crops grown on both owned and leased land are shown in Table 2. The winter wheat and fallow rotation in 2015 has an average net return of $72 per acre on owned land and a $36 per acre on

---

[2] Camelina is more difficult to harvest than wheat and combines must slow down to three miles per hour (as opposed to six mph when harvesting wheat), reducing the number of acres harvested in a day and thus requiring the purchase of an additional combine, or custom hiring the additional harvesting.

Table 1 Changes to hours of use and expected life for tractor, combines, machinery and trucks

| | | Base: wheat and fallow rotation, 2015 | | Wheat (20.3%) and fallow rotation, 2040 | | Wheat (15%) and fallow rotation, 2040 | | Wheat and camelina rotation, 2040 | |
|---|---|---|---|---|---|---|---|---|---|
| Wheat Yield (bu/ac) | | 50 | | 60 | | 57 | | 39 | |
| Camelina Yield (bu/ac) | | – | | | | | | 36 | |
| Yield Increase (bu/ac) | | – | | 10 | | 7 | | 25 | |
| | | Machinery annual hours and expected life | | | | | | | |
| Machine | Size | Hours or miles of annual use | Expected Life (Yrs) | Hours or miles of annual use | Expected Life (Yrs) | Hours or miles of annual use | Expected Life (Yrs) | Hours or miles of annual use | Expected Life (Yrs) |
| Tractor-rubber tracked | 485 hp | 567 | 15.0 | 683 | 12.5 | 677 | 12.6 | 588 | 14.5 |
| Combine | 30' Hillside | 109 | 10.0 | 109 | 10.0 | 109 | 10.0 | 163 | 6.7 |
| Additional Combine | 30' Hillside | NA | NA | NA | NA | NA | NA | 163 | 6.7 |
| Rotary mower | 26' | 167 | 15.0 | 167 | 15.0 | 167 | 15.0 | NA | NA |
| Field sprayer | 90' | 183 | 15.0 | 275 | 10.0 | 275 | 10.0 | 46 | 59.8 |
| Air seeder | 45' | 97 | 15.0 | 97 | 15.0 | 97 | 15.0 | 194 | 7.5 |
| Bank out wagon | 850 bu. capacity | 120 | 20.0 | 144 | 16.6 | 138 | 17.4 | 181 | 13.3 |
| Truck & trailer | Semi, used | 3000 | 20.0 | 3609 | 16.6 | 3450 | 17.4 | 4528 | 13.3 |
| Truck | 2 1/2 ton, older | 2400 | 20.0 | 2887 | 16.6 | 2760 | 17.4 | 3622 | 13.3 |

**Table 2** Per acre returns, total variable cash costs, and net returns for winter wheat and fallow rotations and winter wheat and camelina annual cropping system for crops grown on owned and leased land

| Crops grown on owned land | | | | | | | | |
|---|---|---|---|---|---|---|---|---|
| | 2015 | | 2040 | | 2040 | | 2040 | |
| | Winter wheat | Fallow | Winter wheat (20.3%) | Fallow | Winter wheat (15%) | Fallow | Winter wheat | Camelina |
| Returns | $322 | $0 | $387 | $0 | $370 | $0 | $252 | $270 |
| Total variable cash costs | 118 | 61 | 130 | 71 | 130 | 71 | 135 | 151 |
| Net returns | $204 | ($61) | $257 | ($71) | $240 | ($71) | $116 | $119 |
| Average net returns | $72 | | $93 | | $85 | | $118 | |
| Crops grown on leased land | | | | | | | | |
| | 2015 | | 2040 | | 2040 | | 2040 | |
| | Winter wheat | Fallow | Winter wheat (20.3%) | Fallow | Winter wheat (15%) | Fallow | Winter wheat | Camelina |
| Returns | $215 | $0 | $258 | $0 | $247 | $0 | $168 | $216 |
| Total variable cash costs | 93 | 49 | 105 | 57 | 106 | 57 | 111 | 135 |
| Net returns | $121 | ($49) | $153 | ($57) | $141 | ($57) | $57 | $81 |
| Average net returns | $36 | | $48 | | $42 | | $69 | |

leased land. The low net returns are largely due to the wheat yield of 49.50 bushels per acre. Now consider the impacts of a changing climate, which in this example result in increased wheat yields. When yields are increased 20.3% in 2040 to 59.55 bushels, the net returns increase to $93 per acre on owned land and $48 per acre on leased land; these net returns must also be adjusted to reflect the increase in herbicides and insecticide application costs. We also provide the results for a smaller change in yields due to climatic changes. As expected net returns decrease slightly when wheat yields are increased only 15% relative to the 2015 crop rotation. The net returns are $85 per acre on owned land and $42 per acre on leased land.

To explore some of the tradeoffs that may be present under climate change we incorporate the profitability of changing the cropping system or adapting manage-

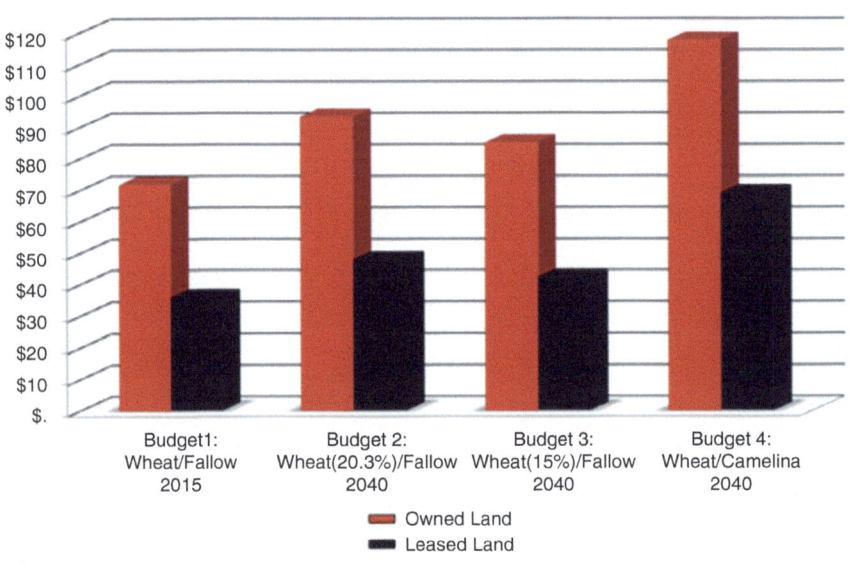

**Fig. 2** *AgBizClimate* output results

ment to new climatic conditions. For this region, research suggests that growers may benefits from climate change when they adapt to an annual cropping system of winter wheat and camelina. The net returns with a winter wheat and camelina rotation are $117 per acre on owned land and $69 per acre with leased land.[3] Figure 2, shows these results as an *AgBizClimate* output. Sensitivity of net returns to output and input prices are available from the authors but not reported in this paper.

As shown in this illustrative example both cropping systems (winter wheat/fallow versus winter wheat/camelina) and cropping arrangements (owned versus

---

[3] Crop leases change in the mid-Columbia region with oilseed crops. The landowner receives 20% of the crop and pays 20% of the fertilizer costs and 100% of the property insurance and taxes. It should also be noted that herbicides are not used in the production of camelina.

leased) will impact net returns. While many alternative cropping systems can be simulated, we provided only the comparison with the winter wheat/camelina and the original system currently used by a majority of the growers in this region. In both the owned and leased situations, both of which are typical of the arrangements in this area, the net returns per acre are higher with the effects of climate change for winter wheat and camelina annual rotation, regardless of whether the crops are grown on owned or leased land.

## 3.3 Profitability of Implementing Investment Strategies

Though we have shown that the winter wheat and camelina rotation has higher average net returns, we do not yet know if it is profitable for an individual producer. In order to switch to an annual cropping system that includes camelina, the producer would need to invest in an additional combine and truck. The profitability of this investment will depend on the timing of the cash flows. An alternative would be to custom hire the harvest of the camelina crop, which eliminates the need for the capital outlay of equipment, but also adds a certain amount of risk due to the uncertainty of the custom operator being available at harvest time. Selecting investments that will improve the financial performance of the business involves two fundamental tasks: (1) economic profitability analysis and (2) financial feasibility analysis. Economic profitability will show if an alternative is economically profitable. However, an investment may not be financially feasible: that is, the cash flows may be insufficient to make the required principal and interest payments (Boehlje and Ehmke 2005). In addition agricultural leases may also change with adaptation strategies as additional inputs and costs are incurred by either the landowner or tenant. The more a tenant or landowner contributes to total costs over the length of a lease, the higher the percentage share of the crop return or annual cash rent payment.

Figure 3 is an *AgBizProfit* output showing the results of a capital investment analysis for the adaptation strategies. Based on a discount rate of 4% and a 7 year analysis, the current wheat and fallow rotation has a net present value (NPV) of $57 per acre. The NPV of the annual cropping system with the purchase of an additional combine and truck is $500/acre. Custom harvesting of the camelina crop results in an NPV of $350 per acre. Therefore, the annual cropping system with the additional equipment purchases is the most profitable strategy. However, if a producer does not have the required cash flow to invest in additional equipment, which is needed for this cropping system, then this change in cropping rotations may not be feasible. The *AgBizFinance* module can be used to determine the feasibility of switching to a camelina rotation.

Conducting an *AgBizFinance* analysis requires a detailed balance sheet, description of current loans, capital leases and cash flows for each enterprise in the farm business. This type of analysis is very specific to a particular farm and difficult to demonstrate and discuss without sufficient data. Therefore an *AgBizFinance* analysis and further discussion is not presented in this paper.

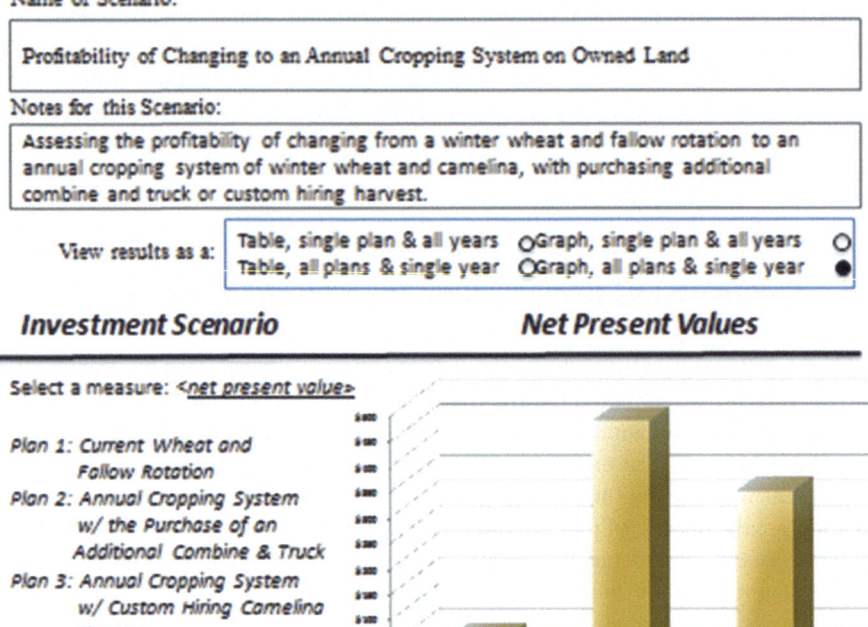

**Fig. 3** *AgBizProfit* results for owned land

## 3.4 Assessing Climate Change Implications for Agricultural Leases

Most of agricultural leases today are based upon what has been done historically or customary for a region. However, as profit margins narrow and climate change impacts yields, production inputs, and crop rotations, there will be a greater focus to base future leases on equitability, where the tenant and landowner are compensated more evenly for their contributions into the lease. Determining the equitability of leases can be explored with a decision support tool such as *AgBizLease*, a module within *AgBiz Logic*. Often times, the net returns on leased land do not equitably compensate the tenant for their financial risk of farming the land. For example, under existing practices, equitable crop leases are established on the percentage of each party's contribution to total costs (Seavert 1999). Using this tool, tenants could review lease terms to determine if current land leases would be equitable in the future. For example, if more insecticides and fungicides are required in future production systems due to a changing climate, those costs could be shared in the same percentages as share of the crop. *AgBizLease* could use the *AgBizClimate* budgets from these analyses to further evaluate the equitability of current lease terms as input costs change, and the resulting sensitivity of net returns.

As shown in Fig. 3, the current crop-share lease is equitable for this winter wheat fallow rotation, however is not profitable for either the tenant or landowner. The accumulated net returns for the tenant and landowner for a ten year lease is −$104 and −$40 per acre. The yields and prices are not sufficient to compensate the tenant for their production inputs and the landowner for their contributions of returns to land, property taxes and both sharing the fertilizer, herbicide and crop insurance costs. However, if this crop-share lease changed to an annual cropping system of winter wheat and camelina with the same sharing of crop and production inputs, both tenant and landowner benefit with $168 and $216 per acre, but not equitably. The *AgBizLease* program calculated an equitable crop-share lease to be 73% of the crop to the tenant and 23% to the landowner. By sharing the crop based on their contributions to this annual cropping system, the tenant would receive $295 per acre and landowner $89 per acre (Fig. 3).

## 4 Assessing Environmental Impacts

*AgBiz Logic* modules are based on the premise that growers maximize net returns over time; the static short run net returns are captured as the difference between revenues and cash costs. Depending upon the scenario, revenues can be defined as revenues associated with selling conventional, market-oriented products or can be expanded to include other services that might be valued by the grower, such as soil carbon, green production, environmental footprint, or other sustainability or risk-management attributes.

To capture the environmental aspects of the production decision, including on-site and off-site impacts, the *AgBizEnvironment* module reflects one of several approaches depending upon whether the environmental impact is considered an input or an output to the production process. Environmental/land quality can be considered as an input into the production process (i.e. soil quality) and thus part of the "natural capital" that impacts growers' net returns. Environmental quality can also be considered as an output of the production process. Way (2015) describes three possible firm-level profit maximization approaches to capture environmental impacts: (1) a conventional approach where environmental quality is reflected in changes in the natural capital variables; (2) the case where changes in environmental characteristics are best reflected using a multiple output production approach; and (3) a constrained profit maximization approach where environmental regulations constrain the choices and production levels of the grower. Each of these approaches requires information on the environmental outcomes from the production processes and/or how these may impact growers' net returns.

The *AgBizEnvironment* module utilizes existing environmental models or calculators to quantify the environmental outcomes and links this information either directly to net returns (if we can construct a shadow price or cost of the outcomes) or provides direct measures of environmental issues of concern such as changes in GHG emissions, soil erosion, carbon soil sequestration and energy usage. Examples include the Environmental Impact Quotient Value (EIQ) formula developed by Cornell University, Cool Farm Tool which measures GHG (carbon dioxide, nitrous oxide, and methane) emissions, COMET-farm which is a whole farm carbon and GHG accounting systems, and the Universal Soil Loss Equation (USLE) calculator and its many variations. Outputs from these models or calculators can be categorized as either an input to the production process and/or an (desirable or undesirable) output from the production process. GHG emissions and soil carbon credits are often characterized as outputs, although soil carbon can also be an input to the quality of the natural capital; pesticide use, soil erosion, and soil carbon are considered both production inputs and outputs. Table 3 provides an overview of these environmental simulation tools available within *AgBizEnvironment*, their outputs, and their applicability in producer-decision support frameworks.

Using the *AgBizEnvironment* module and associated environmental calculators, we explored the economic and environmental tradeoffs for switching to a conservation management practice for the winter wheat-fallow rotation. From *AgBizProfit* we calculated the change in farm-level net returns in the mid-Columbia region of switching to no-till (which is a more conservation-oriented, water conserving management practice) from conventional tillage. No-till has lower variable costs and labor requirements given the absence of the tillage operations pre- and post-harvest. However herbicide applications increase under no-till management in order to control weeds that would otherwise be managed with tillage, and equipment (air-seeded) costs increased. Based on research trials, wheat yields in this micro region are essentially the same between the two systems, at about 63 bu./acre. This yield exceeds the 49.5 bu./acre used in the previous example which was estimated from the 2007 Ag Census data. We opted to use the higher research trial yields for the *AgBizEnvironment* since it reflects the conditions in this smaller micro-region (Table 3).

For the baseline scenario, since the yields and revenues were taken to be the same between the two systems, variation in net returns is due to costs. Under this baseline scenario, net returns for no-till exceed the net returns for conventional tillage by approximately $29 per acre, or alternatively the yield advantage from conventional tillage would need to be about 6–7 bu./acre greater than no-till to equalize the net returns (Way 2015). So why do we not see a much larger adoption rate for the no-till management? In part, the answer may reside with combination of risk and expertise. At this point in the software development, *AgBizProfit* does not incorporate risk as it relates to management expertise.

Environmental impacts of concern also could include GHG emissions and possible soil erosion. These impacts were calculated using the COMET-Farm model for calculating changes in nitrous oxide and soil carbon equivalents only and the

**Table 3** Summary of the environmental tools available with *AgBizEnvironment*

| Simulation tool | Environmental factor | Production input or output | Source |
|---|---|---|---|
| Environmental Impact Quotient (EIQ) Value | Pesticides | Both | http://www.nysipm.cornell.edu/publications/eiq/equation.asp |
| Cool Farm Tool (CFT) | Greenhouse gas emissions/Carbon Sequestration | Output | https://www.coolfarmtool.org |
| COMET-Farm | Greenhouse gas emissions/Carbon Sequestration | Output | http://cometfarm.nrel.colostate.edu |
| Universal Soil Loss Equation (USLE) | Soil Erosion | Both | http://www.ars.usda.gov/Research/docs.htm?docid=10626 |

Universal Soil Loss Equation (USLE) for estimating changes in soil erosion. Our preliminary results indicate a net gain of 0.2 tons soil carbon ($CO_2$equv/yr./acre) from the no-till relative to conventional tillage. There is no accounting for carbon dioxide emissions in the COMET-Farm results since this model does not adjust for changes in energy use. COMET-Farm reflects climate and soil models and thus accounts only for the nitrous oxide and soil carbon activity. With respect to soil erosion, the potential average soil loss for conventional tillage is 5.19 tons/acre/year, and for no–till practice the average soil loss is approximately 1.04 tons/acre/yr. Thus no-till is environmentally preferred over conventional tillage in these two dimensions.

It is noted that the long term average soil loss (5.19 tons/acre/year) for the conventional tillage on this farm, with slopes of 7–15% and Walla Walla silt loam soil type, exceeds the tolerable soil loss limit for maintaining productivity (5.0 tons/acre/year). This brings into question the ability of the conventional tillage farm to continue to maintain yields equivalent to the no-till system. Under a multi-year net returns model, we would likely see yields fall relative to a multi-year no-till system and thus the gap in net returns would increase over time.

This example illustrates the approach to quantifying the economic-environmental tradeoffs associated with alternative management practices and lays the groundwork for monitoring changes in soil carbon or other environmental outcomes that could be used in environmental or carbon accounting policies. What remains in future research is to link the climate changes and projected yield changes that are generated through *AgBizClimate* to the environmental outcomes that are generated through *AgBizEnvironment* and integrate with the economic and financial modules for a fully integrated decision-support framework for growers.

## 5 Toward Landscape-Scale Tradeoff Analysis: Linking to the TOA-MD Platform

This section briefly discusses how farm-level data collected with a farm-level software tool such as *AgBiz Logic* could be combined with landscape-scale data to support regional policy analysis using a framework called TOA-MD (Tradeoff Analysis Model for Multi-dimensional Impact Assessment). We briefly describe the TOA-MD model, and discuss its data requirements and how those could be supported by data generated from *AgBiz Logic*. Also see Antle et al. (2016) for further discussion and an example of the use of the TOA-MD model for analysis of climate smart agriculture.

The TOA-MD model[4] was designed to simulate technology adoption and impacts of climate change or changes in other external drivers within a population of heterogeneous farms. The TOA-MD framework is applied to farmers or growers who choose between the production system currently in use, which in this case would be the winter wheat fallow system, and an alternative production system such as annual cropping (winter wheat camelina), with the choice of system based on the distribution of expected economic returns in the regional farm population.

Unlike the *AgBizLogic* platform, TOA-MD is a model of a farm population, not a model of an individual or "representative" farm, and therefore TOA-MD can simulate an adoption rate for a region (i.e., the proportion of farms that would switch to the alternative production system). TOA-MD is based on a statistical description of the population of farms. Accordingly, the fundamental parameters of the model are population statistics – means, variances and correlations of the economic variables in the models and the associated outcome variables of interest. With suitable bio-physical and economic data, these statistical parameters can be estimated with observational data for a production system in use, combined with experimental, modeled or expert data for a new system that is not yet in use and thus not observable.

The analysis of technology adoption and its impacts at the regional scale depends critically on how the effects of the new technology interact with bio-physical and economic conditions faced by farm decision makers. A key element in the TOA-MD analysis is reliable estimates of the effect of the new "technology" (i.e., the changes in the farming system that farmers could adopt) on the farming system's productivity and profitability. This information can come from various sources, including from formal crop and livestock simulation models, from experimental or observational data such as the information that can be obtained from a set of growers using *AgBizLogic*, or from expert judgment.

The TOA-MD model can be used for what Antle et al. (2014) describe as "adoption-based tradeoffs". Adoption-based tradeoffs occur when the adoption rate of a technology changes in response to an economic incentive or other factor affecting technology adoption. An important example of an adoption-based tradeoff is the

---

[4] See http://tradeoffs.oregonstate.edu.

analysis of GHG mitigation through soil carbon sequestration that occurs when farmers are offered a contract to sequester soil carbon (e.g., see Antle and Stoorvogel 2008). In this type of analysis, the prices faced by the farmers for outputs and inputs are held constant, so the observed changes in behavior are induced by the incentive provided to change management in ways that increase the buildup of the soil carbon. The adoption can also be induced from changes in climate that occurs over a longer time frame.

# 6 Data Requirements for the TOA-MD Model and How It Links to Farm-Scale Decision Support Tools

The parameters of the TOA-MD model are the means, variances and co-variances (or correlations) of the economic returns to each production system being represented in the analysis, and these statistical parameters of the other outcomes of interest, e.g., environmental outcomes such as the change in soil carbon. These statistics represent the farm population of interest, thus the data to be used are ideally obtained from a statistically representative sample of the population of farms and collected over a long enough period of time (e.g., multiple growing seasons) so that statistical methods can be used to account for seasonal variation and other factors that could affect the observed outcomes. The data can be grouped into the following categories:

(i) prices, outputs and costs of production of each production activity;
(ii) farm characteristics, including farm size, family size, and non-agricultural income; and
(iii) other relevant environmental or social outcomes.

The conventional way to obtain the farm production data is to conduct a survey, such as the surveys done periodically by government agencies (e.g., agricultural census or other statistical surveys such as the Agricultural Resource Management Survey in the United States or the Farm Accountancy Data Network data collected in European Community countries). There are limitations to these kinds of data. One is that these data are often collected periodically, e.g., the U.S. agricultural census is carried out on 5-year intervals, and then only made available to researchers with a substantial delay of a year or more. Another major limitation is that these data often lack sufficient detail, particularly for management decisions such as fertilizer and chemical use, machinery use, and agricultural labor. A third limitation is that these surveys can be extremely expensive both for respondents (e.g., to complete large elaborate questionnaires) and for organizations collecting the data (e.g., to employ enumerators, data entry workers, quality control specialists, etc.).

A tool like *AgBiz Logic* could be utilized to provide higher quality, more timely data at lower cost. As portrayed in Fig. 1, a data system that linked farm management software to a confidential database could provide near real-time data on management decisions, and do so for a statistically representative "panel" of farm decision makers over time. Moreover, the level of detailed management data utilized by *AgBiz Logic* would provide the needed level of detail for implementation of analysis using a tool such as TOA-MD. Also, users of *AgBiz Logic* would have every incentive to enter accurate information because they would be using this information to make their actual management decisions. Finally, a tool like *AgBiz Logic* provides a user-friendly, efficient way for farmers to enter data, thus substantially reducing the cost of data collection.

Several considerations need to be incorporated to facilitate a linkage between *AgBiz Logic* and the TOA-MD framework. First, a statistically representative group of farms would need to be identified who would agree to use *AgBiz Logic* and allow their data to be used in a landscape scale analysis. This would involve a sampling process similar to identifying a sample of farms for a farm-level economic survey. Second, software would need to be designed to transmit and assemble the individual farm data into a database that could subsequently be used to estimate TOA-MD parameters while maintaining confidentiality of individual producers. Note that data would need to be collected over multiple growing seasons in most cases to account for crop rotations and other dynamic aspects of the farming system. Farm household characteristic data could be collected as a part of *AgBiz Logic*, or could be collected using a separate survey instrument. Environmental and social outcome data collection would need to be tailored to the specific type of variable. For example, measurement of soil organic matter could require infield soil sampling and laboratory analysis, possibly combined with modeling, or the use of specialized sensors.

In addition it is important to project from current biophysical and socioeconomic conditions into plausible future conditions. This is currently being done on a global scale using new scenario concepts called "Representative Concentration Pathways" and "Shared Socio-Economic Pathways." To translate these future pathways into ones with more detail needed for agricultural assessments, "Representative Agricultural Pathways" are being developed (Valdivia et al. 2015). The data acquired through tools such as *AgBiz Logic* can be combined with these future projections to implement regional integrated assessments using the new methods developed by the Agricultural Model Inter-comparison and Improvement Project (Antle et al. 2015).

## 7 Conclusions

The use of a decision support tools such as *AgBiz Logic* can provide farmers better information on the relative impacts of adapting to a change as reflected in changes in future climate conditions, changes in future policies, prices, and costs or changes in terms of lease arrangements. It can also be used by researchers to understand how decisions about new programs, management options, technologies and varieties may impact a producer's net returns and ultimately his/her choices with respect to adoption of alternative management practices or cropping systems. By

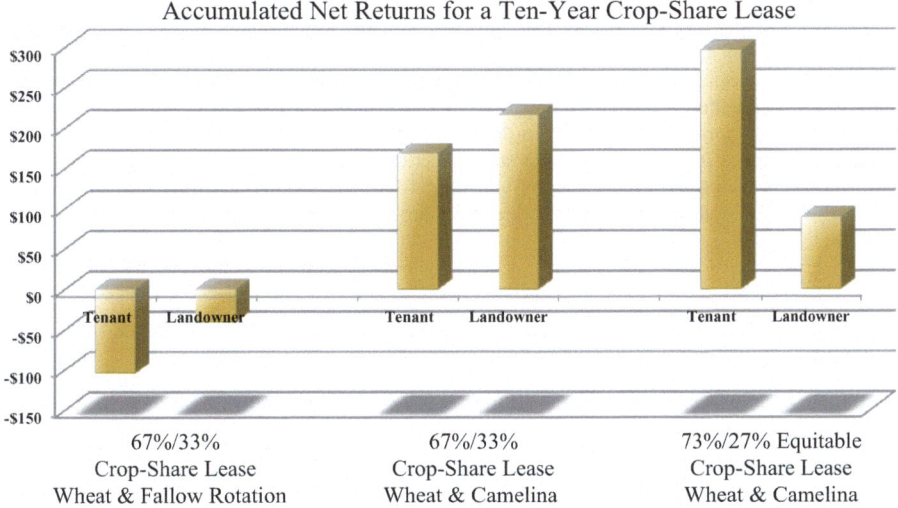

**Fig. 4** *AgBizLease*: results when crop-share leases for a wheat and fallow rotation change to an annual cropping system

incorporating both climate change and environmental outcomes, these decision tools can be used to evaluate climate smart options at the farm-scale.

The examples in this paper illustrate how an integrative decision support tool that is properly fine-tuned for the specific applications can better inform growers and land owners of how changes in climate will impact their operations and their environmental outcomes. *AgBizClimate* was used to show the impacts of climate change to wheat production. *AgBizProfit* was used to show adaptation strategies to an annual cropping system. *AgBizFinance* can be used to show the feasibility of purchasing additional equipment to farm the annual cropping system. *AgBizLease* showed how changing to an annual cropping system also changes the sharing of the crop, and *AgBizEnvironment* showed the tradeoffs of economic returns to environmental impacts (Fig. 4).

A software tool like *AgBiz Logic* could also be utilized to provide higher quality, more timely data for landscape-scale and regional technology assessment. As portrayed in Fig. 1, a data system that linked farm management software to a confidential database could provide near real-time data on management decisions, and do so for a statistically representative "panel" of farm decision makers over time. Moreover, the level of detailed management data utilized by *AgBiz Logic* would provide the needed level of detail for implementation of analyses using a tool such as TOA-MD. Users of *AgBiz Logic* would have every incentive to enter accurate information because they would be using this information to make changes to future management decisions. Finally, a tool like *AgBiz Logic* provides a user-friendly efficient way for farmers to enter data, thus substantially reducing the cost of data collection.

**Acknowledgements** This material is based upon work supported by the National Institute of Food and Agriculture, U.S. Department of Agriculture, under award numbers 2011-68002-30191, 2014-51181-22384 and 2012-38420-30208 (Regional Approaches to Climate Change - Pacific Northwest Agriculture; Developing a Sustainable Biofuels System in the PNW: Economic, Policy and Commercialization Analysis; National Needs Graduate and Postgraduate Fellowship Grants Program (NNF) - Graduate Education in the Economics of Mitigating and Adapting to Climate Change:Evaluating Tradeoffs, Resiliency and Uncertainty using an Interdisciplinary Platform), The Northwest Climate Hub, the Agricultural Model Intercomparison and Improvement Project (AgMIP), and Oregon Agricultural Experiment Station.

## Appendix A: How *AgBiz Logic* Works and Its Web-Based Presence

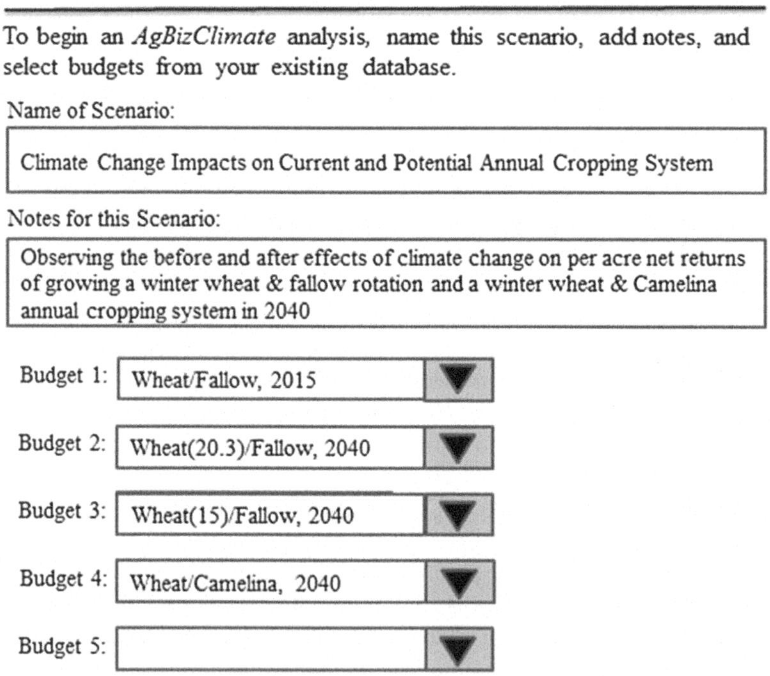

To begin an *AgBizClimate* analysis, name this scenario, add notes, and select budgets from your existing database.

Name of Scenario:

Climate Change Impacts on Current and Potential Annual Cropping System

Notes for this Scenario:

Observing the before and after effects of climate change on per acre net returns of growing a winter wheat & fallow rotation and a winter wheat & Camelina annual cropping system in 2040

Budget 1: Wheat/Fallow, 2015

Budget 2: Wheat(20.3)/Fallow, 2040

Budget 3: Wheat(15)/Fallow, 2040

Budget 4: Wheat/Camelina, 2040

Budget 5:

**Fig. 5** Naming a scenario, inserting notes for a scenario and selectin *ABL* budgets

| Select a state where the crops or livestock enterprises are located: | Select a weather station nearest your crops in this scenario: |
|---|---|
|  | 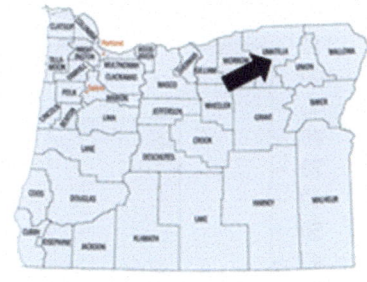 |

**Fig. 6** Selecting Oregon and Umatilla county as the state and county with the closer weather station to crops grown

WHEAT yields will most likely be impacted by climate change. Select the 3 most important weather variables from the list below that will impact yields or quality of the crop in this scenario.

- ○ Seasonal mean temperature
- ○ Number of days above freezing
- ● Number of nights below freezing
- ○ Number of warm nights
- ○ Number of consecutive extremely hot days
- ○ Number of consecutive extremely cold days
- ● Accumulated growing degree days
- ○ Accumulated chilling hours
- ○ 24-hour temperature range (night v. day)
- ○ Number of consecutive wet days
- ○ Number of consecutive dry days
- ● Accumulated seasonal precipitation
- ○ Snowpack

**Fig. 7** Weather variables that will likely impact yields or quality of products for crop and livestock enterprises

Based on your selected weather variables and weather station, the following are projected impacts from climate change.

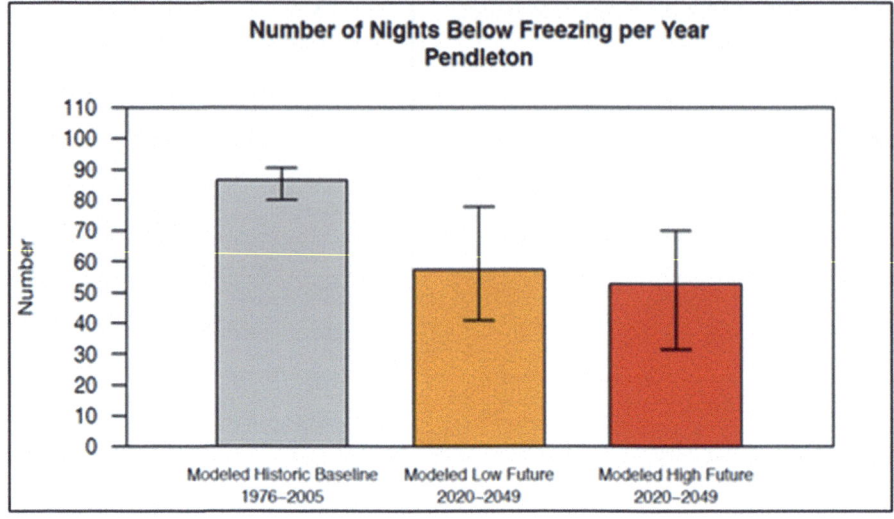

Based on this information, How do YOU think these climate changes will affect your **WHEAT** yields?: 20 % ▲ Change

**Fig. 8** Weather variables that will likely impact yields or quality of products for crop and livestock enterprises

Based on your selected weather variables and weather station, the following are projected impacts from climate change.

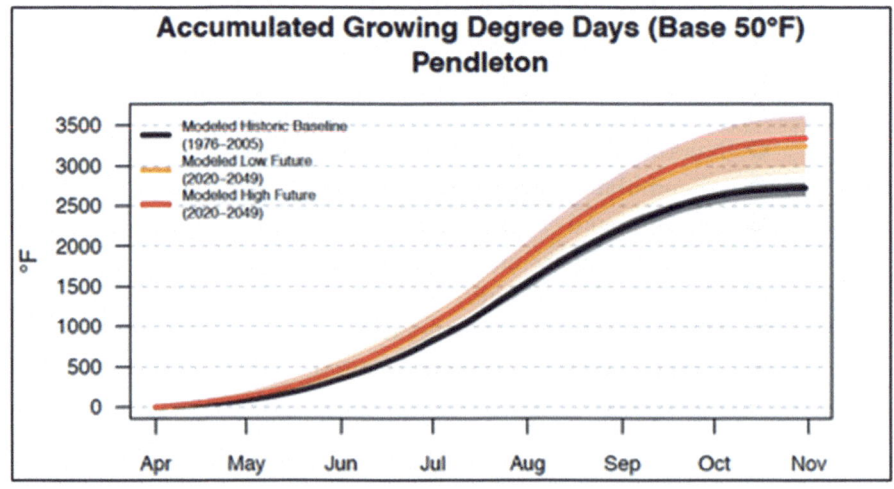

Based on this information, How do YOU think these climate changes will affect your **WHEAT** yields?: 15 % ▲ Change

**Fig. 9** Weather variables that will likely impact yields or quality of products for crop livestock enterprises

Based on your selected weather variables and weather station, the following are projected impacts from climate change.

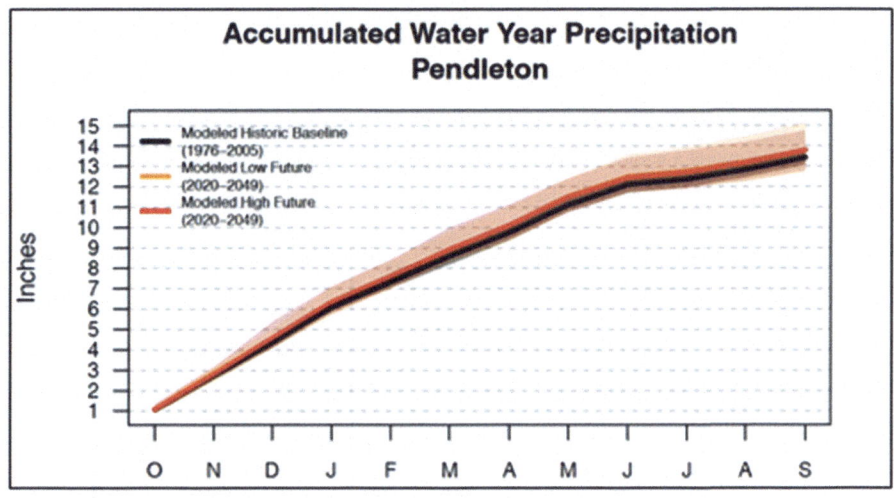

Based on this information, How do YOU think these climate changes will affect your **WHEAT** yields?: 25% ▲ Change

**Fig. 10** Weather variables that will likely impact yields or quality of products for crop and livestock enterprises

Below are estimates of how yields for crops in this scenario may change on average by the 2040s based on crop models, grower focus groups and your estimates from each weather variable. "Your Changes" to yields will be used in this analysis.

|  | Winter Wheat Owned | Wheat Wheat Leased |
|---|---|---|
| Crop Modeling | +20.3% | +20.3% |
| Grower Focus Groups | +15.0% | +15.0% |
| Weather Var. 1 | +20.0% | +20.0% |
| Weather Var. 2 | +15.0% | +15.0% |
| Weather Var. 3 | +25.0% | +25.0% |
| **Your Changes** | +20.3% | +20.3% |

**Fig. 11** Weather variables that will likely impact yields or quality of products for crop and livestock enterprises

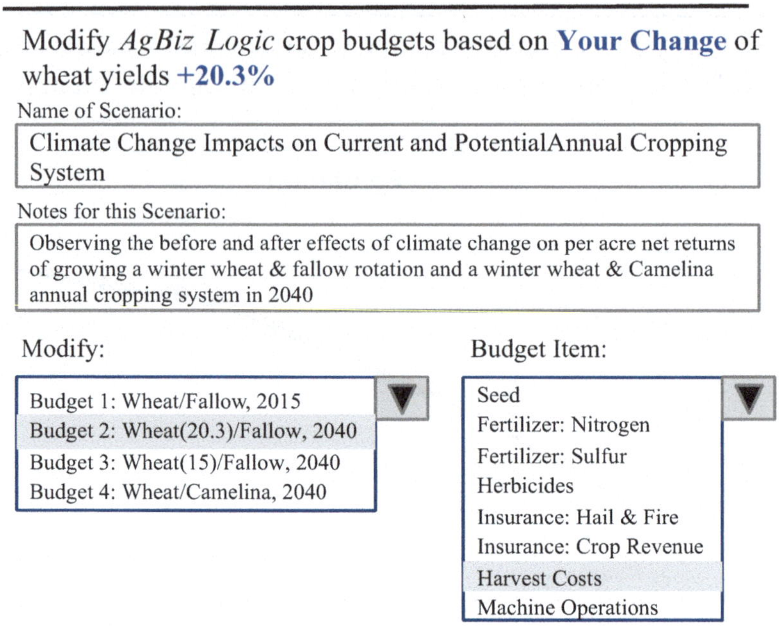

Fig. 12 Modifying 2015 crop budgets for 2040 production

# References

Antle, J. M and Stoorvogel, J. J. 2008. Agricultural carbon sequestration, poverty and sustainability. Environment and Development Economics 13: 327–352.
Antle, J.M., J.J. Stoorvogel, R.O. Valdivia. 2014. New Parsimonious Simulation Methods and Tools to Assess Future Food and Environmental Security of Farm Populations. Phil. Trans. R. Soc. B 2014 369, 20120280.
Antle, J. M., R.O. Valdivia, K.J. Boote, S. Janssen, J.W. Jones, C.H. Porter, C. Rosenzweig, A.C. Ruane, and P.J. Thorburn. 2015. AgMIP's Trans-disciplinary Agricultural Systems Approach to Regional Integrated Assessment of Climate Impact, Vulnerability and Adaptation. C. Rosenzweig and D. Hillel, eds. Handbook of Climate Change and Agroecosystems: The Agricultural Model Intercomparison and Improvement Project Integrated Crop and Economic Assessments, Part 1. London: Imperial College Press.
Antle, J., S. Homann-Kee Tui, K. Descheemaeker, P. Masikate, R. Valdivia. "Using AgMIP Regional Integrated Assessment Methods to Evaluate Climate Impact, Adaptation, Vulnerability and Resilience in Agricultural Systems." D. Zilberman, L. Lipper, N. McCarthy, S. Asfaw, G. Branca, editors. Climate Smart Agriculture - Building Resilience to Climate Change. Elsevier. In preparation, anticipated publication 2016.
Boehlje, Michael and Cole Ehmke, Capital Investment Analysis and Project Assessment, Purdue Extension EC-731. 2005. https://www.extension.purdue.edu/extmedia/ec/ec-731.pdf
COMET Farm. (n.d.) What is COMET-Farm.. Retrieved from http://cometfarm.nrel.colostate.edu
Cool Farm Alliance. (n.d.) Cool Farm Tool.. Retrieved from https://www.coolfarmtool.org
Cornell University. A Method to Measure the Environmental Impact of Pesticides-The EIQ Equation. Retrieved from http://www.nysipm.cornell.edu/publications/eiq/equation.asp

Seavert, C. F., 1999. Negotiating New Lease Arrangements with the Transition to Direct Seed Intensive Cropping Systems. PNW Direct Seed Conference, Spokane, WA. http://pnwsteep.wsu.edu/directseed/conf99/DSPRSEA.htm#Table1

Seavert, C., Steven Petrie, and Sandy Macnab. 2012. Enterprise Budget: Wheat (Winter) Following Fallow, Direct Seed, 12-18 inch Precipitation Zone, North Central Region. AEB 0036. http://arec.oregonstate.edu/oaeb/files/pdf/AEB0036.pdf

United States Department of Agriculture (USDA) Agricultural Research Service (ARS). (n.d.) Universal Soil Loss Equation (USLE).. http://www.ars.usda.gov/Research/docs.htm?docid=10626

United States Department of Agriculture. 2007. 2007 Census of Agriculture, State and County Reports.. http://www.agcensus.usda.gov/Publications/2007/Full_Report/Volume_1,_Chapter_2_County_Level/

U.S. Government Accountability Office. 2014. Climate Change: USDA's Ongoing Efforts Can Be Enhanced with Better Metrics and More Relevant Information for FarmersGAO-14-755: Published: Sep 16, 2014. Publicly Released: Oct 16, 2014. http://www.gao.gov/products/GAO-14-755

Valdivia, R.O., J.M. Antle, C. Rosenzweig, A.C. Ruane, J. Vervoort, M. Ashfaq, I. Hathie, S. Homann-Kee Tui, R. Mulwa, C. Nhemachena, P. Ponnusamy, H. Rasnayaka and H. Singh. 2015. Representative Agricultural Pathways and Scenarios for Regional Integrated Assessment of Climate Change Impact, Vulnerability and Adaptation. C. Rosenzweig and D. Hillel, eds. Handbook of Climate Change and Agroecosystems: The Agricultural Model Intercomparison and Improvement Project Integrated Crop and Economic Assessments, Part 1. London: Imperial College Press.

Way, Jenna. 2015. Linking Farm Profits and Environmental Quality Outcomes for Different On-Farm Conservation Practices. A Project submitted to Oregon State University in partial fulfillment of the requirements for the degree of Master of Science.

**Open Access** This chapter is distributed under the terms of the Creative Commons Attribution-NonCommercial-ShareAlike 3.0 IGO license (https://creativecommons.org/licenses/by-nc-sa/3.0/igo/), which permits any noncommercial use, duplication, adaptation, distribution, and reproduction in any medium or format, as long as you give appropriate credit to the Food and Agriculture Organization of the United Nations (FAO), provide a link to the Creative Commons license and indicate if changes were made. If you remix, transform, or build upon this book or a part thereof, you must distribute your contributions under the same license as the original. Any dispute related to the use of the works of the FAO that cannot be settled amicably shall be submitted to arbitration pursuant to the UNCITRAL rules. The use of the FAO's name for any purpose other than for attribution, and the use of the FAO's logo, shall be subject to a separate written license agreement between the FAO and the user and is not authorized as part of this CC-IGO license. Note that the link provided above includes additional terms and conditions of the license.

The images or other third party material in this chapter are included in the chapter's Creative Commons license, unless indicated otherwise in a credit line to the material. If material is not included in the chapter's Creative Commons license and your intended use is not permitted by statutory regulation or exceeds the permitted use, you will need to obtain permission directly from the copyright holder.

# Part III
# Case Studies: Policy Response to Improving Adaptation and Adaptive Capacity

# Can Insurance Help Manage Climate Risk and Food Insecurity? Evidence from the Pastoral Regions of East Africa

Michael R. Carter, Sarah A. Janzen, and Quentin Stoeffler

**Abstract** Can insurance cost-effectively mitigate the increasingly deleterious impacts of climate risk on poverty and food insecurity? The theory reviewed in this chapter suggests an affirmative answer if well-designed insurance contracts can be implemented and priced at a reasonable level despite the uncertainties that attend climate change. Evidence from the IBLI index insurance project in the pastoral regions in East Africa suggest that these practical difficulties can be overcome and that insurance can have the impacts that underlay the positive theoretical evaluation. At the same time, continuing analysis of the IBLI experience suggests that much remains to be done if quality index insurance contracts are to be scaled up and sustained. We conclude that insurance is not an easy, off-the-shelf solution to the problem of climate risk and food insecurity. Creativity in the technical and institutional design of contracts is still required, as are efforts to forge the more effective public-private partnerships needed to price insurance at levels that will allow insurance to fulfill its potential as part of an integrated approach to social protection and food security in an era of climate change.

There is ample evidence that climate shocks create and sustain poverty and food insecurity in rural regions of the developing world. There is also ample evidence that climate change is increasing the frequency and severity of climate shocks. Together these pieces of evidence in turn provoke the question: Can insurance

M.R. Carter (✉)
NBER, Department of Agricultural and Resource Economics and the Giannini Foundation, University of California, Davis, Davis, CA, USA
e-mail: mrcarter@ucdavis.edu

S.A. Janzen
Department of Economics, Montana State University, Bozeman, MT, USA
e-mail: sarah.janzen@montana.edu

Q. Stoeffler
Department of Economics, Istanbul Technical University, Istanbul, Turkey
e-mail: stoeffler@itu.edu.tr

cost-effectively mitigate the increasingly deleterious impacts of climate risk on poverty and food insecurity?

Two inter-related claims suggest an affirmative answer to this question:

1. After a shock is realized (*ex post*), insurance payments should help families maintain their economic assets (physical and human) and their long-term economic viability. In simpler terms, insurance should help families avoid a (potentially inter-generational) poverty trap.
2. Because it increases *ex post* security, insurance should also have an *ex ante* effect through increasing the expected level and certainty of returns to investment. This *ex ante* 'risk reduction dividend' should allow more families to escape poverty and food insecurity.

Taken together these two arguments suggest that insurance can be a cost-effective instrument to address food insecurity in the face of climate change. As opposed to a policy that simply treats the casualties of climate shocks with, say, food aid transfers, an integrated policy that includes an insurance element may reduce the total required social protection expenditures by addressing the causes, not just the symptoms, of food insecurity. Such an integrated policy cost effective if it allows more more households to maintain and achieve economic viability so that they can take care of their own needs.

The goal of this paper is to interrogate these claims and reflect on obstacles that may limit the efficacy of insurance as an instrument to manage climate risk. To do this, we proceed in several stages. First, in Sect. 1, we use recent theoretical modeling to explore the relative cost effectiveness of insurance as a device to manage the food insecurity induced by climate change. This modeling exercise assumes that:

- A contract can be designed that offers quality protection to inured individuals (*i.e.*, insurance payouts correlate well with household losses) and avoids the problems of moral hazard and adverse selection that can undercut the commercial sustainability of insurance;
- Households understand and trust the insurance and make purchase decisions based on a standard model of economic rationality; and,
- Insurance is commercially priced at the same proportionate levels observed in US crop insurance markets (128% of the actuarially fair price).

Under these assumptions, we find that while the logic outlined above holds and that integrated social protection, which employs an insurance element, can be a part of smart public policy, especially in the face or climate change. We do find that the relative benefits of an integrated social protection begins to weaken as climate change worsens and insurance itself becomes increasingly expensive.

While the theoretical case for insurance-augmented integrated social protection is clear, can it work in practice–that is, can the three conditions assumed by the theoretical analysis be met in practice? To provide insight into this question, we then turn to a specific case study–livestock insurance in the pastoral regions of northern Kenya and southern Ethiopia–to consider the practical barriers that limit the feasibility of insurance as a mechanism to help manage increasing climate risk.

Section 2 first shows how satellite-based index insurance has been developed to overcome the most pressing barriers to using insurance for managing risk among low wealth, spatially disperse rural households. Empirical impact evaluations of the Kenya and Ethiopia programs generally support the *ex post* and the *ex ante* insurance impacts outlined above.

While this evidence from the pastoral regions of East Africa is promising, even in this area the expansion and sustainability of the insurance contract remains fundamentally challenged by a number of issues, including contract quality, demand and pricing. After putting forward a framework for thinking about the factors that limit the quality of index insurance, Sect. 3 reviews new evidence on the quality of the East African insurance contracts and considers possible future steps for improving their quality. Section 4 then summarizes our findings concerning whether insurance can in practice play a useful role in managing climate risk and food insecurity.

# 1 The Logic of Insurance as a Device to Mitigate the Impacts of Climate Change on Food Insecurity

In an earlier paper, Ikegami et al. (forthcoming) identify what might be termed a social protection paradox. They compare two social protection scenarios.

In the first scenario, which mimics the targeting of conventional social protection programs, a fixed government budget is used to bring all poor households up to the poverty line, or as close to the poverty line as the budget permits. This conventional scenario is purely progressive in the sense that larger transfers go to poorer households. In contrast, a second scenario considered by these authors–which they term a triage policy–is not purely progressive. Instead, the fixed government budget is first allocated to the vulnerable non-poor to keep them from falling below a critical asset threshold, thereby stemming their descent into long-term poverty. These transfers to the vulnerable non-poor are contingent transfers that are only made if an unfavorable shock occurs and threatens the vulnerable with economic collapse. After the contingent needs of the vulnerable are met through these transfers, any remaining budget is then allocated progressively to the poor, again moving all poor households as close to the poverty line as possible.

To compare the effectiveness of these two social protection schemes in managing poverty, Ikegami et al. forthcoming employ a dynamic simulation model, similar to the model developed below. In their model, shocks are realized and individuals optimally choose current consumption and the amount of assets to carry forward to generate future income. Based on household asset and consumption levels, an omniscient government then allocates its budget in accordance with its social protection policy regime. Results are derived for both the standard and the triage regimes. Ikegami et al. forthcoming find that while the extent and depth of poverty are lower in the short term under the conventional needs-based approach, those results are reversed in the medium and long terms. In other words, the poor are paradoxically better off in the medium term despite less social assistance being allocated to them and more social assistance targeted to vulnerable but non-poor households.

The reason behind this paradoxical reversal is that when aid is concentrated solely on the neediest and not the vulnerable non-poor, then the number of aid-eligible poor people slowly swells over time, diluting the resources available for each poor individual. In contrast, transfers to the vulnerable both prevent them from falling below the threshold (and becoming poor) and allow them to successfully build up assets and eventually move away from the threshold and the vulnerability that it implies. Over time, under the triage policy an increasingly large share of the social protection resources become allocable to the poor whose ranks have not grown. We might anticipate that this social protection paradox revealed by Ikegami et al. forthcoming will only become larger in the face of climate change.

Building on this work, Janzen et al. (2015) ask whether or not the contingent transfers envisioned in the Ikegami et al. forthcoming triage policy can be implemented via an insurance contract. Implementing these transfers as an insurance contract would have two advantages. First, it may be able to rely on self-selection, obviating the need for the government to monitor needs and issue payments.[1] Second, having an insurance contract available could also offer a benefit to non-vulnerable households, including poorer households. To the extent that these latter households pay a portion of the insurance cost, they would be provisioning a portion of their own social protection.

While this logic may seem compelling, prior theoretical studies have suggested that insurance could actually increase the likelihood of collapse by vulnerable house- holds.[2] However, these other studies ask what happens if vulnerable households are forced to purchase insurance. In contrast to these other theoretical analyses, Janzen et al. (2015) allow individuals to optimally decide and how much insurance to purchase. This difference is subtle but important as Janzen et al. (2015) find that the most vulnerable households optimally purchase only minimal insurance unless it is subsidized. These same households quickly switch to full insurance as soon as they successfully accumulate a small amount of additional productive assets.

Using their model, Janzen et al. (2015) go on to show that the discounted present value of a hybrid policy (which subsidizes insurance and makes cash transfers to close the poverty gap for all poor households) is less than the cost of a conventional transfer program that simply closes the poverty gap for all poor households. After briefly reviewing the Janzen et al. (2015) model, this section then extends their analysis to consider the relative cost effectiveness of an insurance-based hybrid social protection scheme in the face of different climate change scenarios.

---

[1] The Ikegami et al. (forthcoming) policy assumes an omniscient government that can observe shocks and issue precisely the transfer required to protect vulnerable households from slipping into a poverty trap.

[2] See Chantarat et al. (2010) and Kovacevic and Pflug (2011).

## 1.1 Theoretical Model of the Ex Post and Ex Ante Impacts of Insurance on Poverty

Janzen et al. (2015) analyze the following dynamic model of a house- hold optimally allocating its resources across consumption, accumulation of assets that generate income through a risky production process, and purchase of an insurance contract that protects the household against asset losses:

$$\max_{c_t, 0 \leq I_t \leq A_t} \quad E_{\theta, \varepsilon} \sum_{t=0}^{\infty} u(c_t)$$

subject to: (1)

$$c_t + pI_t \leq A_t + f(A_t)$$
$$f(A_t) = \max[F^H(A_t), F^L(A_t)]$$
$$A_{t+1} = (A_t + f(A_t) - c_t)(1 - \theta_{t-1} - \varepsilon_{t+1}) + (\delta(\theta_{t+1}) - p)I_t$$
$$\delta(\theta_{t+1}) = \max((\theta_{t+1} - s), 0)$$
$$A_t \geq 0$$

The first constraint restricts current spending (consumption plus insurance purchases) to cash on hand (current assets plus income). As shown in the second constraint, the model assumes that assets are productive ($f(A_t)$) and that the households have access to both a high and low production technology, $F^H(A_t)$ and $F^L(A_t)$, respectively. Fixed costs associated with the high technology make it the preferred technology only for households above a minimal asset threshold. As has been demonstrated elsewhere, this non-convexity in the production function can lead to multiple equilibria and a poverty trap. Households with assets above a critical threshold level will strive to reach to a higher, non-poor equilibrium level of asset holdings and consumption. Those who begin with assets below that level (or whom shocks push below that level), will settle down at a lower level of asset holding typified by lower consumption and a poor standard of living.

Assets are subject to stochastic shocks (or depreciation). The random variable, $\theta_{t+1} \geq 0$ is a covariant shock and $\varepsilon_{t+1} \geq 0$ is an idiosyncratic shock.[3] Both shocks are exogenous and realized after decision-making in the current period ($t$), but before decision-making in the next period ($t + 1$) occurs. While these risks affect all households, they play an especially important role for households in the vicinity of the critical asset threshold. Because a shock can send households in this vicinity into a downward spiral to the low level equilibrium, we will refer to these households as the 'vulnerable.'

A unit of insurance can be purchased at a price $p$ and the insurance payout is based on the realized covariant shock according to the linear indemnity schedule:

---

[3] The distinction between these two stochastic elements will become important later when we consider feasible insurance mechanisms in the next section.

$$\delta(\theta_t) = \max((\theta_t) - s), 0), \tag{2}$$

where $s$ is the contractually determined depreciation rate above which insurance indemnity payments begin. Note that this insurance mechanism is akin to an index insurance mechanism as it only pays based on common or covariant shocks and does not provide protection against idiosyncratic shocks.

The third constraint is the equation of motion for asset dynamics: period $t$ cash on hand that is not consumed by the household or destroyed by nature is carried forward as assets in period $t + 1$. Finally, the non-negativity restriction on assets reflects the model's assumption that households cannot borrow. This assumption implies that consumption cannot be greater than current production and assets, but it does not preclude saving for the future.

Figure 1 presents some of the key results from the Janzen et al. (2015) analysis of this dynamic model. The horizontal axis represents time periods ("years") in the dynamic model. The vertical axis measures the headcount poverty rate for a stylized economy under three scenarios: An autarky scenario in which no insurance contracts are made available; A market-based insurance scenario in which insurance costs 120% of its actuarially fair price; and, A targeted insurance subsidy scenario in which the government pays half of the commercial insurance premium for all households that hold assets less than the level required to generate an average income equal to 150% of the poverty line. In all cases, the simulation assumes that households behave optimally based on the price of insurance and the dynamic choice problem displayed above.

As can be seen from Fig. 1, under the autarchy scenario with no insurance, headcount poverty steadily increases over time by about 25%, rising from 40% to 50% of the population. Under the targeted insurance subsidy scheme, there is an initial uptick in consumption poverty from 40% to 50%. This initial rise reflects the decision of vulnerable or near poor households to consume at levels below the poverty line in order to invest and (or) purchase insurance. However, over the longer-term, when insurance is partially subsidized for less well-off households, consumption poverty eventually falls to about 15% of the population, as opposed to the 50% level that occurs when there is no insurance market. This long-term drop in consumption poverty when insurance is available and subsidized reflects the fact that a significant fraction of the vulnerable ultimately escape the poverty trap. In contrast, without insurance, more of these vulnerable households fail and swell the ranks of the income poor. When an asset insurance market simply exists, but contracts are not subsidized, the impacts on poverty dynamics are qualitatively similar to the impacts of subsidized insurance, but quantitatively, the impacts are roughly two-thirds the magnitude of the impacts of subsidized insurance. This smaller impact occurs because the risk reduction dividend effects are smaller when insurance is more costly.[4]

---

[4] Janzen et al. (2015) discuss in detail how the price of insurance changes optimal insurance purchase and asset investment decisions.

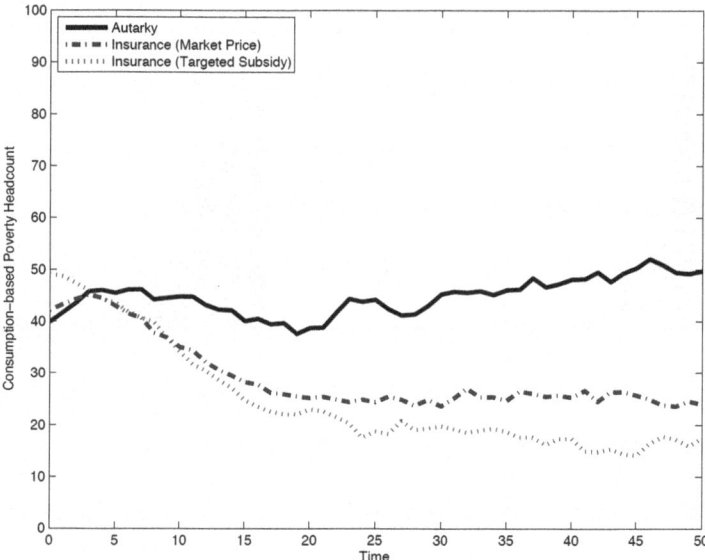

**Fig. 1** Consumption poverty headcount (Source: Janzen et al. (2015))

To gauge the cost-effectiveness of insurance subsidies from a public finance perspective, Janzen et al. (2015) sum the cost of all required cash transfer payments and add to that amount the cost of targeted insurance subsidies. Their analysis reveals an intertemporal tradeoff. The cost of transfers cum insurance subsidies is initially quite high, but over time total social protection costs are higher under the scheme that only provides cash transfers. Achieving the lower long-term poverty measures afforded by insurance subsidies costs more money in the short- term, but leads to substantial long term savings. Using a 5% discount rate the net present value of the two public expenditure streams over the 50 year time horizon of the simulation are 16% lower under the targeted subsidy scheme. Note of course that the public expenditures are only a portion of the full cost of social protection under the insurance scheme as individuals are in some sense privately provisioning a portion of the cost of their own "social" protection.

## 1.2 Analysis of Climate Change Scenarios

The analysis reported in Janzen et al. (2015) assumes a baseline risk scenario that is roughly calibrated to the climate conditions of the pastoral regions of East Africa circa the year 2000. In order to explore the effectiveness of the insurance cum social protection scenario explored by Janzen et al. (2015), we took their model and slowly increased the frequency and severity of the covariant shocks. Figure 2a shows the

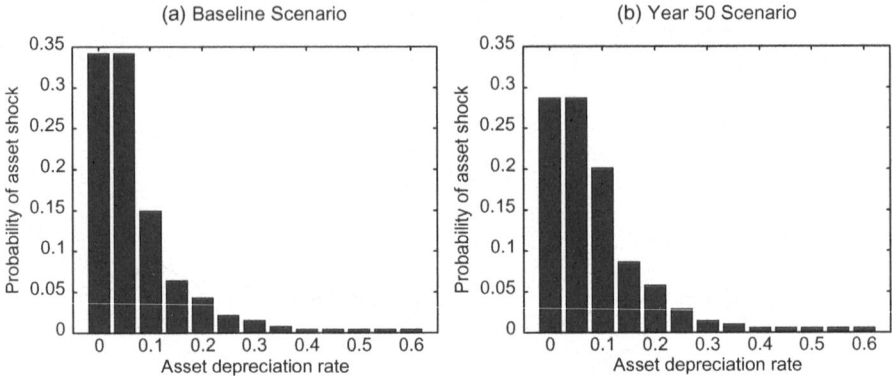

**Fig. 2** Climate change scenarios

baseline scenario on which these results are based.[5] Over a 50 year simulation scenario, we then allowed the climate to worsen every decade. Figure 2 shows the distribution of shocks assumed to exist in the final decade of the simulation. The analysis assumes that individuals are fully informed about the less favorable climate and adjust their behavior accordingly. The cost of insurance is also re-priced with every shift in climate, raising its costs, and the cost of the associated targeted insurance subsidies.

Figure 3 explores the costs of using subsidized insurance as part of a social protection package that seeks to eliminate poverty by transferring to every indigent household the amount of money necessary to lift them to a level of consumption achievable at the poverty line. The vertical axis measures the percentage change in government expenditures relative to the the year-zero transfers that would be required to close the poverty gap for all households under the alternative social protection policies. Results are again shown for three policy scenarios (autarkic risk management; unsubsidized insurance; and, subsidized insurance for poor and vulnerable households). For ease of comparison, we also include the social protection cost trajectories for a given policy both with and without climate change.

As can be seen, as climate change kicks in at year 10 of the simulation, the costs of cash transfers needed to close the poverty gap for all poor households begins to skyrocket above the costs absent climate change. Interestingly, even though insurance becomes increasingly expensive, it manages to hold steady the total cost of social protection (insurance and cash transfers) across the first 3 decades of climate change. This result attains in part because during the first decade of the simulation, many households are able to escape vulnerability and accumulate sufficient assets such that they are no longer eligible for insurance subsidies.

However, when the fourth round of climate change kicks in at year 40 of the simulation, the total costs of social protection begin to accelerate. The hybrid social protection continues to be cost-effective public policy, but as risk rises to an ever higher level, even the hybrid policy begins to loose its effectiveness in absolute terms.

---

[5] The risk levels at baseline in the simulations that follow are similar, but not directly comparable

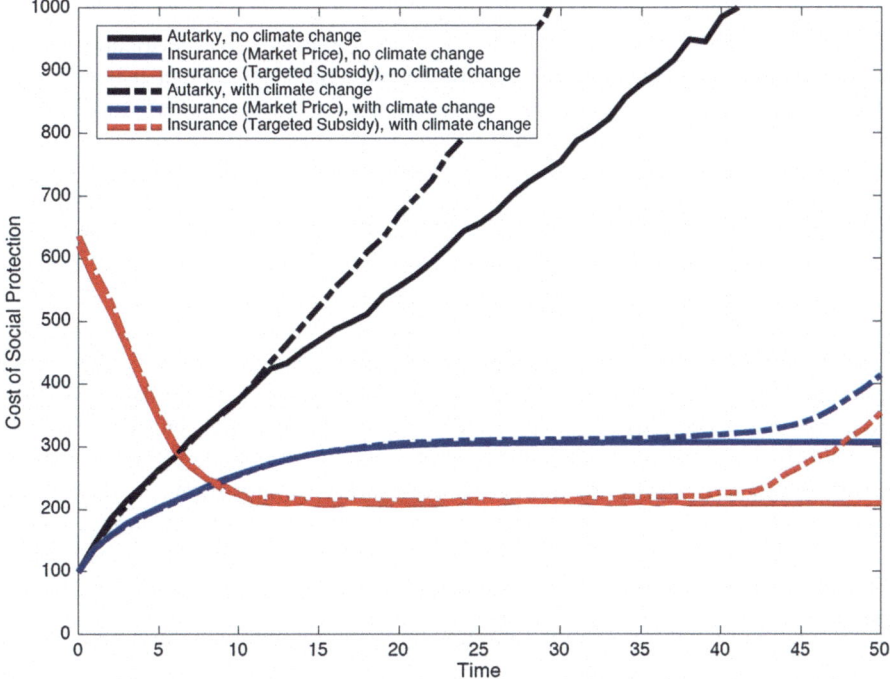

**Fig. 3** Cost of social protection

## 2 Index Insurance as a Solution: Livestock Insurance in the Pastoral Regions of East Africa

Section 1 employed abstract modeling techniques to consider the public finance case for insurance as a mechanism to offset the negative impacts of climate change on poverty and food insecurity. While it is relatively easy to implement an insurance policy in a theoretical model, a key question is whether it is possible to implement an insurance scheme in the real world that offers quality insurance protection, while keeping administrative costs, moral hazard and adverse selection in line.

Conventional agricultural insurance, which requires field visits to verify loss claims by individual households, has a dismal record when applied to small-scale rural house- holds, especially those located in isolated areas. In a study of a conventional insurance program established with heavy subsidies for the small-farm sector in Ecuador, Carter et al. (2014) find that the costs associated with a single loss verification visit may exceed $400. Given that the total annual premium associated with the typical small scale farmer is less than $100, it is easy to see why the business case for individual insurance evaporates. Cutting corners on loss verification is an open invitation to morally hazardous behavior. Moreover, given that it is not cost effective to individually rate the loss probabilities for each and every small-scale farmer, conventional insurance is also subject to problems of adverse selection in which those households most likely to experience a loss are also most likely to buy the insurance. As summarized by Hazell and Valdes (1985) and Hazell (2006), the net result of these problems has been loss ratios well in excess of 100%, implying that the insurance cannot be financially sustained.

Against this backdrop, index insurance appears as a promising, cost-effective solution. Under index insurance, loss verification is not required because payouts are based on an index. For agricultural insurance the index might be yields measured directly or predicted by satellite-based biomass growth indicators for an insurance zone.[6] The index is meant to be highly correlated with, but not identical to, the losses experienced by individual farmers. In principal, index insurance should eliminate problems of high transactions costs, moral hazard and adverse selection. However, its key advantage is also its achilles heel. If the insurance index is only weakly correlated with farmer losses (as Clarke et al. (2012) show in the case of rainfall insurance in India), then index insurance is more similar to a lottery ticket than an insurance contract. Lottery tickets are as likely to pay out when farmers have good crops as when they have bad crops, meaning that lottery ticket 'insurance' is likely to destabilize farmer income by perversely transferring money from bad to good states of the world.

If index insurance is to be part of the solution to helping manage climate risk, then the challenge is clearly to design an insurance index that is sufficiently well correlated with farmer losses such that it offers real *ex post* protection and thereby incentivizes *ex ante* investment such that the risk reduction dividend is gained. The remainder of this section focusses on one of the better researched index insurance projects, the IBLI (index-based livestock insurance) program in the semi-arid pastoral zones of northern Kenya an southern Ethiopia.

## 2.1 Designing the IBLI Index Insurance Contract

As detailed by Chantarat et al. (2013), the IBLI project began with the notion that satellite measures of vegetative growth, which had been in use for some time as part of famine early warning systems, might provide a reliable measure of forage availability for pastoral households. This measure was then transformed into an

---

[6] Because the index is the same for all households in the insurance zone, it does not matter in terms of payout probabilities whether high or loss risk producers select into purchasing the insurance, eliminating the adverse selection problem (assuming that the insurance is priced correctly for each zone). Moreover, as long as the zone is large enough, then moral hazard problems also disappear as no single farmer can influence the index by her actions.

index of predicted livestock mortality losses experienced by pastoral households in drought years.

Figure 4 displays "NDVI" maps for the original IBLI insurance zones in the Marsabit District of Northern Kenya. NDVI (or the Normalized Difference Vegetation Index) measures the intensity of light reflected from the earth's surface in different spectral bands. NDVI is essentially a 'greeness' measure that follows a regular cycle as rains come and forage crops grow. The maps displayed in Fig. 4 are based on a pixel size of 8 km by 8 km–that is, each square of this size receives its own unique NDVI reading on a daily basis as the satellite passes overhead.[7] The plot on the left shows a year with normal conditions, whereas the plot on the right shows a year where drought pressure was severe and livestock losses were high.

While NDVI can clearly distinguish drought from non-drought years, the insurance quality question swings on how well economic losses experienced by pastoralist households can be explained by the NDVI measure. To answer this question, Chantarat et al. (2013) assembled historical data on livestock losses and estimated a non-linear response function that maps NDVI signals into observed livestock mortality losses. Figure 5 gives a sense of the predictive accuracy of this mapping for one of the insurance zones in Marsabit District. Using out-of-sample prediction tests, Chantarat et al. (2013) report that based on the estimated response function and the historical distribution of NDVI, households would have been correctly indemnified 75% of the time when they experienced severe mortality losses (those in excess of 30%). The level of predictive accuracy falls to 60% when losses are 30% or less.

While imperfect, the predictive accuracy of the IBLI mortality was sufficiently high that a pilot project was launched in 2009.[8] While often hampered by implementation problems, the IBLI contract continues to date. Originally rolled out as a randomized controlled trail, the IBLI case study provides an excellent opportunity to learn, not just if index insurance can be implemented, but if it also delivers the expected *ex post* and *ex ante* effects that motivate the use of index insurance as a cost-effective device to help mitigate the costs of climate change. We turn now to consider some of that evidence.

## 2.2 Impacts of the IBLI Contract on Ex Post *Coping* and Ex Ante *Investment*

Severe drought in northern Kenya in 2011 resulted in high rates of livestock mortality in the IBLI pilot zone, with mortality estimates ranging from 25% to 50%. In accordance with the contract, all insured households received indemnity

---

[7] The current version of IBLI operates with much smaller grids based on changes in satellites and satellite technology.

[8] More recent work by Barré et al. (2016) proposes specific quality measures and a safe minimum standard for contract quality.

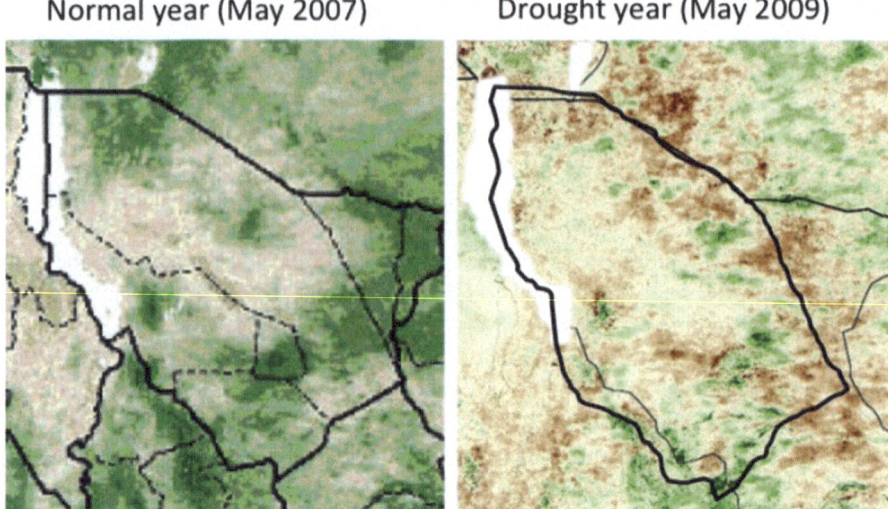

**Fig. 4** Satellite-based NDVI measures of forage availability

payments in October 2011. These payments coincided with the round 3 survey of IBLI study households. While the coincidence of the survey and the payments made it impossible to observe the short run impacts of the payments on coping strategies, households were asked what their coping strategies had been the third quarter of 2011 (the period immediately preceding the payouts, but well into the period of drought losses) and what they anticipated their coping strategies would be in the fourth quarter of 2011. Janzen and Carter (2013) use this data to study the impacts of insurance on families' ability to maintain their assets and food security during and after the severe drought. They achieve causal identification of impacts by exploiting randomly distributed inducements for households to actually purchase the insurance.

The first half of Table 1 summarizes the results of the Janzen and Carter (2017) analysis. The table reports the estimated percentage point reduction in the indicated coping strategy caused by insurance. For example, when pooling all households together, insurance causes 25% point reduction in the probability that the household relies on meal reduction to cope with the drought in the immediate post- payout period.

The first column of the table displays the estimated average impacts of insurance. Looking at the post-payout period, we see that on average insured households reduce anticipated reliance on meal reductions by 25% points and anticipated reliance on livestock sales by 36% points. Looking at the quarter 3, immediate pre-payout figures, we see–perhaps surprisingly–that insurance reduced by 20% points households' reliance on meal reduction. This decrease presumably reflects households' anticipation of the impending insurance payments, which allowed them to reduce hoarding of available food and other stocks.

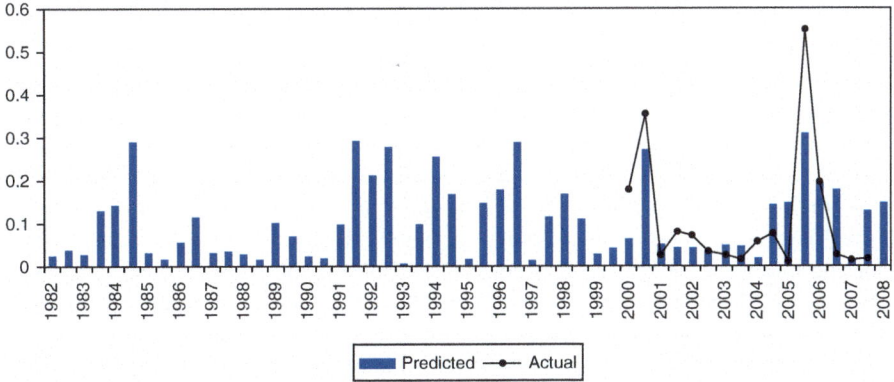

**Fig. 5** Predicted versus actual mortality losses

**Table 1** Causal Impacts of Insurance.

|  | All | Poor | Non-poor |
|---|---|---|---|
| *Ex Ante Risk Management Strategies* | | | |
| Reduce Meals | −20% points | −30% points | – |
| Sell Livestock | – | – | – |
| *Ex Post Risk Coping Strategies* | | | |
| Reduce Meals | −25% points | −43% points | – |
| Sell Livestock | −36% points | – | 64% points |
| *Overall Welfare* | | | |
| Income | +3% | +1% | – |
| MUAC scores | +1 s.d. | – | – |
| *Investment* | | | |
| Expenditures on Livestock | +72% | – | – |

Sources: Janzen and Carter (2017); Jensen et al. (2014a); Jensen et al. (2016)

While these average effects are impressive, looking beyond the averages tells a richer and perhaps more compelling story. As discussed by Janzen and Carter (2017), poverty trap theory (and other theoretical perspectives) suggest that poorer house- holds will confront shocks by holding onto productive assets and destabilizing consumption. While this 'asset-smoothing' behavior reflects an understandable effort to avoid falling into a long-term poverty trap, its impacts on the next generation's human capital are potentially large.[9] At the same time, wealthier households would be expected to respond *ex post* to a shock by selling assets and smoothing consumption.

Motivated by these theoretical propositions, Janzen and Carter (2017) use threshold estimation techniques to test for the presence of a critical asset threshold around

---

[9] See the analysis in Carter and Janzen (2015) for an effort to model these consequences as well as references to other empirical literature that documents this asset smoothing behavior.

which coping behavior switches between asset and consumption smoothing. This estimated threshold is used to distinguish between the poor and non-poor in Table 1. the results are striking. The average post-payout results disguise a strongly heterogenous pattern of insurance impacts. The decrease in meal reductions as a coping strategy is driven almost entirely by poorer households below the threshold, whereas the reduced reliance on livestock sales is driven almost entirely by households above the estimated threshold. These estimates tell an interesting story about the impact of insurance on *ex post* coping strategies. It appears to equally help both poor and non-poor (or at least less poor) households avoid costly coping strategies with potentially deleterious long-term consequences. But the mechanism through which insurance achieves this end is distinctive across the two sub-populations.

The second half of Table 1 reports the results of two additional impact evaluations that take advantage of rich panel data collected for the evaluation of IBLI. Both studies (Jensen et al. 2014b, 2016) also use randomly distributed premium discount coupons to instrument for IBLI purchases. Jensen et al. (2014b) show that insured households demonstrate improved child health (as measured by MUAC) and increased income per adult equivalent. An examination of production strategies also finds that house- holds with IBLI coverage reduce herd sizes and invest more heavily in health and veterinary services for their remaining herd, which is associated with increased milk productivity (and milk income) within the herd. Without explicitly estimating a threshold (as in Janzen and Carter (2017)), Jensen et al. (2016) also reveal heterogeneous impacts, at least for income:[10] the impact on income is significant only for the poorest households. These changes signal the kind of ex ante investment impacts discussed in the introduction, complementing the ex post impact findings of of Janzen and Carter (2017).

## 3 Limitations to Index Insurance as a Solution for Climate Change and Food Insecurity

While the economic case for index insurance as a smart response to managing climate risk and food insecurity is well developed, and while the IBLI project itself has shown that workable contracts can be devised that deliver the anticipated *ex ante* and *ex post* benefits of insurance, it remains far from clear whether index insurance can be scaled and operate as an essential part of the solution to the problem of climate change and food insecurity. Two of the fundamental challenges that may prevent index insurance from reaching its potential are:

---

[10] Jensen et al. (2014a) find no statistically significant difference in impacts for income, MUAC, or investment in their original analysis. They do find a larger impact in milk productivity among poor households, which may partially explain the heterogenous income results revealed in the latter study.

1. *Demand*: Similar to other settings, Jensen et al. (2014b) found that poorer households (in this case, smaller herds) are less likely to purchase IBLI coverage, that liquidity plays an important role in the purchase decision, and that demand is price sensitive. In the model presented in Section 1, Janzen et al. (2015) find that the most vulnerable households, despite having the most to gain from insurance, also have a high opportunity cost of insurance that may inhibit demand for an otherwise valuable product.
2. *Pricing*: A variety of factors have tended to push the price of index insurance contracts in developing country agriculture–including the IBLI project–to levels well in excess of 150% of the actuarially fair price.[11] Small project size is clearly a problem (as many insurance companies do not see it worth their while to participate in these markets), as are thin data problems which makes insurers have imprecise estimates of loss probabilities. Carter (2013) suggests that insurance pricing seems to reflect an 'uncertainty loading,' meaning an extra mark-up that charged when data are of mixed quality and loss probabilities uncertain. Solution to these problems may ultimately require a mixed private- public reinsurance model to keep the price of insurance in the range that it is rational to buy it.

While these challenges are clearly important, in the remainder of this section, we focus on a third, equally important challenge–that of providing scalable high quality contracts. While the IBLI contract was designed with much more care and attention to the ability of the index to adequately cover losses (see Section 2 above), even the IBLI contract shows signs of quality slippage as more data and experience become available. This section analyzes these challenges and suggests a way forward to address them and make IBLI an efficient instrument that protects Kenyan herders from the threat represented by climate change.

## 3.1 The Quality Challenge to Index Insurance

Unlike conventional insurance, index insurance includes a remaining uninsured "basis risk": a farmer or herder may encounter losses when the index does not trigger, or that the index may trigger when she does not have any loss. In the model above, this element was captured with the idiosyncratic risk component. Losses triggered in the model by idiosyncratic shocks were not compensated in the model. It is now widely recognized that basis risk may prevent index insurance to achieve its promise of delivering affordable protection to poor households (Miranda and Farrin 2012; Jensen and Barrett 2015). Clarke (2016) shows that because of basis risk, the most risk averse households may not be interested in purchasing index

---

[11] The actuarially fair price of an insurance contract is the price that is just equal to the expected indemnity payments to the farmers. Clearly the price must be marked up in excess of that amount in order to cover administrative costs, cost of capital, etc. However, a price that is, say, 150% of the actuarially fair price means that the farmer (or whoever is paying the insurance premium) is paying $1.50 for every $1.00 of protection for the farmers.

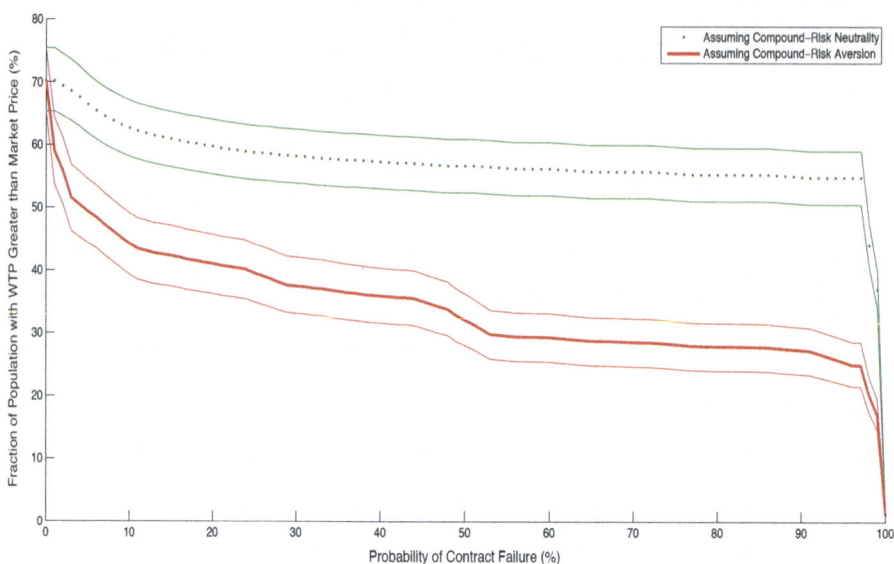

**Fig. 6** Impact of basis risk on willingness to pay for index insurance (Source: Elabed and Carter (2015))

insurance products. Indeed, if they have losses, pay a premium, and fail to receive insurance premiums, they end up in a worse situation than without insurance.

Basis risk may be an even bigger problem than work like Clarke (2016) suggests. Elabed and Carter (2015) use a field experiment in Mali to show that behavioral fac- tors related to basis risk further affect insurance demand. Specifically they show that people dislike the uncertainty of insurance payments, which, added to the original uncertainty of shocks, creates a "compound risk aversion" (the aversion to the combination of two uncertain events) among some households. This behavioral reaction generates a drop in insurance demand from 60% approximately for compound-risk neutral individuals, to only 35% of the population when compound-risk aversion is taken into consideration (Fig. 6).

While the necessity to reduce basis risk is now well acknowledged, there exists a debate regarding its exact definition, which harms efforts to increase overall index insurance quality. For example, there is a disagreement on whether basis risk should measure rainfall index correlation with farmers' rainfall shocks (i.e. accuracy of the index as a rainfall predictor) or its accuracy as a predictor of farmers' overall losses overall quality of the protection). Clearly it is the latter that matters from the farm- er's perspective and that will influence her insurance purchase decision. A mis- placed focus on accuracy of the index as a predictor of, say, rainfall, can lead to inappropriate index insurance products, which trigger payments when rainfalls are low in a given region rather than when farmers have actual losses, as rainfalls in a given region and actual individual losses are, at best, imperfectly correlated. Before analyzing the different sources of low quality of protection, let us step back and examine the objectives of index insurance.

For households, a good insurance means an insurance which improves their well- being by protecting their consumption and assets (see Barré et al. 2016). In addition, the quality of insurance as a development instrument stems from its ability to foster investments and reallocation of resources– and thus generate higher income– by removing risk. In other words, an insurance product needs to be evaluated based on its efficiency in stabilizing highly volatile income streams for poor farmers or herders. As a consequence, an index insurance product should be carefully analyzed to determine if its expected payments are actually correlated with households' losses, or if the insurance rather acts as a weather derivative–or even worse: as a lottery ticket (Jensen et al. 2014b; Barré et al. 2016). In India, Clarke et al. (2012) have shown that insurance payments actually correlates poorly with farmers' low yield events (Fig. 7).

The inadequacy of indemnity payments, observed in India and other settings, raises the issue of index insurance quality. Several sources of errors lead to low levels of index insurance quality. As shown in Fig. 8, for products which aim at covering all types of shocks, these sources of error relate:

- *Design risk* occurs when an insurance index is poorly correlated with *average* losses in the insurance zone covered by the index; and,
- *Idiosyncratic risk* occurs when the individual's losses differ from the average losses in her insurance zone.

In the theoretical model presented in Section 1, the insurance contract exhibited idiosyncratic, but not design risk.

The red line shows the point estimate for an Epanechnikov kernel with a bandwidth of 0.8. The green lines show the 95% confidence intervals for the point esti-

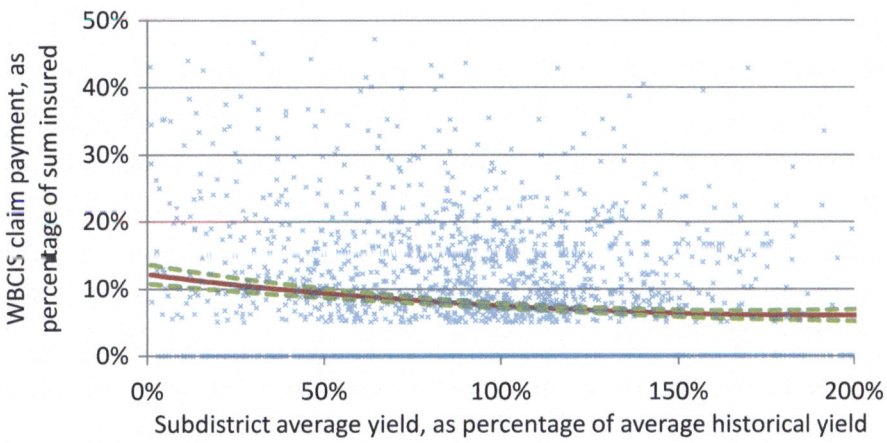

**Fig. 7** Relationship between average yields and insurance payments in India (*Source*: Clarke et al. (2012))

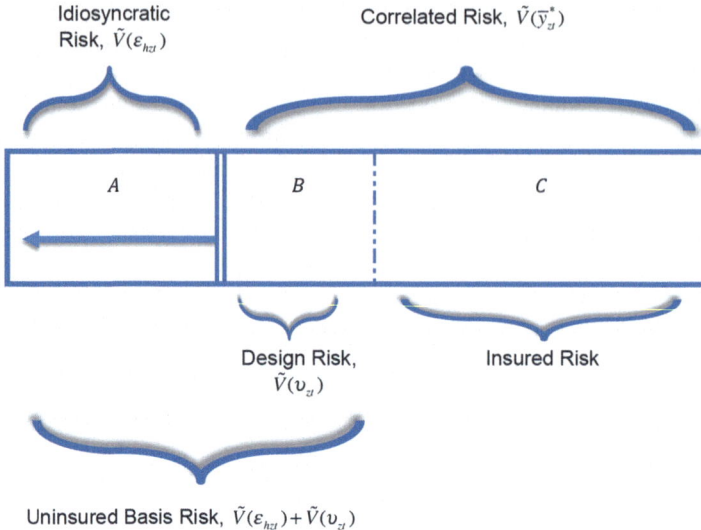

**Fig. 8** Insured and uninsured risk under index insurance (Source: Elabed et al. (2013))

mate. The blue dots represent the scatter plot of claim payments for the respective district yield levels.

Design risk emerges from prediction errors embedded in the index. The average loss within a defined geographic zone can be measured by indices based on several methods: crop cutting, satellite information, weather stations, etc. The contract formula then maps the index into payouts (and, implicitly, losses). Both the index and the mapping necessarily include some errors, which can be limited by using good indices and good insurance designs, but will not be eradicated.

However, even if design risk can be eliminated by improving even further the predictive power of the index, there typically remains some uninsured risk at the individual level. Pure idiosyncratic risk may induce households to encounter agricultural losses. For instance, a single farm's crop may suffer damage from idiosyncratic factors such as animal damage. Local communities often have some informal risk management strategies to cope with such type of pure idiosyncratic shocks when other villagers are not affected. Nevertheless, idiosyncratic risk diminishes the overall protection provided to farmers or herders.

The relative magnitude of both design and idiosyncratic risks are both influenced by the nature of the contract and its geographic scale. In terms of Fig. 8, how much risk appears as idiosyncratic and how much appears as correlated depends on the geographic scale of the index. As the geographic zone covered by a single index increases in size, household losses will correlate less well with the insurance index. For example, a weather-based index that covers households within 30 kilometers of the weather station will track outcomes worse than an index that covers households within 1 kilometer of the weather station. Similarly, an area yield index at the level of a state or province will cover individual farmer losses less well than an index where yields are measured at the level of each municipality or village. However in practice, reducing the geographic scale of the index too much leads to issues related to moral hazard, i.e. the fear that households may become able to manipulate the index.

Finally, for products which do not aim to cover all types of shocks (such as insurance products based on a rainfall index), an additional source of low quality arise from uncovered covariate risks (e.g., locusts, tsunamis). This type of error is related to the traditional distinction between single-peril and multiple-peril insurance products, but the difference is not as clear in the case of index insurance: satellite-based products such as IBLI, for instance, are supposed to cover all types of shocks related to lack of forage- including increase in livestock diseases- but cannot detect shocks which are not related to the ground vegetation- such as a new epidemic affecting well-fed livestock. These uncovered covariate risks further decrease the quality of the protection offered to poor households. Of course, households may be still interested in affordable index insurance products which only protects from one type of shock (*e.g.* drought), but the overall protection provided by this type of product has to be carefully analyzed and put in perspective with the price of the product and the probability that a farmer is made worse off with the insurance than without it.[12]

The lack of a strong negative correlation between the insurance indemnities and income shocks due to yield losses will result in a low demand for the insurance product (Clarke 2016; Smith and Watts 2009). Low correlation will not only fail to protect farmers, but eventually seriously damage livelihoods, because poor households pay high premiums to purchase protection, and plan on being protected when making investment decisions. Thus, a detailed analysis of the sources of errors needs to be conducted before implementing an index-based insurance and after its implementation, in order to rule out low quality products and pave the road for future product improvements. While this type of analysis is rarely undertaken in practice, IBLI is one of the most studied index insurance programs, and its quality has been closely scrutinized before and after implementation.

## 3.2 IBLI's Quality Effort and Remaining Weaknesses

IBLI's initial design considered carefully the above quality challenges, employing the available data. Indeed, as summarized in Sect. 2 above, Chantarat et al. (2013) conducted a rigorous ex-ante analysis intended to design the best performing index insurance product in the Kenyan ASALs. However, ex-post analyses have been less optimistic regarding IBLI's index performance in terms of basis risk and contract quality. Jensen et al. (2014a, b) and have investigated IBLI's performance using data collected between 2009 and 2012 (4 years, eight rainy seasons). This dataset was

---

[12] Note that if farmer pays for an insurance that only covers a sub-set of rainfall events, and then she suffers an uncovered pest invasion, she is actually worse off then if she had the invasion but not purchased the insurance. Clarke (2016) discusses these issues in detail.

employed for the impact evaluation of the IBLI pilot and includes detailed information on livestock dynamics, which can be used to assess the actual protection offered to herders over the period.

Jensen et al. (2014a) decomposes basis risk in several ways. First, by considering livestock surviving rates, the authors show that outcomes for insured households do not stochastically dominate outcomes for uninsured households. Actually, as expected, the insurance contract reduces the mean survival rate (taking into account insurance payments) but reduces skewness of the survival rate distribution. Simulations based on a constant relative risk aversion (CRRA) utility function shows that most households are actually better-off with the insurance at the commercial premium rate, but the benefits vary across locations and households.

To unpack these results, the authors decompose uncovered risks between design risk (the IBLI index was a poor predictor of average losses) and idiosyncratic risk (the individual suffered a worse loss than her neighbors on average did). At the aggregate level, design risk is relatively low since IBLI reduces covariate risk by about 62.8%. However, when individual idiosyncratic risk is added, IBLI only covers between 23.3% and 37.7% of the total risk. Note that at the individual level, the precision of the index when covariate losses are above the strike point is much higher, between 43.1% and 78.6%, which is closer to the objective, but still unsatisfying in some districts. Moreover, covering covariate shocks is arguably a first priority, as households may have informal insurance mechanisms when they receive adverse idiosyncratic shocks (Mobarak and Rosenzweig 2012).[13] Overall, these results call for caution when assessing insurance ex-ante, given that ex-post quality may be lower than expected based on ex-ante, out-of-sample predictions.[14]

An analysis of the consequence of basis risk on insurance demand was further performed by Jensen et al. (2014a). First, basis risk may deter insurance purchase. Second, while index insurance avoids moral hazard issues and individual-level adverse selection, it leaves some room for spatiotemporal adverse selection: households can buy insurance when they anticipate a bad climatic season in a given location, or not buy insurance if they expect a good climatic season in that location. Indeed, households may have an idea of the future season based on their information at the time of the insurance sale, as forage is affected by previous seasons and by the current season early rains. Thus, pastoralists can buy more insurance when they anticipate a bad climatic event– while on the other hand, price tends not to adjust to changing conditions.

---

[13] The complementarity of informal and formal insurance is not straightforward, and depends on the structure of the informal networks and of the index insurance, a point reinforced by Boucher and Delpierre (2014).

[14] The difference between ex-ante and ex-post assessments is striking. Factors explaining this mismatch may include: the use of an out-of-sample prediction which was never used in the design process (thus avoiding overfitting better); the application to a different time period (which was not available at the time of the contract design); the use of more detailed household data; and the computation of mortality rates and basis risk in a different manner.

The analysis shows that price, liquidity and social relationships have a strong impact on index insurance demand. In addition, both basis risk and special adverse selection play a major role. In particular, households in districts with high idiosyncratic risk (which cannot possibly be covered by the index insurance) are much less likely to purchase the IBLI product compared to households living in districts with a higher share of covariate risk. Design risk, on the other hand, plays a much smaller role in diminishing demand by about 1% only, compared to idiosyncratic risk, which explains about 30% of the demand.[15] This conclusion is relatively pessimistic regarding IBLI's potential, as contract design can only address inherent basis risk by lowering the geographic scale of the index. In pastoral regions, where individual households may seasonally migrate across large spaces, there are natural limits to how much a forage index like IBLI can be downscaled.

There are, of course, additional challenges to index insurance quality.[16] However, these issues of basis risk relate directly to the core economic value of the insurance product. If an index insurance does not pay pastoralists when they have losses, it does not matter how precisely it is priced, how efficiently it is implemented, and whether demand is low or high: households are not protected.[17] Index insurance products offer imperfect protection by definition, but efforts have to be made to provide the highest quality of protection as possible. Fortunately, there are several improvements that IBLI has realized in the last year or plans on including, which can improve household protection in several manners.

### 3.3 The Way Forward

Since the introduction of IBLI pilot project in 2009, the program has introduced some improvements and is planning further changes based on recent studies which it conducted. As the project has developed, we learned a lot about the strengths and weaknesses of IBLI. New ex-post data have become available at the household

---

[15] Note that design risk is difficult to measure with a short panel and a limited number of observations, as insured catastrophic losses are rare events by definition.

[16] These challenges relate to contract pricing and implementation (Chantarat et al. 2013), and non-price factors such as trust and liquidity (Jensen et al. 2014b), among others. Climate change also intensifies these challenges, as it creates some short-term uncertainties around future payments (Carter 2013) and may lead to very high premiums if climatic conditions deteriorate in the long-run (Collier et al. 2009; Carter and Janzen 2015).

[17] Of course, for households with full information, demand should be a good indication of the value of an insurance products. However, even for households who understand the product sold, the value of an insurance is difficult to assess ex-ante (Clarke and Wren-Lewis 2013). In addition, households do not always understand very well the insurance product, given the complexity of some index insurance schemes, the low levels of literacy in some contexts, and the poor quality of some marketing/information campaigns. For that reason, implementation of index insurance projects should focus on the quality of the protection offered rather than on the demand for these products only.

level, as well as longer term satellite information. IBLI has also expanded in scale in four districts in Northern Kenya and one district in neighboring Southern Ethiopia. This combination of factors has brought new opportunities and challenges. While IBLI has already operated some modifications since the studies mentioned above, further studies are planned to help continue improving the product design and the protection it provides to herders.

Notably, the program has evolved from an asset replacement mechanism to an asset protection philosophy. From an economic point of view, it is more efficient to intervene early and protect households' productive assets, rather than compensating them after the received a shock and possibly employed other costly coping strategies (Janzen and Carter 2017). In addition, as the project extended to geographic areas where livestock mortality data were lacking (in particular Southern Ethiopia), IBLI had to rely exclusively on NDVI data. Thus, payments would be triggered when NDVI data indicate a deterioration of the climatic conditions.

This move towards early payments have been accompanied by improvements of the product design. Since 2013, in order to limit spatiotemporal adverse selection, IBLI has started to disaggregate more the index, so that households located in different locations receive appropriate (different) insurance contracts. At this disaggregated scale, a larger share of shocks should be considered as covariate risk by the index, and as such reduce the effect of idiosyncratic risk (Jensen et al. 2014b).

Additional analyses have been conducted to further improve index quality. Vrieling et al. (2014) have investigated the possibility to combine remote sensing indices over longer periods in order to increase the predictive power of IBLI's formula. Based on newly constructed remote sensing from 1981 to 2011, the authors show how combining remote sensing indices allow a higher predictive power at a highly disaggregated level–i.e., there is still scope for reducing the magnitude of idiosyncratic risk by downscaling the insurance index. On the other hand, Klisch et al. (2015) have realized technical improvements in the computation of the vegetation index which can be used to detect droughts.[18]

Finally, Vrieling et al. (2016) have conducted some work on the temporality of the payments. The initial IBLI designed considered fixed dates for beginning and end of season in each district location. However, Vrieling et al. (2016) show that it is possible to use a phenomenological model to describe the temporality of forage development, based on historical NDVI data in each location. This change offers the potential to predict more accurately livestock mortality in each district, but also to provide payments one to three months earlier to pastoralists. These early payments could allow pastoralists to protect their herd by buying forage, water or medicine for instance, and prevent other shocks associated with low levels of forage such as animal diseases.

Additional research is required, however, on the relationship between insurance quality and temporality of payments. If early payments do not compromise the cor-

---

[18] These improvements regard the smoothing and filtering of satellite data, the modelling of uncertainty, the spatial and temporal aggregation of satellite data, and the timing of satellite data acquisition and processing.

relation between insurance payments and household's losses, then they are clearly valuable. However, there may be some trade-offs between early protection and accurate protection. Future work will analyze these trade-offs, as well as measure how the identified improvements in satellite indices computations translate into higher index insurance quality for herders.

## 4 Conclusions

We began this paper with the question:

Can insurance cost-effectively mitigate the increasingly deleterious impacts of climate risk on poverty and food insecurity?

The answer, it seems is both yes and no. Theory suggests that if quality insurance coverage can be delivered and the expected *ex post* and *ex ante* impacts take place, then the answer should be yes. Indeed, research on the Index-based Livestock Insurance (IBLI) pilot project in Kenya indicate that these conditions can be met giving further power to the likelihood of a yes answer.

And yet, even within the generally positive environment of the IBLI project, there is ample evidence of the limitations to index insurance. Demand has often been tepid and unstable. Outreach and administration costs have been high. Pricing by a private insurance industry made nervous by climate change has pushed costs up. Finally, the effective quality of the IBLI contact has been scrutinized and found wanting. Efforts to scale the IBLI contract to nearby pastoral regions has proven challenging.

While efforts are underway to respond to these challenges, their breadth and depth make clear that index insurance is not a sliver bullet that can be pulled off the shelf and used to mitigate the food insecurity and other consequences of climate change. Skeptics might suggest that these challenges are insurmountable. Others– and we count ourselves among them–remain undeterred given the evidence that index insurance can be a valuable instrument if these problems can just be solved. Doing so will require continued creativity, piloting and evaluation to see if indeed these not inconsequential challenges can be overcome.

## References

Barré, T. et al. 2016. "Assessing index insurance: conceptual approach and empirical illustration from Burkina Faso." Unpublished.

Barré, T., Q. Stoeffler, and M. Carter. 2016. "Assessing index insurance: conceptual approach and empirical illustration from Burkina Faso." Unpublished.

Barrett, C.B., M.R. Carter, and M. Ikegami. 2013. "Poverty Traps and Social Pro- tection." Unpublished.

Boucher, S., and M. Delpierre. 2014. "The impact of index-based insurance on infor- mal risk-sharing arrangements." Working paper, CEPS/INSTEAD.

Carter, M. 2013. "Sharing the Risk and the Uncertainty: Public-Private Reinsurance Partnerships for Viable Agricultural Insurance Markets." *I4 Index Insurance Innovation Initiative Brief 1*.

Carter, M.R., S. Boucher, and M.J. Castillo. 2014. "Index Insurance: Innovative Financial Technology to Break the Cycle of Risk and Rural Poverty in Ecuador." Working paper, I4, Index Insurance Innovation Initiative.

Carter, M.R., and Janzen, S. 2015. "Social Protection in the Face of Climate Change: Targeting Principles and Financing Mechanisms." *World Bank Policy Research Working Paper* WPS7476.

Chantarat, S., A. Mude, C. Barrett, and C. Turvey. 2010. "The Performance of Index Based Livestock Insurance in the Presence of a Poverty Trap." Unpublished.

Chantarat, S., A.G. Mude, C.B. Barrett, and M.R. Carter. 2013. "Designing index- based livestock insurance for managing asset risk in northern Kenya." *Journal of Risk and Insurance* 80:205–237.

Clarke D. 2016. A theory of rational demand for index insurance. Am Econ J Microecon 8(1):283–306,

Clarke, D., O. Mahul, K.N. Rao, and N. Verma. 2012. "Weather based crop insurance in India." *World Bank Policy Research Working Paper* , pp. .

Collier, B., J. Skees, and B. Barnett. 2009. "Weather index insurance and climate change: opportunities and challenges in lower income countries." *The Geneva Papers on Risk and Insurance-Issues and Practice* 34:401–424.

Elabed, G., M.F. Bellemare, M.R. Carter, and C. Guirkinger. 2013. "Managing basis risk with multiscale index insurance." *Agricultural Economics* 44:419–431.

Elabed, G., and M. Carter. 2017. "Ex-ante impacts of agricultural insurance: Evidence from a field experiment in Mali," working paper.

Hazell, P.B.R. 2006. "The appropriate role of agricultureal insurance in developing countries." *Journal of International Development* 4:567–581.

Hazell, P.B.R., and A. Valdes. 1985. Crop insurance for agricultural development: Issues and experiences,. Baltimore, Maryland, USA: John Hopkins University Press, International Food Policy Research Institute, Washington, DC USA.

Ikegami, M., Carter, M.R., Barrett, C.B. and Janzen, S. (forthcoming). "Poverty Traps and the Social Protection Paradox," in C.B. Barrett, M.R. Carter and J. Chavas The Economics of Asset Accumulation and Poverty Traps (Chicago: University of Chicago Press).

Janzen, S.A., and Cartern, M.R.. 2017. "After the drought: The impact of microinsur- ance on consumption smoothing and asset protection." NBER Working Paper No. 19702.

Janzen, S.A., M.R. Carter, and M. Ikegami. 2015. "Valuing Asset Insurance in the Presence of Poverty Traps," working paper.

Jensen, N., M. Ikegami, and A. Mude. 2016. "Integrating social protection strategies for improved impact: A comparative evaluation of cash transfers and index insurance in Kenya." Unpublished.

Jensen, N.D., and C.B. Barrett. 2015. "Agricultural Index Insurance for Sub-Saharan African Development.", pp. .

Jensen, N.D., C.B. Barrett, and A. Mude. 2014. "Index Insurance and Cash Transfers: A Comparative Analysis from Northern Kenya." *Available at SSRN 2547660*.

Jensen, N.D., A. Mude, and C.B. Barrett. 2014. "How basis risk and spatiotemporal adverse selection influence demand for index insurance: Evidence from northern Kenya." *Available at SSRN 2475187*.

Klisch, A., C. Atzberger, and L. Luminari. 2015. "Satellite-based drought monitoring in Kenya in an operational setting." *The International Archives of Photogrammetry, Remote Sensing and Spatial Information Sciences* 40:433.

Kovacevic, R., and G.C. Pflug. 2011. "Does Insurance Help to Escape the Poverty Trap? A Ruin Theoretic Approach." *The Journal of Risk and Insurance* 78:1003–1028.

Miranda, M.J., and K. Farrin. 2012. "Index insurance for developing countries." *Applied Economic Perspectives and Policy* 34:391–427.

Mobarak, A.M., and M.R. Rosenzweig. 2012. "Selling formal insurance to the informally insured," working paper.

Smith, V., and M. Watts. 2009. "Index based agricultural insurance in developing countries: Feasibility, scalability and sustainability."

Vrieling, A., M. Meroni, A.G. Mude, S. Chantarat, C.C. Ummenhofer, and K.C. de Bie. 2016. "Early assessment of seasonal forage availability for mitigating the impact of drought on East African pastoralists." *Remote Sensing of Environment* 174:44–55.

Vrieling, A., M. Meroni, A. Shee, A.G. Mude, J. Woodard, C.K. de Bie, and F. Rem- bold. 2014. "Historical extension of operational NDVI products for livestock insur- ance in Kenya." *International Journal of Applied Earth Observation and Geoin- formation* 28:238–251.

**Open Access** This chapter is distributed under the terms of the Creative Commons Attribution-NonCommercial-ShareAlike 3.0 IGO license (https://creativecommons.org/licenses/by-nc-sa/3.0/igo/), which permits any noncommercial use, duplication, adaptation, distribution, and reproduction in any medium or format, as long as you give appropriate credit to the Food and Agriculture Organization of the United Nations (FAO), provide a link to the Creative Commons license and indicate if changes were made. If you remix, transform, or build upon this book or a part thereof, you must distribute your contributions under the same license as the original. Any dispute related to the use of the works of the FAO that cannot be settled amicably shall be submitted to arbitration pursuant to the UNCITRAL rules. The use of the FAO's name for any purpose other than for attribution, and the use of the FAO's logo, shall be subject to a separate written license agreement between the FAO and the user and is not authorized as part of this CC-IGO license. Note that the link provided above includes additional terms and conditions of the license.

The images or other third party material in this chapter are included in the chapter's Creative Commons license, unless indicated otherwise in a credit line to the material. If material is not included in the chapter's Creative Commons license and your intended use is not permitted by statutory regulation or exceeds the permitted use, you will need to obtain permission directly from the copyright holder.

# Can Cash Transfer Programmes Promote Household Resilience? Cross-Country Evidence from Sub-Saharan Africa

Solomon Asfaw and Benjamin Davis

**Abstract** Several new initiatives of cash transfer programmes have recently emerged in sub-Saharan Africa, and most target poor rural households dependent on subsistence agriculture. This paper synthesizes the key findings of From Protection to Production Project (PtoP) of FAO and discusses the role of cash transfer programmes risk management tool to increase resilience in sub-Saharan Africa. Results show that such programmes have important implications for household resilience. Although the impacts on risk management are less uniform, the cash transfer programmes seem to strengthen community ties (via increased giving and receiving of transfers) and allow households to save and pay off debts, and decrease the need to rely on adverse risk coping mechanisms. One important finding related to climate change, as illustrated by the Zambia case, is that households receiving cash transfers suffered much less from weather shocks, with poorest households as the biggest gains, and food security increased, although differing across countries. The paper concludes that social protection programmes could be more effective as safety nets by explicitly accounting for climate risk in their design and implementation.

**JEL Classification** I38 • Q01 • Q18

## 1 Introduction

Almost three quarters of economically active rural populations in sub-Saharan Africa (SSA) are smallholder farmers, making them important players in national agricultural development plans. Thus agricultural development that contributes to

S. Asfaw (✉)
FAO of the UN, Rome, Italy
e-mail: solomon.asfaw@fao.org

B. Davis
Food and Agricultural Organization (FAO) of the United Nations,
Viale delle Terme di Caracalla, 00153 Rome, Italy

increasing the productivity, profitability and sustainability of smallholder farming is critical for reducing poverty and improving food security and nutrition. Agriculture in SSA, however, is increasingly exposed to a variety of risks and uncertainties, including market risk, production risks, climate variability, pest and disease outbreaks and windstorms, and institutional risks (Antonaci et al. 2012). The main premise is that by providing a steady and predictable source of income, cash transfer programmes can enhance household and community level resilience by improving human capital, facilitating changes in productive activities by relaxing liquidity constraints, improving natural resource management, and improving the ability to respond to and cope with exogenous shocks (e.g., Handa et al. 2015; Asfaw et al. 2012). The ultimate aim is to strengthen and improve resilience for rural producers to allow them to prevent future fluctuations in consumption and move to the next welfare level (Antonaci et al. 2012).

Government strategies for managing agricultural risks at the household or community level have taken different forms in different countries, but are generally classified into three groups:

1. mitigation/adaptation activities designed to reduce the likelihood of an adverse event or reduce the severity of actual losses. Risk mitigation options are numerous and varied (e.g., irrigation, use of resistant seeds, improved early warning systems, and adoption of better agronomic practices);
2. risk transfer, such as commercial insurance and hedging; and
3. resilience-improving mechanisms to withstand and cope with events ex ante.

Examples of these government strategies include social safety net programmes, buffer funds, savings, strategic reserves, contingent financing, insurance, etc. There are many definitions of resilience in the literature but the common thread in all definitions is the notion that resiliency reflects an ability to successfully manage or withstand a shock or stress (e.g., Alinovi et al. 2010).

Unlike in other parts of the world, most farmers in SSA have no access to government or market-based risk management tools; when they do, government programmes or private sector initiatives to manage price and production instability are often insufficient. Moreover, social protection programmes are seldom institutionalized, and are rarely used as risk management instruments to address food and nutrition insecurity. However, an increasing number of African governments over the last 15 years have launched social protection programmes including cash transfers, workfare and public works programmes and in-kind safety nets.

Cash transfer programmes in African countries have tended to be unconditional (where regular and predictable transfers of money are given directly to beneficiary households without conditions or labour requirements) rather than conditional (more common in Latin America and which require recipients to meet certain conditions, such as using basic health services or sending their children to school). Most of these programmes seek to reduce poverty and vulnerability by improving food consumption, nutritional and health status and school attendance. There is robust evidence from numerous countries (especially within Latin America and increasingly SSA) that cash transfers have leveraged sizeable gains in access to health and education services, as measured by increases in school enrolment (particularly for girls) and use of health services (particularly preventative health, and health monitoring for children and pregnant women) (e.g., Fiszbein and Schady 2009; Davis et al. 2012).

Building on the existing literature, this paper synthesizes the key findings of the From Protection to Production Project (PtoP) of FAO, which studies the impact of cash transfer programmes on household economic decision-making. The cash transfer programmes studied here are government-run cash transfer programmes in SSA. The paper is organized as follows. First we examine cross-country results to test their magnitude and distribution (i.e. heterogeneity) of impact on productivity and economic indicators, and the implications of these impacts on resilience. We will also explore the underlying programme design and implementation features that mediated the impacts. Section 2 provides an overview of the evolution of social cash transfer (SCT) programmes in SSA while Sect. 3 presents the conceptual framework on the linkages between cash transfers and economic impacts and resilience. Section 4 presents the impact evaluation design and data collection methods. Section 5 presents a synthesis of key cross-country findings, while Sec. 6 ends with a short conclusion and policy implications.

## 2 Overview of selected SCT Programmes in SSA

SCTs launched by African governments over the past 10 years have provided assistance to the elderly and to households that are ultra-poor, labour-constrained, and/or caring for orphans and vulnerable children. Typically, ministries of social development manage the programmes. The main types of social protection instruments used in African countries include cash transfers, workfare and public works programmes, and in-kind safety nets.

Workfare and public works programmes supply temporary employment to recipients able to contribute their labour in return for benefits, at the same time creating public goods in the form of new infrastructure, making improvements to existing infrastructure, or performing and delivering services (Del Ninno et al. 2009). In-kind safety nets (e.g. food aid, supplementary feeding and school feeding schemes, etc.) help recipients to access food, health care, education, and other basic goods and services. Other, more common instruments in parts of Southern Africa include social insurance schemes – primarily social pensions and health insurance.

Some of the African social protection instruments implemented during the last decade include the Kenyan Cash Transfer for Orphans and Vulnerable Children (CT-OVC), the Malawi SCTP, Mozambique's Programa de Subsidios de Alimentos (PSA), Ethiopia's PSNP, the Livelihood Empowerment Against Poverty (LEAP) programme in Ghana, the CGP in Lesotho, South Africa's Child Support Grant and Old Age Pensions, Rwanda's Vision 2020 Umurenge Programme, Burkina Faso's nationwide school feeding scheme under the Burkinabé Response to Improve Girls' Chances to Succeed (BRIGHT) integrated programme, Zambia's CGP and the

Zimbabwe SCT. Several other countries, including Uganda, Tanzania and Liberia, have also pursued safety net programmes (Asfaw et al. 2012). Our study focuses on the programmes described in the remainder of this section.

The Lesotho CGP provides an unconditional cash transfer to poor and vulnerable households. The primary objective of the CGP is to improve the living standards of OVC including nutrition and health status and increased school enrolment (Pellerano et al. 2012). The CGP is targeted at poor households with children, including child-headed households. As of the end of 2013 the programme reached approximately 20,000 households and 50,000 children (Pellerano et al. 2014). The Kenyan CT-OVC is the Government's flagship social protection programme, reaching over 130,000 households and 250,000 OVC across the country as of the end of 2011 (Asfaw et al. 2012). In Ethiopia, the cash transfer programme initiated by Tigray regional state and UNICEF aimed to improve the quality of lives of OVCs, elderly and persons with disabilities as well as to enhance their access to essential social welfare services such as health care and education via access to schools in two selected woredas (districts) (Berhane et al. 2015).

The Malawi SCTP was initiated in 2006 in the pilot district of Mchinji, providing small cash grants to ultra-poor, labour-constrained households. The SCTP objectives included reducing poverty and hunger in vulnerable households and increasing child school enrolment. By March 2015 the SCTP covered 100,000 beneficiary households and had gone to full scale in 10 districts, and the Government of Malawi expects to have enrolled over 175,000 households by the end of 2015. The programme was fully executed by the Government of Malawi through the District Councils by Social Welfare Officers (Handa et al. 2015).

The Ghanaian LEAP programme provides cash and health insurance to extremely poor households to improve short-term poverty and encourage long-term human capital development. LEAP started a trial phase in 2008 and began expanding gradually in 2009 and 2010, currently reaching over 70,000 households with an annual expenditure of approximately USD 20 million (Handa et al. 2014). The programme is fully funded from the Government of Ghana's general revenues, and is the of its National Social Protection Strategy's flagship programme. The LEAP programme operates in all 10 regions of rural Ghana. Within regions, districts are selected for inclusion based on the national poverty map; within districts, local DSW offices choose communities based on their knowledge of relative rates of deprivation (Handa and Park 2012).

In 2010, Zambia's Ministry of Community Development and Social Services (MCDSS) began implementing its own CGP in the three districts (Kalabo, Kaputa, and Shongombo) with the highest rates of mortality, morbidity, stunting, and wasting among children under 5. The CGP includes all households with a child under five years of age. Eligible households receive 55 kwacha a month (equivalent to USD 12) irrespective of household size, an amount considered sufficient to purchase one meal a day for everyone in the household for one month. The goal of the programme is to reduce extreme poverty and the intergenerational transfer of poverty, and as of March 2014 the programme reached 20,000 ultra-poor households (Daidone et al. 2014a).

Our impact evaluations focus on measuring the primary objectives of these programmes, including food security, health, and nutritional and educational status, particularly of children. Most programmes are located in some kind of social ministry, administered by professionals with backgrounds in the social sciences, including economists with specialization in the social sectors. The impact evaluations are most often implemented by research institutions and consulting firms with specializations in these social sectors.

## 3 Role of Cash Transfer for Building Resilience: Review of Selected Evidence

The potential benefits of cash transfer programmes are built around the premise that the provision of regular and predictable cash transfers to very poor households, in the context of missing or thin markets, has the potential to both generate economic and productive impacts at the household level (e.g., Handa et al. 2015; Asfaw et al. 2012; Covarrubias et al. 2012; Boone et al. 2013). In rural areas most beneficiaries depend on subsistence agriculture and live in places where markets for financial services (such as credit and insurance), labour, goods and inputs are lacking or do not function well. Cash transfers often represent a dominant share of household income, and can be expected to help households in overcoming the obstacles that block their access to credit or cash. This, in turn, can increase productive and other income-generating investments, influence beneficiaries' role in social networks, increase access to markets, improving the ability to deal with exogenous shocks, and strengthen household and community level resilience (Asfaw et al. 2012).

The predominant view from the literature is that social protection, including cash transfer programmes, may protect beneficiaries from shocks, reduce use of negative coping strategies that undermine longer-term livelihood sustainability, and reduce household risk adversity towards more profitable, yet more risky, activities. One group of empirical literature investigates the impact of social protection on recovery from shocks. Evidence shows that a public works programme in India reduced income fluctuations, while a public works programme in Ethiopia protected households from the negative effects of crop damage on child growth. Nonetheless, although a food-for-work programme in Ethiopia increased risk sharing within treated villages, it also reduced households' capabilities to manage idiosyncratic crop shocks – perhaps as a result of food aid crowding out informal insurance, and subsequently leaving beneficiaries inadequately insured to manage idiosyncratic risk (Dercon and Krishnan 2003). Conditional cash transfers (CCTs) in Latin America also facilitated recovery from shocks; some of the positive effects include reduced child labour in Nicaragua, protection of consumption for coffee farmers in Nicaragua and Honduras during global price drops, income diversification in Brazil and the decline in school dropouts in Mexico.

A second group of empirical studies looks at the impact of social protection on adverse coping strategies. The evidence generally shows a reduction in the use of adverse coping strategies that deplete household assets. One study finds that

Ethiopia's PSNP dissuaded 60% of beneficiaries from engaging in distress sales during a drought (Devereux et al. 2005). The Michinji Malawi Social Cash Transfer pilot scheme reduced begging for food or money by 14%, and reduced school dropout rates by 37% (Covarrubias et al. 2012). In Ghana and Kenya, the LEAP and CT-OVC programmes reduced child labour, distress asset sales and indebtedness. The impact on risk coping behaviour is also influenced by gender and programme design. In the Mchinji pilot scheme, children in female-headed households benefitted from the social cash transfer programme via a decline in non-household wage labour and an increase in participation in household chores, whereas children in male-headed households only experienced a decline in school absenteeism. Yet, these gender-specific outcomes are also a reflection of the constraints facing the households, as female-headed households are also single-guardian households that face challenges in balancing domestic work with income-generating activities (Covarrubias et al. 2012). In addition, cash and in-kind transfers may increase social capital and strengthen informal safety nets and risk-sharing arrangements, provided that appropriate mechanisms and an enabling environment are created.

A third group of studies shows that SCT programmes can have impacts on household decision-making over labour supply, the accumulation of productive assets and productive activities, which would subsequently have implications for resilience. Todd et al. (2010) and Gertler et al. (2012) found that the Mexican PROGRESA programme led to increased land use, livestock ownership, crop production, agricultural expenditures and a greater likelihood of operating a microenterprise. From their analysis of a conditional cash transfer (CCT) programme in Paraguay Soares et al. (2010) found that beneficiary households invested between 45–50% more in agricultural production and that the programme also increased the probability that households would acquire livestock by 6%. Martinez (2004) found that the BONOSOL pension programme in Bolivia had positive impacts on animal ownership, expenditures on farm inputs, and crop output, although the specific choice of investment differed according to the gender of the beneficiary. In contrast, Maluccio (2010) found that the Red de Proteccion Social (RPS) programme in Nicaragua had muted impacts on the acquisition of farm implements and no impact on livestock or land ownership. With respect to SSA, Covarrubias et al. (2012) and Boone et al. (2013) found that the Malawi SCT Programme (SCTP) led to increased investment in agricultural assets, including crop implements and livestock and increased satisfaction of household consumption by own production. Gilligan et al. (2009) found that Ethiopian households with access to both the Productive Safety Net Programme (PSNP) as well as complementary packages of agricultural support were more likely to be food secure, to borrow for productive purposes, use improved agricultural technologies, and operate their own nonfarm business activities. In a later study, Berhane et al. (2011) found that the PSNP led to a significant improvement in food security status for those that had participated in the programme for 5 years versus those who only received 1 year of benefits. Moreover, those households that participated in the PNSP as well as the complementary programmes had significantly higher grain production and fertilizer use. However, beneficiaries did not experience faster asset growth (livestock, land or farm implements) as a result of the programmes (Gilligan et al. 2009).

## 4 Methodology

### 4.1 Programme Evaluation Design and Data

The core of the quantitative analysis for the Lesotho, Malawi, Zambia and Kenya studies was an experimental design impact evaluation. In Ghana and Ethiopia the evaluation designs were quasi-experimental. Table 1 summaries the key evaluation design features of the cash transfer programmes.

In Lesotho, participation in the programme was randomized at the level of the electoral district (ED). First, all 96 EDs in four community councils were paired based on a range of characteristics, with 40 pairs randomly selected for this survey. Within each selected ED, two villages (or clusters of villages) were selected, and in every cluster a random sample of 20 households were selected. Baseline survey data was collected followed by public meetings with a lottery to assign EDs (both sampled and non-sampled) to either treatment or control groups. Selecting the treatment ED after baseline survey helped to avoid anticipation effects (Pellerano et al. 2012). The baseline household survey was carried out in 2011 prior to distribution of cash transfers; a follow up panel survey took place in 2013. A total of 3102 households were surveyed; 1531 programme eligible households (766 treatment and 765 control) were used for impact evaluation analysis, with remaining 1571 programme ineligible households used for analysis of targeting and spillover effects. The baseline analysis report (Pellerano et al. 2012) shows that randomization was quite successful.

**Table 1** Core evaluation designs

| Country | Design | Level of randomization or matching | N | Ineligibles sampled? |
|---|---|---|---|---|
| Ethiopia | Non-experimental (PSM and IPW) | Household level within a village | 3351 | Yes |
| Ghana | Propensity Score Matching (IPW) | Household and Region | 1504 | No |
| Kenya | Social experiment with PSM and IPW | Location | 2234 | No |
| Lesotho | Social experiment | Electoral District | 2150 | Yes |
| Malawi | Social experiment | Village Cluster | 3200 | Yes |
| Zambia | Social experiment | Community Welfare Assistance Committee | 2519 | No |

All studies are longitudinal with a baseline and at least one post-intervention follow-up. N refers to households sampled at follow-up
Source: Davis and Handa (2015)

In Kenya's CT-OVC, the impact evaluation utilized a randomized cluster longitudinal design, with the baseline quantitative survey fieldwork carried out in mid-2007. Within each district, two locations were chosen randomly to receive intervention and two were selected as controls (Ward et al. 2010). This method of randomization was not as robust as in the case of Lesotho due to the fewer units over which the randomization took place. Approximately 2750 households were surveyed in seven districts (namely, Nairobi, Kwale, Garissa, Homa Baye, Migori, Kisumu and Suba). Two-thirds of households were assigned to the treatment group. These households were re-interviewed (first round) two years later, between May and July 2009, in order to assess the impact of the programme on key welfare indicators (Ward et al. 2010). The re-interview success rate was approximately 83%. The second round follow up study was conducted between May and August 2011 with a more detailed economic activity module (including wage labour, self-employment, crop and livestock activities, etc.) to capture potential investment and productive activity benefits of the programme on families. For the household level analysis, we relied on data collected at the baseline (2007) and the second round follow up in 2011, with a sample of 1811 households. However it is important to point out that for many of the outcome variables of interest to the PtoP project, we have only one data point (i.e. no baseline).

In Zambia the baseline survey was carried out in September–October 2010, with follow ups in 2012 and 2013. Communities were randomly assigned to treatment group (incorporated into the programme in December 2010) or control (to be brought into the programme at the end of 2013). Baseline data collection began prior to group assignment. The study includes 2515 households (1228 treatment and 1287 control). Analysis of the baseline data shows that randomization appears to have worked well; greater detail on the randomization process can be found in Seidenfeld and Handa (2011).

In Malawi, baseline data was collected in 2013 and a follow up survey 17 months later in 2014 (Handa et al. 2014). Treatment and control groups each represent about half of communities sampled. The sample is divided between Salima and Mangochi districts which count, respectively, 2192 and 2160 households. Of these households 1775 and 1756, respectively, meet the eligibility criteria. The longitudinal impact evaluation includes 3531 eligible households and 821 ineligible households at baseline.

In Ethiopia, the impact evaluation design was non-experimental; it follows a longitudinal design, with a baseline household survey conducted in mid-2012, followed by separate monitoring surveys, and finally a 24 month follow-up in 2014. The evaluation sample includes three groups of households: treatment beneficiaries, control households, and ineligible households. The development of ranking lists of eligible households based on meeting targeting criteria was a vital component. Treatment and control households were both selected from the list of eligible households. The sample comprises 3664 households at baseline, of which 1629 were beneficiaries and 1589 were control households. In addition 446 sample households were randomly selected for the study from households who were non-eligible to receive support from the programme either because they were less poor and/or because of the presence of able-bodied members. Attrition between baseline (May–August 2012) and endline (2014) was 8.7% or 4.36% per year (Brehane et al. 2012).

The Ghanaian LEAP programme impact evaluation takes advantage of a nationally representative household survey implemented during the first quarter of 2012. It focuses on 7 districts across 3 regions (Brong Ahafo, Central, Volta). The initial treatment sample of 700 households were randomly drawn from the group of 13,500 households that were selected into the programme in the second half of 2009. Households were interviewed prior to indication of selection to lower anticipation effect. The baseline survey instrument was a reduced version of the national household survey instrument, and the national survey sample and the treatment household sample were surveyed at the same time by ISSER. The strategy was to draw the control households from the national survey using PSM techniques. A comparison group of 'matched' households were selected from the ISSER sample and re-interviewed 2 years later, in March–April 2012, along with LEAP beneficiaries to measure changes in outcomes across treatment and comparison groups (Handa and Park 2012).

## 4.2 Analytical Methods

In PtoP project impact evaluation, we seek to answer the question: "How would cash transfer beneficiaries have fared in the absence of the programme?" The identification of the counterfactual is the organizing principle of an impact evaluation as it is impossible to observe a household both participating in the programme and not. The goal is to compare participants with non-participants who are as similar as possible except for receiving the programme in order to measure the differential impact of the intervention. The "with" data are observed in a household survey that records outcomes for recipients of the intervention. The "without" data, however, are fundamentally unobserved since a household cannot be both a participant and a non-participant of the same programme (see Asfaw et al. 2012 for detail).

However, the outcomes of non-beneficiaries may still differ systematically from what the outcomes of participants would have been without the programme, producing selection bias in the estimated impacts. This bias may derive from differences in observable characteristics (e.g., location, demographic composition, access to infrastructure, wealth, etc.) or unobservable characteristics (e.g., natural ability, willingness to work, etc.). Some observable and unobservable characteristics do not vary with time (such as natural ability) while others may vary (such as skills). Furthermore the existence of unobservables correlated with both the outcome of interest and the programme intervention can result in additional bias (i.e., omitted variables).

The validity of experimental estimators relies on the assumption that the control group units are not affected by the programme; this is also referred to as the Stable Unit Treatment Value Assumption (SUTVA) (Rubin 1980; Djebbari and Hassine 2011). However control households can be affected through market interactions and informal transaction and risk sharing (which is also known as non-market interaction). Depending on the nature of the design and the availability of data, different analytical models can be used to estimate the impact of the programme.

Towards this end, two approaches (i.e. a difference-in-difference (DD) estimator and a single difference approach combined with inverse probability weighting and propensity score matching) were used in most of the evaluations, depending on the nature of the design and availability of data (see Asfaw et al. 2012 for detail). When baseline data are not available, as is the case for some of our outcome variables in some countries, the single difference method was applied. When panel data were available with pre- and post-intervention information, which is the case with most of the countries, a DD approach was used. By taking the difference in outcomes for the treatment group before and after receiving the cash transfer, and subtracting the difference in outcomes for the control group before and after the cash transfer was disbursed, DD is able to control for pre-treatment differences between the two groups, and in particular the time invariant unobservable factors that cannot be accounted for otherwise (Wooldridge 2002).

The key assumption is that differences between treated and control households remain constant throughout the duration of the project. If prior outcomes incorporate transitory shocks that differ for treatment and comparison households, DD estimation interprets such shocks as representing a stable difference, and estimates will contain a transitory component that does not represent the true programme effect. When differences between treatment and control groups exist at baseline, the DD estimator with conditioning variables has the advantage of minimizing the standard errors as long as the effects are unrelated to the treatment and are constant over time (Wooldridge 2002). Control variables are most easily introduced by turning to a regression framework which is convenient for the DD, or by combining DD with propensity score matching or DD with inverse probability weighting (DD-IPW).

All estimators presented above assume the cash transfer impact is constant, irrespective of who receives it. The mean impact of a programme or policy based on this assumption is a concise and convenient way of evaluating impacts. Heckman et al. (1997) justify this approach if researchers and policy makers believe that (a) total output increases total welfare and (b) detrimental effects of the programme or policy on certain parts of the population are not important or are offset by transfers—either through an overarching social welfare function or from family members or social networks.

Overall mean impacts are most helpful when complemented with measurements of distributional impact. Even if the mean programme effect were significant, whether the programme had a significant beneficial or detrimental effect might vary across the distribution of targeted households (Khandker et al. 2010). For example, the impact on poorer households as compared to wealthier households is particularly interesting in the context of programmes that aim to alleviate poverty.

There are a number of ways to present the distributional impacts of a cash transfer programme. For example, one could divide the sample of households and individuals into different demographic groups (e.g., by gender or age cohort), perform separate analysis on each group, and see if estimated impacts are different. Interacting the treatment with different household socioeconomic characteristics is another way to capture differences in programme effects, although adding too many interaction terms in the same regression can lead to issues with multicollinearity (Khandker et al. 2010). Another way to present distributional impacts of cash transfer programmes is by using a quintile regression approach to assess the magnitude of impact for each strata of households. Simply investigating changes in the mean programme effect, even across different socioeconomic or demographic groups, may not be sufficient when the entire shape of the distribution changes significantly.

## 5 Results and Discussion

In this section, we synthesize key findings from the PtoP impact evaluation reports and discuss the results over three broad groups of outcome variables linked to household resilience: risk management including climate change, investment in livelihood activities and food security. We focus on the quantitative studies and where applicable we supplement the comparative analysis with results from the qualitative evidence that report on similar outcomes. The results discussed are taken from the following references: Asfaw et al. (2014, 2015a, b, 2016), Daidone et al. (2014a, b), AIR (2013), Handa et al. (2014) and Pellerano et al. (2014).

### 5.1 Can Cash Transfer Promote Ex-Post Risk Management?

By providing a reliable income stream, cash transfer programmes improve risk management by poor rural households. An extra source of income can help households provide for school fees and discourage the need for children to drop-out to work on farms. The transfers flowing in and out of households can also change, and households may engage more in social networks through increased giving and so perhaps be able to rely on these networks in the future. Households can also use that money to pay off debts, purchase on credit, or save the cash. Table 2 presents the cross-country summary of the impact of social cash transfers on risk coping strategies, access to credit, community relations, savings, and debt payments.

Beneficiary households were found to have relied less on risk coping mechanisms thanks to cash transfers. Asfaw et al. (2015b) found households in Malawi to shift away from undesirable ganyu labor as a result of the SCTP. Handa et al. (2015) also found that the SCTP reduced paid work outside the home for children aged 10–17. In the face of negative shocks, use of the cash transfers emerged as the pri-

**Table 2** Synthesis of key findings

| | Ghana | Kenya | Lesotho | Malawi | Zambia | Ethiopia |
|---|---|---|---|---|---|---|
| Ability to manage risk | | | | | | |
| Risk coping mechanisms | + | N/E | +++ | ++ | + | ++ |
| Savings | + | N/E | − | N/A | ++ | N/A |
| Purchase on credit | + | NS | NS | − | NS | 0 |
| Debt payment | ++ | N/E | − | ++ | + | N/E |
| Provide transfer | − | N/E | + | NS | N/E | − |
| Receive transfer | + | N/E | + | − | N/E | NS |
| Remittance receipt | + | N/E | − | N/E | N/E | N/E |
| Agricultural asset | | | | | | |
| Agricultural tools | N/E | + | + | ++ | +++ | 0 |
| Livestock ownership | N/E | ++ | + | +++ | +++ | 0 |
| Crop and livestock production and marketing | | | | | | |
| Agricultural inputs | 0 | − | ++ | ++ | +++ | 0 |
| Livestock inputs | N/A | 0 | 0 | N/E | NS | − |
| Land use | N/E | N/E | NS | N/E | ++ | N/E |
| Agricultural output | N/E | NS | ++ | ++ | ++ | ++ |
| Crop sales | N/E | N/E | 0 | ++ | ++ | 0 |
| Livestock by-products | N/E | N/E | + | N/A | N/A | 0 |
| Non-farm enterprise (NFE) | NS | 0 | − | 0 | +++ | 0 |
| Household welfare | | | | | | |
| Food security | +++ | N/A | +++ | +++ | +++ | +++ |
| Consumption | NS | +++ | + | +++ | +++ | ++ |
| Dietary diversity | 0 | +++ | NS | N/E | ++ | + |
| Home consumption of crop production | N/E | +++ | N/E | NS | + | N/E |

Note: *N/A* not available, *N/E* not estimated, *NS* no shift, 0 overall mixed shift. + = significant positive impact; and − = significant negative impact. One, two or three '+' or '−' signs refer to the level of the impact

mary coping mechanism for a quarter of the negative shocks among SCTP beneficiary households, and there are declines in ganyu labor and in the use of savings as coping mechanisms. The authors also found a smaller percentage of households engaging in coping mechanisms for negative shocks, particularly for the poorest households (Handa et al. 2015). In Ethiopia, the SCTPP reduced the number of hours per day children were engaged in household activities. In particular, children aged 6–12 in beneficiary households worked fewer hours per day on the family farm and across all other activities compared to those in control households (Asfaw et al. 2015a). However, the impact was more mixed in Lesotho: while boys 13–17 may have seen a reduction in engagement in paid work outside the house, girls have seen an increase due to the CGP (Pellerano et al. 2014). Pellerano et al. (2014) found a reduction in the levels of engagement in occasional and irregular occupations among adults, noting the results to indicate that the cash support effectively worked as a safety net preventing households from depending on low paid and precarious occupations. The authors also found CGP beneficiaries to be less likely to send children to live elsewhere by 6 pp., send children to work by 3 pp., take children out of school by 8 pp., and reduce spending on health by 7 pp. as a response to shocks within 12 months previous to the survey.

The decreased need to engage in negative risk coping mechanisms as a result of cash transfers was also shown through increases in enrolment and other educational outcomes for children. Handa et al. (2015) found that children aged 6–17 increased their net enrolment by 12 pp. as a result of the SCTP in Malawi, with slightly stronger impacts considering primary and secondary school-aged children separately. The authors also found the dropout rate to have dropped for primary school-aged children by 4 pp. and temporary withdrawal (missing more than two consecutive weeks of instruction at any time in the past 12 months) to have decreased by 5 pp. By the endline in Ethiopia, Berhane et al. (2015) found the SCTPP to have raised enrolment by around 6 pp. pp. in Hintalo-Wajirat, with the effect for girls particularly strong (13 pppp). Instead of having to take time out of school to earn extra income, children were more readily participating in school thanks to the SCTPP. In Ghana, the LEAP programme reduced the likelihood of school-aged children (5–17) missing any school by 8 pppp and also reduced the chance of missing an entire week by 5 pppp (Handa et al. 2014). Among younger children smaller households appeared to be more protective, with a larger impact on missing any school in smaller households. However, the significant impact on enrolment is entirely driven by larger households. Handa et al. (2014) also found the impact on secondary enrolment for children aged 13–17 to be similar to estimates for South Africa's Child Support Grant (6 pp) and Kenya's CT-OVC (8 pp). While there were mixed results for engagement in paid work with the Lesotho CGP, the programme increased the proportion of children aged 6–19 enrolled in school by 5 pp., with a larger impact on older boys aged 13–17 (Pellerano et al. 2014). AIR (2013) noted that children living in a CGP beneficiary household in Zambia were 1 pp. more likely to ever enroll in school and 2 pp. more likely to enroll on time, for every year less of education their mother has. The authors attribute this effect to the CGP enabling or motivating mothers who did not enroll children in school at baseline to change their actions and start enrolling their children in school.

Cash transfer programmes were found to strengthen community ties through various channels, while the impact on private transfers was mixed. In Lesotho, the CGP had a significant impact in strengthening the reciprocity arrangements around food sharing in treatment villages. Both the proportion of households receiving and the proportion providing in-kind help in the form of food increased as a consequence of the programme. The impact is strong and significant, 15 and 18 pp. respectively, and the magnitude is larger for households with no labour capacity (Daidone et al. 2014b). Handa et al. (2014) found a positive impact on the value of gifts received and the amount of credit extended to others in Ghana. Meanwhile, in Malawi Asfaw et al. (2015b) found SCTP beneficiary households to be 4% points less likely to receive a transfer. In Ethiopia, Asfaw et al. (2015a) found increases in social capital and subjective belief of individuals' quality of life and control. Treated households were more likely to agree with additional support to poor people, have fewer problems with neighbors, and, similarly, agree that people residing in their community are basically honest and trustworthy. Other opinions of life satisfaction and ability to achieve success marked higher among male-headed beneficiary households compared to male-headed control households. However, there were no impacts observed in either receipt or giving of private transfers in Ethiopia.

Beneficiary households were also found to use proceeds from cash transfer programmes to pay off debts. In Ghana, Handa et al. (2014) observed beneficiary households saving more and being more likely to repay debt; smaller beneficiary households also reduced their likelihood of holding a loan by 9 pp. The authors also found a corresponding significant impact on the amount paid off of 19 pp. of adult equivalent consumption. In Malawi, households overall, and female-headed households and large farm households in particular, reduced debt from previous loans due to the SCTP. Male-headed households and large farm households were also less likely to still owe money for previously contracted loans (Asfaw et al. 2015b). Daidone et al. (2014a) also found larger households to pay down loans as a result of the CGP in Zambia.

## 5.2 Can Cash Transfer Contribute to Managing Climate Risk?

Climate change poses severe threats to households' wellbeing across the world, particularly in low-income countries where poor households are often exposed to different sources of risk. Adoption of risk management strategies, such as the promotion of social safety nets, are becoming gradually more relevant for improving the households' abilities to manage climate risk. Given the high incidence of climate shocks in Zambia, we also would like to present the findings of Asfaw et al. (2016) who shed light on how households respond to the CGP cash transfer in a context of weather instability. Asfaw et al. (2016) conducted additional analysis by merging the Zambia CGP impact evaluation data with rainfall data obtained from the Africa Rainfall Climatology v.2 (ARC2) (1983–2012).[1] They assessed whether regular and unconditional small cash payments (via the CGP) helped mitigate the negative effects of climate variability, protect and improve smallholders' livelihoods and ensure food security and nutrition.[2] The authors also investigated how the CGP and climate variability affect households on different quintile of the welfare and food security dimensions.

Asfaw et al. (2016) found the CGP to increase total/food and non-food expenditure, which implies the treatment increases households' welfare. As a result of an

---

[1] Dekads (i.e. 10 days) at 0.1° covering the period 1983–2012 at ward level.
[2] The outcome variables in the study included total expenditure, food/non-food expenditure, daily caloric intake and dietary diversity index.

increase in food expenditure, both quantity and quality of food consumed responded positively to CGP receipt, implying that households benefitted from the CGP in terms of food security and nutrition. With regards to the effect of climatic variables on welfare and food security, results from Asfaw et al. (2016) show that overall, households in areas that experienced lower than average rainfall had lower levels of daily caloric intake as well as food and non-food expenditures, and this effect was most pronounced for the poorest households in the sample. A possible explanation could be that the decline in rainfall had an initial negative impact on agriculture, livestock production and other water-intensive activities. The decline in volume of production thus affected households' purchasing power, forcing them to improve their coping mechanisms.

This study also finds strong evidence that cash transfer programmes have a mitigating role against the negative effects of climate shocks. Households that participated in the CGP had much lower negative effects of the weather shock, with poorest households gaining the most. This indicates the potential of social protection to support food access for households exposed to climate risk. However, the analysis also indicates that while participation in the CGP is beneficial in mitigating negative effects of climate shocks on food security, it is not sufficient to fully overcome these effects. Thus it is important to ensure that SCTs are well aligned with other forms of livelihood programmes and climate risk management, including disaster risk reduction activities. This result confirms the findings of authors like Eriksen et al. (2005), who found a positive relationship between the ability of people to draw on extra sources of income and the ability to withstand droughts in Tanzania and Kenya, with respect to those who were not.

## 5.3 Potential of Cash Transfer to Promote Ex-Ante Risk Management

Cash transfers contribute to ex-ante risk management by increasing household adaptive capacity through accumulation of productive assets, increased crop and livestock production and productivity, and linkages with output markets. We look at various dimensions of the productive process in order to ascertain whether households were found to have increased spending in livelihood activities, including crop production, crop input use and asset building. Given that agriculture represents the primary economic activity of the households studied, investment in agricultural assets and increases in crop production prove critical for strengthening livelihoods and ex-ante risk management. Households can also enhance their resilience by diversifying into different income streams, such as non-farm enterprises. Table 2 presents the cross-country summary of the impact of SCTs on investment in livelihood activities.

### 5.3.1 Impacts on Accumulation of Productive Assets

Beneficiary households overall (and larger sized households in particular) in Zambia owned more axes and hoes, and were more likely to own hammers, shovels, and ploughs as a result of the cash transfer programme (Daidone et al. 2014a). Beneficiary households in Kenya were more likely to own troughs, and male-headed households were also more likely to own machetes and sickles (Asfaw et al. 2014). In Lesotho, Daidone et al. (2014b) found the CGP to increase the use and purchase of scotch-carts. In Malawi, beneficiary households overall, both female and male-headed households, and large farm households owned more agricultural implements (Asfaw et al. 2015b). Handa et al. (2015) also found the SCTP to increase crop production and agricultural assets (sickles in particular). In terms of agricultural asset ownership, beneficiary households in Hintalo-Wajirat were 6 pp. and 7 pp. more likely to own plows and imported sickles, respectively (over baseline shares of 47% and 41%). In contrast, beneficiary households in Abi Adi were less likely to own those agricultural implements. In terms of number owned, there were more negative effects throughout (Asfaw et al. 2015a). However, Berhane et al. (2015) found the SCTPP in Ethiopia to increase a constructed farm productive assets index by 2% in Hintalo-Wajirat.

Cash transfers also led to increased livestock ownership in SSA, particularly of smaller animals. Both small and large beneficiary households in Zambia increased livestock ownership, but the impacts were stronger for large households (Daidone et al. 2014a). Smaller households and female-headed households in Kenya increased their ownership of small livestock (such as sheep and goats) compared to control households. For smaller households, there was about a 15 pp. increase in ownership of small livestock compared to control households, while female-headed households receiving the transfer increased their ownership by 6 pp. (Asfaw et al. 2014). Daidone et al. (2014b) also found the cash transfer in Lesotho to have increased the proportion of households owning pigs by about 8 pp. as well as to have increased the number of pigs owned by 0.1 pp. Whether by number of livestock owned or livestock ownership, SCTP beneficiaries in Malawi faced increases on livestock (also noted by Handa et al. (2015)), such as on chickens, goats and sheep, and pigs (Asfaw et al. 2015b). Meanwhile, in Ethiopia Asfaw et al. (2015a) found the impact on livestock ownership to be more mixed, depending particularly on the area in which the transfer was given. Berhane et al. (2015) found the SCTPP in Ethiopia to increase the likelihood that households own any form of livestock by 7% in Hintalo-Wajirat, with the increase largely driven by the increase in poultry ownership.

### 5.3.2 Impacts on Crop Production and Productivity

The cash transfer programmes evaluated generally led to increased crop production and productivity. Aggregating all crop output by value, the GCP in Zambia increased the value of all crops harvested by ZMK 146, approximately a 50% increase from baseline, with a larger value increase for smaller households at ZMK 182.

Beneficiary households increased their crop production marketing by 12 pp. and also increased their average value of sales (Daidone et al. 2014a). Production of maize, the main staple commodity, increased in CGP households in Lesotho by around 39 kg more than the control group, and more so for households with more available household labour. Sorghum production increased by around 10 kg, with a larger impact in severely constrained households, likely because sorghum requires less labour as compared to other major crops. Furthermore, results on home gardening were consistently larger for unconstrained and moderately labour-constrained households compared to households with no adult members fit to work (Daidone et al. 2014b). In Malawi, beneficiary households increased groundnut production and productivity, with fewer and mixed impacts on other crops. Medium farm households and male-headed households also increased maize yields. Ultimately, both male-headed households and medium farm households increased the value of crop production as a result of the SCTP. Households were more likely to sell any crop, and the value of crop sold increased for female-headed households, small farm households, and medium farm households (although it decreased for large farm households) (Asfaw et al. 2015b). In Ethiopia, Asfaw et al. (2015a) found households to have decreased their yield of sorghum but to have increased sorghum yields, particularly in Hintalo-Wajirat and among male-headed households. Ultimately, beneficiary households increased the total value of their crop production by 18%. For the Kenya CT-OVC, Asfaw et al. (2014) found little impact of the programme on crop production. However, there was an impact on the proportion of food consumption coming from own production, particularly for smaller-sized households and female-headed households. The average treatment effect for the share of consumption from home produced dairy and eggs was 20 pp. for smaller households and 15 pp. for female-headed households.

Increased crop production and productivity for beneficiary households also came through increases in land and crop input use. The CGP in Zambia increased the amount of operated land by about 34% from baseline, and 18 pp. more households spent money on inputs, from a baseline share of 23%. This increase in money spent on inputs was particularly relevant for smaller households (22 pp), and included spending on seeds, fertilizer and hired labour. The increase of 14 pp. in the proportion of small households purchasing seeds is equivalent to more than a doubling in the share of households. Small beneficiary households spent ZMK 42 more on crop inputs than the corresponding control households, including ZMK 15 on hired labour, amounting to three times the value of the baseline mean for overall spending, and four times for hired labour (Daidone et al. 2014a). The CGP in Lesotho significantly increased the share of beneficiary households using pesticides (8 pp), especially those who are labour-unconstrained and who are also more likely to purchase pesticides as a result of receiving the CGP. Households purchased seeds more often (7 pp), although there was no statistically significant change in the intensity of purchase (Daidone et al. 2014b). In Malawi, household expenditure on organic fertilizer increased by MWK 158 (from a baseline of MWK 245). Increases on organic fertilizer expenditure also were found at the disaggregated levels (aside from medium farm households, which faced no increase) and at expenditure-per-acre

(Asfaw et al. 2015b). An increase in the likelihood of chemical fertilizer use is also found among male-headed households. In the case of the Ethiopia SCTPP, female-headed beneficiary households were 4 pp. more likely to practice a soil and water conservation technique on their land, a noticeable increase on their baseline mean of 14%. Female-headed households were also 3 pp. more likely to hire labour for farm work from a low baseline mean of 5% (Asfaw et al. 2015a).

### 5.3.3 Impacts on Non-farm Enterprises

On non-farm enterprises cash transfer programmes were found to have mixed results. In Zambia, non-farm work increased by 20 days overall among beneficiaries and non-farm enterprise by 1.6 days (AIR 2013). Cash beneficiary households participated more often in non-farm enterprises in Kenya if they were female-headed, but less so if they were male-headed; otherwise, there was no impact recoded for the overall sample (Asfaw et al. 2014). In Malawi, results on non-farm enterprise labor were mixed, where beneficiary households were less likely to engage in charcoal/firewood enterprises but were more likely to engage in petty trade enterprises (Asfaw et al. 2015b). In Ethiopia (Asfaw et al. 2015a) and in Ghana (Handa et al. 2014) there were no impacts found on the overall level on the likelihood that households participated more or less often in non-farm enterprises. Pellerano et al. (2014) found a reduction in the proportion of households with an enterprise in operation in the 30 days prior to the survey, but noted that the reduction was mainly driven by households engaging less frequently in home brewing, which is generally small scale and a livelihood strategy of last resort.

## 5.4 Can Cash Transfer Promote Resilience by Enhancing Food Security?

Households consistently more able to consume an adequate amount of food and a more diverse basket are necessarily more resilient and less food insecure than otherwise similar households. Depending on the availability of data across the different countries, we collected the impacts of cash transfer programmes on consumption, dietary diversity and subjective food security indicators. Table 2 presents the cross-country summary of the impact of social cash transfers on food security, consumption and diet diversity.

### 5.4.1 Impact on Food Security

As expected, the studied cash transfer programmes unambiguously increased the food security of beneficiary households. The CGP in Zambia increased the percentage of households eating two or more meals per day by 8 pp. as well as the number of households that were not severely food insecure by 18 pp., (AIR 2013). The share of households consuming from part of their harvest also increased by 6 pp., which came from increased groundnut and rice consumption of home production (Daidone et al. 2014a). In Lesotho, Pellerano et al. (2014) found the CGP to reduce the number of months that households experienced shortages of food and decrease the proportion of households not having enough food to meet their needs at least for one month in the previous 12 months. Food security also increased in Malawi thanks to the cash transfer programme: households overall, for example, were 11 pp. less likely to worry whether they would have enough food in the past seven days. The SCTP also allowed households to eat more meals per day, with effects observed for households at all levels except for large farm households. Medium farm households also increased the number of months that last year's maize harvest lasted (Asfaw et al. 2015b). In Ethiopia, there was a reduction on the number of months with problems satisfying food needs in the overall sample and among male-headed households. There was no impact on number of months in the last 12 months that the household ran out of home-grown food, but there were increases on both the number of times a day children ate in the household and the number of times adults ate in the household. Compared to control households, beneficiary households were also less likely to suffer a shortage of food to eat during the last rainy season as a result of the SCTPP. With regards to measures of last resort, beneficiary households reduced their likelihood of consuming seed stock during the last week, compared to control households (Asfaw et al. 2015a).

### 5.4.2 Impact on Consumption Expenditure

Cash transfers also enabled households to better meet their consumption needs. In Zambia, the programme significantly increased food spending, with the largest share going to cereals, followed by meats, including poultry and fish, followed by fats such as cooking oil and then sugars (AIR 2013). The share of households consuming from part of their harvest also increased by 6 pp., which came from increased groundnut and rice consumption of home production (Daidone et al. 2014a). In Lesotho, Pellerano et al. (2014) detected a statistically significant CGP effect on food expenditure and total consumption when controlling for covariates, including differences in prices across locations, but at low levels of significance. In Kenya, although there was no significant impact on consumption expenditure of cereals and legumes, there was an increase for food spending on dairy and eggs. The programme had no effect on spending on most of the food consumption categories for larger households but it had large increases on three of the outcomes (dairy and eggs, meat and fish and fruit) for smaller households. The programme had larger and positive impacts on female-headed households compared to male-headed households, as in the case of the share of consumption from home produced dairy and eggs. Treated households in Kenya also appeared to consume more animal products, as well as other foods, from their own production compared to control households (Asfaw et al. 2014). In Malawi, there were increases at all levels of daily per capita calories consumed, with those increases in calories coming from food purchases; aside from a decrease for male-headed households, there are no impacts on calories coming from own production. Such results suggest that households are likely using the cash to buy food directly, although calories coming from own production may take more time to see impacts. For both extremely-poor and non-extremely poor household, the pattern holds up: increases in calories consumed come from purchases rather than from own production, with decreases in calories consumed coming from gifts and other sources (Asfaw et al. 2015b). Berhane et al. (2015) found the SCTPP in Ethiopia to reduce the food gap, increase the availability of calories, and to reduce seasonal fluctuations in children's food consumption (Berhane et al. 2015). Meanwhile, Handa et al. (2014) found in Ghana that there was no overall change in food consumption between treated and control households.

### 5.4.3 Impact on Dietary Diversity

There is also some evidence of improved dietary diversity as a result of cash transfer programmes. There was a clear shift away from roots and tubers (primarily cassava) and toward protein (dairy, meats), indicating a possible improvement in dietary diversity among CGP recipients in Zambia (AIR 2013). In smaller households, the impact of the CGP on food expenditures was concentrated on cereals (where 45% of the impact on food is derived) followed by meat (15%), fats (14%), and pulses (13%). Among larger households, the impact of the grant on food expenditures is driven by meats (32%) and then cereals (30%). In the end, food expenditures increase for both groups of households as a result of the cash transfer programme (Daidone et al. 2014a). In Kenya, the results showed no significant impact on consumption expenditure of cereals and legumes. However there was about a 12 pp. increase for food spending on dairy and eggs. The programme had no effect on spending on most of the food consumption categories for households with larger number of members but it had large, positive, and significant effects on three of the outcomes (dairy and eggs, meat and fish and fruit) for smaller sized households. The programme typically had larger and positive impacts on female-headed households compared to male-headed households, such as on consumption of animal products. Treated households also appear to have consumed more animal products, as well as other foods, from their own production compared to control households. Dairy and eggs consumption from own production increased by about 13 pp. for beneficiary households, and the impact on other types of foods was about 4 pp. The average treatment effect for the share of consumption from home produced dairy and eggs was 20 pp. for smaller households and 15 pp. for female-headed households (Asfaw et al. 2014). In Ethiopia, results from Asfaw et al. (2015a) showed an increase in household consumption of oils and fats, sweets, and spices, condiments, and beverages as a result of the SCTPP. This was mixed with reductions in household consumption of fruits and meats. Berhane et al. (2015) found the SCTPP to have improved diet quality, as measured by the Dietary Diversity Index, in both May 2012 and May 2014 by 13% and 12% respectively. In Ghana, although there was no overall change in food consumption between treated and control households, Handa et al. (2014) found a significant decline in starches and meats and an increase in fats and food eaten out. Smaller households also faced a decline in alcohol and tobacco consumption. Among Lesotho CGP beneficiaries, the increased spending on dairy and eggs (as well as meat/fish and fruit for smaller households) did not translate into an impact on dietary diversity (Pellerano et al. 2014).

## 6 Conclusions and Implications

The analysis of impact evaluation studies show that cash transfer programmes overall have important implications for household resilience. By providing a steady and predictable source of income, cash transfer programmes can build human capital and improve food security and potentially strengthen households' ability to respond to and cope with exogenous shocks, and allow them to diversity and strengthen their livelihoods to prevent future fluctuations in consumption. Many of the programmes studied increased investment in agricultural inputs and assets, including farm implements and livestock. Beneficiaries in the studied country programmes generally increased crop production and value of crop production. Although differing across countries, food security indicators revealed increases in the proportion of households being food secure as a result of cash transfer programmes. This too was met by increases in consumption and dietary diversity. Although the impacts on risk management are less uniform, the cash transfer programmes seem to strengthen community ties (via increased giving and receiving of transfers) allow households to save and pay off debts, and decrease the need to rely on adverse risk coping mechanisms. Finally, the case study of the CGP in Zambia demonstrates the potential for cash transfers to help poor households manage climate risk. Not only was CGP receipt associated with increases in total/food and non-food expenditure, and subsequently the quantity and quality of food consumed, but the CGP was also found to benefit households even when they were facing climate shocks. The CGP's climate mitigating effect is particularly evident for households at the lowest quintiles of the distribution, meaning that the CGP better protects poorer households against climate variability than richer households. Thus cash transfers can improve poor households' resilience for an uncertain future in terms of climate change.

The differences in impacts across countries can be attributed to a variety of factors, including the availability of labour given the demographic profile of beneficiary households, the relative distribution of productive assets, the local economic context, the relevance of messaging and soft conditions on spending and the regularity and predictability of the transfers themselves. In the case of LEAP in Ghana, irregular payments may have prevented households from increasing consumption, as consumption is driven by permanent income. Instead, the lumpy flow of cash seems to have promoted declines in the number of households with outstanding loans and increases in the number of households with savings. In Ethiopia, the SCTPP targeted households that were particularly made up with either the elderly or youth, which may explain why beneficiary households did not face increases in labour supply or on other dimensions of agricultural production. The amount offered through the Ethiopia SCTPP as a percentage of per capita income is also not as high compared to cash transfer programmes that have found widespread impacts.

Cash transfers can be more than just social assistance; not only can they help vulnerable households avoid the worst effects of severe deprivation, they can also contribute to economic and social development. Since cash transfer programmes impact the livelihoods of households, articulation with other sectoral development programmes in a coordinated rural development strategy could lead to synergies and greater overall impact. Complementary measures to maximize the positive spillover effects of the income multiplier generated by the cash transfer programme should be targeted not only at cash transfer beneficiary households, but also at ineligible households that provide many of the goods and services in the local economy. However, the potential productive impact of the cash transfer is sensitive to implementation, and delays and irregularities in payment can reduce its effectiveness in terms of helping households invest and manage risk.

Existing social protection programmes rarely takes into account climate risk in their design and implementation. Being poverty reduction instruments, social safety-net interventions tend to target mainly economic (wealth and income) criteria. Including environmental risks and vulnerabilities as targeting criteria could help improve the effectiveness of safety nets as risk-coping instruments. This could be done by developing maps of poverty and climate change vulnerability hotspots or by ensuring effective linkage between social protection management and information and early warning systems. Public works programmes, including productive safety nets, can be designed in ways that simultaneously contribute to increasing household incomes, engaging communities in climate-smart agriculture and generating 'green jobs' in areas such as waste management, reforestation and soil conservation.

## References

Alinovi, L., d'Errico, M., Mane, E. & Romano, D. 2010. Livelihoods strategies and household resilience to food insecurity: an empirical analysis to Kenya. European Report on Development. Available at erd.eui.eu/publications/erd-2010-publications/background-papers/livehoodsstrategies-and-household-resilience-to-food-insecurity.

American Institutes for Research (AIR). (2013). 24-Month Impact Report for the Child Grant Programme. Washington DC, USA.

Antonaci, L., Demeke, M. and Soumare, M.S. (2012). Integrating risk management tools and policies into CAADP: Options and Challenges. FAO-NEPAD policy brief.

Asfaw, S., Covarrubias, K., Davis, B., Dewbre, J., Djebbari, H., Romeo, A. and Winters, P. (2012). Analytical Framework for Evaluating the Productive Impact of Cash Transfer Programmes on Household Behaviour: Methodological Guidelines for the From Protection to Production Project. Paper prepared for the From Protection to Production project. Rome, UN Food and Agriculture Organization.

Asfaw, S., Davis, B., Dewbre, J., Handa, S. and Winters, P. (2014). Cash transfer programme, productive activities and labour supply: Evidence from randomized experiment in Kenya. Journal of Development Studies, 50(8):1172–1196.

Asfaw, S., Carraro, A. and Davis, B. (2016). The Role of Cash Transfers to Manage Climate Risk. The Case of Zambia. ESA working paper, forthcoming.

Asfaw, S., Pickmans, R., Alfani, F. and Davis, B. (2015a). Productive Impact of Ethiopia's Social Cash Transfer Pilot Programme, PtoP project report, FAO, Rome

Asfaw, S., Pickmans, R., & Davis, B. (2015b). Productive Impacts of Malawi's Social Cash Transfer Programme – Midline Report. PtoP project report, forthcoming, FAO, Rome.

Berhane, G., Hoddinott, J., Kumar, N. and Taffesse, A.S., (2011). The impact of Ethiopia's productive safety nets and household asset building programme: 2006–2010. IFPRI, Washington DC. USA.

Berhane, G., Devereux, S., Hoddinott, J., Nega Tegebu, F., Roelen, K., and Schwab, B. (2015). "Evaluation of the Social Cash Transfers Pilot Programme Tigray Region, Ethiopia." *Endline Report, IFPRI, Washington DC. USA.*

Boone, R, Covarrubias. K., Davis. B. and Winters, P. (2013) Cash Transfer Programs and Agricultural Production: The Case of Malawi. Agricultural Economics, 44:365–378.

Covarrubias, K., Davis B. and Winters, P., (2012). "From Protection to Production: Productive Impacts of the Malawi Social Cash Transfer Scheme." *Journal of Development Effectiveness*, 4:1, 50–77.

Daidone, S., Davis, B., Dewbre, J. & Covarrubias, K., (2014a). Lesotho Child Grants Programme: 24-month impact report on productive activities and labour allocation PtoP project report, FAO, Rome

Daidone, S., Davis, B., Dewbre, J., Gonzalez-Flores, M., Handa, S., Seidenfeld, D. & Tembo, G. (2014b). Zambia's Child Grant Programme: 24-month impact report on productive activities and labour allocation PtoP Project Report, FAO, Rome.

Davis, B., Gaarder, M., Handa S. and Yablonski, J., (2012). "Evaluating the impact of cash transfer programs in Sub Saharan Africa: an introduction to the special issue." *Journal of Development Effectiveness* 4(1): 1–8

Davis, B. and Handa, S. (2015). How much do programmes pay? Transfer size in selected national cash transfer programmes in Africa. The Transfer Project Research Brief 2015-09. Chapel Hill, NC: Carolina Population Center, UNC-Chapel Hill.

Del Ninno, C., Subbarao, K., & Milazzo, A. (2009). How to make public works work: A review of the experiences. World Bank, Human Development Network. Washington D.C.: World Bank.

Dercon, S., and Krishnan, P. (2003). Food Aid and In-formal Insurance. Center for the Study of African Economies Working Paper Series 2003-01.

Devereux, S., Marshall, J., MacAskill, J. and Pelham, L. (2005) 'Making Cash Count: Lessons from cash transfer schemes in east and southern Africa for supporting the most vulnerable children and households,' London and Brighton: Save the Children UK and the Institute of Development Studies.

Djebbari, H. and Hassine, N.B., (2011). Methodologies to analyse the local economy impact of SCTs. Report prepared for UNICEF-ESARO.

Eriksen, S.H., Brown, Kelly, P.M., The dynamics of vulnerability: locating coping strategies in Kenya and Tanzania - The Geographical Journal, Vol. 171, No. 4,2005, pp. 287–305

Fiszbein, A. and Schady, N. (2009) Conditional Cash Transfers for Attacking Present and Future Poverty, with Ferreira, F.H.G., Grosh, M., Kelleher, N., Olinto, P., and Skoufias, E., The World Bank Policy Research Report, 2009, Chapters 2, 5.

Gertler, P., Martinez, S. and Rubio-Codina, M. (2012) Investing Cash Transfers to Raise Long Term Living Standards. *American Economic Journal: Applied Economics*, 4(1), pp. 164–92.

Gilligan, D., Hoddinott J. and Taffesse, A. (2009) The Impact of Ethiopia's Productive Safety Net Program and Its Linkages. *Journal of Development Studies* 45(10), pp. 1684–1706.

Handa, S. and Park, M. (2012). Livelihood Empowerment against Poverty Program. Report, University of North Carolina at Chapel Hill.ae Report

Handa, S., Angeles, G., Abdoulayi, S., Mvula, P., Tsoka, M. et al. (2015) Malawi Social Cash Transfer Programme: Midline Impact Evaluation Report. Working Paper. Midline report, University of North Carolina, Chapel Hill, USA.

Handa, S., Park, M., Osei Darko, R., Osei-Akoto, I., Davis, B. & Daidone, S. (2014) Livelihood Empowerment Against Poverty (LEAP) Program - Impact evaluation report, Carolina Population Center, University of North Carolina.

Heckman, J.J., Ichimura, H. and Todd, P.E., (1997). "Matching as an econometric evaluation estimator: evidence from evaluating a job training program." Review of Economic Studies, 64, pp. 605–654.

Khandker, R.K., Koolwal, G.B., and Samad, H.A., (2010). *Handbook on impact evaluation: quantitative methods and practices*. The World Bank, Washington DC., USA.

Maluccio, J. (2010). The Impact of Conditional Cash Transfers in Nicaragua on Consumption, Productive Investments and Labor Allocation. *Journal of Development Studies*, 46 (1),%. 14–38.

Martinez, S. (2004) Pensions, Poverty and Household Investments in Bolivia. Doctoral dissertation, University of California, USA.

Pellerano, L., Hurrell, A., Kardan, A., Barca, V., Hove, F., Beazley, R., Modise, B., MacAuslan, I., Dodd, S. and Crawfurd, L., (2012). CGP impact evaluation: targeting and baseline evaluation report. OPM, January 10.

Pellerano, L., Moratti, M., Jakobsen, M., Bajgar, M.and Barca, V., (2014). Child Grants Programme Impact Evaluation: Follow-up Report. Oxford Policy Management (OPM), Oxford – April 2014

Rubin, D. B., (1980). Discussion of "Randomization Analysis of Experimental Data in the Fisher Randomization Test" *Journal of the American Statistical Association*, 75, 591–593.

Seidenfeld, D. and Handa, S. (2011). Zambia's Child Grant Program: Baseline Report, American Institutes for Research, Washington, DC., November.

Soares, F. V., Ribas, R. P., Hirata, G. I. (2010) The impact evaluation of a rural CCT programme on outcomes beyond health and education. *Journal of Development Effectiveness*, 2(1),%. 138–157.

Todd, J, Winters, P. and Hertz, T. (2010) Conditional Cash Transfers and Agricultural Production: Lessons from the Oportunidades Experience in Mexico. *Journal of Development Studies*, 46(1),%. 39–67.

Ward, P., Hurrell, A., Visram, A., Riemenschneider, N., Pellerano, L. and MacAuslan, I. (2010). Kenya CT-OVC programme operational and impact evaluation 2007–2009. Oxford Policy Management, Oxford, UK.

Wooldridge, J.M. (2002) Econometric analysis of cross-section and panel data. Cambridge, MA: The MIT Press.

**Open Access** This chapter is distributed under the terms of the Creative Commons Attribution-NonCommercial-ShareAlike 3.0 IGO license (https://creativecommons.org/licenses/by-nc-sa/3.0/igo/), which permits any noncommercial use, duplication, adaptation, distribution, and reproduction in any medium or format, as long as you give appropriate credit to the Food and Agriculture Organization of the United Nations (FAO), provide a link to the Creative Commons license and indicate if changes were made. If you remix, transform, or build upon this book or a part thereof, you must distribute your contributions under the same license as the original. Any dispute related to the use of the works of the FAO that cannot be settled amicably shall be submitted to arbitration pursuant to the UNCITRAL rules. The use of the FAO's name for any purpose other than for attribution, and the use of the FAO's logo, shall be subject to a separate written license agreement between the FAO and the user and is not authorized as part of this CC-IGO license. Note that the link provided above includes additional terms and conditions of the license.

The images or other third party material in this chapter are included in the chapter's Creative Commons license, unless indicated otherwise in a credit line to the material. If material is not included in the chapter's Creative Commons license and your intended use is not permitted by statutory regulation or exceeds the permitted use, you will need to obtain permission directly from the copyright holder.

# Input Subsidy Programs and Climate Smart Agriculture: Current Realities and Future Potential

Tom S. Jayne, Nicholas J. Sitko, Nicole M. Mason, and David Skole

**Abstract** The achievement of Climate Smart Agriculture (CSA) goals in Africa will require widespread farmer adoption of practices and technologies that promote resilience and system-wide collective action to promote *ex ante* climate risk management activities and *ex post* coping strategies. Leveraging public sector resources is critical to achieve goals at scale. This study examines the scope for input subsidy programs (ISPs) to contribute to achieving CSA objectives in Africa. Available evidence to date suggests that in most cases ISPs have had either no effect on or have reduced SSA smallholders' use of potentially CSA practices. However, recent innovations in ISPs may promote some climate smart objectives by contributing to system-level *ex-ante* risk management. In particular, restricted voucher systems for improved seed types that utilize private sector distribution supply chains may prove capable of promoting CSA goals. Generally, moving from systems that prescribe a fixed input packet to a flexible system with a range of input choices holds promise, but fixed systems still hold some benefits. Conditional ISPs would require improved monitoring and compliance as well as defining practices with clearly measurable productivity benefits vis-à-vis CSA goals. The potential of ISPs to achieve widespread CSA benefits must address these challenges and be evaluated against benefits of investments in irrigation, physical infrastructure, and public agricultural research and extension, which may generate higher comprehensive social benefits.

---

T.S. Jayne (✉) • N.J. Sitko • N.M. Mason
Department of Agricultural, Food and Resource Economics, Michigan State University, East Lansing, MI, USA
e-mail: jayne@msu.edu; sitkoni1@msu.edu; masonn@msu.edu

D. Skole
Department of Forestry, Michigan State University, East Lansing, MI, USA
e-mail: skole@msu.edu

## 1 Introduction

There is growing global recognition of the urgent need to identify and implement strategies that make food systems more resilient in the face of increasing climate variability. Nowhere is this more evident than in Sub-Saharan Africa.[1] Because the majority of Africans' livelihoods and agrifood systems rely on rainfed farming, Africa is one of the world's regions most vulnerable to climate change. The Intergovernmental Panel on Climate Change concluded that "climate change is expected to have widespread impacts on African society and Africans' interaction with the natural environment" (IPCC 2014, p. 812).

Climate smart agriculture (CSA) has emerged as an approach to enhance the resilience of farm systems to the effects of climate change. CSA is defined by three principle objectives (FAO 2013):

1. sustainably increasing agricultural productivity and incomes;
2. adapting and building resilience to climate change, and;
3. reducing and/or removing greenhouse gases emissions, where possible.

In Africa and other predominantly agrarian regions, there is particular interest in identifying strategies to encourage farmers to adopt practices and technologies that enable more resilient, sustainable and productive farms, while at the same time identifying system-wide collective action to promote a wide range of *ex ante* risk management activities and *ex post* coping strategies. Given the scope and scale of these requirements, leveraging public sector resources is critical.

Input subsidy programs (ISPs) provide a potentially useful means to encourage system-wide coordination and farmer behaviours that raise agricultural productivity and contribute to resilience objectives in Africa, while potentially mitigating the agricultural sector's contribution to GHG emissions. ISPs vary in their distribution modalities and targeting requirements, but generally share the common attributes of providing inorganic fertilizer, and in some countries, improved seeds, to farmers at below-market prices. Many African governments currently devote a large share of their agricultural sector and national budgets to ISPs. The region spends just over US$1.0 billion each year on ISPs (Jayne and Rashid 2013; Jayne et al. forthcoming). A major challenge to enabling ISPs to promote CSA outcomes stems from the major opportunity costs they entail in terms of foregone public spending on other core CSA investments such as irrigation, agricultural R&D, and extension services that could potentially promote CSA practices more effectively per dollar invested than ISPs. However, there is clearly scope for market-smart ISPs to improve smallholder farmers' access to climate smart technologies and overall resilience. This paper assesses the feasibility of leveraging public investments in ISPs to promote adoption of CSA practices and technologies by African farmers.

The paper is organized as follows. Section 2 begins by defining CSA in the context of African smallholder farming systems. Section 3 briefly examines the range

---

[1] Hereafter "Africa".

|  |  | Impact pathway | |
| --- | --- | --- | --- |
|  |  | Household-level | System-wide level |
| Type of strategy | *Ex ante* risk management: promoting resilience and reducing vulnerability | Section 4 | Section 5 |
| | *Ex post* coping strategies: relieving impacts of climate shocks after they have occurred | Section 6 | Section 7 |

**Fig. 1** Various dimensions of how input subsidy programs might contribute to climate smart agriculture

of ISP implementation modalities and approaches in Africa. In Sects. 4, 5, 6, and 7, we adopt the 2×2 matrix framework of Lipper and Zilberman (forthcoming) to consider how ISPs may promote resilience of farming systems in the face of climate shocks through *ex ante* risk management strategies, and how ISPs might be designed to mitigate the effects of climate shocks through *ex post* coping strategies. These impact pathways are evaluated across household/farm level and responses at the system-wide/government level (Fig. 1). Section 4 focuses on household-level *ex ante* risk management strategies. Section 5 focuses on system-wide *ex ante* risk management strategies. Section 6 examines the ability of ISPs to support household-level *ex post* responses to climate shocks. Section 7 examines system-wide *ex post* strategies. Section 8 summarizes our findings and discusses potential implications for ISP policies and programs.

## 2 Defining Climate Smart Agriculture

Although not clearly defined in the academic literature, the term "climate smart agriculture" (CSA) has gained prominence as an emergent agricultural development paradigm (Engel and Muller 2016). The UN Food and Agricultural Organization (FAO), the principle architect of CSA, defines it as an approach that "sustainably increases productivity and resilience (adaptation), reduces/removes GHGs (mitigation), and enhances achievement of national food security and development goals" (FAO 2010, p. ii; FAO 2013). CSA is therefore largely defined by its intended outcomes rather than by a set of specific practices or approaches (Kaczan et al. 2013).

CSA shares many objectives and guiding principles with green economy and sustainable development approaches, including a prioritization of food security and a desire to preserve natural resources. It is also closely linked to the concept of sustainable intensification (SI) (FAO 2013; Campbell et al. 2014). In many cases, SI constitutes a subset of practices that are potentially climate smart under certain current and future climatic conditions. As the FAO Sourcebook on CSA (2013) states, CSA extends these concepts through "a more forward looking dimension, more concern about future potential changes and the need to be prepared for them" (p. 30). Thus, CSA is not a set of new agricultural practices or a new agricultural system. Instead, it is understood as a new approach to guide necessary changes to agricultural systems in order to jointly address challenges of food security and climate change (Lipper et al. 2014; Branca et al. 2011; FAO 2013; Grainger-Jones 2011).

Proponents of CSA emphasize several hallmarks of its approach. First, CSA focuses on risks throughout the food system, with a particular emphasis placed on *ex ante* risks to smallholders resulting from the interaction of changing climate with existing livelihood vulnerabilities (McCarthy et al. 2011; Meinzen-Dick et al. 2013; Grainger-Jones 2011; World Bank 2011). Second, elevating the visibility of emergent risks that smallholders face offers opportunities to focus strategically on practices and technologies that offer multiple benefits in the areas of climate change adaptation, mitigation, and food security. Finally, by linking climate change adaptation and mitigation to smallholder production practices, CSA creates opportunities to link smallholders to previously unavailable sources of support, including climate finance (Meinzen-Dick et al. 2013; Grainger-Jones 2011).

There are a number of SI practices that are often linked to CSA objectives. These include: minimum soil disturbance (zero or minimum tillage); crop rotation and intercropping, particularly with legumes; mulching; crop residue retention; cover cropping; agro-forestry; water management, including irrigation and drainage; integrated soil nutrient management, including efficient use of mineral fertilizer in combination with organic sources; and use of high quality, well-adapted seed varieties. In many cases, these are not new practices, but adoption rates in Africa remain low or sub-optimal (Branca et al. 2011). For the purpose of this paper we will refer to these practices collectively as SI practices, recognizing that they are also closely linked to CSA objectives.

## 3 ISP Implementation Modalities and CSA in Africa

Following the implementation of structural adjustment programs, spending on ISPs in Africa declined substantially. Yet, in the wake of the global food price spike of 2007/2008 and based on the apparent success of Malawi's subsidy program, Africa has seen a resurgence of ISPs. According to Jayne and Rashid (2013), by 2011 ten African countries spent over $1.05 billion on ISPs, or roughly 28.6% of these countries' total public agricultural expenditures.

The majority of new ISPs in Africa focus on subsidizing improved seed and inorganic fertilizers for staple cereal production by smallholder farmers. A few also provide subsidies for small grains and legumes. Variations in ISP design are most notable in terms of: (i) the extent to which the private sector is utilized to distribute inputs, (ii) the range of inputs available to farmers, and (iii) the socio-economic characteristics of the target beneficiaries.

The distribution system and flexibility of input choices for farmers have important implications for their climate smartness. Most ISPs utilize closed voucher systems, where farmers redeem coupons for a prescribed input packet from government-run or designated outlets, or direct delivery systems, where government or contractors deliver prescribed input packets. These types of systems tend to limit farmers' choice of inputs, are rarely attentive to agro-ecological and livelihood variations across space, crowd out private sector participation, and are frequently characterized by elite capture of inputs (Ricker-Gilbert et al. 2011; Mason and Ricker-Gilbert 2013; Pan and Christiaensen 2012; Mason et al. 2013; Lunduka et al. 2013). Such systems, like those in Zambia and Malawi, tend to undermine the development of private sector market channels, encourage mono-cropping and incentivize the production of crops in regions where they are poorly suited (Mason et al. 2013; Lunduka et al. 2013; Levine 2015). These outcomes are clearly contrary to the goals of CSA.

Recently, however, countries have begun to take tentative steps toward implementing more flexible, open voucher systems for ISPs in order to address some of these shortcomings. In Zambia for example, an electronic voucher system was piloted on a limited scale in 2015/2016, where farmers redeem vouchers with registered private sector dealers for a wide range of inputs. These systems can lower ISPs fiscal cost to government, encourage private investments in input supply systems and extension, and allow farmers to choose appropriate inputs (Sitko et al. 2012). These outcomes are decidedly more climate smart than the dominant model.

However, trade-offs exist between the relative flexibility of an ISP and the promotion of particular technologies or farm practices that may be climate smart. For example, open voucher systems may be less effective for promoting the adoption of seed varieties that are drought, heat, or flood tolerant, as there is no way to ensure that farmers will choose these seed types with a completely open voucher. More closed voucher systems may be more appropriate for encouraging the use of particular technologies. Similarly, closed voucher programs may help private seed firms to forecast demand for seed types, such as legume seeds, which is notoriously difficult to predict from year to year. By providing clarity on the effective demand for particular inputs, closed vouchers systems may prove useful to help overcome input supply constraints that hinder the adoption of certain potential SI and CSA practices, such as legume intercropping and rotations.

## 4 Can ISPs Promote Household-Level *Ex Ante* Risk Management?

Having reviewed in general terms how ISPs are implemented and potential linkages to SI and CSA practices, we now examine specific strategies that may foster more climate resilient and productive smallholder farm systems. The sorts of SI and CSA management practices we examine include tillage method, intercropping and rotations, the use of manures and residue retention, and agro-forestry, *inter alia*. More broadly, we explore the potential relationship between ISPs and practices that can potentially improve soil characteristics and stabilize yields in the context of climate variability.

## 4.1 Review of Evidence to Date

The evidence base remains thin but the weight of the available evidence suggests that ***ISPs have had either no effect on or have reduced African smallholders' use of CSA practices***. Empirical evidence across many case studies shows mixed results for many CSA practices considered. In addition, studies show the difficulties posed by delivery mechanisms that provide inputs too late for effective and efficient use by farmers. Finally, the absence of robust agricultural extension services in many African countries makes the diffusion and implementation of CSA practices even more challenging.

More specifically, evidence suggests that ISPs did not affect Ghanaian farmers' investment in soil and water conservation, broadly defined (Vondolia et al. 2012), nor did they affect Malawian or Zambian smallholders' use of manure (Holden and Lunduka 2010, 2012; Levine 2015). And while Malawi's ISP had no statistically significant effect on intercropping (Holden and Lunduka 2010), Zambia's ISP has reduced intercropping in general, but not intercropping involving legumes (Levine 2015). Moreover, Zambia's ISP has negatively affected crop rotation and fallowing (ibid; Mason et al. 2013). The program has contributed to continuous cultivation of mono-cropped maize over time and within seasons, which leave smallholders more vulnerable to climate shocks – the antithesis of CSA. ISPs may increase maize yields in the short run except during extreme weather conditions (see Holden and Lunduka 2010; Mason et al. 2013; Chibwana et al. 2014; Mason et al. 2015; among many others). However, if results similar to Zambia are obtained elsewhere, these yield gains could be coming at the cost of lower soil organic matter and higher soil acidity, both of which will result in lower yields and fertilizer use efficiency in the medium to long run (Marenya and Barrett 2009; Burke 2012).

Empirical evidence on the effects of ISPs on crop diversification is mixed. For example, while Chibwana et al. (2012) and Mason et al. (2013) find that ISPs in Malawi and Zambia, respectively, incentivize households to devote a greater share of their cropped area to maize, other studies from Malawi suggest the opposite (Holden and Lunduka 2010; Karamba 2013) or that ISPs have no statistically significant effect on crop diversification (Karamba 2013). Most likely, the effects of ISPs depend on the range of inputs provided. ISPs that focus less on a specific crop and support a broader range of alternative crops, in particular legumes that add biomass and moisture retention to soil, may generate better outcomes with respect to crop diversification and soil fertility, responsiveness of crops to inorganic fertilizer and other benefits (Snapp et al. 2010).

While ISPs may contribute to sustainable productivity growth by maximizing fertilizer to crop output efficiency, their track record has been disappointing. Jayne et al. (forthcoming) conclude that most African governments to date have focused more on increasing African farmers' use of fertilizer than on providing support for its efficient use.

Another feature of many ISPs that is decidedly *not* climate smart is perennial late delivery of subsidized fertilizer and seeds to beneficiary farmers (Xu et al. 2009; Lunduka et al. 2013; Mason et al. 2013; Namonje-Kapembwa et al. 2015). Late delivery is particularly common when ISP inputs are disseminated through dedicated ISP distribution systems that largely sideline existing input distribution networks. This is how fertilizer for Malawi's ISP and both fertilizer and seed for Zambia's ISP were distributed until 2014/15 and 2015/16, respectively, when each country started piloting agrodealer-based voucher redemption systems (Logistics Unit 2015; ZMAL 2015a; b). Late delivery of ISP inputs results in late planting and/or late fertilizer application, reducing yields and leaving beneficiary households more vulnerable to climate shocks (Xu et al. 2009; Namonje-Kapembwa et al. 2015; Arslan et al. 2015).

Most public agricultural extension systems are seriously under-provisioned to perform their multiple mandates of providing new management advice to farmers, learning from their efforts and difficulties of implementation and liaising with adaptive research systems to generate and disseminate new productive and sustainable practices, including SI practices. Some African public extensions are virtually defunct. Therefore, it should not be surprising that despite heavy spending on ISPs, their impacts on crop yields have been smaller than anticipated (ibid). In Zambia and Malawi, for example, a one-kilogram increase in subsidized fertilizer raises smallholder households' maize output by an average of only 1.88 kg and 1.65 kg, respectively (Mason et al. 2013; Ricker-Gilbert et al. 2011). This low crop yield response to fertilizer is a major reason for the relatively low benefit-cost ratios of the ISPs in Malawi (1.08) and Zambia (0.92) (Jayne et al. 2017).

In response to some of these limitations, many ISPs are currently transforming to more flexible, private-sector, inclusive systems. This creates possibilities for ISPs to be restructured in ways that incentivize farmers to adopt particular SI practices and also bring about system-wide changes that promote resilience. The remainder of this section examines this potential of ISPs, however the discussion is largely conjectural given the limited evidence that ISPs as implemented to date have achieved such benefits.

## 4.2 Looking Forward: Can ISPs Contribute to Climate Smart Farm Management Practices?

A handful of *ex ante* analyses have explored how ISPs might compare to other programs to promote farmers' use of practices that may be climate smart. For example, Marenya et al. (2012) use 30-year crop simulation models for maize, rice, and sorghum calibrated for several districts in Kenya, Malawi, and Uganda to compare changes in the net present value (NPV) of adopting various soil fertility management (SFM) strategies under two sets of policy regimes: a 50% fertilizer subsidy and carbon credits priced at $4, $8, or $12 per metric ton of carbon sequestered in the soil. The SFM strategies considered include various combinations of inorganic (N) fertilizer, animal manure, and crop residue retention – practices that may be 'climate smart' in some contexts. Their results suggest that carbon credits, especially when priced at $8 or $12/mt, produce larger NPV increases than the 50% fertilizer subsidy. While carbon markets are virtually non-existent in Africa, this analysis suggests monetary incentives play an important role in stimulating adoption of climate smart practices. This leaves room for ISPs to deliver monetary incentives to such ends. Yet, this in turn requires that extension systems are capable of delivering appropriate management information and that adoption is effectively monitored, which seems very challenging.

In later work, Marenya et al. (2014) use choice experiments to measure Malawian smallholder farmers' preferences for various hypothetical policy incentives to adopt soil conservation practices, namely minimum tillage with legume intercropping: cash payments, two different types of index-based crop insurance contracts, and fertilizer subsidies.[2] Results suggest that most farmers preferred fertilizer subsidies to cash payments or crop insurance. In addition, farmers generally preferred cash payments to crop insurance, even when the expected payout from the crop insurance was higher than the cash payment. We must be careful, however, in generalizing these results, as they are specific to the choice sets used in the experiments. For example, the expressed preference of fertilizer subsidy over cash payments is likely driven by the fact that cash payment options (ranging from MK 800 to MK 2000) were lower compared to fertilizer subsidy (MK 2000) because of the expected yield gains with fertilizer. Even still, both cases suggest that under the right conditions some combination of conditional subsidy or conditional cash payment can incentivize adoption of farm management practices. Whether or not this leads to a permanent behavioral change, or whether public entities are capable of monitoring adherence to the conditions, remains an open question.

Finally, there is the question about whether raising crop productivity through inorganic fertilizer use might reduce the rate at which forests are converted into farmland and therefore reduce the agricultural sector's contribution to GHG emissions. Recent evidence has begun to question the logic that agricultural productivity growth can arrest rapid farm area expansion and thus conserve the world's forests and grasslands (Hertel 2011; Robertson and Swinton 2005; Byerlee et al. 2014). Instead, a generally positive area response to improved profit incentives is likely to create new pressures for further area expansion and conversion of forest and grasslands to farmland. Policy incentives could play a potential role here. In theory, ISPs could be structured in such a way as to oblige beneficiaries to reduce or maintain the amount of area under cultivation. However, it is not clear whether such

---

[2] Farmers also had the option to decline the soil conservation incentives in favor of continuing 'traditional' practices, which in the context of the choice experiments were defined as not using chemical fertilizer or the soil conservation practices.

rules would impose unreasonable demands on food insecure rural households or whether they could be adequately monitored or enforced.

In summary, while ISPs can be theoretically structured in ways that promote farm-level management changes, the oversight, enforcement, and extension costs needed to make this work are high, and may increase the already substantial opportunity costs of large public expenditures on ISPs.

## 4.3  How Confident Are We That We Know Which Farming Practices Contribute to CSA and SI?

As the development community understandably pushes hard to make progress in helping African farmers, there are major risks of overgeneralization about what kinds of farming practices really contribute to *ex ante* risk management and *ex post* coping strategies. Africa is heterogeneous with respect to its climate conditions, soil types, market access conditions, and factor price ratios. Some parts of Africa are still land abundant; labor and capital may be binding constraints in such areas. Other agricultural areas of Africa are densely populated, facing land pressures and rising land prices. In some of these areas, labor is relatively abundant and hence labor-intensive CSA practices may hold some potential to be scaled-up and incentivized through ISPs. However, in areas with good market access conditions and proximity to urban areas, economic transformation processes are bidding up labor wages and making it difficult for farmers to adopt labor-intensive CSA practices unless they also provide high returns to labor. The heterogeneous conditions of farming systems in Africa warrant great caution against overgeneralization in promoting technologies through ISPs or on their own based on blanket recommendations across wide domains.

As an example, minimizing soil disturbance through no or minimum tillage (MT)[3] strategies are frequently promoted in Africa as a means to mitigate soil erosion, increase soil water retention capacity, and to slow the rate of soil organic carbon (SOC) decomposition, and thus achieve yield growth and stability (Branca et al. 2011; Chivenge et al. 2007). However, yield and soil quality effects of MT practices vary substantially depending on soil type and association of MT with other land management practices, namely crop residue retention and incorporation. Several studies have shown that MT practices lead to an accumulation of SOC in the surface layers of soil (0–10 cm), rather than in the *root* zone (Sisti et al. 2004; Chivenge et al. 2007; Carter and Rennie 1982; Hernanz et al. 2002; Doran 1980). Carter and Rennie (1982) find that microbial biomass and potential mineralizable carbon and nitrogen are high in surface soils where MT is practiced. Conversely, these soil properties are higher in lower soil depths when conventional tillage (CT) is applied. The magnitude and location of the SOC pool are important for yield growth and

---

[3] In this section we present evidence on both zero and minimum tillage methods, which we will refer to broadly as minimum tillage (MT).

stabilization. As Lal (2006) shows, every 1 mt/ha increase in the SOC pool in the root zone is associated with a 30–300 kg/ha increase in maize yields and a 10–50 kg/ha increase in rice yields. Improving SOC pool in the root zone can simultaneously enhance soil's water retention capacity (Mbagwu 1991; Fernández-Ugalde et al. 2009), increase its cation exchange capacity, and thus nutrient retention (Carter et al. 1992), and improve soil aggregation and susceptibility to erosion (Lal 2006; Paul et al. 2013). Thus, further development of MT technologies may be needed to achieve its potential benefits.

Another potential limitation of MT is that without associated investments in crop residue retention and/or crop rotation, fields tilled using MT frequently experience no yield improvement (Hernanz et al. 2002) or in some cases a dramatic drop in yield relative to CT (Rusinamhodzi et al. 2011; Raimbault and Vyn 1991; Paul et al. 2013). When MT practices are applied in conjunction with crop residue retention, legume rotation, and/or nitrogen fertilizer application, the yield effects of MT tend to be higher than those achieved through CT, but again this is highly dependent on prevailing agro-ecological conditions (Raimbault and Vyn 1991; Govaerts et al. 2005; Dalal et al. 1991; Triplett et al. 1968).

As discussed in Section 3, ISPs in the region are not designed to cope with the high level of regional and farm level heterogeneity in input needs and management requirements. Significant region-specific modifications in the composition of ISP inputsm coupled with region-specific farm management promotion strategies will be required for ISPs to contribute meaningfully to CSA goals, which in turn implies significant modification in the logistical design, implementation and cost of ISPs.

A more obvious way in which ISPs can influence overall productivity is through the injection of greater levels of nitrogen (N) into African soils, where nitrogen is often the limiting nutrient factor (Snapp et al. 2010). Rusinamhodzi et al. (2011) in their summary of evidence on conservation agriculture shows that in 73% of the field studies, high levels of nitrogen fertilizer were required to achieve improved yields under these practices. However, recent advances in soil science and agronomy research show that massive nitrogen (N) injections may not be economically feasible for farmers or be social welfare raising without farmer adoption of complementary soil management practices that allow N to be efficiently utilized by plants (Snyder et al. 2009). Thus, the challenge for large-scale programs, such as ISPs, is promoting carbon management practices together with nitrogen to achieve high nitrogen efficiency (Tittonell and Giller 2013). Paul et al. (2013) demonstrate that without sufficient biomass production (often stimulated by inorganic fertilizer application) SI practices of MT and residue retention do not have an effect on yield stability or SOC. Thus, an ongoing challenge is maintaining a large enough N pool in soils containing little organic carbon, which increases N leaching and gaseous loss pathways, adversely affecting CSA goals (Drinkwater and Snapp 2007). Unfortunately, large-scale efforts to promote SI practices that build up soil organic carbon are largely absent from government programs, are largely untested over the wide range of soil types and agro-ecologies found in the region, and are sometimes discounted by some as not being viable from the standpoint of low-resource farmers.

These several examples underscore the lack of consensus within the crop science community about what viable CSA and SI packages appropriate for heterogeneous smallholder agricultural systems should look like. In addition, there is a great deal of uncertainty over how climate will change in the region over the coming decades (Powlson et al. 2016). For these reasons, we conclude that African governments and the development community need an improved empirical evidence base that establishes the practices that actually promote CSA and SI objectives under the wide range of diverse and uncertain farming conditions found in the region. A precondition for making progress on this front is much greater public expenditure on agricultural R&D and adaptive research across the various economic/biophysical micro-climates. While necessary, increased public funding to agricultural R&D is not sufficient. But without a better evidence base on how practices perform under various conditions, the risk is that ISPs may be misguided in choosing which practices to promote.

## 5 Can ISPs Promote System-Wide Ex Ante Risk Management?

This section examines the potential of ISPs to encourage system-wide changes in agricultural value chains that promote resilience to risks associated with climate variability. Due to their scale, ISPs may have capacity to influence the broader systems within which farmers operate and thereby influence farmer behavior both directly as well as indirectly through system-wide changes. We identify three potential areas where these system-wide effects are most evident.

### 5.1 Potential Opportunities

First, as mentioned earlier, by expanding and stabilizing the demand for specified input types and quantities, ISPs can potentially help to overcome some of the persistent risks to commercial legume seed multiplication in the region. Ensuring adequate supplies of these seeds on the market is critical to achieving crop diversification, organic nitrogen fixation, and rotations. However, this potential benefit is mitigated by the trend, among donors and governments, to move toward more open voucher systems. Thus, in many ways there are important trade-offs to consider when promoting particular ISP distribution modalities. While open vouchers are desirable from a farmer choice perspective, restricted-choice vouchers for particular inputs, such as legume seeds, may be necessary to support system-wide improvements in legume seed supply chains. Restricted-choice vouchers may be justified in some instances where there are major beneficial externalities associated with promoting certain inputs and where the social benefits of doing so may greatly outweigh the short-term financial benefits from the perspective of individual farmers. The two approaches may be combined; for example, farmers could be provided an open voucher in addition to a restricted-choice voucher for legume seed. Similar system-wide benefits may accrue by using ISPs to create farmer demand for specific drought-tolerant seed varieties or soil amendments such as lime or inoculants, which are currently not widely used by farmers.

A second way in which ISPs may promote system-wide CSA resilience is through promoting "market-smart" private investments, which could increase private investments in input supply chains and extension services. By encouraging private sector input supply chain development, market-friendly ISPs can foster improved input access conditions for farmers, thus over time making them less dependent on public input supply systems. Private input systems are potentially less prone than public systems to delivery challenges associated with logistical and financial constraints (Jayne and Rashid 2013). There is clear potential for ISPs to promote system-wide investments that are both climate-smart and market-smart and synergistic in their promotion of community resilience to climate variability.

Finally, the move toward digital platforms for delivering ISPs, such as electronic vouchers ('e-vouchers'), create opportunities to use ISPs as delivery mechanisms for other sorts of products, such as weather indexed insurance. This requires that ISP farmer registries collect a wide range of information on beneficiaries, including geographic location and bank information. With this sort of information, ISPs can defray the screening costs of identifying farmers and managing insurance pay-outs when necessary.

## 5.2  Potential Challenges

Unfortunately, some aspects of ISPs may work against climate change mitigation even as they promote resilience objectives. ISPs increase the quantities of fertilizer manufactured and used in the agricultural production process (holding all other factors constant) and therefore ISP proposals that include increased fertilizer use must account for the additional GHG emissions. Inorganic fertilizer use contributes to GHG emissions both through the soil chemical and biological processes and through the production of synthetic fertilizer. According to a recent estimate, 56% of global non-carbon dioxide GHG emissions occur from agricultural production, and roughly 12% of agricultural GHG emissions occur from fertilizer use (IPCC 2014). The additional contribution to GHG emissions caused by the manufacturing of synthetic fertilizer is also significant (see Appendix 1). Thus, the net impact of ISPs on GHG emissions will depend on the effectiveness with which ISPs can be used to promote adoption of CSA practices that raise soil organic carbon, sequester carbon and depress the rate of forest conversion to farmland and offset the adverse effects of increased fertilizer use on GHG emissions. The empirical evidence on these issues is weak and more detailed research is needed. Appendix 1 provides some empirical estimates of the increased GHG emissions caused from additional use of synthetic fertilizers.

Moreover, there is the issue of opportunity costs. Nationwide ISPs tend to be expensive, and they can bid away scarce public funds that could otherwise be used to buffer communities from the effects of climate variability (e.g., irrigation, agricultural research and extension systems, weather insurance, etc.) or to support *ex post* coping responses (e.g., disaster relief programs). In Africa, where irrigation only accounts for 4% of arable land (You et al. 2012) and where there is huge unmet potential for irrigation expansion, ISPs would seemingly compete against public investment in water control and other *ex ante* risk management strategies. Future research is again needed to determine whether smart ISPs may be structured in ways that leverage private sector investments in CSA inputs and services and produce benefits that outweight those generated from other proven types of public investments in agriculture.

## 6 Can ISPs Promote Household-Level *Ex Post* Coping Mechanisms?

There may be limited potential for ISPs' ability to improve the *ex post* capacity of farm households to cope with shocks. Expenditures on ISPs occur before growing season weather outcomes are known. The greatest productivity boost from ISPs occurs in favorable weather years, and vulnerability to climate shocks is quite low during these periods. Vulnerability is of course greatest in extreme weather years. Unfortunately, fertilizer application typically contributes little to crop production growth during such years, and does nothing to stabilize crop yields in the face of extreme weather conditions. This inverse temporal correlation between years of great vulnerability to climate shocks and the payoffs from fertilizer application suggest that ISPs may have limited potential as *ex post* coping mechanisms at least for the period of time until the next harvest, generally 6–9 months later.

However, ISPs are frequently scaled-up in the year following a severe weather event as part of drought-recovery strategies. In such cases, ISPs act as tools to support smallholder households to acquire improved inputs and reengage in production following a severe contraction in farm income, and to potentially re-stock depleted resources that were expended during the crisis to smooth consumption. ISPs can also theoretically be used to help farmers replant crops that failed to survive due to late or false onset rains. Yet, in both cases this would require considerable budgetary flexibility and rapid implementation capacity on the part of governments. In addition, because of the annual crop production cycle characterizing most of the region, it may take time at least 6–9 months after a harvest failure before ISPs could contribute benefits to recipients in the form expanded crop output in the next season.

## 7 Can ISPs Promote System-Wide *Ex Post* Coping Potential?

In their current form, ISPs tend to be costly and therefore compete directly for scarce public sector resources with other CSA risk coping and response strategies, such as disaster risk management plans, rapid repair of damaged infrastructure, emergency feeding, etc. However, ISPs that increase access to weather insurance may help farmers avoid some forms of asset and resource depletion common after a weather shock. In addition, well-targeted ISPs may enable farmers to recover more quickly following extreme weather events. In these ways, ISPs do offer some potential avenues for timely response mechanisms following adverse weather shocks.

## 8 Summary and Implications for ISPs

In almost all countries where they have been implemented, ISPs have clearly promoted national grain production, at least in the years they were implemented. ISPs have a more checkered track record in terms of their impact on farm-level productivity, commercial input market development, and farm management behaviors that promote SI. Longstanding efforts to encourage policy makers to use "market smart" criteria have been disappointing, which has impeded the benefit-cost ratios of ISPs (Jayne and Rashid 2013; Jayne et al. 2017). It may be unrealistic at least in the near future to expect that political economy issues that have impeded efforts to make ISPs more effective can be easily overcome. But given that ISPs are likely to continue, and often account for a large share of public expenditures to agriculture, it may be worth the effort to encourage ISP reforms in ways that contribute to SI practices and CSA objectives.

This study has considered potential avenues of ISP impact on CSA objectives in terms of a time dimension – *ex ante* risk management strategies vs. *ex post* coping strategies – and at different levels of intervention – household-level behavioral change vs. system-wide changes. Using this conceptual lens we find that ISPs hold some potential to influence farmer behavior with respect to *ex ante* risk management strategies, such as the adoption of sustainable land management techniques, private investment in small-scale irrigation, use of drought-, heat-, and saline-resistant crop varieties, use of hardier livestock breeds, and diversifying land and labor activities. Achieving these ends through ISPs is highly dependent on the existence of coordinated investments in both public extension services and research and development, along with monitoring systems. However, the cost of each component will require much greater public budgets devoted to agriculture to achieve the complementary approach needed.

Where ISPs may provide even greater opportunities to promote CSA objectives is through supporting *ex ante* risk management strategies at the system-wide level. Well-designed ISPs may improve seed system performance for legumes and other improved varieties, as well as serving to link farmers to insurance systems. However, trade-offs exist between market development objectives of new ISPs and some of the system-wide constraints to CSA, such as legume seed supply constraints. For ISPs to improve legume seed supplies or access to particular climate improved seed varieties they may need to promote these through restricted-choice vouchers, in addition to or instead of the flexible vouchers being widely promoted in the region. Managing these trade-offs is important for achieving greater system wide benefits through ISPs.

ISP's ability to improve household-level *ex post* coping mechanisms will likely be through support of post-disaster asset accumulation and reengagement with productive agriculture. Yet these outcomes, again, depend on effective public sector performance, particularly in terms of targeting the most affected households and regions.

In summary, ISPs may serve several catalytic functions at a system-level, which can support CSA objectives. However, ISPs can achieve little without the sorts of coordinated public and private investments in areas such as site specific adaptive research and extension, which are necessary to turn potential CSA practices into profitable and adoptable farm management strategies. Indeed, it is currently not possible to point to many, if any, new practices appropriate for smallholder African systems that are tried, tested, and can be confidently promoted as practices that promote CSA, are profitable, and feasible for farmers to adopt. Promoting certain technologies prematurely will lead to high levels of dis-adoption, disillusionment, and difficulties in getting farmers to participate in future programs.

Based on this analysis we propose the following as potential focal areas for improving the climate "smartness" of ISPs in Africa:

- *Support greater concentration of ISPs on legume and climate improved cereal crops:* Many ISPs currently focus primarily on staple cereal crops and inorganic fertilizers. For ISPs to have a more system-wide effect on cropping systems and management practices, seed system constraints for other crops must be addressed. ISPs can serve a catalytic role in this respect.
- *Develop detailed farm registries for ISP beneficiaries:* Detailed registries, that include geo-spatial information, are necessary to delivery support services such as weather insurance to farmers and to track adherence to targeting criteria.
- *Explore the potential for using ISPs to overcome CSA farm management adoption constraints*, bearing in mind that:

   There is limited consensus on what practices are most effective for heterogeneous smallholder systems, and;
   Extension advice and monitoring capacity remains very thin in most of Africa.

- *Support systems to improve timing of input distribution through ISPs:* ISPs chronically deliver fertilizer late (Xu et al. 2009; Namonje et al. 2015; Snapp et al. 2014). Late delivery reduces yields and crop response to fertilizer. This unfavorably affects the ratio of crop output to GHG emissions.
- *Improve targeting capacity of ISPs:* ISPs must more effectively target farmers who can use fertilizer profitably but are not already using it (or using it well below levels considered to be profit-maximizing). This will reduce crowding out of commercial demand and contribute to increased fertilizer use. In addition, effective targeting following a disaster can help support ISPs to support ex post household recovery efforts.
- *Use extension systems and information and communications technologies (ICTs) to show farmers how the use of fertilizer from ISPs and/or commercially obtained*

*fertilizer can become more profitable when complementary SI/CSA practices are adopted.*
- *Promote more secure land tenure/property rights (e.g., through registration or land certification):* land tenure security is important for encouraging the adoption of SI/CSA practices that improve productivity, sustainable land management, and increased use of commercially purchased fertilizer (Lawry et al. 2014; Sitko et al. 2014). Efforts to promote secure land tenure rights are a complement, not necessarily a substitute, for ISPs in promoting CSA, but the cost-effectiveness of both may be different and justify different levels of budget support.

## *8.1 Unresolved Issues for Future Research*

Key knowledge gaps include understanding why farmers are not adopting CSA practices or are subsequently dis-adopting them (which could then point to potential interventions to overcome these constraints); determining which practices are profitable for whom and under what conditions; understanding the interactions between CSA practices and ISP inputs (e.g., do selected CSA practices increase fertilizer use efficiency?); identifying cost-effective, enforceable, and scalable ways to implement a potential CSA precondition requirement for ISPs; and comparing the cost-effectiveness of such a requirement to that of other approaches to promote CSA. Given the very mixed results of ISPs, the rampant elite capture and diversion of inputs intended for the programs, and the high price tag and opportunity cost of ISPs in general and in relation to other programs and investments to develop and stimulate uptake of CSA technologies (see Jayne and Rashid 2013; Lunduka et al. 2013; Mason et al. 2013; among many others), linking CSA promotion to ISPs may be a risky proposition.

## *8.2 Concluding Remarks*

There are three overarching challenges to be addressed for ISPs to effectively contribute to CSA objectives. First is the limited understanding of workable approaches for internalizing the externalities associated with GHG-emitting land management decisions of millions of resource-poor farmers in developing countries. This is a problem for social scientists to resolve by developing ways for carbon markets to be linked to smallholders in Africa and that can provide farmers monetary incentives for the adoption of particular GHG mitigating practices, may be a viable strategy for achieving widespread farm management change, but much remains to be worked out before viable programs could be implemented in most of sub-Saharan Africa.

The second challenge is the currently limited on-shelf technologies and management know-how to improve smallholder yield stability and growth in the face of increasing climate variability. Most on-shelf technologies and practices being promoted as being "climate smart" appear to help at the margin, but cannot be relied upon to meaningfully stabilize harvests in the face of major droughts or floods or to arrest the degree of distress migration often associated with it. More effective water and soil fertility management techniques appropriate for the situation of low-resource farmers are needed, and this will requires significantly increased investment in localized, adaptive research for the wide range of smallholder farming systems in sub-Saharan Africa. This is a challenge both for the scientific research community and for policy makers to make the necessary long-term funding commitments to adaptive agricultural research and development programs.

The third challenge is the near absence of effective bi-directional learning and extension systems to help farmers profitably adopt and adapt proven farm management practices. This again presents challenges for policy makers to make the necessary long-term funding commitments and to social scientists to design extension systems that effectively link scientists and farmers disaggregated by particular agro-ecologies and degrees of resource constraints.

Addressing these three challenges is a tall order. For this reason, we believe that much greater progress is needed in each of these three areas before it could be practical or effective to try to use ISPs as a vehicle to make agriculture more climate-smart. This conclusion is not meant to stifle progress where progress can be made, but is rather to point out the scope of the challenges before us. It will take time for the proposals made here to generate meaningful impacts. This is why there is no time to waste in getting started.

## Appendix 1: Estimating the Contribution of Increased Fertilizer Use to Greenhouse Gas Emissions

African countries contribute to climate change through emissions of greenhouse gases from agriculture, forestry and land use (AFoLU). As much as one third of all emissions globally are from AFoLU, but in many African countries these emission sources constitute the major components of their national GHG inventories, rather than the industrial or energy sectors. For instance, in Malawi 80% of national GHG emissions are from forestry and agriculture, although the absolute contribution to global greenhouse gas emissions is tiny. As a result of the Paris Agreements of the United Nations Framework Convention on Climate Change (UNFCCC) African countries are developing means and measures to mitigate these emissions through actions in the AFoLU sectors, including reducing emissions from deforestation and forest degradation, conservation of carbon stocks in forests and agricultural soils, improved management of agricultural waste and other interventions. In spite of actions to reduce emissions, agriculture and forestry will surely be impacted by climate change. As such, many African countries are taking a broad view and are also implementing adaptation strategies.

National climate action strategies are being developed by all African Countries through the process of the Nationally Determined Contributions, or NDC, which is the main reporting instrument that is the focal point for each country's international commitments. Climate Smart Agriculture (CSA) is being viewed as one model for adaptation. This model focuses on developing interventions in traditional practices that can increase resilience of agricultural systems to adverse effects of climate change and which can be promulgated at the national level and applied locally at farm scale. One compelling intervention under the CSA model is the national subsidy programs for inorganic fertilizers. Increasing the availability and application of chemical fertilizers is seen as a means to increase crop productivity and provide enhanced fertility to nutrient-poor soils, and buffer adverse effects of drought and other climate impacts.

However, at the same time that these measures provide apparent benefits from an adaptation point of view, the use of inorganic fertilizers also increases GHG emissions in agricultural soils, particularly for non-carbon GHGs such as nitrous oxide ($N_2O$). Using estimation methods defined by the Intergovernmental Panel on Climate Change (IPCC 2006), the FAO (FAO 2014) has published estimates of national emissions from agricultural inputs for many African countries. GHG emissions from the application of synthetic fertilizers has increased 25% between 2000 and 2014, from 16,000 GgCO2e to 20,000 GgCO2e, representing about 3% of the total emissions from all agricultural practices, including land clearing. However there is considerable variation across Africa, with a trend toward higher proportional emissions from fertilizers in poorer countries. For instance, in Nigeria where other inputs and energy contributed more to agriculture than in most countries, only about 1.2% of the total emissions from agriculture are attributed to fertilizer applications on soils in 2012, while in Malawi as much as 18% of total agricultural emissions are attributed to fertilizer applications in 2012. In Zambia the proportion is 4%, while in Kenya it is 2% for 2012.

For the most part these are relatively low emissions compared to other components of the agriculture production system; however subsidy programs are expected to raise fertilizer use, particularly for poorer countries such as Malawi. These emissions of GHG, especially non-carbon GHG such as $N_2O$, represent the negative impacts of measures involving increased use of fertilizer to improve resilience of agricultural soils and plant productivity. Thus, interventions that may have positive influence on adaptation may have outcomes that negatively offset gains in mitigation efforts. For instance, annual emission rates of GHG from fertilizer use in agriculture in Malawi is approximately equivalent to protecting 500 hectares of Miombo woodland from deforestation. The exact magnitude of the offset depends on a complex array of factors that are not being studied, including the type of fertilizer used, fertilizer application rates and timing, influence of episodic events that may be changing with climate changes such as severe rain events, soil conditions and land management.

Most studies, and the IPCC (2006), estimate N emission factors for $N_2O$ to be between 1% and 3% of the nitrogen nutrient in fertilizers. Thus, we can estimate the approximate GHG emissions associated with the application of fertilizer under subsidy programs. We assume an application of 300,000 metric tons of fertilizer, of which half is in the form of urea with 50% N and half in the form of inorganic NPK with 30% N. This would equate to roughly 45,000 metric tons of N from NPK fertilizer and 75,000 metric tons of N from urea. Using IPCC emission factors for $N_2O$ emissions this would result in 1200–3600 metric tons of $N_2O$ per ton of N, which when converted to units of nitrous oxide (multiplied by 44/28) and then to carbon dioxide equivalents using a greenhouse warming potential (GWP) of 300 would be 565,714–1,697,143 metric tons of $CO_2$ equivalent ($CO_2e$) greenhouse gas emission. Using IPCC emission factors for urea, we estimate an additional 30,000 metric tons of CO2e. Thus, the total emissions from the application of 300,000 tons of fertilizer of the type we used to make our estimate would be 595,714–1,727,143 metric tons $CO_2e$ per year.

The contributions of inorganic fertilizer to adaption and agricultural resilience would come at a cost to efforts to mitigate emissions from deforestation and degradation; the additional emissions from fertilizer applications would be a significant new emission source and would counter efforts to mitigate emissions in the AFoLU sector.

These estimates are for field applications of inorganic fertilizers. The demand for fertilizer would stimulate production of fertilizers and this production system also produces GHGs, mostly from the large use of energy which are typically from fossil fuels. Although most carbon GHG accounting methods do not attribute production emissions to the end-use emissions, and keep these accounts separate, for the sake of illustration we estimate the additional contribution of producing and transporting 300,000 t of inorganic fertilizer. Several studies suggest an emission factor for fertilizer production to be 2.5–5.67 metric tons of $CO_2e$ per metric ton of fertilizer produced (Kool et al. 2012). Thus, a basic estimate of the magnitude of the emissions associated with the 300,000 additional tons of fertilizer production would be 750,000–1,701,000 metric tons of $CO_2e$.

Combining both agricultural field emissions with emissions associated with production, we estimate that 300,000 tons of additional fertilizer manufacture and use would result in GHG emissions of between 1,345,714 and 3,428,143 metric tons of $CO_2$ equivalent. Approximately 55% of these emissions are attributed to the industrial production of fertilizers (which we believe are conservative estimates). These estimates would represent an increase in fertilizer emission of approximately 10%, and would represent an emission that counter offsets approximately 120,000 to 300,000 hectares of reforestation in mitigation projects.

# References

Arslan, A., N. McCarthy, L. Lipper, S. Asfaw, A. Cattaneo, and M. Kokwe. 2015. "Climate Smart Agriculture? Assessing the Adaptation Implications in Zambia." Journal of Agricultural Economics doi: 10.1111/1477-9552.12107

Branca, G., McCarthy, N., Lipper, L., & Jolejole, M. C. (2011). Climate-smart agriculture: a synthesis of empirical evidence of food security and mitigation benefits from improved cropland management. Mitigation of climate change in agriculture series, 3, 1–42.

Burke, W.J. 2012. "Determinants of Maize Yield Response to Fertilizer Application in Zambia: Implications for Strategies to Promote Smallholder Productivity." PhD dissertation, Michigan State University.

Byerlee, D., J. Stephenson, and N. Villoria. 2014. Does intensification slow crop land expansion or encourage deforestation? Global Food Security (3), 92–98.

Campbell, B.M., P. Thornton, R. Zougmoré, P. van Asten, and L. Lipper. 2014. "Sustainable Intensification: What is its Role in Climate Smart Agriculture?" Current Opinion in Environmental Sustainability 8: 39–43.

Carter, M. R., & Rennie, D. A. (1982). Changes in soil quality under zero tillage farming systems: distribution of microbial biomass and mineralizable C and N potentials. Canadian Journal of Soil Science, 62(4), 587–597.

Carter, D. C., D. Harris, J. B. Youngquist, and N. Persaud. "Soil properties, crop water use and cereal yields in Botswana after additions of mulch and manure." Field Crops Research 30, no. 1–2 (1992): 97–109.

Chibwana, C., M. Fisher, and G. Shively. 2012. "Cropland Allocation Effects of Agricultural Input Subsidies in Malawi." World Development 40(1):124–133.

Chibwana, C., G. Shively, M. Fisher, and C. Jumbe. 2014. "Measuring the Impacts of Malawi's Farm Input Subsidy Programme." African Journal of Agriculture and Resource Economics 9(2):132–147.

Chivenge, P. P., Murwira, H. K., Giller, K. E., Mapfumo, P., & Six, J. (2007). Long-term impact of reduced tillage and residue management on soil carbon stabilization: Implications for conservation agriculture on contrasting soils.Soil and Tillage Research, 94(2), 328–337.

Dalal, R. C., Henderson, P. A., & Glasby, J. M. (1991). Organic matter and microbial biomass in a vertisol after 20 yr of zero-tillage. Soil Biology and Biochemistry 23(5), 435–441.

Doran, J. W. (1980). Soil microbial and biochemical changes associated with reduced tillage. Soil Science Society of America Journal, 44(4), 765–771.

Drinkwater, L. E., & Snapp, S. S. (2007). Nutrients in agroecosystems: rethinking the management paradigm. Advances in Agronomy, 92, 163–186.

Engel, S., & Muller, A. (2016). Payments for Environmental Services to Promote Climate-Smart Agriculture? Potential and Challenges. Potential and Challenges (January 2 2016).

Food and Agricultural Organization. 2010. Climate-Smart Agriculture: Policies, Practices and Financing for Food Security, Adaptation and Mitigation, Rome., http://www.fao.org/docrep/013/i1881e/i1881e00.pdf

FAO. 2013. Climate-Smart Agriculture Sourcebook. Rome, Italy: FAO.

FAO 2014. Agriculture, Forestry and Other Land Use Emissions by Sources and Removals by Sinks. Working Paper ESS/14-02. Food and Agriculture Organization, United Nations, Rome. Wood, S. and A. Cowie 2004. A Review of Greenhouse Gas Emission Factors for Fertilizer Production, Research and Development Division, State Forests of New South Wales. Cooperative Research Centre for Greenhouse Accounting. For IEA Bioenergy Task 38

Grainger-Jones, E. (2011). Climate-smart smallholder agriculture: What's different. IFAD Occasional paper, 3.

Fernández-Ugalde, O., Virto, I., Bescansa, P., Imaz, M. J., Enrique, A., & Karlen, D. L. (2009). No-tillage improvement of soil physical quality in calcareous, degradation-prone, semiarid soils. Soil and Tillage Research,106(1) 29–35.

Govaerts, B., Sayre, K. D., & Deckers, J. (2005). Stable high yields with zero tillage and permanent bed planting?. Field crops research, 94(1), 33–42.

Hernanz, J. L., López, R., Navarrete, L., & Sanchez-Giron, V. (2002). Long-term effects of tillage systems and rotations on soil structural stability and organic carbon stratification in semiarid central Spain. Soil and Tillage Research, 66(2), 129–141.

Hertel, T. W. (2011). The global supply and demand for agricultural land in 2050: A perfect storm in the making?. American Journal of Agricultural Economics, 93(2), 259–275.

Holden, S., and R. Lunduka. 2010. "Too Poor to be Efficient? Impacts of the Targeted Fertilizer Subsidy Programme in Malawi on Farm Plot Level Input Use, Crop Choice and

Land Productivity." Noragric Report No. 55, Department of International Environment and Development Studies, Norwegian University of Life Sciences, Ås, Norway.

Holden, S., and R. Lunduka. 2012. "Do Fertilizer Subsidies Crowd Out Organic Manures? The Case of Malawi." Agricultural Economics 43(3):303–314.

IPCC (Intergovernmental Panel on Climate Change) 2006. 2006 IPCC Guidelines for National Greenhouse Gas Inventories, Volume 4: Agriculture, Forestry and Other Land Use. UNFCCC, Geneva.

IPCC (Intergovernmental Panel on Climate Change) 2014. Mitigation of Climate Change. Contribution of Working Group III to the Fifth Assessment Report of the Intergovernmental Panel on Climate Change [Edenhofer, O., R. Pichs-Madruga, Y. Sokona, E. Farahani, S. Kadner, K. Seyboth, A. Adler, I. Baum, S. Brunner, P. Eickemeier, B. Kriemann, J. Savolainen, S. Schlömer, C. von Stechow, T. Zwickel and J.C. Minx (eds.)]. Cambridge University Press, Cambridge, United Kingdom and New York, NY, USA.

Jayne, T.S., and S. Rashid. 2013. "Input Subsidy Programs in Sub-Saharan Africa: A Synthesis of Recent Evidence." Agricultural Economics 44(6):547–562.

Jayne, T., N. Mason, W. Burke, and J. Ariga. Forthcoming. Input Subsidy Programs in Africa: A Review of Recent Experience. Policy Brief, Food Security Group, Michigan State University, East Lansing.

Karamba, R.W. 2013. "Input Subsidies and Their Effect on Cropland Allocation, Agricultural Productivity, and Child Nutrition: Evidence from Malawi." PhD dissertation, American University.

Kaczan, D., Arslan, A., & Lipper, L. (2013). Climate-smart agriculture. A review of current practice of agroforestry and conservation agriculture in Malawi and Zambia ESA working paper, (13-07).

Kool, A., M. Marinussen, H. Blonk. 2012. LCI Data for the Calculation Tool Feedprint for Greenhouse Gas Emissions of feed production and Utilization: GHG Emissions of N,P and K Fertilizer Production, Blonk Consultants, Gouda, Netherlands.

Lal, R. (2006). Enhancing crop yields in the developing countries through restoration of the soil organic carbon pool in agricultural lands. Land Degradation & Development, 17(2), 197–209.

Lawry, S., Samii, C., Hall, R., Leopold, A., Hornby, D., & Mtero, F. (2014). The impact of land property rights interventions on investment and agricultural productivity in developing countries: a systematic review.Campbell Systematic Reviews, 10(1).

Levine, N.K. 2015. "Do Input Subsidies Crowd In or Crowd Out Other Soil Fertility Management Practices? Evidence from Zambia." MS Plan B Paper, Michigan State University. Available at http://web2.msue.msu.edu/afreTheses/fulltext/N.%20Kendra%20Levine-%20Final%20Plan%20B%20Paper.pdf.

Lipper, L. and D. Zilberman. Forthcoming. Climate Smart Agriculture: Introduction, in D. Zilberman, L. Lipper, S. Asfaw, D. Cattaneo (eds) FAO book on climate change.

Lipper, L., Thornton, P., Campbell, B. M., Baedeker, T., Braimoh, A., Bwalya, M., and Hottle, R. (2014). Climate-smart agriculture for food security. Nature Climate Change, 4(12), 1068–1072.

Logistics Unit. 2015. Final Report on the Implementation of the Agricultural Inputs Subsidy Programme 2014–15. Lilongwe, Malawi: Logistics Unit.

Lunduka, R., J. Ricker-Gilbert, and M. Fisher. 2013. "What are the Farm-Level Impacts of Malawi's Farm Input Subsidy Program? A Critical Review." Agricultural Economics 44(6):563–579.

Marenya, P., and C. Barrett. 2009. "State-Conditional Fertilizer Yield Response on Western Kenyan Farms." Am. J. Agr. Econ. 91(4):991–1006.

Marenya, P., E. Nkonya, W. Xiong, J. Deustua, and E. Kato. 2012. "Which policy would work better for improved soil fertility management in sub-Saharan Africa, feritilzer subsidies or carbon credits?" Agricultural Systems 110: 162–172.

Marenya, P., V. H. Smith, and E. Nkonya. 2014. "Relative preferences for soil conservation incentives among smallholder farmers: evidence from Malawi." American Journal of Agricultural Economics 96 (3): 690–710.

Mason, N. M., Wineman, A., Kirimi, L., & Mather, D. (2015). The effects of Kenya's 'smarter' input subsidy program on crop production, incomes, and poverty. Tegemeo Institute Policy Brief 11.

Mason, N. M., & Ricker-Gilbert, J. (2013). Disrupting demand for commercial seed: Input subsidies in Malawi and Zambia. World Development, 45, 75–91.

Mason, N.M., T.S. Jayne, and R. Mofya-Mukuka. 2013. "Zambia's Input Subsidy Programs." Agricultural Economics 44(6):613–628.

Mbagwu, J. S. (1991). Mulching an ultisol in southern Nigeria: effects on physical properties and maize and cowpea yields. Journal of the Science of Food and Agriculture, 57(4), 517–526.

McCarthy, N., Lipper, L., & Branca, G. (2011). Climate-smart agriculture: smallholder adoption and implications for climate change adaptation and mitigation. Mitigation of Climate Change in Agriculture Working Paper, 3, 1–37.

Meinzen-Dick, R., Bernier, Q., & Haglund, E. (2013). The Six 'ins' of Climate-Smart Agriculture: Inclusive Institutions for Information, Innovation, Investment and Insurance (No. 114). International Food Policy Research Institute (IFPRI).

Namonje-Kapembwa, T., T.S. Jayne, and R. Black. 2015. "Does Late Delivery of Subsidized Fertilizer Affect Smallholder Maize Productivity and Production?" Selected paper presented at the Agricultural and Applied Economics Association and Western Agricultural Economics Association Annual Meeting, San Francisco, CA 26–28 July.

Pan, L., and L. Christiaensen. 2012. Who is Vouching for the Input Voucher? Decentralized Targeting and Elite Capture in Tanzania. World Development 40(8):1619–1633.

Paul, B. K., Vanlauwe, B., Ayuke, F., Gassner, A., Hoogmoed, M., Hurisso, T. T., & Pulleman, M. M. (2013). Medium-term impact of tillage and residue management on soil aggregate stability, soil carbon and crop productivity. Agriculture, ecosystems & environment, 164, 14–22.

Powlson, D. S., Stirling, C. M., Thierfelder, C., White, R. P., & Jat, M. L. (2016). Does conservation agriculture deliver climate change mitigation through soil carbon sequestration in tropical agro-ecosystems? Agriculture, Ecosystems & Environment 220, 164–174.

Raimbault, B. A., & Vyn, T. J. (1991). Crop rotation and tillage effects on corn growth and soil structural stability. Agronomy Journal, 83(6), 979–985.

Ricker-Gilbert, J., Jayne, T. S., & Chirwa, E. (2011). Subsidies and crowding out: A double-hurdle model of fertilizer demand in Malawi. American Journal of Agricultural Economics, aaq122.

Robertson, G. P., & Swinton, S. M. (2005). Reconciling agricultural productivity and environmental integrity: a grand challenge for agriculture. Frontiers in Ecology and the Environment, 3(1), 38–46.

Rusinamhodzi, L., Corbeels, M., van Wijk, M. T., Rufino, M. C., Nyamangara, J., & Giller, K. E. (2011). A meta-analysis of long-term effects of conservation agriculture on maize grain yield under rainfed conditions. agronomy for sustainable development, 31(4), 657–673.

Sisti, C. P., dos Santos, H. P., Kohhann, R., Alves, B. J., Urquiaga, S., & Boddey, R. M. (2004). Change in carbon and nitrogen stocks in soil under 13 years of conventional or zero tillage in southern Brazil. Soil and tillage research, 76(1), 39–58.

Sitko, N. J., Chamberlin, J., & Hichaambwa, M. (2014). Does smallholder land titling facilitate agricultural growth?: An analysis of the determinants and effects of smallholder land titling in Zambia. World Development, 64, 791–802.

Sitko, N. J., Bwalya, R., Kamwanga, J., & Wamulume, M. (2012). Assessing the feasibility of implementing the Farmer Input Support Programme (FISP) through an electronic voucher system in Zambia (No. 123210). Michigan State University, Department of Agricultural, Food, and Resource Economics.

Snapp, S., M. Blackie, R. Gilbert, R. Bezner-Kerr, G. Kanyama-Phiri. 2010. Biodiversity can support a greener revolution in Africa. Proceedings of National Academy of Science, 107 20840–20845. Doi: 10.1073/pnas.1007199107.

Snapp, S., Jayne, T.S., Mhango, W., Ricker-Gilbert, J., & Benson, T. (2014). "Maize yield response to nitrogen in Malawi's smallholder production systems." Working Paper No. 9, Malawi Strategy Support Program. International Food Policy Research Institute.

Snyder, C.S., T.W. Bruulsema, T.L. Jensen, P.E. Fixen. 2009. Review of greenhouse gas emissions from crop production systems and fertilizer management effects, Agriculture, Ecosystems & Environment 133, (3–4): 247–266

Tittonell, P., & Giller, K. E. (2013). When yield gaps are poverty traps: the paradigm of ecological intensification in African smallholder agriculture. Field Crops Research, 143, 76–90.

Triplett, G. B., Van Doren, D. M., & Schmidt, B. L. (1968). Effect of corn (Zea mays L.) stover mulch on no-tillage corn yield and water infiltration.Agronomy Journal, 60(2) 236–239.

Vondolia, G.K., H. Eggert, and J. Stage. 2012. "Nudging Boserup? The Impact of Fertilizer Subsidies on Investment in Soil and Water Conservation." Discussion Paper No. 12–08, Environment for Development and Resources for the Future, Washington, DC.

World Bank. 2011. Policy brief: Opportunities and challenges for climate-smart agriculture in Africa. Washington, D.C.: World Bank.

Xu, Z., Z. Guan, T.S. Jayne, and R. Black. 2009. "Factors Influencing the Profitability of Fertilizer Use on Maize in Zambia." Agricultural Economics 40(4):437–446.

You, L., Ringler, C., Wood-Sichra, U., Robertson, R., Wood, S., Zhou, T., Nelson, G. 2012. What Is the Irrigation Potential for Africa? A Combined Biophysical and Socioeconomic Approach. Food Policy 36, 770–782.

ZMAL (Zambia Ministry of Agriculture and Livestock). 2015a. Farmer Input Support Programme Implementation Manual 2015/16 Agricultural Season. Lusaka, Zambia: ZMAL.

ZMAL. 2015b. Farmer Input Support Programme Electronic Voucher Implementation Manual 2015/16 Agricultural Season. Lusaka, Zambia: ZMAL.

**Open Access** This chapter is distributed under the terms of the Creative Commons Attribution-NonCommercial-ShareAlike 3.0 IGO license (https://creativecommons.org/licenses/by-nc-sa/3.0/igo/), which permits any noncommercial use, duplication, adaptation, distribution, and reproduction in any medium or format, as long as you give appropriate credit to the Food and Agriculture Organization of the United Nations (FAO), provide a link to the Creative Commons license and indicate if changes were made. If you remix, transform, or build upon this book or a part thereof, you must distribute your contributions under the same license as the original. Any dispute related to the use of the works of the FAO that cannot be settled amicably shall be submitted to arbitration pursuant to the UNCITRAL rules. The use of the FAO's name for any purpose other than for attribution, and the use of the FAO's logo, shall be subject to a separate written license agreement between the FAO and the user and is not authorized as part of this CC-IGO license. Note that the link provided above includes additional terms and conditions of the license.

The images or other third party material in this chapter are included in the chapter's Creative Commons license, unless indicated otherwise in a credit line to the material. If material is not included in the chapter's Creative Commons license and your intended use is not permitted by statutory regulation or exceeds the permitted use, you will need to obtain permission directly from the copyright holder.

# Part IV
# Case Studies: System Level Response to Improving Adaptation and Adaptive Capacity

# Robust Decision Making for a Climate-Resilient Development of the Agricultural Sector in Nigeria

Valentina Mereu, Monia Santini, Raffaello Cervigni, Benedicte Augeard, Francesco Bosello, E. Scoccimarro, Donatella Spano, and Riccardo Valentini

**Abstract** Adaptation options that work reasonably well across an entire range of potential outcomes are shown to be preferable in a context of deep uncertainty. This is because robust practices that are expected to perform satisfactorily across the full range of possible future conditions, are preferable to those that are the best ones, but just in one specific scenario. Thus, using a Robust Decision Making Approach in Nigerian agriculture may increase resilience to climate change. To illustrate, the expansion of irrigation might be considered as a complementary strategy to conservation techniques and a shift in sowing/planting dates to enhance resilience of agriculture. However, given large capital expenditures, irrigation must consider climate trends and variability. Using historical climate records is insufficient to size capacity and can result in "regrets" when the investment is undersized/oversized, if the climate turns out to be drier/wetter than expected. Rather utilizing multiple climate outcomes to make decisions will decrease "regrets." This chapter summarizes the main results from a study titled "Toward climate-resilient development in Nigeria" funded by the Word Bank (See Cervigni et al. 2013).

---

V. Mereu (✉) • D. Spano
Euro-Mediterranean Center on Climate Change, Lecce, Italy

Department of Science for Nature and Environmental Resources, University of Sassari, via Enrico de Nicola, 9, Sassari 07100, DC, Italy
e-mail: valentina.mereu@cmcc.it

M. Santini • F. Bosello • D. Spano • R. Valentini • E. Scoccimarro
Euro-Mediterranean Center on Climate Change, Lecce, Italy
e-mail: enrico.scoccimarro@cmcc.it

R. Cervigni
Environment and Natural Resources Global Practice, Africa Region, The World Bank, Washington, DC, USA

B. Augeard
The French National Agency for Water and Aquatic Environments, Vincennes, France

# 1 Introduction

The agricultural sector plays a strategic role for the Nigerian economy, as it contributes to more than 40% of the GDP and accounts for about 65–70% of employment (Yakubu and Akanegbu 2015). Cereals such as maize, sorghum, millet and rice, and tubers as cassava and yam, account for 70% of the production of the agricultural sector in 2013 (FAOSTAT; FAO 2015). Cassava and Yam, with a production of about 53 and 40 million tons respectively (FAO 2015), are the leader crops for the Nigerian economy. Cassava, especially, plays an essential role for food security due to its efficiency in producing carbohydrates, its high flexibility with respect to the timing of planting and harvesting, and its tolerance to drought and to poor soils. Maize and Sorghum are currently the most important cereal food crops in Nigeria either in terms of production or in terms of harvested area (FAO 2015). Other important cereals are Millet, mainly cultivated in the north of the country, and Rice, which is cultivated in all of the Agro-Ecological Zones (AEZs) of Nigeria. Rice production has emerged as the fastest growing sub-sector and the most required commodity in the Nigerian food basket.

Rainfed lowland rice is the predominant production system, accounting for nearly 50% of total rice growing area in Nigeria. Overall, 30% of the production is rainfed upland rice, while just 16% is high yielding irrigated rice. Other production systems make up the remaining 4% (from USAID MARKETS 2009a). Cultivated lands in Nigeria occupied about 44.7% of land area in 2011, with 37.3% and 7.4% consisting of arable lands and permanent crops, respectively (FAO 2015). About two-thirds of the cropped areas are located in the north, with the rest about equally distributed between the center (Middle Belt) and the south. With irrigation accounting for less than 1% of cultivated area (FAO 2015), the rainfall regime highly affects the national crop production. Cultivation calendars and cropping patterns are different in the north and south, largely reflecting differences in precipitation regimes across the country.

Farming systems are mainly (80–90%) smallholder-based, with limited access to pesticides, fertilizers, hybrid seeds, irrigation, and other productive resources. Its farming production systems are inefficient, causing a regular shortfall in national domestic production and a need to import food that accounts for about 10% of overall national imports. Moreover, recent climate patterns (e.g. NIMET's 100-year database or Lebel and Ali 2009) adversely affected national crop production, causing serious implications for food security, public health and the economy of the country. Existing studies on Nigeria (Adejuwon 2005; Odekunle 2004) show that, in general, frequent crop failures and decreases in agricultural productivity are observed as a consequence of climate variability. Nigeria is listed by FAO (AQUASTAT-FAO 2005) among the nations that are technically unable to meet their food needs from rainfed production at a low level of inputs.

In this context, high priority is being posed by Government policies to increase agricultural productivity in order to reduce poverty, increase food security and diversify economy away from oil (NPC 2004; NSSP 2010). One of the options to sustain this goal is represented by irrigation development. Given the limited size of effectively irrigated areas, the contribution of irrigated agriculture to total crop production is quite small at 0.9% and 2.3% of the total national agricultural production of grains – rice and wheat – and vegetables, respectively. According to the International Commission Irrigation and Drainage (ICID[1]) three main types of irrigation schemes are developed in Nigeria: (i) public irrigation schemes, which are under government control; (ii) the farmer-owned and operated irrigation schemes that receive assistance from government in the form of subsidies and training; and (iii) residual flood plains, where no government aid is supplied, that are based on traditional irrigation practices.

Nigeria is considered one of the African countries with the largest potential for irrigation expansion (World Bank 2010). However, as precipitation highly differs across the AEZs, the potential to improve yields by irrigation is highly variable, and a strategic balance between rainfed and irrigated production has to be achieved to ensure effective management of water resources.

The Nigerian government is pursuing several policies that encourage a viable structure of public and private irrigation with a balanced set of small-, medium- and large-scale irrigated production. In addition to rehabilitation and expansion of existing public schemes, the Master Plan for Irrigation and Dam Development proposes the construction of new dams and irrigation schemes to improve the overall infrastructure of the irrigated sub-sector. About 156 km$^3$ of water is exploitable per year from superficial and groundwater resources; currently, only 5% (8 km$^3$) is effectively withdrawn (FAO 2016). According to projections made in the National Water Resources Master Plan (NWRMP) produced by the Japan International Cooperation Agency (JICA 1995), incremental water storage of 2 km$^3$ per year will be required between 2012 and 2020 to meet the increasing water demand from the three competing sectors: agriculture (69%), energy (10%), and domestic use (21%).

Since the vulnerability of the agricultural sector to current climate shocks and resource availability is likely to be exacerbated under future environmental change, achieving food, energy and water security in Nigeria will become more and more challenging. Previous works have addressed the analysis of climate change impacts in Sub-Saharan Africa, highlighting high differences in yield projections across different AEZs (Lobell et al. 2008; Seo et al. 2008a; Thornton et al. 2009; Roudier et al. 2011; Webber et al. 2014), due to differences in climate data, emission scenarios and the modelling approach in simulating crop yield (Roudier et al. 2011). The majority of studies are based on a statistical modelling approach (Parry et al 2004; Lobell et al. 2008; Seo et al. 2008a; Schlenker and Lobell 2010), which however assume stability of the relation between crop and weather. Accordingly, this methodology has a rather limited explanatory power, and is unsuitable for extrapolation outside the range of observed conditions within which it was developed (Challinor et al. 2009; Müller et al. 2011; Rosenzweig et al. 2013).

A minority of studies were conversely based on dynamical simulation of climate change impacts by applying more complex mechanistic process-based crop models.

---

[1] http://www.icid.org/cp_nigeria.html.

These are able to consider both linear and nonlinear crop response to weather variation (Semenov and Porter 1995). According to the available studies, climate change impacts are highly differentiated across specific crops and cropping systems (Mereu et al. 2015; Webber et al. 2014; Roudier et al. 2011), which are characterized by different capacities to adapt to modified climatic conditions and by different strategies implemented to cope with these threats. According to the IPCC AR5 (2014) adaptation strategies for African agriculture can be technological (e.g., stress-tolerant crop varieties, irrigation, enhanced observation/monitoring systems) and agronomic adaptation responses (e.g., agroforestry, conservation agriculture). Seo et al. (2008b, c) point out the need for a careful selection of these measures given the specificity of AEZs and the uncertainty related to climate scenarios.

Conservation agriculture and other land, water and crop management practices are "soft" candidates to reduce climate change impacts on crops and improve the sustainability of agricultural systems. Expansion of irrigation is considered as a complementary strategy. Even so, as irrigation entails large costs and upfront investment, it is crucial to size it adequately by selecting the investment strategies that minimize the risk of misjudgments across multiple climate outcomes and reduce regrets.

This chapter proposes a Robust Decision Making Approach (RDMA) to increase the resilience of Nigerian agricultural sector to climate change and variability. It starts from the analysis of the short- to mid-term risks (2020–2050) posed by climate change to the agricultural sector and it is applied to help in reducing the risks of maladaptation (Daron 2015). In other words, it helps decision makers in identifying and choosing the most suitable adaptation options in a context of deep uncertainty, by favoring those options that will work reasonably well across that entire range of potential outcomes. An important point to consider is that the strategies which are robust, i.e. those are expected to perform satisfactorily across the full range of possible future conditions, are preferable to those that are the best ones, but just in one specific scenario, remain highly sensitive to changes, and may perform very poorly under an alternative, but equally probable, scenario (Lempert et al. 2004, 2006; Wilby and Dessai 2010).

Thus, applying RDMA is one way to cope with uncertainty in future outlooks. Other approaches are adaptive management (i.e. selecting a strategy that can be modified to achieve better performance as one learns more about the issues and how the future is unfolding) and scenario planning (comparing how well alternative policy decisions perform under different plausible future conditions). We chose RMDA building based on the comparative work of Lempert and Collins (2007) concluding that it is preferable to adaptive management when, as in the present case, the decision time scales are such that immediate incremental adaptation would not possible when new information becomes available since investments have already been implemented and infrastructure realized.

## 2 Methodological Approach

Before applying RDMA to support adaptation decisions in irrigation, climate change impacts were quantified using different well-established process based models. Specifically, the analysis includes the following steps and can be represented by the flowchart in Fig. 1:

1. the establishment of a reference development scenario (baseline) that, assuming no-climate change, is the basis for assessing climate change impacts;
2. the definition of a range of possible future climate outcomes to explore the uncertainties related to climate models;
3. the evaluation of climate impacts at the Agro-Ecological Zones (AEZ), watersheds and country-policy level, according to the specific impact investigated;
4. the testing of adaptation strategies and the application of a RDMA to support adaptation decisions in irrigation development.

More details on methods and tools applied are reported in the published report "Toward climate-resilient development in Nigeria" (Cervigni et al. 2013).

### 2.1 Climate Projections and Their Uncertainty

The high resolution Regional Climate Model (RCM) COSMO-CLM at about 8 km$^2$ of resolution (Rockel et al. 2008) was applied to simulate climate trends from 1971 to 2065 under A1B emission scenario and using the boundary conditions of the General Circulation Model (GCM) CMCC-MED (about 80 km of horizontal resolution, Scoccimarro et al. 2011). According to the validation with observed climate along the historical period, the RCM was bias-corrected for the whole simulated period (Cervigni et al. 2013 – Chap. 4 and Appendix B).

To take into account the uncertainty on future climate outcomes nine GCMs simulations taking part of the CMIP3 experiment plus those from the CMCC-MED GCM, were used to "perturb" the RCM results along the period 2006–2065 and maintain high resolution. The GCMs chosen for the simulations were thus: HadCM3, CGCM_2.3.2, CNRM_CM3, CSIRO_Mk3.5, CCSM3, MIROC3.2, GFDL_cm2.1, ECHAM5, FGOALS, and CMCC-MED. The approach to perturb RCM outputs using the variability of global simulations (Buishand and Lenderink 2004) was applied to temperature and precipitation fields (Cervigni et al. 2013 – Chap. 4 and Appendix B).

Such climate simulation ensemble was used to drive the impact assessment described herein comparing impact model outcomes in the short and medium term periods (2006–2035 and 2036–2065, respectively), with the historical baseline (1976–2005). According to the multiple components of the analysis, and their dependence on climate variables suffering from different uncertainty degree in the future (e.g. higher for precipitations than for temperature), the full range of models

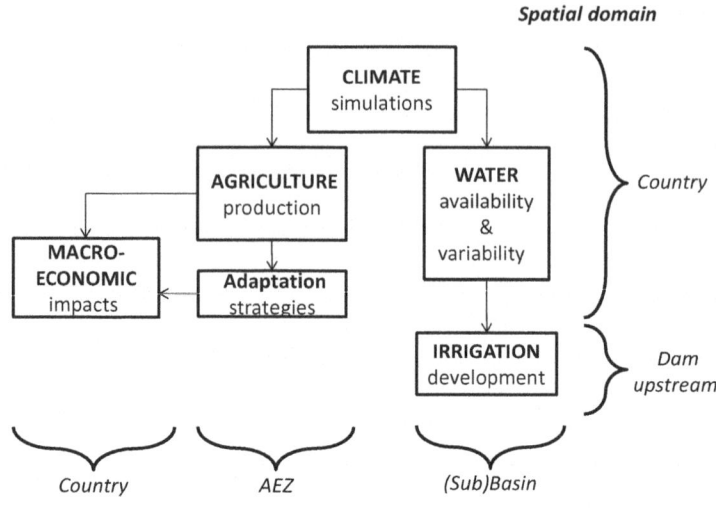

**Fig. 1** Flowchart of the conducted assessment, and spatial levels (coverage, aggregation) of analysis

(the RCM and the 10 perturbations) or their member suggesting the most extreme impacts were used to well represent the uncertainty range of possible climate outcomes.

## 2.2 Crop Modeling: Impacts on Yields

The software DSSAT-CSM, Decision Support System for Agrotechnology Transfer – Cropping System Model (http://dssat.net/; Jones et al. 2003; Hoogenboom et al. 2012) was applied to analyze the impacts of climate change and possible adaptation strategies for the most important staple food crops in Nigeria: sorghum, millet, maize, rice, cassava and yam. The DSSAT-CSM simulates growth, development and yield of a crop growing on a uniform area of land under prescribed or simulated management as well as the modifications in soil, water, carbon, and nitrogen exchanges that take place under the cropping system over time.

Multiple combinations of soil and climate conditions were considered for the different AEZs of Nigeria (Fig. 2), in which specific crop management options, as growing periods and/or crop varieties cultivated (long or medium growing season) were set according to literature (USAID MARKETS 2009b and 2010; ICS-Nigeria reports). The methodology addresses individual crops, considering crop varieties and management systems representative for each AEZ.

For impact analysis on crop yields, simulation results using a sub-ensemble consisting of RCM simulation and its five most extremes and significant

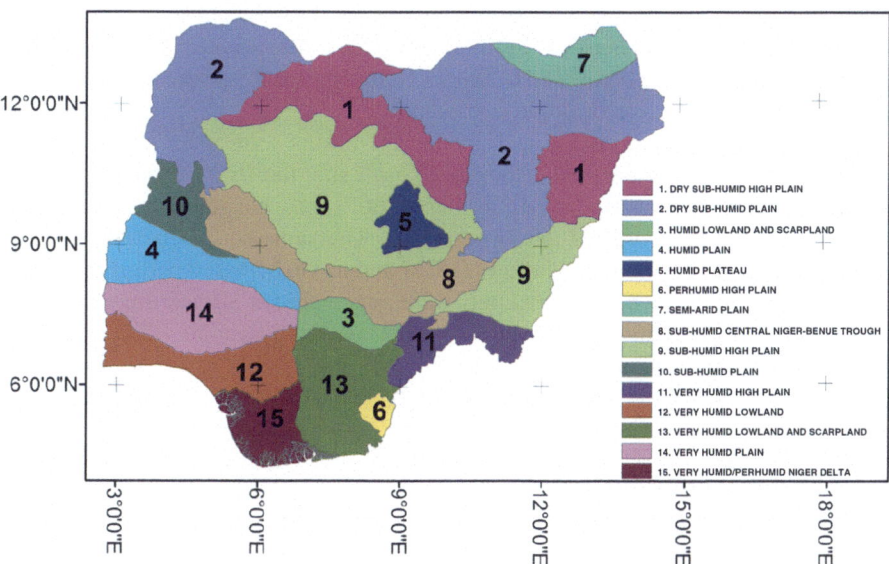

**Fig. 2** Map of Agro-Ecological Zones of Nigeria, considered in this study (From Cervigni et al. 2013)

GCM-perturbations, in terms of climate change projections, were used. Simulations were performed assuming both constant $CO_2$ concentration (380 ppm) and transient $CO_2$ concentration (consistent with the A1B emission scenario). Yield was simulated in both rainfed and irrigated conditions.

The climate impact assessment was made by comparing the yields obtained with the weather data for the reference period 1976–2005 (baseline) and those obtainable under future modified climate conditions in the short- and medium-term periods (Cervigni et al. 2013; Chap. 5 and Appendix C).

## 2.3 Hydrological Modeling: Impacts on Water Availability

An analysis on the spatiotemporal availability of water resources for each of the eight Hydrological Areas (HAs) in Nigeria was also conducted in order to estimate irrigation potential at both existing and planned locations (small and large infrastructures) in selected watersheds.

The GIS version of the SWAT model (ArcSWAT)[2] was applied to evaluate climate risk on water resources. SWAT is a well assessed tool and literature offers good support to its calibration and validation also for the area of interest (Schuol and Abbaspour 2006; Schuol et al. 2008). After modeling the river network through

---

[2] http://swatmodel.tamu.edu/software/arcswat.

the Digital Elevation Model, 893 basins were extracted for the physical-based semi-distributed hydrological analysis. Further, layers of 234 soil types, 16 land covers, and 5 slope classes were combined to extract Hydrological Response Units (HRUs), assumed to have similar hydrological response.

Hydrological simulations for each of the 893 basins were made using the full ensemble of climate projections as input. In each site, the RCM simulated inflow during the historical period (baseline 1976–2005) was bias-corrected based on available historical record for the same period. The same set of coefficients was used to correct all the simulated inflows (RCM and its GCM-based perturbations) in the future period of 2006–2065. Outputs were aggregated at 30-year intervals. The short- and medium-term periods were compared with the baseline (Cervigni et al. 2013, Chap. 5 and Appendix F).

## 2.4 Macro-economic Analysis

The effects of climate-induced yield changes on macroeconomic outcomes (e.g. volume and composition of GDP, imports/exports, etc.) were evaluated by inputting into a Computable General Equilibrium model (ICES) the climate change impacts on agricultural production derived from crop yield analysis. A preliminary step was the construction of a future reference scenario, capturing plausible economic development in Nigeria up to the year of 2050 (Table 1).

This reference scenario is the counterfactual "no climate change", on top of which the impacts of climate change on crop productivity were imposed, and against which the consequent GDP and sectoral performance of the economic system were evaluated.

Assumptions for irrigation, consistent with the Master Plan for Irrigation and Dam Development (but delayed by 5 years), are that in 2025 roughly 5% of Nigerian agriculture (2.1 million hectares) will be irrigated, to reach 25% of total agricultural land in 2050 (11 million hectares). The assumption made here is that future yields will be, in relative terms, as vulnerable as current ones to climate shocks, so that the deviations from current yields obtained from crop modeling can be applied to future yields as well. The rationale is that yield increases in the reference "no climate change" scenario will be achieved largely through irrigation expansion and through management practices that are suited for current climate, but not necessarily to the warmer and more erratic climate of the future. In particular, it is assumed that the uptake of sustainable land management options will be minimal.

Because of the structure of the Social Accounting Matrix (SAM) used in the ICES model, the disaggregation used for crops and zones is as follows. Rice, cassava and yam are modeled individually; millet, sorghum and maize are modeled as a single aggregated crop class, labeled "other cereal crops". Spatially, six global agro-ecological zones were used for the analysis, finding a correspondence with the ones used for the crop modeling.

Table 1 Macroeconomic assumptions for the "no climate change" reference scenario

| Period | Average GDP growth rate (%) |
|---|---|
| 2010–2020 | 9.0 |
| 2021–2030 | 8.4 |
| 2031–2040 | 6.0 |
| 2041–2050 | 4.3 |
| 2010–2025[a] | 9.0 |
| 2025–2050[a] | 5.7 |

|  | Vision 20:2020 | Model simulation |
|---|---|---|
| **A. Sector shares in total value added in 2025** | | |
| Agriculture | 21% | 23% |
| Manufacturing | 18% | 17% |
| Mining | 15% | 21% |
| Services | 46% | 39% |
| **B. Agricultural productivity growth** | | |
| 2010–18 | 3-fold | 2.5-fold |
| 2010–25 | 6-fold | 5.3-fold |
| 2010–50 | NA | 19-fold |

Source: Cervigni et al. (2013)
[a]These rates have been calculated assuming that Nigerian Vision 20:2020 objectives (http://www.nationalplanning.gov.ng/index.php/national-plans/nv20-2020) are achieved with 5-year periods

The exercise was performed under different climate simulations, representing the variability of yield changes – and correspondingly of macro-economic impacts – across climate outcomes corresponding, on average, to the least and the most pessimistic scenario of yield change (Cervigni et al. 2013, Chap. 5 and Appendix I).

## 2.5 Adaptation Strategies in Agriculture

After the assessment of the impacts on crop yield, a set of select farming practices was tested to analyze their potential to offset, across the different AEZs, time horizons (2020 and 2050) and crops, the negative impacts of climate change on yields (Cervigni et al. 2013, Chap. 6 and Appendix C). These adaptation strategies were selected among the most common and suitable farming practices. For rainfed areas, the shift of the sowing/planting dates, conservation/organic agriculture practices and use of inorganic fertilizers were included in the analysis. For irrigated crops, the analysis focused on yield improvements that could be achieved by modifying planting/sowing dates.

In the case of the shift in planting date, for each crop, the simulations were conducted adjusting the sowing/planting period 1 month earlier and 1 month later with respect to the traditional cultivation calendar. In terms of conservation agriculture, the analysis focused on nutrient management, and evaluated the use of manure and residues (manure 1 and residues 1) to complement current nutrient provision; or replace them (manure 2). Finally, additional use of inorganic fertilizers was investigated, at a lower (fertilizer 1) and medium intensity (fertilizer 2).

To address climate model uncertainty, climate data from RCM model and two extreme perturbations (NCAR_CCSM3 and GFDL_cm2.1) were considered. The results were analyzed at AEZ and country level. For each crop, only the AEZs where the crops are mostly diffused are considered in the aggregation at Country level.

The approach selected for undertaking the evaluation of the different adaptation strategies is the "regrets" analysis. The "regrets" of adopting each option were expressed as the percent gap in yield improvement between the option being examined and the best performing option under each of the three climate projections; next, the maximum regret was calculated for each option, across the three climate models; and finally, the "mini-max" adaptation option was identified, i.e., for each combination of crop and AEZ, as the one that minimizes the maximum regrets across climate models.

Successively, an evolution (in 2020 and 2050) of cropping patterns at the level of AEZs was defined using information from the macro-economic model. Moreover, the land area to which the "mini-max" adaptation options should be applied to eliminate as much as possible of the "production gap" between the reference and three climate change scenarios were evaluated.

## 2.6 Costs of Adaptation Options

As an additional experiment, the aggregate costs and benefits of the adaptation strategies identified were explored to investigate if they could be worthwhile in economic terms (Cervigni et al. 2013; Chap. 6). Costs include the direct outlays associated to expanding irrigation and promoting improved farming practices in rainfed areas. In addition to direct outlays, there are also opportunity costs of diverting productive capital, which in the absence of climate change would have been allocated to other development priorities. The benefits are given by the value of the additional output that can be produced once the adaptation measures are in place.

To evaluate the net effect, the macro-economic model was run without negative climate change impacts on yields, as these effects are fully offset by adaptation. At the same time, the model run included a decrease in the annual capital stock, in an amount given by the extra expenditure on adaptation. The metric used to assess the net effect is the terminal value of GDP in 2050, with adaptation, and without.

## 2.7 RDMA for Irrigation Infrastructures

When moving attention to the adaptation strategies for irrigation, it is crucial to consider that uncertainty in future precipitation makes it difficult to project how much water will be available in the future for storage. In case of a changing climate, a given storage design based on historical data can receive less/more water than expected and produce less/more benefit than projected. Climate change impact must therefore be considered in the design of new projects of water storages and irrigation infrastructure development, in order to minimize under- or over-design.

RDMA guiding the selection and design of future irrigation schemes can allow a decision maker to:

1. prioritize the schemes where the area of overdesign risk is smaller than the area of missed opportunity;
2. extend the irrigation area design if the risk of missed opportunity is large; and
3. design the storage facilities conservatively or favor crops that are less sensitive to failures of water supply if the area of overdesign risk is large. Adapting the design to a future climate change has a certain adaptation cost, which is the extra capital cost of building storage or irrigated area; the cost becomes negative if less storage or area is built compared to the historical climate. The benefit is the extra revenue obtained from selling more irrigated crops.

To evaluate what investment decisions on irrigation development are robust under a wide range of climatic outcomes, hydrological modeling results have been used to illustrate the practicability of RDMA for planning irrigation development (Cervigni et al. 2013; Chap. 6 and Appendix J).

The study focused on 18 planned dam sites to identify design options that could minimizes the regrets over a range of possible future climate outlooks. The regrets are defined as the difference in economic return between the chosen option ("no foresight") and the best possible option calculated for each scenario ("perfect foresight"). The Net Present Value (NPV) is the metric used to estimate the value of the different investment decisions.

Monthly data inflows from the hydrological analysis at dam level allowed calculating storage-yield curves (SYCs) for the respective upstream basin, indicating the firm basin yield produced from a given level of storage or, alternatively, storage capacity needed to provide a given basin yield. SYCs were built according to the Sequent Peak Algorithm (SPA; Thomas and Burden 1963) designed for studying reservoir capacity.

The analysis was based on a comparison between SYC referring to the baseline (1976–2005), and 30-year future periods (2006–2035 and 2036–2065), simulated under the whole ensemble of climate projections. Changes in the SYCs for the future simulated flows show the combined effect of predicted changes in flow magnitude and inter-annual variability.

The optimization was carried out with respect to two decision variables: the amount of stored water and the irrigated area. Then, if the purpose of the dam is to irrigate a targeted area, the decision should be made on the amount of storage. If the dam is already built or there are constraints on the storage size, the decision should be made with regard to the irrigated area.

Eleven "perfect foresight" storages were calculated to generate enough yield to provide water to the irrigated area under each climate scenario. Then, the storage of the "no-foresight" case (under current climate) is used to estimate the area for irrigation under each scenario. The difference in storage cost and irrigation revenues between the "perfect foresight" case and the "no-foresight" case corresponds to the regrets. The robust storage option is obtained by adjusting the storage of the "no-foresight" storage in order to minimize the average and the maximum regrets under all climate scenarios. Robust decision making on irrigated area can be estimated following a similar method, but the storage is assumed to be fixed, while the irrigated area is optimized to minimize regrets.

The case study sites were selected in accordance to Government plans to develop irrigation, as reflected in the Master Plan for Irrigation and Dam Development (2009–2020); and using the following criteria: (i) the main basins where new irrigation development is planned should be represented; (ii) the number of sites in each HA should be proportional to the area planned for irrigation development in the HA; (iii) catchment size should be larger than 100 km$^3$ (so that sub-basins are representative of the whole catchment behavior); (iv) lack of dam upstream; and (v) dry and wet future climates should be represented. A small-scale irrigation dam in the northern dry HA was added. The analysis purports to illustrate the policy significance of the RDM approach but should not be considered as an assessment of the technical or financial feasibility of the design solutions investigated, which would require more detailed investigation.

## 3 Results and Discussion

### 3.1 Climate Projections and Their Uncertainty

The simulated air surface temperature averaged over Nigeria shows a strong increasing trend up to 1–2 °C in 2050 compared to the present average temperature, with the highest increases in the North. In the short-term future (2020), the entire country is predicted to experience a moderate surface air temperature increase.

The precipitation time series averaged over Nigeria for the period of 1976–2065 shows no significant trends associated to most of GCM-based perturbations; only the data perturbed through the GFDL model shows a significant negative trend. The model results for precipitation were summarized by defining four classes of risk/conditions at hydrographic sub-basin level: wetting risk, drying risk, stable, uncertain. A given sub-basin is considered "stable" if most climate models (i.e. those falling within the range of the 1st to the 99th percentiles of the ensemble) agree that future rainfall will not be larger (smaller) than 15% (−15%) of historical values. Sub-basins are considered exposed to "dry risks" if the 1st percentile is less than −15% and the 99th percentile if less than 15%, to "wet risk" when the 99th percentile of changes is larger than 15% but the 1st percentile is more than −15%; and are considered uncertain when both a decline larger than −15% and an increase larger than 15% are considered possible.

Cervigni et al. (2013, Chap. 4 – Map 4.2) found that around 2020, 53% of the country's area is expected to be under wetter conditions, 10% under lower rain availability, 35% stable, and the remaining 2% present high uncertainty across precipitation projections. In 2050, 41% of the country is projected to be under wetter conditions 14% under drier conditions, 20% stable, and the area subject to uncertainty increases to 25%. More evident clusters of drying areas in the short- and medium-term are concentrated in the SE plateau and along the SW littoral, the stable areas in the center and along the central and eastern coastal zones, wetting areas in the north with evident uncertainty mainly in the medium-term period.

## 3.2 Impact Analysis on Crop Yields

Climate change impacts on crop yields are expected to be considerably variable over AEZs and crop types. The differences among crops are related to the specific crop sensitivity to modified climatic conditions as well as to crop spatial distribution and crop calendars. The impacts tend to increase from short- to medium-term period. Results are aggregated across AEZs, to develop impacts at the level of individual crops, and across crops, to produce results at the level of AEZs, using base-year information on production shares and value added to define weights used for aggregating. Only the results based in a fixed $CO_2$ concentration are reported here. The full set of results, including increases in $CO_2$ atmosphere concentration, is reported in Cervigni et al. (2013).

In terms of impacts at the level of crops, the results show medium term (2050) yield reductions, with negative median values for all crops in 2050 (Fig. 3b). However, yam, millet and cassava exhibit uncertainty, particularly in 2020 (Fig. 3a), where the median across climate models indicate the possibility of moderate yield increases (in the order of 3–6% or less). In 2050, the consensus across models is higher, with 70% of the model pointing to a decrease in yields. Rice appears to be the most vulnerable crop in both periods, with yield decline of 7% in 2020 and 25% in 2050.

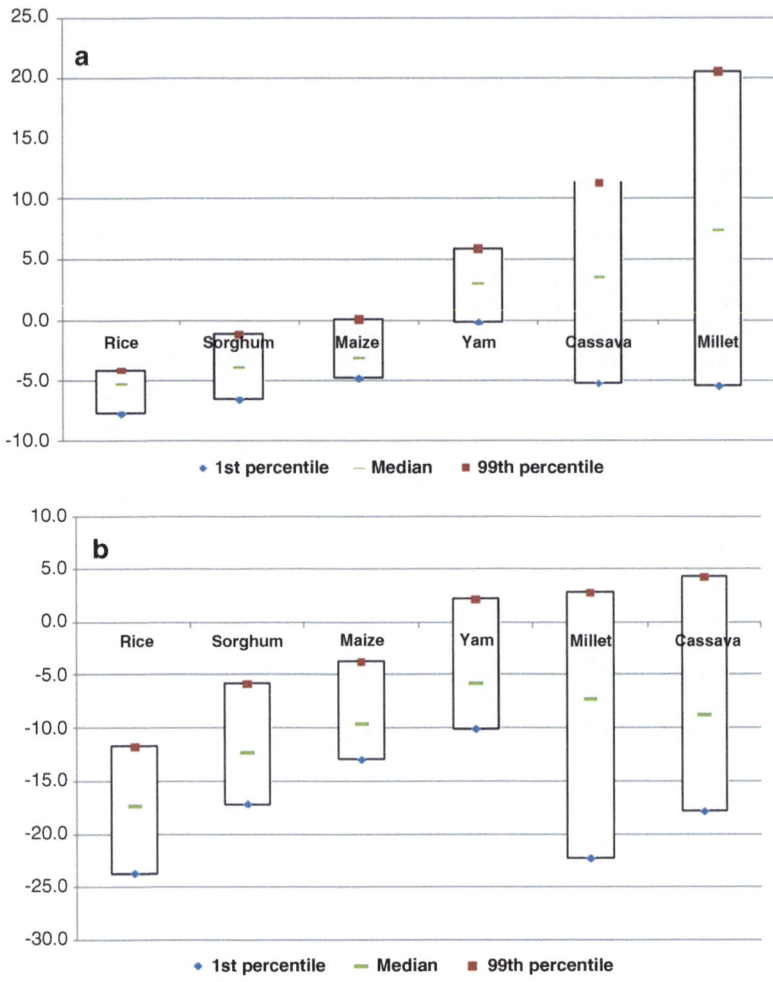

**Fig. 3** Aggregate percent change in crop yields for 2020 (**a**) and 2050 (**b**) (From Cervigni et al. 2013)

Temperature change is likely to be the major driver of yield shocks, rather than water content (this is consistent with other studies such as Lobell et al. 2008 and Lobell and Burke 2010), particularly in presence of less clear signals of precipitation changes. Temperature increase affects crop growth by shortening the crop-growing period and reducing the amount of biomass accumulation. This produces a decrease in crop yield, even if crops are not under water stress conditions.

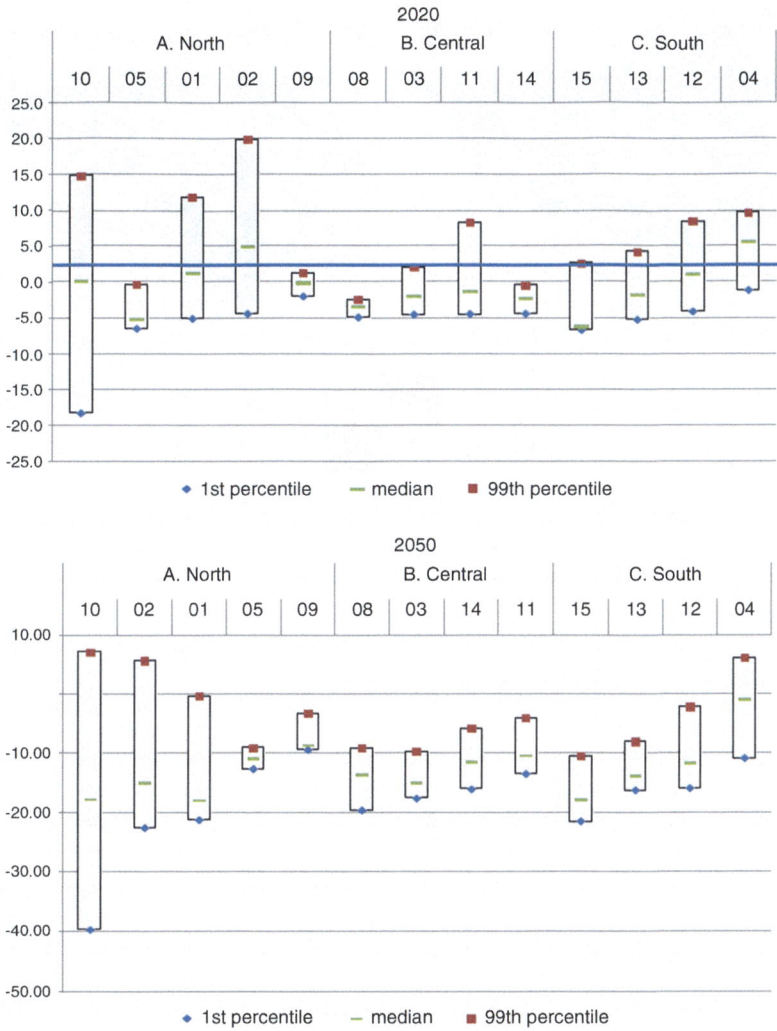

**Fig. 4** Aggregate percent change in crop yields by AEZ (2020 and 2050) (From Cervigni et al. 2013)

In terms of impacts at the AEZ level, the Northern area (Fig. 3) appears more subject to risks of large declines (close to 20% and 40% in 2020 and 2050, respectively), but shows also larger uncertainty. Despite the significant amount of variability across space, by 2050 the likelihood of aggregate yield decline appears stronger in all zones, as indicated by the negative median values observed in Fig. 4.

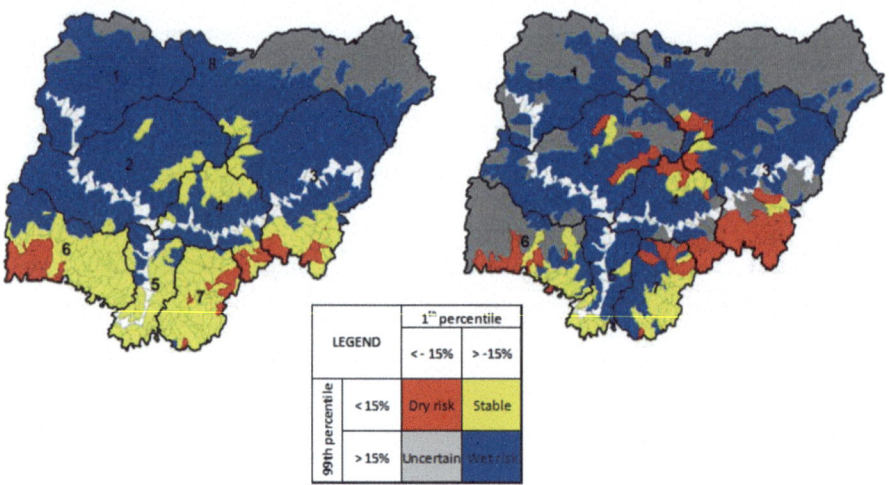

**Fig. 5** Distribution of classes of risk for water flows in 2020 and 2050 vs. 1990. Discretized spatial units are hydrographic sub-basins, while numbered units are Hydrological Areas (From Cervigni et al. 2013)

## 3.3 Water Availability Impact Analysis

The hydrological modeling tools were used to convert changes in climate variables (temperature, precipitation) into changes in water flows, and thus changes in water potentially available for storage to sustain multiple uses. Using the same risk classes defined for the analysis of rainfall changes to summarize the consensus among climate models, it was found (Fig. 5) that, by 2020, 62% of the country is expected to be under wetter conditions, 4% under dry risks, 23% stable, and the remaining 11% are characterized by uncertainty. In 2050, there is still a significant part of the country projected to become wetter (although decreasing from 62% to 49% of land areas); the share of areas under dry risks increases from 4% to 10% (accounting however for 17% of historical runoff). The share of stable sub-basins decreases to 8% of total land areas; while uncertainty increases considerably to 33% of the total.

It is noteworthy that there is a high uncertainty for the arid/hyper-arid regions in the northeast. Except for the central high plateau, the majority of the central and northern parts of Nigeria are expected to experiences an increasing availability of water resources, although the uncertainty for 2050 is more pronounced. The results for central area, SE mountains, and SW littoral indicate a general drying trend in the short and medium-term. Further, while flow is projected to increase up to 200% in some cases, the weighted average of increases is only about 33%, because the largest increases of flow are projected to take place in relatively drier basins. It is only for basins in the bottom 30% of the flow distribution that flow is projected to increase by more than 30%. These changes in water flows are likely to have significant effects on the reliability of irrigation systems, which is determined by magnitude (average) and variability of inflow.

## 3.4 Macro-economic Impacts

The crop model analysis projects a decline in crop production, growing with time and particularly significant by 2050 for the "other cereals" aggregated class, which, unlike the other crops, is in the order of 9.6% even in the most optimistic climate scenario. Low case scenario declines are high also for Rice (−8%). Overall, the outcomes project: (i) an increase in domestic crop prices (particularly severe in the case of rice) suggesting a more rigid demand, and (ii) significant changes in food trade patterns, with net imports increasing in the case of rice and the "cereal crops" to offset the projected decline in domestic production.

Rice and cereals constitute the large majority of agricultural imports in Nigeria in the baseline (35% rice and 46% cereals in 2050). Accordingly, the general equilibrium adjustment to the overall decline in production (occurring for all crops in 2050) consists in meeting demand where possible via an increase in imports, which is higher for crops with relatively lower import prices in the baseline (such as rice and other cereals). The combined effect of changes in production, prices and imports turns into an overall reduction in GDP compared to the no-climate change reference scenario, which by 2050 varies between 3% and 4.5% (Fig. 6), depending on the climate model. These results should probably be considered as a conservative, lower bound estimate of macro-economic impacts of climate change.

## 3.5 Adaptation Options in the Agriculture and Water Sectors

It is likely that an efficient adaptation strategy for the agricultural sector in Nigeria requires a combination of expansion in irrigated areas and improved management practices for rainfed crops, allocated accorded to the considerations discussed in this paper. Several factors will contribute to determining the ultimate outcome, including relative costs, resource availability, the institutional context, etc. This section presents analyses of options that can be deployed in rainfed areas and to what extent they could counter the overall impact of climate change on production, and at what cost.

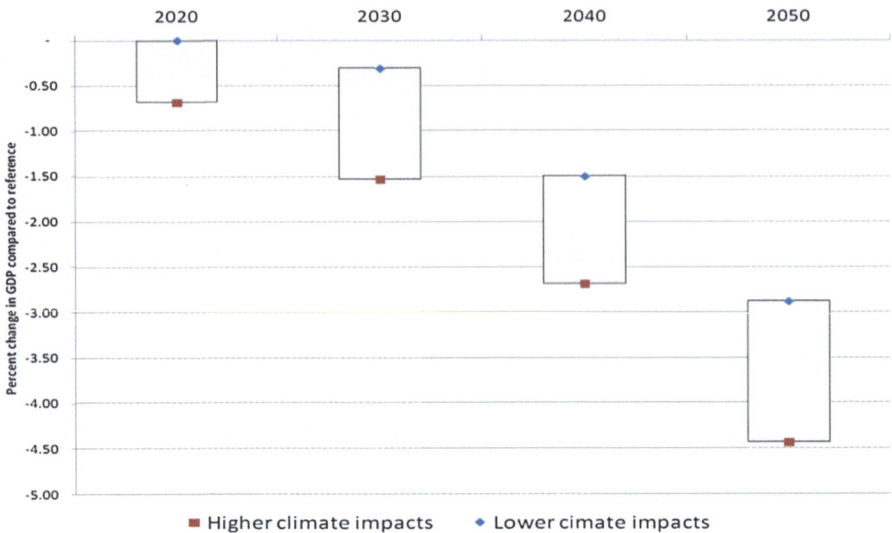

**Fig. 6** Deviation of GDP from the no-climate change reference scenario (From Cervigni et al. 2013)

### 3.5.1 Adaptation Through Sustainable Land Management Practices

The adaptation options tested (Table 2) appear to perform well, both in the short-term (2020) and medium-term (2050), improving yields (compared to a no-adaptation case) from 20% (e.g. changes in sowing/planting dates) to 90% (e.g. residues and other nutrient management options) of the cases, depending on crop, time horizon, climate model and AEZ considered (Figs. 7 and 8).

The use of residues and "manure 1", at worst, performs slightly less than the no-adaptation case; in the best cases, they deliver yields 30% higher. Change in planting dates can produce significant improvements (in excess of 20%), but in some crops and zones they can actually result in a further yield decline. The wide range of variability in the performance of the options points to the need of further evaluating the suitability of different adaptation options to different crops and AEZs under conditions of climate uncertainty.

Results of the regret analysis (Fig. 9) shows that "Manure 2", "Manure 1" and "Residues" are the best performing options, accounting for 75% of total mini-max options. It is important to note that besides increasing nutrient availability, these options increase soil fertility in a broader sense: through improvement of soil characteristics, of soil water retention and thus availability; and through reducing nutrient losses by runoff and leaching.

The optimal mix of adaptation options is highly crop- and location-specific (Fig. 10): e.g., the mini-max strategy for Cassava is "Manure 2" in 90% and "Manure 1" in 10% of the AEZs; while in the case of Rice, the strategy is to adopt "Manure 1" in 75%, "Fertilizer 2" in 17%, and "Residues" in 8% of the AEZs.

**Table 2** Adaptation options tested

| Group | Adaptation option | Description | Benefits | Constraints |
|---|---|---|---|---|
| *Rain-fed areas* | | | | |
| Change in planting/sowing dates | Plus 1 month<br>Minus 1 month | Shift the sowing/planting date 1 month before and 1 month after the ordinary sowing/planting date. | It may allow avoiding very hot and/or dry periods. It does not imply cost for farmers and can be immediately put in place, if the results are positive.<br>In some agro-ecological subzones (AESZs) and for some crops (cereals), yields have increased 20–30%, depending on crop and AESZ. | Farmers need extensive training and access to skilled advisory services.<br>Results are highly variable depending on the crop and the cultivar. |
| Inorganic fertilization | Fertilizer 1<br>Fertilizer 2 | Increase by 30% (fertilizer 1) and by 60% (fertilizer 2) over the ordinary fertilization amount. | Yields increase up to 20–30% for cereals and yams, and up to 40% for cassava. | Relatively high cost of fertilizers; farmers need access to skilled advisory services.<br>There may be an impact on the environment. |
| Conservation agriculture | Manure 1<br>Manure 2<br>Residue | Application of manure (manure 1) or residues from crop production (residue) to complement baseline nutrient management; complete substitution of inorganic fertilization with manure (manure 2). | Yields increase up to 25% for sorghum and millet, up to 35% for rice, and up to 50% for maize and cassava. | Farmers need extensive training and access to skilled advisory services.<br>There may be a relatively high up-front cost for the purchase or application of manure and residues. |
| *Irrigated areas* | | | | |
| Combining shift in growing period and irrigation | | Shift the sowing/planting date 1 month before and 1 month after the traditional date, in addition to irrigation practice. | Yields increase for cassava and yams, and there is a positive synergy between irrigation and the shift in growing period. | Farmers need extensive training and access to skilled advisory services. |

From Cervigni et al. (2013)

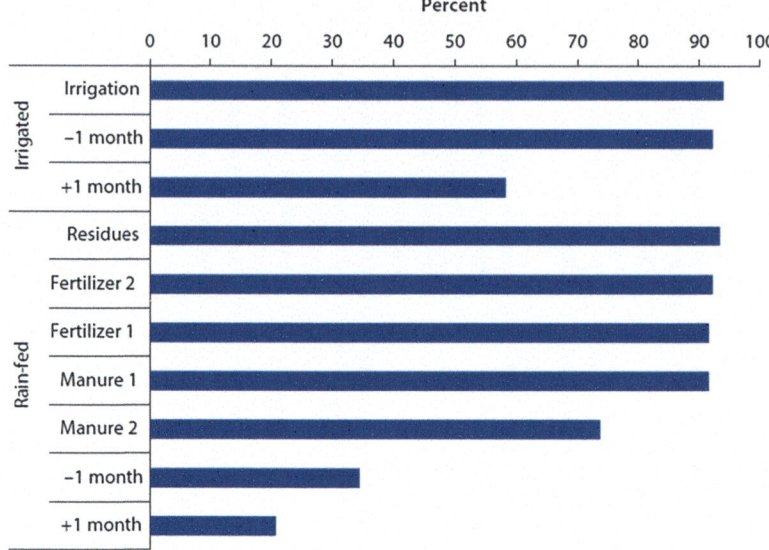

**Fig. 7** Safety ratio of the adaptation options 2020 (From Cervigni et al. 2013)

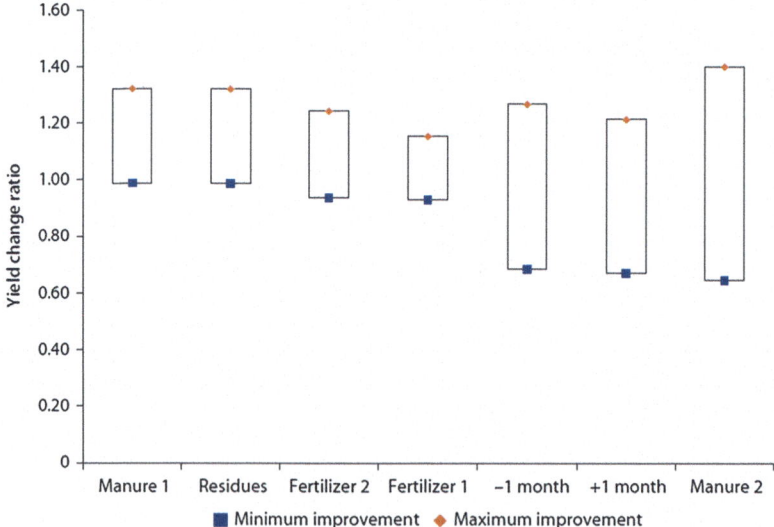

**Fig. 8** Adaptation options: maximum and minimum yield improvement (From Cervigni et al. 2013)

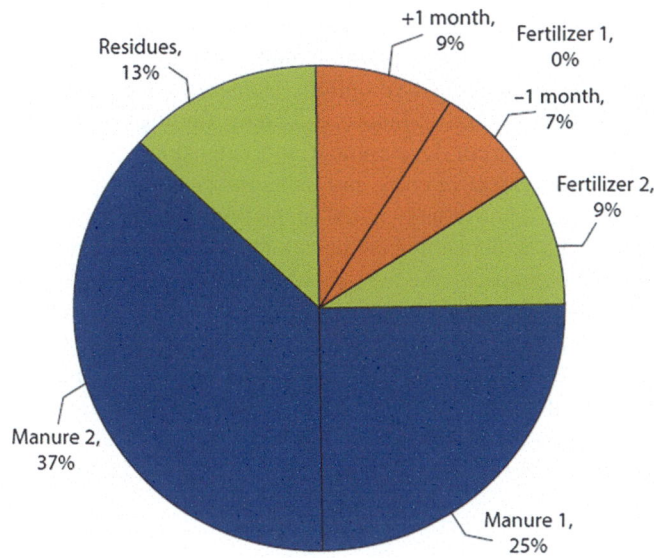

**Fig. 9** Mini-max adaptation options for rainfed areas (From Cervigni et al. 2013)

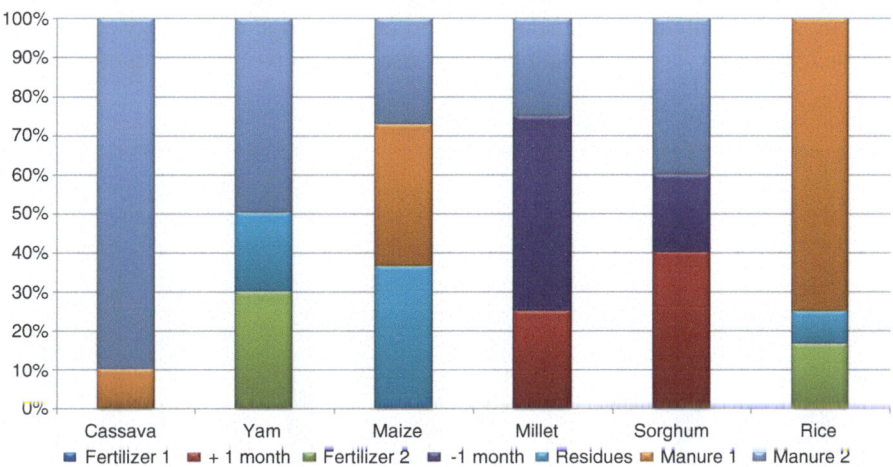

**Fig. 10** Composition of mini-max adaptation strategies across rainfed crops (From Cervigni et al. 2013)

Similarly, at the level of AEZ (Fig. 11), the mini-max adaptation strategy in AEZ 10 entails the adoption of a single option, namely "manure 1", whereas in the case of AEZ 11, the strategy includes five options, namely "−1 month", "+ 1 month", "Fertilizer 2", "Residues", and "Manure 2". These findings highlight the importance of stepping up research, development and extension services, to enable the identification and deployment of crop- and location-specific adaptation options.

Our analysis (Table 3) finds that by 2020 adaptation should be applied to a total of 0.6 to 1.1 million hectares (depending on the climate model considered); by 2050, due to more severe climate impacts, the area should increase to 14–18 million hectares. While in 2020 the mini-max adaptation options succeed (with the exception of millet in one climate model) in fully offsetting climate impacts, a residual gap remains in 2050, ranging from 1% to 22%, depending on crops and climate models (Table 4). Taking into account the yield differential over time between rainfed and irrigated conditions, the remaining production gap could be filled by expanding irrigation in the medium term (2050) to between 1.5 and 1.7 million hectares (Table 5).

### 3.5.2 Costs of Adaptation

Our results (Table 6) also indicate that adaptation is effective at reducing the net GDP loss, provided that unit costs can be kept in check.

In the "low unit cost" case, the terminal year loss in GDP is always lower with adaptation than without; the benefit-cost ratio of adaptation ranges between 1.2 to almost 2. However, under the high unit cost case, the proposed adaptation strategy

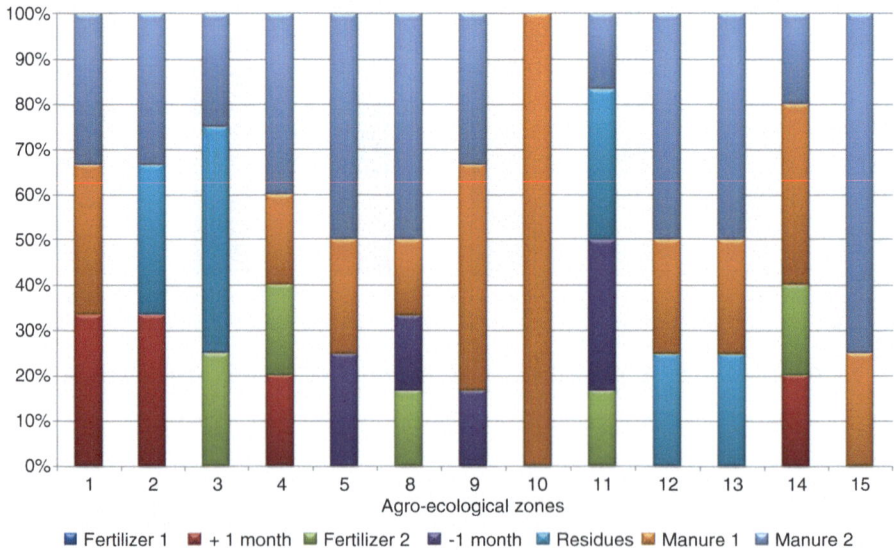

**Fig. 11** Composition of mini-max adaptation strategies across agro-ecological zones (From Cervigni et al. 2013)

**Table 3** Applying mini-max rainfed adaptation options by year and climate model

|  | 2020 | | | 2050 | | |
|---|---|---|---|---|---|---|
| Crops | NCAR | GFDL | RCM | NCAR | GFDL | RCM |
| Cassava | 0.00 | 0.00 | 0.22 | 0.00 | 0.23 | 2.06 |
| Maize | 0.07 | 0.33 | 0.18 | 3.84 | 4.05 | 4.05 |
| Millet | 0.00 | 0.00 | 0.27 | 3.01 | 3.16 | 3.16 |
| Rice | 0.17 | 0.10 | 0.13 | 2.29 | 2.63 | 2.63 |
| Sorghum | 0.36 | 0.34 | 0.29 | 4.01 | 4.42 | 4.42 |
| Yams | 0.00 | 0.00 | 0.02 | 0.00 | 1.66 | 1.66 |
| Total | 0.59 | 0.77 | 1.11 | 13.15 | 16.15 | 17.98 |

Source: Cervigni et al. (2013)
In hectares and millions

**Table 4** Production gap eliminated by mini-max rainfed options, by year and climate model

|  | 2020 | | | 2050 | | |
|---|---|---|---|---|---|---|
| Crops | NCAR | GFDL | RCM | NCAR | GFDL | RCM |
| Cassava | n.a. | n.a. | 100 | n.a. | 100 | 92.2 |
| Maize | 100 | 100 | 100 | 100 | 99.9 | 99.1 |
| Millet | n.a. | n.a. | 95.1 | 100 | 82.6 | 78.3 |
| Rice | 100 | 100 | 100 | 100 | 89.2 | 89.0 |
| Sorghum | 100 | 100 | 100 | 100 | 94.0 | 93.9 |
| Yams | n.a. | n.a. | 100 | n.a. | 97.4 | 92.3 |

Source: Cervigni et al. (2013)
In percent

**Table 5** Area of adaptation application by climate model

|  | 2020 | | | 2050 | | |
|---|---|---|---|---|---|---|
| Areas | NCAR | GFDL | RCM | NCAR | GFDL | RCM |
| Farm practices in rain-fed areas | 0.59 | 0.77 | 1.11 | 14.26 | 16.15 | 17.98 |
| Additional irrigation | 0.00 | 0.00 | 0.02 | 0.00 | 1.49 | 1.67 |
| Total | 0.59 | 0.77 | 1.13 | 14.26 | 17.65 | 19.65 |

Source: Cervigni et al. (2013)
In hectares and millions

**Table 6** Aggregate costs and benefits of adaptation

| Variables | NCAR | GDFL | RCM |
|---|---|---|---|
| GDP loss induced by climate change in 2050 | 2.9% | 3.6% | 4.5% |
| GDP loss induced by adaptation in 2050: | | | |
|    Low unit cost case | 2.3% | 2.6% | 2.3% |
|    High unit cost case | 15.5% | 14.3% | 12.7% |
| Benefit cost ratio: | | | |
|    Low unit cost case | 1.26 | 1.38 | 1.96 |
|    High unit cost case | 0.19 | 0.25 | 0.35 |

Source: Cervigni et al. (2013)

is no longer attractive, with the opportunity cost of capital diverted to adaptation far exceeding the benefit in terms of recovered production. The benefit-cost ratio is consistently less than 1 under all climate scenarios. These findings underscore the importance of supporting adaptation with measures to control the unit costs of investments in irrigation and sustainable land management practices, which appear to be consistently quite higher in Nigeria than in comparator countries in Africa.

### 3.5.3 Robust Decision Making Approach for Irrigation Infrastructure

The impact of adapting the design of reservoir or irrigation area to a wetter or dryer climate is quantified by calculating the avoided regrets. The regrets of using historical climate as a basis for planning and design of irrigation are typically between 10% (storage optimization, minimum average regrets) and 40% (irrigated area optimization, minimum maximum regrets) of the investment cost. Results of the analysis show that these regrets can be greatly reduced by optimizing the design of irrigation schemes. On average, the regrets decrease 30–50% depending on the type of optimization. Moreover, the results vary greatly among case studies, with up to 90% of the regrets that can be avoided in some locations.

Different classes of avoided regrets were defined based on their value compared to the investment cost. Optimizing the design has a high (low) impact if the avoided regrets exceed 20% (are less than 5%) of the investment cost, while the impact is moderate if the avoided regrets are between 5% and 20% of the investment cost. Results show that, in about half of the case studies, taking into account climate change in the design has a moderate to high impact, whichever optimization method is considered. Results obtained by optimizing the storage and the irrigated area optimization are illustrated on maps in Figs. 12 and 13.

The reduction in regrets exceeded 50% of the investment cost in two case studies in the northern part of the country. In these areas, the climate is projected to be much wetter than the historical scenario for all the perturbed models, as shown by the mean annual runoff and the storage-yield curves. Therefore, there is a strong incentive to build smaller dams to irrigate a given area (or larger irrigated area for a given storage). Nevertheless, these results should be taken with caution because of the significant uncertainties in climate models and the hydrological model, and should be completed with additional ensemble members (e.g. emission scenarios, climate models, hydrological model parameterization).

## 4 Conclusions and Recommendations

The results of this analysis indicate that in Nigeria the significance of climate shifts will increase in the medium term (2036–2065) compared to the short term (2006–2035). On average, temperatures in Nigeria will rise from 1 to 2 °C, with the north more affected than the south. Projected changes in the amount and seasonal

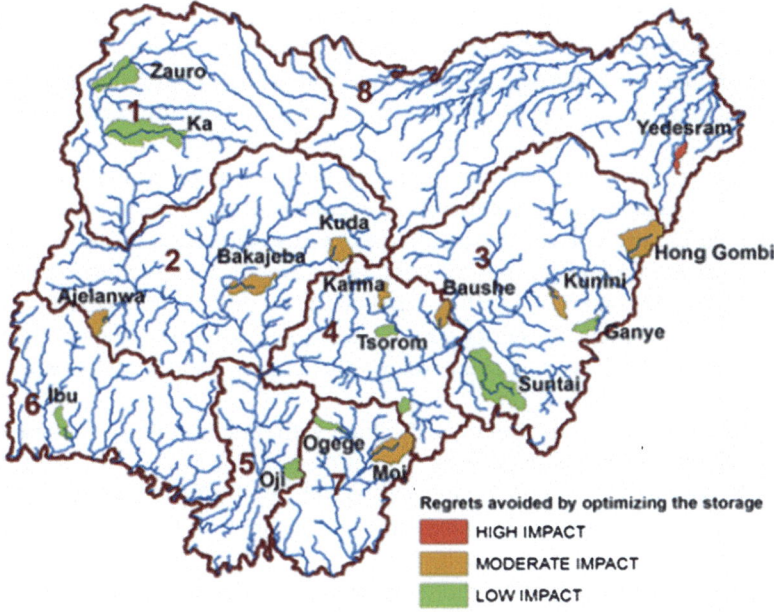

**Fig. 12** Regrets avoided by optimizing the storage (From Cervigni et al. 2013). Note: Low impact: decrease in regrets is less than 5% of the investment cost; moderate impact: between 5% and 20%; high impact: more than 20%

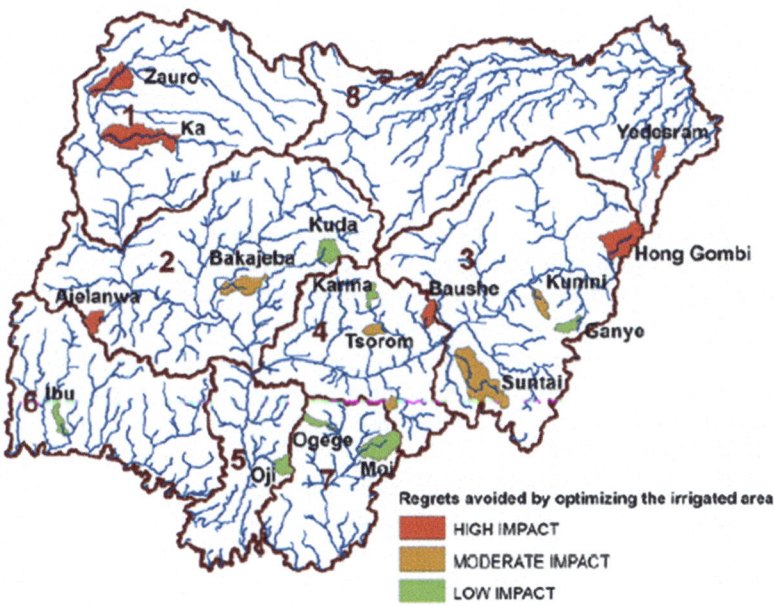

**Fig. 13** Regrets avoided by optimizing the irrigated area (From Cervigni et al. 2013). Note: Low impact: decrease in regrets is less than 5% of the investment cost; moderate impact: between 5% and 20%; high impact: more than 20%

distribution of rainfall are quite uncertain, with no clear agreement among climate models on whether rainfall would rise or fall.

The combination of changes in temperature and precipitation shows biophysical impacts that can have significant consequences for the agriculture and related water sector. The likely negative impacts of climate change on rainfed agriculture and the increased uncertainty about water resources available in the future make it essential to consider climate change into agricultural sector planning.

Indeed agriculture will mainly be affected by loss of yields for the main crops (cassava, millet, yam, maize, sorghum, and rice), even if precipitation increases in several parts of the country. The effects are fairly clear in the longer term but somehow more ambiguous in the shorter term (2020) when, according to more than half of the climate models, cassava, and perhaps other crops, might actually experience an increase in productivity.

The projected decline in rainfed yields along with projected rises in temperature might ultimately reduce food security. It is projected that half of Nigeria's agro-ecological zones will be food insecure by 2020 and 75% by 2050 unless their diminishing local food production is complemented by improved in-country trade or more imports.

Impacts on water resources are more uncertain, but it looks very likely that availability of water for storage and use will be different from the past. In particular, our analysis suggests that, by 2050, in only 23% of the country the hydrological regime will remain stable. In the rest of the country, the hydrology of the future will be very different than today, with 50% of the country expected to have higher runoff than the historical average, 10% of the country projected to be exposed to drier conditions, and 33% of total land area will be uncertain as climate models disagree so much that is difficult to define where runoff will increase or decrease.

The decline in crop yields will have significant consequences also for the national economy, by 2050 reducing GDP (compared to the no-climate change scenario) by up to 4.5%. Climate change is also projected to increase net import of various crops, particularly rice and other cereals.

The major policy implication of our analysis is that ignoring the effects of climate change in the design of agriculture policies, programs and projects would have dire consequences on the sector's development prospects, and indeed on the country's overall growth. At the same time, because of large uncertainties on the magnitude, speed and, in the case of precipitation, even direction of change, there is no silver bullet to consider in the design of climate change adaptation interventions. In fact, selecting the wrong adaptation response to climate change may have costs as large as not adapting at all. In the case of the rainfed agriculture, the adoption of certain adaptation technologies (e.g. the shift in sowing date) may turn out be ill-suited for some crops or agro-ecological regions, and result in a net *decline* in yields, rather than reducing climate change impacts. Similarly, development of irrigation schemes may lead to wrong-sizing of the amount of storage or irrigated area, both if climate change is ignored, *and* if a single scenario of climate change is arbitrarily selected (instead of considering the full range of possible outcomes).

Admittedly, addressing head-on the challenge of uncertainty in designing adaptation responses to climate change requires investments in developing the human and institutional capacity required to assess the full spectrum of development outcomes of any given project. In that sense, there is a trade-off between rapidity (and political expediency) of adaptation response, and their longer term effectiveness and ability to minimize risks and regrets. It is easier to come up with a package of interventions that might only look at one end, rather than the full spectrum of possible climate rather outcomes; and it may put the country in a favorable position to gain access to bilateral and multilateral sources of climate finance. However, our analysis suggests that there may be considerable risks at stake, both for the country (which will not achieve the intended development benefits); and for the international donor community, which may not get the expected adaptation value for taxpayer money.

Our analysis suggests there is a wide range of land and water management practices that can offset or even reverse the effects of climate change on crops, and can do so in a robust way, i.e. improving yields, compared to the no-adaptation case, over a wide range of future climate scenarios. These practices include elements of conservation agriculture (e.g., integrated soil fertility management, water harvesting, and agroforestry). Other options are shifts in sowing/planting dates, crop rotation, minimum or no tillage, and restoration of degraded pasture.

A combination of robust sustainable land management practices for 14–18 million hectares (ha) of rainfed areas and 1.5–1.7 million additional irrigated ha might fully offset medium-term climate change impacts on agriculture. At low unit costs, this adaptation package has a benefit-cost ratio exceeding 1 in all climate scenarios considered.

Similarly, on irrigation, application of a robust decision making approach can assist in building climate resilience into investments. Testing the use of the approach on to 18 planned irrigation schemes, this work finds that the regrets for not including climate change in the design can be as high as 40% of investment costs; and that by selecting the investment strategy that minimizes regrets across multiple climate outcomes, they can be reduced by 30–50% on average, and up to 90% in some locations.

Finally, an important challenge for policy is that action on adaptation may be perceived as having benefits too differed in time (i.e. too far past the time of action). Nevertheless, there are at least three reasons why the Government may act now to deal with climate change. First, many actions that will strengthen longer-term climate resilience will also help reduce the vulnerability to current climate swings. Second, investment decisions that will be taken in the near future on long-lived infrastructure, such as irrigation schemes, will determine how resilient these investments will be to the harsher climate of the future. To avoid locking the sector in a state of future climate vulnerability, it is essential to carefully evaluate the implications of alternative planning and design options overs a wide range of future climate scenarios. Third, building the knowledge, capacity, institutions and policies needed to deal with the climate of the future takes time. The longer Nigeria delays action, the less time it will have to get ready, and the more it will have to resort to reactive practices rather than prevention.

The actions that Nigeria could consider to enhance its overall ability to plan and implement climate-resilient development could be organized around the three areas:

1. consolidate and harmonize policies and legislation to effectively integrate climate change considerations into sector planning and development;
2. develop practical knowledge on climate resilience practices and technologies to define and prioritize, across space and crops, opportunities for adopting "triple-win" agricultural options (higher yields, higher climate resilience, reduced carbon emissions) and solutions on the ground that farmers can adopt;
3. promote investments and resource mobilization.

Enhancing the climate resilience of the economy is likely to be a major undertaking that no individual institution can accomplish on its own. Considering that States and LGAs control a large share of public spending in many of the highly climate vulnerable sectors, the Federal Government may want to establish strategic partnerships with the States to optimize the planning and implementation of adaptation efforts across levels of government and budgetary lines.

## References

Adejuwon JO (2005) Food crop production in Nigeria. I. Present effects of climate variability. Clim Res 30:53–60.
Buishand TA, and Lenderink G (2004) Estimation of Future Discharges of the River Rhine in the SWURVE Project. KNMI, Technical Report, Royal Netherlands Meteorological Institute, De Bilt, Netherlands.
Cervigni R, Valentini R, Santini M (2013) Toward climate-resilient development in Nigeria. Directions in development. World Bank, Washington.
Challinor AJ, Ewert F, Arnold S et al. (2009) Crops and climate change: progress, trends, and challenges in simulating impacts and informing adaptation. J Exp Bot. 60(10):2775–2789.
Daron J. (2015) Challenges in using a Robust Decision Making approach to guide climate change adaptation in South Africa. Climatic Change 132:459–473. DOI 10.1007/s10584-014-1242-9.
FAO (2005) Irrigation in Africa in figures: AQUASTAT Survey-2005. FAO Water Report 29 (with CD ROM). Rome.
FAO (2016) AQUASTAT website. Food and Agriculture Organization of the United Nations (FAO). Website accessed on [2016/05/07].
FAO (2015) FAOSTAT.. http://faostat.fao.org/
Hoogenboom G, Jones JW, Wilkens PW, Porter CH et al. (2012) Decision Support System for Agrotechnology Transfer (DSSAT) version 4.5. University of Hawaii, Honolulu.
ICS-Nigeria, Information and Communication Support for Agricultural Growth in Nigeria. http://www.icsnigeria.org. Accessed 23 June 2011.
IPCC (2014) Summary for policymakers. In: Climate Change 2014: Impacts, Adaptation, and Vulnerability. Part A: Global and Sectoral Aspects. Contribution of Working Group II to the Fifth Assessment Report of the Intergovernmental Panel on Climate Change [Field CB, Barros VR, Dokken DJ, Mach KJ, Mastrandrea MD, Bilir TE, Chatterjee M, Ebi KL, Estrada YO, Genova RC, Girma B, Kissel ES, Levy AN, MacCracken S, Mastrandrea PR, and White LL (eds.)]. Cambridge University Press, Cambridge, United Kingdom and New York, NY, USA, pp. 1–32.
JICA (1995) The Study on the National Water Master Plan. Sector Report Vol 2. Report prepared for the Federal Ministry of Water Resources and Rural Development. Nigeria.

Jones JW, Hoogenboom G, Porter CH et al. (2003) The DSSATcropping system model. Eur J Agron 18:235–265.
Lebel T., Ali A. (2009) Recent trends in the Central and Western Sahel rainfall regime (1990–2007). Journal of Hydrology, 375:52–64.
Lempert RJ, Nakicenovic N, Sarewitz D, Schlesinger M (2004) Characterizing Climate-Change Uncertainties for Decision-Makers. An Editorial Essay. Climatic Change, 65(1):1–9.
Lempert RJ, Groves DG, Popper SW, Bankes SC (2006) A General, Analytic Method for Generating Robust Strategies and Narrative Scenarios. Management Science, 52(4):514–528.
Lempert RJ, Collins MT (2007) Managing the Risk of Uncertain Threshold Responses: Comparison of Robust, Optimum and Precautionary Approaches. Risk Analysis 27(4):1009–1026.
Lobell DB, Burke MB, Tebaldi C et al. (2008) Prioritizing climate change adaptation needs for food security in 2030. Science 319:607–610.
Lobell DB, Burke MB (2010) On the use of statistical models to predict crop yield responses to climate change. Agric. Forest Meteorol. 150:1443–1452.
Mereu V, Carboni G, Gallo A, Cervigni R, Spano D (2015) Impact of climate change on staple food crop production in Nigeria. Climatic Change 132(2):321–336.
Müller C, CramerW, Hare WL et al (2011) Climate change risks for African agriculture. Proc Natl Acad Sci USA 108:4313–4315.
NPC – National Planning Commission (2004) Meeting everyone"s needs: National Economic Empowerment and Development Strategy. Abuja, Nigeria: National Planning Commission.
NSSP – Nigeria Strategy Support Program (2010) Background Paper No. NSSP 011, January 2010. Available at: http://nssp.ifpri.info/2010/02/09/%e2%80%9coptions-for-enhancing-agricultural-productivity-in-nigeria%e2%80%9d/. [Accessed on 08 May 2016].
Odekunle TO (2004) Rainfall and the length of the growing season in Nigeria. Int J Climatol 24:467–79.
Parry M, Rosenzweig C, Iglesias A, Livermore M et al (2004) Effects of climate change on global food production under SRES emissions and socio-economic scenarios. Glob Environ Chang 14:53–67.
Rockel B, Will A, Hense A (2008) The regional Climate Model COSMO-CLM (CCLM). Meteorologische Zeitschrift 17:347–348.
Rosenzweig C, Jones JW, Hatfield JL et al (2013) The Agricultural Model Intercomparison and Improvement Project (AgMIP): protocols and pilot studies. Agric For Meteorol 170:166–182.
Roudier P, Sultan B, Quirion P, Berg A (2011) The impact of future climate change on West African crop yields: what does the recent literature say? Global Environmental Change 21:1073–1083.
Schlenker W, Lobell DB (2010) Robust negative impacts of climate change on African agriculture. Environ Res Lett, 5:1–8.
Schuol J, Abbaspour KC (2006) Calibration and uncertainty issues of a hydrological model (SWAT) applied to West Africa. Advances in Geosciences, 9:137–143.
Schuol J, Abbaspour KC, Sarinivasan R, Yang H (2008) Estimation of freshwater availability in the West African Sub-continent using the SWAT hydrologic model. J Hydrol, 352:30–49.
Scoccimarro E, Gualdi S, Bellucci A et al (2011) Effects of tropical cyclones on ocean heat transport in a high resolution coupled general circulation model. Journal of Climate 24:4368–4384.
Semenov MA, Porter JR (1995) Climatic variability and the modelling of crop yields. Agric For Meteorol, 73:265–283.
Seo N, Mendelsohn R, Dinar A, Hassan R, Kurukulasuriya P. (2008a) A Ricardian Analysis of the Distribution of Climate Change Impacts on Agriculture Across Agro-Ecological Zones in Africa. World Bank Policy Research Working Paper 4599.
Seo N, Mendelsohn R, Dinar A, Kurukulasuriya P, Hassan R (2008b) Differential Adaptation Strategies to Climate Change in African Cropland by Agro-Ecological Zones. World Bank Policy Research Working Paper 4600.
Seo N, Mendelsohn R, Dinar A, Kulukulasuriya P., Hassan R (2008c) Long-Term Differential Adaptation by Selection of Farm Types Across Agro Ecological Zones in Africa. World Bank Policy Research Working Paper 4602.

Thomas HA, Burden RP (1963) Operations research in water quality management. Division of Engineering and Applied Physics, Harvard University.

Thornton PK, Jones PG, Alagarswamy G, Andresen J (2009) Spatial variation of crop yield response to climate change in East Africa. Glob Environ Chang 19:54–65.

USAID MARKETS (2009a) Package of practices for sorghum production.. http://www.nigeriamarkets.org/files/Sorghum_Pop_English_July_2009.pdf. Accessed 01 May 2011.

USAID MARKETS (2009b) Package of practices for rice production.. http://www.nigeriamarkets.org/files/Rice_Pop_English_June_2009.pdf. Accessed 13 December 2011.

USAID MARKETS (2010) Package of practices for maize production.. http://www.nigeriamarkets.org/files/Maize_Pop_2010_English_final.pdf. Accessed 01 May 2011.

Webber H, Gaiser T, Ewert F (2014) What role can crop models play in supporting climate change adaptation decisions to enhance food security in Sub-Saharan Africa? Agricultural Systems 127:161–177.

Wilby RL, Dessai S (2010) Robust adaptation to climate change. Weather, 65(7):180–185.

World Bank (2010) Africa Infrastructure: a time for transformation. Washington DC. World Bank.

Yakubu MM, Akanegbu BN (2015) The impact of international trade on economic growth in Nigeria: 1981 – 2012. European Journal of Business, Economics and Accountancy, 3(6): 26–36.

**Open Access** This chapter is distributed under the terms of the Creative Commons Attribution-NonCommercial-ShareAlike 3.0 IGO license (https://creativecommons.org/licenses/by-nc-sa/3.0/igo/), which permits any noncommercial use, duplication, adaptation, distribution, and reproduction in any medium or format, as long as you give appropriate credit to the Food and Agriculture Organization of the United Nations (FAO), provide a link to the Creative Commons license and indicate if changes were made. If you remix, transform, or build upon this book or a part thereof, you must distribute your contributions under the same license as the original. Any dispute related to the use of the works of the FAO that cannot be settled amicably shall be submitted to arbitration pursuant to the UNCITRAL rules. The use of the FAO's name for any purpose other than for attribution, and the use of the FAO's logo, shall be subject to a separate written license agreement between the FAO and the user and is not authorized as part of this CC-IGO license. Note that the link provided above includes additional terms and conditions of the license.

The images or other third party material in this chapter are included in the chapter's Creative Commons license, unless indicated otherwise in a credit line to the material. If material is not included in the chapter's Creative Commons license and your intended use is not permitted by statutory regulation or exceeds the permitted use, you will need to obtain permission directly from the copyright holder.

# Using AgMIP Regional Integrated Assessment Methods to Evaluate Vulnerability, Resilience and Adaptive Capacity for Climate Smart Agricultural Systems

John M. Antle, Sabine Homann-KeeTui, Katrien Descheemaeker, Patricia Masikati, and Roberto O. Valdivia

**Abstract** The predicted effects of climate change call for a multi-dimensional method to assess the performance of various agricultural systems across economic, environmental and social dimensions. Climate smart agriculture (CSA) recognizes that the three goals of climate adaptation, mitigation and resilience must be integrated into the framework of a sustainable agricultural system. However, current methods to determine a systems' ability to achieve CSA goals are lacking. This paper presents a new simulation-based method based on the Regional Integrated Assessment (RIA) methods developed by the Agricultural Model Inter-comparison and Improvement Project (AgMIP) for climate impact assessment. This method combines available data, field- and stakeholder-based surveys, biophysical and economic models, and future climate and socio-economic scenarios. It features an integrated farm and household approach and accounts for heterogeneity across biophysical and socioeconomic variables as well as temporal variability of climate indicators. This method allows for assessment of the technologies and practices of an agricultural system to achieve the three goals of CSA. The case study of a mixed crop livestock system in western Zimbabwe is highlighted as a typical smallholder agricultural systems in Africa.

---

J.M. Antle (✉)
College of Agricultural Sciences, Oregon State University, Corvallis, OR, USA
e-mail: john.antle@oregonstate.edu

R.O. Valdivia
Department of Applied Economics, Corvallis, OR, USA

S. Homann-KeeTui
International Crops Research Institute for the Semi-Arid Tropics, Bulawayo, Zimbabwe

K. Descheemaeker
Wageningen University, Wageningen, Netherlands

P. Masikati
World Agroforestry Centre, Lusaka, Zambia

© FAO 2018
L. Lipper et al. (eds.), *Climate Smart Agriculture*, Natural Resource Management and Policy 52, DOI 10.1007/978-3-319-61194-5_14

## 1 Introduction

One of the most important challenges for agricultural researchers is to evaluate the potential adoption and impact of agricultural technologies. Early research focused on economic impacts, but the search for more sustainable systems has shown the need for multi-dimensional assessments that consider agricultural system performance in economic, environmental and social dimensions and the inevitable tradeoffs among those dimensions (Antle 2011; Antle et al. 2014). The emerging reality of climate change means that the search for sustainable systems must also consider vulnerability to climate change, which may include increasing frequency and magnitude of climate extremes. The recent calls for "climate smart" agriculture recognize that climate adaptation, mitigation and resilience must be integrated into the broader agenda of developing sustainable agricultural systems.

As Lipper et al. (2014) emphasize, climate-smart agriculture (CSA) is an approach for transforming and reorienting agricultural systems to support food security under climate change. Part of that process of re-orientation is to evaluate the performance of existing farming systems, and possible modifications of those systems, under a changing climate as well as with other changes (e.g., policy and technology) that may affect agricultural system performance and farm household well-being. Various elements of climate-smart agricultural systems have been identified, and a number of metrics can be utilized to evaluate systems for climate-smart attributes (Rosenzweig et al. 2015 and Rosenzweig et al. 2016).

Evaluating technologies for their performance in the multiple dimensions of sustainability poses major conceptual, analytical and data challenges: evaluating the farming system and farm household as an integrated unit, rather than individual production activities; linking the farming system to the other environmental and social outcomes that it may impact, including greenhouse gas emissions and carbon sequestration; and evaluating performance in more extreme and possibly variable climate conditions. Furthermore, there is a need to assess the usefulness of prospective changes in production systems that are not yet in widespread use, as well as the use of existing or new technologies under future climate and socio-economic conditions (Antle et al. 2015a).

The goal of this article is to describe and demonstrate the use of new simulation-based methods to evaluate the potential for currently available or prospective agricultural systems to achieve the goals of CSA. The motivation for this approach is the fact that conventional field experiments and *ex post* assessments are not appropriate tools to evaluate agricultural system performance in changing and uncertain climatic conditions and future socio-economic conditions. The approach presented here combines the available data, including observational data from field experiments and from surveys of actual farming system performance, with biophysical and economic models and future climate and socio-economic scenarios. These models become the "laboratory" in which simulation experiments are conducted to explore the performance of agricultural systems under the range of conditions considered relevant by stakeholders and scientists. An important feature and strength of this method is that it relies on input from stakeholders and thus provides a process to effectively engage stakeholders in the development and evaluation of technological options (Valdivia et al. 2015).

The approach we present is based on the Regional Integrated Assessment (RIA) methods developed by the Agricultural Model Inter-comparison and Improvement Project (AgMIP) for climate impact assessment (Antle et al. 2015b; AgMIP 2015). In this chapter, we first describe some of the key features of smallholder farming systems typical in many parts of Africa as well as other parts of the world, focusing in particular on the smallholder systems that involve rainfed crops and livestock and that are particularly vulnerable to climate and other changes and also have limited capacity to adapt to such changes. Next we provide an overview of the AgMIP methods for technology impact assessment, and discuss how they can be used for CSA assessments of vulnerability, resilience and adaptive capacity. We illustrate the application of these methods with a case study of crop-livestock systems in Zimbabwe. We conclude with a discussion of the strengths and limitations of these methods, and how they could be improved to be more useful for CSA.

## 2 Key Features of Crop-Livestock Systems: Implications for Modeling

To motivate the discussion of methods to follow, we first describe key features of typical smallholder agricultural systems in Africa, using the example of mixed crop livestock systems found in the Nkayi district of western Zimbabwe. Crop production is rainfed, and average annual rainfall ranges from 450 to 650 mm, making the system vulnerable to erratic rainfall with a drought frequency of one in every 5 years. Long-term average maximum and minimum temperatures are 26.9 and 13.4 °C, respectively. The soils vary from inherently infertile deep Kalahari sands, which are mainly nitrogen- and phosphorus-deficient, to clay and clay loams that are also nutrient-deficient due to continuous cropping without soil replenishment. Farmers use mainly a mono-cereal cropping system with addition of low amounts of inorganic and organic soil amendments. Natural pasture provides the main feed for livestock, and biomass availability is seasonal. During the wet season feed quantity and quality is appreciable, while during the dry season there is low biomass of poor quality. The natural pastures are mainly composed of savannah woodlands, with various grass species (Homann et al. 2007; Masikati et al. 2015).

As in many parts of Africa, mixed crop–livestock production systems are dominant in Nkayi. These farming systems are mainly based on maize, with smaller portions of sorghum, groundnuts, and cowpeas as staple crops, combined with the use of communal range lands, fallow land, and crop residues for livestock production (Fig. 1). Household livestock holdings vary from a few to 40 head per household of cattle, donkeys, and goats. Livestock offer opportunities for risk spreading, farm diversification, and intensification, and provide significant livelihood benefits

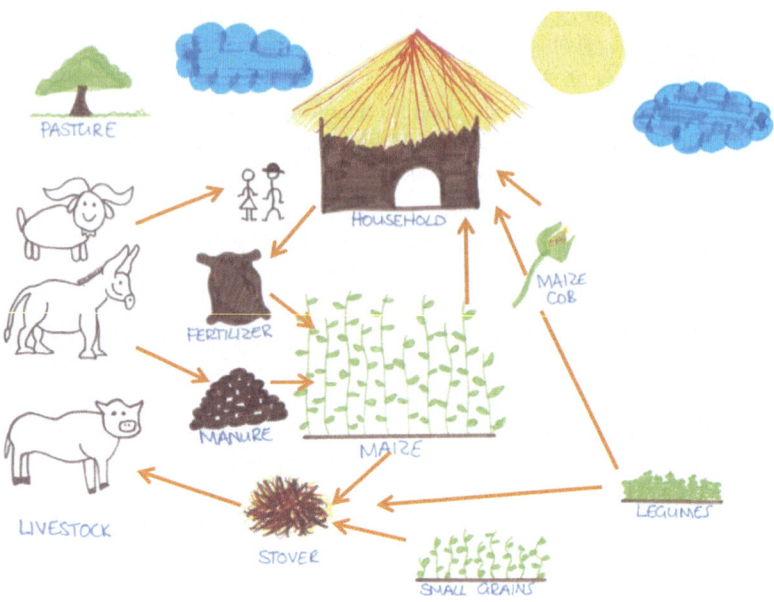

**Fig. 1** Mixed crop livestock farming systems, provider of food and livelihoods, and most common form of land use, affected by climate change in semi-arid Zimbabwe (Figure 2 of Masikati et al. 2015)

(Bossio 2009; Williams et al. 2002). Animals complement cropping activities through the provision of manure for soil fertility maintenance, draft power for cultivation, transport, cash, and food, while crop residues are used as adjuncts to dry-season feed. These systems evolve in response to various interrelated drivers, such as increased demographic pressure along with higher incomes earned by the urban populations, which results in a growing demand for crop and livestock products with the development of local and urban markets (Homann-KeeTui et al. 2013). This increased demand for crop and livestock products could benefit small-scale farmers as they gain access to markets, if they are able to intensify and diversify production in a sustainable way. These diverse income sources could reduce risk and increase resilience of farmers.

Another key characteristic of crop-livestock systems in many regions of Africa is low productivity due to a combination of factors that include unfavorable climatic conditions, poor and depleted soils, environmental degradation, and low level of capital endowment that leads to limited uptake of improved technologies, as well as adverse policies (Kandji et al. 2006; Morton 2007; World Bank Report 2009). Climate variability and change stressors, superimposed on the many structural problems in smallholders farming systems where there is not much support nor adequate adaptation strategies, can exacerbate food insecurity and increase vulnerability (Kandji et al. 2006; Morton 2007).

These characteristics of smallholder farming systems suggest that assessment models need the following features and capabilities.

**Integrated farm and household approach** A whole-farm approach is needed to represent all possible adaptation and mitigation options, including crop-livestock interactions and nutrient cycling, effects of specialization and diversification, and scale effects. In addition, a whole farm and household approach is needed to represent all components of the household's income, including both on-farm and off-farm income sources and employment opportunities. The household approach is also needed to represent economic vulnerability and resilience, for example, off-farm income may be impacted differently than farm income by climate change.

**Bio-physical and socio-economic heterogeneity** Analysis must account for the heterogeneity that is often high in farm household populations, in terms of soil conditions and climate, as well as differences in farm and herd size, behavioral differences due to the farm decision makers' knowledge and experience, the age, gender and health of the farm household members, and location and access to markets, capital and information.

**Temporal variation and system dynamics** Temporal variation in inputs and outputs of these systems has important effects on system performance and human well-being. For example, a key element of food security is the stability of food availability over the annual cycle. Adaptation, mitigation and resilience all involve change over time that can be thought of as investment or dis-investment in natural capital (e.g., soil fertility), physical capital (tools, machinery and structures, as well as livestock), human capital (farm family members' health, education and knowledge), and social capital (social networks and relationships). Resilience involves the capacity of a system to withstand a shock or disruption and naturally involves an understanding of system dynamics.

## 3 AgMIP Regional Integrated Assessment Methods

AgMIP has developed a methodology for RIA of climate change impact, adaptation, mitigation and vulnerability, and thus provides a framework for CSA assessment. The approach is designed to quantify indicators of system performance deemed to be relevant by both stakeholders and scientists, and then conduct simulation experiments to evaluate how system performance responds to climate and other changes, including system changes for climate adaptation and mitigation. These methods can be used in various ways to support technology development, e.g., to facilitate the targeting of agricultural interventions to farm types, for design and impact assessment of context specific safety-net, food security or market oriented intervention packages.

Based on discussions with stakeholders and the research assessment literature, a number of key indicators were identified to assess impact, vulnerability, mitigation and adaptation. These indicators are also relevant to the evaluation of CSA.

- Physical quantities and value of principal agricultural products, at the farm household level and aggregated to the regional or population level.
- Net value of single agricultural commodities as well as entire farms
- Average household per-capita income or wealth.
- The headcount poverty rate in the population (i.e., the proportion of households below the poverty line) and other poverty measures such as the poverty gap (i.e., the degree to which individuals are below the poverty line).
- Food security indicators, including capability to buy an adequate diet, per-capita food consumption, calories and other nutrient intake, dietary diversity indicators, and impacts on children such as stunting or mortality.
- Environmental indicators, including soil fertility, soil erosion, and indicators of greenhouse gas emissions and mitigation.
- Vulnerability, defined as the proportion of households that may be adversely affected by climate change. Losses can be measured in economic terms or in other dimensions of well-being such as health.
- Resilience, defined as the capability of a system to minimize the magnitude of adverse impacts or enhance positive effects towards greater adaptive capacity.

The foundation of the AgMIP RIA approach is the design of the simulation experiments that are used to evaluate climate impacts and the effects of system adaptations. There are many possible simulation experiments that can be carried out. Working with various stakeholders, AgMIP has identified four "core" research questions for regional integrated assessments. Figure 2 illustrates these Core Questions described below. Note that climate change can have either negative (left figure) or positive (right figure) effects without adaptation, and in a given population of farm households some may experience negative effects and some may experience positive. Effective climate adaptations will reduce negative effects or enhance positive effects. Another key element of Fig. 2 is that the climate assessment is carried out in the context of a plausible future state of the world (i.e., the non-climate biophysical and socio-economic future conditions) embodied in a "representative agricultural pathway" or RAP. As we discuss further below, the AgMIP RIA method includes the development of RAPs with inputs from scientists as well as stakeholders.

The four core questions are defined as follows:

**Core Question 1: What is the sensitivity of current agricultural production systems to climate change?** This question addresses the isolated impacts of a change in climate assuming that the production system does not change from its current state. It is useful as a baseline for comparison with other combinations of technology and states of the world.

**Core Question 2: What are the effects of adaptation in the current state of the world?** This question is one often raised by stakeholders: what is the value of

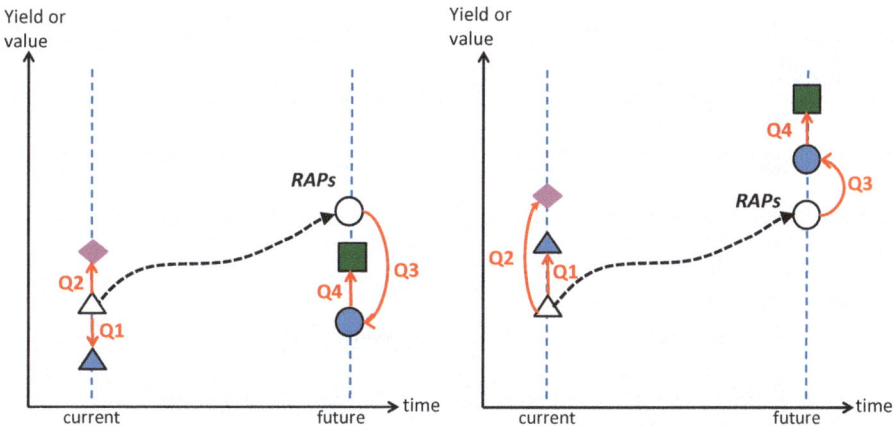

**Fig. 2** Overview of core climate assessment questions and the production system states that are simulated. The *dashed black line* represents the evolution of the production system in response to development in the agricultural sector that would occur without climate change, or independently of climate change, as defined by a Representative Agricultural Pathway (*RAP*). *Arrows* illustrate effects associated with the four core questions described in the text (Source: adapted from Antle et al. 2015b)

adapting today's agricultural systems to climate changes that may be occurring now and in the near future?

**Core Question 3: What is the impact of climate change on future agricultural production systems?** This question evaluates the isolated role of climate impacts on a future production system, which will differ from the current production system due to development in the agricultural sector not directly motivated by climate changes.

**Core Question 4: What are the benefits of climate change adaptations?** This question analyzes the benefit of potential adaptation options in the production system of the future, which may offset climate vulnerabilities or enhance positive effects identified in Core Question 3 above.

The AgMIP RIA methodology is designed to enable research teams, in collaboration with stakeholders, to answer each of these core questions. Figure 3 provides an overview of the approach. As noted in the previous section, an integrated whole-farm and household modeling approach is needed for CSA. Accordingly, the AgMIP approach to RIA is built on the concept of the farm household and the farming system that it uses. The foundation of the AgMIP approach is the characterization of the existing farming system, typically by developing "cartoons" or system diagrams (see Fig. 1, and Fig. 3b). The research team uses this characterization of the current systems to identify the key system components, and the corresponding data and models that will be needed to implement the RIA analysis.

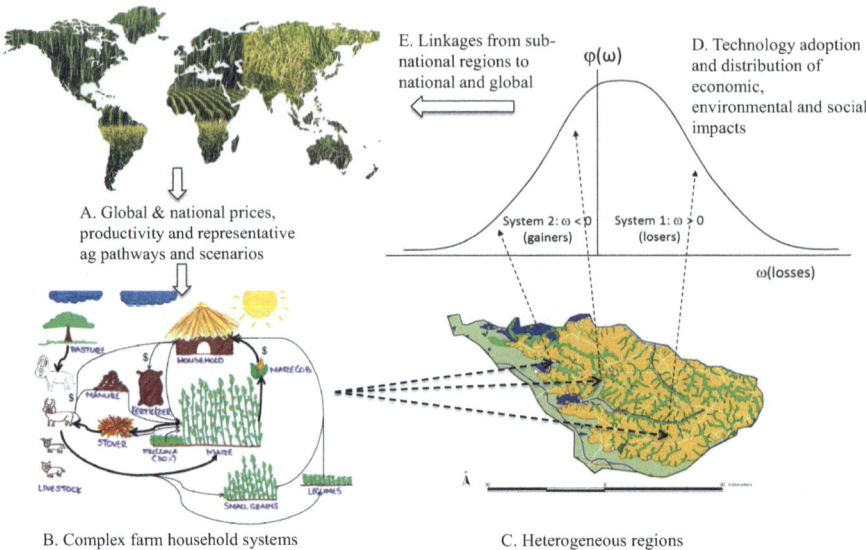

**Fig. 3** AgMIP Regional Integrated Assessment approach simulates climate change impact, vulnerability and adaptation through climate data, bio-physical simulation models and economic models representing a population of heterogeneous farm household systems. (**a**) RAPS together with global and national price, productivity and land use projections define the bio-physical and socio-economic environment in which (**b**) complex farm household systems operate in heterogeneous regions (**c**). Analysis of technology adoption and impact assessment is implemented in these heterogeneous farm household populations (**d**). This regional analysis may feed back to the country and global scales (**e**) (Source: Antle et al. 2015a)

In the AgMIP RIA methodology, the heterogeneous response to climate change derives from the productivity impacts of climate change incorporated in the model through crop and livestock simulation models, as well as the socio-economic heterogeneity in the farm household system due to variations in farm size, household size, and non-farm income. As explained in detail in the AgMIP RIA Handbook (AgMIP 2015), the AgMIP method uses crop and livestock model simulations to project the effects of climate change on the productivity of a system. In this method a yield under a changed climate is approximated as $y^c = r^c \cdot y^o$ where $y^o$ is an observed yield and $r^c$ is a simulated relative yield calculated as $r^c = y^{sc}/y^{so}$, where $y^{sc}$ is the simulated yield under the changed condition, and $y^{so}$ is the simulated yield under the observed condition. This procedure is used rather than directly using $y^{sc}$ as an estimate of $y^c$ to account for the fact that simulated yields do not incorporate all the factors affecting observed yields and thus tend to be biased. If this bias is (approximately) proportional and equal for both $y^{sc}$ and $y^{so}$ then it will cancel out. In cases where process-based models are not available for a crop or livestock species, assumptions for yield impacts are included in scenarios based on expert judgment and other available data such as behavior of similar species or studies of analog climates.

For analysis of adaptations, a similar method is used to assess how the existing system could be changed. These changes can range from management of the existing production activities, changes in the land or other resources allocated to those activities, as well as the introduction of new activities or the elimination of activities. Also, changes in the farm household's labor allocation between production activities, and between agricultural and non-agricultural activities can be considered. These characterization of the existing and prospective farming systems also helps to develop future socio-economic pathways (i.e., Representative Agricultural Pathways, see below) by identifying the "external" or "driving" variables that define the bio-physical and socio-economic conditions in which the analysis is conducted. For example, if the analysis is being designed for a future period, it is likely that prices received or paid by the farmers will be different. It is also likely that characteristics of the farm household population will change, such as the farm size distribution, non-agricultural income and household size.

## 3.1 Quantifying Vulnerability

The AgMIP RIA methods are designed to assess vulnerability of farm households to climate change. We define a climate as a probability distribution of weather events that occur at a specific place and during a defined period of time. A change in climate is a change in the probability distribution of weather events. These changes are often described in terms of the mean temperature over a period of time such as a day, month or year, but can also be changes in temperature extremes, the variability of weather events, and other aspects such as rainfall amount and intensity and wind velocity.

Impacts of climate change are quantified as gains and losses in economic well-being (e.g., farm income or per capita income) or other metrics of well-being (e.g., changes in health or environmental quality). In this framework, some or all individuals may gain or lose from a change, and we say the losers are *vulnerable to loss from climate change*. The AgMIP RIA methodology is designed to quantify the proportion of the population that are losers, as well as the magnitude of loss. It is important to note, however, that in a heterogeneous population there are typically some gainers and some losers, and thus the net impact may be positive or negative.

The AgMIP RIA method is designed to quantify climate vulnerability by modeling a heterogeneous population of farm households rather than modeling a "representative" or average or typical farm. This approach begins with the representation of impacts on the farm household using the concept of economic gains and losses (other metrics of impact can be also used depending on available data, e.g., the impact on health of household members). As Fig. 3 shows, the AgMIP RIA approach uses a statistical representation of the farming system in a heterogeneous region or population to quantify the distribution of gains and losses, e.g., due to climate change. Figure 4 illustrates this idea with two loss distributions. The area under the distribution on the positive side of zero is the proportion of losers and is the measure

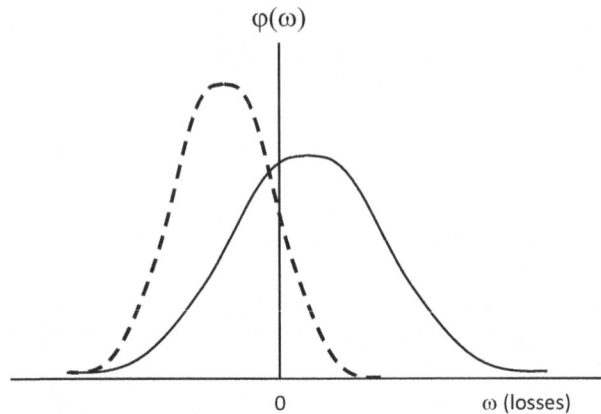

**Fig. 4** Vulnerability Assessment Using the Distribution of Losses Associated with Climate Change. The area under the distribution on the positive side of zero is the proportion of losers and a measure of vulnerability. Here the solid distribution represents a system for which the average loss is positive and there are more losers than gainers. The *dashed* distribution represents a system with more gainers than losers. The goal of climate adaptation is to shift the distribution leftward

of vulnerability. The solid distribution in Fig. 4 represents a system for which the average loss is positive and there are more losers than gainers. Note, however, that even in this case there are some gainers.

The goal of analysis for CSA is to improve the performance of farming systems. In the context of vulnerability analysis, this means reducing the number of losers (the vulnerable) and increasing the gainers from any perturbation of the system, be it climate change or any other change. The dashed distribution in Fig. 4 represents a system that is less vulnerable to climate change, and has more gainers than losers. Note that in this case, even though gainers outnumber losers, there are still some losers. It is also important to note that both the mean and the dispersion of the distribution of gains and losses matters to the measurement of vulnerability. Indeed, the dispersion (i.e., variance) of the distribution of losses represents the heterogeneity of the impacts of climate change on the population. In the AgMIP RIA methodology, this heterogeneous response to climate change derives from the productivity impacts of climate change incorporated in the model through crop and livestock simulation models (see discussion below), as well as the socio-economic heterogeneity in the farm household system due to variations in farm size, household size, and non-farm income. The areas under the distributions on the positive side in Fig. 4 represent the proportion of vulnerable farm households. The AgMIP RIA methodology also provides the capability to simulate the magnitude of impacts on the vulnerable members of the population, as well as the impact on those that gain, and the net or aggregate impact in the population.

## 3.2 Quantifying Resilience

Resilience has been defined in a number of ways in the scientific literature. In ecology, resilience is defined as the capacity of a system to maintain its form and function in response to a shock or disruption (Folke 2006; Nelson et al. 2007). In economic terms, resilience can be defined as the capacity to restore or maintain economic values, such as farm income (Antle et al. 2006; Antle and Capalbo 2010), or to minimize the loss from an adverse disruption or "disaster" over the time it takes for a system to return to its "normal" state (Hallegate 2014). Resilience to climate change can also be defined more broadly as the capacity to cope with change and minimize losses from change and enhance possible benefits of change, and thus can incorporate longer-term responses through adaptation (Malone 2009).

The definition of resilience as the capacity to withstand disruptions refers to the properties of a given system's performance, and is most relevant to analysis of relatively short-term events such as a storm or drought where it can be expected that the system will return to its normal state. In contrast, the capability to adapt or respond by making purposeful changes in a system seems most relevant to longer-term permanent changes in climate, and can include adaptations that are designed to improve the capability to withstand shocks or disruptions. Clearly, both concepts of resilience – the ability to minimize the effects of temporary shocks and disruptions, as well as the capacity to cope with the long-term shifts in weather patterns associated with climate change – are relevant to analysis of agricultural system performance.

The AgMIP RIA framework illustrated in Figs. 2 and 3 can be used to quantify resilience using the various indicators identified above. As noted above, vulnerability is measured as the proportion of farm households that experience a loss over a specified period of time. Loss can be measured in economic terms as reduced income or loss of the capitalized value of income plus assets, and also in non-economic terms such as reduced health or degraded environmental conditions. To see how resilience can be quantified, define the minimum possible loss for a given system as $Loss_{min}$ and define the realized loss as $Loss$. This minimum loss can be measured in various ways depending on the context. For example, it could be the loss that would be incurred if the best coping actions are undertaken as soon as possible and as effectively as possible. A resilience indicator can be calculated as 100 ($Loss_{min}$ / $Loss$), similar to what Hallegate (2014) defines as "microeconomic resilience". Thus, if a system can achieve the minimum possible loss its resilience is 100%, and otherwise its resilience is less than 100%.

This measure of resilience fits the situation where there is a loss, whereas with climate change and other types of change there can be net aggregate gains in some cases, and even when there are losers, there are also likely to be some gainers. To accommodate both gains and losses, we adopt the convention that resilience is 100% for gainers. Letting $v$ be the percent of vulnerable population, the resilience indicator for the population of gainers and losers is then calculated as $100(1 - v) + v \, Loss_{min} / Loss$.

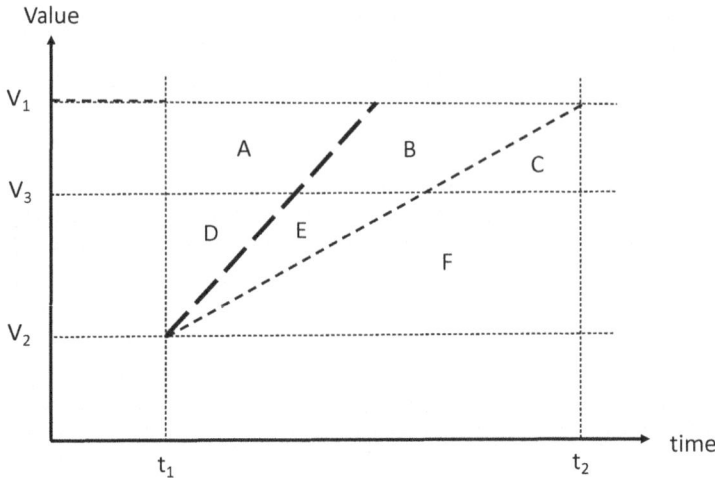

**Fig. 5** Analysis of Resilience to Temporary Disruptions and Long-term Change. See the text for explanation

This definition of resilience makes sense for a temporary change or disruption that a system can fully recover from, such as a seasonal drought followed by normal weather. However, if there are long-term changes, such as climate change, then the minimum loss would grow over time and the ratio $Loss_{min} / Loss$ would be undefined. A solution to this problem is to measure the losses over a finite time period relevant to decision making for making technology investment decisions, so that the minimum loss and actual loss are both bounded.

Figure 5 provides a stylized graphical representation of how resilience can be quantified for a temporary disruption as well as for a permanent change, over a specified time horizon from time $t_1$ to time $t_2$. In the analysis of a temporary disruption, the system provides a value $V_1$ before the disruption occurs at $t_1$. The disruption lowers the system performance to $V_2$, and the system then recovers along some path from $V_2$ back to $V_1$ (the path is shown as linear in Fig. 5, but more generally may be nonlinear). Suppose we are comparing two different systems, one more resilient than the other. The heavy dashed line in Fig. 5 indicates the system with the most rapid recovery possible, and thus $Loss_{min}$ equals area (A + D) and its resilience is 100% The less resilient system recovers along the path indicated by the lighter dashed line, so the loss is area (A + B + D + E), and the system resilience is calculated as 100 (A + D)/ (A + B + D + E) < 100%.

The analysis of resilience to a long-term change in climate is somewhat different than the case of a temporary disturbance in several respects. In response to long-term changes we expect systems to be adapted to climate change to some degree. There are three types of adaptations that can be expected to occur and can overlap at different scales. First, there are the kinds of changes in management that farmers can undertake within the existing system, such as changes in planting dates and reallocation of land and other resources among existing crops and livestock activities, or reallocation of their time among farm and non-farm activities. These types of adaptations have been called "autonomous or incremental adaptations." Second, there are adaptations that require investments external to the farm, such as investments in research and development of new technologies, such as improved crop varieties, or diversification and risk management options, sometimes referred to as "planned or systems adaptations." Third, transformational adaptation requires more fundamental changes in production systems, institutional arrangements, priorities for investment, and norms and behaviour (Kates et al. 2012). Zimbabwe is among the countries where transformational adaptation is recommended, to shift the systems towards more livestock-oriented and diversified systems with drought-tolerant food and feed crops, and development of the associated value chains (Rippke et al. 2016; Rickards and Howden 2012).

As illustrated in Fig. 2, the system currently in use would follow a path over time from the value indicated by the white triangle to the blue circle, whereas a system better adapted to the future climate would achieve a higher level of performance indicated by the green square. However, it is not clear from this diagram at what point in time along this path adaptations take place. One might assume that autonomous adaptations occur more-or-less continuously as farmers learn about climate changes and how to adapt management, whereas planned adaptations could occur in more discrete steps, e.g., as new crop varieties are developed and released.

The complexity of the progression of adaptation over time creates a major challenge for the analysis of adaptation. Given the difficulty analysts face in knowing how adaptations would evolve over time, the approach we adopt here is to treat each adapted system as if it were to become available at a discrete point in time, and that its effectiveness increases over time up to its maximum, depending on the characteristics of the technology and the capacity of farmers to acquire and use it successfully.

Following this approach, in Fig. 5 we can interpret $V_1$ as the performance of the current system in the future period without climate change (i.e., as the value represented by the white circle in Fig. 2). $V_2$ represents the value the same system would achieve with climate change (i.e., the blue circle in Fig. 2), and $V_3$ represents the value that an adapted system can achieve (i.e., the green square in Fig. 2). We can now interpret the heavy dashed line as a more rapid adoption pathway for the adapted technology, and the lighter dashed line as a less-rapid adoption pathway. Thus, under the rapid adaptation scenario, the loss due to climate change from $t_1$ to $t_2$ would be equal to area (A + B + C + D) which we could interpret as $Loss_{min}$ and corresponding to a resilience measure of 100%. Under the slower adaptation pathway, the loss would be (A + B + C + D + E), implying a resilience of 100 (A + B + C + D)/ (A + B + C + D + E) < 100%. The resilience of the unadapted system would be lower, and equal to 100 (A + B + C + D)/ (A + B + C + D + E + F).

## 3.3 Representing Future Socio-economic Conditions

In a climate change analysis, it is necessary to distinguish between three basic factors affecting the expected value of a production system: the production methods used (i.e., the system technology); the physical environment in which the system is operated, including soils and climate; and the economic and social environment in which the system is operated, i.e., the socio-economic setting. In the AgMIP RIA methodology, the non-climate bio-physical conditions and socio-economic conditions are embodied in a Representative Agricultural Pathway, or RAP (Valdivia et al. 2015). RAPs are qualitative storylines that can be translated into model parameters such as farm and household size, prices and costs of production, and policy. Following the four core climate impact assessment questions discussed above, the model can be set up with appropriate combinations of parameters to represent the corresponding technologies, climates, and socio-economic conditions.

As indicated in Fig. 2, the analysis of Core Questions 3 and 4 is carried out under plausible future conditions defined by Representative Agricultural Pathways. To project the average level of productivity into the future that would occur with ongoing technological advancements (not associated with climate change or adaptation), the AgMIP methodology utilizes the technology trend and price projections developed for global economic models (e.g., see Nelson et al. 2013), together with the assessment of technology trends made by research teams in the development of regional RAPs.

## 3.4 Defining and Quantifying Adaptation

The goal of adaptation analysis is to improve the performance of farming systems, e.g., to reduce vulnerability as illustrated in Fig. 4. The relative yield concept discussed above for modeling climate productivity impacts can also be applied to quantify the effects of an adaptation on a crop yield. Let a yield for an adapted system (say, a change in planting date) be $y^a = r^a \cdot y^o$ where $y^o$ is an observed yield and $r^a$ is a simulated relative yield calculated as $r^a = y^{sa}/y^{so}$, where $y^{sa}$ is the simulated yield under the adapted management, and $y^{so}$ is the simulated yield under the non-adapted (observed) management. This method can be applied under any climate conditions. Thus, for projecting yield with climate change and adapted management, we have $y^{ac} = r^a \cdot y^c = r^a \cdot r^c \cdot y^o$.

As we discussed above, the analysis of climate impact and adaptation must be carried out under future socio-economic conditions defined by a RAP. By definition, the RAP represents changes in socio-economic conditions that would occur without climate change. Therefore, any changes in crop or livestock systems and productivity described in a RAP cannot be a climate adaptation. Changes defined as a climate adaptation must, by definition, be changes that would occur in response to changes in climate, given any other changes that would have occurred regardless of climate change. The "simulation experiments" carried out for a climate adaptation analysis are designed to show the effect of climate adaptation *holding all else constant*, including any changes in productivity that would have occurred without climate change.

## 4 Assessing Crop-Livestock System Adaptations in Zimbabwe for CSA

In this section we summarize results from a recent study of the crop-livestock systems described in Fig. 1 and Section 1 that used the AgMIP integrated assessment approach to evaluate the climate vulnerability and benefits of adaptation strategies in these systems for multiple climate change scenarios (Masikati et al. 2015). Data from climate projections and RAPs were combined with soils and weather data and farm survey data to parameterize crop, livestock and economic simulation models to simulate the performance of systems under future socio-economic conditions with climate change. Next these models were used to simulate the performance of the systems with three adaptations that could improve crop and livestock productivity: applying higher levels of N fertilizer with micro-dosing; producing maize with recommended N fertilizer application rates; and with maize being grown in a rotation with mucuna.

To illustrate the use of the AgMIP RIA methods, here we report crop and livestock modeling results using averages over projections from five mid-century climate models that were run with a high emissions scenario (referred to by climate modelers as Representative Concentration Pathway 8.5), together with a business as usual Representative Agricultural Pathway for mid-century. We evaluate the economic impacts of the driest climate scenario on the crop-livestock system of Nkayi, Zimbabwe without adaptation, and with the following package of adaptations designed for resource-limited households.

- Adoption of long duration maize varieties instead of short duration varieties, with grain yield increases between 8% and 18%, and residue increases between 5% and 11%.
- Converting 1/3 of the maize land to maize-mucuna rotation, 30% of the mucuna biomass left on the fields as inorganic fertilizer for subsequent maize. 70% fed to cattle or available for sale.
- Application of micro-dosing (17 kg N/ha) on 1/3 of the maize field, second year after the maize mucuna rotation.

It is important to emphasize that the results reported here are for a single scenario to illustrate how the AgMIP methods can be used for CSA analysis. Each of the components of the analysis is uncertain, and to represent that uncertainty a more complete analysis would utilize multiple climate projections and multiple socio-economic scenarios and model components. Also, we emphasize that by interpreting these changes as climate adaptations, it is assumed that these changes would not have been adopted in order to achieve the productivity gains already embodied in the Representative Agricultural Pathway.

## 4.1 Climate Projections

For the climate scenario used, temperatures are projected to increase across the whole region of southern Africa. Changes range from large increases inland (above 3 °C in southwestern Botswana and surrounding areas) to smaller increases in coastal areas. Rainfall projections are less certain; rainy seasons are likely to start later and there are indications that rainfall will decrease over most of southern Africa, particularly over the western and central regions. Here we present results using one model that shows a mean temperature increase of about 3 °C and a mean rainfall decrease of about 0.6 mm/day over October–March, compared to the current average of about 3.4 mm/day.

## 4.2 Crop Models

The Agricultural Production Systems Simulator (APSIM) (Keating et al. 2003) was used to assess the impacts of climate change on crop production. The model was calibrated for maize and the forage legume, mucuna (mucuna pruriens) using on-farm experimental data obtained from the ICRISAT research work under different projects in Nkayi district (Masikati 2011; Homann-KeeTui et al. 2013). APSIM Results were judged satisfactory with observed mean maize grain yield of 1115 kg/ha and simulated of 1185 kg/ha. However, the model had a tendency to over-predict maize biomass with mean observed yield of 2460 kg/ha and simulated of 3385 kg/ha. For mucuna biomass results were satisfactory with mean observed yields of 4263 kg/ha and simulated of 4224 kg/ha.

The model was also evaluated for its ability to simulate maize grain yield variability across farming households. The model showed capacity to simulate the middle yield range from the farming households but did not perform so well for the lower and higher yields (Masikati et al. 2015). To offset the models' effects on projected future yields, the simulated yields were bias corrected before doing the economic analyses; the biomass yields were also adjusted before they were used for livestock simulations.

## 4.3 Livestock Model

Household-level livestock production was modeled with LIVSIM (LIVestock SIMulator, Rufino et al. 2009). The LIVSIM model was earlier calibrated for Zimbabwean conditions and the Mashona breed, for which it is also used here (Rufino et al. 2008; Rufino et al. 2011). LIVSIM simulates production with a monthly time step, based on breed-specific genetic potential and feed intake, taking into account specific rules for herd management. The impact of climate change and the various adaptation strategies on livestock production was predicted based solely on simulated changes in on-farm feed production resulting from the crop model runs. Livestock rely on community rangelands during the whole year and in the dry season, crop residues constitute an important feedbase component (Masikati 2011). However, the feed quality of the crop residues and of the dried grasses in the rangeland is low and also the risk of low crop production during dry years is relatively high. Therefore, feed gaps in the dry season are common, leading to important inefficiencies in the livestock component of the system. Hence grass and on-farm feed production and composition change with climate, and the effects of these changes on livestock were simulated with LIVSIM for climate change under current practices and for the adaptation strategies. The effects of increased crop residue availability in the fertilizer adaptation strategies and of higher-quality feed in the mucuna strategy were investigated. However, potential changes in rangeland productivity and direct effects of temperature on animal performance were not taken into account in this study.

## 4.4 Economic Model

AgMIP is using the Tradeoff Analysis model for Multi-Dimensional impact assessment (TOA-MD) to implement the economic analysis component of the RIA methodology. The TOA-MD model is a parsimonious, generic model for analysis of technology adoption and impact assessment, and ecosystem services analysis. Further details on the impact assessment aspects of the model are provided in Antle (2011) and Antle et al. (2014). The model software and the data used in various studies are available to researchers with documentation and self-guided learning modules at http://tradcoffs.oregonstate.edu.

There are several features of this model that make it appropriate for assessment of technologies for climate impact assessment as well as analysis of technologies for CSA. First, TOA-MD represents the whole farm production system which can be composed of (as appropriate) a crop sub-system containing multiple crops, a livestock subsystem with multiple livestock species, an aquaculture sub-system with multiple species, and the farm household (characterized by the number of family members and the amount of off-farm income). Second, TOA-MD is a model of a farm population, not a model of an individual or "representative" farm. Accordingly, the TOA-MD model is designed to quantify vulnerability and resilience using gains and losses as discussed above. With suitable bio-physical and economic data, these statistical parameters can be estimated for an observable production system. Using the methods described in the AgMIP Regional Integrated Assessment Handbook (AgMIP 2015), model parameters under climate change, without or with adaptation, can be calculated, and the model can be used to evaluate the four Core Questions identified in Fig. 2.

The TOA-MD model was parameterized using household surveys conducted in 2011 with 160 farmers interviewed in 8 villages that provided data on farm, herd and household size, off-farm income, revenues from crops and livestock, and the costs of production. In addition, 8 focus group discussions, one per each village surveyed, assessed agricultural output and input prices, perceived as normal prices during the observation year, not peak prices (Homann-KeeTui et al. 2013, 2015). For the calculation of net returns, monetary values of the crop (grain and residues) and livestock (sale, draft power, manure, milk) outputs were estimated with observed values or at opportunity cost, with internally used crop and livestock outputs factored in as costs under the respective activities, taking into account the local user practices. For the analysis presented here, the farm households were stratified into three groups according to livestock ownership as the locally most important wealth criterion (none; 1–8 cattle, or more than 8 cattle).

A Representative Agricultural Pathway (RAP) was developed with stakeholder collaboration to project the current systems into the future. In this analysis, the optimistic assumption was made that Zimbabwe will move out of 15 years of economic crisis towards positive economic development. Acknowledging the challenges and time required for institutional change, pro-active governance and investments, conservative projections were made for future productivity trends and prices. The pathway used was based on growth through market-oriented crop and livestock production, as government seeks to promote agricultural production and restore investor confidence. Severe liquidity constraints however restrict public and private investments. Limited employment opportunities in urban areas reduce rural-urban migration. An exogenous yield increase of 40% was assumed for maize as the predominant crop, and 35% increase for small grains and legumes. Fodder crops were only recently introduced and no market exists, and no increase was assumed. Productivity increases of 35% for cattle and 25% for small stock offtake was assumed, made possible by reducing mortality and improving livestock quality, and also modest 10% increases milk, manure and draft power production were assumed.

It was also assumed that international product prices are not fully transmitted to the national and local markets. Price increases for grain and live animal sales was assumed to be 10% from 2005 to 2050, and a 5% increase for the other products that are usually not traded. Input prices tend to remain high with 10% price growth. Input subsidies are assumed to be limited to vulnerable households during recovery and rehabilitation.

## 4.5 Impact of Climate Impact and Adaptation on Crop and Livestock Productivity

The mean of the crop model simulations showed projected crop yield losses under the current farming practices were modest, in the range of 7–9%, although some climate model projections were much higher or lower (Fig. 6). Crop systems in Nkayi are low

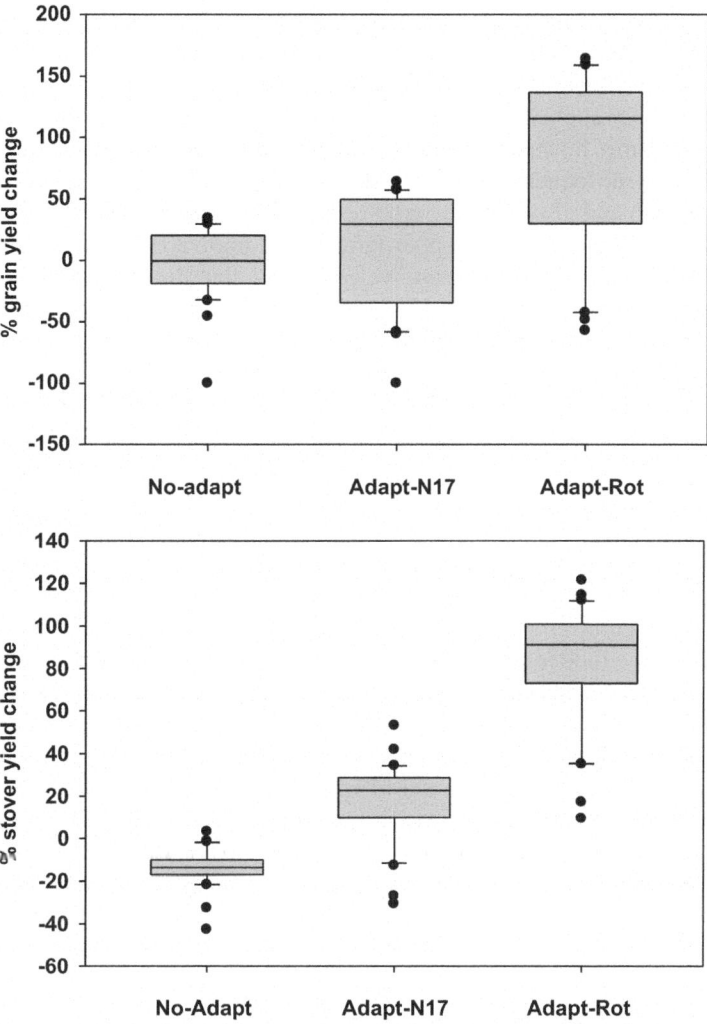

**Fig. 6** Boxplots showing average percent maize grain and stover yield change in Nkayi district, Zimbabwe, under current farmer practice (no-adapt) and different adaptation strategies (Adapt-N17 = microdosing at 17 kg N/ha and Adapt-Rot = maize-mucuna rotation system. The percent change under adapted scenarios is calculated with respect to the non-adapted scenario under climate change while for the non adapted scenario yield change is relative to current practice under current climate

input systems where average yields are around 500–700 kg/ha. Temperature thresholds for maize in the APSIM model are greater than 30 °C (Hatfield et al. 2011; Hatfield and Prueger 2015), and current average maximum temperature during the growing season is about 27 °C, so higher average temperatures of 2–3 °C do not substantially affect crop production unless there are more extreme events in a particular growing season. The simulations show average impacts on yields are small but some larger positive and negative outcomes can also be expected (Fig. 6).

Because the average reductions predicted by the crop models are relatively small, the use of soil amendments as adaptation strategies can more than offset the negative impact of climate change, with mean yield gains ranging between 20% and 80% (Fig. 6). The use of organic amendments such as legume residues and low inorganic fertilizer application show higher yield variability as compared to the no-adaptation scenario, however average yields under adapted management are greater than 2 t/ha. The subsequent maize crop after mucuna would benefit from biological nitrogen fixation and also from the crop residues that are applied. Such adaptation strategies would benefit resource-poor farmers to improve main staple crop yields with minimal external inputs. Again, we emphasize that the analysis assumes that these changes in management would not be made as part of the ongoing improvement in practices that is represented in the RAP.

Impacts of climate change and adaptation packages on livestock productivity were assessed through changes in feed quantity and quality. Reduced grass growth due to climate change lowered feed intake from the rangelands by 10% and 50% in the rainy season and dry season respectively. Climate change reduced on-farm maize stover yield by on average 15%, further aggravating the dry season feed gaps that are characteristic for the mixed crop-livestock systems in semi-arid areas. The adaptation package helped offsetting the adverse effects of climate change on fodder availability by increasing the fodder quantity through fertilizer input and rotations with legume crops. The diversification with legume grain and fodder crops also improved the fodder quality, primarily through higher protein content.

Climate change resulted in a 35–39% and 30–35% reduction of annual milk production for households with small and large herds respectively (Fig. 7). Offtake was roughly halved by climate change (Fig. 7) and with lower feed availability resulting in underfed animals, mortality rates rose by 8% and 14% for households with small and large herds respectively. With the adaptation package, on-farm feed quantity and quality was improved, resulting in milk production at roughly the same level that was obtained without climate change. The offtake was brought back to about 80% and 90% of the offtake in the current climate for households with small and large herds respectively.

## 4.6 Economic Analysis: Climate Impact, Adaptation, Vulnerability and Resilience

Table 2 summarizes the results of the economic analysis of climate change impact for the farm population in Nkayi stratified by cattle ownership. We compare climate change impact without adaptation and with the adaptation package (comprised of

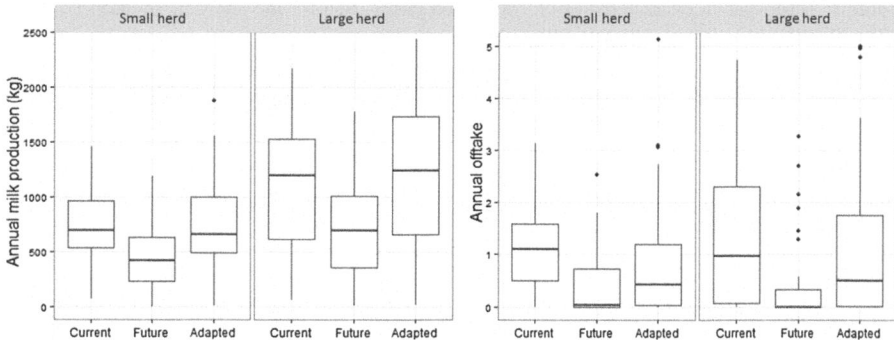

**Fig. 7** Annual milk production and offtake per farm in the current and future climate without adaptation package and with the adaptation package (long duration maize varieties, allocation of land to a maize-mucuna rotation, mucuna biomass left on the fields as inorganic fertilizer for subsequent maize, and use of micro-dosing of N on maize) for households with small and large herds

the elements identified above: long duration maize varieties, allocation of land to a maize-mucuna rotation, mucuna biomass left on the fields as inorganic fertilizer for subsequent maize, and use of micro-dosing of N on maize in the maize-mucuna rotation). We emphasize that these results are based on a single scenario comprised of one climate model projection, one crop model and livestock model, and one socio-economic scenario, to illustrate the type of analysis that can be done. More generally, it is important to consider the uncertainties in each component by utilizing a range of scenarios and model assumptions.

Without adaptation, Table 2 shows that vulnerability to loss from climate change ranges from 45% of the farm households without cattle, to 61% and 71% of households with small and large herds. The households with cattle are more vulnerable because, as discussed above, the main adverse impact shown by the crop and livestock model simulations is on livestock feed availability and livestock productivity. These losses range from 25% to 57% of mean farm net returns before climate change, and thus represent a substantial loss for the vulnerable households, and correspond to losses of 11–16% of per capita income. However, some farms gain, and these gains range from 28% to 34% of mean returns before climate change. These gains are attributed to the heterogeneity in the bio-physical and economic conditions that exist. For example, in any given year, rainfall varies across the landscape with some areas drier and some wetter, with corresponding variation in crop and forage productivity. The net impacts aggregated across all farms are small for farms without livestock (about +3%), but much larger and negative for farms with large herds (−23%). It is important to recognize that even though the losses are a larger percent of farm income for the farms with cattle, the farms without cattle are much poorer. Thus, with climate change the negatively impacted farms without cattle will be in an even worse condition than before climate change and much poorer than the farms with cattle.

Table 2 shows that farms without cattle are very likely to adopt the adaptations being considered, with adoption rates about 96% in the rapid adaptation scenario and over 75% in the scenario of a transitional adaptation in which the benefits are

**Table 1** Base system characteristics of 160 mixed farms used for the analysis, by farm type, in Nkayi district

| Variables | Units | 0 cattle Mean | 1–8 cattle Mean | >8 cattle Mean | Total Mean | Std. Dev. |
|---|---|---|---|---|---|---|
| Proportion in community | % | 42.5 | 38.1 | 19.4 | n.a. | n.a. |
| Household members | people | 5.9 | 6.9 | 7.4 | 6.6 | 2.5 |
| Proportion of female headed households | % | 27.9 | 31.1 | 22.6 | 28.1 | n.a. |
| Net returns maize | US$/farm | 60 | 162 | 63 | 100 | 121 |
| Net returns other crops | US$/farm | 31 | 62 | 35 | 44 | 53 |
| Net returns cattle | US$/farm | 0 | 472 | 1347 | 443 | 586 |
| Net returns other livestock | US$/farm | 9 | 19 | 15 | 14 | 29 |
| Off-farm income | US$/farm | 220 | 300 | 294 | 265 | 217 |
| Farms with maize | % | 98.5 | 100.0 | 100.0 | 100.0 | 0.1 |
| Maize area | Ha | 1.1 | 1.4 | 1.8 | 1.3 | 0.8 |
| Maize grain yield | kg/ha | 497 | 826 | 675 | 657 | 531 |
| Farms with small grains | % | 23.5 | 32.8 | 41.9 | 30.6 | 46.2 |
| Small grain area | Ha | 0.7 | 0.7 | 1.0 | 0.8 | 0.8 |
| Small grain yield | kg/ha | 393 | 726 | 327 | 512 | 622 |
| Farms with legumes | % | 33.8 | 49.2 | 48.4 | 42.5 | 49.6 |
| Legume area | ha | 0.4 | 0.4 | 0.5 | 0.4 | 0.3 |
| Legume yields | kg/ha | 452 | 722 | 388 | 557 | 541 |
| Cattle[a] | TLU | 0 | 5.4 | 13.9 | 4.7 | 4.7 |
| Other livestock[a] | TLU | 0.3 | 0.5 | 1.6 | 0.6 | 0.9 |

[a]Herd size: Cattle = 1.14 Tropical Livestock Unit (TLU), donkeys = 0.5 TLU, goats and sheep =0.11 TLU

realized more gradually over 10 years. Farms without livestock would gain more (as a percent of their base system farm income) than farms with cattle, but do not necessarily gain more in absolute terms because the farms without livestock have much lower incomes (Table 1). The relatively smaller impact of climate change and greater benefit from adaptation for farms without livestock is because these adaptations improve crop productivity more than livestock productivity (Fig. 6). The adaptations have substantial impacts on per capita incomes, more than doubling the farm incomes of the poorest households.

For analysis of resilience, we considered two versions of the adaptation scenarios, a transitional case in which adaptation takes 10 years for farmers to realize the full benefits of the practices (e.g., due to a gradual dissemination of the technology and information), and a rapid case in which farmers realize the full benefits immediately. Recall that we defined resilience as the degree to which a system can be adapted to minimize the losses of climate change. In the analysis presented in Table 2, we interpret the rapid adaptation as the smallest possible loss, so its resilience is 100%, and we evaluate the no-adaptation case and the transitional adaptation case relative to the

**Table 2** Future (2050) farming system vulnerability and resilience, and net economic impacts of climate change, for crop-livestock systems in Nkayi, Zimbabwe, for no adaptation, transitional adaptation and rapid adaptation scenarios, hot dry GCM (all values are percent)

| Stratum | Adaptation | Vulnerability | Climate impact on net returns | | | | Adoption of adaptations | |
|---|---|---|---|---|---|---|---|---|
| | | | Gains | Losses | Net impact | Resilience | Adoption rate | Adopter gain |
| No cattle | None | 45 | 28 | −25 | 3 | 91 | n.a. | n.a. |
| No cattle | Transition | 18 | 73 | −32 | 41 | 93 | 75 | 60.5 |
| No cattle | Rapid | 1 | 139 | −20 | 119 | 100 | 96 | 136 |
| Small herd | None | 61 | 32 | −41 | −9 | 79 | n.a. | n.a. |
| Small herd | Transition | 39 | 42 | −33 | 9 | 93 | 80 | 20 |
| Small herd | Rapid | 25 | 51 | −27 | 24 | 100 | 98 | 51 |
| Large herd | None | 71 | 34 | −57 | −23 | 79 | n.a. | n.a. |
| Large herd | Transition | 46 | 47 | −42 | 5 | 98 | 64 | 43 |
| Large herd | Rapid | 42 | 48 | −40 | 8 | 100 | 80 | 87 |

Note: Transitional adaptation occurs over 10 growing seasons. Rapid adaptation occurs in the first growing season. Gains, Losses, Net Impact and Adopter Gain are percent of base system net returns.

rapid adoption case. The analysis considers the benefits over a 10-year period using a discount rate of 10%.

With these assumptions, the no-adaptation scenario gives the farms without cattle a resilience of 91%, somewhat higher than the resilience of the systems with cattle (79%). With transitional adaptation, the farms without livestock improve from 91% to 93%, whereas the farms with livestock improve from 79% to 93% (small herds) and 98% (large herds). Table 2 also shows that with rapid adaptation more farmers would adopt and the benefits would be much larger, especially for the small farms without livestock. This analysis illustrates the potential benefits of enhancing the adaptive capability of farmers, enabling them to substantially reduce vulnerability and enhance resilience when effective adaptation options are available.

## 5 Conclusions

In this chapter we described and demonstrated the use of new simulation-based technology impact assessment methods, developed by AgMIP, to evaluate the potential for currently available or prospective agricultural systems to achieve the goals of CSA. We described methods used to quantify the vulnerability and resilience of agricultural systems, two key elements of CSA. We used a case study of crop-livestock systems in Zimbabwe to illustrate how these methods can be used to evaluate alternative management practices for climate smart agriculture.

Our analysis of the Zimbabwe case illustrates the potential for these methods to test the usefulness of specific modifications to raise incomes, reduce vulnerability to climate change and to enhance resilience. While we must caution against generalizing from this single example, we do think that it illustrates the potential importance of making improved technologies available but also the role that adaptive capacity will play. This example also serves to demonstrate why it is important to clearly define the "simulation experiment," i.e., the conditions under which climate impacts and adaptations are being evaluated. In this example, it was assumed that there would be relatively little change in productivity over time, and that a package of improved practices that we called "climate adaptations" could provide higher incomes for many of the farmers. However, one could ask why these improvements are considered "climate adaptations" and what changes in the institutional or policy environment would be needed to facilitate their use. Thus, for a meaningful analysis of CSA, or climate adaptation more generally, these policy dimensions of the story must be addressed. Otherwise, the type of analysis we have presented here risks overstating the potential for adaptations to offset the potentially adverse effects of climate change.

Although we have not discussed mitigation of greenhouse gases in this chapter, it is important to note that the framework presented here can also incorporate greenhouse gas emissions as part of a technology assessment. Examples of how this modeling framework can be used for that purpose are presented in a number of publications, including Antle and Stoorvogel (2008). However, it should be noted that accurate quantification of greenhouse gas emissions, including changes in soil carbon, nitrous oxide emissions from soils, and methane emissions from livestock, is data-intensive and requires the use of complex models. Alternatively, estimates of average rates of emissions under alternative practices could be used. This is an area in need of further research.

Another area that clearly needs additional research is the incorporation of livestock herd dynamics and the interaction of crop and livestock systems. This is particularly important for smallholder farm households whose livelihoods and well-being depend on livestock both as a source of food and income as well as an asset that can be used to cope with climate variability and extremes. Further work on the role of livestock and crop-livestock systems in the context of climate smart agriculture is clearly warranted.

# References

Agricultural Model Inter-comparison and Improvement Project (AgMIP) *Guide for Regional Integrated Assessments: Handbook of Methods and Procedures, Version 6.0* (2015). http://agmip.org.

Antle J (2011) Parsimonious Multi-dimensional Impact Assessment. Amer. J. Agr. Econ. 93(5):1292–1311. doi:10.1093/ajae/aar052

Antle JM, SM Capalbo (2010) Adaptation of Agricultural and Food Systems to Climate Change: An Economic and Policy Perspective. Applied Economic Perspectives and Policy 32:386–416.

Antle JM, JJ Stoorvogel, RO Valdivia (2006) Multiple Equilibria, Soil Conservation Investments, and the Resilience of Agricultural Systems. Environment and Development Economics 11(4):477–492.

Antle JM, J Stoorvogel, R Valdivia (2014) New Parsimonious Simulation Methods and Tools to Assess Future Food and Environmental Security of Farm Populations. Philosophical Transactions of the Royal Society B. doi: 369:20120280

Antle JM, RO Valdivia, KJ Boote et al (2015b) AgMIP's Trans-disciplinary Agricultural Systems Approach to Regional Integrated Assessment of Climate Impact, Vulnerability and Adaptation. In: Rosenzweig C, D Hillel (eds) Handbook of Climate Change and Agroecosystems: The Agricultural Model Intercomparison and Improvement Project Integrated Crop and Economic Assessments, Part 1. Imperial College Press, London.

Bossio D (2009) Livestock and water: understanding the context based on the 'Compressive Assessment of Water Management in Agriculture'. The Rangeland Journal 31 (2):179–186.

Folke, C. (2006). Resilience: The emergence of a perspective for social-ecological systems analyses. Global Environmental Change, 16, 253–267. doi:10.1016/j.gloenvcha.2006.04.002

Hallegate, S. 2014. Economic Resilience: Definition and Measurement. Policy Research Working Paper 6852, The World Bank.

Hatfield, J.L., K.J. Boote, B.A. Kimball, L.H. Ziska, R.C. Izaurralde, D. Ort, A.M. Thomson, and D.W. Wolfe.2011. Climate impacts on agriculture: Implications for crop production. Agron. J. 103:351–370. doi:10.2134/agronj2010.0303

Hatfield, J. L. and Prueger, J.H. 2015. Temperature extremes: Effect on plant growth and development. Weather and climate extremes. 10. A. 4-10. doi: org/10.1016/j.wace.2015.08.001

Homann-KeeTui, S., D. Valbuena, P. Masikati, K. Descheemaeker, J. Nyamangara, L. Claessens, O. Erenstein, A. van Rooyen, D. Nkomboni. 2015. Economic trade-offs of biomass use in crop-livestock systems: Exploring more sustainable options in semi-arid Zimbabwe. Agricultural Systems 134:48–60.

Homann, S., van Rooyen, A., Moyo, T., Nengomasha, Z., 2007 Goat production and marketing: Baseline information for semi-arid Zimbabwe. International Crops Research Institute for the Semi-Arid Tropics, pp84

Homann-KeeTui, S., Bandason, E., Maute, F., Nkomboni, D., Mpofu, N., Tanganyika, J., Van Rooyen, A. F., Gondwe, T., Dias, P., Ncube, S., Moyo, S., Hendricks, S., Nisrane, F. 2013. Optimizing Livelihood and Environmental Benefits from Crop Residues in Smallholder Crop-Livestock Systems in Southern Africa. ICRISAT Socio-economics Discussion Paper Series. Series Paper Number 11.

Kandji, T. S., Verchot, L., and Mackensen, J. 2006. *Climate change and variability in Southern Africa: Impacts and adaptation in the agricultural sector*, World Agroforestry Centre and United Nations Environment Programme Report, p. 42.

Kates, R.W., Travis, W.R., Wilbanks, T.J. 2012. Transformational adaptation when incremental adaptations toclimate change are insufficient. PNAS, 109;19, 7156–7161

Keating, BA., P.S. Carberry, G.L. Hammer, M.E. Probert, M.J. Robertson, D. Holzworth. 2003. An overview of APSIM, a model designed for farming systems simulation. *European Journal of Agronomy* 18: 267–288.

Lipper, L. P. Thornton, B.M. Campbell, T. Baedeker, A. Braimoh, M. Bwalya, P. Caron, A. Cattaneo, D. Garrity, K. Henry, R. Hottle, L. Jackson, A. Jarvis, F. Kossam, W. Mann, N. McCarthy, A. Meybeck, H. Neufeldt, T. Remington, P. Thi Sen, R. Sessa, R. Shula, A. Tibu, E.F. Torquebiau. 2014. Climate Smart Agriculture for Food Security. *Nature Climate Change* **4**, 1068–1072.doi:10.1038/nclimate2437.

Malone, E.L. 2009. Vulnerability and Resilience in the Face of Climate Change: Current Research and Needs for Population Information. Battelle Pacific Northwest Division Richland,

Washington 99352. http://www.globalchange.umd.edu/data/publications/Resilience_and_Climate_Change.pdf. Accessed December 6 2015.

Masikati, P., S. Homann-KeeTui, K. Descheemaeker, O. Crespo, S. Walker, C.J. Lennard, L. Claessens, A.C. Gama, S. Famba, A.F. van Rooyen, and R.O. Valdivia. 2015. Crop–Livestock Intensification in the Face of Climate Change: Exploring Opportunities to Reduce Risk and Increase Resilience in Southern Africa by Using an Integrated Multi-modeling Approach. C. Rosenzweig and D. Hillel, eds. *Handbook of Climate Change and Agroecosystems: The Agricultural Model Intercomparison and Improvement Project Integrated Crop and Economic Assessments, Part 2*. London: Imperial College Press.

Masikati, P., 2011. Improving the Water Productivity of Integrated Crop-livestock Systems in the Semi-arid Tropics of Zimbabwe: Ex-ante Analysis Using Simulation Modeling. Ph.D. Thesis, ZEF, Bonn.

Morton, J. F. 2007. The impact of climate change on smallholder and subsistence agriculture, *PNAS*, **104**(50), 19680–19685.

Nelson, G. C.; Valin, H.; Sands, R. D.; Havlik, P.; Ahammad, H.; Deryng, D.; Elliott, J.; Fujimori, S.; Hasegawa, T.; Heyhoe, E.; Kyle, P.; Lampe, M. V.; Lotze-Campen, H.; d'Cros, D. M.; van Meijl, H.; van der Mensbrugghe, D.; Muller, C.; Popp, A.; Robertson, R.; Robinson, S.; Schmid, E.; Schmitz, C.; Tabeau, A. & Willenbockel, D. 2013. 'Climate change effects on agriculture: Economic responses to biophysical shocks', *Proceedings of the National Academy of Sciences of the United States of America* doi:10.1073/pnas.1222465110, 1–6.

Nelson, Donald R., W. Neil Adger, and Katrina Brown. 2007. "Adaptation to Environmental Change: Contributions of a Resilience Framework." Annual Review of Environment and Resources 32:395–419. http://eprints.icrisat.ac.in/4245/1/AnnualReviewofEnvResources_32_395-419_2007.pdf

Rickards, L. and Howden, S. M. 2012. Transformational adaptation: agriculture and climate change. Crop & Pasture Science, **63**, 240–250., http://dx.doi.org/10.1071/CP11172

Rippke, U., Ramirez-Villegas, J., Jarvis, A., Vermeulen, S.J., Parker, L., Mer, F., Diekkrüger, B., Challinor, A.J. Howden, M. 2016. Timescales of transformational climate change adaptation in sub-Saharan African agriculture. Nature Climate Change Letters. | DOI: 10.1038/NCLIMATE2947

Rosenzweig, C., J.W. Jones, J.L. Hatfield, J.M. Antle, A.C. Ruane and C.Z. Mutter. (2015). The Agricultural Model Intercomparison and Improvement Project: Phase I Activities by a Global Community of Science. C. Rosenzweig and D. Hillel, eds. *Handbook of Climate Change and Agroecosystems: The Agricultural Model Intercomparison and Improvement Project Integrated Crop and Economic Assessments, Part 1*. London: Imperial College Press.

Rosenzweig, C., A. Arslan, F. Matteoli, M. Ngugi, and T. Rosenstock. 2016. KAG Sub-Group on Integrated Planning and Monitoring for Climate-Smart Agriculture. (in preparation).

Rufino, M.C., J. Dury, P. Tittonell, M.T. Van Wijk, S. Zingore, M. Herrero and K.E. Giller. 2008. Collective management of feed resources at village scale and the productivity of different farm types in a smallholder community of North East Zimbabwe. Submitted to Agric. Syst.

Rufino, M.C., Herrero, M., van Wijk, M.T., Hemerik, L., de Ridder, N., Giller, K.E., 2009. Lifetime productivity of dairycows in smallholder farming systems of the highlands of Central Kenya. Animal 3, 1044–1056

Rufino, M.C., Dury, J., Tittonell, P., van Wijk, M.T., Herrero, M., Zingore, S., Mapfumo, P. and Giller, K.E. (2011). Competing use of organic resources village-level interactions between farm types and climate variability in a communal area of NE Zimbabwe, *Agric. Syst.* **104**, 2011, 175–190.

Valdivia, R.O., J.M. Antle, C. Rosenzweig, A.C. Ruane, J. Vervoort, M. Ashfaq, I. Hathie, S. Homann-KeeTui, R. Mulwa, C. Nhemachena, P. Ponnusamy, H. Rasnayaka and H. Singh. (2015). Representative Agricultural Pathways and Scenarios for Regional Integrated Assessment of Climate Change Impact, Vulnerability and Adaptation. C. Rosenzweig and D. Hillel, eds. *Handbook of Climate Change and Agroecosystems: The Agricultural Model Intercomparison and Improvement Project Integrated Crop and Economic Assessments, Part 1*. London: Imperial College Press.

Williams, T. O., Thornton, P., Fernandez-Rivera, S., (2002). Trends and prospects for livestock systems in the semi-arid tropics of Sub-Saharan Africa. In Targeting Agricultural Research for Development in Semi-Arid Tropics of Sub-Saharan Africa. Proceedings held at International Center for Research in Agroforestry Nairobi, Kenya from 1 to 3 July 2002, pp155–172

World Bank (2009). Making development climate resilient: A World Bank strategy for Sub-Saharan Africa, Report No. 46947 — AFR.

**Open Access** This chapter is distributed under the terms of the Creative Commons Attribution-NonCommercial-ShareAlike 3.0 IGO license (https://creativecommons.org/licenses/by-nc-sa/3.0/igo/), which permits any noncommercial use, duplication, adaptation, distribution, and reproduction in any medium or format, as long as you give appropriate credit to the Food and Agriculture Organization of the United Nations (FAO), provide a link to the Creative Commons license and indicate if changes were made. If you remix, transform, or build upon this book or a part thereof, you must distribute your contributions under the same license as the original. Any dispute related to the use of the works of the FAO that cannot be settled amicably shall be submitted to arbitration pursuant to the UNCITRAL rules. The use of the FAO's name for any purpose other than for attribution, and the use of the FAO's logo, shall be subject to a separate written license agreement between the FAO and the user and is not authorized as part of this CC-IGO license. Note that the link provided above includes additional terms and conditions of the license.

The images or other third party material in this chapter are included in the chapter's Creative Commons license, unless indicated otherwise in a credit line to the material. If material is not included in the chapter's Creative Commons license and your intended use is not permitted by statutory regulation or exceeds the permitted use, you will need to obtain permission directly from the copyright holder.

# Climate Smart Food Supply Chains in Developing Countries in an Era of Rapid Dual Change in Agrifood Systems and the Climate

Thomas Reardon and David Zilberman

**Abstract** Food supply chains are essential to food security in developing regions where today the great majority of food consumed is purchased from rural-urban, rural-rural, and urban-rural supply chains. Disrupting those supply chains means disrupting food security. Yet short-term climate shocks and long-term climate change threaten to cause that disruption. This chapter does four things: (1) analyzes the types and determinants of vulnerabilities of food supply chains to climate shocks and change; (2) considers how those vulnerabilities are conditioned by urbanization, diet change, and rapid transformation of food systems; (3) discusses how supply chain actors, from farmers to processors and distributors and input suppliers, invest in mitigation of the risks of these shocks and reduction of their vulnerabilities; (4) discusses policy implications and lays out an agenda for research for climate smart food supply chains in developing regions.

## 1 Introduction

The literature on Climate Smart Agriculture (CSA) has mostly ignored that farming of a given product, like fish, fruit or milk, occurs within a complex supply chain. Even CSA literature on "food systems" (such as FAO's book on this theme, Elbehri (ed.) Elbehri 2015) tends to focus on biophysical dimensions of climate change in farm areas, not all the way along supply chains. This chapter aims at that gap.

The supply chain has a complex dendritic cluster structure composed of three dimensions:

---

T. Reardon (✉)
Department of Agricultural, Food and Resource Economics, Michigan State University, East Lansing, MI, USA
e-mail: reardon@msu.edu

D. Zilberman
Department of Agriculture and Resource Economics, University of California Berkeley, Berkeley, CA, USA

(a) Four main segments of the product's supply chain, with flow from upstream to downstream: (1) farm input supply chains, which are "upstream" of farms; (2) the farm segment, which forms approximately half of the full costs and value added of food supply chains in developing countries (Reardon 2015); (3) the "midstream" segment consisting of processing, wholesale, and transport; and (4) the "downstream" segment consisting of retailing (stores and restaurants).
(b) Supply chains of innovation (R&D) supply innovations for the technologies, institutions, and organizations of each segment.
(c) Input supply chains upstream of each segment provide inputs to it such as equipment to fertilizer manufacturers and fuel to transporters.

This three-dimensional cluster is important to food security and livelihoods as well as vulnerable to climate shocks. These complex supply chains are important for food security because they provide 90% of urban consumers' food in Africa and Asia (the other 10% from imports) (Reardon 2016). Further, our research shows that 50–80% of rural diets (in value terms) in Africa and Asia come from purchased food. We also found that nearly 100% of rural households in Asia, and 98% in Africa, buy food and thus depend on the supply chain for food security. With respect to rural livelihoods, supply chains provide supply inputs to farmers and households depend on them as sales conduit to urban areas, the main markets in developing countries. Moreover, the off-farm components of the supply chains, such as transport, commerce and processing and handling, are key sources of employment in rural areas for a majority of rural households, and of cash for farm investments.

Finally, while conventional wisdom tends to see domestic food supply chains in developing regions as traditional and stagnant, in fact they have transformed greatly. The transformation has involved moving from traditional short fragmented chains to transitional and modern forms in developing regions. This is discussed in more detail in Sect. 2. Supply chains have grown massively in volume and length as the urban areas served rapidly expand and reach out into rural areas. Supply chains have transformed rapidly in structure – such as the rise of supermarkets and large processors, and in product composition – with a concurrent shift in diets toward processed foods, and non-grain products such as milk, meat, fish, fruits, and vegetables (Tschirley et al. 2015). They have also transformed rapidly in conduct – with basic changes in food processing and transport technology, and with the rise of standards and contracts (Swinnen 2007).

A crucial point is that short-term climate shocks and long-term change can heavily affect not just the farm segment of the above complex supply chains, but also the segments upstream and downstream from the supply chain's "four legged chair", as well as the input supply chains to all those segments. These shocks and changes can be challenging – even fundamentally disruptive – to these complex chains, endangering food security and livelihoods for both rural and urban households. For these reasons, to develop CSA systems we need to understand all three dimensions of food supply chains, how the three off-farm segments respond to climate change and shocks, how farmers do and should react to these responses, and then what type of policies are needed to take into account the complete supply chain.

Moreover, climate smart supply chains must consider that the short run and long run impacts of climate change are different, and thus so are short-term and long-term impact of and response to climate change. In the shorter run, we may observe increased probabilities of extreme climatic events such as typhoons and droughts, which shock agriculture and disrupt supply chains (Kleindorfer and Saad 2005). These require risk management and climate shock coping innovations in supply chains. For most agrifood companies, the short-term climate shocks dominate their climate-related concerns as they are issues of immediate business survival; all but the largest companies are forced to have short-run planning horizons.

In the longer run, climate change may affect the configuration of agricultural supply chains. The changes include migration of weather patterns from the equator towards the poles, melting snow and ice, and rising sea levels (Nicholls and Cazenave 2010). This may require that supply chains adapt by shifting supply zones and innovating structurally to new configurations. These will appear as short-term shock adaptations over time; but they will also appear as long-term planning issues for the largest companies and governments. Adaptation may appear in the form of introduction and adoption of new innovations, changes in land use and trade patterns, migration, increased reliance on insurance, and increased investment to enhance resilience of farming operations as well as distribution and processing (Zilberman et al. 2012). Each of these activities has significant impacts on the organization of agriculture and agricultural supply chains over time.

A research agenda on the impact of climate change on simple and complex agricultural supply chains is essential to comprehend its impacts and in developing policies and mechanisms to address them. A literature has emerged on managing risks of disruption in nonfood manufactures supply chains (e.g., Oke and Gopalakrishnan 2009), but there has been little on this topic for food supply chains, let alone with climate crises and change as the cause of the disruption.

Moreover, climate changes manifested in both short-term crises and long-term changes represent a new and complex set of shocks that call on researchers to alter the way supply chains are studied; that is, to date, supply chain contexts have been taken to be either static or only slowly changing, and hanging along one or other particular dimension, such as urbanization or market liberalization. By contrast, climate change represents fundamental increases in unanticipated risk as well as basic changes in the agroecological contexts of the supply chain. In addition, climate change will result in new innovations that are likely to lead to modifications and changes in supply chains (Du et al. 2016). These will require an agenda of research on new methods and models of understanding these changes.

The next section presents the main patterns of transformation of supply chains. It will be followed by two sections discussing the short and long-term climate changes, their impacts on supply chains, and the measures post-harvest actors take to mitigate these impacts. In the conclusion we offer initial implications of these changes for the climate change debate and propose policy implications and a research agenda.

## 2 Background: Up-to-Date Vision of Transforming Supply Chains in Africa and Asia as Basis for Assessment of Climate Impacts

To assess the challenges of climate change and shocks on food supply chains, it is crucial to have a realistic and up-to-date vision of what food supply chains have become in Africa and Asia. The transformation of complex supply chains has taken place at somewhat different speeds and extent over products and zones as well as countries and regions. Farmers relying on traditional supply chains (short, fragmented, and with low dependency on infrastructure and equipment) are paradoxically exposed to more income risk from local climate shocks as their overall livelihood is less commercialized and more dependent on the vicissitudes and low purchasing power of the local economy. At the same time, traditional supply chains are less exposed to climate shocks that can occur along geographically longer and more complex supply chains.

Moreover, while modern supply chains have more sources of vulnerability than traditional chains, the modern chains also have potentially more means to escape from and mitigate climate vulnerability. For example, longer chains afford much greater chances to increase overall income and manage risk through market engagement. Altogether, whether the farmer who depends on traditional or modern supply chains is exposed to lower overall risk is an empirical question that will vary by place and crop. It can be addressed with respect both to the degree of transformation (modernization) of the supply chains and the degree to which post-harvest actors in the supply chain invest in climate-mitigating technologies and institutions.

At the essence of transformation is a system that is local, grain focused, small scale, spot market, and labor or land intensive to a system with geographically long supply chains, a diversified product portfolio, coordination institutions such as standards and contracts, varying degrees of disintermediation and consolidation, and increasingly capital-intensive technology used in each of the segments. In general the transformation has developed around two broad axes. The first is structural, including change in the number of segments and sub-segments of agrifood value chains (VCs), and the degree of concentration and ownership of capital (public versus private, domestic versus foreign) per segment. The second is behavioral, including how actors per segment buy, make, and sell, and the choices made of technology, institutions (like standards and contracts) and organizations (like vertical and horizontal integration and coordination). This transformation is led by a rapidly urbanizing population, domestically sourced food in Africa and Asia and increasing food purchases by rural populations.

First, urbanization has been very rapid, and the urban share in national food markets is now dominant, or nearly so. In (developing) Eastern and Southern Africa (ESA), 30% of the population is urban but represents 40% of national food consumption, and roughly 50% of food market purchases. In West Africa, those shares are roughly 40%, 50%, and 60%, and in China and Southeast Asia, roughly 45%, 55%, and 65%, respectively. These numbers are surprising in national debates in

Africa and Asia partly because, as we perceive it, the extent and importance of urbanization has not yet entered national food security debates, and because the image of large rural populations dominating food needs is a persistent vestige of the situation only a decade or two ago when the urban market was a small niche market. Further, urban consumption volumes have become massively larger. Rural-urban food supply chains have expanded 800% in Africa, 300% in South Asia, 1000% in Southeast Asia over the past three decades. The key implication of this for climate shock vulnerability is that the length of supply chains is growing as urban areas source further afield.

Second, the great majority of food supply in Africa and Asia is from domestic sources; food imports are only about 10% of total food consumption in Africa and Asia. Driven by food security issues, we focus mainly on domestic supply chains. The implication for the climate change debate is that the great majority of food supply is vulnerable to climate shocks internal to particular developing countries. Still, international supply chains that provide the small share (10%) of the domestic food supply from imports, or convey the 5–10% of farm incomes that come from exports, are vulnerable to climate shocks on long maritime passages as well as to policy shocks such as export bans.

Third, rural households have shifted from subsistence farming to depending more on food purchases. In ESA, the share represents 45% of total rural food expenditures (meaning 45% of their food consumption comes from purchases and 55% from own-farm production) and in Asia 60–70% of total rural food expenditures.

These first three points imply that climate vulnerability issues along a supply chain can be divided directionally, namely: (1) rural to urban ("rural-urban") supply chains; (2) rural to rural ("rural-rural"); (3) urban to rural ("urban-rural"). To date the vast majority of research on supply chains has been international or rural-urban. We focus here on domestic rural-urban supply chains and leave to future analysis (based on field research currently being conducted in Africa) to elucidate the transformations of the other two types of supply chains and explore to what extent they have climate vulnerabilities and strategies that differ from rural to urban chains.

Rural-urban food supply chains are transforming rapidly, consolidating, and undergoing technological change. Transformation is occurring in terms of growth in supply chain volume, rapid diversification beyond grains, and increasing delivery of processed foods. For example, in ESA non-grains in urban household total food expenditure (TFE) is 66%, while rural is 61%. This implies more perishable goods and thus increased vulnerability to transport and storage conditions, which affect food safety and food loss. Processed foods in Asia constitute 73% of TFE in urban, and 60% in rural TFE; in ESA 56% of urban TFE and 29% of rural TFE. Processing is vulnerable to energy and climate shocks in two primary ways. First is availability of energy supply due to cost, reliability of the grid, and transportation routes. Second is delays in transportation that result in spoilage. In the past, drying, salting, and pickling reduced vulnerability to delays, but the shares of these technologies are declining, even in developing countries, as fast transport chains and cold storage, which are more vulnerable, take over.

Post-harvest consolidation is occurring in two ways. Supply chains are becoming "intermediationally shorter", which implies consolidation over actors transporting food. This is increasing the scale of transport vehicles and wholesale/logistics and further concentrating transport. In retail, consolidation is occurring with the rapid rise of supermarkets, processing, wholesale/logistics, and agricultural input sectors. These changes are driven, in part, by a rise in the share of foreign direct investment (FDI), or "multinationalization". Both changes imply a concentration of the effects of climate change over fewer actors. Finally, technological change is resulting in a greater reliance on energy production with more dependence on equipment, as shown through an increase in the capital-labor ratio.

## 3 Short-Term Climate Change Impacts on Transforming Food Supply Chains

### 3.1 Impacts on Food Supply Chains from Short-Term Climate Change

Short-term climate shocks increase climate vulnerability and can be measured at various points along the supply chain. Climate-shock vulnerability points are called "hotspots" in the energy or food safety or phytosanitary literature (Giorgi 2006). Hotspots occur both in segments themselves (such as cold storage points, dry storage points, processing points, farming points, and input delivery paths to farms) as well as in sub-segments or individual operation points (such as mountain feeder roads to main highways) and input ingress points (such as water canals for farms or fuel, or electricity delivery interfaces). Vulnerability at each hotspot is dependent on the type of shock and attributes of a given segment. Further, the points need not be directly in the supply chain, but rather in secondary supply chains that feed into the product supply chain.

Examples of short-term climate shocks are floods or hillside rock avalanches on highways, tidal wave or typhoon destruction of sea or river ports or disruption of energy or fuel sources. These changes can disrupt or stop the product or input flow, especially along longer supply chains. For example, large poultry production and processing in Thailand by CP Foods relies on grain imports from the United States and imports chicken parts to China and Russia. A stoppage of operation in one of the facilities may disrupt production throughout the system, and may be very costly. Along a domestic supply chain, poultry production for urban consumption in Bangladesh or Nigeria relies on feed ingredient shipments from grain and cassava zones to peri-urban feed and poultry production facilities, which are vulnerable to road flood-outs and political strife (Liverpool-Tasie et al. 2016 for Nigeria).

A key point is that vulnerability of a supply chain often increases with the number and nature of hotspots. Further, the number and nature of the hotspots are in turn functions of the structure, conduct, and performance of the supply chain. We can categorize these conditioning factors, which are elements of the transformation of the supply chain, as follows.

The first determinant of a hotspot in the supply chain is the physical infrastructure affecting production risk in the supply area. The irrigation and drainage and flood control infrastructure upstream in the farming area is a crucial conditioner of the impact of drought and flooding shocks. This kind of private and public infrastructure is present far more in Asia, particularly East and Southeast Asia and in some zones of South Asia, and far less in Africa (Rosegrant et al. 2009). This discrepancy highlights the relative vulnerabilities by geography.

The second factor is the geographic distance along the supply chain. Longer geographic distance to the farm zone, and/or longer "lead time" from the assemblage and first stage processing and the final processing and demand points, increase vulnerability to climate shocks. There is however a trade-off between the vulnerability this implies and the diversification of urban food supply sources that long supply chains afford, which could reduce vulnerability to some degree. Even so, the rapid urbanization in both Africa and Asia is resulting in longer supply chains with increased climate vulnerability.

The third factor is the degree of product perishability. The greater the perishability of the product, and thus the need for fast delivery and/or cold storage, the greater the vulnerability to climate shock. This factor again increases climate vulnerability in Africa and Asia as the diet transformation has brought a huge surge in the demand for perishables.

A fourth factor is physical intensity in a given segment (e.g. irrigation, farm equipment, cold storage, delivery trucks). The robustness of physical capital is a key element in the vulnerability of supply chains to climate shocks. An example is the widespread damage to flimsy bamboo greenhouses on Java during unexpectedly virulent storms in the past few years. There is a general tendency for the capital/labor ratio to rise in food supply chains as one moves from traditional to transitional to modern chains, which increases vulnerability. That tendency is for three reasons: (1) the labor market tends to tighten with urbanization and physical capital substitutes for labor; (2) physical capital enables supply chain managers to reduce vulnerability by off-setting climate-imposed costs with economies of scale, and reducing transport times with larger vehicles and inter-modal facilities, and increased cost competition in commoditizing supply chains further drives this investment; and (3) increased quality competition in modernizing supply chains increases equipment needs to achieve quality and safety attributes from suppliers to meet buyer requirements and standards. Growing dependence of suppliers and buyers on "asset-specific investments" may increase incentives to protect these assets from climate shocks (such as by investments in flood control).

A fifth factor is the location specificity of production or intermediation. Vulnerability to climate change decreases with more interchangeable places to produce a crop or handle it logistically. Location specificity, as a special case, can be linked to asset-specificity in that buyers depend on, are perhaps "locked into," sourcing from a farm zone or intermediation point due to specialized resources, firms or farms. This in a sense "holds hostage" the supply chain to these locations and thus to climate shocks they undergo. The "lock in" may run both ways – suppliers may be dependent on specific buyers in order to make profitable the specific investments they have made for that relationship. Moreover, asset specificity tends to be correlated with the product being a "differentiated product" instead of a commodity competing only on cost.

In a situation where there is a confluence of location and asset specificity and product differentiation, suppliers and buyers may have a strong incentive to invest in climate shock mitigation to protect the mutually profitable linkage. However, climate shocks may reach a level that requires too high an investment in mitigation for the linkage to be profitable, at which point the buyer or supplier would back away from this linkage. For example, a buyer who requires a high level of food safety (and thus low pesticide use), may break away from a given zone when climatic changes increase insect density to the point where more pesticide use is required to have acceptable fruit cosmetic quality, and thus make it uneconomic to rely on that zone.

Further, supply chain networks such as a supermarket chain source from several different zones (such as occurs in Mexico for tomatoes, see Reardon et al. 2007) over the year in order to smooth product supply inter-seasonally. While inter-season average vulnerability may remain low, periodic shocks due to climate or violence may increase dependence (such as in the North-South maize supply to feed mills for chicken and fish in Southern Nigeria; see Liverpool-Tasie et al. 2016). Sixth, more concentrated (as defined by industrial organizational terms) segments of the supply chain may either increase or decrease vulnerability to climate shocks. On the one hand, concentrating a process in a single large firm rather than in many small firms could make the process more risky (such as happened in the US in 1993 when the beef supply of the large chain Jack in the Box was tainted by E. coli from a single source and then infected the many points of supply). However, large companies have the means to make the "threshold investments" needed to mitigate or cope with a climate shock, as discussed in the next sub-section.

Finally, a seventh factor is variation over time in one location and over locations in the exposure to climate risk, controlling for the nature and occurrence of the hotspots per se. This acts as a magnifier and complement to the above six determinants of whether a point in the supply chain is a hotspot.

In sum, the determinants of hotspots described above, namely physical infrastructure to reduce production risk in supply zones, geographic length of the supply chain, perishability of the product, intensity and robustness of physical capital, asset specificity cum location specificity, concentration, and exposure to climate risk) generate a large number of "hotspots" in developing country food supply chains, before and after the farm gate. They also vary enormously over locations and products and the degree of transformation of supply chains. That implies that solutions to climate risk for supply chains will need to be highly differentiated and adapted to varying circumstances.

Moreover, these determinants are present in all directions of supply chains, including rural-urban chains, urban-rural, and rural-rural. While research on this is still in its infancy, we surmise that rural-urban and urban-rural supply chains, compared with rural-rural, will tend to have better infrastructure, be as long, involve more perishable products, and be more concentrated and asset-specific than rural-rural supply chains. This difference likely arises because rural-rural chains move more grains and tubers and shelf-stable vegetables like potatoes, while rural-urban and urban-rural, which include cities as origins or destinations, are more varied in product terms and more transformed in industrial organization terms.

### 3.2 Impacts on Supply Chain Structure/Conduct/Performance of Short-Term Climate Shocks from Strategic Responses of Supply Chain Actors

Enterprises in any segment of the supply chain, including input firms, farms, processors, and distributors, can be said to maximize utility under constraints. Utility derives from the level and stability of profits, which are a function of costs, product quality and safety (the latter two being in turn a function of requirements derived from the governance of the supply chain, such as the degree to which standards are imposed). Constraints are a function of assets, including productive assets and human capital, which can be private, collective, or public.

Within the constrained optimization framework, a firm (such as an urban retailer or processor, or an urban or rural wholesaler) has to decide on the design of the supply chain used to source inputs and market outputs. Du et al. (forthcoming) decompose the "optimal supply chain choice of the innovator" to six detailed choices: (1) production quantity given capital constraints and market conditions; (2) in-house versus purchased supplies (upstream this means deciding how much feedstock to grow vs. purchase from other farmers, midstream is inventory levels, and downstream is creation of marketing services in-house or outsourced); (3) for purchased supplies, whether to buy through contracts or spot market arrangements; (4) when using contracts, what terms and conditions to include; (5) for in-house production, what technology to use; and (6) how the degree of monopsony and monopoly, and government regulations that affect market power, change the choices made for these five considerations. These basic questions form the basis from which a supply chain is designed. The vulnerability or resilience to climate shocks are derived from the nature of the supply chain (controlling for the climate shock) which in turn is formed by design decisions of firms using them.

All else equal, a short-term climate shock reduces profit for these firms. To attenuate profit loss, firms or farms need to innovate and make investments to manage risks *ex ante* or cope with shocks *ex post*, at a type and level appropriate for the nature of risk. We follow a long literature on investment and call these "threshold investments" (Hubbard 1994). Typically, a firm or farm would make the threshold investment itself to mitigate the effects of a shock. At times, a mitigation measure taken by a single firm provides external economies to firms around it (or up or downstream from it). An example could be a firm constructing a culvert that diverts flood water not just from it but also from those physically downstream from it.

Moreover, the needed threshold investments (and returns to these investments) will be conditioned by the sources of vulnerability related to the seven determinants of hotspots discussed above. We surmise that there is a greater possibility for threshold investments to reduce risks on some determinants of hotspots (such as physical infrastructure to reduce production risks) than others (like intensity and robustness of physical capital). In addition, we expect the risk mitigation strategy of a firm in an area of very low density of physical capital or non-robust physical capital to be different than that of one in area of high density of capital and high robustness. There are also mutual externalities of items of capital stock in a given area; for example, if a sea wall is fragile or flimsy a mitigation investment in flood control canals next to it would be ineffective. By contrast, there could be a positive externality where pond lining reinforcement is undertaken in an aquaculture area bordering the sea where strong sea walls have been erected.

A key point is that not all zones, firms, and farms will be able to make the needed threshold investments. The challenge is exacerbated by the need for *ex ante* investments – implying an investment, credit, and planning horizon foreign to small firms and farms. This can create a kind of "poverty trap" (Carter and Barrett 2006) caused by climate shocks and accompanied by exclusion of certain zones, firms and farm strata. This can lead to a concentration of the segments of supply chains, such as when large processing firms gain market share after a shock. It can lead either to concentration of zones where the product is produced, or a shift toward new zones (similar to what can happen in long-term climate change discussed below).

The threshold investments cum strategies of managing risk from short-term climate shocks or coping fall into several categories. A major distinction is between large, transnational companies and smaller, domestic companies. For example, firms and farms may need to temporarily or permanently switch away from supplying zone or intermediate point. This of course is done constantly in international trade, such as the example of a US fruit processing firm recently shifting from Mexico to China to Argentina as costs changed. Some large companies do the same in large domestic markets, such as Charoen Pokhpand (CP) building compartmentalization of its supply chains in Asia to allow switching from one source zone to another after a climate shock. International sourcing also diminishes climate shock risk by having a more diversified network of suppliers with low degree of correlated exposure to climatic risks.

Such investments are less easy for most domestic sourcing, which we noted is 90% of the food supply of Africa and Asia. The challenges can be substantial for several reasons. First, there may be no cost-effective sourcing alternative in the short run, either in terms of switching from long distance to "local" sourcing, or switching to another zone. This difficulty may be more acute for urban-rural and rural-rural supply chains as the web of transport routes and the economic sourcing distances for rural consumers may be more limited for these supply chains. By contrast, rural-urban supply chains utilize a more extensive web of transport links including large highways, radiating from and to a large city.

Second, another zone might be available but lack prior requisite investments to meet the buyer's requirements. An example is the requirement by most European retailers for perishables suppliers in developing countries to have GLOBALGAP certification. This would involve "asset specificity" of investments, often substantial, by suppliers in a given zone (and typically by larger producers). If the buyer suddenly had to switch zones, it may well not be able to find the qualified suppliers. Again, this challenge might be more acute for the rural-rural and urban-rural supply chains than for the urban-rural chains, but the issue is present for all three depending on the product and the degree of transformation of the market.

A similar challenge might go for a range of post-harvest transport and processing facilities that would be needed to source. Moreover, a large buyer with standards needs to provide an ongoing incentive for suppliers to make investments in the requisite quality and so on. If the buyer is seen to be risky as a client, farmers, processors, and distributors will shy away from making needed relation-specific investments for that client. The buyer would need to maintain a minimum of demand from that zone or set of suppliers to maintain the incentive.

Third, the business management literature references the need to reduce lead time and "increase agility" to avoid risks or cope with shocks (Ponomarov and Holcomb 2009). This involves investing in alternative arrangements to existing suppliers or supply routes and systems, all of which are costly. For example, CP built "redundant ports" for rice supply from Thailand to its foreign markets, building several ports along rivers to provide alternatives in the case of a typhoon or tidal wave. With the growing need for these investments in the face of increased climate shocks, market concentration in larger firms will likely increase.

As a consequence of the above challenges, firms and farms may make induced innovations in "climate proofing" or "climate adapting" their equipment and processes. Firm-level investments might include energy saving or less energy dependent equipment (e.g. larger equipment), larger and more vehicles, and more rapid transport (to reduce inventories "held hostage" to climate shocks). Firms may also invest in enhanced storage through driers and dehumidifiers or stronger storage (for example[1] investment by a cocoa cooperative in typhoon-proofed cocoa containers in Vanuatu), and increased access to information flows for better "supply chain intelligence" as well as purchase insurance policies, where available. Finally, firm-level investments may seek to enhance supply chain-level efficiencies. At the government- and community-level, investments could seek to reinforce and/or build deepwater/off-shore ports (as in Indonesia, Shanghai, Rabobank), increase resilience in urban logistics, and seek to improve arrangements between governments for facilitation of shipping and supply (such as Hangzhou government did with Heilongjiang). Finally, an improved regulatory environment could further induce the private investments noted above, and create incentives and capacity for these investments.

---

[1] Personal communication Randy Stringer, Professor at University of Adelaide, July 2016.

## 4 Long-Term Climate Change Impacts on Transforming Food Supply Chains: Challenges and Strategies

### 4.1 Supply Chains and Melting Snow and Ice

Climate change is increasing the likelihood and rate of melting snow and ice, which may have permanent effects on the economics of agricultural production in many regions. These changes in seasonal water availability patterns may result in floods and disrupt patterns of farm production. Melting snow and ice may change patterns of availability of water to irrigated agriculture. In locations such as close to the Himalayan mountains, there may be more floods during the rainy season and less water for irrigation during the dry seasons (Xu et al. 2009).

Intermediaries may suffer because flooding may harm infrastructure, including both storage facilities and roads, as well as affect the availability of supply. The risk of floods may necessitate moving processing and storage facilities, and may require added investment in transportation. Changes in patterns of farm production and the availability of food supplies may change procurement strategies of intermediaries as well as prompt them to invest in agricultural production in regions less vulnerable to these effects. Similar to mitigation measures done for short-term climate shock risks, some of the implications of the melting of snowcaps and ice can be mitigated by construction of dams or new storage facilities to protect against the increased flooding and to store water during the dry seasons (Xie and Zilberman 2016).

The response needed to large scale long-term climate change is of a far greater scale, and much greater investment requirements than are the mitigation measures made for short-term shocks discussed in the previous section. Thus there will be a need for public sector support; assembling resources for such grand investments can be politically challenging. Countries with superior governance system will be able to adapt more effectively to these long-term changes. Because many of these changes supersede national borders, for example Himalayan ice melt affects many countries, there is a growing role for multilateral organizations and international agreements. There will also be many opportunities for the private sector, at times in concert with the public sector, to intervene by investing in water projects. In some cases, organizations that have the financial capacity and creative ability to modify water patterns may become new important players in agricultural resource management. These water projects may include dams, hydropower facilities and other investments that will enhance agricultural productivity and provide a new source of value for the existing entities.

### 4.2 Supply Chain and Migrating Weather

There are many possible effects of migrating weather patterns on agriculture. Migrating weather will impact farm-level production and consideration of where to locate new production. With increasing knowledge of the evolution of changing weather patterns, agile firms may be able to exploit differences in impact across space by making strategic investments in land, processing facilities, and equipment. However, the gap between agile and non-agile firms and farms will be exacerbated. Finally, public research will be needed to support development and dissemination of technologies to adapt to, mitigate and slow migrating weather impacts.

One possible outcome is migration of crop production associated with specific weather. For example, production of certain dry wine varieties requires specific weather patterns. Increased heat may increase sugar content and may harm the ability to maintain wine quality. One solution may be to relocate production of grapes to another region (e.g. from California to Oregon). The production of wine also involves processing and shipping of grapes by wineries. Wineries have invested a lot in infrastructure and have recognized brand names. For instance, some of the reputation of wine is location-specific (e.g. high quality Bordeaux wine is produced only in Bordeaux). Thus, migration of weather may lead to migration of infrastructure and changes in regional and brand reputation, but also provide opportunities for other brands to grow or shift.

This will be a special challenge for denomination by locality/terroir, for example of cheeses and wines. Response by growers will vary. Growers in areas with warming weather may adapt their practices on-site to maintain location and quality, while other growers will shift location (such as wine production moving from Napa to Oregon). Further, with a migration of supply location, there may be a decline in terroir branding and a shift to marketing by variety (e.g., Cabernet) rather than by region, and an increasing importance of brand rather than location. This can induce further concentration in formerly location-bound industries as companies with good R&D, branding and scale make alliances with growers in developing countries for contracted production of intermediate inputs based on detailed specifications. Of course to some extent this already occurs in commodity olive oil or wine, such as with Italian producers buying olives and grapes, first processed, from Eastern Europe and North Africa. This is also part of a larger trend where food industry companies source commodities (cheap bulk intermediate inputs) and market differentiated products, such as Smithfield Foods does in Europe by sourcing cheap pork from Eastern Europe and marketing quality branded products in France and Spain.

Large organizations that are aware of the differential impact of weather patterns that worsen productivity of certain regions (e.g. southern China), while increasing it in other regions (e.g. northern China), may invest and hold land resources to later build infrastructure for new agricultural production. Such behavior requires the ability to predict spatial differences in the evolution of climate change over time; that ability is still limited. However, as our understanding of patterns of climate change develop, we are likely to see more speculative investments in regions that may benefit from climate change. For instance, the projected water depletion/shortage in the Middle East is leading investors from those countries to buy land swaths in well-watered regions, such as in sub-humid Africa.

Given technological change in agricultural production processing and transport, weather migration may prompt more rapid transition to a modernized agricultural system as older facilities are retired. In this case, adaptation to climate change will have the unintended consequence of modernization – even vice versa, where modernization leads to adaptation. However, the extent to which this occurs depends on the ability to attract financial capital to areas most suited to expansion. It is quite likely that better managed, informed and more agile players, big or small, will be the ones that take advantage of these new opportunities.

Moreover, climate change may exacerbate differences between agile and non-agile actors. One of the main unintended consequences of climate change is an increasing of the gap between more traditional and less mobile farming communities with more entrepreneurial, mobile groups. This means that policies that assist in relocation and provide access to new opportunities may help overcome some of the negative distributional effects of adaptation to climate change.

Similarly, climate change may cause migration of farm workers from areas that suffer from worsening conditions. Migration is a difficult process and one may expect to see the emergence of networks of labor contractors that will enable movement of labor across regions. These types of labor movements are sometimes associated with illicit activities and human rights violations, and thus may require regulation and policy interventions, but the migration itself may provide better outcomes for people who live in regions that suffer from climate change.

Finally, one of the important challenges of public research is to develop and disseminate technologies that will slow the impacts of migrating weather. Even relatively small changes in temperature may have significant impact that require adaptation (Di Falco and Veronesi 2014). Moreover, slightly higher temperatures may increase vulnerability to pests and reduces chill days, which are required for blooming of some tree crops. Addressing these changes may require significant science-based adaptation. This may include new varieties better suited to changing agro-climatic conditions as well as practices to decrease the negative side effects of warming that may include new pests and shorter tree bloom. While government in developed countries may engage in supporting this type of research and development, in some developing countries the private sector may be engaged in pursuing appropriate technologies to assure availability of inputs. For example, multinationals which depend on the production of cacao, rubber, and other tropical crops, may engage in enhancing the capacity of producers to withstand the impacts of a changing climate. At the same time, these organizations may also encourage investment in production of these crops in new regions.

## 5 Conclusions and Agenda

In this chapter we emphasized several key points. First, it is important to analyze climate short-term shocks and long-term change on the full food supply chain (inputs, farms, processing, distribution). The farm is just one segment of the chain and accounts for only about half its costs and value added. The supply chain as a whole is important to food security and livelihoods (as employment) in both rural and urban areas. We identified three types of supply chains as important to rural and urban areas: rural to urban, urban to rural, and rural to rural. It is important to analyze all three in a dynamic context. Climate change and supply chains are dynamic phenomena. Analysis, and especially development of policies to affect the adaptation of all elements of supply chains to climate change, should take into account that supply chains are evolving and therefore be based on expectations of their future, rather than present, form.

Second, it is crucial to approach the analysis of climate shocks on supply chains with a clear view of the complexity of a given supply chain as an interdependent set of segments and sub-segments. Climate shocks upstream in the supply chain can disrupt a wide complex of midstream and downstream activities; a shock such as a flood in an intermediate area, which may impact assemblage and transport, can then block the sale of surplus from the rural area and ingress of input supply chains to farmers. These impacts could also block or delay supply to urban areas, which now constitute the majority of food consumption and markets in Africa and Asia, and rural areas, which now depend to a large extent on food supply purchases.

Third, it is important to analyze climate change impacts on supply chains from the viewpoint of "hot spots" of vulnerability along the chains, both before and after the farm gate. We identified seven determinants of these hot spots: physical infrastructure to reduce production risk in supply zones, geographic length of the supply chain, perishability of the product, intensity and robustness of physical capital, asset specificity cum location specificity, concentration, and exposure to climate risk. They vary enormously over locations and products and the degree of transformation of supply chains. This implies that solutions to climate risk for supply chains will need to be highly differentiated and adapted to the varying circumstances.

Fourth, it is important to view climate shocks, and strategies to mitigate them, from the point of view of (1) strategic supply chain design choices by actors along the supply chain, of sourcing and marketing systems, geography, institutions, and organization; and (2) threshold investments by actors (firms and farms) along all supply chains. It is thus crucial to understand the incentives and capacity of the actors in the segments of the supply chain, alongside the vulnerability of the segments in the case of insufficient or untimely incentive (or risk itself) or incapacity to make the needed investments. Moreover, it is probable that many small scale farms and firms will not be able to make the needed adjustments and investments and may fail because of climate shocks and ensuing supply chain adaptations undertaken by the leaders of the segments of the supply chains.

The above four points suggest a research agenda examining several dimensions of the climate change-supply chain interaction.

First, applied field research should study supply chains and understand their structure, conduct, and performance, and the variants of a given product's supply chains, geographically and by degree of transformation (traditional, transitional, modern).

Second, applied field research should analyze the vulnerabilities (potential and realized disruptions) of the supply chain by segment and by vector of impact, such as intermediate point flooding, energy constraints from stymied fuel supply chains, droughts in farming areas, and so on.

Third, a study of the actors' strategies and constraints in both their design of and behavior in the supply chains should be done, with a particular application to understanding their choices and threshold investments to reduce their vulnerability to (*ex ante*) or cope with (*ex post*) supply chain disruptions due to climate shocks and changes.

Fourth, the research on innovation systems and public R&D policies associated with climate change should be expanded to take into account all the components of the supply chain. It should consider allocation of efforts between public-private interaction in R&D activities throughout the supply chain, and the policies that can affect them.

Fifth, the research on climate change should emphasize policy and infrastructure investment constraints in the context of supply chains and identify potential areas to improve incentives and capacity of firms and farms and to facilitate public sector actions to make the needed climate adaptations.

# References

Carter, MR and CB Barrett. 2006. "The economics of poverty traps and persistent poverty: An asset-based approach," The Journal of Development Studies, 42(2): 178–199.

Di Falco, S., and M. Veronesi. 2014. "Climatic anomalies and conflicts: the role of tenure security on land disputes." Paper prepared for presentation at the EAAE 2014 Congress 'Agri-Food and Rural Innovations for Healthier Societies', August 26 to 29, Ljubljana, Slovenia.

Du, X., L. Lu, T. Reardon, and D. Zilberman. Forthcoming. 2016. "The economics of agricultural supply chain design: A portfolio selection approach," *American Journal of Agricultural Economics*, 98(5): 1377–1388; http://dx.doi.org/10.1093/ajae/aaw074

Elbehri, A. (ed.) 2015. Climate Change and Food Systems: Global assessments and implications for food security and trade. Rome: FAO.

Giorgi, F. (2006). Climate change hot-spots. *Geophysical research letters*, *33*(8).

Hubbard, R. G. (1994). Investment under uncertainty: keeping one's options open. *Journal of Economic Literature*, *32*(4), 1816–1831.

Kleindorfer, P.R. and G.H. Saad. 2005. "Managing Disruption Risks in Supply Chains," *Production and Operations Management*, 14(1), Spring: 53–68.

Liverpool-Tasie, L., Adjognon, S., & Reardon, T. (2016). Transformation of the food system in Nigeria and female participation in the Non-Farm Economy (NFE). In *2016 Annual Meeting, July 31-August 2, 2016, Boston, Massachusetts* (No. 236277). Agricultural and Applied Economics Association.

Nicholls, R.J. and A. Cazenave. 2010. "Sea-level rise and its impact on coastal zones," *Science*, 328, 18 June: 1517–1520.

Oke, A. and M. Gopalakrishnan. 2009. "Managing disruptions in supply chains: A case study of a retail supply chain," *International Journal of Production Economics*, 118(1), March: 168–174.

Ponomarov, S. Y. and M.C. Holcomb. 2009. Understanding the concept of supply chain resilience. *The International Journal of Logistics Management*, 20(1), 124–143.

Reardon, T. 2015. "The Hidden Middle: The Quiet Revolution in the Midstream of Agrifood Value Chains in Developing Countries," *Oxford Review of Economic Policy*, 31(1), Spring: 45–63.

Reardon, T. 2016. *Growing Food for Growing Cities: Transforming Food Systems in an Urbanizing World*. Chicago: The Chicago Council on Global Affairs. April.

Reardon, T. J.A. Berdegué, F. Echánove, R. Cook, N. Tucker, A. Martínez, R. Medina, M. Aguirre, R. Hernández, F. Balsevich 2007. *Supermarkets and Horticultural Development in Mexico:*

*Synthesis of Findings and Recommendations to USAID and GOM*, Report submitted by MSU to USAID/Mexico and USDA/Washington, August.
Rosegrant, M.W., C. Ringler, and T. Zhu. 2009. "Water for agriculture: Maintaining food security under growing scarcity," *Annual Review of Environment and Resources*, 34: 205–22.
Swinnen, JFM (ed.). 2007. *Global supply chains, standards, and the poor*. CABI Press.
Tschirley, D., T. Reardon, M. Dolislager, and J. Snyder. 2015. "The Rise of a Middle Class in Urban and Rural East and Southern Africa: Implications for Food System Transformation," *Journal of International Development*, 27(5), July: 628–646.
Xie, Y. and D. Zilberman. 2016. "Theoretical implications of institutional, environmental, and technological changes for capacity choices of water projects," *Water Resources and Economics*, 13: 19–29.
Xu, J., R.E. Grumbine, A. Shrestha, M. Eriksson, X. Yang, Y. Wang, and A. Wilkes. 2009. "The melting Himalayas: Cascading effects of climate change on water, biodiversity, and livelihoods," *Conservation Biology*, 23(3): 520–530.
Zilberman, D., Zhao, J., & Heiman, A. (2012). Adoption versus adaptation, with emphasis on climate change. *Annu. Rev. Resour. Econ.*, 4(1), 27–53.

**Open Access** This chapter is distributed under the terms of the Creative Commons Attribution-NonCommercial-ShareAlike 3.0 IGO license (https://creativecommons.org/licenses/by-nc-sa/3.0/igo/), which permits any noncommercial use, duplication, adaptation, distribution, and reproduction in any medium or format, as long as you give appropriate credit to the Food and Agriculture Organization of the United Nations (FAO), provide a link to the Creative Commons license and indicate if changes were made. If you remix, transform, or build upon this book or a part thereof, you must distribute your contributions under the same license as the original. Any dispute related to the use of the works of the FAO that cannot be settled amicably shall be submitted to arbitration pursuant to the UNCITRAL rules. The use of the FAO's name for any purpose other than for attribution, and the use of the FAO's logo, shall be subject to a separate written license agreement between the FAO and the user and is not authorized as part of this CC-IGO license. Note that the link provided above includes additional terms and conditions of the license.

The images or other third party material in this chapter are included in the chapter's Creative Commons license, unless indicated otherwise in a credit line to the material. If material is not included in the chapter's Creative Commons license and your intended use is not permitted by statutory regulation or exceeds the permitted use, you will need to obtain permission directly from the copyright holder.

# The Adoption of Climate Smart Agriculture: The Role of Information and Insurance Under Climate Change

Jamie Mullins, Joshua Graff Zivin, Andrea Cattaneo, Adriana Paolantonio, and Romina Cavatassi

**Abstract** Climate change adds to the existing challenges in improving crop productivity and welfare for smallholder agricultural households by affecting the mean and variability of weather conditions and the frequency of extreme weather events. In the face of such growing uncertainty, agricultural practices of small landholders need to be adapted to better manage the changing risk structures. Since government risk management programs may complement or substitute for farmer adaptation, this chapter examines how a range of institutional interventions might assist, obstruct, channel, or change smallholder agricultural adaptation to climate change. Taken together, our results underscore the importance of the informational role of the agriculture extension, suggest that insurance can lead to significant changes in farmer planting and land management decisions, and show how information about changing conditions and insurance can be complimentary in driving changes in farmer behavior.

## 1 Introduction

Climate change adds to the existing challenges in improving crop productivity and welfare for smallholder agricultural households by affecting the mean and variability of weather conditions and the frequency of extreme weather events. In the face of such growing uncertainty, agricultural practices of small landholders need to be

---

J. Mullins (✉)
Department of Resource Economics, University of Massachusetts Amherst,
Amherst, MA, USA
e mail: jmullins@umass.edu

J.G. Zivin
School of Global Policy and Strategy, University of California San Diego, San Diego, CA, USA

A. Cattaneo
FAO of the UN, Rome, Italy

A. Paolantonio • R. Cavatassi
International Fund for Agriculture Development (IFAD), Rome, Italy

adapted to better manage the changing risk structures. Since government risk management programs may complement or substitute for farmer adaptation (Anton et al. 2013), this chapter examines how a range of institutional interventions might assist, obstruct, channel, or change smallholder agricultural adaptation to climate change.

Our analysis begins with a stylized conceptual model from which we build a series of simulations based on empirical data from smallholder agriculture households in Malawi. We proceed by analysing three climate change scenarios, looking at the spectrum of farmer responses as a function of extension information provision, weather index insurance, and the interaction of the two institutions.

Our approach grapples with three distinct dimensions of uncertainty central to understanding how the policies of an institutional actor might affect smallholder agricultural adaptation to climate change. First, uncertainty about farmers' perceived risks and their degree and direction of adaptation response to climate change is addressed through the implementation of an empirically founded expected-utility-optimization framework which accounts for farmer risk preferences and the role of weather conditions and yield variability in adaptation decisions. Second, we address uncertainty about the quantitative impacts of climate change on the variability of yields and production risks through a regression analysis linking weather conditions and yields across a range of crops and conservation techniques. Finally the wide range of possible policy options is narrowed through a focus on the effects of two program types: information provision regarding likely changes in weather conditions under climate change and weather indexed insurance coverage.

The basis of the analysis in this chapter is that climate change affects the distribution of weather conditions during the growing season, which in turn impacts yields under a given set of management practices.[1] Changes in yield distributions ultimately alter expected farmer incomes, and thus planting and management decisions. In our simulations, farmers can adopt adaptation strategies along two distinct dimensions. First, farmers can change cropping decisions between staple and cash crops and amongst crop types within these categories. Second, farmers can make changes in land management practices through the adoption of Climate Smart Agricultural (CSA) techniques (e.g. Kassie et al. 2008; Rosenzweig and Binswanger 1993; Heltberg and Tarp 2002; Deressa and Hassan 2010). CSA practices that are considered in the simulations include intercropping of staple (maize) and cash (legumes) crops, as well as the improvement of soil water-holding capacity by adding crop residues or manure, and/or by adopting conservation tillage in response to changes in water availability (Smith and Olesen 2010). Investments in soil-water holding capacity (SWC) may be a particularly important adaptive response in light of recent research that finds a positive correlation between rainfall variability and the selection of SWC type practices (Arslan et al. 2013).

---

[1] A necessary limitation of our simulations is that they rely upon data from the 2009-2010 growing season and thus cannot attend to new seed varieties or cultivation practices that may arise in the face of climate change.

This chapter focuses on how smallholder adaptation to changing conditions under climate change might be affected by government risk management interventions. While Mendelsohn (2010) finds that farmers without insurance have a strong incentive to adapt to climate change, Skees et al. (1999) show that the assumption of risk by farmers may stymie farmer investment in certain adaptation strategies. Collier et al. (2009) underscore the importance of specific policy design features in impacting behaviour. For example, traditional agricultural insurance (which makes an indemnity payment when the farm incurs a verifiable production loss) can help to manage production risk but may diminish incentives to adapt to climate change. Conversely, area-yield insurance and weather index insurance (as we examine in this chapter) approaches can minimize these moral hazard concerns since indemnities are paid independently of the actual loss incurred by a policyholder. Of course, all risk management policies will change the framework under which farmers make production decisions. Deepening our understanding of how institutional policies impact farmer decisions under climate change is of critical importance for well-designed climate adaptation strategies now and in the future.

The following questions anchor our analyses as we build on previous work examining risk management under climate change (Collier et al. 2009; Heltberg et al. 2009, Anton et al. 2013).

1. Can policy makers assist in risk management without steering farmers away from beneficial adaptation?
2. How do insurance and information programs impact farmer behaviours and might these two policy approaches interact in their effects on farmer decisions?
3. How can policy makers decide between interventions when the information about how various instruments would perform under an increasingly variable climate is very limited?

The contribution of this chapter is to address – in the context of smallholder agriculture in Malawi – the risk and the uncertainties introduced by climate change and the role of perceptions regarding this uncertainty in shaping farmer decisions and the appropriate risk management instruments to improve smallholder welfare.

## 2 Conceptual Model

In this section, we develop a basic model of smallholder agricultural management when yields are stochastic and farmers are risk averse. We begin with the assumption that farmers are growing a single staple crop on a fixed plot of land. Farmers maximize their expected utility from profits by choosing agricultural inputs, $x$, and techniques, $\phi$. The vector **x** will include a range of purchased agricultural inputs, such as fertilizer, pesticides, herbicides, and seed. The variable $\phi$ will correspond to the labour requirements of the dominant agricultural technique used to cultivate the crop. In this model, possible techniques include a variety of CSA practices as well as more chemically-intensive ones. The key distinction between inputs and technique is that the former is assumed to impact expected yield while the latter is assumed to impact the volatility of yield.[2] Without loss of generality, we define $\phi$ as the intensivity with which the chosen technique reduces yield volatility.

In particular, agricultural yield on land of given quality is equal to $f(x) + 2(1 - g(\phi))$ θ, where θ is a stochastic weather variable with an expected value of zero and variance $\sigma^2$ (Just and Pope 1978).[3] Expected yield f is assumed to be increasing in inputs at a decreasing rate, i.e. f'(x) > 0, f''(x) < 0. The function g can be thought of as a measure of protection against weather volatility, such that 1-g is a measure of weather sensitivity (Graff-Zivin and Lipper 2008). Protection is assumed to be increasing in technique at a decreasing rate, i.e. g'($\phi$) > 0, g''($\phi$) < 0. Let p represent the market price per unit of agricultural output. For simplicity, we will also assume that this price represents the per unit value of agricultural output consumed by the farmer, which is tantamount to assuming that all farmers have market access and that food production levels always exceed the subsistence demands of the household.

Revenue can thus be expressed as $R = pf(x) - 2p(1 - g(\phi))\sigma^2$. Taking a second-order Taylor-Series approximation of EU(R) yields the following expression:

$$EU(R) \approx pf(x) - rp(1 - g(\phi))\sigma^2, \qquad (1)$$

where r is the Arrow-Pratt measure of risk aversion. Utility from agricultural revenues is increasing in average yield and decreasing in the variability of yields. This type of utility function is frequently used in finance (Markowitz 1987) and can be viewed as a special case of the more general class of mean-variance utility functions. The properties of these utility functions and their consistency with expected utility theory are discussed in great detail elsewhere (Meyer 1987).

Turning to costs, several differences between inputs and technique are worth highlighting. First, inputs require market purchases early in the growing season that only pay dividends at harvest. As such, limited savings and the imperfect credit markets that are commonplace in developing countries may play an important role in input purchases. On the other hand, technique will generally be 'purchased' with household labor. Since technique does not require an initial cash outlay, credit constraints should be immaterial. In particular, we let λ represent the costs of credit, which can be viewed as the shadow value on a credit constraint. A larger λ represents dearer credit and thus raises the effective costs of input purchases while leaving the costs of technique unaffected.

Second, the nature of costs for the $x$ and $\phi$ choice variables also differ, independent of cash flow concerns. While the costs of inputs are based on market prices net of any subsidies, the costs of technique are a bit more complicated. This complication arises because we would like to allow for the possibility that technique can be

---

[2] As will be made clear below, technique can potentially impact long-term expected yields. Since these benefits will accrue with a considerable delay, they are best reflected in an appropriately discounted cost function.

[3] The assumption of additive risk can be relaxed during simulations.

mean yield augmenting in the long-run. For example, several studies suggest that conservation agriculture can increase expected yield after a 3–5 year period of ecosystem disequilibrium (see Graff, Zivin and Lipper 2008). Rather than model this as part of f(), which would require more explicit assumptions regarding the timing of those benefits, we include them in the 'effective' costs of technique. In particular, we assume that the costs of technique will include the direct costs of its application net of the present discounted value of any future yield benefits. As such, the per-unit costs of technique will be a function of discount rates $\delta$.

We denote the costs of inputs as $c_x$ and the costs of technique as $c_\varphi(\delta)$, with the usual assumption regarding the convexity of costs, such that the cost of technique are increasing in discount rates at an increasing rate, e.g. $c_\varphi' > 0$ and $c_\varphi'' > 0$. Moreover, we introduce the terms $(1-s_x)$ and $(1-s_\varphi)$ to denote targeted government subsidies for inputs and technique, respectively. Suppressing the expected utility notation, the objective of the farmer is to maximize the expected utility of profits, which can be expressed as follows:

$$\pi = pf(x) - p(1-g(\phi))r\sigma^2 - (1-s_x)\lambda c_x x - (1-s_\phi)c_\phi(\delta)\phi \quad (2)$$

The first order conditions imply:

$$p\frac{\partial f}{\partial x} - (1-s_x)\lambda c_x = 0 \quad (3)$$

$$p\frac{\partial g}{\partial \phi}r\sigma^2 - (1-s_\phi)c_\phi(\delta) = 0. \quad (4)$$

Inputs and technique will be chosen such that the marginal benefits from each will be equal to its marginal cost, net of subsidies. In the case of inputs, these costs will also depend on borrowing costs as measured by $\lambda$. The marginal benefits from inputs are due to expected yield augmentation. The marginal benefits from technique are due to protection from yield volatility.

## 2.1 Inputs, Technique, Insurance, and Diversification

This basic framework can be generalized to expand the portfolio of farmer investment options by introducing the possibility of insurance coverage, $\psi$, and crop diversification, D. Insurance could play an important role in this setting going forward, as climate change is expected to increase yield volatility considerably. Since credible documentation of individual farmer yield losses is likely prohibitively expensive and/or infeasible in the developing country context, we assume that insurance contracts are written based on 'local' realizations of weather. Of course, yield volatility depends on weather, among other things, so one can view this insurance contract as one that partially indemnifies households against agricultural risk. Moreover, since it is based on weather rather than experienced yields that will also depend on a host of farmer behaviors, it eliminates very practical concerns about moral hazard.

In particular, we view insurance as a state-contingent contract, where farmers receive a payment Z that depends on the probability of a given weather realization and thus the variance of weather, and the amount of insurance coverage purchased. This insurance is distinct from the type of 'insurance' purchased through the use of technique since insurance shrinks downside risk while technique decreases both downside and upside risk by compressing volatility. More formally, $Z(\psi, \sigma^2)$, where the payout Z is increasing in coverage and volatility at a decreasing rate. Similar to agricultural inputs, this contract is purchased at the beginning of the growing season in return for protection in the future, so the costs of credit will play a role in the purchase decision. Insurance costs are increasing and convex with the volatility of weather, reflecting the additional costs of provision by insurers.

It is interesting to note that while the value of insurance (or for that matter technique) to farmers depends on perceived volatility, the premiums are expected to depend on actual volatility as understood by insurers.[4] To formalize the notion of this wedge between perceptions and actual, we introduce the term $m$ such that the true volatility $\sigma_T^2 = \sigma^2 / m^2$, with $0 < m < 1$. When $m=0$ farmers believe weather to be non-stochastic. When $m=1$ they have a perfect estimate of volatility. All cases in between correspond to the case where farmers underestimate the realization of weather by a fixed proportion equal to $m$. As with inputs and technique, we allow the government to subsidize the purchase of insurance, such that the 'effective' cost of purchase can be expressed as: $(1 - s_\psi)\lambda c_\psi(\sigma^2/m^2)$.[5]

Our approach to modelling diversification is highly stylized to maintain a focus on the core tradeoffs associated with pursuing this strategy rather than the specifics of alternative crops. In particular, we assume that diversification helps protect farmers against revenue volatility in much the same way as technique, i.e. we assume g is increasing in diversification at a decreasing rate. The costs of diversification depend on the net expected revenue reductions associated with planting it instead of the staple crop; simply denoted by $c_D$. Since these costs are only realized at harvest time, credit is not a concern for this strategy. Allowing subsidies for diversification strategies, denoted $s_D$, we can rewrite the farmers expected profit function as follows:

$$\pi = pf(x) - p(1 - g(\phi, D))r\sigma^2 + Z(\psi, \sigma^2) - \lambda(1 - s_x)c_x x$$

---

[4] One notable exception is the case where insurance markets are not competitive, since insurers will be able to set prices, at least partly, based on farmer perceptions as embodied in their willingness to pay for insurance.

[5] We also note that government safety nets can be viewed as a special case of insurance that is offered at fixed coverage levels with zero direct cost to the farmer.

$$-(1-s_\phi)c_\phi(\delta)\phi - \lambda(1-s_\psi)c_\psi(\sigma^2/m^2)\psi - (1-s_D)c_D D \quad (5)$$

This yields the following FOCS:

$$p\frac{\partial f}{\partial x} - \lambda(1-s_x)c_x = 0 \quad (6)$$

$$p\frac{\partial g}{\partial \phi}r\sigma^2 - (1-s_\phi)c_\phi(\delta) = 0 \quad (7)$$

$$\frac{\partial Z}{\partial \psi} - \lambda(1-s_\psi)c_\psi(\sigma^2/m^2) = 0 \quad (8)$$

$$p\frac{\partial g}{\partial D}r\sigma^2 - (1-s_D)c_D = 0 \quad (9)$$

Here again we see that investments are made such that the marginal benefit of those investments is equal to the marginal costs of those investments. Since the role played by agricultural inputs is independent of the other investment activities – it only affects expected yield – optimal input usage is identical to that found in our simpler case. The introduction of diversification, which competes with technique to shape effective risk exposure, makes the role of policy levers more complicated. Since insurance contracts are written on weather rather than agricultural yield, optimal coverage is orthogonal to the other risk management strategies. Although our simulations do not address credit constraints, it is worth noting that input usage and insurance purchases will depend upon the state of credit markets, while technique and diversification eschew such concerns.[6]

## 2.2 The Impacts of Climate Changes: Weather Volatility and Extension

As noted earlier, uncertainty about weather and attendant yield volatility are expected to increase under climate change. While volatility has no impact on input usage, its impacts on technique and diversification are straightforward. Greater volatility leads to greater perceived volatility (except in the special case where m = 0) and thus increases the returns to protection from yield risk. How much additional investment is made in each will depend on the curvature of the risk protection function g in technique and diversification spaces.

---

[6] If investments can be differentially collateralized or credit is targeted toward particular actions, credit constraints can differ for each type of expenditure.

In contrast, the impact of uncertainty on the purchase of insurance is ambiguous. The net effect will depend on the relative curvatures of the payout and cost function. It will also depend on the wedge between actual and perceived uncertainty since expected benefits are based on farmer perceptions but the price of insurance will be driven by the true underlying risk. The more farmers underestimate the risk (as m approaches zero) the larger the first term in brackets and the more likely insurance will be decreasing in risk. Put another way, the more farmers misjudge risk the more they will undervalue insurance relative to its costs and the less likely they are to purchase it.

$$\frac{dx}{d\sigma^2} = 0 \tag{10}$$

$$\frac{d\phi}{d\sigma^2} = p\frac{\partial^2 f}{\partial x^2} \cdot \frac{\partial^2 z}{\partial \psi^2} \left[ \left( p\frac{\partial g}{\partial \phi} r \right) \left( p\frac{\partial^2 g}{\partial D^2} r\sigma^2 \right) - \left( p\frac{\partial^2 g}{\partial \phi \partial D} r\sigma^2 \right) \left( p\frac{\partial g}{\partial D} r \right) \right] / |H| > 0 \tag{11}$$

$$\frac{d\psi}{d\sigma^2} = \left[ (1-s_\psi)\lambda \frac{\partial c_\psi}{\partial \sigma^2} \frac{1}{m^2} - \frac{\partial^2 z}{\partial \psi \partial \sigma^2} \right] / \frac{\partial^2 z}{\partial \psi^2} \begin{matrix}<\\>\end{matrix} 0 \tag{12}$$

$$\frac{dD}{d\sigma^2} = p\frac{\partial^2 f}{\partial x^2} \cdot \frac{\partial^2 z}{\partial \psi^2} \left[ \left( p\frac{\partial g}{\partial D} r \right) \left( p\frac{\partial^2 g}{\partial \phi^2} r\sigma^2 \right) - \left( p\frac{\partial^2 g}{\partial \phi \partial D} r\sigma^2 \right) \left( p\frac{\partial g}{\partial \phi} r \right) \right] / |H| > 0 \tag{13}$$

While we have not yet formally modeled policies to expand agricultural extension, nearly all of the comparative statics described above could be influenced by it. If, for example, an increase in extension efforts helps farmers understand that appropriate fertilizer applications can increase their yields, then this is tantamount to a change in the function f to the farmer. Similarly, if extension provides farmers with new information about diversification opportunities or new agricultural techniques, this translates into a change in the function g from the farmer's perspective. Since f and g feature prominently in all expressions above, extension of this sort will influence optimal decision making as well as the responsiveness of optimal decision making to changes in other policies and parameters.[7]

One such parameter that deserves particular attention is misperceptions regarding weather volatility. In particular, it is possible that extension could make targeted efforts to help farmers better understand weather and help them update their heuristics under a changing climate. This is, in fact, one of the risk management interventions we will examine via simulation in later sections.

---

[7] The impacts of extension could also be linearly approximated by modeling them as changes in the 'effective' costs of inputs, technique, insurance, and diversification. In this case, the impacts of extension will be entirely analogous to the earlier analysis on subsidies. Whether such an approximation is a reasonable one remains an empirical question.

Formally, we can view extension efforts to increase farmer understanding of weather conditions as an effort to increase the parameter $m$. In this case, it is straightforward to show that all of the risk reducing activities – technique, diversification, and insurance – are increasing in $m$ and thus increasing in extension (or other informational) activities that move farmer priors closer to 'actual' distributions under a changing a climate. Letting $\sigma_T^2$ denote true weather volatility (as opposed to perceived volatility) the specific relationships are as follows:

$$\frac{dx}{dm} = 0 \tag{14}$$

$$\frac{d\phi}{dm} = p\frac{\partial^2 f}{\partial x^2} \cdot \frac{\partial^2 z}{\partial \psi^2} \left[ \left(2p\frac{\partial g}{\partial \phi}rm\sigma_T^2\right)\left(p\frac{\partial^2 g}{\partial D^2}r\sigma_T^2\right) \right.$$
$$\left. -\left(p\frac{\partial^2 g}{\partial \phi \partial D}r\sigma_T^2\right)\left(2p\frac{\partial g}{\partial D}rm\sigma_T^2\right) \right]/|H| > 0 \tag{15}$$

$$\frac{d\psi}{dm} = -\frac{\partial^2 z}{\partial \psi \partial m} / \frac{\partial^2 z}{\partial \psi^2} > 0 \tag{16}$$

$$\frac{dD}{dm} = p\frac{\partial^2 f}{\partial x^2} \cdot \frac{\partial^2 z}{\partial \psi^2} \left[ \left(2p\frac{\partial g}{\partial D}rm\sigma_T^2\right)\left(p\frac{\partial^2 g}{\partial \phi^2}r\sigma_T^2\right) \right.$$
$$\left. -\left(p\frac{\partial^2 g}{\partial \phi \partial D}r\sigma_T^2\right)\left(2p\frac{\partial g}{\partial \phi}rm\sigma_T^2\right) \right]/|H| > 0 \tag{17}$$

The impacts of these policy instruments on farmer welfare can be obtained by plugging the relevant relationships back into the expected profit function, defined in (5). Heterogeneity with respect to time or risk preferences can be similarly explored.

Of particular note are the predictions of Eqs. 15 and 17, which suggest that better information regarding higher weather volatility ought to lead to increased use of CSA techniques and diversification crops. These are outcomes that will be examined directly as part of the simulations in the following sections.

## 3 The Simulation Framework

While the conceptual model highlights a number of policy tools that can be used to influence farmer choices under climate change, we will limit our empirical attention to those policies that are most directly tied to the increased weather volatility that is expected under climate change. In particular, we simulate the impacts of insurance and extension policies on cropping patterns and farmer welfare under zero, modest, and more severe climate change scenarios. Simulated crop choices are based on estimated agricultural production functions for smallholder farmers in Malawi as well as assumed constraints regarding the cultivation of staple crops for subsistence purposes. Following a brief description of the Malawian agricultural context, we will explain the simulation approach, assumptions, and results.

## 3.1 Institutional Context in Malawi Relevant to the Empirical Application

There are a number of institutions that serve farmers in Malawi, including extension and other sources of agricultural information, credit sources, input and output markets, farmers unions, and social safety net programs. The density and quality of these institutions should increase farm productivity and the ability of farm household members to manage shocks to income, contributing to greater and more stable livelihoods.

- In this context, access to **credit**, extension services, and safety nets are of particular relevance to this paper since these three institutional avenues are central to managing agricultural risk. In terms of access to credit, in the 2010 LSMS-ISA household survey, just 16% of all households accessed some form of credit, from both formal and informal sources, indicating that access to credit is quite constrained. This is further supported by the fact that among those accessing credit, 57% of loans came from neighbors/relatives/friends.
- In terms of **extension services**, despite a relatively large numbers of communities with agriculture extension officers, in 2010, information from the household survey indicates that just 21% of households received any extension advice in the Northern region, followed by 18% and 12% in Central and Southern regions, respectively. Beyond the limited reach of extension, Nkonya et al. (2015) also report that in Sub-Saharan Africa when extension advice is received it fails to provide advice on adaptation to climate change.
- Finally, concerning the Malawi Social Action Fund (MASAF), which provides a **safety net** to vulnerable households, in the 2010 survey 28% of villages surveyed had a MASAF program.

Without going into the detail of the functioning of these institutions, the picture that emerges from these statistics is one that highlights the limited access to information, credit, and safety nets for Malawian farmers. The challenges of managing risk faced by agricultural households are therefore numerous. The application presented here tries to provide new insights that would allow focusing potential efforts by policymakers interested in addressing agricultural risk management issues in Malawi.

Concerning insurance, in 2005, the World Bank, in close collaboration with Malawi's National Association of Small Farmers (NASFAM), developed an index-based crop insurance program, which led to 892 groundnut farmers purchasing weather-based crop insurance policies. During the 2006/2007 cropping season, the pilot expanded to 1710 farmers, with the inclusion of coverage for maize. A positive effect of the program was that, as the crop insurance contracts mitigated the weather risk associated with lending, local banks came forward to offer loans to insured farmers. However, what emerged from this pilot was that index-based weather insurance is not a panacea, since farmers face a broad spectrum of risk beyond just weather risk (Bryla and Syroka 2009). Furthermore, to be effective index-based weather insurance contracts require reliable, timely, and high quality weather data with a long historical record. More importantly from an institutional perspective, an improved enabling legal and regulatory framework is necessary for the expansion of any weather index insurance in Malawi. These challenges, combined with the often limited understanding of insurance, can lead to low adoption of insurance. We are aware of these challenges, and here we discuss weather index insurance as one possible tool in a portfolio of risk management options, as indicated by the theoretical model presented in the previous section.

## 3.2 Background Information on Malawi for the Empirical Application

Agriculture is the mainstay of the economy of Malawi accounting for about 34% of GDP, 85% of the labour force and 83% of foreign exchange earnings (Mucavele 2007). Smallholders account for 78% of the cultivated land and generate about 75% of Malawi's total agricultural output, indicating the predominance of the smallholder agricultural sector (Chirwa and Quinion 2005; Tchale 2009). Malawi is densely populated, with 84% of farmers practicing rainfed agriculture only, and more than 72% of the smallholder farms having an area of less than one hectare. Such conditions already make food self-sufficiency at the household level difficult, and the predicted impacts of climate change in Malawi are expected to primarily impact smallholder, rain dependent farmers (Denning et al. 2009).

The principal crops grown in Malawi are maize, tea, sugarcane, groundnut, cotton, wheat, coffee, rice and pulses. A significant feature of Malawi's agriculture is the dominance of maize in farming systems. It is estimated that more than 70% of the arable land is allocated to maize production (GoM 2006). According to Dorward et al. (2008), the share of farmers growing maize varies from 93% to 99% in the country's main regions. Although agriculture and maize are clearly very important to the livelihoods of most Malawians, their overall productivity performance raises serious concerns about long-term viability. The factors that are commonly cited as underlying low crop productivity include weather variability, declining soil fertility, limited use of improved agricultural technologies and sustainable land management practices, low/poor agricultural extension services, market failures, and underdevelopment and poorly maintained infrastructure (World Bank 2010).

Of relevance to agricultural risk management in Malawi, the yield of crops is limited to differing degrees by water availability and temperature depending on the agro-ecological zone (see Fig. 1). A synthesis of climate data by the United Nations Development Program (McSweeney et al. 2012) indicated that in the period 1960 to 2006, mean annual temperature in Malawi increased by 0.9 °C. This increase in temperature has been concentrated during the rainy summer season (December – February), and is expected to increase further. Long term rainfall trends are difficult to characterize due to the highly varied inter-annual rainfall pattern in Malawi, though such variability is expected to increase under climate change (McSweeney et al. 2012).

## 3.3 Data and Estimated Production Functions

We now turn our attention to simulations of smallholder Malawian farmer planting decisions and outcomes under a number of climate change and policy intervention scenarios. The relationships between input usage and yields for each crop and CSA technique, are estimated separately using multiple regressions with data from the Third Integrated Household Survey (IHS3 2012), which was conducted from March 2010 to March 2011 and implemented by the Malawian National Statistical Office (NSO) in collaboration with the World Bank. From this dataset we rely on information from ~7800 Malawian rural households covering ~18,500 individual plots cultivated during the 2009–2010 agricultural season. While such estimates are made for all four agro-ecological zones (AEZs) in Malawi, this investigation focuses on Tropical Warm/Semiarid AEZ for which the most data are available (nearly 9000 plot observations). Crop specific production functions are estimated by regressing logged plot level yields on logged input usages and weather conditions. The use of logged values of yields and inputs in a linear framework is equivalent to assuming a Cobb-Douglas production function with a translog structure. As weather variables enter linearly (i.e.- not logged), they are treated as TFP shifters. The resulting estimated regression equations serve as the production functions for later simulation of farmer outcomes under various weather, price, information, and restriction scenarios. Table 1 presents the coefficient estimates from the 2009–2010 data for the Tropical Warm/Semiarid AEZ. These coefficients define the production functions used in the simulation of farmer planting decisions and outcomes.

Crop specific functions for variation in yields are also estimated through linear regressions of the standard deviation of yields between plots within the 768 Enumeration Areas on measures of the level and variation of rainfall and temperatures during the 2009–2010 growing season. The resulting estimated equations (one for each crop type) serve to simulate the variation in yields under scenario specific conditions.

As outlined in Table 1, agricultural inputs included in the estimation of production functions include seed quantity, fertilizer usage, days of labor, and land area of the plot. Weather conditions, which are used to estimate both the production and yield variation functions, include mean and standard deviation of temperatures and

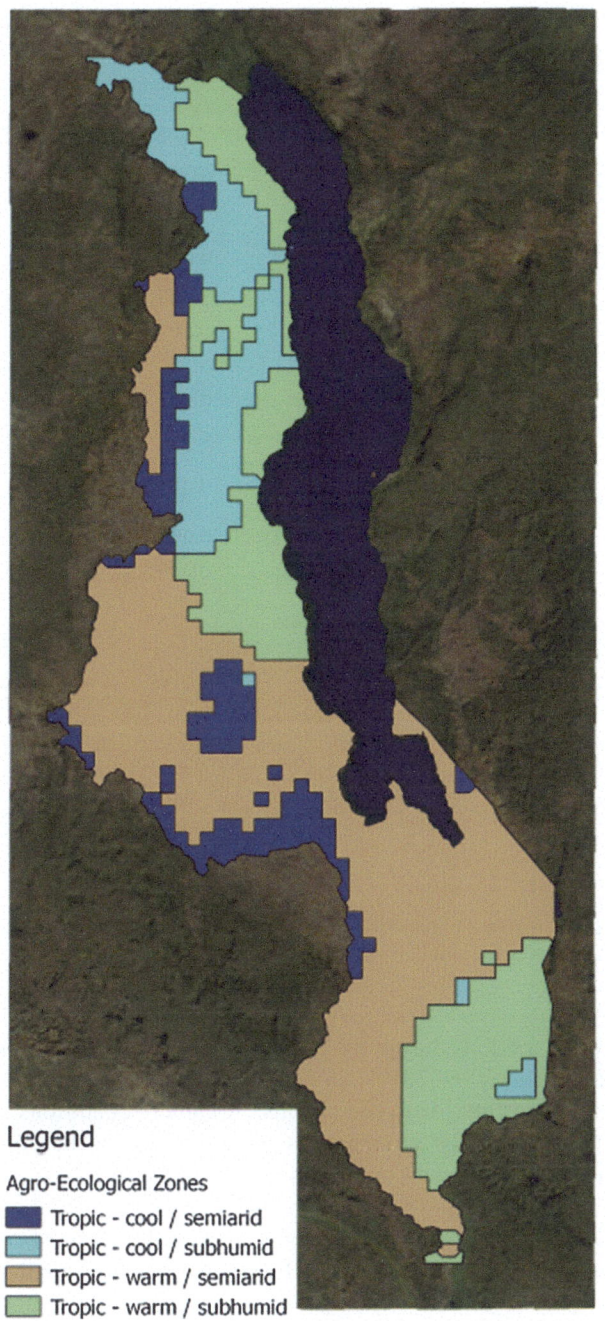

Fig. 1 Malawi agro-ecological zones

**Table 1** Coefficient estimates for production function by crop – dependent variable is logged yield

|  | Maize local | Maize hybrid | Groundnut Chalimbana | Groundnut CG7 | Beans | Pigeonpea (nandolo) |
|---|---|---|---|---|---|---|
| Seed Quantity – Logged (kg) | 0.0288 (0.02) | 0.0403 (0.02) | 0.0745 (0.03) | 0.062 (0.02) | 0.18 (0.06) | 0.164 (0.03) |
| Fertilizer Usage – Logged (kg) | 0.0411 (0.00) | 0.0453 (0.00) | −0.0153 (0.01) | 0.00131 (0.01) | −0.0265 (0.02) | 0.00514 (0.01) |
| Labor Days – Logged (days) | 0.161 (0.04) | 0.0985 (0.04) | 0.0715 (0.06) | 0.175 (0.06) | 0.0347 (0.14) | 0.25 (0.05) |
| Cultivated Area – Logged (hectares) | 0.491 (0.04) | 0.436 (0.04) | 0.296 (0.06) | 0.55 (0.06) | 0.263 (0.14) | 0.196 (0.06) |
| Mean Temperature (10-day mean) | −0.0702 (0.03) | −0.17 (0.03) | −0.0151 (0.06) | 0.0546 (0.06) | 0.0837 (0.11) | 0.246 (0.07) |
| SD Temperature | −0.858 (0.20) | −0.243 (0.20) | −0.0946 (0.32) | −0.155 (0.31) | 0.253 (0.90) | 0.928 (0.28) |
| Mean Precipitation (mm/10-day period) | 0.0178 (0.01) | 0.036 (0.01) | 0.0383 (0.02) | 0.0314 (0.01) | 0.0771 (0.03) | −0.0511 (0.01) |
| SD Precipitation | −0.0186 (0.01) | −0.0573 (0.01) | −0.0342 (0.02) | −0.0327 (0.02) | −0.0654 (0.03) | −0.0239 (0.02) |
| Constant | 8.386 (0.91) | 10.68 (0.93) | 4.816 (1.55) | 4.068 (1.59) | 0.806 (2.15) | −1.42 (1.67) |

Notes: Standard errors reported in parenthesis. Significance of estimates is not taken into account when applying estimates in the simulation.

rainfall over 10-day periods during the growing season, and are observed for each enumeration area. The 2009–2010 rainy season therefore serves as the basis for defining the relationships between inputs, weather conditions, and yields. Additionally, the 2009–2010 growing season serves as the baseline period for weather conditions and all prices used in the simulations.

It is important to note also that the direct reliance of the model on data for the estimation of yield functions and input usages restricts the scope of crops and agricultural techniques that are considered in the simulations to those that are in wide use during the 2009–2010 Malawian growing season and that, more in general, characterize Malawi agricultural production. In particular, neither crop varieties nor cultivation practices that are particularly adapted to varying conditions under climate change are considered in the simulations because no basis for modelling the relevant relationships between inputs and outputs exists in the data, nor information on crop varieties. Practically, this approach assumes that crop and technique availability doesn't change in the simulated future and thus implicitly limits the scope of extension activities (when extension is considered) to the provision of information regarding growing conditions.[8]

## 3.4 Simulation Model Assumptions

Following the estimation of the production and yield variation functions using the 2009–2010 data, the simulation of farmer decisions and resulting outcomes for a future growing season are undertaken in two distinct stages. In the first, a representative farmer is faced with a planting decision based on known input prices, anticipated weather conditions, and known relationships between inputs, weather, and yields.[9] This information, along with anticipated output prices is used by the farmer to maximize expected utility through decisions about which crops to plant and what, if any, CSA techniques to use. In the second stage, farmer outcomes are simulated based on crop and CSA choices and scenario specific weather conditions. The degree to which farmer expectation of weather conditions align (or not) with realized conditions serves as a measure of farmer information regarding climate change. Changes in the level of farmer "informedness" are the means through which extension informational programs can impact simulated farmer cropping choices and outcomes.

The representative farmer must choose between local and hybrid maize as a staple crop, and may also plant a cash crop for diversification purposes. The simulated diversification crops are all legumes and include Chalimbana Groundnuts, CG7 Groundnuts, Beans, and Pigeon Peas. The farmer is restricted to planting a minimum share of the chosen staple crop in order to ensure subsistence (which is not given an explicit utility or profit value in the simulations), and can choose up to one diversification crop to plant in addition to the staple (thus, planting 100% staple crop is always an option).

For any combination of staple and diversification crop, the farmer also selects whether and which CSA techniques to apply to the growing of the staple crop. Specifically, the farmer chooses between soil and water conservation (SWC) techniques, legume intercropping, or both in these simulations. Each CSA technique modulates the impact of inputs and weather on yields of the staple (but not the

---

[8] It is worth noting that maize utilization in Malawi is largely linked to the fertilizers input subsidy program (FISP) which accounts for a limited range of varieties distributed but even accounting for varietal diversity the main distinction would still be linked to local versus hybrid maize utilization.

[9] Input and output prices, as well as the production and yield variance functions are not altered in any of the scenarios considered in this chapter, but are instead held fixed as observed in the 2009-2010 growing season.

diversification crop, if any) in ways that are assumed to be understood (and thus taken account of) by the farmers.

As in our conceptual model, farmers are assumed to have a mean-variance utility function in net profits. They choose the crop mix (up to one staple and one diversification crop) and CSA technique usage by maximizing expected utility given anticipated weather and price conditions. As noted, farmers are allowed at most one diversification crop and, for simplicity, we limit our analysis to crop shares in 10% increments. In the second stage of the simulation, net profits and total utility are calculated (using the same mean-variance utility function) using scenario-specific weather conditions.

Simulated utility levels – both for anticipated utility in stage 1 and realized utility in stage 2 – are simply the sum of the simulated net profit minus the simulated variance of revenues times the coefficient of absolute risk aversion. This mean-variance utility function is laid out explicitly below:

$$U = p_1 \widehat{yield_1} + p_2 \widehat{yield_2} - IP \cdot I$$
$$-ARA^* \left[ p_1^2 \, Var\left(\widehat{yield_1}\right) + p_2^2 \, Var\left(\widehat{yield_2}\right) + 2\, p_1\, p_2 \operatorname{cov}\left(yield_1, yield_2\right) \right]$$

Above we see the simulated yield levels for the staple, $\widehat{yield_1}$, and diversification crop, $\widehat{yield_2}$, multiplied by their respective output prices, $p_1$ and $p_2$. From this simulated revenue, the dot product of the vectors of input prices, $IP$, and input usages, $I$, is subtracted to yield net profits. The second line of the equation is the variance portion of the utility function, discounted by the coefficient of absolute risk aversion. The variance of revenues (equivalent to the variance of profits since input prices are non-stochastic) is simply the simulated variance of yields from the two crops, $Var(\widehat{yield_1})$ and $Var(\widehat{yield_2})$, each multiplied by the square of their respective output prices, plus the covariance correction term $2\, p_1\, p_2\, Cov(yield_1, yield_2)$.[10] The between-crop covariance term is estimated directly from yields in the 2009–2010 data.

The representative farmer is simulated making planting decisions for a single average sized plot of 0.74 hectares and is assumed to apply mean input levels for each crop planted and CSA technique utilized. Labor costs for different crop choices and CSA usages are included in the cost calculation used by the farmer for planting decisions but are omitted from the simulation of realized utility, as most labor is provided without monetary cost (by family, friends, or for an in-kind payment). Finally, a coefficient of Absolute Risk Aversion of 0.00016 is assumed, which implies a coefficient of Relative Risk Aversion of approximately 1.5 for the representative farmer. The modelled level of risk aversion is informed by the estimated

---

[10] This summing procedure is simply following the rules for adding variances, namely:

$$Var(aX + bY) = a^2 \, Var(X) + b^2 \, Var(Y) + 2ab \operatorname{cov}(X, Y)$$

risk aversion parameters of De Brauw & Eozenou (2014) for Mozambican farmers, while taking into account the lower average incomes of Malawians.

## 3.5 Climate Scenarios

Three climate change scenarios are considered in these simulations. These include a "No Climate Change" scenario under which weather conditions remain at baseline, that is as observed in the 2009–2010 rainy season, a "Mid-line Climate Change" scenario under which mean temperature, standard deviation of temperature and standard deviation of rainfall are all increased by 10% from baseline, and a "High Climate Change" scenario under which the levels of these three weather variables are increased by 20% from baseline. Due to the uncertainty of the effects of climate change on rainfall levels in Malawi, we do not simulate changes in mean rainfall as part of our climate change scenarios.[11]

Observed price levels in the 2009–2010 data are used for both inputs and outputs under all three climate change scenarios, thus the general equilibrium effects of climate change on market prices are not considered by this analysis.[12]

## 4 Simulation Results

As described earlier, we will simulate the impacts of insurance and extension under a variety of climate change scenarios. For the purposes of simulation, the function of the extension will be limited to providing farmers with information about changing weather conditions due to climate change. This is akin to extension activities only impacting $m$ in the conceptual model. While it is likely that extension services would be much broader in practice, the simulation of such effects is left for later work. Since the effectiveness of these two policy instruments will be inter-connected when extension is influencing farmer perceptions about climate change and thus the returns to insurance acquisition, we also present some stylized simulations where both are implemented simultaneously. Throughout, we contemplate two distinct assumptions regarding constraints on staple crop cultivation for subsistence purposes – a 50% and a 70% requirement – in part to illustrate the importance of crop diversification as a potential response to increased weather volatility and also to

---

[11] See McSweeney et al. (2012) for more information on the anticipated impacts of climate change on Malawi.

[12] Given the high proportion of subsistence farmers in Malawi, increased output prices due to increased scarcity under climate change are likely to be detrimental on net, and thus farmer outcomes simulated in a general equilibrium framework would likely be associated with lower levels of overall utility than those presented here.

demonstrate the additional value of information for farmers that are less constrained – by subsistence requirements or otherwise – in their planting decisions.

## 4.1 Insurance

Insurance in this context is assumed to be rainfall index insurance with a predetermined payout amount that is varied in certain simulations to model different levels of insurance coverage. Payouts are received if rainfall is below a pre-specified level, fixed in our simulations at the 30th percentile of the rainfall distribution at baseline. Universal participation in the rainfall insurance program is assumed when the program is available, and premiums are assumed to be zero (or covered by the government or other outside institution).[13]

We begin by looking at the impacts of insurance coverage on farmer decisions and outcomes. Panels a & b of Fig. 2 report the results of simulations in which the payout amount for rainfall insurance is varied between zero and 6000 MWK (which is slightly above 100% of expected net profits under baseline conditions) for each of the three climate change scenarios with a 70% staple requirement (Fig. 2a) and a 50% staple requirement (Fig. 2b). As the level of payout increases we see the simulated average total utility rise in all three climate change scenarios under both staple constraints. These simulations assume farmers are unaware of the changes to weather conditions under the climate change scenarios, and thus we observe no differences in crop choice or CSA usage between scenarios. The changes in weather conditions do however affect farmer outcomes as illustrated by the lower utilities simulated under the Mid-line and High Climate Change scenarios. The greater the difference between farmer-anticipated and realized weather conditions, the larger the loss of utility to farmers.

It is worth noting that under more extreme climate change scenarios, the variance of rainfall (but not the mean) increases. This slightly increases the likelihood of payout at the 30th percentile of baseline rainfall (as well as at all other rainfall trigger levels below the 50th percentile), but this change is not significant enough to be easily distinguished in the presented figures as the effects of climate change on production greatly outweigh the effects on the probability of insurance payout. Nonetheless, farmer outcomes improve slightly more under the Mid-line Climate Change scenario than under the No Climate Change scenario and under the High Climate Change scenario compared to the Mid-line Climate Change scenario because of the increase in likelihood of a payout.

Comparing outcomes under the more and less restrictive staple requirements, we see higher levels of diversification when the staple requirement is relaxed, but that greater diversification into a cash crop (in this case beans) opens the farmer up to greater harm under climate change.

---

[13] Mapping this insurance policy and subsequent simulations into the conceptual model involves setting $c_\psi = 0$ and varying $\psi$ exogenously.

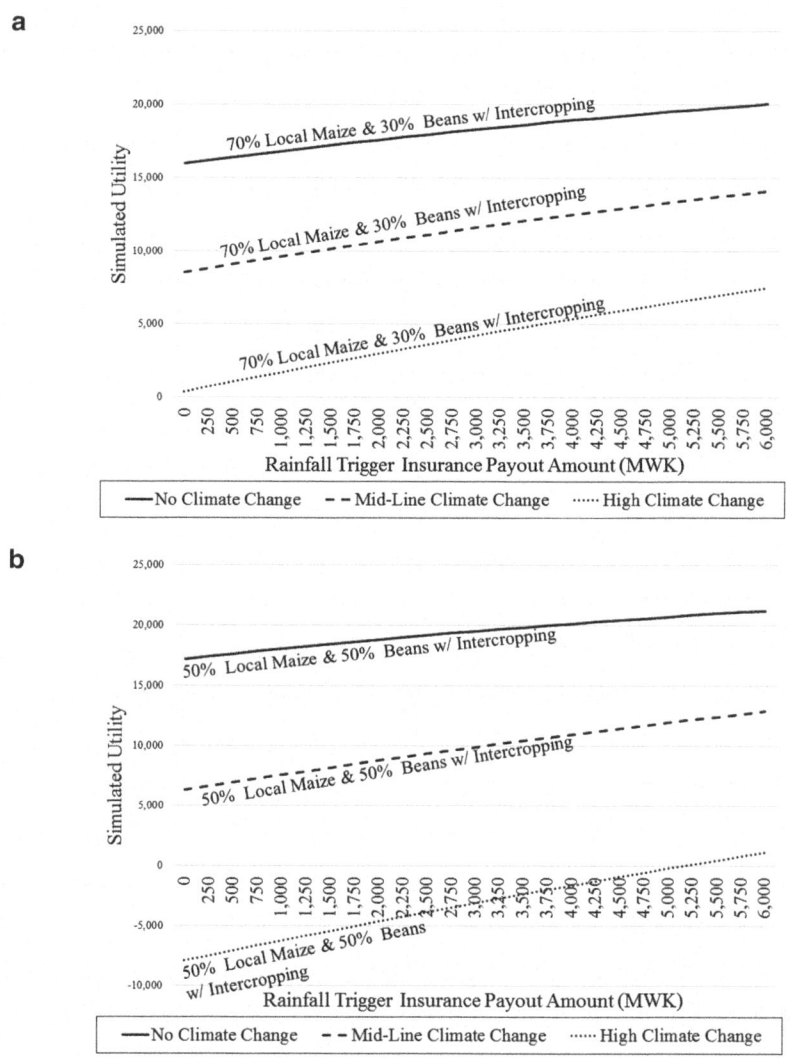

**Fig. 2** (**a**) Simulated Utility by Insurance Payout, Unanticipated Climate Change – 70% Staple Requirement. *Notes:* Simulations are based on coefficient estimates and baseline parameter values from the Tropical Warm/Semiarid Agro-ecological Zone, which is the AEZ in Malawi for which the most data are available. Simulated profits under baseline conditions are equal to 5934 MWK, thus the maximum insurance payout simulated here amounts to a full replacement of baseline profits. All utility levels are normalized via the addition of 30,000 units. Crop Choice and CSA usage does not differ between scenarios because climate change is unanticipated. (**b**) Simulated Utility vs. Insurance Payout, Unanticipated Climate Change – 50% Staple Requirement. *Notes:* Simulations are based on coefficient estimates and baseline parameter values from the Tropical Warm/Semiarid Agro-ecological Zone, which is the AEZ in Malawi for which the most data are available. Simulated profits under baseline conditions are equal to 5152 MWK, thus the maximum insurance payout simulated here amounts to more than a full replacement of baseline profits. All utility levels are normalized via the addition of 30,000 units. Crop Choice and CSA usage does not differ between scenarios because climate change is unanticipated

In Fig. 3, the payout amount of the rainfall trigger insurance is again varied (and the trigger level is still held fixed at the 30th percentile of baseline rainfall), only this time farmers are informed about the changes in weather conditions under the climate change scenarios and can adjust their planting decisions accordingly. This allows farmers to adopt additional CSA techniques in the face of harsher weather conditions, and also to switch diversification crop from Beans to Groundnut CG7 (which is specifically noted for its drought tolerance, see Subrahmanyam et al. 2000). These adaptations on the part of the farmer lead to utility outcomes under climate change that are much more similar to the baseline outcomes than those achieved when changes in weather conditions were unanticipated. As weather variation increases, that is, as we move from the No Climate Change scenario to the Mid-line and on to the High Climate Change scenario, we see planting decisions moving toward greater adoption of CSA techniques, consistent with Eq. 11 in the conceptual model as well as the results of Arslan et al. (2013).[14]

Comparing Panels a & b in Figs. 2 and 3, we see again that the relaxation of staple requirements leads to further diversification and poorer outcomes under climate change. These results suggest that farmers that are currently somewhat better off (and thus are less constrained by subsistence requirements to plant a staple crop) are more susceptible to harm under unanticipated climate change. While information regarding climate change (i.e.- when the changes in weather conditions are anticipated) improves outcomes for all farmers under the climate change scenarios, this improvement is most dramatic when staple requirements are less stringent, suggesting a higher value of information for less constrained farmers. Put another way, without information on climate change, shifts in weather conditions have greater potential to harm farmers that are less constrained. Without good information on climate change, this effect would tend to increase subsistence constraints in successive years as farmers that began with more flexibility will tend to face greater harm from unanticipated changes in weather conditions. Finally, it is notable that better information regarding weather conditions (that is comparing Figs. 2 and 3) leads to additional uptake of CSA techniques, providing a concrete example of increased farmer adaptive behavior in the face of climate change following a risk management intervention.

## 4.2 Extension and Information Provision

Given the results in Figs. 2 and 3, we now turn to a more direct examination of how more information about changing weather conditions under climate change might impact farmer choices and outcomes. Panels a and b of Fig. 4 demonstrate the results of bringing farmer expectations regarding weather conditions closer in line

---

[14] We do not however see increasing diversification in response to growing weather variability as predicted by Equation 13. Likely reasons for this are discussed in Section IV.B below in the context of better information regarding variability.

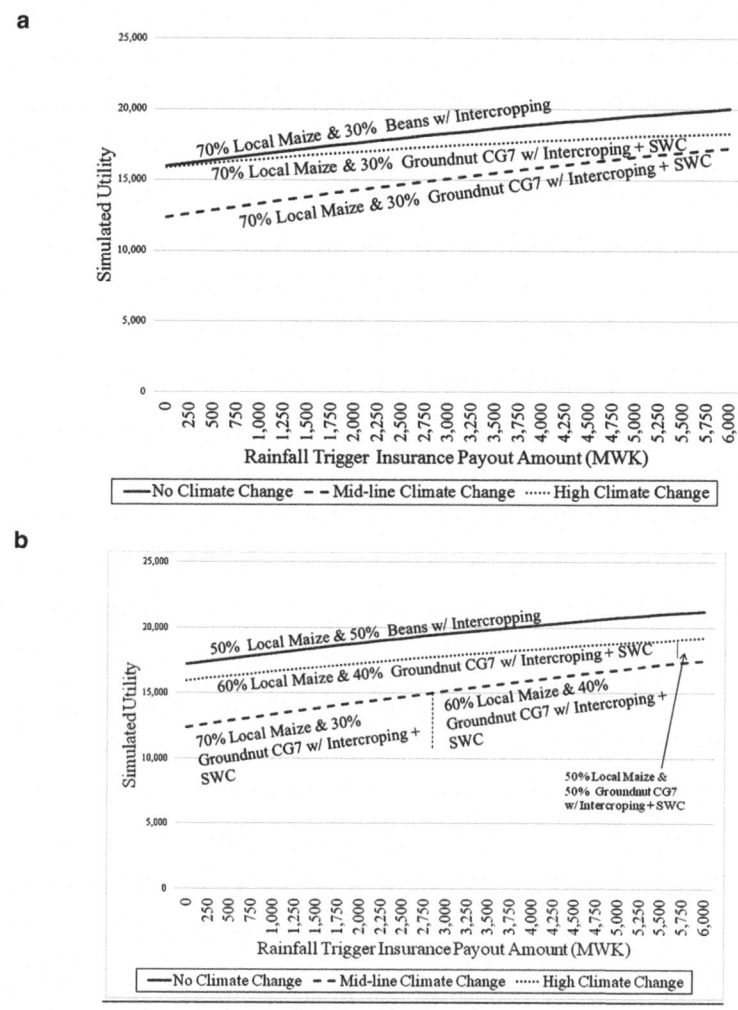

**Fig. 3** (**a**) Simulated Utility by Insurance Payout, Anticipated Climate Change – 70% Staple Requirement. *Notes:* Simulations are based on coefficient estimates and baseline parameter values from the Tropical Warm/Semiarid Agro-ecological Zone, which is the AEZ in Malawi for which the most data are available. Simulated profits under baseline conditions are equal to 5934 MWK, thus the maximum insurance payout simulated here amounts to a full replacement of baseline profits. All utility levels are normalized via the addition of 30,000 units. (**b**) Simulated Utility by Insurance Payout, Anticipated Climate Change – 50% Staple Requirement. *Notes:* Simulations are based on coefficient estimates and baseline parameter values from the Tropical Warm/Semiarid Agro-ecological Zone, which is the AEZ in Malawi for which the most data are available. Simulated profits under baseline conditions are equal to 5152 MWK, thus the maximum insurance payout simulated here amounts to more than a full replacement of baseline profits. All utility levels are normalized via the addition of 30,000 units

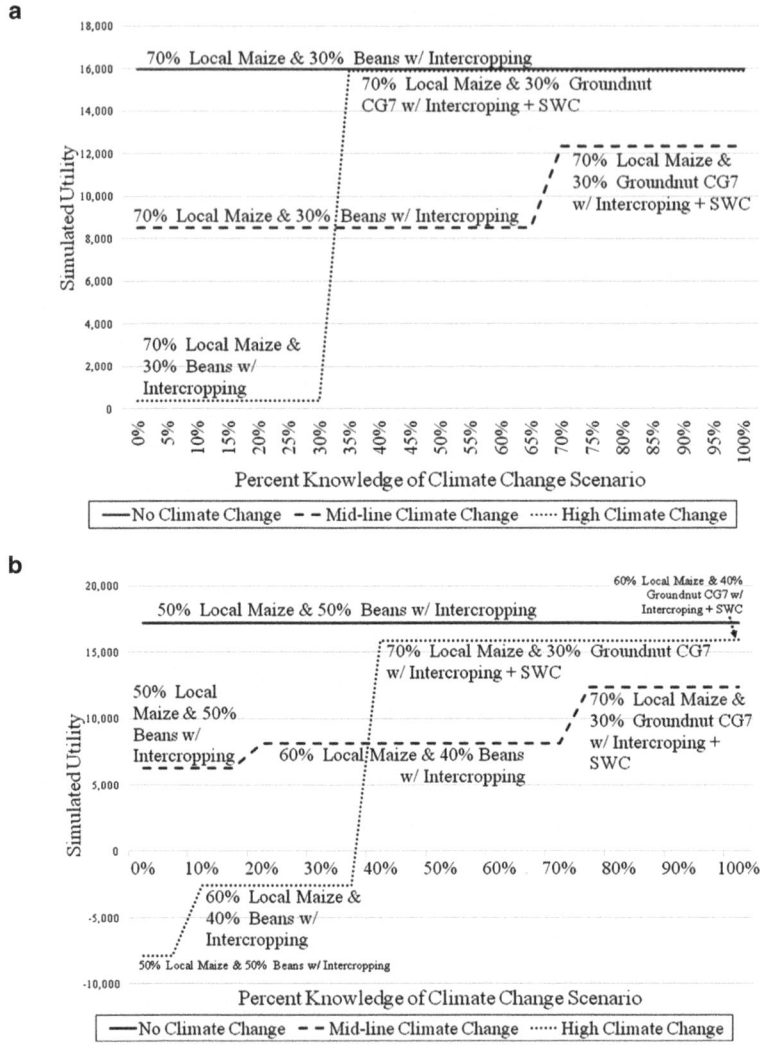

**Fig. 4** (**a**) Simulated Utility by Informedness Regarding Climate Change – 70% Staple Requirement. *Notes:* Simulations are based on coefficient estimates and baseline parameter values from the Tropical Warm/Semiarid Agro-ecological Zone, which is the AEZ in Malawi for which the most data are available. All utility levels are normalized via the addition of 30,000 units. Crop Choice and CSA usage does not change under the No Climate Change scenario because weather conditions conform to farmer's expectations. (**b**) Simulated Utility by Informedness Regarding Climate Change – 50% Staple Requirement. *Notes:* Simulations are based on coefficient estimates and baseline parameter values from the Tropical Warm/Semiarid Agro-ecological Zone, which is the AEZ in Malawi for which the most data are available. All utility levels are normalized via the addition of 30,000 units. Crop Choice and CSA usage does not change under the No Climate Change scenario because weather conditions conform to farmer's expectations

with the scenario conditions that drive yields. Partial information may capture either incomplete penetration of information provision (i.e. some share of perfectly informed farmers make for a representative farmer that is partially informed), imperfect information regarding the climate change scenario that farmers are encountering, or some combination of the two. However, since we simulate outcomes for a single "representative farmer", rather than all farmers on average, the simulations reflect an improving quality of information, such that the information the farmer relies on increasingly reflects the true conditions of the climate scenario that will determine yield outcomes. Over time, the smooth evolution of farmer expectations toward conditions under climate change could arise from straightforward Bayesian updating.

Under the No Climate Change scenario in Panels a & b of Fig. 4, we see that increased information has no effect on planting decisions or outcomes because there is no deviation between farmers' baseline expectations and realized conditions (in effect, farmers are fully informed at baseline). However, when conditions do deviate from past levels – as they do under the Mid-Line and High Climate Change scenarios – we see that more information does drive different crop and CSA usage decisions. That is to say that farmer decisions change when farmer expectations about conditions deviate from baseline to the degree that another crop choice/CSA combination yields higher total utility. Specifically, as informedness regarding changing weather conditions increases, we see the adoption of SWC techniques – in addition to legume intercropping – and the planting of CG7 Groundnuts, which are high yielding and better suited to the weather conditions under climate change than Beans. Importantly we see that farmer outcomes improve as they are provided with additional information, and that the value of information increases as realized weather conditions deviate further from baseline expectations (that is, under scenarios in which climate change is more extreme).

Bringing expectations regarding weather conditions in line with the new realities under climate change is akin to increasing $m$ in the conceptual model. In response we see increased CSA usage as predicted by Eq. 15, but we generally see a fall in the level of the diversification crop planted. This apparent contradiction with the predictions of Eq. 17 is likely explained by better yields of local maize under climate change conditions relative to the cash crops. In our simplified conceptual model, diversification only reduces yield variability, but in our empirical context it can also lower the yields of cash crops relative to the staple, potentially increasing the level of staple planted.

Returning to the simulation results, we again see that the loosening of staple requirements weakly increases the usage of diversification crops under all scenarios. Additionally, farmers move away from baseline planting behaviors and achieve higher profits/utility with less information when they have a wider range of crop combination possibilities under the less restrictive staple requirements. This illustrates that farmers are better able to make use of partial information on climate change when they are less constrained by subsistence requirements. This point is particularly important given that perfect information on future weather conditions cannot be provided in the real world, so any information will necessarily be partial information.

## 4.3 Insurance and Extension

Tables 2 and 3 examine the potential impacts of insurance and information provided in concert on farmer decisions under the Mid-line and High Climate Change scenarios respectively. Results are not presented for the No Climate Change scenario as information regarding climate change is of no value in that case. Moreover, we only present results under the 50% staple requirement in order to focus attention on the effects of changes in information and insurance coverage rather than the staple constraint. Simulated utility levels are not presented in these tables, but utility levels weakly increase as the levels of information and insurance payout increase (that is as we move toward the bottom right of each table). It is worth noting that the top and bottom rows of Tables 2 and 3 correspond to the dashed and dotted lines in Figs. 2b and 3b, while the first columns of the tables correspond to the dashed and dotted lines respectively in Fig. 4b.

Tables 2 and 3 show, without exception, that the amount of land dedicated to cash crops weakly increases as the level of insurance coverage increases. This finding is consistent with the conclusions of Collier et al. (2009), who argue that weather index insurance can, if appropriately designed, be used to facilitate farmer adaptation to climate change. These results suggest that government or donor assistance could be justified, and it should focus on funding the start-up costs of developing weather insurance markets and addressing the catastrophic layer of risk.

Turning to the effects of increased information regarding climate change, we see consistent switching from Beans as a cash crop to Groundnut CG7 which is better adapted to the climate change impacted weather conditions. Similarly, we see the wider adoption of SWC techniques as better information on the extent of climate change is made available to the farmers. Both these characterizations hold across insurance coverage levels, and suggest greater adaptation in the face of greater anticipated change, no matter the level of insurance coverage.

It is also worth noting that in a number of cases where insurance payouts are high and climate change expectations are moderate, hybrid maize will be planted rather than the local maize that is more typical. It would appear that these cases represent scenarios when the farmer, relieved of some downside risk by high insurance coverage, seeks to take advantage of the upside potential of hybrid maize. This response proves ex post problematic since hybrid maize is more sensitive to weather variability. Once the full extent of the changes in weather conditions due to climate change are revealed, the farmer returns to more conservative cropping choices and the disincentivizing impacts of insurance coverage on adaptation disappear. These results, however, should be interpreted in light of the data limitation of the study as well as the characteristics of Malawian agriculture. As already pointed out, the IHS3

**Table 2** Crop choice and conservation technique usage by level of information regarding climate change scenario and insurance payout level: mid-line climate change – 50% staple requirement

| | | Rainfall trigger insurance payout amount | | | | | | |
|---|---|---|---|---|---|---|---|---|
| | | 0 MWK | 1000 MWK | 2000 MWK | 3000 MWK | 4000 MWK | 5000 MWK | 6000 MWK |
| Climate change scenario | 0% | 50% local maize & 50% beans w/ intercropping | 50% local maize & 50% beans w/ intercropping | 50% local maize & 50% beans w/ intercropping | 50% local maize & 50% beans w/ intercropping | 50% local maize & 50% beans w/ intercropping | 50% local maize & 50% beans w/ intercropping | 50% local maize & 50% beans w/ intercropping |
| | 25% | 60% local maize & 40% beans w/ intercropping | 60% local maize & 40% beans w/ intercropping | 60% local maize & 40% beans w/ intercropping | 50% local maize & 50% beans w/ intercropping | 50% local maize & 50% beans w/ intercropping | 50% local maize & 50% beans w/ intercropping | 50% local maize & 50% beans w/ intercropping |
| | 50% | 60% local maize & 40% beans w/ intercropping | 60% local maize & 40% beans w/ intercropping | 60% local maize & 40% beans w/ intercropping | 60% local maize & 40% beans w/ intercropping | 60% local maize & 40% beans w/ intercropping | 60% local maize & 40% beans w/ intercropping | 50% hybrid maize & 50% groundnut CG7 w/ intercroping + SWC |
| | 75% | 70% local maize & 30% groundnut CG7 w/ intercroping + SWC | 70% local maize & 30% groundnut CG7 w/ intercroping + SWC | 70% local maize & 30% groundnut CG7 w/ intercroping + SWC | 50% hybrid maize & 50% groundnut CG7 w/ intercroping + SWC | 50% hybrid maize & 50% groundnut CG7 w/ intercroping + SWC | 50% hybrid maize & 50% groundnut CG7 w/ intercroping + SWC | 50% hybrid maize & 50% groundnut CG7 w/ intercroping + SWC |
| | 100% | 70% local maize & 30% groundnut CG7 w/ intercroping + SWC | 70% local maize & 30% groundnut CG7 w/ intercroping + SWC | 70% local maize & 30% groundnut CG7 w/ intercroping + SWC | 60% local maize & 40% groundnut CG7 w/ intercroping + SWC | 60% local maize & 40% groundnut CG7 w/ intercroping + SWC | 60% local maize & 40% groundnut CG7 w/ intercroping + SWC | 60% local maize & 40% groundnut CG7 w/ intercroping + SWC |

*Notes*: MWK stands for Malawian Kwacha, the local currency in our empirical context. Simulations are based on coefficient estimates and baseline parameter values from the Tropical Warm/Semiarid Agro-ecological Zone, which is the AEZ in Malawi for which the most data are available. Simulated profits under baseline conditions are equal to 5152 MWK. thus the maximum insurance payout simulated here amounts to more than a full replacement of baseline profits. Percent knowledge regarding the relevant climate change scenario represents a share of the linear distance between the baseline conditions and actual conditions under the climate change scenario at which farmers anticipate conditions will be when the planting decision is made

**Table 3** Crop choice and conservation technique usage by level of information regarding climate change scenario and insurance payout level: high climate change – 50% staple requirement

| | | Rainfall trigger insurance payout amount | | | | | | |
|---|---|---|---|---|---|---|---|---|
| | | 0MWK | 1000 MWK | 2000 MWK | 3000 MWK | 4000 MWK | 5000 MWK | 6000 MWK |
| Climate change scenario | 0% | 50% local maize & 50% beans w/ intercropping | 50% local maize & 50% beans w/ intercropping | 50% local maize & 50% beans w/ intercropping | 50% local maize & 50% beans w/ intercropping | 50% local maize & 50% beans w/ intercropping | 50% local Maize m 50% beans w/ intercropping | 50% local maize & 50% beans w/ intercropping |
| | 25% | 60% local maize & 40% beans w/ intercropping | 60% local maize & 40% beans w/ intercropping | 60% local maize & 0% beans w/ intercropping | 60% local maize & 40% beans w/ intercropping | 60% local maize & 40% beans w/ intercropping | 60% local maize & 40% beans w/ intercropping | 50% hybrid maize & 50% groundnut CG7 w/ intercroping + SWC |
| | 550% | 70% local maize & 30% groundnut CG7 w/ Intercroping + SWC | 70% local maize & 30% groundnut CG7 w/ intercroping + SWC | 70% local maize & 30% groundnut CG7 w/ intercroping + SWC | 60% local maize & 40% groundnut CG7 w/ intercroping + SWC | 60% local maize & 40% groundnut CG7 w/ intercroping + SWC | 60% local maize & 40% groundnut CG7 w/ intercroping + SWC | 60% local maize & 40% groundnut CG7 w/ intercroping + SWC |
| | 75% | 70% local maize & 30% groundnut CG7 w/ intercroping + SWC | 70% local maize & 30% groundnut CG7 w/ intercroping + SWC | 60% local maize & 40% groundnut CG7 w/ intercroping + SWC | 60% local maize & 40% groundnut CG7 w/ intercroping + SWC | 60% local maize & 40% groundnut CG7 w/ intercroping + SWC | 60% local maize & 40% groundnut CG7 w/ intercroping + SWC | 60% local maize & 40% groundnut CG7 w/ intercroping + SWC |
| | 100% | 60% local maize & 40% groundnut CG7 w/ intercroping + SWC | 60% Local Maize & 40% Groundnut CG7 w/ Intercroping + SWC | 60% local maize & 40% groundnut CG7 w/ intercroping + SWC | 60% local maize & 40% groundnut CG7 w/ intercroping + SWC | 60% local maize & 40% groundnut CG7 w/ intercroping + SWC | 60% local maize & 40% groundnut CG7 w/ intercroping + SWC | 50% local maize & 50% groundnut CG7 w/ intercroping + SWC |

*Notes:* MWK stands for Malawian Kwaclia, the local currency in our empirical context. Simulations are based on coefficient estimates and baseline parameter values from the Tropical Warm/Semiarid Agro-ecological Zone, which is the AEZ in Malawi for which the most data are available. Simulated profits under baseline conditions are equal to 5152 MWK. Thus the maximum insurance payout simulated here amounts to more than a full replacement of baseline profits. Percent knowledge regarding the relevant climate change scenario represents a share of the linear distance between the baseline conditions and actual conditions under the climate change scenario at which farmers anticipate conditions will be when the planting decision is made data used to model the relationships between input usage and maize yields do not allow to further distinguish between specific hybrid varieties including those that can be specifically adapted to climate change conditions. Moreover, Malawi in not a country of origin for the crop, which implies that genetic diversity is rather low compared to traditional maize domestication countries.[15]

## 5 Conclusions and Policy Implications

This chapter ventures into key support services that are explicitly addressed and contemplated in the Agriculture Sector Wide Approach (ASWAp) of Malawi- the national policy program of the country- namely: (1) technology generation and dissemination (whereby a key role is precisely identified for weather forecasting) and (2) institutional strengthening (including insurance) and capacity building.

The conceptual model built was also driven by results of an evidence base project that has been conducted in Malawi between 2012 and 2015 (FAO and GoM 2015). Results of the study indicate that:

(a) weather variability is a key factor determining which strategies will work across different locations in Malawi for agricultural practices, types of crops and diversification strategies suggesting explicitly that "using weather data in planning any agricultural and food security intervention" would be highly advisable.
(b) Improving communication of information and tailoring extension services to local conditions (including weather variability) is likely to increase adoption rates of different crops and agricultural practices as well as farm incomes across the country, therefore a stronger investment should be made to strengthen extension based service.
(c) In terms of risk management instruments available to farmers, no insurance exists in the country and as such insurance schemes and simulations could be examined in more depth as part of the agricultural risk management portfolio of options provided by policymakers.

As a result, the chapter built an empirical model, which aimed at advancing the state of knowledge on the options and choices between diversification and land management practices, through the presence or absence of institutional support provided by insurance and extension in the form of awareness of climate scenarios. Different potential welfare outcomes for agricultural households are, hence, investigated and examined as a result of the model. A third key institution, access to credit, is indirectly addressed through the implication of analysis conducted and results obtained.

---

[15] We nevertheless recognize room for improvement in our analysis as additional information may become available from the new wave of the IHS (IHS4) that the World Bank is currently implementing in Malawi.

The conceptual model developed highlights that the interaction, in addressing risk, between diversification and land management complicates the role of policy levers and their impact. The model simulates the impacts of weather index insurance and extension under a range of climate change scenarios for two levels of staple requirements.

The empirical application, which presents results for farmers in the tropical warm/semi-arid AEZ of Malawi, builds on the conceptual model by estimating production functions and yield variation functions for different crops, and then simulating the outcome of farmer decisions.

As a first result, the crucial role played by extension, although in this model simply limited to climatic scenarios, is confirmed by the simulations, indicating that more information on climatic variables and their impact on yields can drive farmers to choose different crops, as well as different and more sustainable land management practices (SLM). It is interesting to note that among the SLM the main role is played by Soil and Water Conservation structures, confirming findings reported by FAO and GoM (2015), which suggested that "in areas where there is high and increasing variability of rainfall and higher aridity, the evidence indicates that sustainable land management practices such as soil and water conservation, legume rotation or intercropping and agroforestry (fertilizer tree systems) are more productive than conventional practices".

The important implication of this finding is that farmer welfare outcomes, driven by diversification of crop and adoption of SLM, improve as they are provided with additional information, and that the value of information increases as realized weather conditions deviate further from baseline expectations (that is, under scenarios in which climate change is more extreme). These results highlight how the value of information is higher for farmers that are less restricted in their planting choices, since they have a broader scope to adapt, suggesting important implications also with regard to access to other seed crops via credit.

Comparing outcomes under the more and less restrictive staple requirements, we see higher levels of diversification when the staple requirement is relaxed, but that greater diversification into a cash crop opens the farmer up to greater losses under climate change when this is not anticipated, suggesting that farmers that are currently somewhat better off (and thus are less constrained by subsistence requirements to plant a staple crop) are more susceptible to unanticipated climate change. While information regarding climate change (i.e. when the changes in weather conditions are anticipated) improves outcomes for all farmers under the climate change scenarios, this improvement is most dramatic when staple requirements are less stringent.

Moving to the role of insurance, it is important to note that the insurance instrument we analyzed is triggered by rainfall level and not by realized losses, as such the insurance parameters tend to not affect the cropping and land management practices adopted. This is important to avoid inhibiting adaptation measures. However, this may not always be the case in practice since diversification and management practices may differ from insurance in the way they affect a risk profile. Insurance will exclusively reduce downside risk whereas diversification and land management practices may reduce both downside and upside risk. Indeed we observe that in the case where farmers anticipate climate change, as the insurance payout amount increases there is a switch in the level of diversification under the more pronounced climate change scenario. Interestingly this effect is in the direction of greater diversification towards cash crops. This result is in line with literature that claims that a lack of access to insurance leads to a lower likelihood of farmers adopting new technologies (Feder et al. 1985; Antle and Crissman 1990). It is also confirmed by results from Asfaw et al. (2015), which suggest that policy interventions as well as insurance and credit scheme need to be prioritized taking households exposure to climatic risk into account and enabling farmers to pursue choices and diversify their portfolio of choices, for crop and income, so to reduce their vulnerability to poverty. This is suggested in our case, through the mechanism in play such that, as insurance reduces downside risk, farmers have an incentive to invest in higher risk and higher returns activities.

Last but not least, our simulations further suggest that extension and weather index insurance are complementary in the Malawian context, both leading to greater levels of adaptation and improved farmer welfare.

Farming is a risky enterprise and one that will only become riskier under climate change. While our analyses have highlighted the important role that extension and insurance can play in better managing that risk, limited financial resources will require governments to carefully weigh the costs and benefits of each strategy in the design of national or subnational policies. Although we did not explore it here, extension, in the form of information on climate change impacts, is likely to affect the budgetary outlays for any subsidized weather index insurance by helping in its design. The general conclusion is therefore that priority should be given to providing accurate and useful weather and climate information to farmers, as well as clear explanation of its implications in terms of adaptation options. Insurance, although not an adaptation strategy per se, can help in the adaptation process if appropriately designed to minimize the moral hazard that may attend insurance schemes that incentivize additional risk taking.

# References

Antle, J. M. and C. C. Crissman (1990), Risk, Efficiency, and the Adoption of Modern Crop Varieties: Evidence from the Philippines. *Economic Development and Cultural Change*, 38(3): 517-537.
Antón, J. A. Cattaneo, S. Kimura, J. Lankoski (2013) Agricultural risk management policies under climate uncertainty. *Global Environmental Change*, Available online 10 September 2013., http://dx.doi.org/10.1016/j.gloenvcha.2013.08.007
Arslan, A., McCarthy, N., Lipper, L., Asfaw, S. and Cattaneo, A. (2013). Adoption and intensity of adoption of conservation farming practices in Zambia. *Agriculture, Ecosystems and Environment,* In Press, Available online 1 October 2013.
Asfaw, S., Mc Carthy, N., Paolantonio, A., Cavatassi, R., Amare, M., Lipper, L. (2015), "Livelihood diversification and vulnerability to poverty in rural Malawi". FAO-ESA Working Paper No. 15-02, August 2015.

Bryla, E. and J. Syroka (2009) *Micro- and meso-level weather risk management : deficit rainfall in Malawi*. Experiential briefing note. Washington DC ; World Bank.

Chirwa, P. and Quinion, A. (2005). Impact of soil fertility replenishment agroforestry technology adoption on the livelihoods and food security of smallholder farmers in central and southern Malawi, in Sharma, P and Abrol, V (eds.) *'Crop Production Technologies'*, InTech, Rijeka, Croatia.

Collier, B., J. R. Skees and B. J. Barnett (2009), "Weather Index Insurance and Climate Change: Opportunities and Challenges in Lower Income Countries." *Geneva Papers on Risk and Insurance – Issues and Practice* 34:401-424., July.

De Brauw, Alan, and Patrick Eozenou. "Measuring risk attitudes among Mozambican farmers." *Journal of Development Economics* 111 (2014): 61-74.

Denning G, Kabambe P, Sanchez P, Malik A, Flor R, Harawa, R, Nkhoma, P, Zamba, C, Banda, C, Magombo, C, Keating, M, Wangila, J and Sachs, J (2009). Input subsidies to improve smallholder maize productivity in Malawi: Toward an African green revolution. *PLoS Biology*, 7(1): 2-10.

Deressa, T.T. and Hassan, R.H. (2010). Economic Impact of Climate Change on Crop Production in Ethiopia: Evidence from Cross-Section Measures. *Journal of African Economies* 18(4):529-554.

Dorward, A., Chirwa, E., Boughton D., Crawford, E., Jayne, T., Slater, R., Kelly, V., and Tsoka, M., (2008). Towards Smart Subsidies in Agriculture? Lessons from Recent Experience in Malawi, Natural Resource Perspectives 116: Overseas Development Institute, London, UK

Feder, G., R. Just, and D. Zilberman (1985), Adoption of Agricultural Innovations in Developing Countries: A Survey. *Economic Development and Cultural Change*, 33(2): 255-298.

Food and Agriculture Organization of the United Nations (FAO) and Government of Malawi (GoM) (2015), A Strategic Framework for Climate Smart Agriculture in Malawi, unpublished.

Government of Malawi (GoM) (2006). Malawi growth and development strategy 2006-2011: Ministry of Economic Planning and Development: Lilongwe, Malawi.

Graff-Zivin, J., & Lipper, L. (2008). Poverty, risk, and the supply of soil carbon sequestration. *Environment and Development Economics* 13(03), 353-373.

Heltberg, R., P. B. Siegel, and S. L. Jorgensen. 2009. "Addressing human vulnerability to climate change: Toward a 'no-regrets' approach." *Global Environmental Change* 19(1): 89-99.

Heltberg, R. and Tarp, F. (2002). Agricultural Supply Response and Poverty in Mozambique. *Food Policy* 27: 103-124.

IHS3 (2012). Household socio-economic characteristics report. National statistical office, Lilongwe, Malawi.

Just, R. E., & Pope, R. D. (1978). Stochastic specification of production functions and economic implications. *Journal of Econometrics*, 7(1), 67-86.

Kassie, M., Pender, J., Yesuf, M., Kohlin, G., Bluffstone, R. A. and Mulugeta, E. (2008). Estimating returns to soil conservation adoption in the northern Ethiopian highlands, *Agricultural economics*, 38: 213–232.

Markowitz H. (1987). *Mean-Variance Analysis In Portfolio Choice And Capital Markets*, Blackwell, Cambridge., Mass. and Oxford.

McSweeney, C., New, M. & Lizcano, G. 2012. UNDP Climate Change Country Profiles: Malawi.. Available: http://country-profiles.geog.ox.ac.uk/

Mendelsohn, R. (2010), Agriculture and economic adaptation to agriculture, COM/TAD/CA/ENV/EPOC(2010)40/REV1.

Meyer, J. (1987). Two-moment decision models and expected utility maximization. *The American Economic Review*, 421-430.

Mucavele, F. G. (2007), True Contribution of Agriculture to Economic Growth and Poverty Reduction: Malawi, Mozambique and Zambia Synthesis Report. Food, Agriculture and Natural Resource Policy Analysis Network.

Nkonya E., F. Place, E. Kato, and M. Mwanjololo. 2015. Climate Risk Management Through Sustainable Land Management in Sub-Saharan Africa. In R. Lal B. Singh, D. Mwaseba,

D. Kraybill, D. Hansen and L. Eik (eds.), Sustainable Intensification to Advance Food Security and Enhance Climate Resilience in Africa, Springer International Publishing Switzerland. Page 75112. DOI 10.1007/978-3-319-09360-4_5. pp 665

Rosenzweig, M.R. and Binswanger, H.P. (1993). Wealth, weather risk and the composition and profitability of agricultural investments. *Economic Journal* 103(416) : 56-78.

Skees, J., P. Hazell, and M. Miranda (1999), New Approaches to Crop Yield Insurance in Developing Countries. EPTD Discussion Paper No. 55. Washington, DC: International Food Policy Research Institute.

Smith, P., and J. E. Olesen (2010), "Synergies between the mitigation of, and adaptation to, climate change in agriculture", Journal of Agricultural Science Cambridge 148: 543-552.

Subrahmanyam, P and Merwe, P J A Van der and Chiyembekeza, A J and Ngulube, S and Freeman, H A (2000) *Groundnut Variety CG 7: A Boost to Malawian Agriculture.* International Arachis Newsletter 20. pp. 33-35.

Tchale, H. (2009), "The efficiency of smallholder agriculture in Malawi", African Journal of Agricultural and Resource Economics, 3(2): 101-121.

World Bank (2010). Social dimensions of climate change: equity and vulnerability in a warming world. World Bank, Washington DC

**Open Access** This chapter is distributed under the terms of the Creative Commons Attribution-NonCommercial-ShareAlike 3.0 IGO license (https://creativecommons.org/licenses/by-nc-sa/3.0/igo/), which permits any noncommercial use, duplication, adaptation, distribution, and reproduction in any medium or format, as long as you give appropriate credit to the Food and Agriculture Organization of the United Nations (FAO), provide a link to the Creative Commons license and indicate if changes were made. If you remix, transform, or build upon this book or a part thereof, you must distribute your contributions under the same license as the original. Any dispute related to the use of the works of the FAO that cannot be settled amicably shall be submitted to arbitration pursuant to the UNCITRAL rules. The use of the FAO's name for any purpose other than for attribution, and the use of the FAO's logo, shall be subject to a separate written license agreement between the FAO and the user and is not authorized as part of this CC-IGO license. Note that the link provided above includes additional terms and conditions of the license.

The images or other third party material in this chapter are included in the chapter's Creative Commons license, unless indicated otherwise in a credit line to the material. If material is not included in the chapter's Creative Commons license and your intended use is not permitted by statutory regulation or exceeds the permitted use, you will need to obtain permission directly from the copyright holder.

# A Qualitative Evaluation of CSA Options in Mixed Crop-Livestock Systems in Developing Countries

Philip K. Thornton, Todd Rosenstock, Wiebke Förch, Christine Lamanna, Patrick Bell, Ben Henderson, and Mario Herrero

**Abstract** The mixed crop-livestock systems of the developing world will become increasingly important for meeting the food security challenges of the coming decades. The synergies and trade-offs between food security, adaptation, and mitigation objectives are not well studied, however. Comprehensive evaluations of the costs and benefits, and the synergies and trade-offs, of different options in developing-country mixed systems do not exist as yet. Here we summarise what we know about the climate smartness of different alternatives in the mixed crop-livestock systems in developing countries, based on published literature supplemented by a survey of experts. We discuss constraints to the uptake of different interventions and the potential for their adoption, and highlight some of the technical and policy implications of current knowledge and knowledge gaps.

---

P.K. Thornton (✉)
CGIAR Research Program on Climate Change, Agriculture and Food Security (CCAFS), ILRI, Nairobi, Kenya
e-mail: p.thornton@cgiar.org

T. Rosenstock • C. Lamanna
World Agroforestry Centre, Nairobi, Kenya

W. Förch
Deutsche Gesellschaft für Internationale Zusammenarbeit (GIZ) GmbH, Private Bag X12 (Village), Gaboron, Botswana

P. Bell
Ohio State University, Columbus, USA

B. Henderson • M. Herrero
Commonwealth Scientific and Industrial Research Organization (CSIRO), Clayton South, Australia

# 1 Background and Methods

Mixed crop-livestock systems, in which crops and livestock are raised on the same farm, are the backbone of smallholder production in the developing countries of the tropics (Herrero et al. 2010). It is estimated that they cover 2.5 billion hectares of land globally, of which 1.1 billion hectares are rainfed arable lands, 0.2 billion hectares are irrigated croplands, and 1.2 billion hectares are grasslands (de Haan et al. 1997). Mixed crop-livestock systems produce over 90 per cent of the world's milk supply and 80 per cent of the meat from ruminants (Herrero et al. 2013). They occur in nearly all agro-ecological zones in developing countries, with an enormous variety of climatic and soil conditions. The location of the mixed systems in the global tropics and subtropics is shown in Fig. 1. The mixed systems are those in which more than 10% of the dry matter fed to animals comes from crop by-products or stubble, or more than 10% of the total value of production comes from non-livestock farming activities (Seré and Steinfeld 1996). Rather than break the mixed systems down further in terms of whether they are rainfed or irrigated and on the basis of temperature and length of growing period (LGP), as in Robinson et al. (2011), Fig. 1 uses the breakdown in Herrero et al. (2009) on the basis of whether the mixed systems are "extensive", with lower agroecological potential (LGP < 180 days per year), or "intensifying", with higher agroecological potential (LGP >= 180 days per year) coupled with better access to urban markets (<8 hours' travel time to urban centres with a population > 250,000).

In both Latin America and sub-Saharan Africa (SSA) the great majority of the mixed systems are rain-fed. In Asia, a large proportion of the mixed systems are irrigated. The mixed systems extend to the tropical highlands of Latin America, East and southern Africa and northern Asia. In well-integrated crop-livestock systems, livestock provide draft power to cultivate the land and manure to fertilize the soil, and crop residues are a key feed resource for livestock. These mixed systems currently provide most of the staples consumed by many millions of poor people in the global tropics: between 41 and 86 per cent of the maize, rice, sorghum and millet, and 75 per cent of the milk and 60 per cent of the meat (Herrero et al. 2010). The mixed systems will be critically important for future food security too. Human population may peak in Asia and Latin America soon after 2050, but growth is projected in Africa until well into the twenty-second century, and some of this growth will occur not only in urban areas but also in the rural-based mixed systems, where more than 60% of people already live (Herrero et al. 2010).

The justification for integrating crop and livestock activities is that crop (or livestock) production can produce resources that can be used to benefit livestock (or crop) production, leading to greater farm efficiency, productivity or sustainability (Sumberg 2003). Optimal interactions between different operations on the farm can increase farmer's incomes, as well as system-wide resilience and environmental sustainability (Descheemaeker et al. 2010). With limited access to agricultural inputs, combining crops with livestock also offers complementary benefits to each that would otherwise require external inputs to maintain. These resources can be in

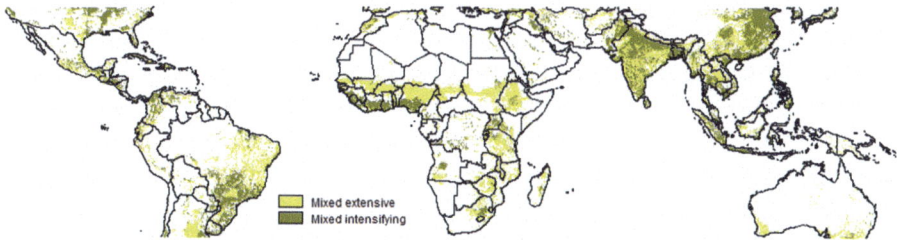

**Fig. 1** Mixed crop-livestock systems in the tropics and subtropics (from Herrero et al. 2009). Mixed systems (M): those in which >10% of the dry matter fed to animals comes from crop by-products or stubble, or >10% of the total value of production comes from non-livestock farming activities. Original classification of Seré and Steinfeld (1995). Mixed systems broken down into "extensive", LGP < 180 days per year (lower agronomic potential), and "mixed intensifying", LGP > 180 days per year (higher agronomic potential) plus better market access (<8 h travel time to an urban centre with >250,000 people)

the form of feed biomass such as crop residues, animal manure, draught power, and cash. Resource-poor farmers depend directly on the food production system for livelihood security, and thus, mixed systems offer key livelihood diversification options, as smallholders in developing countries aim to minimise risk associated with agricultural production, liquidity constraints, high transaction costs, that can all result in income and consumption fluctuations (Dercon 1996; Davies et al. 2009; Barrett et al. 2001).

The future of mixed crop-livestock systems in developing countries is a subject of ongoing debate. On the one hand, they have been seen as one stage in an evolutionary process of intensification via increasing human population pressure on a relatively fixed land resource (Boserup 1965). Intensification dynamics may lead to land consolidation and the exiting of some producers from agriculture altogether (e.g., Australia and North America); or it may lead to exchanges and market-mediated interactions between different producers who may be widely separated geographically (e.g., parts of Asia). On the other hand, structural constraints may continue to impede both adoption of intensification technology and land consolidation in smallholder farming (Waithaka et al. 2006; Fritz et al. 2015). Possibilities for sustainably intensifying production and productivity in many parts of SSA (in particular) are likely to remain severely constrained well into the future. The impacts of climate change on smallholder mixed systems will constitute an additional, and in places severe, challenge in the future (Thornton and Herrero 2015).

Despite the adaptation challenge, the mixed systems could play a critical role in mitigating greenhouse gases from the agriculture, forestry and land-use sectors. Mixed crop-livestock systems are a considerable source of greenhouse gas (GHG) emissions, accounting for 63% of the emissions from ruminants globally (Herrero et al. 2013). Even so, the emissions intensities (the amount of greenhouse gases emitted per kg of meat or milk) of the mixed systems are 24–37% lower than those of grazing systems in Africa (Herrero et al. 2013), mostly because of the higher-quality diets of ruminants in the former compared with the latter systems. The mixed systems also provide 15% of the nitrogen inputs for crop production via manure amendments (Liu et al. 2010). Carbon sequestration in soils and biomass provides another mitigation opportunity in the mixed systems (Seebauer 2014).

The mixed systems have considerable potential for addressing the three pillars of Climate Smart Agriculture (CSA), namely production for food security, adaptation, and mitigation. The synergies and trade-offs between these three, however, are not well studied, particularly in the mixed systems regarding the effects of climate change on livestock and on crop-livestock interactions in a smallholder context (Thornton and Herrero 2015). Quantifying the baseline situation as well as the effects of different alternatives on the various dimensions of climate smartness in different contexts are needed before robust statements can be made about what is and what is not "climate smarter" than current practice.

Mixed crop-livestock systems offer a wide range of possibilities for adapting to climate change and mitigating the contribution of crop and livestock production to GHG emissions. This is in large part the result of the interactions between crop and livestock enterprises that may be able to be exploited to raise productivity and increase resource use efficiency, increasing household incomes and securing availability and access to food (Thornton and Herrero 2015). Integration of crops and livestock can reduce resource depletion and environmental fluxes to the atmosphere and hydrosphere, it can result in more diversified landscapes that favour biodiversity, and it can increase the flexibility of the farming system to manage socio-economic and climate variability (Lemaire et al. 2014). Integration also reduces the risk of smallholders who are often vulnerable not only to crop failure and climate change, but also to other risks such as agricultural trade risk, and food price risk, as well as health and demographic risks (Devereux 2001).

Mixed farming systems have various characteristics that may be advantageous in some situations and disadvantageous in others (and sometimes both in the same situation) (van Keulen and Schiere 2004). For example, the use of draught power allows larger areas of land to be cultivated and it allows more rapid planting when conditions are appropriate. On the other hand, this may mean that extra labour (often women's) is required for weeding. On a mixed farm, crop residues can be mulched, thereby helping to control weeds and conserve water; and they are an alternative source of low-quality roughage for livestock. But again, feeding crop residues may compete with other uses of such material, such as mulching, construction, and nutrient cycling. A major constraint to increased crop-livestock integration is that it can be complex to operate and manage (van Keulen and Schiere 2004; Russelle et al. 2007). Nonetheless, this integration is critical for smallholders in order to increase livelihood security while reducing vulnerability to food insecurity, as well as to climate change.

Comprehensive evaluations of the costs and benefits, and the synergies and trade-offs, of different options in developing-country mixed systems do not exist as yet. The question this chapter seeks to answer is, what can presently be said about the climate smartness of different alternatives in the mixed crop-livestock systems in developing countries, from both a technical and an institutional perspective? We build on the listing in FAO (2013) of crop and livestock management interventions that may be able to deliver multiple benefits (food security and improved climate change mitigation and adaptation) in different situations. These span the range of crop and grazing land management, water management, and livestock management, and include options related to food storage and processing, insurance, and use of weather information. Many of these alternatives have far wider geographic applicability and are not limited just to the mixed systems of the developing world.

The methods used here to evaluate how farm-level CSA management practices and technologies affect food production, adaptive capacity and climate change mitigation in mixed farming systems are based on the protocol of Rosenstock et al. (2016), supplemented by a survey of experts. We evaluated their responses through an informal survey. CSA experts were asked to identify the effects of each intervention on indicators of CSA (as in Rosenstock et al. 2016) in relation to food production (e.g., yield and income effects), resilience (e.g., effects on quality of soil resources, resource use efficiency, labour requirements), and mitigation (e.g., effects on emissions and emission intensities). Additional, the survey gathered information regarding key climate risks that each potential CSA practice addresses as well and identified socioeconomic conditions that enhance the practice. The results from this survey were averaged to determine whether the practice had a positive (+), negative (−), or undetermined (+/−) impact on the key CSA indicators noted above, such as carbon sequestration.

The next section contains descriptions and brief evaluations of the CSA interventions. Section 3 contains brief discussions of constraints to the uptake of these interventions and the potential for their adoption. In Sect. 4 we highlight some of the technical and policy implications of current knowledge as well as knowledge gaps concerning CSA interventions in the mixed crop-livestock systems of developing countries.

## 2 CSA Interventions in the Mixed Systems

Climate-smart options for mixed crop-livestock system vary widely in their potential impacts on agricultural productivity, climate change resilience, and GHG mitigation (Table 1). While experts agree that most options will improve productivity, impacts on resilience and mitigation are particularly variable. This variability is due in part to context specificity in the effect of a particular intervention. For some of the interventions, the strength of evidence to support the assessments is very limited. In the following subsections, we unpack the potential trade-offs, context-specificity, and constraints to adoption for each CSA option for mixed crop-livestock systems.

### 2.1 Changing Crop Varieties

Decades of research has gone into developing crop varieties that can improve agricultural productivity and resilience by increasing yield, reducing the time for crops to mature, increasing tolerance to stresses such as drought, salinity, pests, and disease, and improving the nutritional quality of crops. Without such innovations, it is

**Table 1** Climate-smart options available to smallholders in mixed crop-livestock systems in developing countries: potential impacts and strength of evidence. Scoring based on authors' assessment of the articles found in a systematic review of CSA (described in Rosenstock et al. 2016), supplemented with a survey of nine experts through an informal survey

| Options | Potential impacts | | | Strength of evidence | Selected examples |
| --- | --- | --- | --- | --- | --- |
| | Prod. | Res. | Mit. | | |
| Change crop varieties | + | +/− | +/− | *** | Krouma 2010, Kumar et al. 2008, Kamara et al. 2003 |
| Change crops | + | + | +/− | * | Sauerborn et al. 2000 |
| Crop residue management | +/− | + | − | ** | Liu et al. 2003, Mrabet 2000, Obalum et al. 2011, Omer et al. 1997, Sissoko et al. 2013 |
| Crop management | + | +/− | +/− | * | Wang et al. 2006, Borgemeister et al. 1998 |
| Nutrient management | + | + | + | *** | Surekha et al. 2010, Szilas et al. 2007, Torres et al. 1995, Witt et al. 2000, Yadav and Tarafdar 2012 |
| Soil management | + | + | +/− | ** | Kywe et al. 2008, Yang et al. 2010, Yusuf et al. 2009, Zougmore et al. 2000, Suriyakup et al. 2007 |
| Change livestock breed | + | + | + | * | Thornton and Herrero 2010 |
| Manure management | + | +/− | +/− | * | Rabary et al. 2008, Salako et al. 2007, Srinivasarao et al. 2012, Taddesse et al. 2003 |
| Change livestock species | + | +/− | +/− | * | Limited information; discussed by Hoffmann 2010; FAO 2013 |
| Improved feeding | + | +/− | +/− | ** | Akinlade et al. 2003, Akinleye et al. 2012, Barman and Rai 2008, Kaitho et al. 1998, Lallo and Garcia 1994; Thornton and Herrero 2010 |
| Grazing management | + | + | +/− | ** | Bozkurt and Kaya 2011, Moyo et al. 2011, Mattiauda et al. 2013, Ma et al. 2014 |
| Alter integration within the system | + | + | + | * | Tuwei et al. 2003, Kaitho et al. 1998 |
| Water use efficiency and management | + | + | +/− | ** | Kipkorir et al. 2002, Li et al. 2004, Mahmoodi 2008, Mailhol et al. 2004, Speelman et al. 2008 |
| Food storage | + | + | + | * | Sadfi et al. 2002, Haile 2006, Ilboudo et al. 2010, Koona et al. 2007 |
| Food processing | + | +/− | | * | Mahmutoğlu et al. 1996 |
| Use of weather information | + | + | +/− | − | Hansen et al. 2011 |
| Weather-index insurance | + | +/− | +/− | * | Cole et al. 2012 |

The results from this survey were averaged to determine whether the practice had a positive (+), negative (−), or undetermined (+/−) impact on the key CSA indicators. Potential impacts (prod = production, res = resilience, mit = mitigation): + = positive, − = negative, +/− = uncertain. Strength of evidence: ***= confident, **likely, *poor, − speculation

thought that crop yield in developing countries would be 20–24% lower than current levels, 6–8% more children would be undernourished, and per capita calorie consumption would be 14% lower than current levels (Evanson and Gollin 2003). Adaptation strategies such as improved varieties may reduce projected yield losses under climate change, particularly among rice and wheat in the tropics (Challinor et al. 2014). High yielding varieties can improve the food self-sufficiency of smallholders and increase income without needing to cultivate extra land. Drought-tolerant varieties have helped to stabilize yields, particularly of cereal crops in rain-fed systems (La Rovere et al. 2014). As drought, pest and disease outbreaks, and water salinization become more common with climate change and increasing demands on natural resources, changing crop varieties will continue to be among the first lines of defence for improving productivity and resilience in mixed crop-livestock systems. However, research on crop improvement and resilience has been limited to staple grains for the most part. Within mixed systems, a diverse number of crops including feed and forage species as well as trees or fodder shrubs contribute to the resilience of the system. More attention is needed to understand how the climate resilience of non-traditional products that contribute to smallholder health and nutrition and overall system performance can be enhanced.

Adoption rates of improved varieties and seeds in areas where those seeds are available and awareness is high can be as much as 85% among smallholder farmers (e.g., see Kyazze and Kristjanson 2011). That study showed that high-yielding varieties have the greatest appeal among smallholder farmers, followed by tolerance to drought and pests. However, recent evidence shows that very few farmers actually have access to improved crop varieties or improved seeds in the developing world. In SSA, 68–97% of seed grown by smallholder farmers comes from informal sources (i.e. seed saving, friends and relatives) and local markets (McGuire and Sperling 2015). Thus a primary barrier to adoption of improved varieties is availability of seeds (Westermann et al. 2015).

## 2.2 Changing Crops

Under climate change, the suitable area for cultivation of most staple crops in the tropics is likely to both shift and decrease, requiring farmers to adopt transformative types of adaptation, such as switching crops (Vermeulen et al. 2013). Maize, beans, banana, and finger millet, staple crops in much of SSA, could experience reduction in suitable areas for cropping by 30–50% (Ramirez-Villegas and Thornton 2015). Changing from less suitable crops to those more suitable in future climates is an effective strategy for maintaining productivity and increasing resilience to climate change. While many studies have looked at climate impacts on staples, information on the likely impacts of climate change on forages such as Napier grass that are typically used in mixed systems is practically non-existent. In areas that are projected to see improvements in crop suitability, such as a relaxation of current cold temperature constraints in parts of the tropical highlands in East Africa, for example, mixed crop-livestock farmers may be able to capitalise by planting crops appropriate to the changing climatic conditions.

While changing crops is a more substantial alteration to a mixed crop-livestock system than simply changing varieties, adoption rates of new crops and switching crops can still be quite high compared with other management practices. In Rakai, Uganda, for example, more than a quarter of surveyed smallholder farmers had introduced a new crop in the last 10 years, whereas more than a third of households had also stopped growing a crop that was no longer seen as profitable or suitable (Kyazze and Kristjanson 2011). However, in many cases the potential to change crop species will depend on the familiarity of farmers with the new species as well as cultural preferences. Barring potentially catastrophic losses (such as the introduction of maize lethal necrosis diseases in Kenya in 2013), the transition to new crops is likely to be a gradual and relatively slow process.

## 2.3 Crop Residue Management

Crop residue management practices determine the destination and use of stover and other crop byproducts. Some effective residue management solutions retain plant residues and practices that minimally disturb the soil. In addition to potential increases in soil organic carbon and subsequently increased water infiltration and storage within the soil, effective crop residue management can dramatically decrease soil erosion through the protection of the soil surface from rainfall (Lal 1997). Such practices can include minimum or no-tillage, cover cropping, and the addition of mulch. Minimum tillage practices limit disturbance of the soil and therefore protect the soil structure from degradation. Additionally, limiting tillage can decrease soil crust formation. Both of these factors contribute to enhancing water infiltration into the soil and subsequently increase water productivity of agroecosystems (Rockström et al. 2009). Cover cropping includes the growing of typically a non-harvested or partially harvested crop either in a crop rotation or in the non-main growing season. Cover cropping with leguminous crops can be very beneficial to typically low-fertility and highly weathered soils common in smallholder systems (Snapp et al. 2005). Similar to both minimum tillage and cover cropping, mulching can increase soil aggregation (Mulumba and Lal 2008), and thus soil physical quality. In addition, the use of mulching also protects soils from direct impact by rainfall, greatly reducing nutrients and organic matter lost through soil erosion (Barton et al. 2004).

Minimum tillage practices must be adapted to local conditions and must contain strong incentives for farmer adoption. A study in Central Kenya found that profitability and yield depend on the soil fertility status (low, medium, high), with neither tillage nor crop residue retention practices being profitable (Guto et al. 2012). While cover crops offer great potential, there are costs that must be weighed by potential adopters. Cover cropping can potentially interfere with subsequent crops by using finite soil water, they can decrease soil warming, subsequently inhibiting seed germination, and increasing the direct cost and production risks to farmers (Snapp et al.

2005). Current practices of grazing livestock on harvested fields (and other free grazing practices) would need to be addressed at the same time, and there may also be implications for women's labour requirements, for example. Increased soil degradation and subsequent loss of crop yield can result from this practice (Udo et al. 2011). Many smallholder agroecosystems already have a high demand for crop biomass for feed and fuel. Areas where mulching has higher adoption potential are those where increased biomass production is high enough to meet feed, fuel, and mulching requirements (Valbuena et al. 2012).

## 2.4 Crop Management

Crop management techniques within the perspective of climate change range widely and include practices such as modifying planting date and multicropping with multiple crops and varieties. As the world climate system changes, local weather patterns will become more unpredictable. In addition to accessing available weather forecasting information, farmers will need to adjust planting seasons accordingly. Changes in planting dates can have profound impacts on farm productivity. A study in Zimbabwe found that delayed planting results in a 32% loss of grain yield (Shumba et al. 1992). However, in order for some farmers to effectively plant earlier might require adjusting cultivation practices. In the same study, Shumba et al. (1992) reported that earlier planting was only feasible with the use of select pesticides and minimum tillage techniques. Multicropping involves the growing of multiple crops within the same growing season—and can include intercropping (within the same field at the same time) with both leguminous and non-leguminous crops and trees (agroforestry). Intercropping—the planting of two or more crops on the same field within one season—has profound effects on the ability of smallholder farmers to reduce risk. Crops in intercropping systems typically access different soil water and nutrient resources, have difference water requirements, and have varying growth and maturity rates, all of which reduce the risk of total crop failure (and the associated risk of food insecurity) due to erratic or decreased precipitation (Ghosh et al. 2006). An extensive analysis reported that monocropping—the most common agricultural practice in Africa—is the most susceptible to the negative effects of climate change (Nhemachena and Rashid 2008).

While changing planting date for many crops in some areas might be very simple, a study of the Nile Basin in Ethiopia indicated that lack of access to weather information and extension services is a formidable constraint to changing planting dates (Deressa et al. 2009). Even with access to these services, farmers will require time to test planting dates before adoption, or more likely, adjustments might need to be made on a season-by-season basis. Additionally, changes in planting and harvest dates might require changes in cultivation practices as well as changes in market systems. In some situations, labour availability may become an issue – for instance, when children are in school and cannot help with weeding. With respect to intercropping, determining the proper crop combinations and intercropping type requires both local knowledge and evaluation. The use of intercropping systems can also increase labour demands as some intercropped plants will require varying weeding, applications, and harvest times (Rusinamhodzi et al. 2012).

## 2.5 Nutrient Management

Smallholders manage complex nutrient cycles on mixed crop-livestock farms (Tittonell et al. 2009) offering multiple opportunities to become more climate-smart. Producers control the distribution of nutrients through the same means as mono-specific growers and ranchers such as the application of inorganic and organic fertilizers and composts, growing trees, recycling of wastes, and improving animal diets which all have known benefits for improving productivity, water and nutrient use efficiency, and reducing GHG intensity of production (Kimaro et al. 2015; Barton et al. 2004; Zingore et al. 2007). A key feature of nutrient management in mixed farming is that farmers transfer nutrient-rich materials – manure, residues, feeds – between production activities. Technological change for any specific subcomponent of the system, therefore, has cascading affects across the farm because of concomitant changes in nutrient availability (van Wijk et al. 2009). The consequence is that individual management changes can create either trade-offs or synergies not only within, but also among, farm subcomponents and products. For example, conservation agriculture is often promoted in mixed crop-livestock systems to help maintain soil chemical and physical properties amongst other CSA-relevant goals (for example, water-use efficiency and soil carbon sequestration). However, crop residues in mixed systems are typically fed to livestock, often serving as a vital feed resource during periods of low supply (Giller et al. 2015). Thus, conserving crop residues for fertility may reduce nutrients available for other subcomponents of the system.

At this time, much is known about nutrient dynamics of individual subcomponents and entire mixed systems (Abegaz et al. 2007); however, less is understood about how to optimize the various subcomponents to meet multiple objectives (Groot et al. 2012). For example, recycling of manure nutrients back to crop fields is one of the most often cited interventions to improve nutrient management in mixed systems. Closing the nutrient cycle in this way has the potential to increase crop yields (including feed byproducts) and farm output while reducing GHG emissions from stored manures. In practice, however, the efficiency of this practice to preserve the nutrient composition of the manure is highly subject to handling and storage conditions and transfer time, with farmer practice having a significant impact on the final fertilizer value of the material (Rufino et al. 2006). Farmer practice is subject to available resources, materials and labour, and as such utilization of manure nutrients may be impractical when put up against other competing goals of the household. Similar practical challenges obstruct implementation of other nutrient management options; and the use of human waste comes with its own challenges relating to health and cultural acceptability. Mixed system farmers have the opportunity to improve feeding on farms, typically by supplying high protein feeds. Higher protein diets tend to increase productivity of livestock through improved digestibility and intake of crude protein (Bekele et al. 2013) and decrease emissions intensity from milk and meat production (Barton et al. 2004). However, the potential to plant legume species is often constrained by factors as varied as seed availability, access to knowledge, and land rights (Franzel et al. 2014).

## 2.6 Soil Management

Managing soil resources for climate-related risks often involves increasing soil physical quality while maintaining or improving soil fertility status. Soil physical characteristics important for climate change adaptation include increased soil organic carbon and soil aggregation, and enhancing these properties can lead to increased water infiltration into the soil and subsequently soil water storage for plant use. Additionally, management of soil fertility within smallholder agroecosystems is especially important as climate change is expected to negatively affect soil fertility and the mineral nutrition contained within plants (St Clair and Lynch 2010). These important aspects of soil quality are managed through effective use of crop rotations, leguminous plants, and livestock density management. The use of crop rotations decreases disease incidence, suppresses weed infestation, and can enhance nutrient cycling when leguminous plants are used (Mureithi et al. 2003). Leguminous plants and trees can be effectively incorporated into smallholder agroecosystems through intercropping, relay cropping, and planting boundaries. The nitrogen-fixing capabilities of leguminous plants can increase soil fertility of smallholder soils as well as provide important nutrients to smallholder farmers (Kerr et al. 2007). Livestock stocking management is less straightforward, however. While the determination of livestock density varies by environment and livestock type, Taddesse et al. (2003) reported that medium-stocking intensity can lead to higher species richness compared with both a high-stocking intensity and the non-grazed control, as well as resulting in less soil compaction than the high-stocking intensity treatment. These results may not hold in other situations because of the diverse conditions found in smallholder livestock keeping systems.

While each of these practices represents possible techniques to effectively manage soil resources, each practice must be assessed to identify possible constraints or drawbacks. For example, a study in Tanzania found that adoption of leguminous crop rotations was negatively affected by longer distances from houses to farm plots, smaller plot sizes, and poor fertility soils (Kassie et al. 2013). While leguminous plants offer many benefits to smallholder farmers, farmers are not likely to adopt this practice unless there are clear market returns (Snapp et al. 2002). The effects of livestock grazing management on soil quality is affected by many geographic-specific factors including soil type and topography. Precipitation can also exacerbate the effect of livestock grazing on compaction during heavy rainfall events (Ghosh et al. 2006). Additionally, stocking intensity must be managed in such a way that sufficient crop residue is returned to the soil to maintain nutrient cycling and soil physical quality (de Faccio Carvalho et al. 2010). The ways in which different soil management interventions interact at the systems level in helping to meet food security objectives remain to be elucidated (Hurni et al. 2015).

## 2.7 Changes in Livestock Breed

The local breeds of cattle that are raised in the developing world are generally well-adapted to their environments in terms of disease resistance, heat tolerance and nutritional demand. Their productivity is often low, however, and the emissions intensity of production (the amount of GHG emissions produced per kilogram of milk and meat) can be high. The utilisation of more productive animals is one strategy that can lead to higher productivity and reduced emissions intensity. Livestock populations exhibit natural genetic variation, and selection within breeds of farm livestock may produce genetic changes in the range 1–3% per year in trait(s) of interest (Smith 1984). Attempts to utilize this genetic variation to breed reduced-emissions cattle, for instance, are inconclusive as yet. Within-breed selection often poses challenges in developing countries because appropriate infrastructure such as performance recording and genetic evaluation schemes are often lacking. Cross-breeding is usually more feasible, and can deliver simultaneous adaptation, food security and mitigation benefits. Locally-adapted breeds can be utilised that are tolerant to heat, poor nutrition and parasites and diseases, and these traits can be transferred to crossbred animals. Cross-breeding coupled with diet intensification can lead to substantial efficiency gains in livestock production and methane output. Crossbred cattle, for example, can easily produce more than double the amount of milk and meat, compared with local breeds (Galukande et al. 2013). Widespread uptake could result in fewer but larger, more productive animals being kept, which would have positive consequences for incomes, methane production and land use. The adoption potential of cross-bred cattle is high: adoption rates of crossbred dairy animals of 29% have been observed in Kenya (Muriuki and Thorpe 2006). The benefits on production are substantial, and the mitigation potential is positive, though relatively modest; for the mixed systems of the tropics and subtropics it is estimated at about 6 Mt. $CO_2$-eq per year (Thornton and Herrero 2010).

There are significant issues associated with the feasibility of widespread adoption of crossbred animals, however. The adoption rate of crossbreds in Kenya is atypical of developing countries as a whole. There are several reasons for this. Larger, more productive animals need more and higher-quality feed and water, which may have substantial impacts on land and labour resources at the household level. For example, women collect water for animals in many African households when it is not immediately available. Adoption of crossbreds may therefore increase work burden on women. Crossbreds also require some capital investment, and smallholders may have no access to viable lines of credit. A key constraint seems to be an adequate understanding of the objectives and attitudes of smallholders; smallholders have often found breeding programs to be unsuitable, unprofitable, or impossible to implement – this applies to small ruminants as well as to large (Kosgey et al. 2006). In addition, the impacts of an increasingly variable climate on crossbred animal performance may increase household risk in ways that are unacceptable. Some East African livestock keepers, for example, generally prefer dealing with indigenous breeds, especially during times of severe drought, as smaller animals can be physically handled in ways that become impossible with heavier animals (BurnSilver 2009).

## 2.8 Manure Management

The utilisation of livestock manure to add nutrients back to the soil is one of the key crop-livestock interactions in mixed farming systems. Manure when used as a soil amendment can benefit the soil, resulting in crop production and resilience benefits for smallholders via increased nutrient supply to crops and improved soil structure and water holding capacity, for example. Manure has well-documented impacts on soil chemical and physical properties. For example, Srinivasarao et al. (2012) showed a positive interaction between the application of manure and mineral fertilizer on carbon stocks in the soil in semiarid regions of India, with beneficial effects on crop yield stability. Taddesse et al. (2003) demonstrated positive impacts of manure application in the Ethiopian highlands on pasture biomass production, species richness and water infiltration rates. The GHG emissions dimension associated with manure is complex. When stored, manure can release significant amounts of nitrous oxide and methane. Nitrous oxide and other GHGs are also released when manure is applied to the land (Smith et al. 2008). In tropical mixed farming systems, the opportunities for manure management, treatment and storage are often quite limited, although there may be opportunities in zero-grazing smallholder dairy systems, for example (FAO 2013). In more extensive systems, manure has to be collected from the field, usually once it has dried and methane emissions are negligible (Smith et al. 2008). Various options exist to modify GHG emissions in the production, storage and application of manure. Improved livestock diets and the use of certain feed additives can substantially reduce methane emissions from enteric fermentation and manure storage (FAO 2013). Storage emissions can be reduced by composting the manure or by covering manure heaps; and manure can be digested anaerobically to produce methane as an energy source, for example (Smith et al. 2008). Generally, however, manure storage under anaerobic conditions is only viable in the highly intensive livestock production systems, and anaerobic digestion technology is unlikely to be applicable in smallholder mixed systems for the foreseeable future. Emissions during and after the application of manure to the field can be reduced by rapid incorporation of the manure into the soil (FAO 2013).

These manure management options can all contribute to increased productivity, but the synergies and trade-offs in relation to household resilience and mitigation benefits in different contexts and production systems are not well studied. Their applicability in relatively low-input mixed farming systems is likely to remain limited (FAO 2013), as the investment costs, labour demands and technical know-how will be beyond the reach of the great majority of smallholders. Some recent studies indicate that there is potential for communal biogas digesters to improve soil fertility in the developing world (see, for example, Smith et al. 2014), but the constraints of unaffordability, water scarcity, inappropriate technology and lack of technical capacity may be insuperable without considerable public sector investment (Mwakaje 2008). The conditions under which such interventions are climate smarter are still largely unknown.

## 2.9 Changes in Livestock Species

The substitution of one species of livestock for another is one strategy that livestock farmers can use to increase their resilience to climatic and economic shocks. There are various mechanisms by which this can occur: risk can be spread by having a more diverse species portfolio, and for a farm with small stock, it will often be easier to shift between small stock species than between larger, less "liquid" stock. The last several decades have seen species substitution in several parts of Africa, as a result of long- and/or short-term climate and vegetation changes. In parts of the Sahel, dromedaries have replaced cattle and goats have replaced sheep, in the wake of the droughts of the 1980s (Hoffmann 2010). In Ethiopia, smallholders are adopting goats and sheep rather than cattle in response to market opportunities: there is strong urban demand for meat, it is easier to sell small animals, and profits accrue more quickly and are generally less risky. Traditional cattle keepers in parts of northern Kenya and southern Ethiopia have adopted camels as part of their livelihood strategy as a result of drought, cattle raiding and epizootics. More widespread adoption of camels and goats in the drylands of Africa is now being observed in many other places – unlike cattle and sheep, browsers feed on shrubs and trees, and browse may be a relatively plentiful feed resource even in situations where herbaceous feed availability is declining. Livestock species substitution may also arise from considerations of GHG emissions, given that there are considerable differences in emissions and emission intensities between ruminant livestock production systems and monogastric systems producing chickens and pigs, for example (Hoffmann 2010).

Livestock species substitution will no doubt continue to occur, and it is clear that these substitutions can deliver various benefits: enhancing resilience, maintaining or increasing productivity in the face of shocks, and mitigating GHG emissions. There is little evidence, however, of how the synergies and trade-offs may play out in the mixed crop-livestock systems, particularly through time: while there may be long-term benefits of species substitution, there are likely to be short-term costs and challenges associated with species switching that smallholders may be unwilling or unable to address (FAO 2013). The challenges revolve around the capital outlays involved, and the lack of technical know-how needed to manage unfamiliar livestock species.

## 2.10 Improved Feeding

Interventions that target improved feed resources can result in faster animal growth rates, higher milk production, earlier age at first calving, and increased incomes. Better nutrition can also increase fertility rates and reduce mortality rates of calves and mature animals, thus improving animal and herd performance and system resilience to climatic shocks. For cattle, such interventions may include the use of improved pasture and agroforestry species and the use of nutritious diet supplements. Feed availability for ruminants can be a major constraint in the mixed systems of the tropics during the dry season. The options available to smallholders include higher-digestibility crop residues, diet supplementation with grain, small areas of planted legumes ("fodder banks"), the leaves of certain agroforestry species, and grass species that can be planted on field boundaries or in rehabilitated gullies (with added erosion control benefits). These kinds of supplements can substantially increase productivity per animal while also increasing resilience by making substantial impacts on income. For example, the feeding of 1 kg of *Leucaena leucocephala* leaves per animal per day can nearly triple milk yields and live-weight gains (Thornton and Herrero 2010). At the same time, because these supplements improve the diet of ruminant livestock, the amount of methane produced by the animal per kilogram of meat and milk produced is substantially reduced (Bryan et al. 2013). There may also be soil carbon sequestration benefits from planting trees and deep-rooted pasture species. For example, planting *Leucaena* trees on farms increases carbon sequestration in the soil, possibly by up to 38 tonnes of carbon per ha (Albrecht and Kandji 2003). In many regions, crop residues (stover) are a critical feed resource; increases in stover digestibility of 10 percentage points are well within the range of variation in digestibility that has been observed in sorghum, for example (Blümmel and Reddy 2006). Such genetically improved dual-purpose crops (food and feed), both cereals and legumes, are widely grown in some parts of the tropics.

Improving the diets of ruminants is one of the most direct and effective ways of increasing productivity and incomes, while mitigating GHGs at the same time. Mixed crop-livestock system diets are often complex and amenable to modification. Widespread application of the different options above is plausible in many situations. Adoption rates of up to 43% for genetically improved dual-purpose crops have been observed in some parts of West Africa, though lower adoption rates are more usual (Kristjanson et al. 2002).

There may be constraints at the local level, however: diet intensification may require additional household labour, and the availability of appropriate planting material may be inadequate, for example. In addition, some of these alternatives require appropriate technical capacity to manage them as well as some cash investment. Some also require land, although sometimes competition for land can be avoided: in an example from Ethiopia, degraded land is given to female headed households or landless youth, who thus get a chance to produce small stock for sale. Overall, the above constraints may not pose insuperable barriers to the continuing uptake of climate-smarter feeding practices in the future.

## 2.11 Grazing Management

Native grasses in rangelands and mixed systems are often of relatively low digestibility. The productivity of pastures can be increased through adding nitrogen and phosphorus fertilizers, adjusting the frequency and severity of grazing, changing plant composition, and utilizing irrigation. Improving pasture productivity offers a readily available means of increasing livestock production, particularly in the humid/sub-humid tropics. Substantial improvements in livestock productivity and soil carbon sequestration are possible, as well as reductions in enteric emission intensities, by replacing natural vegetation with deep-rooted pasture species. For example, in Latin America, *Brachiaria* grasses have been widely adopted; animal productivity can be increased by 5–10 times compared with animals subsisting on diets of native savanna vegetation. In Brazil, where about 99 million hectares have been planted, annual benefits are about US$4 billion. In the humid-subhumid livestock of systems of Latin America, the total mitigation potential of improved pastures such as *Brachiaria* is estimated to be 44 Mt. $CO_2$-eq (Thornton and Herrero 2010; Rao et al. 2014). However, while such practices will generally improve pasture quality and animal performance, they will not always reduce GHG emissions. For example, Henderson et al. (2015) found that while the inclusion of legumes in animal diets improved livestock productivity, the nitrogen emissions from sown legumes exceeded soil carbon sequestration benefits in most grasslands. Similarly, the addition of nitrogen fertilizer in a grazing system may reduce methane emissions but increase nitrous oxide emissions (FAO 2013). A third way in which grazing management may deliver productivity, mitigation and adaptation benefits is by balancing and adapting grazing pressure on land, though the effects are highly dependent on the context, such as plant species and soil and climatic conditions, for instance (Smith et al. 2008). Bozkurt and Kaya (2011) reported substantially improved grazing performance of beef cattle on upland rangeland conditions in Turkey from rotational grazing compared with set stocking, while Moyo et al. (2011) found no benefit in animal performance using rotational grazing schemes in the communal areas of Zimbabwe without controlling stocking rates in relation to the season's rainfall. In colder conditions in the Chinese steppe, Ma et al. (2014) found pronounced effects of grazing intensity and grazing period on sheep and grassland productivity, with deferred spring grazing combined with higher stocking rates in summer and relatively low stocking rates in autumn found to be a sustainable grazing strategy for these conditions. Any grazing management that enhances the quality and digestibility of the forage potentially improves livestock productivity and reduces the intensity of GHG emissions in the same way as for diet intensification.

There are considerable constraints associated with these grazing management options in the smallholder mixed systems of the tropics, however. First, managed pasture systems will require considerable investment costs (for fencing, watering points) and additional labour (FAO 2013). Second, such systems require high levels of technical capacity to operate and maintain. As noted above, the adoption rates of improved pastures in humid-subhumid Latin America have been high, but in smallholder mixed systems of SSA, adoption rates have been considerably lower for a range of reasons (see Sumberg (2002) in relation to fodder legumes). There may also be governance issues: replacing free grazing systems with cut-and-curry systems (as is happening in parts of Ethiopia, for example) may benefit pasture and animal productivity, but it requires changes in community bylaws and the development of mechanisms that can enforce the rules for zero grazing. For the arid and semi-arid systems in the tropics and subtropics, in general, there are far fewer opportunities for feasible grazing management options.

## 2.12 Alter Integration Within the System

Various options are available to smallholders in mixed systems involving changes to the proportion of crops to livestock and additions or subtractions to the enterprises that farmers engage in. Such changes can directly and indirectly affect the integration of the different elements in the farming system with respect to its resources of feed, manure, draft power and labour, and cash. Integrated crop-livestock systems offer some buffering capacity in relation to adaptation, with mitigation and resilience benefits too (Thornton and Herrero 2015). In many places smallholders are continually reassessing their activities, and risk reduction is often much more important than productivity increases per se (Kraaijvanger and Veldkamp 2015). In dry spells, farmers may reduce their investment in crops or even stop planting altogether and focus instead on livestock production (Thomas et al. 2007). Others may increase off-farm income in poor seasons via trading or some other business activity (Thornton et al. 2007). Remittances form an important source of income in some regions that can be invested in climate smarter activities (Deshingkar 2012). Such measures may help households to adapt and manage risk, though they may not necessarily deliver productivity and mitigation benefits directly, particularly in the short term (FAO 2013), though it could be argued that off-farm income invested in natural resource management-based alternatives may deliver such benefits in time. In the medium and longer term, smallholders may undertake more permanent (or semi-permanent) farming system transitions.

In marginal areas of southern Africa, reductions in length of growing period and increased rainfall variability are tending to push farmers to convert from mixed crop-livestock systems to rangeland-based systems, as farmers find growing crops too risky in marginal environments (Thornton and Herrero 2015). On the other hand, agricultural system transitions in some of the marginal areas of East Africa are operating the other way round: in recent years, the traditionally pastoral Pokot people of semi-arid north-western Kenya have started engaging in opportunistic cropping using residual moisture in dry river beds as a means of diversifying their livelihood options in the face of increasing rainfall variability and conflict over resources (Rufino et al. 2013). The addition of trees and shrubs to mixed farming systems can have well-documented benefits on animal production (Kaitho et al. 1998; Tuwei et al. 2003) as well as on mitigation, as outlined in Sect. 2.10 above.

Options that alter the integration of enterprises within mixed systems may deliver multiple benefits, although it is likely that there will be some tradeoffs that have to be made in the short term with respect to mitigation, productivity and food security (FAO 2013). There is still limited information currently that quantifies what these tradeoffs are in different contexts (e.g. Tschakert 2007), and given the prevalence of smallholder mixed systems in the tropics and subtropics, this warrants considerable attention (Thornton and Herrero 2015). At the same time, any change towards climate-smarter agriculture needs to have direct, short-term financial benefits for farmers, otherwise adoption is not likely to occur. In addition to potential short-term losses associated with these tradeoffs, there may be other obstacles to smallholder farmers making what may be quite radical changes to their farming and livelihood systems, related to cash availability and the technical know-how that new or unfamiliar crops or livestock species may require. There may be cultural constraints to their adoption as well. Lack of information, or of adequately packaged and communicated information, concerning likely seasonal weather conditions or longer-term climatic trends and economic conditions may also act as barriers to famers' being willing to make substantial changes to their production and livelihood systems (FAO 2013).

## 2.13 Water Use Efficiency and Management

Improving water use efficiency and water management on mixed farms is arguably the most important and high potential improvement for farmers to be climate-smart. An assessment of more than 60 economic studies of various management practices ranging from alley cropping to tillage and fertilizer indicates that water management strategies increase net returns and purchasing power parity of households much more than any other and perhaps presents the only viable pathway to help transition smallholder farmers out of poverty (Harris and Orr 2014). Without a doubt, the ability to supply water, mitigate the impacts of variable rainfall on crops, pasture and animals, and extend growing seasons has significant impacts on smallholder livelihoods, increasing yields and economic returns (Burney and Naylor 2012; Kurwakumire et al. 2014; Thierfelder and Wall 2009; Gebrehiwot et al. 2015). As an alternative to establishing irrigation schemes, more passive water harvesting techniques can equally yield big gains for smallholders. Small-scale water harvesting can include practices such as digging zai pits for individual plants and constructing ditches, terraces or stone lines to direct water to where it is needed. Simple techniques conserve soil moisture and improve productivity of most crops (Amede et al. 2011; Zougmoré et al. 2004). Water harvesting is often already a locally adapted measure and there are well known examples such as the Fanya-juu terraces for vegetable and staple production and chaco dams to increase water availability for cattle and other livestock in East Africa. Large-scale investments in soil and water conservation in northern Ethiopia, combined with collective action and conducive policy environments, has transformed semiarid, degraded lands into productive farming systems that are far less prone to droughts, thus transforming smallholder livelihoods and food security (Walraevens et al. 2015).

The promise of water management and increasing water use efficiency for improving livelihoods, especially under more variable weather conditions, has led to calls for this to be a priority investment (Burney et al. 2013; Rockström and Falkenmark 2015). Will water management transform smallholder mixed systems? Like other technologies, adoption of improved water management is significantly constrained by social, economic and environmental factors. In some cases, the labour hours required to dig channels and planting basins as such outweigh the perceived benefits or the labour is simply not available at the time of peak demand (Drechsel et al. 2005). This may often require community investment and collective action, and associated policy change and institutional mobilisation (Mengistu 2014). In addition to high labour demands, farmers in the highlands of Ethiopia are often reluctant to construct stone terraces in their fields due to the pest harbouring effects, as crop losses may outweigh yield gains (Teshome et al. 2014). These factors can reduce the attractiveness of water harvesting to producers. Furthermore, water management typically requires investments, capital for technologies such as pumps or boreholes or time for building terraces. In many cases, farmers are hesitant to make such investments without appropriate land rights (Lanckriet et al. 2015). Zimbabwe, for example, saw very low levels of adoption of key water saving technologies in the arid and semi-arid zones throughout the late twentieth century due to political instability and insecure tenure rules (Nyamadzawo et al. 2013). Thus, while the potential of water management for smallholder productivity is significant, so are the challenges; greater attention is needed to build the enabling environment for adoption than to develop new technologies.

## 2.14 Food Storage

The significance of food losses for smallholder farmers in Africa, including in mixed systems, is categorically different than in the developed world. Consumer waste, responsible for 95–115 kg food per person per year in developed countries (FAO 2011), is typically not a serious problem in developing countries or more specifically in crop-livestock systems. In contrast, food losses in SSA occur during the postharvest phases where due to a lack of information on harvesting techniques, storage facilities, and pests and diseases cause losses at a near equivalent amount (30–40%) to that of consumer waste in developed countries (Affognon et al. 2015). For example, postharvest losses of grains in Tanzania occur in the field (15%), during processing (13–20%), and during storage (15–25%) (Abass et al. 2014). Postharvest losses can be reduced using existing low-cost technologies and methods, many of which have been adopted rapidly in Asia, but are not widely used in SSA. Baoua et al. (2012) show that any number of techniques ranging from simple mixing of cowpea grain with ash to more advanced and costly storage in hermetically sealed plastic bags significantly reduce pest infestation, by more than 50%. Though the appropriate strategy to reduce losses needs to be tailored to the enterprise (resources available, market orientation, and commodity), an ample number of approaches are already available, even for small-scale producers, such as harvesting in the morning and separating out pest infected produce, and general principles to develop best practices are known for crops (Kitinoja and Kader 2003).

Storage of highly perishable animal products, milk and meat, as well as of higher-value vegetables and fruit, present unique challenges in resource limited and small-scale producer environments and have received markedly less attention. But gaps in knowledge should not discourage promotion of postharvest interventions, gains in food availability due to better storage practices at even modest levels of loss reduction (for example 10–15%) anywhere on the farm would have cascading impacts on food and nutrition security, adaptive capacity and the climate, though it is difficult to predict by precisely how much.

Many factors contribute to postharvest loss including mechanical injuries, water stress, physiological disorders, temperature, humidity, wind, marketing systems, regulations, a lack of tools, and equipment of information; many of these are recalcitrant problems obstructing agricultural development more generally. However, given that few other interventions offer the immediate ability to increase food availability by such a margin in such a short period, it is troubling how little effort is being directed toward solving this issue compared with increasing production, especially when the latter will become even less tenable under climate change.

## 2.15 Food Processing

Like improved postharvest storage methods, food processing presents an opportunity to extend the shelf-life of perishable farm products. Food processing, however, adds an additional layer of utility; it provides a mechanism for smallholders to add value to products at the farm gate. In mixed systems, farmers typically have potential to create fermented milk products, dried meat products as well as creating derivatives from crop products. By reducing the speed of food degradation, food processing increases or at least maintains the level of consumable farm output. Food processing also typically generates value-addition and/or an extra product that can be sold into the market, facilitating livelihood diversification by creating an alternative revenue stream. Improved longevity of production and increased marketability may make smallholders less susceptible to the annual cycles of food insecurity and less vulnerable to shifting weather patterns. Smallholder participation and integration into markets cannot be taken as a foregone conclusion, however. A link between food processing and GHG emissions can also be drawn. Similar to other postharvest methods that preserve food, increased food availability may decrease production-related emissions, assuming that demand and output remain constant. When processing requires energy and facilitates off-farm transport, it is important to consider the full lifecycle emissions of the product to understand the net climate impacts of production.

## 2.16 Use of Weather Information

Smallholders in rainfed mixed systems are vulnerable to weather variability both between seasons and within a season. They deal with this variability in several ways, usually building on long experience. The uncertainty associated with rainfall variability can be reduced through the use of weather information and climate advisories, enabling smallholders to better manage risks and take advantage of favourable climate conditions when they occur (Hansen et al. 2011). Reducing smallholders' vulnerability to current climate risk is often seen as one of the most appropriate entry points into future adaptation, given that climate change may most often be experienced as changes in the frequency and severity of extreme events. The provision of appropriate weather information and associated advisories can help smallholders make more informed decisions regarding the management of their crops and livestock, leading to increased productivity. The effective use of weather information may also be able to contribute to resilience by helping smallholders better manage the negative impacts of weather-related risks in poor seasons while taking greater advantage of better-than-average seasons. Use of weather information may also contribute to GHG emissions mitigation in some situations – for example, by better matching the use of fertilizer and other crop and pasture production inputs with prevailing weather conditions.

Climate services for agriculture are being scaled up in several developing countries. For example, some 560,000 rural households in Senegal now have access to climate information services via rural radio, provided by journalists trained to understand and communicate climate information in local languages and in an interactive format to engage listeners (Ndiaye et al. 2013). In this and other cases, demand for weather information is clearly driven by farmers. There is much less evidence as to how such weather information is being used, however, and the extent to which its use contributes to increased resilience and productivity (and any mitigation co-benefits). Robust impact assessment of the use of weather information and its effects on development outcomes (in addition to climate smartness) in developing country situations is sorely needed. There are several important constraints to the use of climate services, which include bridging the gap between the content, scale, format and lead-time that farmers need and the information that is routinely available (Hansen et al. 2011); ensuring that the information produced is credible, and that it can be understood and appropriately acted upon; and in ways that do not disadvantage economically and socially marginalized groups. One approach, based on combining climate information with participatory farm planning and budgeting tools, is showing promise in helping to overcome some of these constraints (Dorward et al. 2015) in pilot studies in Tanzania and elsewhere.

## 2.17 Weather-Index Insurance

Agricultural insurance is one approach to managing weather-related risks; it normally relies on direct measurement of the loss or damage suffered by each farmer, which can be costly and time consuming. An alternative is index-based insurance that uses a weather index (e.g., amount of rainfall in a specified period) to determine payouts for the hazard insured. Index-based insurance for crops is often based on rainfall received at a particular meteorological station, with thresholds set for making lump-sum or incremental payouts to those insured. In remote areas, another approach is to use an index based on satellite imagery of vegetation ground cover as a proxy for fodder availability to insure livestock keepers against drought (Chantarat et al. 2013). Index insurance is often coupled with access to credit, allowing farmers to invest in improved practices that can increase productivity and food security, even in adverse weather conditions. In many parts of the global tropics, rainfall is highly variable, and many smallholders inevitably experience livestock loss and crop yield reductions if not total crop failure. Index insurance can make a substantial contribution to smallholders' resilience.

Agricultural insurance is being applied in a range of situations in the developing world. In India, for example, national index insurance programmes, linked to agricultural credit provision and enabled with strong government support, have reached more than 30 million farmers. The Agriculture and Climate Risk Enterprise (ACRE) program in East Africa now reaches nearly 200,000 farmers with bundled index insurance, agricultural credit and farm inputs (Greatrex et al. 2015). Index insurance may have few direct mitigation co-benefits, but smallholders may be able to enhance carbon sequestration or reduce GHG emissions via the management decisions they make as a result of being insured.

Since the 1990s, there has been considerable debate about the potential uses of index-based insurance to manage weather risks in agriculture. In addition to the challenge of basis risk, questions have been raised as to its general scalability (Hazell et al. 2010). There is also a substantial challenge in reconciling simplicity, transparency and efficiency in weather-index insurance programs: they are often complicated instruments needing outreach, education and extension, and the building of trust through time. A key challenge is that the current evidence base as to the impacts of weather-index insurance is weak; when applied at scale in different contexts, the tangible and sustainable impacts on poverty and food security are not yet clear. Nor is it clear whether changes in farmers' production practices tend to increase or decrease farm-level income risk. There may be equity issues too: provision of weather-index insurance to some may exacerbate the losses of segments of society that cannot purchase insurance (Miranda and Farrin 2012). As for climate services, robust impact assessments of weather-index insurance and its relative climate smartness are greatly needed.

## 3 Adoption Constraints and the Potential for Uptake of CSA Interventions

As shown in the previous section, a wide range of options exists for mixed crop-livestock farmers in developing countries, and many of these have positive impacts on at least one or two of the three CSA pillars, and some on all three. The evidence base is mixed, however: the scientific literature for some of these options is scanty, and the survey results of expert opinion clearly show that local context can have an over-riding influence on whether particular practices are positive or negative in any particular situation, given that some 40% of the impacts shown in Table 2 are adjudged to be uncertain. One key message from this analysis is that broad-brush targeting of CSA interventions is apparently not appropriate, from a technical standpoint, given that the impacts are often not clear and/or highly context-specific. The technical potential of CSA interventions in developing country agriculture is going to remain difficult to estimate for some time to come.

Independent of context, common elements can be identified that are important to facilitate the adoption of CSA in developing countries, while these tend to be similar to those that characterise the adoption of other types of sustainable agricultural development or natural resource management strategies. In light of the limited capacity of smallholders to bear risk, they tend to select farm portfolios that stabilise income flows and consumption (Barrett et al. 2001). Under climate change, this ability is determined by high-level factors such as the need for conducive enabling policy environments and public investment, the assurance of peace and security, stable macro-economic conditions, functioning markets and appropriate incentives (or the development of these, including financial, labour, land and input markets), as well as the ability and willingness of farmers to invest their own human, social, natural and physical capitals (Westermann et al. 2015; Ehui and Pender 2005). Socio-cultural traditions, including structural social inequalities, marginalisation of specific groups and gender relations, local institutions (that include informal rules and regulations) that guide resource use, and the division of labour and household decision making, all play a key role in determining whether climate smarter practices are feasible in specific locations.

With respect to agricultural technology adoption and uptake in general, many of the CSA interventions discussed in Sect. 2 have different constraints. These are laid out in Table 2 by intervention, for the following constraints:

- Investment cost: the upfront infrastructural and/or technological costs that farmers may have to make before some types of intervention can be implemented, such as fencing material or irrigation equipment.
- Input/operating cost: these are the recurring costs of inputs needed, including labour, fertilizer or hybrid seed.
- Risk: certain technologies in some situations (e.g., higher levels of purchased inputs in places with high rainfall variability) may have unintended impacts on production or income variability, which can severely constraint adoption.

**Table 2** Constraints to the widespread adoption of climate-smart options (Table 1 and Sect. 3) available to smallholders in mixed crop-livestock systems in developing countries

| Option | Constraint | | | | | | | | | |
|---|---|---|---|---|---|---|---|---|---|---|
| | Investment Cost | Input/operating cost | Risk | Access to technology | Technical know-how | Temporal trade-offs | CSA trade-offs | Information | Acceptability | State of evidence base |
| 2.1 Change crop varieties | | * | | ** | | | | * | | |
| 2.2 Change crops | | * | * | * | * | | | * | * | |
| 2.3 Crop residue management | | * | * | | | ** | | ** | ** | |
| 2.4 Crop management | | * | * | | | | | | * | |
| 2.5 Nutrient management | | ** | | | * | * | * | | | |
| 2.6 Soil management | * | * | | | * | * | * | | | |
| 2.7 Change livestock breed | ** | * | * | * | ** | * | * | * | ** | * |
| 2.8 Manure management | *(*) | | | * | ** | | | ** | * | ** |
| 2.9 Change livestock species | ** | * | * | * | ** | * | | ** | ** | * |
| 2.10 Improved feeding | * | ** | | * | * | | * | * | * | |
| 2.11 Grazing management | ** | * | | * | ** | * | * | ** | * | |

| | | | | | | | | | |
|---|---|---|---|---|---|---|---|---|---|
| 2.12 Alter system integration | * | | ** | * | ** | * | ** | ** | ** |
| 2.13 Water use efficiency / mgmt | ** | ** | | * | * | * | | ** | |
| 2.14 Food storage | * | | | * | | | * | * | ** |
| 2.15 Food processing | * | * | | * | | ? | | * | ** |
| 2.16 Use of weather information | | | * | * | * | *? | * | * | ** |
| 2.17 Weather-index insurance | * | | ** | ** | * | *? | ** | * | ** |

Importance of constraint: **major, *moderate, ? unknown and/or highly context-specific. Authors' evaluation

- Access to technology: adoption may well be constrained in situations where smallholders have limited physical access to the technology (e.g. seeds of improved varieties of crops or pastures).
- Technical know-how: some interventions require high levels of technical knowledge about how to implement and manage the option, and this may act as a powerful deterrent to adoption.
- Temporal trade-offs: sometimes trade-offs may need to be made in the short term to realise medium- or longer-term benefits (e.g., losing access to a piece of land while waiting for certain cash crops to produce harvestable yield), and farmers may not have the wherewithal to wait for these benefits to materialise.
- CSA trade-offs: some interventions in some situations may involve trade-offs between the CSA pillars (production, resilience and mitigation objectives); productivity-enhancing technology may increase resilience by improving household cash flow, but may increase GHG emissions or emission intensities at the same time (e.g., adding nitrogen fertilizer under some circumstances).
- Information: some interventions have recurring informational needs such as seasonal weather forecasts.
- Acceptability: some CSA interventions may go against socio-cultural norms, directly affecting a technology's acceptability in a community (e.g., practices that may affect communal grazing governance in a location, or weak land tenure arrangements affecting the acceptability of investment).
- State of evidence base: insufficient evidence to be able to make robust statements about the relative climate smartness of different alternatives in differing contexts may indirectly constrain their uptake.

Table 2 demonstrates clearly that all interventions are associated with some constraints that may affect adoption in different circumstances. Despite the constraints, all of these interventions may be suitable in some circumstances, but identifying those circumstances may not be straightforward. This is a serious knowledge gap. The scale of the agricultural production and food security challenge in the coming decades is known well enough: by 2030, population may be 8.5 billion, with still-rapid growth in SSA in particular (UNPD 2015). Much of the food production needed will be produced by smallholder mixed farmers, whose numbers are projected to increase from about 560 million today to some 750 million by 2030, mostly in SSA and Asia (Campbell and Thornton 2014). Many of these current and future smallholders will have to become adopters of climate-smart interventions if future food demand is to be satisfied in sustainable ways. Currently, there is only limited information concerning the potential uptake of CSA interventions at scale, in terms of geographic or other domains. A highly indicative analysis is shown in Box 1 for SSA, as a simple example; much more robust and detailed information than is contained in Box 1 would be of considerable value in helping to target research-for-development initiatives to overcome the key adoption barriers in particular places and to prioritise investments in CSA.

**Box 1 Towards prioritising investments in CSA: sub-Saharan Africa as an example**

One preliminary step towards generating the information needed to prioritise investments in CSA is identifying those locations where different interventions may be profitable for smallholders, feasible given their biophysical, informational and socio-economic constraints, and socio-culturally acceptable. As an illustration, we mapped the 17 interventions outlined in Sect. 2 to spatial domains in sub-Saharan Africa based on the mixed system classification shown in Table 1. We used the potential impacts of the intervention from Table 1 and the nature of the constraints to adoption from Table 2, and then subjectively evaluated the suitability of each intervention as zero, low, medium or high in each system. One way to evaluated suitability is in relation to potential adoption rates. To date, adoption rates of agricultural technology in SSA have not often exceeded 30% over one or two decades (see, for example, a discussion in Thornton and Herrero (2010)). Accordingly, we used potential adoption rates of 5% (low suitability) 15% (medium suitability) and 30% (high suitability), nominally for the period to 2030, for the 17 CSA interventions in Table 1. For each intervention, we calculated the size of the rural area and the current number of rural people in each system, crudely multiplied by the associated adoption rate, and summed these to give a highly approximate indication of the relative size of the "suitability domain" (in terms of size and rural population) for each intervention. Results are shown in the table below. Improved feeding and altering the enterprise balance may be suitable over relatively large areas and for large numbers of people living in the rural areas, not all of whom are engaged in agriculture, of course (Lowder et al. 2014). Food storage, grazing management and changes in livestock species (particularly large to small ruminants, or ruminants to non-ruminants, for example) are also options with relatively large domains, according to this analysis. The results for food storage are noteworthy; this intervention appears to have solid CSA benefits (particularly related to increased food availability), and considerable effort and resources might well be warranted to increase the uptake of simple food storage technology and the availability of appropriate information.

There are many problems with this particular analysis: to name just three, the subjective nature of the suitability index, the fact that potential adoption rates are likely to be context- and intervention-specific, and the lack of specificity as to what the exact intervention actually is in each category (for instance, "improved feeding" is a broad term covering many different types of intervention). Nevertheless, this type of broad-brush analysis, if done on a global basis in relation to specific interventions and with as much quantifiable information as possible, could be very helpful in prioritising investments in CSA over the next few years (Table B1).

## 4 Conclusions

The analysis presented here is largely qualitative, based on a systematic review protocol coupled with a survey of experts. We recognise this as a weakness, but as noted in Sect. 1, at present we lack comprehensive information on the costs, benefits, synergies and trade-offs of many of the interventions examined. This is partly because the current state of science for CSA in the mixed systems in developing countries is sparse. There are gaps in our understanding of some of the key biophysical and socioeconomic interactions at the farm level, and work remains to be done before we can inform agricultural development planning for food security in the face of climate change, particularly at the household level, with the accuracy scientists typically strive for.

At the same time, we do not lack analytical tools and methods that could be used for quantitative priority setting to help allocate the resources needed to stimulate the widespread adoption of CSA. To overcome the dearth of field-based evidence on CSA practices and their interactions, modelling tools for the *ex ante* evaluation of these practices will be particularly useful in these early stages of CSA programming. Process-based models such as APSIM (Keating et al. 2003) and IAT (Lisson et al. 2010) can further our understanding of key biophysical interactions under a range CSA management options in the absence of empirical field results (Rigolot et al. 2016). The outputs of these models can in turn be used to help specify the biophysical relationships in bio-economic models suited to the *ex ante* assessment of CSA practices. Mathematical programming techniques can be used to construct bio-economic models that are well-equipped to evaluate CSA practices and help rank practices based on their economic viability in the presence of risk. Their strength lies in their flexibility to incorporate multiple interactions, such as those characterised by CSA, as well as flexibility to include a variety of constraints (Hazell and Norton 1986), including many of those identified in Table 2. Their weakness is in their generally normative nature, as farmers do not tend to behave as optimally as these tools suggest, due in part to various non-economic and non-biophysical considerations that affect farmer decision making. However, recent developments in the growing field of positive mathematical programming have considerably improved the reliability of these models to more accurately simulate farmer behaviour (Mérel and Howitt 2014; Qureshi et al. 2013). Given that the success of CSA practices is highly context-dependent, the usefulness of *ex ante* analyses will have to explicitly account for the heterogeneity of farms and adoption impacts within rural populations and landscapes. This will in turn depend on adequate representation of farm populations in household survey data coupled with spatial data on farming systems, especially when assessing the potential for adoption at regional scales. Naturally, there is no substitute for field-based research and *ex post* analyses of the adoption CSA practices and their economic impacts. As more field and survey-based data accrue over time, these *ex post* analyses can run in parallel with and complement *ex ante* analyses, further building the evidence base for CSA practices and policies.

**Table B1** Agricultural system domains where climate-smart options (Table 1 and Sect. 2) for smallholders in mixed crop-livestock systems in sub-Saharan Africa may be suitable. Relative suitability: 0, not suitable; 1 (low), 5% potential adoption; 2 (medium) 15% potential adoption; 3 (high), 30% potential adoption. EM, extensive mixed systems; IM, intensifying mixed systems (From Herrero et al. 2009; see Fig. 1). Population data from CIESIN (2005). Suitability ratings are the authors' own estimates.

| Option | "Suitability" EM | IM | Total area (km² million) | Total rural population (million 2000) |
|---|---|---|---|---|
| 2.1 Change crop varieties | 1 | 3 | 0.67 | 60.62 |
| 2.2 Change crops | 2 | 3 | 1.12 | 85.78 |
| 2.3 Crop residue management | 0 | 1 | 0.07 | 8.01 |
| 2.4 Crop management | 1 | 2 | 0.45 | 36.60 |
| 2.5 Nutrient management | 1 | 2 | 0.45 | 36.60 |
| 2.6 Soil management | 1 | 2 | 0.45 | 36.60 |
| 2.7 Change livestock breed | 2 | 3 | 1.12 | 85.78 |
| 2.8 Manure management | 2 | 2 | 0.91 | 61.76 |
| 2.9 Change livestock species | 3 | 2 | 1.59 | 99.50 |
| 2.10 Improved feeding | 3 | 3 | 1.81 | 123.52 |
| 2.11 Grazing management | 3 | 2 | 1.59 | 99.50 |
| 2.12 Alter integration between crops and livestock | 3 | 3 | 1.81 | 123.52 |
| 2.13 Water use efficiency | 2 | 1 | 0.76 | 45.75 |
| 2.14 Food storage | 3 | 2 | 1.59 | 99.50 |
| 2.15 Food processing | 1 | 2 | 0.45 | 36.60 |
| 2.16 Weather information | 3 | 1 | 1.45 | 83.49 |
| 2.17 Weather-index insurance | 2 | 2 | 0.91 | 61.76 |

Despite the limitations of the analysis conducted here, some conclusions can be drawn. First, from a technical perspective, there appear to exist no "silver bullets" for achieving climate-smart mixed systems. While this echoes the conclusions of the semi-quantitative analysis in Thornton and Herrero (2014), here we looked at a much wider range of possible interventions than was done there. Triple wins undoubtedly exist, but technical recommendations over broad domains that will work in all or even most circumstances may not be appropriate. Second, from an adoption perspective, a range of different constraints exist that may impede the widespread adoption of all these innovations. These may be to do with investment and/or running costs and access to technology and knowledge of how to implement it, as well as social acceptability and local governance issues. In different contexts, these may conspire to prevent the incremental and transformational shifts that may be needed to result in more climate smart agriculture in many places. Third, for some of the interventions evaluated, there are significant trade-offs between meeting shorter-term food production or food security objectives and longer-term resilience objectives. This applies particularly to crop residue management and altering the integration of crops and livestock within the system, but also to several other interventions (nutrient, soil, water management; grazing management; changing livestock species and breeds; and use of weather information and weather-index insurance). These temporal trade-offs may be difficult to resolve in many local contexts, and the triple wins involving these interventions will sometimes be elusive.

Despite some key knowledge gaps, the lack of silver bullets, the constraints to adoption, and the trade-offs that may arise between shorter- and longer-term objectives at the household level, much is being done. As noted above, more comprehensive information could help target interventions more effectively and precisely, but in many situations, there is already appropriate information to enable no-regret interventions to be suggested – those that already fit in well within current farming practices and do not significantly increase labour demands and household risk, for example. Impacts of adoption of CSA interventions are already appearing (e.g., Nyasimi et al. 2014) and countries such as Myanmar and Cambodia are developing national agricultural strategies around CSA (Hom et al. 2015; CCAFS 2016).

Evidence is also accumulating of the kinds of approaches that can support the scaling up of CSA interventions. Multi-stakeholder platforms and policy making networks are key, especially if paired with capacity enhancement, learning, and innovative approaches to support decision making of farmers (Westermann et al. 2015). Modern information and communications technology offers efficient and cost-effective ways to disseminate and collect information at massive scale, as well as an infrastructure for developing and utilising new and diverse partnerships (with the private sector, for example). A certain level of local engagement may still usually be needed, paying attention to farmers' needs and their own situations (Westermann et al. 2015).

**Acknowledgements** This work was partially supported by the CGIAR Program on Climate Change, Agriculture and Food Security (CCAFS). PKT acknowledges the support of a 2015–2016 CSIRO McMaster Research Fellowship.

# References

Abass, A.B., Ndunguru, G., Mamiro, P., Alenkhe, B., Mlingi, N. and Bekunda, M. 2014. Post-harvest food losses in a maize-based farming system of semi-arid savannah area of Tanzania. Journal of Stored Products Research, 57, pp.49–57.

Abegaz, A., van Keulen, H., Haile, M. and Oosting, S.J. 2007. Nutrient dynamics on smallholder farms in Teghane, Northern Highlands of Ethiopia. In: Advances in integrated soil fertility management in sub-Saharan Africa: Challenges and opportunities (pp. 365–378). Springer Netherlands.

Affognon, H., Mutungi, C., Sanginga, P. and Borgemeister, C. 2015. Unpacking postharvest losses in sub-Saharan Africa: a meta-analysis. World Development, 66, pp.49–68.

Akinlade, J. A., J.W. Smith, A. Larbi, I.O. Adekunle, A. A. Taiwo, and A. A. Busari. 2003. Impact of Forage Legume Hays Derived from Intercrop as Dry Season Feed Supplements for Lactating Bunaji Cows and N'dama Beef Cattle. J. Appl. Anim. Res. 24(2): 185–191.

Akinleye, A.O., V. Kumar, H.P.S. Makkar, and K. Becker. 2012. Jatropha platyphylla kernel meal as feed ingredient for Nile tilapia ( Oreochromis niloticus L .): growth , nutrient utilization and blood parameters. J. Anim. Physiol. Anim. Nutr. (Berl). 96: 119–129.

Albrecht A., Kandji S.T. 2003. Carbon sequestration in tropical agroforestry systems. Agriculture, Ecosystems and Environment 99 15–27.

Amede, T., Menza, M. and Awlachew, S.B. 2011. Zai Improves Nutrient and Water Productivity in the Ethiopian Highlands. Experimental Agriculture 47, 7–20.

Baoua, I. B., Amadou, L.. Margam, V., Murdock, L. L. 2012. Comparative evaluation of six storage methods for postharvest preservation of cowpea grain. Journal of Stored Products Research 49 171–175.

Barman K, Rai SN 2008. Utilization of tanniniferous feeds. 4. Effect of supplementation of Acacia nilotica pods on nutrient utilization and extent of tannin degradation in cattle. Indian Journal of Animal Sciences 78(2): 191–196.

Barrett CB, Reardon T, Webb P 2001. Nonfarm income diversification and household livelihood strategies in rural Africa: concepts, dynamics, and policy implications. Food Policy 26, 315–331.

Barton, A.P., M.A. Fullen, D.J. Mitchell, T.J. Hocking, L. Liu, Z. Wu Bo, Y. Zheng, and Z.Y. Xia. 2004. Effects of soil conservation measures on erosion rates and crop productivity on subtropical Ultisols in Yunnan Province, China. Agric. Ecosyst. Environ. 104(2): 343–357.

Bekele W, Melaku S, Mekasha Y 2013. Effect of substitution of concentrate mix with Sesbania sesban on feed intake, digestibility, body weight change, and carcass parameters of Arsi-Bale sheep fed a basal diet of native grass hay, Trop Anim Health Prod. 45(8): 1677–1685.

Blümmel M., Reddy B.V.S. 2006. Stover fodder quality traits for dual-purpose sorghum genetic improvement. SAT 2(1), online at http://ejournal.icrisat.org/cropimprovement/ v2i1/v2i1stoverfodder.pdf.

Borgemeister, C., C. Adda, M. Sétamou, K. Hell, B. Djomamou, R.H. Markham, and K.F. Cardwell 1998. Timing of harvest in maize: Effects on post harvest losses due to insects and fungi in central Benin, with particular reference to Prostephanus truncatus (Horn) (Coleoptera: Bostrichidae). Agric. Ecosyst. Environ. 69(3): 233–242.

Boserup, E. 1965. The condition of agricultural growth: the economics of agrarian change under population pressure. London: Aldine Publishing Company.

Bozkurt Y, Kaya I 2011. Effect of two different grazing systems on the performance of beef cattle grazing on hilly rangeland conditions. J. Appl. Anim. Res. 39(2): 94–96.

Bryan E, Ringler C, Okoba B, Koo J, Herrero M, Silvestri S 2013. Can agriculture support climate change adaptation, greenhouse gas mitigation and rural livelihoods? Insights from Kenya. Climatic Change 118(2) 151–165.

Burney J A, Naylor RL 2012. Smallholder Irrigation as a Poverty Alleviation Tool in Sub-Saharan Africa. World Development 40(1) 110–123.

Burney, J. A, Naylor, R.L., Postel, S.L. 2013. The case for distributed irrigation as a development priority in sub-Saharan Africa. Proceedings of the National Academy of Sciences of the United States of America 110(31), pp.12513–7.

BurnSilver SB 2009. Pathways of continuity and change: Maasai livelihoods in Amboseli, Kajiado District, Kenya. In Staying Maasai? (pp. 161–207). Springer, New York.

Campbell B, Thornton PK 2014. How many farmers in 2030 and how many will ardopt climate resilient innovations? CCAFS Info Note. CGIAR Research Program on Climate Change, Agriculture and Food Security (CCAFS), Copenhagen.

CCAFS 2016. CCAFS Scenarios Guide the Cambodian Climate Change Priorities Action Plan for Agriculture. Outcome Case. Copenhagen, Denmark: CGIAR Research Program on Climate Change, Agriculture and Food Security (CCAFS).

Challinor AJ, Watson J, Lobell DB, Howden SM, Smith DR, Chhetri N 2014. A meta-analysis of crop yield under climate change and adaptation. Nature Climate Change 4 (4) 287–291.

Chantarat S, Mude AG, Barrett CB, Carter MR 2013. Designing index-based livestock insurance for managing asset risk in northern Kenya. The Journal of Risk and Insurance 80 (1) 205–237.

Center for International Earth Science Information Network (CIESIN), Columbia University; and Centro Internacional de Agricultura Tropical (CIAT) 2005. Gridded Population of the World Version 3 (GPWv3): Population Grids. Palisades, NY: Socioeconomic Data and Applications Center (SEDAC), Columbia University. Available at http://sedac.ciesin.columbia.edu/gpw.

Cole S, GG Bastian, S Vyas, C Vendel and D Stein 2012. The effectiveness of index-based microinsurance in helping smallholders manage weather related risks. EPPI-Centre, Social Science Research Unit, Institute of Education, University of London.

Davies M, Guenther B, Leavy J, Mitchell T, Tanner T 2009. Climate Change Adaptation, Disaster Risk Reduction and Social Protection: Complementary Roles in Agriculture and Rural Growth? IDS Working Paper 320, Institute of Development Studies, Sussex.

Descheemaeker K, Amede T, Haileslassie A 2010. Improving water productivity in mixed crop–livestock farming systems of sub-Saharan Africa. Agricultural Water Management 97, 579–586.

Dercon S 1996. Risk, Crop Choice, and Savings: Evidence from Tanzania. Economic Development and Cultural Change 44, 485–513.

Deressa, T.T., R.M. Hassan, C. Ringler, T. Alemu, and M. Yesuf. 2009. Determinants of farmers' choice of adaptation methods to climate change in the Nile Basin of Ethiopia. Glob. Environ. Chang. 19(2): 248–255.

de Faccio Carvalho, P.C., I. Anghinoni, A. de Moraes, E.D. de Souza, R.M. Sulc, C.R. Lang, J.P.C. Flores, M.L. Terra Lopes, J.L.S. da Silva, O. Conte, C. de Lima Wesp, R. Levien, R.S. Fontaneli, and C. Bayer 2010. Managing grazing animals to achieve nutrient cycling and soil improvement in no-till integrated systems. Nutr. Cycl. Agroecosystems 88(2): 259–273.

De Haan C, Steinfeld H, Blackburn H 1997. Livestock and the Environment: Finding a Balance, Eye, Suffolk: WRENmedia.

Deshingkar P 2012. Environmental risk, resilience and migration: implications for natural resource management and agriculture Environ. Res. Lett. 7 015603.

Devereux S 2001. Livelihood insecurity and social protection: a re-emerging issue in rural development. Development Policy Review 19 (4): 507–519.

Dorward P, Clarkson G, Stern R 2015. Participatory Integrated Climate Services for Agriculture (PICSA): Field Manual. Walker Institute, University of Reading. Online at. https://cgspace.cgiar.org/rest/bitstreams/60947/retrieve

Drechsel, P., Olaleye, A., Adeoti, A., Thiombiano, L., Barry, B. and Vohland, K. 2005. Adoption driver and constraints of resource conservation technologies in sub-saharan Africa. Berlin: FAO, IWMI, Humbold Universitaet.

Ehui S, Pender J 2005. Resource degradation, low agricultural productivity, and poverty in sub-Saharan Africa: pathways out of the spiral. Online at http://onlinelibrary.wiley.com/doi/10.1111/j.0169-5150.2004.00026.x/epdf

Evanson RE, Gollin D 2003. Assessing the impact of the Green Revolution 1960 to 2000. Science 300: 758–762.

FAO 2011. Global Food Losses and Food Waste. FAO, Rome, Italy.

FAO 2013. Climate-Smart Agriculture Source Book. FAO, Rome, Italy.

Franzel, S., Carsan, S., Lukuyu, B., Sinja, J. and Wambugu, C. 2014. Fodder trees for improving livestock productivity and smallholder livelihoods in Africa. Current Opinion in Environmental Sustainability, 6, pp.98–103.

Fritz S, See L, McCallum I, You L, Bun A, Albrecht F, Schill C, Perger C, Duerauer M, Havlik P, Mosnier A, Thornton P,Wood-Sichra U, Herrero M, Becker-Reshef I, Justice C, Hansen M,Gong P, Abdel Aziz S, Cipriani A, Cumani R, Cecchi G, Conchedda G, Ferreira S, Gomez A, Haffani M, Kayitakire F, Malanding J, Mueller R, Newby T, Nonguierma A,Olesegun A,Ortner S,Ram R,Rocha J, Schepaschenko D, Schepaschenko S, Terekhov A,Tiangwa A, Vancutsem C, Vintrou E, Wenbin W, van der Velde M, Dunwoody A, Kraxner F, Obersteiner M 2015. Mapping global cropland and field size. Global Change Biology, DOI: 10.1111/gcb.12838

Galukande, E., Mulindwa, H., Wurzinger, H., Roschinsky, M., Mwai, R., Solkner, A.O. 2013. Cross-breeding cattle for milk production in the tropics: achievements, challenges and opportunities. Animal Genetic Resources 52 111–125.

Gebrehiwot NT, Mesfin KA, Nyssen J 2015. Small-scale irrigation: the driver for promoting agricultural production and food security (the case of Tigray Regional State, Northern Ethiopia). Irrigation and Drainage Systems Engineeri006Eg 4 2, http://dx.doi.org/10.4172/2168-9768.1000141

Giller, K.E., Andersson, J.A., Corbeels, M., Kirkegaard, J., Mortensen, D., Erenstein, O. and Vanlauwe, B. 2015. Beyond conservation agriculture. Frontiers in plant science, 6: 870.

Greatrex H, Hansen JW, Garvin S, Diro R, Blakeley S, Le Guen M, Rao KN, Osgood, DE. 2015. Scaling up index insurance for smallholder farmers: Recent evidence and insights. CCAFS Report No. 14. Copenhagen. Available online at: www.ccafs.cgiar.org

Groot JCJ, Oomen GJM, Rossing WAH 2012. Multi-objective optimization and design of farming systems. Agricultural Systems 110, 63–77.

Ghosh, P.K., M. Mohanty, K.K. Bandyopadhyay, D.K. Painuli, and A. K. Misra. 2006. Growth, competition, yield advantage and economics in soybean/pigeonpea intercropping system in semi-arid tropics of India: I. Effect of subsoiling. F. Crop. Res. 96(1): 80–89.

Guto, S.N., P. Pypers, B. Vanlauwe, N. de Ridder, and K.E. Giller. 2012. Socio-ecological niches for minimum tillage and crop-residue retention in continuous maize cropping systems in smallholder farms of central Kenya. Agron. J. 104(1): 188–198.

Haile, A. 2006. On-farm storage studies on sorghum and chickpea in Eritrea. African Journal of Biotechnology 5(17): 1537–1544.

Hansen JW, Mason SJ, Sun L, Tall A. 2011. Review of seasonal climate forecasting for agriculture in sub-Saharan Africa. Experimental Agriculture 47(02): 205–240.

Harris, D. and Orr, A. 2014. Is rainfed agriculture really a pathway from poverty? Agricultural Systems 123, pp.84–96.

Hazell PB, Norton RD 1986. Mathematical Programming for Economic Analysis in Agriculture. New York, Macmillan.

Hazell, P., Poulton, C., Wiggins, S. and Dorward, A. 2010. The future of small farms: trajectories and policy priorities. World Development 38(10) 1349–1361.

Henderson B, Gerber P, Hilinski T, Falcucci A, Ojima DS, Salvatore M, Conant RT 2015. Greenhouse gas mitigation potential of the world's grazing lands: modelling soil carbon and nitrogen fluxes of mitigation practices. Agr Ecosyst Environ 207, 91–100.

Herrero M, Thornton P K, Notenbaert A, Msangi S, Wood S, Kruska R L, Dixon J, Bossio D, van de Steeg J A, Freeman H A, Li X, Rao P P 2009. Drivers of change in crop-livestock systems and their potential impacts on agro-ecosystems services and human well-being to 2030. Study commissioned by the CGIAR Systemwide Livestock Programme (SLP). ILRI, Nairobi, Kenya.

Herrero M, Thornton PK, Notenbaert A, Wood S, Msangi S, Freeman HA, Bossio D, Dixon J, Peters M, van de Steeg J, Lynam J, Parthasarathy Rao P, Macmillan S, Gerard B, McDermott J, Seré C, Rosegrant M 2010. Smart investments in sustainable food production: revisiting mixed crop-livestock systems. Science 327, 822–825.

Herrero M, Havlík P, Valin H, Notenbaert AM, Rufino M, Thornton PK, Blummel M, Weiss F, Obersteiner M 2013. Global livestock systems: biomass use, production, feed efficiencies and greenhouse gas emissions. PNAS 110 (52) 20888–20893.

Hoffmann, I. 2010. Climate change and the characterization, breeding and conservation of animal genetic resources. Animal genetics, 41(s1), pp.32–46.

Hom NH, Htwe NM, Hein Y, Than SM, Kywe M, Htut T. 2015. Myanmar Climate-Smart Agriculture Strategy. Ministry of Agriculture and Irrigation (MOAI). Naypyitaw, Myanmar: CGIAR Research Program on Climate Change, Agriculture and Food Security (CCAFS), International Rice Research Institute (IRRI).

Hurni H, Giger M, Liniger H, Studer RM, Messerli P, Portner B, Schwilch G, Wolfgramm B, Breu T 2015. Soils, agriculture and food security: the interplay between ecosystem functioning and human well-being. Current Opinion in Environmental Sustainability 15 25–34.

Ilboudo, Z., L.C.B. Dabiré, R.C.H. Nébié, I.O. Dicko, S. Dugravot, A. M. Cortesero, A. Sanon 2010. Biological activity and persistence of four essential oils towards the main pest of stored cowpeas, Callosobruchus maculatus (F.) (Coleoptera: Bruchidae). J. Stored Prod. Res. 46(2): 124–128.

Kaitho, R., A. Tegegne, N. Umunna, I. Nsahlai, S. Tamminga, J. Van Bruchem, J. Arts 1998. Effect of *Leucaena* and *Sesbania* supplementation on body growth and scrotal circumference of Ethiopian highland sheep and goats fed teff straw basal diet. Livest. Prod. Sci. 54(2): 173–181.

Kamara, A. Y., A. Menkir, B. Badu-Apraku, and O. Ibikunle. 2003. The influence of drought stress on growth, yield and yield components of selected maize genotypes. J. Agric. Sci. 141(1): 43–50.

Kassie, M., M. Jaleta, B. Shiferaw, F. Mmbando, and M. Mekuria. 2013. Adoption of interrelated sustainable agricultural practices in smallholder systems : Evidence from rural Tanzania. Technol. Forecast. Soc. Chang. 80(3): 525–540.

Keating BA, Carberry PS, Hammer GL, Probert ME, Robertson MJ, Holzworth D, Huth NI, Hargreaves JN, Meinke H, Hochman Z, McLean G 2003. An overview of APSIM, a model designed for farming systems simulation. European Journal of Agronomy 18(3), pp.267–288.

Kerr, K.B., S.S. Snapp, M. Chirwa, L. Shumba, and R. Msachi. 2007. Participatory research on legume diversification with Malawian smallholder farmers for improved human nutrition and soil fertility. Exp. Agric. 43(04): 437–453.

Kimaro, A.A., Mpanda, M., Rioux, J., Aynekulu, E., Shaba, S., Thiong'o, M., Mutuo, P., Abwanda, S., Shepherd, K., Neufeldt, H. and Rosenstock, T.S. 2015. Is conservation agriculture 'climate-smart' for maize farmers in the highlands of Tanzania? Nutrient Cycling in Agroecosystems, pp.1–12.

Kipkorir, E.C., D. Raes, and B. Massawe 2002. Seasonal water production functions and yield response factors for maize and onion in Perkerra, Kenya. Agric. Water Manag. 56(3): 229–240.

Kitinoja, L. & Kader, A.A. 2003. Small-Scale Postharvest Handling Practices: A Manual for Horticultural Crops, Davis, USA: Postharvest Technology Research and Information Centre.

Koona, P., V. Tatchago, and D. Malaa. 2007. Impregnated bags for safer storage of legume grains in West and Central Africa. J. Stored Prod. Res. 43(3): 248–251.

Kosgey, I.S., Baker, R.L., Udo, H.M.J. and Van Arendonk, J.A.M. 2006. Successes and failures of small ruminant breeding programmes in the tropics: a review. Small Ruminant Research, 61(1), pp.13–28.

Kraaijvanger R, Veldkamp T 2015. Rain productivity, fertilizer response and nutrient balance of farming systems in Tigray, Ethiopia: a multi-perspective view in relation to soil fertility degradation. Land Degradation and Development 26, 701–710.

Kristjanson P.M., Tarawali S., Okike I., Singh B.B., Thornton P.K., Manyong V.M., Kruska R.L., Hoogenboom G. 2002. Genetically Improved Dual-Purpose Cowpea: Ex-Ante Assessment of Adoption and Impact in the Dry Savannas of West Africa. International Livestock Research Institute Impact Assessment Series Number 9. ILRI, Nairobi, Kenya.

Krouma, A. 2010. Plant water relations and photosynthetic activity in three Tunisian chickpea (Cicer arietinum L.) genotypes. Turk. J. Agric. For. 34: 257–264.

Kumar, A., J. Bernier, S. Verulkar, H.R. Lafitte, and G.N. Atlin. 2008. Breeding for drought tolerance: Direct selection for yield, response to selection and use of drought-tolerant donors in upland and lowland-adapted populations. F. Crop. Res. 107(3): 221–231.

Kurwakumire, N., Chikowo, R., Mtambanengwe, F., Mapfumo, P., Snapp, S., Johnston, A. and Zingore, S. 2014. Maize productivity and nutrient and water use efficiencies across soil fertility domains on smallholder farms in Zimbabwe. Field Crops Research 164, pp.136–147.

Kyazze FB, Kristjanson P 2011. Summary of Baseline Household Survey Results: Rakai District, South Central Uganda. CGIAR Research Program on Climate Change, Agriculture and Food Security (CCAFS) Copenhagen, Denmark. Available online at: http://ccafs.cgiar.org/resources/baseline-surveys

Kywe, M., M.R. Finckh, and A. Buerkert. 2008. Green Gram rotation effects on Maize growth parameters and soil quality in Myanmar. J. Agric. Rural Dev. Trop. Subtrop. 109(2): 123–137.

La Rovere, R., Abdoulaye, T., Kostandini, G., Guo, Z., Mwangi, W., MacRobert, J. and Dixon, J. 2014. Economic, Production, and Poverty Impacts of Investing in Maize Tolerant to Drought in Africa: An Ex-Ante Assessment. The Journal of Developing Areas 48(1) 199–225.

Lal, R. 1997. Residue management, conservation tillage and soil restoration for mitigating greenhouse effect by CO2-enrichment. Soil Tillage Res. 43(1–2): 81–107

Lallo, C.H.O., and G.W. Garcia. 1994. Poultry by-product meal as a substitute for soybean meal in the diets of growing hair sheep lambs fed whole chopped sugarcane. Small Rumin. Res. 14: 107–114.

Lanckriet S, Derudder B, Naudts J, Bauer H, Deckers J, Haile M, Nyssen J 2015. A political ecology perspective of land degradation in the north Ethiopian highlands. Land Degradation & Development 26, 521–530.

Lemaire G, Franzluebbers A, de Faccio Carvalho PC, Dedieu B 2014. Integrated crop–livestock systems: strategies to achieve synergy between agricultural production and environmental quality. Agriculture, Ecosystems & Environment 190, 4–8.

Li, F.M., P. Wang, J. Wang, and J.Z. Xu. 2004. Effects of irrigation before sowing and plastic film mulching on yield and water uptake of spring wheat in semiarid Loess Plateau of China. Agric. Water Manag. 67(2): 77–88.

Lisson S, MacLeod N, McDonald C, Corfield J, Pengelly B, Wirajaswadi L, Rahman R, Bahar S, Padjung R, Razak N, Puspadi K 2010. A participatory, farming systems approach to improving Bali cattle production in the smallholder crop–livestock systems of Eastern Indonesia. Agric. Syst. 103 (7), 486–497.

Liu, X.J., J.C. Wang, S.H. Lu, F.S. Zhang, X.Z. Zeng, Y.W. Ai, S.B. Peng, and P. Christie. 2003. Effects of non-flooded mulching cultivation on crop yield, nutrient uptake and nutrient balance in rice-wheat cropping systems. F. Crop. Res. 83(3): 297–311.

Liu J, You L, Amini M, Obersteiner M, Herrero M, Zehnder AJB, Yang H 2010. A high-resolution assessment on global nitrogen flows in cropland. PNAS 107(17), 8035–8040.

Lowder SK, Skoet J, Singh S 2014. What do we really know about the number and distribution of farms and family farms worldwide? ESA Working Paper No. 14–02. Rome, FAO.

Ma, L., F. Yuan, H. Liang, and Y. Rong. 2014. The effects of grazing management strategies on the vegetation, diet quality, intake and performance of free grazing sheep. Livest. Sci. 161(1): 185–192.

Mahmoodi, R. 2008. Effects of limited irrigation on root yield and quality of sugar beet (Beta vulgaris L.). African J. Biotechnol. 7(24): 4475–4478.

Mahmutoğlu, T., F. Emír, and Y.B. Saygi. 1996. Sun/solar drying of differently treated grapes and storage stability of dried grapes. J. Food Eng. 29(3–4): 289–300.

Mailhol, J.C., A. Zaïri, A. Slatni, B. Ben Nouma, and H. El Amani. 2004. Analysis of irrigation systems and irrigation strategies for durum wheat in Tunisia. Agric. Water Manag. 70(1): 19–37.

Mattiauda, D. A., S. Tamminga, M.J. Gibb, P. Soca, O. Bentancur, and P. Chilibroste. 2013. Restricting access time at pasture and time of grazing allocation for Holstein dairy cows: Ingestive behaviour, dry matter intake and milk production. Livest. Sci. 152(1): 53–62.

McGuire S, Sperling L 2015. Seed systems smallholder farmers use. Food Security (2015) 1–17.

Mengistu T 2014. Restoring degraded landscapes not a miracle anymore. Online at https://wle.cgiar.org/thrive/2014/11/28/restoring-degraded-landscapes-not-miracle-anymore

Mérel P. and Howitt R 2014. Theory and Application of Positive Mathematical Programming in Agriculture and the Environment. Annu. Rev. Resour. Econ., 6, 451–70.

Miranda, M.J. and Farrin, K. 2012. Index insurance for developing countries. Applied Economic Perspectives and Policy 34(3), 391–427.

Moyo, B., S. Dube, C. Moyo, and E. Nesamvuni. 2011. Heavily stocked 5-paddock rotational grazing effect on cross-bred Afrikaner steer performance and herbaceous vegetation dynamics in a semi-arid veld of Zimbabwe. African J. Agric. Res. 6(10): 2166–2174.

Mrabet, R. 2000. Differential response of wheat to tillage management systems in a semiarid area of Morocco. F. Crop. Res. 66(2): 165–174.

Mulumba, L.N., and R. Lal. 2008. Mulching effects on selected soil physical properties. Soil Tillage Res. 98(1): 106–111.

Mureithi, J.G., Gachene, C.K.K., Ojiem, J. 2003. The role of green manure legumes in smallholder farming systems in Kenya: the legume research network project. Trop. Subtrop. Ecosyst. 1: 57–70.

Muriuki HG, Thorpe W 2006. Smallholder Dairy Production and Marketing in Eastern and Southern Africa: Regional Synthesis. The South-South Workshop on Smallholder Dairy Production and Marketing (International Livestock Research Institute, Nairobi, Kenya).

Mwakaje, A. G. 2008. Dairy farming and biogas use in Rungwe district, South-west Tanzania: A study of opportunities and constraints. Renewable and Sustainable Energy Reviews 12(8) 2240–2252.

Ndiaye O, Moussa AS, Seck M, Zougmore R, Hansen J. 2013. Communicating seasonal forecasts to farmers in Kaffrine, Senegal for better agricultural management. Dublin, Ireland: Irish Aid. http://www.mrfcj.org/pdf/case-studies/2013-04-16-Senegal.pdf

Nhemachena, C., and H. Rashid. 2008. Determinants of African farmers' strategies for adapting to climate change: Multinomial choice analysis. AfJARE 2(1): 83–104.

Nyamadzawo, G., Wuta, M., Nyamangara, J. and Gumbo, D. 2013. Opportunities for optimization of in-field water harvesting to cope with changing climate in semi-arid smallholder farming areas of Zimbabwe. SpringerPlus 2(1), p.100.

Nyasimi M, Amwata D, Hove L, Kinyangi J, Wamukoya G. 2014. Evidence of impact: Climate-smart agriculture in Africa. Wageningen, Netherlands: CGIAR Research Program on Climate Change, Agriculture and Food Security (CCAFS) and the Technical Centre for Agricultural and Rural Cooperation (CTA).

Obalum, E.E., U.C. Amalu, M.E. Obi, and T. Wakatsuki. 2011. Soil Water Balance and Grain Yield of Sorghum Under No-Till Versus Conventional Tillage With Surface Mulch in the Derived Savanna Zone of Southeastern Nigeria. Exp. Agric. 47(01): 89–109.

Omer, M.A., E.M. Elamin, and M.A. Ojmer. 1997. Effect of tillage and contour diking on sorghum establishment and yield on sandy clay soil in Sudan. Soil Tillage Res. 43(3–4): 229–240.

Qureshi M., Whitten S., Franklin B. 2013. Impacts of climate variability on the irrigation sector in the southern Murray-Darling Basin, Australia. Water Resources and Economics, 4, 52–68.

Rabary, B., S. Sall, P. Letourmy, O. Husson, E. Ralambofetra, N. Moussa, and J.L. Chotte. 2008. Effects of living mulches or residue amendments on soil microbial properties in direct seeded cropping systems of Madagascar. Appl. Soil Ecol. 39(2): 236–243.

Ramirez-Villegas J and Thornton PK 2015. Climate change impacts on crop production. Working Paper 119, CCAFS, Copenhagen, Denmark.

Rao IM, Peters M, van der Hoek R, Castro A, Subbarao G, Cadisch G, Rincón A 2014. Tropical forage-based systems for climate-smart livestock production in Latin America. Rural21, online at http://www.rural21.com/english/news/detail/article/tropical-forage-based-systems-for-climate-smart-livestock-production-in-latin-america-00001322/

Rigolot C, De Voil P, Douxchamps S, Prestwidge D, Van Wijk M, Thornton PK, Rodriguez D, Henderson B, Medina D, Herrero M 2016. Adapting smallholder mixed crop-livestock farming systems to climate variability in northern Burkina Faso with crop-livestock interactions. Agricultural Systems, doi:10.1016.j.agsy.2015.12.017

Robinson TP, Thornton PK, Franceschini G, Kruska RL, Chiozza F, Notenbaert A, Cecchi G, Herrero M, Epprecht M, Fritz S, You L, Conchedda G, See L 2011. Global livestock production systems. Rome, Food and Agriculture Organization of the United Nations (FAO) and International Livestock Research Institute (ILRI) 152 pp.

Rockström, J., P. Kaumbutho, J. Mwalley, a. W. Nzabi, M. Temesgen, L. Mawenya, J. Barron, J. Mutua, and S. Damgaard-Larsen. 2009. Conservation farming strategies in East and Southern Africa: Yields and rain water productivity from on-farm action research. Soil Tillage Res. 103(1): 23–32.

Rockström J, Falkenmark M 2015. Increase water harvesting in Africa. Nature 519(7543) 283–5.

Rosenstock TS, Lamanna C, Chesterman S, Bell P, Arslan A, Richards M, Rioux J, Champalle C, Eyrich A-S, English W, Ström H, Madalinska A, McFatridge S, Poultouchidou A, Akinleye AO, Kerr A, Corner-Dolloff C, Zhou W, Lizarazo M, Girvetz EH, Tully KL, Dohn J, Morris KS 2016. The scientific basis of climate-smart agriculture: A systematic review protocol. Working Paper, ICRAF, Nairobi.

Rufino, M.C., Rowe, E.C., Delve, R.J. and Giller, K.E. 2006. Nitrogen cycling efficiencies through resource-poor African crop–livestock systems. Agriculture, ecosystems & environment 112(4), pp.261–282.

Rufino MC, Thornton PK, Ng'ang'a SK, Mutie I, Jones PG, van Wijk MT, Herrero M 2013. Transitions in agro-pastoralist systems of East Africa: impacts on food security and poverty. Agriculture, Ecosystems and Environment 179 215–230.

Rusinamhodzi, L., M. Corbeels, J. Nyamangara, and K.E. Giller. 2012. Maize-grain legume intercropping is an attractive option for ecological intensification that reduces climatic risk for smallholder farmers in central Mozambique. F. Crop. Res. 136: 12–22.

Russelle MP, Entz MH, Franzluebbers AJ 2007. Reconsidering integrated crop–livestock systems in North America. Agronomy Journal 99, 325–334.

Sadfi, N., M. Cherif, M.R. Hajlaoui, and A. Boudabbous. 2002. Biological control of the potato tubers dry rot caused by Fusarium roseum var. sambucinum under greenhouse, field and storage conditions using Bacillus spp. isolates. J. Phytopathol. Zeitschrift 150(11–12): 640–648.

Salako, F.K., P.O. Dada, C.O. Adejuyigbe, M.O. Adedire, O. Martins, C. A. Akwuebu, and O.E. Williams. 2007. Soil strength and maize yield after topsoil removal and application of nutrient amendments on a gravelly Alfisol toposequence. Soil Tillage Res. 94(1): 21–35.

Sauerborn, H., J. Sprich, and H. Mercer-Quarshie. 2000. Crop Rotation to Improve Agricultural Production in Sub-Saharan Africa. J.Agronomy Crop Sci. 184: 56–61.

Seebauer M 2014. Whole farm quantification of GHG emissions within smallholder farms in developing countries. Environ. Res. Lett. 9, 035006 doi:10.1088/1748-9326/9/3/035006

Seré C, Steinfeld H, 1996. World livestock production systems: Current status, issues and trends. FAO Animal Production and Health Paper 127. FAO (Food and Agriculture Organization of the United Nations), Rome, Italy.

Shumba, E.M., S.R. Waddington, and Rukuni M. 1992. The use of tine-tillage, with Atrazine weed control, to permit earlier planting of maize by smallholder farmers in Zimbabwe. Exp. Agric. 28: 443–452.

Sissoko, F., F. Affholder, P. Autfray, J. Wery, and B. Rapidel. 2013. Wet years and farmers' practices may offset the benefits of residue retention on runoff and yield in cotton fields in the Sudan-Sahelian zone. Agric. Water Manag. 119(0): 89–99.

Smith, C. 1984. Rates of genetic change in farm livestock. Res. Dev. Agric. 1, 79–85.

Smith, P., Martino, D., Cai, Z., Gwary, D., Janzen, H., Kumar, P., McCarl, B., Ogle, S., O'Mara, F., Rice, C., Scholes, B., Sirotenko, O., Howden, M., McAllister, T., Pan, G., Romanenkov, V., Schneider, U., Towprayoon, S., Wattenbach M. & Smith, J. 2008. Greenhouse gas mitigation in agriculture. Phil. Trans. R. Soc. B, 363: 789–813.

Smith, J., Abegaz, A., Matthews, R. B., Subedi, M., Orskov, E. R., Tumwesige, V., & Smith, P. 2014. What is the potential for biogas digesters to improve soil fertility and crop production in Sub-Saharan Africa? Biomass and Bioenergy, 70, 58–72.

Snapp, S.S., D.D. Rohrbach, F. Simtowe, and H.A. Freeman. 2002. Sustainable soil management options for Malawi: can smallholder farmers grow more legumes? Agric. Ecosyst. Environ. 91(1–3): 159–174.

Snapp, S.S., Swinton, S.M., Labarta, R., Mutch, D., Black, J.R., Leep, R., Nyiraneza, J. and O'Neil, K. 2005. Evaluating cover crops for benefits, costs and performance within cropping system niches. Agronomy Journal, 97(1), pp.322–332.

Speelman, S., S. Farolfi, S. Perret, L. D'haese, and M. D'haese. 2008. Irrigation Water Value at Small-scale Schemes: Evidence from the North West Province, South Africa. Int. J. Water Resour. Dev. 24(August 2014): 621–633.

Srinivasarao, C., B. Venkateswarlu, A.K. Singh, K.P.R. Vittal, S. Kundu, G.R. Chary, G.N. Gajanan, and B.K. Ramachandrappa. 2012. Critical carbon inputs to maintain soil organic carbon stocks under long-term finger-millet (Eleusine coracana [L.] Gaertn.) cropping on Alfisols in semiarid tropical India. J. Plant Nutr. Soil Sci. 175(5): 681–688.

St Clair, S.B., and J.P. Lynch. 2010. The opening of Pandora's Box: climate change impacts on soil fertility and crop nutrition in developing countries. Plant Soil 335: 101–115.

Sumberg, J. 2002. The logic of fodder legumes in Africa. Food policy 27(3) 285–300.

Sumberg J 2003. Toward a disaggregated view of crop–livestock integration in Western Africa. Land Use Policy 20 253–264.

Surekha, K., P.C. Latha, K. V. Rao, and R.M. Kumar. 2010. Grain Yield, Yield Components, Soil Fertility, and Biological Activity under Organic and Conventional Rice Production Systems. Commun. Soil Sci. Plant Anal. 41(19): 2279–2292.

Suriyakup, P., A. Polthanee, K. Pannangpetch, R. Katawatin, J.C. Mouret, and C. Clermont-Dauphin. 2007. Introducing mungbean as a preceding crop to enhance nitrogen uptake and yield of rainfed rice in the north-east of Thailand. Aust. J. Agric. Res. 58(11): 1059–1067.

Szilas, C., J.M.R. Semoka, and O.K. Borggaard. 2007. Can local Minjingu phosphate rock replace superphosphate on acid soils in Tanzania? Nutr. Cycl. Agroecosystems 77(3): 257–268.

Taddesse, G., D. Peden, A. Abiye, and A. Wagnew. 2003. Effect of Manure on Grazing Lands in Ethiopia, East African Highlands. Mt. Res. Dev. 23(2): 156–160.

Teshome A, de Graaff J, Stroosnijder L 2014. Evaluation of soil and water conservation practices in the north-western Ethiopian highlands using multi-criteria analysis. Frontiers in Environmental Science 260, doi_10.3389fenvs.2014.00060/links/54a0ff020cf257a636021e11.pdf

Thierfelder C, Wall PC 2009. Effects of conservation agriculture techniques on infiltration and soil water content in Zambia and Zimbabwe. Soil and Tillage Research 105(2), pp.217–227.

Thomas, D. S. G., Twyman, C., Osbahr, H., Hewitson, B. 2007. Adaptation to climate change and variability: farmer responses to intra-seasonal precipitation trends in South Africa. Climatic Change, 83(3): 301–322.

Thornton P K, Boone R B, Galvin K A, BurnSilver S B, Waithaka M M, Kuyiah J, Karanja S, González-Estrada E and Herrero M 2007. Coping strategies in livestock-dependent households in East and southern Africa: a synthesis of four case studies. Human Ecology 35 (4), 461–476.

Thornton PK, Herrero M 2010. The potential for reduced methane and carbon dioxide emissions from livestock and pasture management in the tropics. PNAS 107 (46) 19667–19672.

Thornton PK, Herrero M 2014. Climate change adaptation in mixed crop-livestock systems in developing countries. Global Food Security 3, 99–107.

Thornton PK, Herrero M 2015. Adapting to climate change in the mixed crop-livestock farming systems in sub-Saharan Africa. Nature Climate Change 5, 830–836.

Tittonell, P., Van Wijk, M.T., Herrero, M., Rufino, M.C., de Ridder, N. and Giller, K.E. 2009. Beyond resource constraints–Exploring the biophysical feasibility of options for the intensification of smallholder crop-livestock systems in Vihiga district, Kenya. Agricultural Systems 101(1), pp.1–19.

Torres, R.O., R.P. Pareek, J.K. Ladha, and D.P. Garrity. 1995. Stem-nodulating legumes as relay-cropped or intercropped green manures for lowland rice. F. Crop. Res. 42(1): 39–47.

Tschakert P 2007. Environmental services and poverty reduction: Options for smallholders in the Sahel Agricultural Systems 94, 75–86.

Tuwei, P.K., Kang'Ara, J.N.N., Mueller-Harvey, I., Poole, J., Ngugi, F.K. and Stewart, J.L. 2003. Factors affecting biomass production and nutritive value of Calliandra calothyrsus leaf as fodder for ruminants. The Journal of Agricultural Science 141(01), pp.113–127.

Udo, H.M.J., H.A. Aklilu, L.T. Phong, R.H. Bosma, I.G.S. Budisatria, B.R. Patil, T. Samdup, and B.O. Bebe. 2011. Impact of intensification of different types of livestock production in smallholder crop-livestock systems. Livest. Sci. 139(1–2): 22–29.

United Nations Population Division 2015. World Population Prospects: The 2015 Revision. Online at http://esa.un.org/unpd/wpp/Download/Probabilistic/Population/

Valbuena, D., O. Erenstein, S. Homann-Kee Tui, T. Abdoulaye, L. Claessens, A.J. Duncan, B. Gérard, M.C. Rufino, N. Teufel, A. van Rooyen, and M.T. van Wijk. 2012. Conservation Agriculture in mixed crop-livestock systems: Scoping crop residue trade-offs in Sub-Saharan Africa and South Asia. F. Crop. Res. 132: 175–184.

van Keulen H, Schiere H 2004. Crop-livestock systems: old wine in new bottles? In: New directions for a diverse planet, Proceedings of the 4th International Crop Science Congress 26 Sep – 1 Oct 2004, Brisbane, Australia. Published on CDROM, www.cropscience.org.au

van Wijk MT, Tittonell P, Rufino MC, Herrero M, Pacini C, de Ridder N, Giller KE 2009. Identifying key entry-points for strategic management of smallholder farming systems in sub-Saharan Africa using the dynamic farm-scale simulation model NUANCES-FARMSIM. Agricultural Systems 102(1), 89–101.

Vermeulen SJ, Challinor AJ, Thornton PK, Campbell BM, Eriyagama N, Vervoort JM, Kinyangi J, Jarvis A, Läderach P, Ramirez-Villegas J, Nicklin KJ, Hawkins E and Smith DR 2013. Addressing uncertainty in adaptation planning for agriculture. PNAS 110 (21), 8357–8362.

Waithaka MM, Thornton PK, Shepherd KD, Herrero M 2006. Bio-economic evaluation of farmers' perceptions of viable farms in western Kenya. Agricultural Systems 90 243–271.

Walraevens, K., Gebreyohannes Tewolde, T., Amare, K., Hussein, A., Berhane, G., Baert, R., Ronsse, S., Kebede, S., Van Hulle, L., Deckers, J. and Martens, K. 2015. Water Balance Components for Sustainability Assessment of Groundwater-Dependent Agriculture: Example of the Mendae Plain (Tigray, Ethiopia). Land Degradation & Development 26(7), 725–736.

Wang, P., D. W. Zhou, Valentine I 2006. Seed maturity and harvest time effects seed quantity and quality of Hordeum brevisubulatum. Seed Science and Technology 34(1): 125–132.

Westermann O, Thornton P, Förch W. 2015. Reaching more farmers – innovative approaches to scaling up climate smart agriculture. CCAFS Working Paper no. 135. Copenhagen, Denmark: CGIAR Research Program on Climate Change, Agriculture and Food Security (CCAFS).

Witt, C., U. Biker, C.C. Galicia, and J.C.G. Ottow. 2000. Dynamics of soil microbial biomass and nitrogen availability in a flooded rice soil amended with different C and N sources. Biol. Fertil. Soils 30(5–6): 520–527.

Yadav, B.K., and J.C. Tarafdar. 2012. Efficiency of Bacillus coagulans as P biofertilizer to mobilize native soil organic and poorly soluble phosphates and increase crop yield. Arch. Agron. Soil Sci. 58(10): 1099–1115.

Yang, C.H., Q. Chai, and G.B. Huang. 2010. Root distribution and yield responses of wheat/maize intercropping to alternate irrigation in the arid areas of northwest China. Plant, Soil Environ. 56(6): 253–262.

Yusuf, A.A., R.C. Abaidoo, E.N.O. Iwuafor, O.O. Olufajo, and N. Sanginga. 2009. Rotation effects of grain legumes and fallow on maize yield, microbial biomass and chemical properties of an Alfisol in the Nigerian savanna. Agric. Ecosyst. Environ. 129(1–3): 325–331.

Zingore, S., Murwira, H.K., Delve, R.J. and Giller, K.E. 2007. Influence of nutrient management strategies on variability of soil fertility, crop yields and nutrient balances on smallholder farms in Zimbabwe. Agriculture, Ecosystems & Environment 119(1), pp.112–126.

Zougmore, R., F.N. Kambou, K. Ouattara, and S. Guillobez. 2000. Sorghum-cowpea Intercropping: An Effective Technique Against Runoff and Soil Erosion in the Sahel (Saria, Burkina Faso). Arid Soil Res. Rehabil. 14(May 2014): 329–342.

Zougmoré R, Mando A, Stroosnijder L 2004. Effect of soil and water conservation and nutrient management on the soil-plant water balance in semi-arid Burkina Faso. Agricultural Water Management 65 103–120.

**Open Access** This chapter is distributed under the terms of the Creative Commons Attribution-NonCommercial-ShareAlike 3.0 IGO license (https://creativecommons.org/licenses/by-nc-sa/3.0/igo/), which permits any noncommercial use, duplication, adaptation, distribution, and reproduction in any medium or format, as long as you give appropriate credit to the Food and Agriculture Organization of the United Nations (FAO), provide a link to the Creative Commons license and indicate if changes were made. If you remix, transform, or build upon this book or a part thereof, you must distribute your contributions under the same license as the original. Any dispute related to the use of the works of the FAO that cannot be settled amicably shall be submitted to arbitration pursuant to the UNCITRAL rules. The use of the FAO's name for any purpose other than for attribution, and the use of the FAO's logo, shall be subject to a separate written license agreement between the FAO and the user and is not authorized as part of this CC-IGO license. Note that the link provided above includes additional terms and conditions of the license.

The images or other third party material in this chapter are included in the chapter's Creative Commons license, unless indicated otherwise in a credit line to the material. If material is not included in the chapter's Creative Commons license and your intended use is not permitted by statutory regulation or exceeds the permitted use, you will need to obtain permission directly from the copyright holder.

# Identifying Strategies to Enhance the Resilience of Smallholder Farming Systems: Evidence from Zambia

**Oscar Cacho, Adriana Paolantonio, Giacomo Branca, Romina Cavatassi, Aslihan Arslan, and Leslie Lipper**

**Abstract** To support countries implementing CSA solutions, the Economics and Policy Innovations for Climate Smart Agriculture (EPIC) group at FAO uses a methodology based on building a solid evidence base. The knowledge gained from datasets that combine household, geographical and climate data helps design policies that enhance food security and climate resilience while also taking advantage of mitigation opportunities to obtain financing. Appropriate application of CSA principles depends on specific conditions that vary between and within countries. Demographic, environmental, economic and institutional factors are all important determinants of the effectiveness of any particular policy. This chapter builds upon econometric results obtained from previous analyses by developing a conceptual model that introduces the temporal aspects of household vulnerability. The method is based on a factorial design with two vulnerability levels (high and low) and two production methods (conventional or business as usual, and improved agricultural management with high CSA potential). Farms are classified into groups based on cluster analysis of survey data from Zambia. Results provide a baseline consisting of probability distributions of yields, labor use, cash inputs and profit for each of the four combinations of vulnerability level and production system. This is useful for stochastic dominance analysis, but additional work is required to incorporate the temporal aspect of the problem. The chapter identifies data gaps and additional analyses required to capture the spatio-temporal aspects of household vulnerability and adaptive capacity.

---

O. Cacho (✉)
University of New England Business School, Armidale, Australia
e-mail: ocacho@une.edu.au

A. Paolantonio • R. Cavatassi • A. Arslan
International Fund for Agriculture Development (IFAD), Rome, Italy

G. Branca
Department of Economics, University of Tuscia, Viterbo, Italy

L. Lipper
ISPC-CGIAR, Rome, Italy

# 1 Introduction

In its most general definition, resilience is the ability of a system to react or cope with change. More specifically, the concept refers to the ability of a system to respond to shocks (temporary) or more persistent adverse trends (stressors) (Hoddinott 2014). In the context of food security, resilience means being able to achieve or maintain food security in spite of shocks or permanent stressors. This implies reducing the risk of becoming food insecure, increasing adaptive capacity to cope with risks and effectively respond to change over time.[1]

From the standpoint of CSA, of which food security is one key pillar, the importance of understanding resilience arises from the need to address the vulnerability of farm households to climate change, which is determined by a combination of adaptive capacity and exposure to shocks and slower changes (Adger et al. 2004; IPCC 2007a; OECD 2009; IPCC 2014).

A conceptual framework for thinking about resilience is illustrated in Fig. 1. Adaptive capacity is affected by both the internal state of the farm household (education, age, farm area, assets owned, land productivity) and the external state experienced at the local level (technologies available, institutions, policies, infrastructure, markets).

This is a dynamic system where the internal state changes over time depending on the outcomes of household decisions such as the crop mix, input use, production methods and off-farm activities. The outcomes are affected by climate (through yields) and markets (through prices) which are out of the control of the household. For example, a good season combined with strong markets helps build financial capital reducing vulnerability, whereas a string of poor seasons may result in loss of financial or human capital (by the selling of assets or migration of family members to the city), increasing vulnerability of the household.

Both the internal and external states can change over time depending on policies, for example education and extension improve the internal state (human capital), whereas R&D and transport infrastructure improve the external state by providing new technologies and improving access to markets. Climate change affects the internal state indirectly by changing the yield probability distributions, for example due to increasing frequency of dry spells, floods and storms. It can also affect the external state, as in the case of severe storms destroying transport and communication infrastructure.

Individual households make decisions based on the options available to them (Fig. 1), and their actions result in outcomes (i.e. profits) whose probability distribution is determined by both the internal and external state as well as by the changing climate. These influences are represented as dotted lines in Fig. 1. The dynamic aspect of the problem is represented by the solid arrow between outcomes and the

---

[1] HLPE, Climate change and food security. A report by the High Level Panel of Experts on Food Security and Nutrition of the Committee on World Food Security, (FAO, Rome, 2012). http://www.ifpri.org/sites/default/files/HLPE-Report-3-Food_security_and_climate_change-June_2012.pdf.

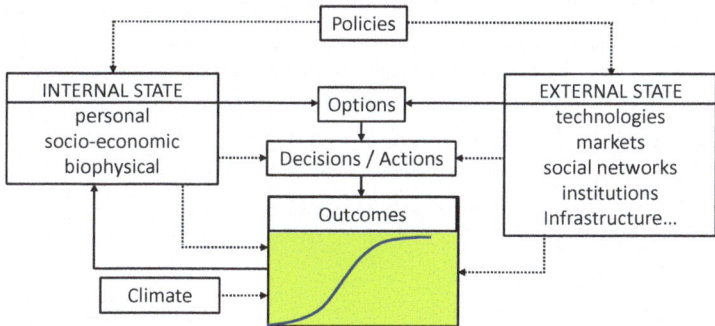

**Fig. 1** Conceptual model illustrating the key relationships of concern in this study

internal state. The outcomes at the end of each growing season will determine whether the household is able to improve its state (i.e. build human and natural capital), thus enhancing its resilience.

The empirical implementation of the model illustrated in Fig. 1 requires a number of relationships to be known for the particular situation of interest. The options available to households depend not only on the technologies that are suitable for the area, but also on their ability to access these technologies through knowledge and investment capital. This suggests that understanding constraints at the household level is a key to assessing vulnerability. A behavioral model is required to understand the decisions taken by households given the constraints they face. The standard approach is to assume utility maximization, where utility is a function of expected profits and risk (Moschini and Hennessy 2001).

The propensity of households to adopt given technology packages, and the probability distributions of outcomes, can be inferred empirically from existing data. Estimating the effects of climate shocks on the shape of these distributions is more difficult as it would require panel data for a number of years involving a range of different climatic conditions. In the absence of these, it may be possible to infer changes in outcome distributions using crop simulation models.

Many CSA practices can increase food production and the adaptive capacity of the food production system, while at the same time reducing net greenhouse gas emissions by capturing carbon in biomass and soils. However, capturing these long-term synergies may entail significant costs in the short term, and other barriers to adoption of CSA may be present, particularly for smallholders (McCarthy et al. 2011).

According to FAO (2011) the pillars of adaptation in agriculture are soil health, water conservation, diversification and local institutions. The Economics and Policy Innovations for Climate Smart Agriculture (EPIC) programe at FAO has been addressing these issues for a number of years, formally grounded on a substantial evidence base that continues to grow (Arslan et al. 2014, 2015; Asfaw et al. 2014). In this study we focus on the first two factors: soil health and water conservation, both of which are related to farming methods involving minimum soil disturbance (MSD). MSD, while contributing to soil health, increases water retention and moisture and is considered as one of the practices with potential to contribute to the CSA pillars. This chapter contributes towards building up an empirical model for the conceptual framework illustrated in Fig. 1 as a useful tool for policy analysis. This paper forms a base from which the temporal aspects of the problem can be addressed through simulation of climate scenarios in future research.

## 2 Data and Methods

The data used in this analysis come from a household survey conducted by EPIC in 2013 to support a detailed cost benefit analysis of crop practices in Zambia, with the purpose of comparing agricultural practices with CSA potential to conventional ones (see Branca et al. 2015). Given the low adoption rate of agricultural practices with CSA potential encountered in the country (Arslan et al. 2014), the need for an ad hoc study emerged to understand the performance of households who adopt the recommended practices as well as related costs and benefits.

The first step required identifying a sample that allowed such comparison, starting with defining what was "conventional" for Zambia as opposed to "alternative practices," whose CSA potential had to be assessed. Initial screening of the farming practices in use in the country was conducted through literature review, key informant interviews and qualitative analysis. The screening allowed identification of the most common farming practices defined as "conventional". Conventional practices were then contrasted with the "alternative practices" identified by compiling a list of various farming practices in different combinations with sustainable land management as a common factor (see Branca et al. 2015).

Households were randomly selected from the population of adopters of "alternative practices", maintaining representativeness of agro-ecologies in different districts, provinces and camps. Households were selected so as to cover enough agricultural camps with adopters of improved practices in a diversity of agro-ecological regions while also ensuring a balanced presence of non-adopters. The final sample included 695 rural households randomly selected within the population of adopters and non-adopters in eight districts of two agro-ecological regions (AER IIa and AER III, see Fig. 2).[2] The data collected include detailed information on household structural characteristics, farming practices adopted, quantities and costs of all inputs (including hired or family labor), yields and marketed returns, and input and output farm-gate prices. This information provides a baseline to study the adaptive capacity of different types of households based on a factorial design whereby we compare two vulnerability levels (high and low) and two production

---

[2] The sample covers the districts of Mumbwa, Chibombo, Katete, Chipata, Chinsali, Mpika, Kalomo, and Choma.

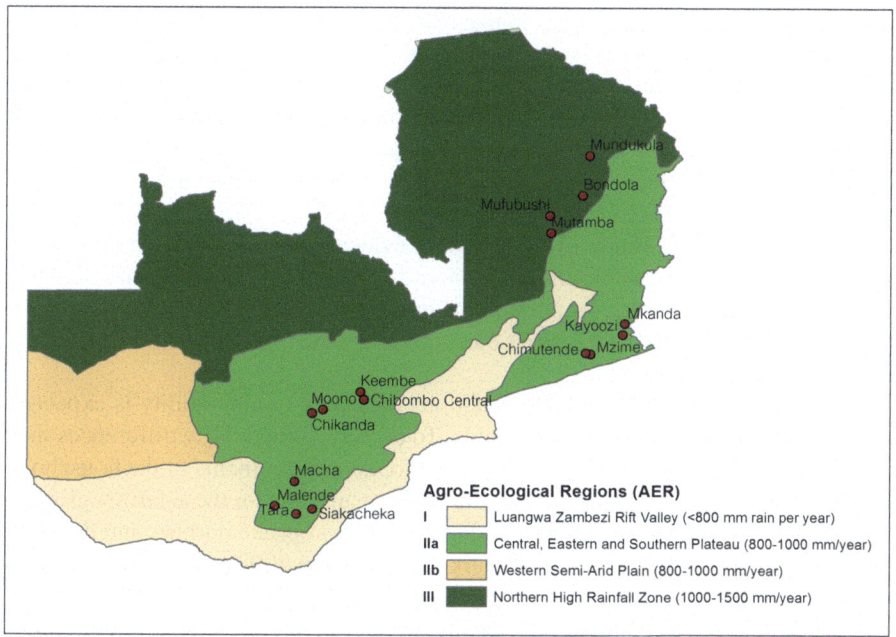

**Fig. 2** Map of study area and sample points (Source: Branca et al. 2015)

methods (conventional and MSD). Farms are classified into groups based on cluster analysis as described later.

The data suggest that a wide range of combinations of practices are being used by farm households in Zambia, and these have been grouped into two main categories based on the tillage method applied: (1) farmers that use conventional (CONV) tillage techniques (including oxen ploughing and hand hoe ploughing, ridging and bunding) as opposed to (2) farmers that adopt sustainable land management practices based on the principle of MSD and water conservation (including planting basins and potholes and ripping with oxen/tractor). Later in the analysis MSD is further split according to its emphasis on labor or capital inputs.

Previous work has shown that MSD generates higher average benefits in drier areas (Branca et al. 2013) and that adoption rates are higher in these areas, especially under high rainfall variability, both of which are conditions that characterize AER I, IIa and IIb in Zambia (Arslan et al. 2014). However, it should be noted that various SLM practices (including MSD, crop rotations with legumes, residue retention and agroforestry) have been primarily promoted in AER IIa, likely due to its proximity to the railway line and to Lusaka and other urban centers. Region IIa has received more assistance from government, NGOs and donor organizations, and is the geographic focus of outgrower schemes and conservation farming. This is also reflected in our sample as MSD fields are found only in AER IIa, which runs east-west through the center of the country on the plateau of the Central, Lusaka, and Eastern Provinces and parts of Western and Southern Provinces. The region is sometimes referred to as Zambia's maize belt, as almost half of all maize produced in the country is grown in twelve of its districts (MAL 2007). AER IIa is also recognized as a vulnerable area. About 41% of Zambian farm households live in this region and are mostly engaged in crop production.[3] The area is characterized by a semi-arid climate, where maize yields are projected to decrease significantly as a result of increased frequency of droughts and hot days and nights based on country-specific climate change models (Kanyanga et al. 2013).

Given the sampling frame of the data and evidence of expected benefits of the practices analyzed here under climate change,[4] we focus our analysis only on AER IIa. Moreover, given the key importance of maize for food and nutrition security in the country (MAL 2007), we restrict our sample to maize producers, resulting in a subset of 487 households.

The heterogeneity of the farm populations means that vulnerability is expected to differ significantly between households. To capture vulnerability differences that are relevant to policy choices, it is convenient to identify segments of the household population with common attributes, and to conduct analysis for these farmer groups. Cluster analysis provides a method to identify the appropriate number and description of farmer typologies (Acosta-Michlik and Espaldon 2008).

We conduct our analysis for two types of smallholder households that were clearly identified based on cluster analysis: (i) smaller farms with few assets (hypothesized to be more vulnerable), and (ii) larger farms with more assets (hypothesized to be less vulnerable). We first conduct analyses of means to detect differences between the probability distributions of the production methods (CONV and MSD) between these two farm types. Variables analyzed include maize yields, labor use, fertilizer use, cash inputs, profits and returns to labor.

Given the baselines obtained from the analysis of household types (low and high vulnerability) and production systems (CONV and MSD) it was clear that there are two distinct types of MSD applications in the sample: one that relies mostly on labor (using hand hoes to dig planting basins/potholes) and another that uses capital (oxen or machinery) for ripping. We denote these groups as MSD-L and MSD-K, respectively. This classification conforms with reports in the literature that find labor requirements for planting basins as one of the main constraints for the adoption of this practice in the region (Baudron et al. 2007; Mazvimavi 2011; Ngoma et al. 2014).[5] No distinction regarding emphasis on capital or labor was identified in the case of CONV, which consisted of a relatively small sample.

---

[3] The statistical surveys conducted by the Ministry of Agriculture and Livestock in collaboration with the Central Statistical Office in 2002/03 show that more than 97% of households residing in AER IIa are engaged in crop production activities.

[4] MSD is effective in keeping soil moisture, therefore it can be expected to be adopted more widely in dry areas that are projected to get even drier – as reported in Arslan et al. (2014).

[5] MSD primarily based on planting basins is the integral part of the Conservation Farming packages that have been heavily promoted in Zambia since 1990's. In recent years there is a shift towards promoting CF based on ripping, which require less labor compared to planting basins.

The analysis concludes by comparing the full probability distributions of key variables between farm clusters and production methods. The key variables are compared in terms of stochastic dominance to determine whether any one practice would be preferred to others independently of the risk aversion level of the decision maker. The chapter concludes by identifying the additional information and analyses that would be required to implement an analytical model such as illustrated in Fig. 1.

## 3 Results and Discussion

**Descriptive analysis** Analysis of unconditional means (Table 1) provides evidence that farms using MSD have significantly higher average yields than farms using conventional till (CONV) in the study area (2101 vs 1675 kg/ha). However, this is accompanied by higher labor requirements (108 vs 80 days/ha) and cash inputs (274 vs 207 $/ha). The amount of fertilizer used by farmers practicing MSD tended to be higher (211 vs 180 kg/ha) but not significantly ($p = 0.12$).

The combination of higher yields and higher input use still results in higher average gross margins under MSD ($160/ha) than under conventional till ($139/ha), but this difference is not statistically significant (Table 1). When the imputed cost of family labor is included in the calculation, profits are quite similar (50 vs 58 $/ha for MSD against CONV) (see also Branca et al. 2015). Return to labor is significantly

**Table 1** Tests of differences in means of key variables between farms using conventional till (CONV) and those using sustainable land management (MSD)

| Variable | | CONV | MSD | Total | |
|---|---|---|---|---|---|
| Number of farms | | 84 | 370 | 454 | p(|T| > |t|) |
| Maize yield** | Mean | 1674.52 | 2101.47 | 2022.47 | |
| (kg/ha) | SE | 170.49 | 82.34 | 74.47 | 0.03 |
| Labor** | Mean | 80.49 | 107.97 | 102.88 | |
| (pd/ha) | SE | 8.41 | 5.46 | 4.74 | 0.01 |
| Fertilizer | Mean | 179.81 | 211.33 | 205.50 | |
| (kg/ha) | SE | 17.39 | 8.79 | 7.86 | 0.11 |
| Cash inputs*** | Mean | 206.85 | 273.57 | 261.22 | |
| ($/ha) | SE | 15.53 | 9.47 | 8.32 | 0.00 |
| Gross margin | Mean | 139.12 | 160.49 | 156.53 | |
| ($/ha) | SE | 32.22 | 14.88 | 13.50 | 0.54 |
| Profit | Mean | 58.54 | 49.67 | 51.31 | |
| ($/ha) | SE | 32.86 | 15.10 | 13.71 | 0.80 |
| Labor productivity* | Mean | 71.63 | 40.64 | 46.37 | |
| (kg maize/pd) | SE | 34.19 | 3.34 | 6.88 | 0.08 |
| Return to labor* | Mean | 6.64 | 2.99 | 3.67 | |
| ($/pd) | SE | 4.23 | 0.48 | 0.87 | 0.10 |

Means are significantly different at $p<0.1$ (*); $p<0.05$ (**); or $p<0.01$ (***)

lower for MSD than for CONV (2.99 vs 6.64 $/pd), corresponding to lower labor productivity (40.6 vs 71.6 kg maize/pd).

Using nationally representative data from 2004 to 2008, Arslan et al. (2014) found that the adoption rate of MSD was quite low, and that it had decreased significantly between the two years. The only province with increased adoption levels was the Eastern province, which is mostly in AER IIa with a high density of projects promoting conservation farming, of which MSD is the main component. Possible reasons for low adoption in general include that farmers face labor and capital constraints, or that they do not perceive MSD to be more profitable than using tillage – at least in the short run during which there may be no significant yield difference until the soil quality is improved, which requires 3–5 years of repeated MSD practice (McCarthy et al. 2011). Although average gross margins and average profits were positive for both systems, they were quite low (Table 1), and a high proportion of farms experienced negative profits, suggesting that the opportunity cost of their labor is lower than the wage rate used in the calculations,[6] perhaps because there are no alternative employment opportunities.

**Cluster analysis** Cluster analysis revealed two distinct groups of farms as described above and illustrated in the dendogram in Fig. 3, consisting of 55 and 45 percent of the sample. There are clear differences in the mean values of variables used to form the clusters (Table 2). Although all the farms in the sample are smallholders, Cluster 1 has larger farms than Cluster 2 (with means of 4.02 ha vs 2.21 ha). Farmers in Cluster 1 tend to be better educated, have more livestock, more wealth and larger households. The difference in wealth is especially obvious, with an average wealth index[7] of 0.64 for Cluster 1 compared to −0.47 for Cluster 2. All household heads are male in Cluster 1, whereas 30 percent of them are female in Cluster 2. These results suggest that farms in Cluster 2 are potentially more vulnerable to shocks, as they have fewer assets to draw from in emergencies (particularly livestock) and have less wealth. This means they are likely to be less resilient than farms in Cluster 1.

Table 3 shows that, on average, Cluster 1 farms have higher maize yields (2172 kg/ha vs 1838 kg/ha) and higher profits (85.69 vs 8.80 $/ha) than Cluster 2 farms. In contrast, Cluster 2 farms use more labor (124 vs 86 pd/ha on average) and less cash inputs (241 vs 277 $/ha), reflecting the presence of cash constraints. This becomes more evident in the distribution analyses presented later. The large difference in profits between clusters (Table 3) reflects the higher reliance on labor experienced by Cluster 2, which combined with lower labor productivity (28.7 vs 60.7 kg maize/pd) results in lower returns to labor (2.18 vs 4.87 $/pd).

Tests of differences between CONV and MSD within each cluster (Table 4) indicate that the patterns observed above for the pooled data are also present within each of the two clusters: MSD produces higher yields on average, but it requires more

---

[6] Labor costs were estimated at the prevailing wage rate in the rural labor market in the study area using rates that differ by farm activity type collected through a Community level questionnaire.

[7] The wealth index is constructed using principal component analysis. It includes the following variables representing key assets owned by the household: number of ploughs, number of harrows, number of cultivators, number of rippers, number of tractors, number of cars, number of bikes.

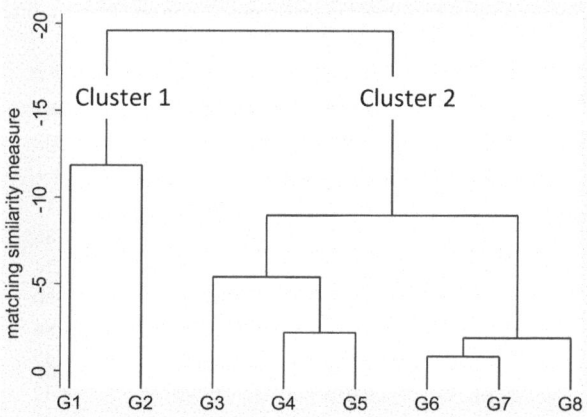

**Fig. 3** Dendogram of cluster analysis

**Table 2** Mean values and standard errors (SE) of variables used in cluster analysis and results of t test of differences between means

|  |  | Cluster 1 | Cluster 2 | Total |  |
|---|---|---|---|---|---|
| Number of farms |  | 251 | 203 | 454 | p(ITI > ItI) |
| Female head*** | Mean | 0.00 | 0.30 | 0.13 |  |
|  | SE | 0.00 | 0.03 | 0.02 | 0.00 |
| Age of head | Mean | 45.71 | 46.00 | 45.84 |  |
|  | SE | 0.78 | 0.91 | 0.59 | 0.81 |
| Average education* | Mean | 7.27 | 6.91 | 7.11 |  |
|  | SE | 0.11 | 0.16 | 0.10 | 0.06 |
| Adults per ha*** | Mean | 1.55 | 2.18 | 1.83 |  |
|  | SE | 0.08 | 0.12 | 0.07 | 0.00 |
| Dependency ratio | Mean | 1.25 | 1.23 | 1.24 |  |
|  | SE | 0.06 | 0.07 | 0.04 | 0.81 |
| Household size*** | Mean | 8.38 | 6.57 | 7.57 |  |
|  | SE | 0.20 | 0.17 | 0.14 | 0.00 |
| Farm size*** | Mean | 4.02 | 2.21 | 3.21 |  |
|  | SE | 0.20 | 0.11 | 0.13 | 0.00 |
| Cattle*** | Mean | 9.56 | 0.62 | 5.56 |  |
|  | SE | 0.87 | 0.18 | 0.53 | 0.00 |
| Goats and sheep*** | Mean | 9.90 | 3.79 | 7.17 |  |
|  | SE | 1.37 | 0.60 | 0.82 | 0.00 |
| Wealth index*** | Mean | 0.64 | −0.47 | 0.15 |  |
|  | SE | 0.07 | 0.03 | 0.05 | 0.00 |

Means are significantly different at p<0.1 (*); p<0.05 (**); or p<0.01(***)

**Table 3** Means of selected variables related to maize production and t test of differences between clusters

| Variable | | Cluster 1 | Cluster 2 | Total | p(|T| > |t|) |
|---|---|---|---|---|---|
| Practicing MSD | Mean | 0.81 | 0.82 | 0.81 | |
| | SE | 0.02 | 0.03 | 0.02 | 0.71 |
| Maize yield** | Mean | 2171.87 | 1837.75 | 2022.47 | |
| (kg/ha) | SE | 105.67 | 102.10 | 74.47 | 0.03 |
| Labor*** | Mean | 85.63 | 124.21 | 102.88 | |
| (pd/ha) | SE | 5.34 | 8.05 | 4.74 | 0.00 |
| Fertilizer | Mean | 215.43 | 193.22 | 205.50 | |
| (kg/ha) | SE | 10.04 | 12.43 | 7.86 | 0.16 |
| Cash inputs** | Mean | 277.25 | 241.41 | 261.22 | |
| ($/ha) | SE | 10.35 | 13.40 | 8.32 | 0.03 |
| Gross margin | Mean | 171.84 | 137.61 | 156.53 | |
| ($/ha) | SE | 19.73 | 17.76 | 13.50 | 0.21 |
| Profit *** | Mean | 85.69 | 8.80 | 51.31 | |
| ($/ha) | SE | 19.80 | 18.09 | 13.71 | 0.01 |
| Labor productivity ** | Mean | 60.69 | 28.68 | 46.37 | |
| (kg maize/pd) | SE | 11.95 | 4.02 | 6.88 | 0.02 |
| Return to labor | Mean | 4.87 | 2.18 | 3.67 | |
| ($/pd) | SE | 1.48 | 0.66 | 0.87 | 0.13 |

Means are significantly different at $p<0.1$ (*); $p<0.05$ (**); or $p<0.01$(***)

labor and cash inputs. As a result, MSD has lower returns to labor, with the lowest return ($2.15/pd) experienced by Cluster 2 farms.

It is difficult to draw general conclusions from the analysis of differences between means presented in Table 4. In some cases there are significant differences between clusters or between production methods, but these differences are not always consistent. This suggests that further partitioning of MSD is required as explained in the Methods section. The remaining analyses distinguish between MSD-L and MSD-K to indicate emphasis on the use of labor or capital respectively.

Table 5 presents average values for the variables of interest, partitioning the data by cluster and by production system. These results show the logic behind distinguishing between MSD practices based on their labor intensity. The average labor required by MSD-L (140 and 174 pd./ha for clusters 1 and 2 respectively) is considerably higher than that required by MSD-K (76 and 99 pd./ha). In fact, the labor used in MSD-K is comparable to that of CONV in both clusters (79 and 83 pd./ha). This indicates the extent to which the availability of capital (oxen in this case) helps overcome labor constraints of adopting MSD. As before, return to labor tends to be higher for CONV than for MSD (Table 5), with the exception of MSD-K in Cluster 2, which is higher than for CONV (2.63 vs 2.34 $/pd).

Figure 4 presents cumulative distribution functions (CDF) for yields, labor and fertilizer use. The left sections of the yield distributions for MSD are to the right of those for CONV in both clusters (Fig. 4a, b), except for the lowest-yielding farms under MSD-K in Cluster 1. The higher labor requirements of MSD identified above

Identifying Strategies to Enhance the Resilience of Smallholder Farming Systems...

**Table 4** Tests of differences in means of key variables between farms using conventional till (CONV) and those using sustainable land management (MSD)

| Variable | Cluster 1 | | Cluster 2 | | Prob > F | | |
| --- | --- | --- | --- | --- | --- | --- | --- |
| | CONV | MSD | CONV | MSD | Cluster | Method | Interaction |
| N | 48 | 203 | 36 | 167 | | | |
| Maize yield | 1925.74 | 2230.07 | 1339.55 | 1945.15 | ** | ** | |
| (kg/ha) | 227.07 | 110.42 | 262.20 | 121.74 | 0.02 | 0.02 | 0.43 |
| Labor | 78.77 | 87.25 | 82.78 | 133.15 | ** | *** | * |
| (pd/ha) | 14.22 | 6.92 | 16.42 | 7.63 | 0.04 | 0.01 | 0.08 |
| Fertilizer | 214.54 | 215.64 | 133.50 | 206.10 | ** | * | * |
| (kg/ha) | 24.06 | 11.70 | 27.78 | 12.90 | 0.03 | 0.07 | 0.08 |
| Cash inputs | 244.03 | 285.11 | 157.28 | 259.54 | *** | *** | |
| ($/ha) | 25.19 | 12.25 | 29.09 | 13.50 | 0.01 | 0.00 | 0.15 |
| Gross margin | 152.90 | 176.32 | 120.74 | 141.24 | | | |
| ($/ha) | 41.56 | 20.21 | 47.99 | 22.28 | 0.34 | 0.53 | 0.97 |
| Profit | 73.99 | 88.46 | 37.96 | 2.51 | * | | |
| ($/ha) | 41.93 | 20.39 | 48.41 | 22.48 | 0.09 | 0.77 | 0.48 |
| Labor productivity | 101.61 | 51.01 | 31.66 | 28.04 | *** | | |
| ($/pd) | 21.00 | 10.21 | 24.25 | 11.26 | 0.01 | 0.13 | 0.19 |
| Return to labor | 9.87 | 3.68 | 2.34 | 2.15 | ** | | |
| (kg maize/pd) | 2.67 | 1.30 | 3.09 | 1.43 | 0.05 | 0.16 | 0.18 |

Means are significantly different at p<0.1 (*); p<0.05 (**); or p<0.01(***)

**Table 5** Mean values of key variables related to maize production by cluster x production system

| | Cluster 1 | | | Cluster 2 | | |
| --- | --- | --- | --- | --- | --- | --- |
| | CONV | MSD-L | MSD-K | CONV | MSD-L | MSD-K |
| N | 48 | 35 | 168 | 36 | 77 | 90 |
| Maize yield | 1926 | 2188 | 2239 | 1340 | 2097 | 1815 |
| Labor | 79 | 140 | 76 | 83 | 174 | 99 |
| Fertilizer | 215 | 197 | 220 | 133 | 187 | 222 |
| Cash inputs | 244 | 250 | 292 | 157 | 233 | 282 |
| Gross margin | 153 | 204 | 171 | 121 | 194 | 96 |
| Profit | 74 | 64 | 94 | 38 | 17 | −10 |
| Labor productivity | 102 | 32 | 55 | 32 | 19 | 36 |
| Return to labor | 9.87 | 3.28 | 3.76 | 2.34 | 1.59 | 2.63 |

in terms of means are also evident when looking at the full distributions (Fig. 4c, d). These differences apply for MSD-L but not for MSD-K, which has similar distributions to CONV in both clusters.

It is interesting to note that the distributions of fertilizer use are very similar in Cluster 1 across all three production systems (Fig. 4e), but in the case of Cluster 2 the distributions for MSD are to the right of those for CONV (Fig. 4f). This is a clear indication of the constraints faced by farmers in this cluster. Many of these farmers

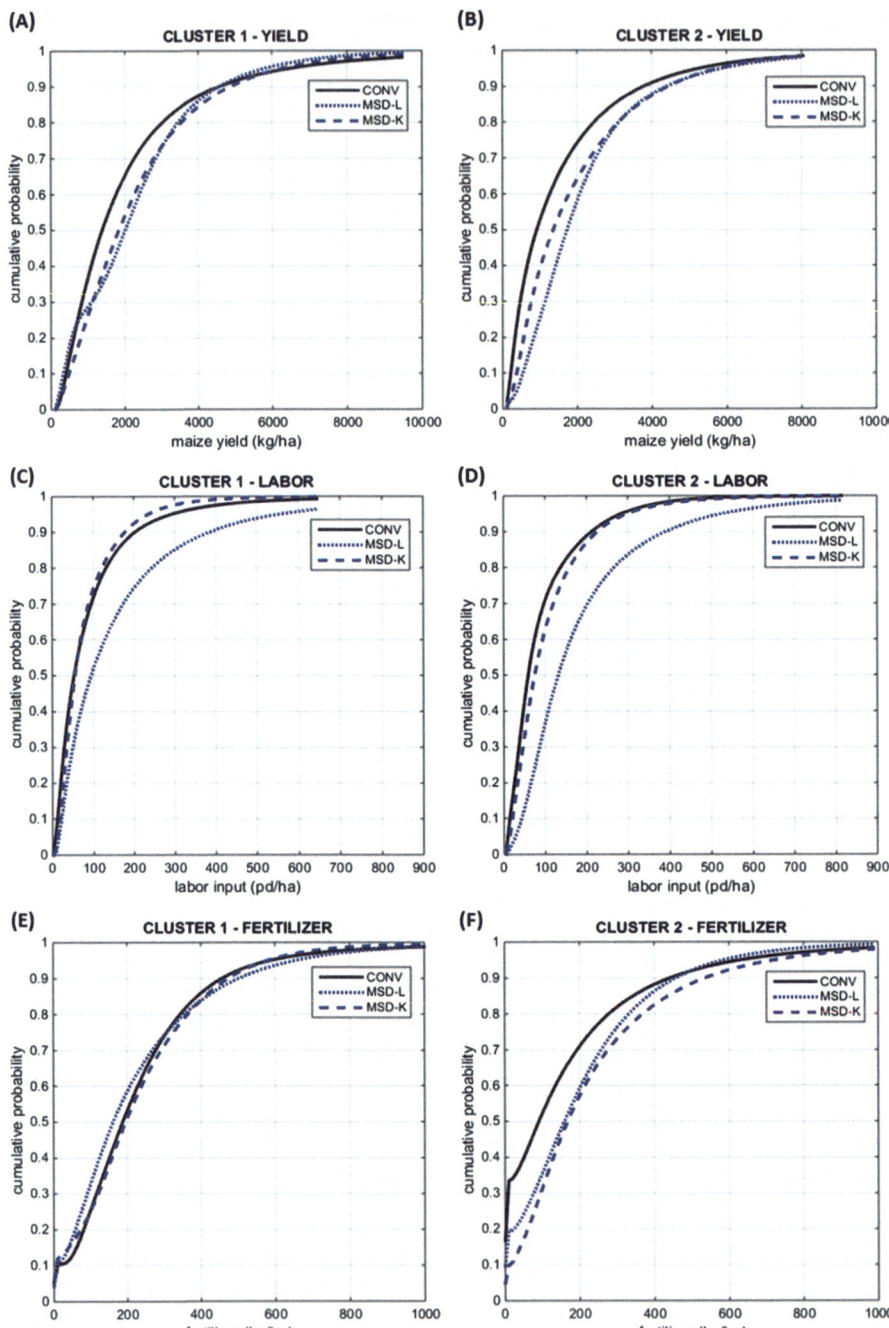

**Fig. 4** Kernel density estimates of cumulative distribution functions for maize yields (**a**, **b**), labor use (**c**, **d**) and fertilizer use (**e**, **f**) for farmers in clusters 1 or 2 and using conventional tillage (CONV) or minimum soil disturbance (MSD-L, MSD-K)

**Table 6** Probability of losses in terms of gross margins and profits by cluster and production method

|  | Cluster 1 | | | Cluster 2 | | |
| --- | --- | --- | --- | --- | --- | --- |
|  | CONV | MSD-L | MSD-K | CONV | MSD-L | MSD-K |
| N | 50 | 181 | 52 | 93 | 96 | 37 |
| P(GM < 0) | 0.32 | 0.26 | 0.32 | 0.30 | 0.20 | 0.37 |
| P(PROFIT < 0) | 0.43 | 0.43 | 0.44 | 0.47 | 0.50 | 0.60 |
| P(GM < $50) | 0.41 | 0.32 | 0.40 | 0.42 | 0.29 | 0.49 |
| P(PROFIT < $50) | 0.54 | 0.50 | 0.52 | 0.58 | 0.58 | 0.69 |

can only afford to apply fertilizer when they participate in MSD promotion programs that provide fertilizer as part of an MSD package described in this paper.

Figure 5 presents cumulative distribution functions for cash inputs, gross margins and profits. It is clear that MSD requires more cash inputs than CONV, and the differences are larger in Cluster 2 (Fig. 5b) than in Cluster 1 (Fig. 5a), once again suggesting the constraints faced by small farmers in adopting MSD. Regarding gross margins, both MSD options dominate CONV in terms of second degree stochastic dominance in the case of Cluster 1, (Fig. 5c).[8] This dominance disappears when expressed in terms of profit (Fig. 5e), which considers the cash value of family labor. In contrast, there is no clear dominance relationship in Cluster 2 in terms of either gross margins (Fig. 5d) or profits (Fig. 5f).

In general, about one-third of farms experienced a loss in terms of gross margins (Table 6), except for the case of MSD-L in Cluster 2, where only about one-fifth of farms experienced a loss. This is an interesting finding that shows that poor farms use family labor to cope with risk.

In both clusters, when the high labor requirements of MSD-L are priced at market rates to calculate profits, there is no clear preference relative to CONV on stochastic dominance grounds.

## 4 Implications and Further Work

From a policy standpoint the main issue arising from this analysis is that small, vulnerable farms are more likely to face labor and cash constraints, which may prevent them from adopting technologies that have the potential to sustainably improve food security and enhance their adaptive capacity, i.e. be climate-smart. Widespread adoption, however, will require policies that address the barriers identified here to provide: (i) improved techniques that are less labor intensive, (ii) improved availability of fertilizers, and (iii) credit to cover the up-front costs of investing in soil health that takes several years to bear fruit.

---

[8] Second degree stochastic dominance occurs when the area under the CDF for MSD is ≥ than the area under the CDF for CONV throughout the distribution (Anderson et al. 1977).

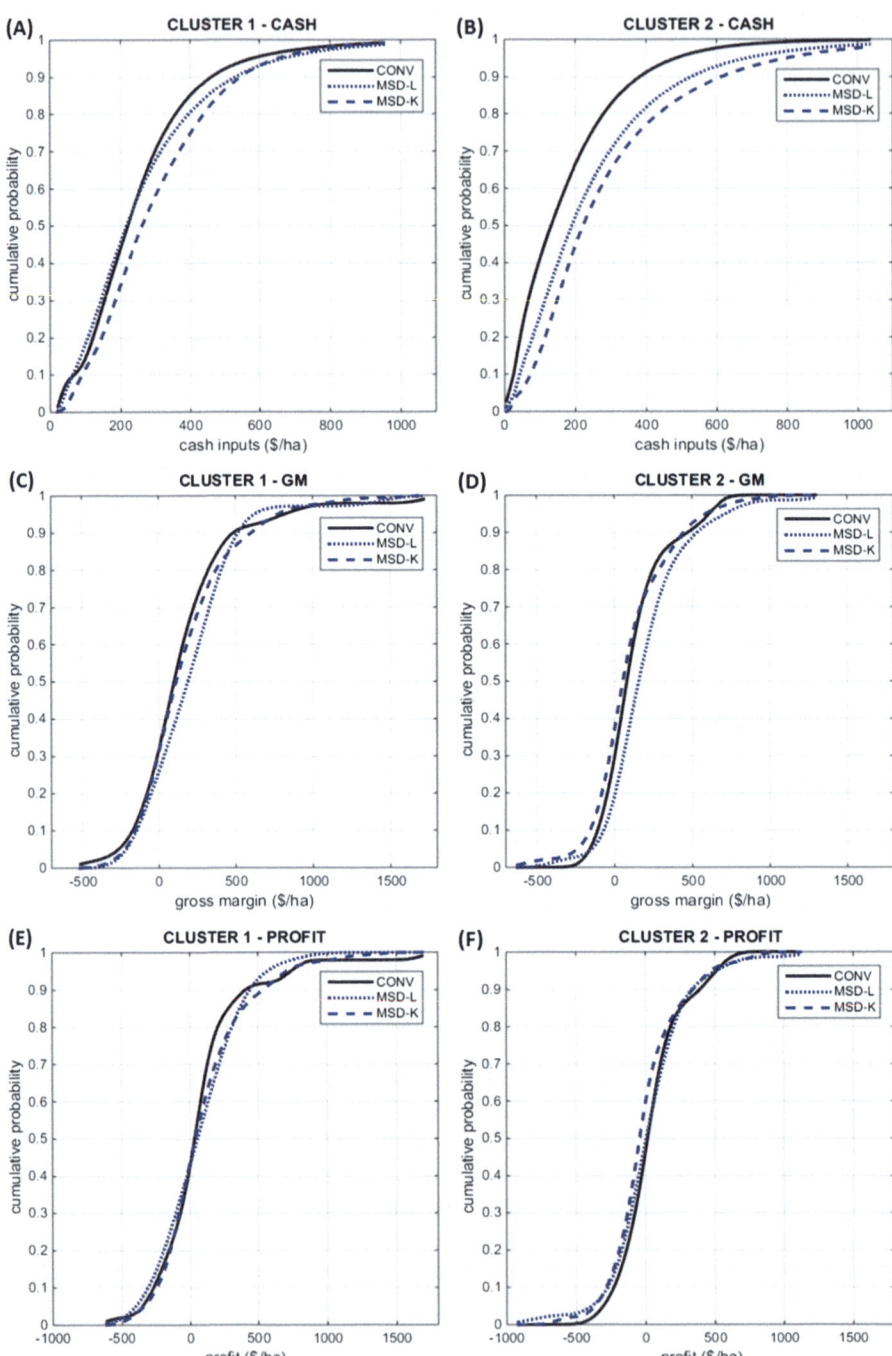

**Fig. 5** Kernel density estimates of cumulative distribution functions of cash inputs (**a**, **b**) gross margins (**c**, **d**) and profits (**e**, **f**) for farmers in clusters 1 or 2 and using conventional tillage (CONV) or minimum soil disturbance (MSD-L, MSD-K)

Some agronomists argue that switching from 'conventional' to MSD technologies increases crop yields after a few years of declining or stable yields (e.g. see Erenstein et al. 2008). Also farmers may need a few years of experience to acquire the additional knowledge and management skills necessary for more diversified operations. Most farmers adopt alternatives gradually. In the sample, an average number of 3–4 years of adoption is recorded, which is generally considered not enough for 'conservative' practices to generate the full expected benefits (Erenstein et al. 2008). Unfortunately not enough observations were available to for a disaggregated analysis by categories of number of years since adoption (e.g. up to 2 years and above 3 years).

The outcome distribution in Fig. 1 can be replaced with actual profit distributions such as Fig. 5e, f, using a different distribution for each combination of vulnerability (high or low) and production method (CONV, MSD-L or MSD-K). These distributions provide a baseline from which the dynamic aspects of the problem may be addressed.

The analyses presented in this chapter provide baselines to identify the most vulnerable farm households based on the whole distribution of the farm households in the sample. The potential contribution of MSD practices to enhanced resilience of households faced with climate change is better understood by focusing on particular segments of the farm population: the most vulnerable households. The distributions of yields and profits illustrated in this paper would shift in response to changes in climate, and the nature of these shifts may differ between CONV and MSD. The hypothesis is that more vulnerable households (Cluster 2) will have lower average yields under uncertain weather events than less vulnerable households (Cluster 1), and that MSD will lessen this negative effect during dry spells.

The expectation that MSD will show its true worth in dry years could not be tested because that source of variation is not included in the data. Studies of adaptation to climate change in Sub-Saharan Africa have found that smallholders are already using a range of strategies to deal with climate variability (Skjeflo 2013), many of them related to sustainable land management. However, evidence also shows that the key variables explaining adoption of these practices are availability of financing and risk management instruments, availability of technical information to enable the adoption process, collective action at the local level, and tenure security (McCarthy et al. 2011). Some of these constraints have been considered in this chapter by focusing on the most vulnerable households, but additional work is needed to estimate changes in the probability distributions of yields and profits caused by alternative policies in the presence of climate change.

The probability distributions derived in this study are useful for stochastic-dominance analysis but they tell only part of the story. The data are for a single cropping season and so do not cover variations in time. To get the full picture we need data on a variety of climate years, including dry and wet years. This can be obtained from panel data or from simulations using crop and livestock production models. These data are required to implement the conceptual model (Fig. 1) proposed in this chapter. Resilience is a dynamic concept implying adjustment through time as climatic, economic and social conditions change. Future empirical work on this topic should focus on introducing alternative climate scenarios and undertaking dynamic analysis by combining econometric results and crop simulation models.

# References

Acosta-Michlik L, Espaldon V. (2008). Assessing vulnerability of selected farming communities in the Philippines based on a behavioural model of agent's adaptation to global environmental change. Global Environmental Change. 18: 554–563.

Adger, W. N., N. Brooks, et al. (2004). New Indicators of Vulnerability and Adaptive Capacity. Technical Report 7. Norwich, University of East Anglia. Tyndall Centre for Climate Change Research.

Anderson, J. R., Dillon, J. L., & Hardaker, J. E. (1977). Agricultural decision analysis. The Iowa State University Press, Ames.

Arslan, A., McCarthy, N., Lipper, L., Asfaw, S. and Cattaneo, A. and Kokwe, M. (2015). Climate Smart Agriculture? Assessing the Adaptation Implications in Zambia. Journal of Agricultural Economics, 66 (3): 753–780.

Arslan, A., McCarthy, N., Lipper, L., Asfaw, S. and Cattaneo, A. (2014). Adoption and intensity of adoption of conservation farming practices in Zambia. Agriculture, Ecosystems & Environment. 187: 72–86.

Asfaw, S., McCarthy, N., Lipper, L., Arslan, A., Cattaneo, A., Kachulu, M. (2014) Climate variability, adaptation strategies and food security in Malawi. ESA Working Paper No. 14–08, FAO, Rome.

Baudron, F., H. M. Mwanza, B. Triomphe, and M. Bwalya (2007). Conservation Agriculture in Zambia: A Case Study of Southern Province. Nairobi: African Conservation Tillage Network, Centre de Coopération Internationale de Recherche Agronomique pour le Développement, and Food and Agriculture Organization of the United Nations.

Branca G, McCarthy N, Lipper L and Jolejole M.C. (2013). Food security, climate change and sustainable land management, a review. Agronomy for sustainable development. 33:635–650.

Branca, G. et al., (2015). Benefit-cost analysis of sustainable land management technologies for CSA in Zambia. Final report. FAO-CSA Project. May.

Erenstein, O., Sayre, K., Wall, P., Dixon, J., & Hellin, J. (2008). Adapting no-tillage agriculture to the conditions of smallholder maize and wheat farmers in the tropics and sub-tropics. No-till Farming Systems. 253–278.

FAO (2011). FAO-Adapt: FAO's Framework programme on Climate Change Adaptation. Rome, FAO.

Hoddinott, John F. (2014). Resilience: A primer. 2020 Conference Brief 8. May 17–19, Addis Ababa, Ethiopia. Washington, D.C.: International Food Policy Research Institute (IFPRI).

IPCC 2014: Summary for policymakers. In: Climate Change 2014: Impacts, Adaptation, and Vulnerability. Part A: Global and Sectoral Aspects. Contribution of Working Group II to the Fifth Assessment Report of the Intergovernmental Panel on Climate Change [Field, C.B., V.R. Barros, D.J. Dokken, K.J. Mach, M.D. Mastrandrea, T.E. Bilir, M. Chatterjee, K.L. Ebi, Y.O. Estrada, R.C. Genova, B. Girma, E.S. Kissel, A.N. Levy, S. MacCracken, P.R. Mastrandrea, and L.L. White (eds.)]. Cambridge University Press, Cambridge, United Kingdom and New York, NY, USA, pp. 1–32.

IPCC (2007a). Contribution of Working Group II to the Fourth Assessment Report of the Intergovernmental Panel on Climate Change Climate Change 2007, Fourth Assessment Report. M. L. Parry, O. F. Canziani, J. P. Palutikof, P. J. van der Linden and C. E. Hanson. Cambridge, United Kingdom and New York , NY, USA, Cambridge University Press.

Kanyanga, J., Thomas, T. S., Hachigonta, S. and Sibanda, L.M. (2013). Zambia. in S. Hachigonta, G. C. Nelson, T. S. Thomas and L. M. Sibanda (eds.), Southern African Agriculture and Climate Change (Washington, D.C.: International Food Policy Research Institute 2013, pp. 225–289).

MAL. 2007. Investment Opportunities in Agriculture, Government of the Republic of Zambia.

Mazvimavi, K. (2011). Socio-economic Analysis of Conservation Agriculture in Southern Africa. FAO Network Paper No. 2. Rome, Italy: Food and Agriculture Organization of the United Nations.

McCarthy, N., Lipper, L., & Branca, G. (2011). Climate-Smart Agriculture: Smallholder Adoption and Implications for Climate Change Adaptation and Mitigation. Mitigation of Climate Change in Agriculture Series (Vol. 4).

Moschini, G. and Hennessy, D.A. (2001). Uncertainty, risk aversion, and risk management for agricultural producers, Handbook of agricultural economics 1: 88–153.

Ngoma, H., Mulenga B. P., Jayne T. S. (2014). What Explains Minimal Usage of Minimum Tillage Practices in Zambia? Evidence from District-representative Data. No 165886, Food Security Collaborative Working Papers, Michigan State University, Department of Agricultural, Food, and Resource Economics.

OECD (2009). Integrating Climate Change Adaptation into Development Co-operation: Policy Guidance, OECD.

Skjeflo, S. (2013). Measuring household vulnerability to climate change—Why markets matter. Global Environmental Change 23(6): 1694–1701.

**Open Access** This chapter is distributed under the terms of the Creative Commons Attribution-NonCommercial-ShareAlike 3.0 IGO license (https://creativecommons.org/licenses/by-nc-sa/3.0/igo/), which permits any noncommercial use, duplication, adaptation, distribution, and reproduction in any medium or format, as long as you give appropriate credit to the Food and Agriculture Organization of the United Nations (FAO), provide a link to the Creative Commons license and indicate if changes were made. If you remix, transform, or build upon this book or a part thereof, you must distribute your contributions under the same license as the original. Any dispute related to the use of the works of the FAO that cannot be settled amicably shall be submitted to arbitration pursuant to the UNCITRAL rules. The use of the FAO's name for any purpose other than for attribution, and the use of the FAO's logo, shall be subject to a separate written license agreement between the FAO and the user and is not authorized as part of this CC-IGO license. Note that the link provided above includes additional terms and conditions of the license.

The images or other third party material in this chapter are included in the chapter's Creative Commons license, unless indicated otherwise in a credit line to the material. If material is not included in the chapter's Creative Commons license and your intended use is not permitted by statutory regulation or exceeds the permitted use, you will need to obtain permission directly from the copyright holder.

# Part V
# Case Studies: Farm Level Response to Improving Adaptation and Adaptive Capacity

# Climate Risk Management through Sustainable Land and Water Management in Sub-Saharan Africa

**Ephraim Nkonya, Jawoo Koo, Edward Kato, and Timothy Johnson**

**Abstract** Weather volatility is increasing, hence the need to build resilience for farmers and the poor, who are affected the most. Using Mali and Nigeria as case study countries, this study shows that climate change may reduce the yield of staple food crops – namely maize, rice, and millet – by 20% in 2050 compared to their levels in 2000. Sustainable land and water management (SLWM) – which includes a combination of organic soil fertility, inorganic fertilizer, and water managements – will more than offset the effect of climate change on yield under the current management practices. Additionally, SLWM is more profitable and could therefore increase household income and address poverty.

Unfortunately, adoption rates of SLWM remain low. Policies and strategies for increasing their adoption includes improvement of market access, enhancing the capacity of agricultural extension service providers to provide advisory services on SLWM, and building an effective carbon market that involves both domestic and international buyers. The recent United Nations Framework Convention on Climate Change (UNFCCC) provides one of the opportunities for reducing climate risks and achieving sustainable agricultural production under climate change.

## 1 Introduction

Building smallholder farmer resilience in sub-Saharan Africa (SSA) is increasingly becoming an important policy agenda due to an increase in frequency and magnitude of shocks and stresses resulting from significant changes in biophysical and socio-economic factors. Food and energy price volatility, economic recession, climate change, and land degradation are the recent major changes that have increased smallholder farmer vulnerability to shocks and stresses (Torero 2015; Nazlioglu and Soytas 2012; Barrett and Constas 2014; Nkonya et al. 2016a). The global food price index increased dramatically in 2007/08 and 2011/12 and have remained relatively higher than the long-term average (Torero 2015). Rainfall variability in SSA is high

E. Nkonya (✉) • J. Koo • E. Kato • T. Johnson
Environment and Production Technology, IFPRI, Washington, DC, USA
e-mail: e.nkonya@cgiar.org

and frequency of hydrological shocks is increasing (Zseleczky and Yosef 2014). The impacts of these shocks on food security and welfare of smallholder farmers in general are enormous. Climate change is predicted to decrease production of major crops in SSA significantly. Maize production – the region's most important crop that account for 13% of cropland area (FAO 2012) – is estimated to decrease by 22% by 2050 – the largest impact among the major crops in SSA (Schlenker and Lobell 2010). Similarly, production of sorghum and millet are each estimated to decrease by 17% (Ibid). IPCC (2007) estimates a 50% reduction in rainfed crop yield due to climate change.

In the last decade, SSA experienced the worst land degradation in the world, accounting for 22% of the total global annual cost of land degradation of about US$300 billion (Nkonya et al. 2016b). In addition to reducing agricultural productivity, land degradation increases production risks – especially for smallholder farmers who do not use greater inputs to mask negative impacts of land degradation (Moussa et al. 2016; Nkonya et al. 2015a).

SSA countries have designed a number of policies and strategies for adaptation to climate change and to address other shocks and stressors. All 51 countries in SSA have ratified the UNFCCC and two thirds have submitted their national adaptation program of action (NAPA) (UNFCCC 2014a). In terms of mitigation, 22 SSA countries have submitted the Nationally Appropriate Mitigation Actions (NAMA) to the UNFCCC (UNFCCC 2014a, 2014b). The NAMAs are voluntary mitigation strategies designed by developing countries. They include technology, financing, and capacity-building that lead to mitigation of greenhouse gas emissions (GHG). In addition to the NAMAs, parties to the UNFCCC were asked to submit country level strategies for reduction of GHG to the 21st Conference of Parties (COP21) in Paris (Höhne et al. 2014). The COP21 GHG emission reduction strategies are known as intended nationally determined contributions (INDC). By December 2015, a total of 47 SSA countries had submitted their INDC (UNFCC 2015a, 2015b).[1] All NAPAs and NAMAs/INDC mention generic land improvement action plans.

In order to design cost-effective and appropriate adaptation and mitigation strategies, policy makers and development partners need empirical evidence of effectiveness of policies and strategies for building resilience and adaptation to climate change. Accordingly, this study addresses the following major research questions:

(i) What are the impacts of climate change on production of staple foods in SSA?
(ii) What are the SLWM practices that could be used to adapt to climate change?
(iii) What is the impact of SLWM practices on production risks in SSA?
(iv) What are the drivers of adoption of SLWM practices?
(v) What are the policy implications for enhancing adaptation to climate change using SLWM practices?

In this study, we define SLWM practices as the use of soils, water, animals, and plants, for the production of ecosystem services in a manner that maintains their long-term productive potential and ecosystem functions (Liniger and Critchley

---
[1] Exceptions are Cote d'Ivoire, Mayotte, Cape Verde, & Reunion.

2007). Given that this definition involves complex processes, we will refer to a management practice as an SLWM when it is better than the common land degrading management practices – which largely includes no external or other organic soil fertility management (OSFM) practices that enhance soil fertility. Our SLWM practice will focus on integrated soil fertility management (ISFM) practice and irrigation. ISFM is a management practice in which appropriate germplasm is used together with judicious amounts of inorganic fertilizer and organic inputs as well as good agronomic practices (Vanlauwe et al. 2015). In addition to increasing soil carbon and thus contributing to mitigation of climate change (Vanlauwe et al. 2015), ISFM and other SLWM reduce downward production risks and increase food security (Kassie et al. 2015).

Using Mali and Nigeria as case study countries, this chapter examines the impacts of climate change on maize, rice, and millet production and risks. Selection of the countries was driven by data availability and their biophysical and socio-economic characteristics. Mali and Nigeria represent a large share of drylands – which are most affected by climate change (Christensen et al. 2007). Nigeria and parts of Kayes and Sikasso regions in Southern Mali are also in sub-humid and humid agroecological zones (Fig. 5). This further enhances the two countries' representativeness of agroclimatic characteristics in SSA.

The section below sets the context of the chapter by discussing the background of the case study countries. The discussion explores the biophysical and socio-economic characteristics of the case study countries relevant to climate change.

## 2 Background of the Case Study Countries

We explore the general socio-economic and biophysical characteristics of the case study countries and compare them with SSA. To put into context the climate risk management, we also discuss risk management and climate change policies.

### 2.1 Socio-Economic and Biophysical Characteristics of Mali and Nigeria

With more than 50% of the population in Mali and Nigeria living below the international poverty line, the 2015 United Nations human development report puts both countries in the low human development group (Table 1). Mali and Nigeria are respectively 179th and 152th countries in the human development index (HDI) ranking of 188 countries (Table 1).[2] Mali's economy is heavily dependent on agriculture as the sector accounts for 42% of the GDP and 75% of the economically

---

[2] HDI is an index of life expectancy, education, and per capita gross income. HDI ranges from 1 to 0. The higher the HDI the higher the human development.

**Table 1** Human development status in the case study countries

| Development indicator | Mali | Nigeria | SSA |
|---|---|---|---|
| HDI 2014 | 0.42 | 0.51 | 0.52 |
| HDI rank | 179 | 152 | |
| Gross National income per capita (US$) | | | |
| Men | 2.195 | 6.585 | 4.148 |
| Women | 961 | 4.052 | 2.626 |
| Percent of population living below | | | |
| National poverty line | 44 | 46 | |
| International poverty line (PPP US$1.25 per day) | 51 | 62 | |
| Agricultural value added as % of GDP | 42 | 20 | 14 |
| Agricultural share (%) of economically active population[a] | 75 | 54 | 58 |
| Agricultural expenditure as % of total public expenditure | 13.4 | 5.2 | 5 |
| Area equipped for irrigation as share of total irrigation potential[b] | 42 | 13 | |
| Ratified UNFCC? | Yes | Yes | |
| Year submitted NAPA/INDC | 2007 | 2015 | |
| Submitted NAMA? | No | No | |
| Savings in a formal financial institution (% of population 15 years or older) | 5 | 24 | 12 |

[a]For Nigeria, (NBS 2012)
[b]AQUASTAT raw data (Available at http://www.fao.org/nr/water/aquastat/main/index.stm)
Sources: Agriculture value as percent of GDP – World Bank (2015); Rest of the data – UNDP (2015)

active population is employed in agriculture (Table 1). For Nigeria, 54% of the 54 million economically active population is employed in agriculture – a sector that contributes 31% of the country's GDP (NBS 2012).[3]

In terms of investment in land-based sectors in general, Mali allocates over 13% of its public expenditure budget to agriculture (Benin and Yu 2012) – which is more than twice the SSA regional average and larger than the Maputo Declaration target of allocating 10% of public expenditure to agriculture (AU 2014). Nigeria's public expenditure budget allocation is about the regional average of 5% and half of the Maputo Declaration target (Ibid).

As stated above, Mali and Nigeria represent well SSA's agroecological zones. The drylands areas in both countries represent a large share of SSA as 54% of SSA land area is in the arid and semi-arid zone (Jahnke 1982) – which is home to 268 million people, 75% of which live in rural areas and are heavily dependent on agriculture (Fabricius et al. 2008). About 51% of Mali's land area is in the hyper-arid zone (Sahara desert) while 23% and 18% is in the Sahelian and Sudan-Guinean zones respectively (RDM 2007). The share of population residing in the Sahelian and Sudan-Guinean zones are respectively 27% and 68% (INS 2009). In Nigeria the

---

[3] The oil sector accounts for 41% of the GDP. The agricultural sector includes crops, livestock, fish, and forestry (NBS 2012).

Sudan Sahelian area – covering the Northeast and Northwest geopolitical zones accounts for 51% of the total area and is home to a third of the country's population (NBS 2012). The humid and subhumid areas in represent about 57% of land area in SSA (Dixon et al. 2001), which is home to over 61% of SSA population (Fabricius et al. 2008).

## 2.2 Risk Management Policies and Irrigation Development

Savings are one of the key strategies for risk management (World Bank 2014). In developing countries, livestock serve as savings and insurance against risks. Only 5% and 24% of the population above 15 years old in Mali and Nigeria, respectively, has savings in a formal banking institution (World Bank 2014). The SSA regional average is 12% indicating that Mali is below and Nigeria is above the regional average. Nigeria represents regional average human development and above average risk management while Mali is below average for both indicators. Livestock accounts for more than 50% of capital held by SSA rural households (Kamuanga et al. 2008). However, the livestock sector's contribution to income is low because of its low productivity (Nkonya et al. 2016b). Accordingly, livestock contributes respectively 15% and 3.3% of Mali and Nigeria GDP (FAO 2005a, 2005b). In both countries, over 90% of the rural households own livestock – suggesting that smallholder farmers use the traditional savings and insurance mechanisms more than the formal instruments. Unfortunately, government investments in enhancing the livestock sector are quite low: the sector receives less than 5% of the public expenditure budget in SSA (Nkonya et al. 2016b).

On climate change adaptation policies, Nigeria submitted its INDC prior to the Paris COP21 in which one of its strategies for adaptation to climate change include climate smart agriculture and reforestation. The country has not yet delivered NAPA or NAMA – suggesting a weak political will to invest in adaptation to climate change. Mali has submitted its NAMA in which SLWM practices are among the adaptation strategies (RDM 2007). However, Mali has has also submitted its INDC with an agriculture-related commitment to increase rice irrigation efficiency to reduce water loss. The INDC also aims at protection of forests and reforestation to enhance carbon mitigation (Ibid).

Irrigation development is an important strategy for climate change adaptation and for enhancing food security in SSA (Burney et al. 2013). This is especially important in the drylands which will be most affected by climate change. Nelson et al. (2009) estimate that about 24% of the US$3 billion annual investment expenditure (as of 2000) required to offset the effect of climate change on nutrition in SSA will be for irrigation development (Fig. 1).

Mali has significantly invested in irrigation as 42% of its irrigation potential is equipped for irrigation (FAO 2005c). However, the country remains highly vulnerable due to the large area being in the drylands and large share of population dependent on agriculture. Only 13% of irrigable area in Nigeria is equipped for irrigation – a

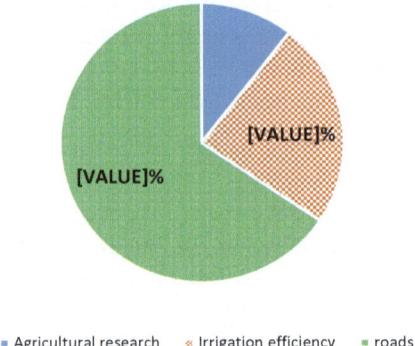

**Fig. 1** Contribution of irrigation, roads, and R&D to total additional annual investment (2000 US$3 billion) required to offset the effects of climate change on nutrition in SSA (Note: Nelson et al. (2009) separate irrigation and road investments into supporting area expansion and yield increase (for roads) and enhancement of water use efficiency (for irrigation). The Percentages reported for irrigation & roads are derived from a sum of the two groups (Source: Extracted from Nelson et al. 2009)

level that puts Nigeria among 24 SSA countries with less than 50% of irrigation potential equipped for irrigation (FAO 2005c). However, Nigeria has invested significantly to support irrigation development in the semi-arid areas as 68% of the irrigated area in Nigeria is located in the semi-arid northern zone (FAO 2005c).

Below, we discuss the methods and data used in this study, in which we show the temporal and spatial scale of the analysis of impact of climate change on food security. In order to draw relevant policy implications and strategies required to enhance adaptation to climate change, we discuss the analytical approaches of the drivers of adoption of SLWM practices and their impacts on climate-related risks.

## 3 Methods and Data

### 3.1 Impact of Climate Change on Food Security

We estimate the impact of climate change on crop productivity in the year 2050 using climate simulation models with different assumptions that lead to optimistic and pessimistic predictions. The National Center for Atmospheric Research (NCAR) predicts greater precipitation (10% increase), while the Commonwealth Scientific and Industrial Research Organization (CSIRO) model predicts a drier climate (2% decrease in 2050) (Nelson et al. 2009).

Additionally, we use a crop simulation model to estimate the impact of SLWM practices on climate-related production risks with and without climate change from the year 2000 to 2050. We also use the same model to estimate carbon sequestration since soil carbon is one of the most important elements determining adaptation and mitigation to climate change (Lal 2004, 2011). We use the DSSAT (Decision Support System for Agrotechnology Transfer) Cropping System Model v4.5 (Hoogenboom et al. 2010; Jones et al. 2003), which combines crop, soil, and weather databases for access by a suite of crop models embodied in one system. The models integrate the effects of crop system components and management options to simulate the states of all the components of the cropping system and the interaction between them. DSSAT crop models are designed on the basis of a systems approach, which provides a framework for users to understand how the overall cropping system and its components function throughout cropping seasons, on a daily basis. The DSSAT model has been widely used in various types of cropping systems all over the world, including low-input subsistence ones in SSA. The model was modified by incorporating a soil organic matter and residue module from the CENTURY model and this combined model, DSSAT-CENTURY, was used in this study, as it was designed to be more suitable for simulating low-input cropping systems and conducting long-term sustainability analyses in SSA (Gijsman et al. 2002).

## 3.2 Drivers of Adoption of SLWM Technologies and their Impact of Climate-Related Production Risks

We estimate the drivers of adoption of SLWM using a Probit model shown below:

$$Y^* = \Phi - 1(Y) = X\beta + \varepsilon,$$

Where $Y^*$ is a latent variable, given by:

$$Y = \begin{cases} 0 \text{ if } Y^* \leq 0 \\ 1 \text{ if } Y^* \geq 1 \end{cases},$$

$\Phi$ is a normally distributed cumulative static with Z-distribution, i.e. $\Phi(Z)\epsilon(0,1)$, X is a vector of covariates of determinants of adoption of SLWM practices and $\beta$ is a vector of the associated coefficients. $X\beta \sim N(0,1)$; $\varepsilon$ is an error term with normal distribution, i.e., $\varepsilon \sim N(0,1)$.

Choice of the elements of the X vector in the empirical model is guided by literature[4] and data availability. Given that some drivers of adoption of SLWM are potentially endogenous, we estimate a reduced form model to determine the robustness of the coefficients. The coefficients reported in the results section show that they were generally robust to statistical errors.

Impacts of SLWM on production risks is estimated using Just-Pope mean-variance model (Just and Pope 1979) – a model that estimates deviation from conditional mean crop yield:

---

[4] Please see Nkonya et al. (2008) and Di Falco (2014) for a review.

$$Y = f(X,C) = p(X,C) + \sqrt{\varphi(X,C)}\, e(\xi)$$

Where Y = yield which is affected by a deterministic production function $P(\bullet)$ and stochastic risk function $\varphi(\bullet)$ with an error term ($e(\xi)$) determined by rainfall and other production risks.

C and X are respectively covariates of land management practices and other covariates, which simultaneously affect $P(\bullet)$ and $\varphi(\bullet)$.

$\dfrac{\partial \operatorname{var}(Y)}{\partial C} > 0$ → Risk-increasing land management practice,

$\dfrac{\partial \operatorname{var}(Y)}{\partial C} < 0$ → Risk-reducing land management practice.

## 3.3 Data

Plot and household level survey data were used from both countries to determine farmers' land management practices and yield. For Mali, the 2004/05 agricultural household survey data were used. The data were nationally representative and included 10,000 households. The agricultural household survey data from Nigeria were collected by IFPRI for impact assessment of a large agricultural project that covered the entire country. A total of 9176 households from all 37 states were surveyed. The 37 states formed the strata and the data were representative at state level. Unfortunately, the data collected in Mali and Nigeria were not the same and the covariates included in each country differ slightly but largely remain comparable on a broader scale.

We use three staple crops – namely maize, rice, and millet, which account for the largest caloric requirements in both countries. The three crops are staple crops in both countries and in total account for 45% and 27% of the harvested area in Mali and Nigeria respectively (FAOSTAT 2013). However, rice consumption in both countries is rising and for the case of Nigeria, the country is the second largest rice importer in the world (after China) (Johnson et al. 2013). Nigeria rice import is worth about US$2 billion per year (Ibid). Through its agriculture transformation agenda policy, the country has embarked on achieving rice self-sufficiency by 2015 (Ibid) – a target that was not achieved.

The major soil fertility management practice scenarios simulated for maize, rice, and millet are given in Table 2. Irrigation is not shown since it is only used for rice and no scenario for rainfed rice is simulated. In all simulations, we assumed no carbon fertilization, since maize and millet are C4 species, which are not significantly affected by carbon fertilization (Leakey 2009). However, carbon fertilization

**Table 2** Soil fertility management scenarios used for crop simulation

| Treatment code | Description of treatment (scenario) | Relevance |
|---|---|---|
| TR0 | Normal practices, all zero inputs, no crop residues left on farm after harvest | Farmer practice as majority of farmers in both countries don't apply any inputs |
| TR1 | 100% Crop residue left on farm after harvest | Farmer practice |
| TR2 | Manure 5 tons/ha +100% Crop residue left on farm after harvest | First level of improved farmer practice |
| TR3 | 40kgN/ha + 1.67 t/ha Manure + 50% Crop residue left on farm after harvest – most likely practice that farmers are likely to afford | About half the recommended application rate for maize and rice |
| TR4 | 80kgN/ha + 100% Crop residue left farm after harvest | Represents government policies that provide fertilizer subsidy |
| TR5 | 80kgN/ha + 5 t/ha Manure +100% Crop residue left on farm after harvest – recommended practices for maize and rice | Recommended soil fertility management practice – Aduayi et al. (2002) |

Source: Authors' review

is likely to increase yield for rice (C3) and this means our estimates for rice under climate change may be underestimated.

### 3.3.1 DSSAT Model Calibration

Calibration of the DSSAT model was achieved through a process of parameter adjustment in the DSSAT default settings so that the final simulations were as close as reasonably possible to data that were reported in the literature as representing farmers' fields. Data for calibration of the DSSAT model were obtained from agricultural research institutes in Mali and Nigeria that focused on soil fertility management practices.

The weather data solar radiation, minimum and maximum temperatures, and rainfall were generated using stochastic functions based on historical weather data obtained from WorldClim http://worldclim.org; Hijmans et al. 2005). For the base climate scenario, the WorldClim current conditions data set, which are an average of 1950 to 2000, and which reports monthly average minimum and maximum temperatures and monthly average precipitation, are used. Precipitation rates and solar radiation data were obtained from NASA's LDAS website (http://ldas.gsfc.nasa.gov). The future rainfall data (2000 to 2050) were obtained from CSIRO (Commonwealth Scientific and Industrial Research Organization) and NCAR. All average climate variables were generated at a 10 km × 10 km grid scale. In order to decrease the simulation workload, only projections under the IPCC (Intergovernmental Panel on Climate Change) scenarios a2 and 2050s (corresponding $CO_2$ concentration of 599 ppm) are used.

Soil profile data were obtained from the FAO harmonized soil profile database.[5] Topographic data were obtained from the HydroSHEDS database – a global topographic database derived from NASA's SRTM (Shuttle Radar Topography Mission) data and contains 90 m hydrologically conditioned digital elevation model (DEM) data.

On water management, farmer management practices are reflected by using rainfed scenarios for maize and millet and irrigation for rice. In Nigeria, 52% of rice production is under lowland flood irrigation and 16% under fully equipped irrigation (Johnson et al. 2013). In Mali, 50% of rice production is under equipped irrigation (Ministère de l'Agriculture (2009) and about 68% of farmers use some form of irrigation for rice production (Dillon 2008). In both countries, maize and millet are almost entirely rainfed.

## 4 Results

### 4.1 Impact of Climate Change on Crop Yield and Food Security Implications

In both countries, maize and rice yields are significantly reduced by climate change. Table 3 shows that between 2000 and 2050, yields of maize and rice are expected to decrease by 3% to 39% depending on the climate change scenario used. Yield of all three staple crops would decrease under both the NCAR and CSIRO models. As expected, yield reduction under CSIRO is greater than is the case under NCAR. Decrease of millet is the lowest – underscoring its resilience in the drylands. The maize and rice yields in both countries have a greater decrease for treatments receiving inorganic fertilizer than those which do not receive the treatment (Tables 3 and 4). This could be due to the higher variability of high input production systems under climate stress. Rainfed millet yield will decrease the least due to its resilience to dry conditions.

The results show an average decrease of about 21% of staple food production – suggesting a reduction of household food security. This is especially high under farmer management practices, which are already lower and will decrease further even without climate change. Additionally, the results show different crop response to climate change and the need to emphasize crop diversification among farmers as one of the strategies for climate risk management.

---

[5] http://www.fao.org/soils-portal/soil-survey/soil-maps-and-databases/harmonized-world-soil-database-v12/en/.

**Table 3** Maize, rice, and millet yield in 2050 under different climate change scenarios, Mali

|  | TR0 | TR1 | TR2 | TR3 | TR4 | TR5 |
|---|---|---|---|---|---|---|
| Yield (tons/ha) | | | | | | |
| **No climate change** | | | | | | |
| Maize[a] | 0.4 | 0.5 | 0.7 | 3.4 | 1.7 | 3.6 |
| Rice[b] | 0.6 | 1.6 | 1.3 | 6.4 | 3.5 | 6.8 |
| Millet[c] | 0.4 | 0.4 | 0.3 | 0.4 | 0.4 | 0.4 |
| **Climate change: NCAR** | | | | | | |
| Maize[a] | 0.34 | 0.48 | 0.63 | 2.72 | 1.43 | 2.87 |
| Rice[b] | 0.45 | 1.19 | 0.91 | 4.65 | 1.77 | 5.01 |
| Millet[c] | 0.39 | 0.38 | 0.32 | 0.37 | 0.32 | 0.39 |
| **Climate change: CSIRO** | | | | | | |
| Maize[a] | 0.38 | 0.50 | 0.69 | 3.03 | 1.58 | 3.25 |
| Rice[b] | 0.67 | 1.48 | 1.07 | 4.93 | 2.04 | 5.20 |
| Millet[c] | 0.40 | 0.39 | 0.34 | 0.42 | 0.40 | 0.50 |
| **Impact of climate change on yield (Percent change)[d]** | | | | | | |
| Maize[a] | −13.3 | −2.7 | −12.0 | −16.3 | −19.3 | −19.1 |
| Rice[b] | −20.2 | −31.0 | −33.9 | −37.3 | −35.6 | −34.0 |
| Millet[c] | 6.3 | 2.1 | −0.8 | 5.7 | −3.6 | −1.9 |

Note: see Table 2 for definition of Treatments TR0-TR5
Sites: [a] Sikasso,
[b] Segou
[c] Cinzana
[d] $\frac{No\_CC - \overline{CC}}{No\_CC} * 100$ where $No\_CC$ = No climate change, $\overline{CC}$ = average yield for NCAR & CSIRO models

## 4.2 How Much Does SLWM Help Reduce Impact of Climate Change on Crop Yield?

We compared the yield of crops with and without SLWM under climate change to determine the level at which SLWM could help reduce the impact of climate change. The impact of SLWM practices on climate adaptation strategies offers some insights on the options that farmers could use to adapt to climate change. For brevity, we only compared TR3, i.e., 40kgN/ha, 1.67 t/ha Manure and 50% crop residue (TR3) – which could be regarded as an ISFM practice since 40kgN/ha is half of the recommended amount of 80kgN/ha (Table 2). We compare T3 with the average yield of farmer practice (TR0 & TR1). Figure 2 shows that SLWM practices are predicted more than double the yield of maize and rice under farmer practice in both countries. This means SLWM could not only offset the negative impact of climate change but could increase yield under farmer practice. The results underscore the importance of promoting SLWM practices as a strategy for addressing climate change.

**Table 4** Maize, rice, and millet yield in 2050 under different climate change scenarios, Nigeria

|  | TR0 | TR1 | TR2 | TR3 | TR4 | TR5 |
|---|---|---|---|---|---|---|
| **Maize** | | | | | | |
| No CC | 1.03 | 1.32 | 1.58 | 4.32 | 3.26 | 4.33 |
| NCAR | 1.0 | 1.2 | 1.4 | 2.6 | 3.4 | 3.4 |
| CSIRO | 0.9 | 1.1 | 1.6 | 2.3 | 3.1 | 3.1 |
| Average | 0.9 | 1.1 | 1.5 | 2.5 | 3.3 | 3.2 |
| **Rice** | | | | | | |
| No CC | 1.12 | 2.79 | 1.98 | 9.49 | 4.33 | 10.24 |
| NCAR | 1.0 | 2.3 | 1.4 | 3.5 | 7.8 | 8.8 |
| CSIRO | 0.9 | 2.2 | 1.6 | 8.1 | 3.5 | 8.9 |
| Average | 0.9 | 2.3 | 1.5 | 7.9 | 3.5 | 8.9 |
| **Millet** | | | | | | |
| No CC | 0.71 | 1.22 | 1.27 | 1.38 | 1.22 | 2.31 |
| NCAR | 0.7 | 1.1 | 1.2 | 1.1 | 1.0 | 1.7 |
| CSIRO | 0.6 | 1.1 | 1.1 | 1.2 | 1.0 | 1.8 |
| **Average** | 0.6 | 1.1 | 1.1 | 1.2 | 1.0 | 1.8 |
| **Impact of climate change on yield (Percent change)[a]** | | | | | | |
| Maize | −8.5 | −13.2 | −5.7 | −24.6 | −23.5 | −25.9 |
| Rice | −18.4 | −19.4 | −23.8 | −16.4 | −18.8 | −13.5 |
| Millet | −9.2 | −9.9 | −10 | −15.2 | −15.5 | −22.7 |

Note: $\dfrac{No\_CC - \overline{CC}}{No\_CC} *100$ where $No\_CC$ = No climate change, $\overline{CC}$ = average yield for NCAR & CSIRO models

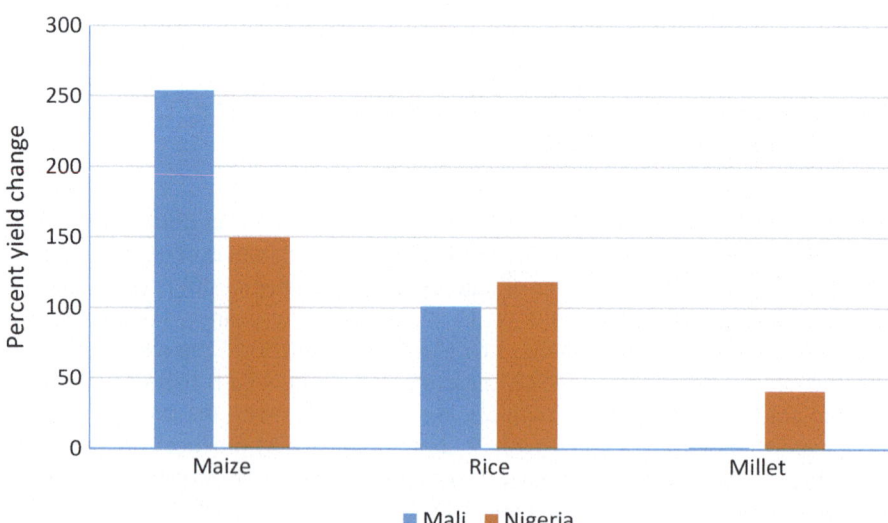

**Fig. 2** Change of crop yield under farmer practice change under climate change due to use of SLWM practices

It is important to examine SLWM adoption rate and drivers of adoption in order to identify the appropriate polices and strategies for enhancing their adoption. The next section addresses these important questions.

$$\Delta y = \frac{y_1 - y_0}{y_0} * 100$$

Where $y_1$ = T3 yield & $y_0$ = average yield of T0 & T1 under climate change (see Table 2 for definition of T0, T1 & T3).

## 4.3 Adoption Rate of SLWM Practices

About 50% of farmers in SSA do not use external inputs such as inorganic fertilizer or organic inputs (Table 5). Adoption of inorganic fertilizer and organic inputs are respectively about 19% and 25% (Table 5). Pender et al. (2009) observed lower adoption rates of external inputs since they observed that only 3% of farmers in SSA use low-cost productivity enhancing management practices – such as organic inputs. The low adoption of organic inputs is especially troubling given that it could be produced by farmers and is crucial in reducing climate-related production risks.

Even though irrigation could increase crop yield by at least 50% (Ringler and Nkonya 2012), its adoption is only 7% (Table 5) – an aspect which illustrates the weak irrigation development in SSA (You et al. 2011). As discussed earlier, irrigation development is one of the key investments required for adaptation to climate change in SSA (Nelson et al. 2009) and its low adoption rate underscores the urgent need for increasing investment and promoting its adoption.

More detailed analysis for the case study countries shows an interesting pattern. About 6% of crop farmers in Mali and 12% in Nigeria use irrigation (Figs. 3 and 4). There is large variation of adoption of irrigation in both countries across

**Table 5** Adoption and profitability of soil fertility management practices in SSA

| Country | ISFM | Fertilizer | Organic inputs | Irrigation | Nothing |
|---|---|---|---|---|---|
| | Adoption (percent) | | | | |
| Mali | 18 | 16 | 39 | 6.0 | 27 |
| Uganda | 0 | 1 | 68 | 0.1 | 31 |
| Kenya | 16 | 17 | 22 | 2.0 | 44 |
| Nigeria | 1 | 23 | 28 | 12.0 | 47 |
| Malawi | 8 | 52 | 3 | 2.3 | 38 |
| Tanzania | 1 | 1 | 3 | 3.6 | 95 |
| Average adoption rate and profit | | | | | |
| Adoption rate (%) | 6.2 | 19.1 | 24.6 | 7.0 | 49.8 |
| Profit (US$/ha/year) | 36.5 | 24.6 | 15.1 | | 10.4 |

Source: Nkonya et al. (2016a)

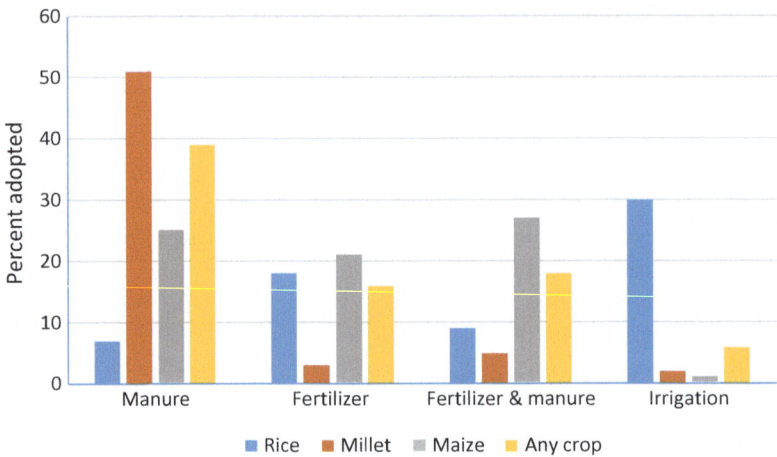

**Fig. 3** Adoption rates of manure, fertilizer, and irrigation in Mali. Source: computed from raw data, Mali agricultural census 2004/05

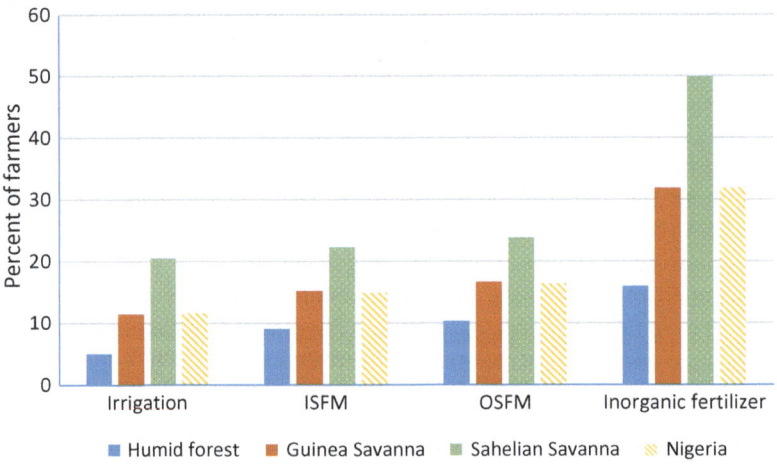

**Fig. 4** Adoption rate of soil fertility management practices, Nigeria

agroecological zones and crops. Drylands account for the largest share of irrigated area. About 30% of rice is irrigated in Mali (Fig. 3), a level that reflects the dominance of rice as an irrigated crop in Africa. About 14% of the area under full or partial control irrigation in Africa is planted with rice (FAO 2005). On spatial distribution, irrigation is concentrated in the drylands in both countries. About 70% of irrigated area in Mali is in the Sahelian zone located in the middle belt (Fig. 5). Likewise, adoption of irrigation is highest in the Sahelian Sudan and Guinea Sudan in Nigeria (Figs. 4 and 5), both of which account for 68% of irrigated area in Nigeria (FAO 2005).

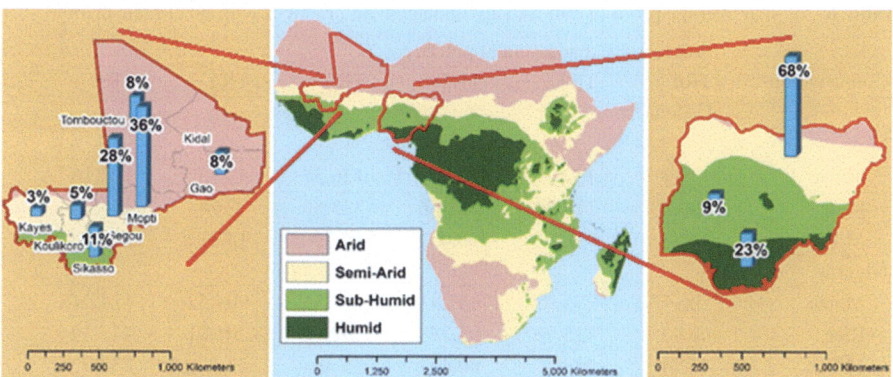

**Fig. 5** Distribution of irrigated area across agroecological zones, Mali & Nigeria (Sources: RDM (2009) and FAO (2005))

Adoption of OSFM is very high in Mali but quite limited in Nigeria. Over 50% of millet farmers in Mali apply manure and 39% of all crop farmers in the country apply manure (Fig. 3). This level is much higher than the adoption rate of inorganic fertilizer – which is only 16%. In Nigeria, only 16% of farmers apply OSFM practices, which includes animal and green manure, agroforestry, and composting. Adoption of inorganic fertilizer is much higher (32%) (Fig. 4). One of the reasons behind such high adoption of inorganic fertilizer could be the high fertilizer subsidy rate and relatively higher income of farmers in Nigeria compared to those in Mali (Table 1). Adoption of ISFM is 18% and 1% in Mali and Nigeria, respectively (Table 5) – and in both cases lower than less profitable practices (Figs. 3 and 4).

Given the adoption patterns of soil fertility management practices discussed above, it is important to analyze the adoption incentives and drivers of their adoption. To better understand the adoption incentives and competitiveness of the land management practices, the section below analyzes profitability of soil fertility management practices in the case study countries. This is followed by analysis of the drivers of adoption of soil fertility management practices, which will be used to draw implications on policies and strategies for increasing their adoption, and consequently enhancing adaptation to climate change.

## 4.4 How Profitable Are the SLWM Practices?

Soil fertility management practices that combine manure and inorganic fertilizer – or ISFM (TR3 & TR5) generally have the highest profit for all crops (Table 6). This is consistent with other studies (e.g. Doraiswamy et al. 2007; Sauer et al. 2007; Nkonya et al. 2016a). The predominant management practices (TR0 & TR1) – regarded in this study as farmer management practices – are least profitable, and are shown to have greater yield variability.

Table 6 50-year average profit of soil fertility management practices with no climate change

| Country/crop | Soil fertility management practices | | | | | |
|---|---|---|---|---|---|---|
|  | TR0 | TR1 | TR2 | TR3 | TR4 | TR5 |
|  | US$/ha | | | | | |
| Mali | | | | | | |
| Maize | 13.34 | 15.72 | 20.05 | 126.57 | 52.53 | 127.16 |
| Rice | 109.53 | 128.09 | 248.66 | 383.39 | 72.74 | 494.63 |
| Millet | 8.88 | 9.65 | 16.05 | 14.72 | 13.21 | 20.90 |
| Nigeria | | | | | | |
| Maize | 206.13 | 295.98 | 451.41 | 881.51 | 904.52 | 1142.46 |
| Rice | 66.93 | 115.78 | 96.38 | 192.71 | 201.01 | 447.39 |
| Millet | 47.74 | 63.87 | 66.43 | 75.38 | 53.52 | 78.09 |

Notes: See Table 2 for definition of TR0-TR5
Average profit for the 50-year average (2000–50), 2015 constant price

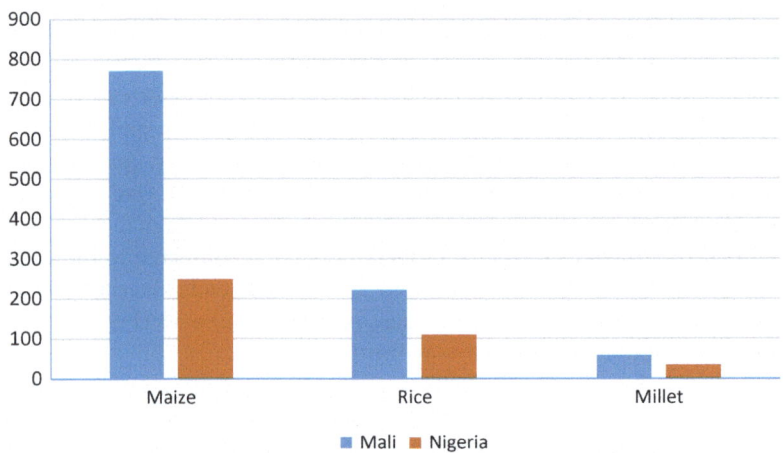

Fig. 6 Change of profit per ha due to a switch from farmer practice to ISFM

If a household switches from the farmer practice (TR0 & TR1) to medium-level ISFM (TR3), their maize and rice profits are expected to more than double in both countries (Fig. 6) – suggesting that adoption of ISFM will simultaneously reduce poverty and production risks and increase food security.

As seen in Table 5, adoption rate of ISFM is low – despite its high returns. There are several reasons that contribute to this pattern and the econometric analysis below will shed light on this. However, a couple of factors need to be examined in detail since they may not be reflected fully in an econometric analysis.

(i) *ISFM and OSFM practices are labor intensive*: In all treatments using manure, labor costs amounted to 50–80% of total production costs. This is a major constraint for OSFM that includes a transfer of biomass – especially under SSA's low mechanization. Ownership of livestock is an important driver given that

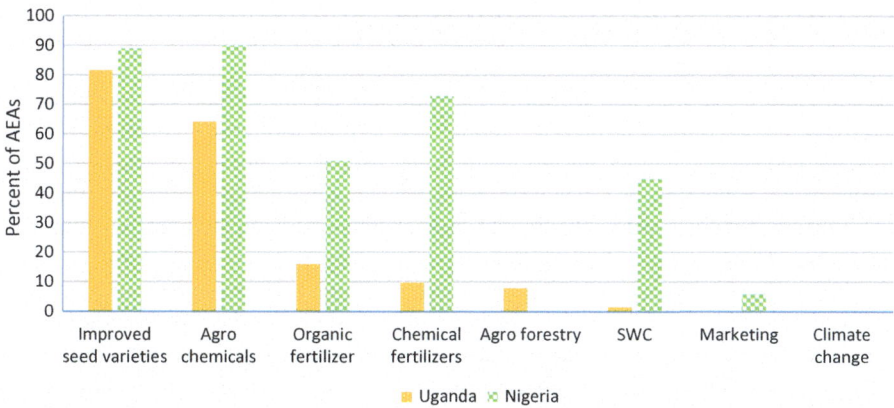

**Fig. 7** Major topics promoted by AEAs in Nigeria and Uganda (Notes: *AEAs* Agricultural extension agents, *SWC* Soil and water conservation practices. Sources: Extracted from: Banful et al. (2010) – Nigeria results; Nkonya et al. (2013) – Uganda results)

there are no marketing mechanisms for organic inputs and that biomass transfer could require animal draft power. This means farmers need to produce their own organic inputs and must have transportation to move animal and household refuse from home to crop plots.

(ii) *Low capacity of agricultural extension agents to provide advisory services on ISFM & adaptation to climate change*: Studies done in Nigeria and Uganda involving agricultural extension agents (AEAs) on topics they promote to farmers showed that improved seeds and agrochemicals are the most important technologies promoted (Fig. 7). Promotion of chemical fertilizers in Nigeria is high but limited in Uganda. Promotion of agroforestry and other organic inputs was quite low in both countries. No AEAs promoted agroforestry in Nigeria though 51% promoted organic fertilizer like manure. In both countries, no AEAs promoted climate change adaptation strategies. The results reflect the low capacity of AEAs to provide advisory services on OSFM and climate change.

Where average profit is given by:

$$\Delta \pi \% = \frac{\pi_{ISFM} - \overline{\pi_c}}{\overline{\pi_c}} * 100$$

Where $\Delta \pi$ = change in crop profit per ha, $\pi_{ISFM}$= Profit with middle-level ISFM (40 kgN/ha + 1.7 tons manure/ha, 50% crop residues, $\overline{\pi_c}$ = Average profit per ha for farmer practice, i.e., TR0 & TR1.

The section below analyzes the drivers of adoption of SLWM by taking into account other factors.

## 4.5 Drivers of Adoption of SLWM Practices

*Human capital* We find that in Mali, older farmers are more likely to use manure and ISFM while younger farmers are more likely to use inorganic fertilizer (Table 7). In Nigeria however, older farmers are more likely to use inorganic fertilizer (Table 8) – a reflection of older farmers' higher income and/or political influence that increase their access to subsidized fertilizer. The results in Mali suggest that for quick wins, fertilizer subsidy programs need to be targeted to younger farmers, who also happen to be poorer and – as observed in Ethiopia by Krishnan and Patnam (2014) – could serve as AEAs to other farmers. As demonstrated by Bandiera and Rasul (2006), and Conley and Udry (2010) young adopters of agricultural production technologies could influence other farmers decision to adopt new technologies through peer influence. The dependence ratio in Mali and Nigeria increases the propensity to use ISFM and inorganic fertilizer. This is likely driven by an attempt by households with a large number of children to increase per unit area production to address family food needs. Household size in Mali and Nigeria increases with adoption of ISFM and inorganic fertilizer in Mali. Family size also increases the propensity to use OSFM in Nigeria. This could be due to high family labor for large households allowing them to adopt labor intensive practices – especially those including biomass transfer.

Secondary and post-secondary education has a negative impact on propensity to adopt irrigation in Nigeria. Contrary to Di Falco (2014), level of formal education has no significant effect on probability to adopt any soil fertility management practice considered in Nigeria. This could be due to a small percent of households of the benchmark group – farmers with no formal education – who only constitute 10% of the sample population. In Mali, primary and secondary education increases adoption of manure. Similarly, secondary and post-secondary education increases adoption of inorganic fertilizer in Mali.

Female-headed households in Mali have greater propensity to adopt ISFM but the converse is the case in Nigeria. The greater likelihood of adoption of ISFM in Mali by female-headed households reflects their greater efficiency in utilizing labor and agricultural investments (Oladeebo and Fajuyigbe 2007). It is also a reflection of the higher and lower adoption rates of organic inputs in Mali and Nigeria respectively. In both countries, female-headed households are less likely to adopt irrigation. This is consistent with van Koppen et al. (2013) who found that the irrigation adoption rate of female headed households in SSA is only two-thirds of the rate of male-headed households.

### 4.5.1 Financial and Physical Capital

As expected, access to credit increases probability to use inorganic fertilizer, ISFM, and irrigation in both countries (Tables 7 and 8). However, access to credit reduces propensity to adopt manure. This could be due to a substitution effect in that farmers

Table 7 Determinants of adoption of SLWM practices, Mali (Probit model)

| Explanatory Variables | Manure | | Fertilizer | | ISFM | | Irrigation | |
|---|---|---|---|---|---|---|---|---|
| | Structural | Reduced | Structural | Reduced | Structural | Reduced | Structural | Reduced |
| | Average marginal effects | | | | | | | |
| **Household level characteristics** | | | | | | | | |
| Age of household head | 0.001* | 0.001** | −0.000* | −0.001*** | 0.0001* | 0.0001 | 0.001 | 0.001 |
| Female headed household | −0.009 | −0.016 | −0.009 | −0.027 | 0.027* | 0.02* | −0.02 | −0.026 |
| Dependence ratio | −0.001* | −0.001** | 0.0001 | 0.001*** | 0.0001 | 0.001*** | 0.0001 | 0.001 |
| Household size | 0.001 | 0.001* | 0.002*** | 0.003*** | 0.001** | 0.003*** | 0.002 | 0.003 |
| **Level of education of household head (cf no formal education)** | | | | | | | | |
| Primary | 0.036* | −0.015 | −0.016 | 0.116** | −0.001 | 0.072 | 0.007 | 0.009 |
| Secondary | 0.085** | 0.066* | 0.057** | 0.053* | −0.003 | −0.019 | 0.02 | 0.031* |
| Post-secondary | 0.008 | 0.034 | 0.114*** | −0.02 | 0.057 | −0.004 | −0.000 | 0.007 |
| **Physical capital** | | | | | | | | |
| Number of cattle owned | 0.021*** | 0.018*** | 0.0001 | 0.015*** | 0.017* | 0.027*** | −0.002* | −0.03*** |
| Farmsize | 0.026** | 0.003*** | 0.011 | 0.001*** | 0.034 | 0.002*** | −0.105* | 0.006 |
| **Plot level characteristics** | | | | | | | | |
| Plot owned by male | 0.003 | 0.011 | 0.072** | 0.099*** | 0.017 | 0.058 | 0.019** | 0.020** |
| Distance home-plot | −0.006*** | −0.007*** | 0.001 | 0.001 | 0.003*** | 0.003*** | 0.000 | 0.000 |
| **Plot topography (cf flat):** | | | | | | | | |
| Plateau | 0.029** | −0.016 | 0.064*** | 0.157*** | 0.046*** | 0.159*** | −0.034*** | −0.041*** |

(continued)

**Table 7** (continued)

| Explanatory Variables | Manure | | Fertilizer | | ISFM | | Irrigation | |
|---|---|---|---|---|---|---|---|---|
| | Structural | Reduced | Structural | Reduced | Structural | Reduced | Structural | Reduced |
| | Average marginal effects | | | | | | | |
| Valley bottom | −0.080*** | −0.113*** | −0.016 | −0.001 | 0.009 | 0.025** | −0.002 | −0.001 |
| Gentle slope | 0.054*** | 0.054*** | 0.029*** | 0.048*** | −0.005 | 0.016 | −0.011*** | −0.020*** |
| Steep slope | −0.051** | −0.051* | 0.003 | 0.003 | −0.07*** | −0.08*** | −0.019** | −0.026*** |
| **Plot tenure/method of acquisition (cf given as gift)** | | | | | | | | |
| Leasehold | 0.134** | 0.133** | −0.038 | −0.078** | 0.047 | −0.004 | 0.011 | 0.015 |
| Sharecropping | −0.127** | −0.131** | −0.079** | −0.102*** | −0.080* | −0.13*** | 0.096*** | 0.129*** |
| Rented | 0.062 | 0.053 | −0.069 | −0.06 | 0.014 | 0.086 | 0.024 | 0.033 |
| Customary tenure | 0.245** | 0.254*** | −0.094 | −0.117** | −0.108* | −0.14*** | 0.623*** | 0.793*** |
| Given by village council | −0.047 | −0.236*** | 0.062*** | 0.333*** | 0.019 | 0.160*** | – | – |
| Other acquisition methods | −0.054 | −0.066 | 0.009 | 0.018 | 0.004 | −0.001 | 0.060** | 0.055** |
| **Access to rural services** | | | | | | | | |
| Extension services | −0.037 | | 0.027 | | 0.004 | | 0.0002 | |
| Roads | −0.006*** | −0.007*** | 0.0001 | 0.001 | −0.003*** | −0.003*** | −0.001* | −0.003* |
| Credit | −0.036*** | | 0.047*** | | 0.05*** | | 0.015*** | |

Note: Marginal effect = percentage change in the probability of adoption due to unit change of corresponding covariate
*, **, & *** respectively mean: associated coefficient is statistically significant at 10%, 5% and 1%.

Table 8 Determinants of adoption of SLWM practices (Probit model), Nigeria

| | Inorganic fertilizer | | OSFM | | ISFM | | Irrigation | |
|---|---|---|---|---|---|---|---|---|
| | Structural | Reduced | Structural | Reduced | Structural | Reduced | Structural | Reduced |
| **Human capital** | Average marginal effects | | | | | | | |
| Level of formal education of household head (cf no formal education) | | | | | | | | |
| Primary | 0.015 | 0.019 | 0.01 | 0.015 | 0.014 | 0.02 | 0.003 | 0.004 |
| Koranic | 0.005 | 0.012 | 0.004 | 0.009 | 0.013 | 0.019 | 0.013 | 0.015 |
| Secondary | 0.019 | 0.024 | −0.014 | −0.01 | −0.004 | 0.001 | −0.037** | −0.035** |
| Post-secondary | 0.013 | 0.018 | −0.025 | −0.017 | −0.023 | −0.014 | −0.05*** | −0.05*** |
| Household size | 0.001 | 0.001 | 0.001*** | 0.001*** | 0.002*** | 0.003*** | 0.01*** | 0.01*** |
| Dependence ratio | −0.001 | 0.001 | 0.004*** | 0.004*** | 0.001*** | 0.001*** | −0.0001 | −0.002 |
| Female headed household | −0.031** | −0.025* | −0.026* | −0.023 | −0.028* | −0.026* | −0.05*** | −0.05** |
| Age, household head | 0.001* | 0.001 | 0.001 | −0.001 | 0.001 | −0.001 | −0.001* | −0.001** |
| Farmer group member | 0.061*** | | 0.009 | | 0.004 | | −0.005 | |
| Non-farm activity | −0.028** | −0.033*** | 0.021* | 0.020* | 0.013 | 0.013 | −0.006 | −0.004 |
| **Physical & financial capital** | | | | | | | | |
| Farm area (ha) | 0.003*** | 0.003*** | 0.001*** | 0.001*** | 0.002*** | 0.002*** | 0.001 | 0.001 |
| TLU | 0.004 | 0.004 | 0.004* | 0.001* | 0.002* | 0.001* | 0.0006 | 0.0001 |
| Remittance | 0.013 | | −0.031** | | −0.028** | | 0.024** | |

(continued)

Table 8 (continued)

| | Inorganic fertilizer | | OSFM | | | ISFM | | | Irrigation | | |
|---|---|---|---|---|---|---|---|---|---|---|---|
| | Structural | Reduced | Structural | Reduced | | Structural | Reduced | | Structural | Reduced | |
| Average marginal effects | | | | | | | | | | | |
| **Access to rural services** | | | | | | | | | | | |
| Distance (km) to | | | | | | | | | | | |
| Road | −0.001* | −0.001 | −0.001 | 0.001 | | −0.001* | 0.001 | | −0.001** | −0.001** | |
| Town | −0.001 | −0.001* | −0.001*** | −0.001*** | | −0.001*** | −0.001*** | | −0.002 | −0.002 | |
| Credit | 0.009*** | | 0.002*** | | | 0.001** | | | 0.05** | | |
| Extension services | −0.025** | | 0.001 | | | −0.001 | | | 0.013 | | |
| **Agroecological zones (cf humid forest)** | | | | | | | | | | | |
| Guinea Savanna | 0.135*** | 0.127*** | 0.095*** | 0.064*** | | 0.11*** | 0.079*** | | 0.098*** | 0.078*** | |
| Sahelian Savanna | 0.146*** | 0.152*** | 0.175*** | 0.17*** | | 0.175*** | 0.170*** | | 0.144*** | 0.145*** | |

Notes: *OSFM* Organic soil fertility management (agroforestry, manure, & green manure)
\*, \*\*, & \*\*\* = Associated coefficient is significant at 0.10, 0.05 & 0.01 level

with access to credit substitute manure with fertilizer. The results suggest the importance of access to credit for adoption of purchased inputs and its role in adaptation to climate change. Remittances reduce the probability to use inorganic fertilizer and ISFM in Nigeria. This could be due to the fact that remittances are received in time of emergencies and not used for purchase of inputs.

Contrary to Di Falco (2014), access to extension services also has no significant impact on propensity to adopt soil fertility management practices in both countries. In fact extension services reduce probability to adopt manure. As discussed earlier, only a small share of AEAs promote OSFM practices (Fig. 7). Consistent with Di Falco (2014) and Barrett and Constas (2014), proximity to roads increases the propensity to irrigation in both countries. This is consistent with the theory that irrigated crops are marketed more than rainfed crops because irrigation involves large investments in equipment and infrastructure and use greater amounts of inputs for irrigated crops (You et al. 2011). Proximity to roads also increases the probability to use manure and ISFM in Mali. In Nigeria, proximity to roads and cities of more than 50,000 residents increases the probability to use all three soil fertility management practices considered. Consistent with Nelson et al. (2009), the results underscore the importance of market access for enhancing adaptation to climate change.

As expected, physical capital endowment (livestock and farm size) increases propensity to use fertilizer and ISFM in both countries but reduces the propensity to adopt irrigation in Mali. The inverse relationship between irrigation adoption and farm size is expected given that farmers in SSA irrigate small farms more than large ones (Domenech and Ringler 2013). The number of livestock owned also increases the probability to use manure in Mali – an aspect that underscores the lack of a market for OSFM that forces farmers to depend on their own production. The number of livestock is also inversely associated with the probability to use irrigation – an aspect that is expected given that pastoralists with large herds of cattle are less likely to be engaged in large investment crop production.

Of specific interest is the topography of plots, which is an indicator for irrigation use, and here analyzed in Mali only.[6] Small scale irrigation is largely done on gentle slope plots because irrigation on steep slopes is difficult and expensive to implement (Nielsen et al. 2015). Accordingly, irrigation is more likely to be done on plots with flatter terrain than on any other topography (Table 7).

## 4.6 Reducing Climate-Related Risks – The Role of Soil Carbon and SLWM Practices

Soil carbon enhances soil moisture conservation and consequently reduces yield variability in areas with low-rainfall and highly variable moisture (Lal 2015; Govaerts et al. 2009; Manna et al. 2005). Consistent with this, our 30-year

---

[6] Plot topography data were not collected in Nigeria.

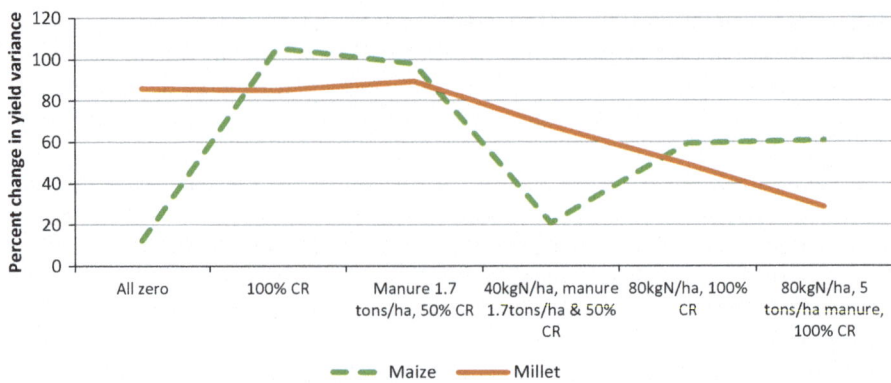

**Fig. 8** Impact of SLWM on maize and millet yield variance – 30 year DSSAT simulation results, Mali

simulation results show that maize and millet yield variance in the dry areas of Mali fell as the amount of soil carbon increased (Fig. 8).

Accordingly, the Just-Pope mean-variance results show that almost all SLWM practices in both countries reduce production risks (Tables 9 and 10) – underscoring their importance in designing appropriate climate change adaptation strategies.

Other variables are also important in reducing climate-related production risks. Specifically, access to roads reduces production risks in both countries further demonstrating the role played by market access in adaptation to climate change. This supports Nelson et al. (2009) findings that two thirds of US$3 billion additional investment required to offset climate change impacts on nutrition in SSA will need to be directed to roads (Fig. 1). Similarly, irrigation is associated with lower production risks in both countries.

Rainfed area also reduces production risks. This could be due to largescale farmers' ability to invest in management practices that could lead to reduction of production risks. The number of livestock is a risk reducing asset in both countries. This underscores the role played by livestock in risk management through organic soil fertility improvement and provision of animal power for biomass transfer.

Post-primary education in Mali reduces production risks – underscoring the key role of human capital in adaptation to climate change. Like the case of adoption of SLWM however, education does not have a significant impact on production risks in Nigeria.

**Table 9** Impact of SLWM practices on climate-related production risks – Just-Pope mean-variance model, Mali

| Explanatory variables | Variance Function, (FGLS) |
|---|---|
| **SLWM practices:** | Ln(crop value XOF)$^2$ |
| Manure | −0.084*** |
| Inorganic fertilizer | −0.131*** |
| ISFM | −0.005*** |
| **Human capital** | |
| (Ln(age of household head) | 0.094*** |
| Male-headed household head | −0.185** |
| Dependence ratio | 0.001** |
| Ln(Family size) | −0.023** |
| Level of education of household head (cf no formal education) | |
| Primary | −0.016 |
| Secondary | −0.605*** |
| Post-secondary | −0.508** |
| **Physical endowment** | |
| TLU | −0.087*** |
| **Access to credit** | 1.58*** |
| **Access to rural services** | |
| Access to extension services | −0.112 |
| Ln(distance to road, km) | −0.030** |
| **Plot level characteristics** | |
| Plot owned by male | 0.373 |
| Ln(Distance (km) – homestead to plot) | −0.019** |
| Slope position (cf flat) | |
| Plateau | −0.023** |
| Valley bottom | −0.078** |
| Gentle slope | 0.003*** |
| Steep slope | −0.026*** |
| Constant | −0.294*** |

Note: *FGLS* feasible generalized least squares
*,**, & *** means associated coefficient is statistically significant at 0.10, 0.05 & 0.01 confidence interval

## 5 Conclusions and Policy Implications

Our estimates show that climate change is predicted to reduce production of staple foods (maize, rice, and millet) by about 20% by 2050 if farmers do not take adaptive strategies. This jeopardizes food security – especially for the poorest farmers who

**Table 10** Impact of SLWM practices on climate-related production risks – Just-Pope mean-variance model, Nigeria

|  | Variance model (FGLS) |
|---|---|
| **SLWM practices** |  |
| **Irrigation** | **−0.009*** |
| Soil bunds | −0.613*** |
| Stone bunds | −0.843*** |
| Mulching | 0.248 |
| Grass strip | −0.592*** |
| Ditches | −0.715*** |
| Ridges | −0.272*** |
| Animal manure | 0.13 |
| Compost | −0.427*** |
| Inorganic fertilizer | −0.139** |
| ISFM | −0.134** |
| **Human capital** |  |
| Ln(Age of household head) | 0.034 |
| Ln(number of adult males) | −0.055 |
| Ln(adult females) | −0.065 |
| Female household head | 0.062 |
| Education of household head (cf no formal education) |  |
| Primary | 0.044 |
| Koranic | 0.114 |
| Secondary | 0.001 |
| Post-secondary | −0.098 |
| **Physical and financial capital** |  |
| Remittance | −0.157*** |
| Ln(value of productive) assets) | −0.023** |
| Ln(TLU) | −0.025* |
| Ln(rainfed area) | −0.177*** |
| **Access to rural services** |  |
| Ln(distance to market, km) | 0.021 |
| Ln(distance to road, km) | −0.032* |
| Agroecological zones (cf Sahelian Savannah) |  |
| Humid forest | 1.101*** |
| Guinea Savannah | 0.192*** |
| Constant | 1.470*** |

Note: *FGLS* feasible generalized least squares
*,**, & *** means associated coefficient is statistically significant at 0.10, 0.05 & 0.01 confidence

heavily depend on rainfed agriculture and who do not use soil carbon-enhancing management practices. Our results show that even though all land management practices considered lead to a lower yield due to climate change, adoption of SLWM practices could completely offset the negative effect of climate change on crop production related to farmer management practices and significantly reduce production risks. Specifically, adoption of SLWM will simultaneously increase crop yield and profit under current farmer practice by at least twofold. This means SLWM could simultaneously increase food security and reduce poverty and climate-related production risks. This is in addition to the off-site benefit of carbon sequestration, which farmers do not consider in their planning. This underscores the importance of promoting SLWM practices to help smallholder farmer adaptation and resilience to climate change and to help SSA countries to achieve their commitment to the UNFCCC 21st Conference of Parties (CO21) to contribute to the reduction of GHG emissions.

The low adoption rate of SLWM calls for major changes in the agricultural development policies and strategies.

The major drivers for adoption of SLWM include access to agricultural extension services, market access, credit, and greater endowment of physical resources. The results underscore the need for increasing access to rural services – especially for farmers in remote areas and poor farmers and female-headed households. Improvement of market access will provide incentives for farmers to use SLWM and other production technologies. Development of market infrastructure could serve multiple purposes of rural poverty reduction and modernization of agriculture. This could be done in conjunction with other rural development and poverty reduction programs. This demonstrates that adaptation to climate change will need to be more holistic and go beyond the traditional approach of compartmentalized development strategies.

There is need for increasing the training of agricultural extension service providers about SLWM and climate change – both of which are relatively new to many older agricultural extension services. Additionally, advisory services on irrigation development and management remain weak. This is especially true for irrigation engineering advisory services, which remain largely confined to large-scale irrigation schemes (Nkonya et al. 2015b). As a result of this and other factors, water loss in irrigation schemes and irrigation systems is more than 50% in Africa (Delaney 2009). Short-term training with specific focus on these important topics will be more effective and practical than long-term training. Additionally, sex of extension agent providers has a large impact on type of advisory services provided and beneficiaries of such services (Takeshima and Edeh 2013; Davis et al. 2012). Our results show that female-headed households are less likely to adopt SLWM. One strategy for increasing their adoption is to recruit more female extension agents who are better able to provide advisory services and SLWM messages to women than male extension agents (Nkonya et al. 2013; Davis et al. 2012; Takeshima and Edeh 2013).

The challenges of adoption of SLWM also includes high labor intensity of practices which involves biomass transfer, limited marketing infrastructure, and production of organic inputs like manure. Promotion of agroforestry is likely to be an amenable practice since it is less labor intensive once it gets established and it simultaneously addresses both lack of markets and production challenges of organic inputs. Unfortunately, current soil fertility management policies gravitate around inorganic fertilizer subsidy. There are no programs that provide incentives for adoption of OSFM practices like agroforestry. Given the multiple benefits of OSFM practices, it is important to consider initiatives that provide incentives for adoption of agroforestry, ISFM, and OSFM practices. For example, it is possible to provide conditional fertilizer subsidies given to beneficiaries who have planted trees in croplands. Such incentives are easy to verify and could serve as a form of payment for ecosystem services since they will increase carbon sequestration. A study in Malawi showed that farmers are highly receptive to conditional fertilizer subsidies given to farmers to plant agroforestry trees (Marenya et al. 2014).

Following UNFCCC's COP21 resolution to include agriculture in the carbon sequestration program, adaptation and mitigation in the agricultural sector is included in 80% of the national INDCs (Richards et al. 2015). This provides a unique opportunity for building carbon markets in SSA by organizing smallholder land users to participate in the carbon market. This could be effectively achieved by organizing them in groups and giving them the mandate to manage their natural resources. Implementing this would require revision of the Decentralization Act in order to give villagers a full mandate to manage their own resources. Efforts to increase economic interest groups and cooperatives would also help smallholder land users to work collectively. Success of carbon markets is greater when both international and domestic buyers are involved. The domestic buyers could include governments. Additionally, experience has shown that the payment for ecosystem services (PES) are successful in countries with strong policies and investment in PES. For example, the Costa Rica constitution sets a framework for rewarding land users who provide significant off-site benefits (Salazar and Chacón 2011). The constitution further states that revenue collected from fossil fuel taxes, water fees, and from donors be allocated to PES (Ibid). The land users also are exempted from paying some local taxes. These incentives have significantly helped to combat deforestation in Costa Rica. This suggests that the governments in SSA need to enhance their policies that enhance incentives of land users to adopt ISFM and OSFM practices.

The impact of climate change on food security and rural development in general are large and require immediate action to offset their effects on the rural poor. The opportunities for addressing climate risks using SLWM are large but they need strong government commitment to exploit them in order to achieve food security and ensure sustainable agricultural development in Africa.

**Acknowledgement** We are grateful to the TerrAfrica World Bank for providing funding for this study. We are also grateful to the farmers and community leaders in Mali and Nigeria who provided data and information used in this study. We thank ministries of agriculture and environment as well research institutions and bureaus of statistics from both countries for providing data and documents. We are indebted to a number of colleagues from the World Bank and participants to various workshops who provided insightful comments to various versions of this paper. We are also grateful to Taoufiq Bennouna, Stephen Danyo, and Florence Richard who managed this study project from the World Bank. The authors take responsibility for all errors and omissions of this report.

# References

Aduayi, E. A., Chude, V. O., Adebusuyi, B. A., & Olayiwola, S. O. (2002). Fertilizer use and management practices for crops in Nigeria. *Federal Ministry of Agriculture and Rural Development Abuja, Nigeria P*, 63–65.

AU (African Union). 2014. Malabo declaration on accelerated agricultural growth and transformation for shared prosperity and improved livelihoods. Online at http://www.au.int/en/content/malabo-26-27-june-2014-decisions-declarations-and-resolution-assembly-union-twenty-third-ord. Accessed on August 25, 2015.

Bandiera, O., & Rasul, I. (2006). "Social networks and technology adoption in northern Mozambique." The Economic Journal, 116(514):869–902.

Banful A.B., E. Nkonya and V. Oboh. 2010. Constraints to Fertilizer Use in Nigeria Insights from Agricultural Extension Service. IFPRI Discussion Paper 01010.

Barrett, C. B., & Constas, M. A. (2014). Toward a theory of resilience for international development applications. Proceedings of the National Academy of Sciences, 111(40), 14625–14630.

Benin, S. and Yu, B. 2012. Complying the Maputo Declaration Target: trends in public agricultural expenditures and implications for pursuit of optimal allocation of public agricultural spending. ReSAKSS Annual Trends and Outlook Report 2012. International Food Policy Research Institute (IFPRI).

Burney, J. A., Naylor, R. L., & Postel, S. L. (2013). The case for distributed irrigation as a development priority in sub-Saharan Africa. Proceedings of the National Academy of Sciences, 110(31), 12513–12517.

Christensen, J. H., B. Hewitson, A. Busuioc, A. Chen, X. Gao, I. Held, R. Jones, et al. 2007. "Regional Climate Projections." In Climate Change 2007: The Physical Science Basis. Contribution of Working Group I to the Fourth Assessment Report of the Intergovernmental Panel on Climate Change. In: Solomon S., D. Qin, M. Manning, Z. Chen, M. Marquis, K. B. Averyt, M. Tignor, et al. (eds), 847–940. Cambridge, UK: Cambridge University Press.

Conley, T. and C. Udry (2010). "Learning about a New Technology." American Economic Review, 100(1):35–69.

Davis, K., Nkonya, E., Kato, E., Mekonnen, D. A., Odendo, M., Miiro, R., & Nkuba, J. (2012). Impact of farmer field schools on agricultural productivity and poverty in East Africa. World Development, 40(2), 402–413.

Delaney S. 2009. Challenges and opportunities for agricultural water management in West and Central Africa: Lessons from IFAD experience. IFAD, Rome.

Di Falco S. 2014. Adaptation to climate change in Sub-Saharan agriculture: assessing the evidence and rethinking the drivers. European Review of Agricultural Economics 41 (3): 405–430.

Dillon, A. (2008). Access to irrigation and the escape from poverty: Evidence from northern. International Food Policy Research Institute Discussion paper 782.

Doraiswamy P, McCarty G, Hunt E, Yost R, Doumbia M, Franzluebbers A (2007) Modeling soil carbon sequestration in agricultural lands of Mali. Agricultural Systems 94(1):63–74

Domenech, L., & Ringler, C. (2013). The impact of irrigation on nutrition, health, and gender: A review paper with insights for Africa south of the Sahara. International Food Policy Research Institute (FPRI) Discussion paper #1428

Fabricius C., A. Ainslie, J. Cloete, C. Shackleton, S. Shackleton, P. Urquhart, J. Gambiza, E. Nel, K. Rowntree, M. Mortimore, J. Ariyo, M. Bila, A. Faye, A. Faye, S. Herrmann, S. Mohammed, H. Seyni, K. Vogt, B. Yamba, S. Herrmann, S. Maddrell, C. Nzioka and I. Bond. 2008. Situation Analysis of Ecosystem Services and Poverty Alleviation in arid and semi-arid Africa. Ecosystem Services for Poverty Alleviation (ESPA) report. Online at http://www.nerc.ac.uk/research/funded/programmes/espa/final-report-africa/ accessed December 22, 2015.

FAO (2005). Irrigation in Africa in figures. AQUASTAT Survey – 2005. FAO Water Report # 29. Online at ftp://ftp.fao.org/agl/aglw/docs/wr29_eng.pdf. Accessed on December 22, 2015.

FAO (Food and Agriculture Organization). 2005a. Mali Livestock sector Brief. FAO, Rome

FAO (Food and Agriculture Organization). 2005b. Nigeria Livestock sector Brief. FAO, Rome

FAO (Food and Agriculture Organization). 2005c. Nigeria Water Report. Water Report 29. Online at http://www.fao.org/nr/water/aquastat/countries_regions/nga/index.stm

FAO (Food and Agriculture Organization). 2012. The State Of Food and Agriculture, 2012. Investing in Agriculture. FAO, Rome.

Gijsman, A. J., Hoogenboom, G., Parton, W. J., & Kerridge, P. C. (2002). Modifying DSSAT crop models for low-input agricultural systems using a soil organic matter–residue module from CENTURY. *Agronomy Journal, 94*(3), 462–474.

FAOSTAT (2013). Agricultural statistics. Online at http://www.fao.org/faostat/en/#data. Accessed on January 12, 2017.

Govaerts, B., N. Verhulst, A. Castellanos-Navarrete, K.D. Sayre, J. Dixon, & L. Dendooven, (2009). Conservation agriculture and soil carbon sequestration: between myth and farmer reality. Critical Reviews in Plant Science, 28(3):97–122.

Hijmans, R.J., S.E. Cameron, J.L. Parra, P.G. Jones and A. Jarvis, 2005. Very high resolution interpolated climate surfaces for global land areas. *International Journal of Climatology* 25: 1965–1978

Höhne, N., Ellermann, C., & Li, L. (2014). Intended Nationally Determined Contributions under the UNFCCC. Ecofys Discussion paper. Cologne: Ecofys. Online at http://mitigationpartnership.net/sites/default/files/u1585/discussion-paper-indcs.pdf. Accessed on August 23, 2015.

Hoogenboom, G., J.W. Jones, P.W. Wilkens, C.H. Porter, K.J. Boote, L.A. Hunt, U. Singh, J.L. Lizaso, J.W. White, O. Uryasev, F.S. Royce, R. Ogoshi, A.J. Gijsman, and G.Y. Tsuji. 2010. Decision Support System for Agrotechnology Transfer (DSSAT) Version 4.5 [CD-ROM]. University of Hawaii, Honolulu, Hawaii.

INS (Institut National de la Statistique du Mali, Mali). 2009. 2009 Census data. Online at http://www.instat-mali.org/ Accessed on August 23, 2015.

IPCC (Intergovernmental Panel on Climate Change). 2007. Summary for policymakers. Climate change 2007: the physical science basis. Working Group I Contribution to IPCC Fourth Assessment Report: Climate Change 2007, Geneva

Jahnke, H.E. (1982). Livestock Production Systems and Livestock Development in Tropical Africa, Kiel.

Johnson, M., Takeshima, H., & Gyimah-Brempong, K. (2013). Assessing the potential and policy alternatives for achieving rice competitiveness and growth in Nigeria. IFPRI Discussion Paper #01301.

Jones, J.W., G. Hoogenboom, C.H. Porter, K.J. Boote, W.D. Batchelor, L.A. Hunt, P.W. Wilkens, U. Singh, A.J. Gijsman, and J.T. Ritchie. 2003. DSSAT Cropping System Model. European Journal of Agronomy 18:235–265.

Just, R., & Pope, R. D. (1979). Production function estimation and related risk considerations. American Journal of Agricultural Economics, 61(2), 276–284.

Kamuanga M.J.B., J. Somda, Y. Sanon, and H. Kagoné. 2008. Livestock and regional market in the Sahel and West Africa Potentials and challenges. Online at http://www.oecd.org/swac/publications/41848366.pdf, accessed August 21, 2015.

Kassie, M., H. Teklewold, P. Marenya, M. Jaleta, & O. Erenstein. 2015. Production risks and food security under alternative technology choices in Malawi: Application of a multinomial endogenous switching regression. Journal of Agricultural Economics, 66(3):640–659.

Krishnan, P., & M. Patnam. (2014). Neighbors and extension agents in Ethiopia: Who matters more for technology adoption? American Journal of Agricultural Economics, 96(1):308–327.

Lal R. 2015. Soil carbon sequestration and aggregation by cover cropping. Journal of Soil And Water Conservation 70(6):329–339.

Lal, R. (2004). Soil carbon sequestration to mitigate climate change. Geoderma, 123(1), 1–22.

Lal, R. (2011). Sequestering carbon in soils of agro-ecosystems. Food Policy, 36, S33-S39.

Leakey, A. D. (2009). Rising atmospheric carbon dioxide concentration and the future of C4 crops for food and fuel. Proceedings of the Royal Society of London B: Biological Sciences, rspb-2008:1–11.

Liniger, H., & Critchley, W. (2007). Where the land is greener. Bern, Switzerland: CTA, FAO, UNEP, CDE on behalf of the World Overview of Conservation Approaches and Technologies (WOCAT).

Manna, M. C., Swarup, A., Wanjari, R. H., Ravankar, H. N., Mishra, B., Saha, M. N., ... & Sarap, P. A. (2005). Long-term effect of fertilizer and manure application on soil organic carbon storage, soil quality and yield sustainability under sub-humid and semi-arid tropical India. Field crops research, 93(2), 264–280.

Marenya, P., Smith, V. H., & Nkonya, E. (2014). Relative Preferences for Soil Conservation Incentives among Smallholder Farmers: Evidence from Malawi. American Journal of Agricultural Economics, 96(3):690–710.

Ministère de l'Agriculture (2009). Bilan de la Campagne Agricole de l'Initiative Riz (2008–2009). Secrétariat Général, Ministère de l'Agriculture, République du Mali. Bamako, Mali.

Moussa, B., Nkonya, E., Meyer, S., Kato, E., Johnson, T., & Hawkins, J. (2016). Economics of land degradation and improvement in Niger. In Economics of Land Degradation and Improvement–A Global Assessment for Sustainable Development (pp. 499–539). Springer International Publishing.

Nazlioglu, S., & Soytas, U. (2012). Oil price, agricultural commodity prices, and the dollar: A panel cointegration and causality analysis. Energy Economics, 34(4), 1098–1104.

NBS (National Bureau of Statistics). 2012. Annual Abstract of Statistics, 2012

Nelson, G. C., Rosegrant, M. W., Koo, J., Robertson, R., Sulser, T., Zhu, T. Ringler, C. Msangi, S. Palazzo, A. Batka, M. Magalhaes, M. Valmonte-Santos, R. Ewing, M. Lee D. (2009). Climate change: Impact on agriculture and costs of adaptation (Vol. 21). Intl Food Policy Research Institute. Washington DC

Nielsen T., F. Schünemann, E. McNulty, M. Zeller, E.M. Nkonya, E. Kato, S. Meyer, W. Anderson, T. Zhu, A. Queface, L. Mapemba. 2015. The food-energy-water security nexus: Definitions, policies, and methods in an application to Malawi and Mozambique. IFPRI Discussion paper #01480 pp. 71.

Dixon, J. A., Gibbon, D. P., & Gulliver, A. (2001). *Farming systems and poverty: improving farmers' livelihoods in a changing world*. Food & Agriculture Organization.

Nkonya E., H. Takeshima, T. Johnson, L. You, H. Xie, M. Adesugba, E. Kato, J. Ogbe, A. Madukwe, and T. Edeh. 2015b. Turning Tragedy into Opportunity: Water Management Solutions for Flood Recession and Dry Season Farming in Nigeria. IFPRI mimeo.

Nkonya E., Kwapong N.A., B. Bashaasha, M. Mangheni and E. Kato. 2013. Effectiveness of pluralistic and demand-driven and versus supply-driven agricultural extension services in Africa: Which reaches more farmers and women? The case of Uganda. IFPRI mimeo.

Nkonya E., T. Johnson, H.Y. Kwon, and E. Kato. 2016a. Economics of land degradation in sub-Saharan Africa In: E. Nkonya, A. Mirzabaev and J. von Braun (eds). Economics of Land Degradation and Improvement – A Global Assessment for Sustainable Development. Springer, New York: 215–260.

Nkonya, E., Anderson, W., Kato, E., Koo, J., Mirzabaev, A., von Braun, J., & Meyer, S. (2016b). Global cost of land degradation. In Economics of Land Degradation and Improvement–A Global Assessment for Sustainable Development (pp. 117–165). Springer International Publishing, New York.

Nkonya, E., Place, F., Kato, E., & Mwanjololo, M. (2015a). Climate risk management through sustainable land management in Sub-Saharan Africa. In Sustainable Intensification to Advance Food Security and Enhance Climate Resilience in Africa (pp. 75–111). Springer International Publishing.

Nkonya, E., J. Pender, E. Kato. 2008. "https://www.cambridge.org/core/journals/environment-and-development-economics/article/who-knows-who-cares-the-determinants-of-enactment-awareness-and-compliance-with-community-natural-resource-management-regulations-in-uganda/011594149EC749785A2614DCBC6600BA" Who knows who cares? Determinants of enactment, awareness and compliance with community natural resource management regulations in Uganda," Environment and Development Economics 13(1),79-109.

Oladeebo, J. O., & Fajuyigbe, A. A. (2007). Technical efficiency of men and women upland rice farmers in Osun State, Nigeria. Journal of Human Ecology, 22(2): 93–100.

Pender, J., Ringler, C., Magalhaes, M., & Place, F. (2009). The role of sustainable land management for climate change adaptation and mitigation in sub-Saharan Africa. TerrAfrica.

RDM (Republique du Mali). 2007. Government Programme d'Action National d'Adaptation aux Changements Climatique.

RDM (Republique du Mali). 2009. National Strategy for the Development of Rice Growing. Online at http://www.jica.go.jp/english/our_work/thematic_issues/agricultural/pdf/mali_en.pdf

Richards M, Bruun TB, Campbell B, Gregersen LE, Huyer S, Kuntze V, Madsen STN, Oldvig MB, Vasileiou I. 2015. How countries plan to address agricultural adaptation and mitigation: An analysis of Intended Nationally Determined Contributions. CCAFS Info Note. Copenhagen, Denmark: CGIAR Research Program on Climate Change, Agriculture and Food Security (CCAFS). http://bit.ly/1Yfsotb

Ringler C. and E. Nkonya. 2012. Sustainable land and water management policies. In Lal R. and B. Stewart (eds). Soil Water and Agronomic Productivity. Advances in soil science. CRC Press Taylor Francis Group, New York: 523–538.

Salazar M.C. and M.P. Chacón. 2011. The case of Costa Rica. In: Greiber T. and Simone Schiele (Eds.). Governance of Ecosystem Services. Gland, Switzerland: IUCN. xii + 140 pp

Sauer, J., Tchale, H., & Wobst, P. (2007). Alternative soil fertility management options in Malawi: an economic analysis. Journal of Sustainable Agriculture, 29(3), 29–53.

Schlenker, W., & Lobell, D. B. (2010). Robust negative impacts of climate change on African agriculture. Environmental Research Letters, 5(1), 014010.

Takeshima, H. and Edeh, H. (2013), Typology of Farm Households and Irrigation Systems: Some Evidence from Nigeria, IFPRI Discussion Paper 01267. International Food Policy Research Institute, Washington D.C.

Torero, M. (2015). Consistency between Theory and Practice in Policy Recommendation by International Organizations for Extreme Price and Extreme Volatility Situations.

UNDP (United Nations Development Program). 2015. Human Development Report 2015. Work for Human Development. UNDP, New York.

UNFCCC. 2015a. Adoption of the Paris Agreement. United Nations, New York.

UNFCCC. 2015b. INDC submissions. Available at http://www4.unfccc.int/submissions/INDC/Submission%20Pages/submissions.aspx. Accessed on August 23, 2015.

United Nations Framework Convention on Climate Change (UNFCCC). 2014a. "National Adaptation Programmes of Action Received by the Secretariat."

United Nations Framework Convention on Climate Change (UNFCCC). 2014b. "Appendix II – Nationally Appropriate Mitigation Actions of Developing Country Parties." Online at http://unfccc.int/meetings/cop_15/copenhagen_accord/items/5265.php. Accessed on August 23, 2015.

van Koppen, B., Hope, L., & Colenbrander, W. (2013). Gender aspects of small-scale private irrigation in Africa International Water Management Institute (IWMI) Working Paper # 1543.

Vanlauwe B., K. Descheemaeker, K. E. Giller, J. Huising, R. Merckx, G. Nziguheba, J. Wendt, and S. Zingore. 2015. Integrated soil fertility management in sub-Saharan Africa: unravelling local adaptation. Soil, 1:491–508.

World Bank. 2014. World Development Report, 2014. Risk and Opportunity Managing Risk for Development. World Bank, Washington DC., 344pp

World Bank. 2015. World Development raw database. Online at http://databank.worldbank.org/data/reports.aspx?source=world-development-indicators. Accessed January 14, 2017.

Zseleczky, L., & Yosef, S. (2014). Are shocks becoming more frequent or intense? Resilience for Food and Nutrition Security, 9.

**Open Access** This chapter is distributed under the terms of the Creative Commons Attribution-NonCommercial-ShareAlike 3.0 IGO license (https://creativecommons.org/licenses/by-nc-sa/3.0/igo/), which permits any noncommercial use, duplication, adaptation, distribution, and reproduction in any medium or format, as long as you give appropriate credit to the Food and Agriculture Organization of the United Nations (FAO), provide a link to the Creative Commons license and indicate if changes were made. If you remix, transform, or build upon this book or a part thereof, you must distribute your contributions under the same license as the original. Any dispute related to the use of the works of the FAO that cannot be settled amicably shall be submitted to arbitration pursuant to the UNCITRAL rules. The use of the FAO's name for any purpose other than for attribution, and the use of the FAO's logo, shall be subject to a separate written license agreement between the FAO and the user and is not authorized as part of this CC-IGO license. Note that the link provided above includes additional terms and conditions of the license.

The images or other third party material in this chapter are included in the chapter's Creative Commons license, unless indicated otherwise in a credit line to the material. If material is not included in the chapter's Creative Commons license and your intended use is not permitted by statutory regulation or exceeds the permitted use, you will need to obtain permission directly from the copyright holder.

# Improving the Resilience of Central Asian Agriculture to Weather Variability and Climate Change

Alisher Mirzabaev

**Abstract** Central Asia is projected to experience significant climate change, combined with increased weather volatility. Agriculture is a key economic sector and a major source of livelihoods for Central Asia's predominantly rural population, especially for the poor. Agricultural production, being sensitive to weather shocks and climate volatility, may be negatively affected by climate change if no adaptive actions are taken. Climate smart technologies could help in strengthening the resilience of agricultural producers in the region to increased weather variability due to climate change. This study identifies the key barriers and opportunities for a wider adoption of climate smart technologies and also evaluates their potential impacts on agricultural revenues of differentiated groups of agricultural producers, with a focus on the poor. Adoption of climate smart agricultural technologies was found to raise farming profits of both poorer and richer households, although these positive impacts may likely to be higher for richer households. The study also shows that policies facilitating improved access to markets and agricultural extension services, as well as higher commercialization of household agricultural output may increase the adoption of climate smart agricultural technologies in the region.

## 1 Introduction

The four countries of Central Asia – Kazakhstan, Kyrgyzstan, Tajikistan and Uzbekistan – are located in arid, semiarid and sub-humid regions. The climate in the region is intrinsically volatile, with often recurring weather shocks, such as droughts, heatwaves, frosts and hails (Gupta et al. 2009). Agriculture is an important sector for the region. Even in richer Kazakhstan, where the share of agriculture is 6% of the Gross Domestic Product (GDP), it employs almost 30% of the labor force. In the rest of the region, the share of agriculture in GDP is as high as 30% in Kyrgyzstan, and in employment as high as 66% in Tajikistan (Mirzabaev 2013). Thus,

A. Mirzabaev (✉)
University of Bonn, Bonn, Germany
e-mail: almir@uni-bonn.de

agriculture is a major source of livelihood, especially for the rural poor. Because agricultural production is sensitive to weather, increased weather variability due to climate change may have a negative impact on agricultural production and farming incomes. Therefore, appropriate actions are needed to dynamically adapt the agricultural practices to changing climatic and weather conditions (Zilberman et al. 2012). In this context, the poorest rural households are more vulnerable to climate change because they have lower adaptive capacities and higher dependence on farming incomes. In fact, Mirzabaev (2013) finds that every 10% decrease in farming incomes due to weather variability in the region is likely to reduce the per capita food consumption of the poorest quartile of households by 5.2%, while a similar decrease in farming incomes would result only in 3.9% decrease in the per capita food expenses of the richest 10% of rural households. Taking this into account, any analysis of adaptation to climate change would be deficient unless it specifically looks into the factors that enable or prevent the poorest agricultural households from adapting to increased weather variability and climate change. Ultimately, major impacts of climate change are expected to be not through aggregate changes, but through their distributional effects (Zilberman et al. 2004).

Despite a decade of strong economic growth, rural incomes remain low in many parts of the region, with related challenges of food insecurity and rural poverty. Adaptive actions are required not only to cope with weather shocks, but also for being resilient enough to successfully overcome the negative impacts of weather shocks and achieve agricultural growth and rural poverty reduction. The adoption of sustainable and climate-smart agricultural technologies (CSATs) could help in increasing such a resilience of agricultural households to climate change (Lipper et al. 2014). This is especially important in the context of significant uncertainties about the direction and magnitudes of climate change impacts in Central Asia. Limited resources require that these adaptive actions are made up of no-regret measures, capable of positively contributing to regional food security, agricultural growth and poverty reduction even with perfect climate change mitigation.

Based on the above problem compounded by regional challenges, the proposed study seeks to answer the following research questions:

1. what are the key barriers and catalysts for the adoption of CSATs in Central Asia, and
2. what may be the distributional effects of the adoption of CSATs on the farming revenues of different categories of agricultural households, with a focus on the poor?

## 2 Literature Review

### 2.1 Climate Change in Central Asia

The regional downscaling of IPCC forecasts for Central Asia (de Pauw 2012) indicates that there may be likely increases in the average annual mean, minimum and maximum temperatures throughout the region, though the temperature increases would be lower in the west of the region near the Caspian Sea, and higher in the north of the region (de Pauw 2012). In general, the annual precipitation may increase in the region, with higher increases in the north of the region, and some slight decreases in the south of the region. Spring and fall precipitations are likely to increase while summer precipitation to decrease. Wetter winters may be more frequent, as well as drier springs, summers and autumns. However, unlike the temperature projections there are big disagreements among different models on the direction and magnitudes of precipitation changes in the region. Warming could increase the water run-off in Central Asia for decades, or even centuries as suggested by Gupta et al. (2009). However, the seasonality of runoff may change, with more runoff in spring and less in summer (ibid). Moreover, Stulina (2008) indicates that forecasts of the flow of the Amudarya and Syrdarya Rivers strongly vary depending on the model. For example, under the Geophysical Fluid Dynamics Laboratory (GFDL) model of the United States' National Oceanic and Atmospheric Administration (NOAA), there may be 1% increase in the average flow of Syrdarya and no change in the flow of Amudarya by 2030. In contrast, using the Canada Climate Change Model (CCCM) may lead to predictions of significant decreases in the flow of both rivers, −28% and −40% for Syrdarya and Amudarya, respectively (ibid.). All in all, the climate change forecasts for Central Asia indicate that temperatures may be rising all across the region. There is no consensus in precipitation and water run-off predictions.

Mirzabaev (2013) estimates the aggregate impacts of climate change on Central Asian agriculture to range between +1.21% and −1.43% of net crop production revenues by 2040. Though small in relative terms, the absolute monetary impact is not negligible, ranging from +180 mln USD annually in the optimistic scenario, to − 210 mln USD annually in the pessimistic scenario relative to 2010 levels, where optimistic and pessimistic scenarios are defined to correspond to B1 (lowest future emission trajectory) and A1FI (highest future emission trajectory) scenarios by IPCC (2007), respectively. However, these aggregate impacts have significant geographic and socio-economic distributional effects, whereby the poorer provinces in Central Asia and poorest agricultural households would be affected more negatively by climate change due to their lower adaptive capacities and higher dependence on agricultural incomes (Mirzabaev 2013). This is also supported by several other studies on the region. Nelson et al. (2010) find that by 2050, climate change may lead to higher rainfed wheat yields in Kazakhstan and Kyrgyzstan (by 0–11%), while in Tajikistan, Turkmenistan and Uzbekistan rainfed wheat yields may decline (by 8–18%). The yields for irrigated wheat may decrease in all countries (by 7–14%), except in Uzbekistan (+1%). Sommer et al. (2013) find that wheat yields may grow on average by +12% across Central Asia, ranging from − 3% to +27%. Bobojonov et al. (2012) estimate that during 2040–2070, the climate change may increase agricultural incomes in northern rainfed areas of Central Asia (in some areas by up 50%), and reduce incomes in the southern irrigated areas, especially under the conditions of water scarcity (in some areas by more than 17%).

As we can see from these studies, major impacts of climate change in Central Asia are likely to be through their negative effects on the poorest agricultural households, while the aggregate effects do not seem to be substantial relative to the overall economy. Therefore, the link between climate change and poverty is vital for responses to climate change in the region. In this regard, climate smart agriculture may help reduce vulnerability by stabilizing or even increasing agricultural production (Meinzen-Dick et al. 2012; Wheeler and von Braun 2013).

## 2.2 The Role of Climate-Smart Agricultural Technologies

Climate smart agriculture is an approach to transform agricultural systems and to support food security under a changing climate by providing context-specific and flexible solutions (Lipper et al. 2014). In general, climate-smart agriculture has three objectives (McCarthy and Brubaker 2014):

1. Increasing agricultural productivity in a sustainable way, and to support equitable increases in farm income, food security and development
2. Strengthening the resilience of agricultural and food security systems towards climate change
3. Reducing greenhouse gas emissions from agriculture (including crops, livestock and fisheries).

Thus, climate-smart agriculture has, in fact, wide-reaching implications beyond narrowly defined climate change and adaptation to it, and covers a broad spectrum of sustainable development objectives. Climate-smart agriculture involves technological, institutional and policy solutions. For example, a crop rotation with nitrogen-fixing crops increases biomass production. Diverse production systems tend to produce more biomass than monocultures (Tilman et al. 2006), which also entails opportunities of additional carbon storage. Improved water management also has impacts on biomass production as it can increase the amount of water in the root zone and therefore enhances carbon sequestration potential (Kimmelshue et al. 1995). Reduced tillage could lead to decreases in carbon losses (Branca et al. 2011). Gupta et al. (2009) indicate several dozen of such climate smart technologies experimented with in Central Asia for the last decade, such as zero tillage, direct seeding, cutback and zigzag irrigation, double cropping, etc. The corresponding economic analyses of these technologies also show that many of them have positive cost-benefit ratios (Pender et al. 2009), i.e. could be used as no-regret options for both adapting to climate change, sustainably managing soil and water resources, and raising farming productivity and incomes.

## 3 Conceptual Framework

The adoption of CSATs, like those listed above, could be highly useful to strengthen the resilience of the agricultural households and improve their capacities to adapt to climate change. In general, adaptation can be considered as all changes an individual or an institution, such as government, makes to adjust to a changing environment (Osberghaus et al. 2010; Seo 2011). However, when faced with slow onset uncertain risks such as climate change, raising public awareness could be necessary for correct attribution of the causes of on-going climatic changes and appropriate reactions to these changes. It also needs to be acknowledged, as suggested by Nhemachena and Hassan (2007) and Mertz et al. (2009), that adaptation measures undertaken by farmers may have other driving forces than actual climate effects. For this reason, adaptation actions are a function of both perceiving the risks associated with the climate change, but are also dependent on personal environmental knowledge and beliefs, as well as personal characteristics such as gender, age, education, etc. (O'Connor et al. 1999).

Adaptation can be classified into two categories: (i) private adaptation and (ii) public adaptation (Mendelsohn 2000). Private adaptation is undertaken by individuals themselves seeking to maximize their utility, while public adaptation is undertaken by governments seeking to achieve a higher public benefit for the entire society (Osberghaus et al. 2010). Adaptation can happen *ex post* or *ex ante* to a climatic shock (Mendelsohn 2000).

The vulnerability of agricultural production to climatic and weather changes is greatly modulated by timely adaptation and coping actions. However, when evaluating uncertain and low probability events individuals may often make decisions based on their intuitive risk judgments, i.e. perceptions, rather than rational expected utility maximization (Tversky and Kahneman 1986), which is influenced by individual's previous experiences, education, age, gender, socio-economic, institutional, cultural and other characteristics. However, perceiving climate change is not by itself sufficient for adapting to it. One of the key incentives for successful adaptation is when agricultural producers do perceive that climate is changing and that this change is affecting their agricultural activities, necessitating them to take appropriate actions to modify their farming practices to better suit the new climate. Households start adapting only when the costs of inaction on the changes that they perceive outweigh the costs of adaptive actions. Even if households perceive certain changes in the climate, they may still be unwilling to incur costs of adapting to these changes if these changes do not pose a sufficiently high level of damage risk, especially since individuals tend to underestimate the occurrence of low probability events (Tversky and Kahneman 1986).

Even when households perceive the changes and are willing to take adaptive actions, they may still be constrained by low adaptive capacities. Households' adaptive capacities, in turn, depend on their resource endowments, specifically, their access to five "capitals": human, natural, financial, social and physical (Chambers and Conway 1992), which largely fashion households' resilience to external shocks, including weather and climate shocks. A major purpose of the analysis would be to

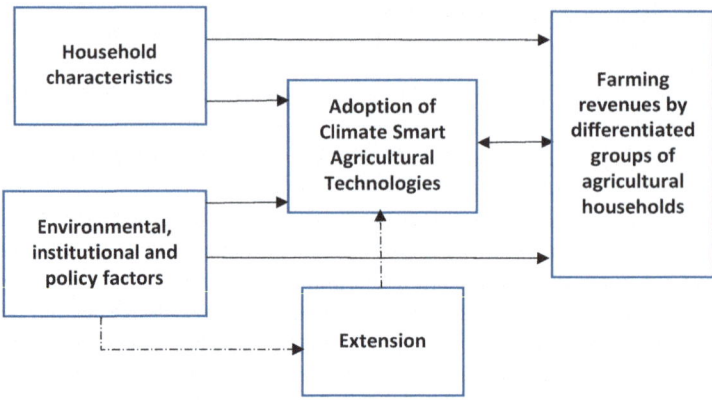

**Fig. 1** The conceptual framework of causal relationships

estimate the impact of the adoption of CSATs by different categories of surveyed households, with a focus on the poorer households.

Following these broad outlines, this study is based on the conceptual framework of causal links shown below (Fig. 1), which also motivated the empirical strategy outlined in the following section.

The conceptual framework indicates that both the adoption of climate smart technologies and farming incomes depend on the characteristics of agricultural households and the environmental, institutional and policy factors affecting the decision making by households. The adoption of CSATs would also affect the farming incomes. However, the relationship is likely to be endogenous, whereby farming incomes of the households would also influence their decisions about the adoption of CSATs. To address this endogeneity, access to agricultural extension services would be used as an instrumental variable. Extension services on CSATs are provided for free to agricultural households in the region (also corroborated by the survey dataset used in this study) by government-run farming associations or non-profit organizations, and therefore, is likely to influence farming incomes only through its impact on the adoption of CSATs, and is not influenced by farming incomes of the households.

## 4 Empirical Framework

### 4.1 Data

The dataset used for this analysis comes from nationally representative agricultural household surveys conducted in the four countries of Central Asia studied in this paper. The survey covers the 2009–2010 cropping season. The multi-stage survey sampling was conducted in a way to ensure representativeness of the survey sample with the overall population of agricultural producers: farmers and household producers, across different agro-ecologies and farming systems in each country. The confidence interval of 95% was used to calculate the sample size. The calculated sample size varied between 380 and 385 respondents between the countries. To compensate for any missing or failed cases, the actual sample size for each country was determined to be 400 respondents, i.e. 1600 respondents in total.

Uzbekistan and Kazakhstan (the larger countries in the dataset) were first divided into major agro-ecological zones – west, south, center and east for Uzbekistan, north, center, west, south and east for Kazakhstan. Then in each zone, one province was randomly selected. In the case of Tajikistan and Kyrgyzstan (the smaller countries) all provinces were selected for further sampling of villages in each of them. The number of respondents was allocated to each province depending on the share of the agro-ecological zone (or province, in the cases of Tajikistan and Kyrgyzstan) in the value of the national agricultural production.

Following this, the total list of villages was obtained for each province selected. The villages in each province were numbered, and the corresponding numbers for the selected villages were randomly drawn (35 villages in Kazakhstan, 22 in Kyrgyzstan, 25 in Tajikistan, 25 in Uzbekistan). The number of respondents per village was evenly distributed within each province. At the village level, the list of all agricultural producers, including household producers, were obtained from the local administrations; agricultural producers were numbered, and then from this numbered list, respondents were randomly selected. Due to civil unrest during most of 2010 in southern Kyrgyzstan, it was impossible to include the three provinces in the south of Kyrgyzstan in the sampling. Similarly, Gorno-Badahshan autonomous province of Tajikistan was also excluded from sampling due its trivial share in agricultural production and population, as well as extremely high surveying costs due to its location in high altitude areas with difficult access (Fig. 2). In summary, in spite of these geographical gaps, the selected samples are well representative of the key areas in the region in terms of their share in the overall agricultural production, population, and different income levels.

## 4.2 Methods

As an initial step, an exploratory analysis of the survey datasets is conducted with the purpose of highlighting the major characteristics of the surveyed households. Then, two-stage regression is run to identify the impact of adoption of climate smart technologies on net farming profits. The purpose of the two-stage procedure is to address the endogeneity between the farming incomes and the adoption of CSATs. In the first stage, the probit model is used to regress the variable representing the adoption of climate smart agricultural practices on a number of explanatory variables including household, farm, climatic and institutional characteristics, also including the instrumental variable: access to extension services. The motivation for

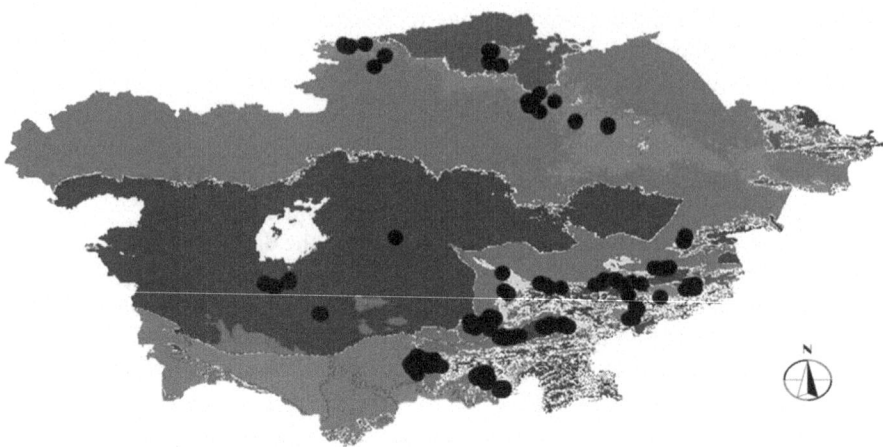

**Fig. 2** Location of surveyed households across agro-ecological zones in Central Asia

using access to extension as instrument is because it affects farming incomes through adoption of CSATs only and households' access to extension is not affected by farming incomes. Extension is usually provided by the governments or by non-profit/donor funded organizations in the region. In the second stage, net household farming profits are regressed using ordinary least squares (OLS) on the same explanatory variables as above (excluding the instrument) and the fitted values of adoption of climate smart technologies from the first stage.

However, to identify the distribution impacts on different categories of households, separate regressions are run for two categories of households. The first group with per capita daily food expenses less than the median for the whole sample (0.83 USD) – named as the "poor", and the group with per capita daily food expenses more than the medium, named as the "non-poor". The econometric model specifications for the first and second stages are given below.

### 4.2.1 The First Stage

$$A = \alpha H + \beta C + \phi F + \delta I + \mu G + \epsilon \qquad (1)$$

where,

$A$ = adoption of CSATs (0-no, 1-yes)
$H$ = a vector of household characteristics
$C$ = a vector of climate variables (temperature and precipitation, etc)
$F$ = a vector of farm characteristics, such as farm size and livestock value.
$I$ = a vector of institutional variables (market access, land tenure, etc)
$G$ = the instrumental variable: access to extension services
$\epsilon$ = error term

### 4.2.2 The Second Stage

$$\pi = \alpha H + \beta C + \phi F + \delta I + \mu fvA + e \qquad (2)$$

where,

$\pi$ = net farm profits
H = a vector of household characteristics
C = a vector of climate variables (temperature and precipitation, etc)
F = a vector of farm characteristics, such as farm size and livestock value.
I = a vector of institutional variables (market access, land tenure, etc)
fvA = fitted values of adoption of climate smart technologies from the first stage
e = error term

## 4.3 Variable Selection

Literature on the adaptation to climate change in agriculture has strong linkages to the previous research on adoption of new technologies by agricultural producers, including under risky decision making contexts (Zilberman et al. 2012). Based on the previous lines of research and earlier work on agricultural adaptation to climate change *per se*, it is hypothesized that there are a number of variables which influence the adoption of CSATs. These variables are grouped into four major categories, following Gbetibouo (2009): (i) household characteristics, corresponding to human dimension of the five "capitals", (ii) farm characteristics (physical capital), (iii) climate-related variables (natural capital), and (iv) institutional variables.

## 4.4 Household Characteristics

**Family size, age, education and gender of the household head** are standard variables used in most adaptation and agricultural technology adoption studies, though there is no firm theoretical consensus on the direction of their impact on adaptation/adoption. In most cases, this is a matter of empirical analysis and can differ from one context to another. Income of the household may have an effect on adaptation as richer households have more resources and relatively greater adaptive capacities making them more likely to adapt. To capture the income status of the households the **value of total household assets** are used.

## 4.5 Farm Characteristics

**Total farm size** is expected to have a positive effect on technology adoption as economies of scale could allow undertaking adaptation measures with scale-sensitive costs. Many rural households in Central Asia keep livestock as one of the key saving and investment strategies, hence **the value of the livestock owned** (different from income status) by the household can be a good indicator of the level of adaptive capacity.

## 4.6 Climatic Characteristics

Higher **frequency of climatic shocks** can provide with more incentives for adaptation. Significance of these variables would also corroborate the intuition that unless Governments encourage farmers for *ex ante* adaptation most of adaptation to climate change could be *ex post*. It is believed that many impacts of climate change would be felt along the **agro-ecological zones**, hence the estimation includes indicators for agro-ecological zones. Higher long-term climate variability (30 years, 1980–2010) in terms of more **variable temperature and precipitation** could necessitate a more adaptive behavior. Finally, the estimation also takes into account **long-term average precipitation and temperature** (30 years, 1980–2010). The climate variables have been compiled for about 400 weather stations across Central Asia. The data come from national meteorological agencies, Williams and Konovalov (2008), NASA's Global Summary of the Day, and other sources. Climate variables from individual weather stations were spatially projected to the digital map of Central Asia using spatial interpolation technique of inverse weighted distance. Following this, corresponding weather variables were extracted for each household using the GPS location of the household.

## 4.7 Institutional Characteristics

**Land tenure** is a potentially important factor influencing farmers' decisions, including those on adapting to climate change (Quan and Dyer 2008). Adaptation to climate change may lead to increased production costs and/or necessitate long-term farm investments. Quan and Dyer (2008) note that secure land tenure arrangements are needed for better climate change adaptation. Farmers in Central Asia may operate several parcels with different tenure arrangements ranging from privately owned to those leased from the State. To measure this in one variable, taking into account different levels of incentives for long-term investments inherent to different land tenure arrangements, the share of privately owned land area in the total farm size is used in the model, even though, admittedly, this variable may not perfectly capture the tenure security. Higher **market access** would normally lead to more adaptation. The **country dummies** are included to implicitly account for other country-specific characteristics that are not included in the models explicitly. The **intensity of nighttime lighting** (DMSP-OLS Nighttime Lights Time Series, NOAA's National Geophysical Data Center, using the data collected by US Air Force Weather Agency) is used **as a proxy for availability of electricity.** More lighting could indicate at economic dynamism of the region and availability of non-farm job opportunities.

## 5 Results and Discussion

The survey responses show that about 62% of the respondents adopted at least one CSAT. The adoption rates among the poorer half of the households are lower than among richer half of the households (Fig. 3).

Many households in the surveyed sample, especially poorer households, report to be constrained in their technology adoption by lack of credit, inputs, water and information (Fig. 4). Major constraints to adaptation that are faced specifically by poorest one third of agricultural households are found to be lack of access to credit and inputs.

Table 1 presents the mean values for some major variables used in the analysis segregated by country. Country-wise in the region, the adoption of climate smart technologies is higher in Uzbekistan and Tajikistan, and the lower in Kyrgyzstan. Agricultural households in Uzbekistan and Tajikistan also report to have much higher access to extension services. There are no major differences among the households in the countries of the region in terms of their demographic characteristics. Farm sizes are the largest in Kazakhstan and lowest in Tajikistan and Kyrgyzstan. In terms of total assets, the households in Kazakhstan are much richer than those in other countries of the region. In general, Table 1 shows that despite considerable similarities across the countries of the region, there are also substantial structural and institutional differences, which need to be taken into account while devising country-specific polices for promoting resilience to climate change. In this paper, the focus is on common patterns across the region and discuss about development policies that could be beneficial across different settings in the region.

Specifically, Table 2 presents the findings on the major determinants of the adoption of CSATs (CSATs) in the region (the first stage of the estimation) and Table 3 presents the estimation of potential impacts of adoption of CSATs among households of two categories (the poorer half of the sample and the richer half) (2nd stage).

The results in the first stage indicate that the selected instrument, extension services, positively influences the adoption of climate smart technologies with statistical significance at 5%. The F-statistic of the excluded instrument is equal to 19.43, which points also at the statistical validity of the instrument. Access to extension would increase the knowledge and information of the households about CSATs and the ways to apply them in their farms, thus allowing for higher adoption of CSATs

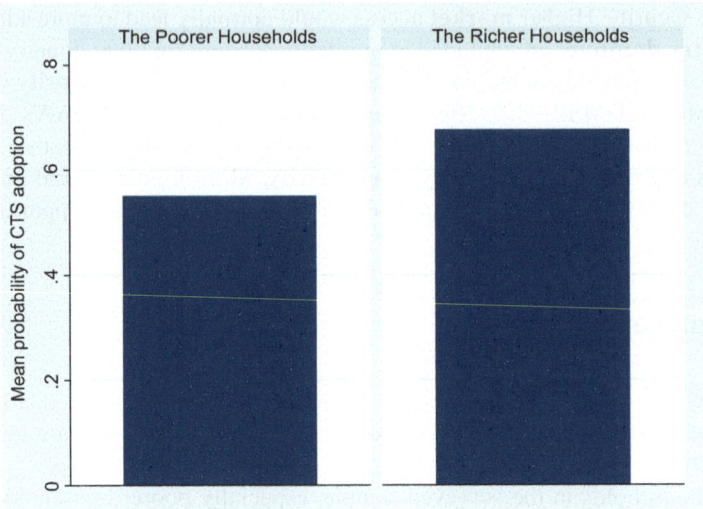

**Fig. 3** CSAT adoption by households according to per capita food consumption (below and above medium food expenses per capita). 0 = no, 1 = yes

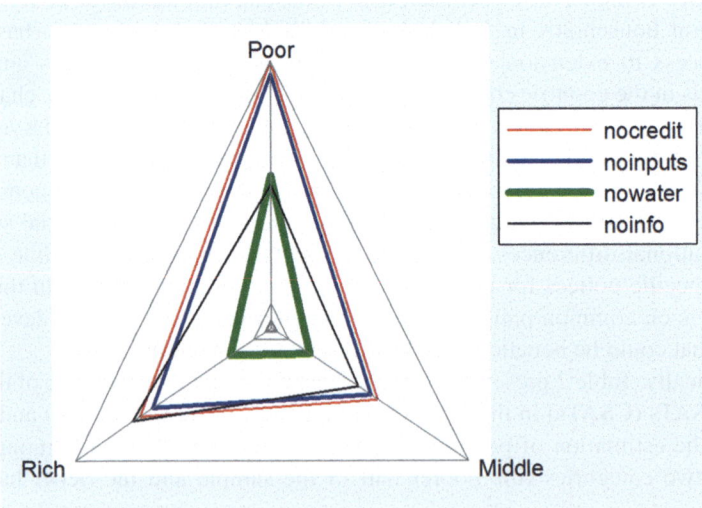

**Fig. 4** Constraints to CSAT adoption by household's economic status (Terciles food expenses per capita)

**Table 1** Mean values of key household, institutional, and environmental characteristics

| Variables | Kazakhstan | Kyrgyzstan | Tajikistan | Uzbekistan |
|---|---|---|---|---|
| Number of climate smart technologies used | 2.8 | 0.2 | 4.4 | 4.9 |
| Household size | 6 | 6 | 8 | 6 |
| Age of household head in years | 51 | 50 | 52 | 47 |
| Length of growing periods in months | 97 | 102 | 131 | 92 |
| Annual precipitation in mm | 402 | 448 | 486 | 289 |
| Annual temperature in degree Celsius | 7.0 | 5.7 | 14.4 | 14.4 |
| The number of weather shocks during the last 5 years | 2.7 | 0.4 | 1.1 | 1.4 |
| Land tenure (0-not private, 1-private) | 0.63 | 0.90 | 0.73 | 0.60 |
| Farm size in hectares | 194 | 5 | 4 | 28 |
| Access to extension (binary) | 0.1 | 0.2 | 0.7 | 0.7 |
| Value of livestock (in USD) | 5255 | 8998 | 869 | 6796 |
| Distance to markets (in minutes) | 133 | 150 | 59 | 75 |
| Value of total assets (in USD) | 83,123 | 20,727 | 7407 | 34,939 |

Source: Mirzabaev (2013)

(Deressa et al. 2009). However, as we have seen in Table 1 and Fig. 4, access to extension services and to information remains inadequate in many parts of the region. Hence, there is a need for public polices and development interventions facilitating greater access to extension among agricultural producers. Other major factors found to be positively affecting the adoption of climate smart technologies are better access to markets and commercialization of the households farming activities (vs. subsistence production), length of growing days and frequency of weather shocks.

Both access to markets and higher commercialization of the produce allow households to increase the profitability of their sales, thus incentivizing and enabling them to make investments into CSATs (von Braun and Kennedy 1994). In the face of higher frequency of weather shocks households might adopt CSATs precisely for increasing the resilience of their production activities against these shocks. The factors which are negatively associated with the adoption are night time lighting intensity, share of the land privately owned. Moreover, the likelihood of adoption of climate smart technologies is lower in more humid areas as compared to arid agro-ecological zones. The night time lighting intensity is used here as a proxy for availability of non-farm jobs. Better access to non-farm jobs could increase the opportunity costs of farm labor, thus making it costlier to adopt labor-intensive CSATs. Whereas the negative impact of the share of land held under private property on adoption is surprising, it is true that this variable used here is an imperfect proxy for land tenure security.

The second stage of the regression shows that the adoption of CSATs has a positive impact on the net farming profits of both poorer and richer households, with higher levels of positive impacts on the net profits of the richer households. The share of land under private property is another major factor positively affecting farm

**Table 2** The 1st stage results. Determinants of CSAT adoption

| Variables | Coef. | Conf. interval |
|---|---|---|
| **Access to extension (binary)** | **0.527\*\*\*** | **(0.294 to 0.761)** |
| Age of household head | 0.0485\* | (−0.00334 to 0.100) |
| Age of household head, squared | −0.000398 | (−0.000892 to 9.55e-05) |
| Education of HH head | −0.0324 | (−0.368 to 0.303) |
| Education of HH head, squared | 0.0134 | (−0.0558 to 0.0826) |
| Gender of HH head (0-female, 1-male) | −0.0545 | (−0.299 to 0.190) |
| Family size | 0.000330 | (−0.0289 to 0.0295) |
| Distance to markets (log) | −0.134\*\* | (−0.249 to −0.0191) |
| Night-time lighting intensity | −0.0146\*\* | (−0.0267 to −0.00241) |
| Total household assets | 3.00e-06 | (−2.40e-06 to 8.41e-06) |
| Livestock value | −3.44e-06 | (−1.12e-05 to 4.35e-06) |
| Farm size (ha) | 0.00183 | (−0.00122 to 0.00488) |
| Aridity level | 0.254 | (−1.353 to 1.861) |
| Length of growing days | 0.0160\*\*\* | (0.00779 to 0.0242) |
| Share of land privately owned | −0.443\*\*\* | (−0.680 to −0.207) |
| Agro-ecological zone (base: arid) | | |
| Semiarid | −2.245\*\*\* | (−3.089 to −1.400) |
| Sub-humid | −2.702\*\*\* | (−3.697 to −1.707) |
| Humid | −1.765\*\* | (−3.186 to −0.345) |
| Subsistence farmer (binary) | 0.499\* | (−0.00377 to 1.001) |
| Mean annual temperature | 0.0314 | (−0.00965 to 0.0725) |
| Annual precipitation | 0.00142\*\* | (0.000262 to 0.00258) |
| Variance of precipitation | −0.00787\*\*\* | (−0.0111 to −0.00466) |
| Variance of temperature | 0.0406 | (−0.119 to 0.200) |
| Number of weather shocks during the last 5 years | 0.0577\*\* | (0.00453 to 0.111) |
| Country dummies (base: Kazakhstan) | | |
| Kyrgyzstan | −1.575\*\*\* | (−2.048 to −1.103) |
| Tajikistan | −0.982\*\*\* | (−1.455 to −0.508) |
| Uzbekistan | 0.955\*\*\* | (0.516 to 1.394) |
| Constant | 0.447 | (−1.306 to 2.201) |
| R-squared | 0.43 | |
| F-Statistic of the excluded instrument | 19.43 | |

\*\*\*$p < 0.01$, \*\*$p < 0.05$, \*$p < 0.1$

profitability for both poorer and richer households. On the other hand, there are several variables with statistically significant effects only under one category of households. There are no variables with statistically significant opposite signs under the two categories of households. Better access to markets, higher livestock assets, and previous experiences with weather shocks seem to be positively related to higher on-farm profitability, especially among the poorer agricultural households.

**Table 3** The 2nd stage results. Potential impacts of CSAT adoption

| Variables | Stage 2, Poorer households, net farming profits (log) Coef. | Stage 2, Richer households, net farming profits (log) Coef. |
|---|---|---|
| **CSAT adoption, fitted values** | 0.205* | 0.531*** |
| Age of household head | −0.000785 | −0.00924 |
| Age of household head, squared | 9.13E-06 | 8.45E-05 |
| Education of HH head | 0.0202 | 0.0138 |
| Education of HH head, squared | −0.00869 | −0.000772 |
| Gender of HH head (0-female, 1-male) | 0.0391 | 0.0477 |
| Family size | −0.000273 | 0.00958 |
| Distance to markets (log) | −0.0293** | 0.0161 |
| Night-time lighting intensity | 0.000417 | 0.00566*** |
| Total household assets | 4.41E-08 | 2.72E-07 |
| Livestock value | 4.01e-06*** | 1.76E-07 |
| Farm size (ha) | −0.000220* | −0.000205 |
| Aridity level | −0.245 | 0.172 |
| Length of growing days | −0.00204* | −0.00264* |
| Share of land privately owned | 0.193*** | 0.201*** |
| Agro-ecological zone (base-arid) | | |
| Semiarid | 0.230*** | 0.115 |
| Sub-humid | 0.393*** | 0.196 |
| Humid | 0.459*** | 0.0538 |
| Subsistence farmer (binary) | 0.0351 | −0.0856 |
| Mean annual temperature | 0.0132*** | 0.00905 |
| Annual precipitation | −0.00014 | −9.04E-05 |
| Variance of precipitation | 0.000254 | 0.00138*** |
| Variance of temperature | −0.0804*** | −0.0545** |
| The number of weather shocks during the last years | 0.0122* | −0.00138 |
| Country dummies (base-Kazakhstan) | | |
| Kyrgyzstan | 0.239** | 0.348*** |
| Tajikistan | 0.125 | 0.0102 |
| Uzbekistan | 0.0219 | −0.00421 |
| Constant | 8.975*** | 8.559*** |
| R-squared | 0.244 | 0.186 |
| Number of observations | 760 | 758 |

***$p < 0.01$, **$p < 0.05$, *$p < 0.1$

## 6 Discussion and Policy Implications

The results presented in this paper indicate that the adoption of CSATs increases the on-farm profitability. In their extensive review of climate smart and sustainable land management technologies (SLM) in the region, Pender et al. (2009) also find that cost-benefit ratios of many SLM technologies such as zero tillage, mulching, improved irrigation techniques, raised bed planting, etc. are positive in the region, often substantially so, thus corroborating the findings of this paper. However, the adoption rates of these technologies remain relatively low. These low adoption rates are often due to various barriers to adoption, as discussed above, such as lack of access to credit, to extension services and to input and output markets. Financial institutions can often be unwilling to extend credit to small-scale farming households with unknown risk profile and lack of collateral to guarantee the credit. In many instance, these farming households only lease their land from the State without the legal entitlement to use their land as collateral for obtaining credit. Government policies could target expanding farmers' legal rights in using their land leased from the State as collateral for obtaining credit. An alternative option would be government-financed soft loan programs to farmers targeting the adoption of new resource-efficient and climate smart technologies. The large scale adoption of conservation tillage practices in Kazakhstan on several millions of hectares was partially found to be facilitated by government subsidies promoting this technology (Kienzler et al. 2012). However, limited public funds may serve as barriers for other such programs at a larger scale, especially in the poorer parts of the region. Furthermore, the overall impact of such soft loan or subsidy programs on poverty reduction may also be reduced by asymmetric bargaining powers and access to credit funds between richer farmers and poorer farmers.

Other more promising areas for catalyzing the adoption of CSATs include providing better access to markets, including through better infrastructure, improving the investment climate for post-harvest processing and moving towards higher liberalization of input and output markets. In some countries of the region, input and output markets, as well as acreage decisions, especially for cotton and wheat crops, are still administratively managed by governments. Although abrupt removal of these regulation could be counter-productive in the short-term, gradual liberalization of the agricultural sector is likely to improve the agricultural profitability and reduce resource misallocations.

## 7 Conclusions

Central Asia is expected to experience a significant climate change in the coming decades, even though there are high uncertainties about the exact magnitudes of these changes. Importantly, previous studies point at important distributional effects of climate change on different categories of rural agricultural households, with more negative impacts on the poor. Given the uncertainties related with climate change, there is a need for such CSATs that would strengthen the resilience of agricultural production against a variety of climatic shocks, at the same time allowing for agricultural productivity growth and rural poverty reduction. This study finds a positive impact of the adoption of CSATs on agricultural revenues, both for poorer and richer households. Despite this potential, the adoption rates of CSATs remain relatively low in the region. The findings show that policy actions targeted towards improving access to markets and agricultural extension services, and higher commercialization of household agricultural production can serve as catalysts for the adoption of CSATs by rural households.

Although there are numerous analyses of the costs and benefits of adoption of CSATs at a farm level, larger scale effects of these adoptions, and social rates of returns from adopting these technologies need yet to be studied in the region. More information on the macroeconomic and social rates of returns from investing into CSATs as well as the extent of transaction costs for the implementation of CSAT programs and initiatives could provide with the necessary evidence base for better informed policies on the promotion of CSATs in Central Asia.

**Acknowledgments** I would like to thank Julia Anna Matz and the anonymous reviewer for their insightful and very helpful comments and suggestions. The data used in this study was collected by International Center for Agricultural Research in the Dry Areas (ICARDA) under a project funded by the Asian Development Bank (ADB).

# References

Bobojonov. I, Sommer, R., Nkonya, E., Kato, E. and Aw-Hassan, A., (2012). Assessment of climate change impact on Central Asian agriculture: Bio-economic farm modeling approach. ICARDA Research report (unpublished). Syria

Branca, G. N. McCarthy, L. Lipper and M.C. Jolejole (2011): Climate Smart Agriculture: A Synthesis of Empirical Evidence of Food Security and Mitigation Benefits for Improved Cropland Management. Mitigation of Climate Change in Agriculture Series 3. Rome, Italy: Food and Agriculture Organization of the United Nations (FAO).

Chambers, R., Conway, G., (1992). Sustainable rural livelihoods : practical concepts for the 21st century. Brighton, England: Institute of Development Studies.

Deressa, T. T., Hassan, R. M., Ringler, C., Alemu, T., & Yesuf, M. (2009). Determinants of farmers' choice of adaptation methods to climate change in the Nile Basin of Ethiopia. Global environmental change, 19(2), 248–255.

de Pauw, E., 2012. Downscaling Basic Climatic Variables for Central Asia. Adapta-tion to Climate Change in Central Asia and People's Republic of China. ICARDA, Aleppo.

Gbetibouo, G.A., (2009). Understanding Farmers' Perceptions and Adaptations to Climate Change and Variability. Discussion Paper 00849. International Food and Policy Research Institute.

Gupta, R., K. Kienzler, C. Martius, A. Mirzabaev, T. Oweis, E. de Pauw, M. Qadir, K. Shideed, R. Sommer, R. Thomas, K. Sayre, C. Carli, A. Saparov, M. Bekenov, S. Sanginov, M. Nepesov, and R. Ikramov (2009) Research Prospectus: A Vision for Sustainable Land Management Research in Central Asia. ICARDA Central Asia and Caucasus Program. Sustainable Agriculture in Central Asia and the Caucasus Series No.1. CGIAR-PFU, Tashkent, Uzbekistan.

Kienzler, K. M., Lamers, J. P. A., McDonald, A., Mirzabaev, A., Ibragimov, N., Egamberdiev, O., ... & Akramkhanov, A. (2012). Conservation agriculture in Central Asia—What do we know and where do we go from here?. Field Crops Research, 132, 95–105.

Kimmelshue, J. E., Gilliam, J. W., & Volk, R. J. (1995). Water management effects on mineralization of soil organic matter and corn residue. Soil Science Society of America Journal, 59(4), 1156–1162.

Lipper, L., P. Thornton, B.M. Campbell, T. Baedeker, A. Braimoh, M. Bwalya, P. Caron, A. Cattaneo, D. Garrity, K. Henry, R. Hottle, L. Jackson, A. Jarvis, F. Kossam, W. Mann, N. McCarthy, Alexandre Meybeck, H. Neufeldt, T. Remington, P.T. Sen, Reuben Sessa, R. Shula, A. Tibu and E. Torquebiau (2014): Climate-smart agriculture for food security. In: Nature Climate Change 2014(4), pp. 1068–1072. doi:10.1038/nclimate2437

McCarthy, N. and J. Brubaker (2014): Climate-Smart Agriculture and Resource Tenure in Sub-Saharan Africa: A Conceptual Framework. Rome, FAO.

Meinzen-Dick, R., Q. Bernier and E. Haglund (2012): The six "ins" of climate-smart agriculture: Inclusive institutions for information, innovation, investment, and insurance. CAPRi Working Paper No. 114. Washington, D.C.: International Food Policy Research Institute. http://dx.doi.org/10.2499/CAPRiWP114

Mendelsohn, R., (2000). Efficient adaptation to climate change. Climatic Change, Vol. 45, 583–600

Mertz, O., Mbow, Ch., Reenberg, A. and Diouf, A., (2009). Farmers' Perception of Climate Change and Agricultural Adaptation Strategies in Rural Sahel, Environmental Management, Vol. 43, 804–816

Mirzabaev, A.. 2013. Climate Volatility and Change in Central Asia: Economic Impacts and Adaptation. Doctoral mthesis at Agricultural Faculty, University of Bonn. urn:nbn:de:hbz:5n-3238

Nelson, G., Rosegrant, M., Palazzo, A., Gray, I., Ingersoll, C., Robertson, R., Tokgoz, S., Zhu, T., Sulser, T., Ringler, C., Msangi, S., and You, L., (2010). Food Security, Farming and Climate Change to 2050: Scenarios, Results, Policy Options', IFPRI Research Monograph, Washington, DC: IFPRI.

Nhemachena, Ch. and Hassan, R. (2007). Micro-Level Analysis of Farmers' Adaptation to Climate Change in Southern Africa. IFPRI Discussion Paper 00714

IPCC, (2007). Climate Change 2007: The Scientific Basis. Contribution of Working Group I to the Fourth Assessment Report of the Intergovernmental Panel on Climate Change, edited by S. Solomon et al., Cambridge Univ. Press, New York

O'Connor, R., Bord, R., and Fisher, A., (1999). Risk Perceptions, General Environmental Beliefs, and Willingness to Address Climate Change. Risk Analysis, Vol.19, No. 3, 461–471

Osberghaus, D., Finkel, E., and Pohl, M., (2010). Individual adaptation to climate change: The role of information and perceived risk. Discussion Paper No. 10–061. Center for European Economic Research.

Pender, J., Mirzabaev, A., & Kato, E. (2009). Economic Analysis of Sustainable Land Management Options in Central Asia. Final report for the ADB. IFPRI/ICARDA, 168.

Quan, Julian and Dyer, Nat (2008) Climate change and land tenure: the implications of climate change for land tenure and land policy. Working Paper. Food and Agriculture Organization of The United Nations, Rome.

Seo, S.N., (2011). An Analysis of Public Adaptation to Climate Change Using Agricultural Water Schemes in South America", Ecological Economics, Vol. 70, No. 4, 825–834

Sommer, R., Glazirina, M., Yuldashev, T., Otarov, A., Ibraeva, M., Martynova, L., ... & De Pauw, E. (2013). Impact of climate change on wheat productivity in Central Asia. Agriculture, ecosystems & environment, 178, 78–99.

Stulina G. (2008). Acting Today, Preparing for Tomorrow, Case study Central Asia. Addressing Water Scarcity and Drought in Central Asia Due to Climate Change. Presentation ICWC, available online at https://www.google.de/url?sa=t&rct=j&q=&esrc=s&source=web&cd=13&ved=0ahUKEwiGjPXYpfLUAhVGDJoKHXZcAQM4ChAWCCswAg&url=http%3A%2F%2Fwww.wkhcca.my%2Fdownload.php%3Ffile%3D318&usg=AFQjCNGMXGu6tk_pv0_WgEi7_MheeQLPFQ, access 05 July 2017.

Tilman, D., P.B. Reich and J. Knops (2006): Biodiversity and ecosystem stability in a decade-long grassland experiment. In: Nature 2006(441), pp. 629–632. doi:10.1038/nature04742.
Tversky, A., Kahneman, D., (1986). Rational Choice and the Framing of Decisions. The Journal of Business, 59(4), S251-S278.
von Braun, J. V., & Kennedy, E. (1994). Agricultural commercialization, economic development, and nutrition. Johns Hopkins University Press.
Wheeler, T., & von Braun, J. (2013). Climate change impacts on global food security. Science, 341(6145), 508–513.
Williams, M. and Konovalov, V. (2008). Central Asia Temperature and Precipitation Data, 1879–2003. Boulder, Colorado: USA National Snow and Ice Data Center. Digital media.
Zilberman, D., Liu, X., Roland-Holst, D., & Sunding, D. (2004). The economics of climate change in agriculture. Mitigation and Adaptation Strategies for Global Change, 9(4), 365–382.
Zilberman, D., Zhao, J., & Heiman, A. (2012). Adoption versus adaptation, with emphasis on climate change. Annu. Rev. Resour. Econ., 4(1), 27–53.

**Open Access** This chapter is distributed under the terms of the Creative Commons Attribution-NonCommercial-ShareAlike 3.0 IGO license (https://creativecommons.org/licenses/by-nc-sa/3.0/igo/), which permits any noncommercial use, duplication, adaptation, distribution, and reproduction in any medium or format, as long as you give appropriate credit to the Food and Agriculture Organization of the United Nations (FAO), provide a link to the Creative Commons license and indicate if changes were made. If you remix, transform, or build upon this book or a part thereof, you must distribute your contributions under the same license as the original. Any dispute related to the use of the works of the FAO that cannot be settled amicably shall be submitted to arbitration pursuant to the UNCITRAL rules. The use of the FAO's name for any purpose other than for attribution, and the use of the FAO's logo, shall be subject to a separate written license agreement between the FAO and the user and is not authorized as part of this CC-IGO license. Note that the link provided above includes additional terms and conditions of the license.

The images or other third party material in this chapter are included in the chapter's Creative Commons license, unless indicated otherwise in a credit line to the material. If material is not included in the chapter's Creative Commons license and your intended use is not permitted by statutory regulation or exceeds the permitted use, you will need to obtain permission directly from the copyright holder.

# Managing Environmental Risk in Presence of Climate Change: The Role of Adaptation in the Nile Basin of Ethiopia

Salvatore Di Falco and Marcella Veronesi

**Abstract** This study investigates the impact of climate change adaptation on farm households' downside risk exposure in the Nile Basin of Ethiopia. The analysis relies on a moment-based specification of the stochastic production function. We use an empirical strategy that accounts for the heterogeneity in the decision on whether to adapt or not, and for unobservable characteristics of farmers and their farm. We find that past adaptation to climate change (i) reduces current downside risk exposure, and subsequently the risk of crop failure; (ii) would have been more beneficial to the non-adapters if they adapted, **in terms of reduction in downside risk exposure; and (iii)** is a successful risk management strategy that makes the adapters more resilient to climatic conditions.

**JEL Classification** D80 • Q18 • Q54

---

This book chapter appeared previously on environmental and resource economics 2014 vol 57(4):553–577.

Previous versions of this paper have been presented at the 2012 Nordic Economic Development conference, the 2012 annual meeting of the Environment for Development Initiative, the 2011 EAERE (European Association of Environmental and Resource Economics) meeting, the 2011 AES (Agricultural Economics Society) Meeting, at the 2011 EAAE (European Association of Agricultural Economics), and at the 2010 Ascona Workshop on "Environmental Decisions: Risks and Uncertainties." We would like to thank session participants for suggestions. We also would like to thank the three anonymous reviewers for their comments and suggestions. All remaining errors and omissions are our own responsibility. Funding from SIDA through the Environment for Development Initiative is gratefully acknowledged.

S. Di Falco (✉)
Department of Economics, University of Geneva, Geneva, Switzerland
e-mail: salvatore.difalco@unige.ch

M. Veronesi
Department of Economics, University of Verona, Verona, Italy

# 1 Introduction

One consequence of climate change in sub Saharan Africa is that farmers will be more exposed to environmental risk. More erratic and scarce rainfall and higher temperature imply that farmers will face greater uncertainty. Ethiopia is a prime example in that rainfall variability and associated drought have been major causes of food shortage and famine. During the last 40 years, Ethiopia has experienced many severe droughts leading to production levels that fell short of basic subsistence levels for many farm households (Relief Society of Tigray, REST and NORAGRIC at the Agricultural University of Norway 1995, p. 137). Harvest failure due to weather events is the most important cause of risk-related hardship of Ethiopian rural households, with adverse effects on farm household consumption and welfare (Dercon 2004, 2005). Climate change is projected to further exacerbate these issues (Parry et al. 2005; Lobell et al. 2008; Schlenker and Lobell 2010; World Bank 2010). Thus, the implementation of adaptation strategies can be very important (Mendelsohn and Dinar 2003; Deressa et al. 2009; Di Falco and Veronesi 2013). For instance, farmers may face drier soil, and therefore they implement investments in soil conservation so that soil moisture may be retained. They can plant trees to procure shading on the soil or utilize irrigation and water harvesting technologies (Kurukulasuriya et al. 2011). They can also simply switch to different crops or activities that are more suited to drier or wetter environmental conditions (Seo and Mendelsohn 2008a).[1]

This paper uses survey data from the Nile basin of Ethiopia (IFPRI 2010) to investigate whether having adapted to climate change, defined as having implemented a set of strategies such as changing crop varieties, adopting water harvesting or soil and water conservation in response to long-term changes temperature and rainfall, affects current environmental risk exposure. In particular, we pose the following questions:

1. Are farm households that in the past implemented climate change adaptation strategies getting benefits in terms of a reduction in current risk exposure?
2. Are there significant differences in risk exposure between farm households that did and those that did not adapt to climate change?
3. Is climate change adaptation a successful risk management strategy that makes the adapters more resilient to current environmental risk?

The Nile basin of Ethiopia provides a relevant area to address these issues for a number of reasons. This is a very large area that covers about 34% of the total geographical area and almost 40% of the population of the entire country (Deressa et al. 2009). Farming is characterized by small-holder subsistence farmers. Farm size is on average quite small (less than one hectare). Production is traditional with plough

---

[1] It can be argued that if the production conditions become too challenging, farmers may see less of a scope for action (i.e., prospects are too gloomy) and be forced out of agriculture and migrate. However, this possibility (along with other non-crop related strategies) have not been observed in the sample used in our study.

and animals' draught power. Labor is the major input in the production process during land preparation, planting, and post-harvest processing. The use of other inputs is extremely limited (Deressa et al. 2009). The region is prone to extreme weather events such as droughts and floods. These have often resulted in crop failure, water shortage, and food insecurity (Di Falco et al. 2011). Drought is characterized by abnormal soil water deficiency. This is due to climatic variability, such as precipitation shortage or increased evapotranspiration (Gadisso 2007). Moreover, a number of papers have looked at either the impact of climate change on productivity or farm revenues (e.g., Deressa and Hassan 2009; Di Falco et al. 2011; Di Falco and Veronesi 2013) as well as the determinants of adaptation (Deressa et al. 2009, 2011; Di Falco et al. 2011).[2] However, the study of the risk implications of adaptation to climate change has been overlooked. This paper aims to fill this gap.

For our purpose, it is important to identify a suitable metric to capture the extent of environmental risk. In a rainfed agricultural production setting, the focus on crop failure seems natural. Avoiding crop failure is indeed the major preoccupation of farmers in Ethiopia. This is captured by the downside risk exposure measured by the skewness of yields. Our analysis relies on a moment-based specification of the stochastic production function (Antle 1983; Antle and Goodger 1984; Chavas 2004). This method has been widely used in the context of risk management in agriculture (Just and Pope 1979; Kim and Chavas 2003; Koundouri et al. 2006). It could be argued that the variance of yields is also a possible measure of risk exposure. However, it should be noted that the variance does not distinguish between unexpected good and bad events. We therefore focus on the skewness in risk analysis, that is we approximate downside risk exposure by the lack of symmetry of crop yield distribution. If the skewness of yield increases and becomes positive, then it means that downside risk exposure decreases, that is the probability of crop failure decreases (Di Falco and Chavas 2009). This approach can thus capture a fuller extent of risk exposure.

We investigate the effects of climate change adaptation on risk exposure in an endogenous switching regression framework.[3] The survey collected information on both farm households that did and did not adapt plus on a very large set of control variables. We take into account that the differences in risk exposure between those farm households that did and those that did not adapt to climate change could be due to unobserved heterogeneity. Not distinguishing between the casual effect of climate change adaptation and the effect of unobserved heterogeneity could lead to misleading policy implications. We account for the endogeneity of the adaptation decision by estimating a simultaneous equations model with endogenous switching by full information maximum likelihood estimation. In addition, we build a

---

[2] There are other very relevant studies addressing similar issues in different countries or at a different scale. The interested reader is referred to Mendelsohn et al. (1994), Gbetibouo and Hassan (2005), Seo and Mendelsohn (2008b), Hassan and Nhemachena (2008), Kurukulasuriya and Mendelsohn (2008), and Seo et al. (2009).

[3] This framework allows for testing the exogeneity hypothesis, in this case correcting for selection bias. It is especially useful when risks vary across categories, but have absolute thresholds.

counterfactual analysis, and compare the expected downside risk exposure under the actual and counterfactual cases of whether the farm household did or did not adapt to climate change. Treatment and heterogeneity effects are calculated to understand the differences in downside risk exposure between farm households that adapted and those that did not adapt.

Key findings of our analysis are that:

1. past adaptation to climate change *decreases* current downside risk exposure, and thereby the risk of crop failure;
2. there are significant and non-negligible differences in risk exposure between adapters and non-adapters;
3. **farm households that did not adapt would benefit the most in terms of reduction in downside risk exposure from adaptation**; and
4. the implementation of adaptation strategies is a successful risk management strategy that makes the adapters more resilient to climatic conditions.

The paper proceeds as follows. Sections 2 and 3 describe the study sites and survey instruments. Section 4 outlines the model and the estimation procedure used. Section 5 presents the results, and section 6 concludes by offering some final remarks and directions for future research.

## 2 Background

Ethiopia's GDP is closely associated with the performance of its rainfed agriculture (Deressa and Hassan 2009). For instance, about 40 percent of national GDP, 90 percent of exports, and 85 percent of employment stem from the agricultural sector (Ministry of Finance and Economic Development, MoFED 2007). The rainfed production environment is characterized by land degradation and very erratic and variable climate. Rainfall variability and associated droughts have been major causes of food shortage and famine in Ethiopia (World Bank 2010). A recent mapping on vulnerability and poverty in Africa listed Ethiopia as one of the most vulnerable countries to climate change with the least capacity to respond (Orindi et al. 2006; Stige et al. 2006).

The success of the agricultural sector is crucially determined by the productivity of small holder farm households. They account for about 95 percent of the national agricultural output, of which about 75 percent is consumed at the household level (World Bank 2006). With low diversified economy and reliance on rain-fed agriculture, Ethiopia's development prospects have been thus associated with climate (Deressa et al. 2009). For instance, the World Bank (2006) reported that catastrophic hydrological events such as droughts and floods have reduced its economic growth by more than a third. The frequency of droughts has increased over the past few decades, especially in the lowlands (Lautze et al. 2003). A 2007 study, undertaken by the national meteorological service (NMS), highlights that the annual minimum temperature has been increasing by about 0.37 degrees Celsius every 10 years over the past 55 years. Rainfall has been more erratic with some areas becoming drier while others becoming relatively wetter. These findings show that climatic variations are already happened in this part of the world. The prospect of further climate change can exacerbate this already difficult situation. Climate change is projected to further reduce agricultural productivity (Rosenzweig and Parry 1994; Parry et al. 2005; Cline 2007). Most climate models converge in forecasting scenarios of increased temperatures for most of Ethiopia (Dinar et al. 2008).

## 3 Survey Design and Data Description

The survey was carried out in the Nile River Basin in Ethiopia in 2005.[4] The household sampling frame was developed to ensure representation for the Nile River Basin at the *woreda* (an administrative division equivalent to a district) level regarding level of rainfall patterns in terms of both annual total and variation. The data used for the sample frame are from the Atlas of the Ethiopian Rural Economy (IFPRI 2010). The survey considered traditional typology of agro-ecological zones in the country (namely, *Dega, Woina Dega, Kolla*, and *Berha*), percent of cultivated land, degree of irrigation activity, average annual rainfall, rainfall variability, and vulnerability (number of food aid dependent population). The sampling frame selected the *woredas* in such a way that each stratum in the sample matched to the proportions for each stratum in the entire Nile basin. The procedure resulted in the inclusion of twenty *woredas*. Random sampling was then used in selecting fifty households from each *woreda*. The final dataset contains comprehensive observations from almost 1000 farms. Information on agricultural practices and production, costs, investments, and revenues as well as tenure security, past shocks, and access to credit were collected.[5] One of the survey instruments was in particular designed to capture farmers' perceptions and understanding on climate change, and their approaches for adaptation. Questions were included to investigate whether farmers have noticed changes in mean temperature and rainfall over the last two decades, and reasons for observed changes. Overall, increased temperature and declining rainfall are the predominant perceptions in our study sites. These perceptions do match with the existing evidence reported in the previous section.

Furthermore, some questions investigated whether farm households made some adjustments in their farming practices in response to long-term changes in mean temperature and rainfall by adopting some particular strategies. Changing crop varieties and adoption of soil and water conservation strategies were major forms of adaptation strategies followed by the farm households in our study sites. These adaptation strategies are mainly yield-related and account for more than 95 per cent of the adaptation strategies followed by the farm households who actually undertook an adaptation strategy. The remaining adaptation strategies accounting for less

---

[4] To our knowledge there has not been a follow up survey yet.
[5] For complete information on the survey, please refer to IFPRI (2010).

than five percent were water harvesting, irrigation, non-yield related strategies such as migration, and shift in farming practice from crop production to livestock herding or other sectors. We use this information from the survey to create the variable *adaptation*. This is equal to 1 if a farm household adopted any of the above strategies, and to 0 otherwise.

As mentioned, detailed production data were collected at different production stages (i.e., land preparation, planting, weeding, harvesting, and post-harvest processing). Most of the sample population is composed of rainfed farms (less than 9 per cent of them have access to irrigation). Ethiopian rural households face high weather and climatic variability. Significant spatial variations exist in agroecological conditions, including topography, soil type, temperature, and soil fertility (Hagos et al. 1999).[6] The farming system in the survey sites is very traditional with plough and animals' draught power. Labor is the major input in the production process during land preparation, planting, and post-harvest processing. Labor inputs were disaggregated as adult male labor, adult female labor, and children labor. The three forms of labor were aggregated as one labor input using adult equivalents.[7]

Monthly rainfall and temperature data were collected from all the meteorological stations in the country for the period 1970–2000. Then, the *Thin Plate Spline* method of spatial interpolation was used to impute the household specific rainfall and temperature values using latitude, longitude, and elevation information of each household. The *Thin Plate Spline* is a physically based two-dimensional interpolation scheme for arbitrarily spaced tabulated data. The Spline surface represents a thin metal sheet that is constrained not to move at the grid points, which ensures that the generated rainfall and temperature data at the weather stations are exactly the same as data at the weather station sites that were used for the interpolation. In our case, the rainfall and temperature data at the weather stations are reproduced by the interpolation for those stations, which ensures the credibility of the method (see Wahba 1990). This method is one of the most commonly used to create spatial climate data sets (e.g., Di Falco et al. 2011; Deressa and Hassan 2009). Its strengths are that it is readily available, relatively easy to apply, and accounts for spatially varying elevation relationships. However, it only simulates elevation relationships and has difficulty handling very sharp spatial gradients, which can be typical of coastal areas. Given that our area of the study is characterized by significant terrain features, and no climatically important coastlines, the choice of the *Thin Spline method* is reasonable (for more details on the properties of this method in comparison to the other methods see Daly 2006).

---

[6] Note that the cross-section and plot level nature of the data does not allow an analysis of the dynamic aspects of farm-level management decisions. Panel data would be required to explore such issues. To our knowledge, there is no climate change survey where the same household has been interviewed in different point in time.

[7] We employed the OECD/EU conversion factor in the literature in developing countries, where adult female and child labor are converted into the adult male labor equivalent with the conversion factors 0.8 and 0.3, respectively.

However, it should be noted that the impact of variations in temperature and rainfall may vary across seasons, and should be taken into account.[8] We therefore investigate the differential impact of the two main rainy seasons in Ethiopia: the long rainy season (*Meher*) and the short rainy season (*Belg*). We do not distinguish between differences in temperatures between seasons because we did not find large differences in average temperature between months in the period 1970–2000. This may be related to the location of Ethiopia near the Equator.

The final sample includes twenty *woredas*, 941 farm households (i.e., on average about forty-seven farm households per *woreda*), and 2801 plots (i.e., on average about three plots per farm household). The scale of the analysis is at the plot-level.[9] The basic descriptive statistics are presented in Table 1, and the definition of the variables in Table A1 of the appendix (Table A2).

## 4 Model of Climate Change Adaptation and Risk Exposure

In this section we specify an econometric model of climate change adaptation and risk exposure. Particular functional forms are chosen to remain within the spirit of previous work in this area (Di Falco et al. 2011). The simplest approach to examine the impact of climate change adaptation on farm households' downside risk exposure would be to include in the risk equation a dummy variable equal to one if the farm household adapted to climate change, and then, to apply ordinary least squares. This approach, however, might yield biased estimates because it assumes that adaptation to climate change is exogenously determined while, in fact, it may be endogenous to other factors. Namely, the decision on whether to adapt or not to climate change is voluntary and may be based on individual self-selection. Farmers that adapted may have systematically different characteristics from the farmers that did not adapt, and they may have decided to adapt based on expected benefits. Unobservable characteristics of farmers and their farm may affect both the adaptation decision and risk exposure, resulting in inconsistent estimates of the effect of adaptation on production risk and risk of crop failure. For example, if only the most skilled or motivated farmers chose to adapt and we fail to control for skills, then we will incur upward bias.

We account for the endogeneity of the adaptation decision by estimating a switching regression model of climate change adaptation and risk exposure with endogenous switching. In particular, we model the climate change adaptation

---

[8] We thank a reviewer for emphasizing this aspect.

[9] Although a total of 48 annual crops were grown in the basin, the first five major annual crops (teff, maize, wheat, barley, and beans) cover 65 per cent of the plots. These are also the crops that constitute the staple foods of the local diet and are relevant in the context of self-subsistence farming. It should be also noted that including the other crops (e.g., perennials) would have implication for the specification of the production technology represented by the production function. We therefore limit the estimation to these primary, annual, crops.

**Table 1** Descriptive statistics

| Variable name | Total sample | | Adapters | | Non-adapters | |
|---|---|---|---|---|---|---|
| | Mean | Std. Dev. | Mean | Std. Dev. | Mean | Std. Dev. |
| Dependent variables | | | | | | |
| Adaptation | 0.690 | 0.463 | 1.000 | 0.000 | 0.000 | 0.000 |
| Skewness | 0.593 | 14.877 | 0.845 | 17.903 | 0.034 | 0.320 |
| Explanatory variables | | | | | | |
| Climatic factors | | | | | | |
| Average temperature | 18.523 | 2.228 | 17.945 | 1.991 | 19.809 | 2.190 |
| Belg rainfall | 257.064 | 146.275 | 224.635 | 135.490 | 329.284 | 143.617 |
| Meher rainfall | 960.439 | 293.511 | 910.282 | 304.337 | 1072.136 | 231.788 |
| Crops varieties | | | | | | |
| Barley | 0.185 | 0.389 | 0.208 | 0.406 | 0.135 | 0.342 |
| Maize | 0.199 | 0.399 | 0.194 | 0.396 | 0.211 | 0.408 |
| Teff | 0.271 | 0.445 | 0.242 | 0.428 | 0.336 | 0.473 |
| Wheat | 0.208 | 0.406 | 0.212 | 0.409 | 0.200 | 0.401 |
| Soil characteristics | | | | | | |
| Highly fertile | 0.280 | 0.449 | 0.257 | 0.437 | 0.333 | 0.472 |
| Infertile | 0.158 | 0.365 | 0.172 | 0.378 | 0.127 | 0.333 |
| No erosion | 0.484 | 0.500 | 0.472 | 0.499 | 0.510 | 0.500 |
| Severe erosion | 0.104 | 0.306 | 0.114 | 0.318 | 0.082 | 0.274 |
| Assets | | | | | | |
| Machinery | 0.019 | 0.136 | 0.024 | 0.153 | 0.007 | 0.085 |
| Animals | 0.874 | 0.332 | 0.887 | 0.317 | 0.845 | 0.362 |
| Inputs | | | | | | |
| Labor | 101.088 | 121.383 | 105.912 | 133.503 | 90.344 | 87.743 |
| Seeds | 115.181 | 148.732 | 125.867 | 163.948 | 91.385 | 103.552 |
| Fertilizers | 60.760 | 176.962 | 62.092 | 177.988 | 57.795 | 174.720 |
| Manure | 198.572 | 832.187 | 254.955 | 952.355 | 73.009 | 438.860 |
| Farm head and farm household characteristics | | | | | | |
| Literacy | 0.489 | 0.500 | 0.524 | 0.500 | 0.414 | 0.493 |
| Male | 0.926 | 0.262 | 0.932 | 0.252 | 0.914 | 0.281 |
| Married | 0.928 | 0.259 | 0.931 | 0.254 | 0.922 | 0.269 |
| Age | 45.740 | 12.548 | 46.267 | 11.914 | 44.566 | 13.790 |
| Household size | 6.603 | 2.189 | 6.765 | 2.136 | 6.243 | 2.261 |
| Off-farm job | 0.249 | 0.433 | 0.286 | 0.452 | 0.169 | 0.375 |
| Relatives | 16.494 | 43.682 | 19.561 | 51.321 | 9.473 | 13.287 |
| Access to credit | 0.259 | 0.438 | 0.306 | 0.461 | 0.156 | 0.363 |
| Flood | 0.172 | 0.378 | 0.217 | 0.412 | 0.074 | 0.261 |
| Drought | 0.443 | 0.497 | 0.565 | 0.496 | 0.171 | 0.376 |
| Information sources | | | | | | |
| Government extension | 0.609 | 0.488 | 0.761 | 0.427 | 0.270 | 0.444 |

(continued)

**Table 1** (continued)

| Variable name | Total sample | | Adapters | | Non-adapters | |
|---|---|---|---|---|---|---|
| | Mean | Std. Dev. | Mean | Std. Dev. | Mean | Std. Dev. |
| Farmer-to-farmer extension | 0.516 | 0.500 | 0.659 | 0.474 | 0.197 | 0.398 |
| Radio information | 0.307 | 0.461 | 0.382 | 0.486 | 0.139 | 0.347 |
| Neighborhood information | 0.316 | 0.465 | 0.321 | 0.467 | 0.305 | 0.461 |
| Climate information | 0.422 | 0.494 | 0.563 | 0.496 | 0.111 | 0.314 |
| Sample size | 2801 | | 1933 | | 868 | |

The sample size refers to the total number of plots. The final total sample includes 20 woredas, 941 farm households, and 2801 plots

decision and its implications in terms of risk exposure in the setting of a two-stage framework.[10] In the first stage, we use a selection model where a representative farm household chooses whether to adapt or not to adapt, while in the second stage we estimate conditional risk exposure functions accounting for the endogenous selection. Finally, we produce selection-corrected predictions of counterfactual downside risk exposure.

*Stage I – Selection Model of Climate Change Adaptation* In the first stage, we use a selection model for climate change adaptation where a representative risk averse farm household $i$ chooses to implement climate change adaptation strategies if the expected utility from adapting $U(\pi_1)$ is greater than the expected utility from not adapting $U(\pi_0)$, i.e., $E[U(\pi_1) - U(\pi_0)] > 0$, where $E$ is the expectation operator based on the subjective distribution of the uncertain variables facing the decision maker, and $U(\bullet)$ is the von Neumann-Morgenstern utility function representing the farm household's preferences under risk. Let $A^*$ be the latent variable that captures the expected benefits from the adaptation choice with respect to not adapting. We specify the latent variable as:

$$A_i^* = \mathbf{z}_i \alpha + \eta_i \text{ with } A_i = \begin{cases} 1 & \text{if } A_i^* > 0 \\ 0 & \text{otherwise} \end{cases}, \quad (1)$$

that is farm household $i$ will choose to adapt ($A_i = 1$) through the implementation of some strategy or set of strategies in response to long term changes in mean temperature and rainfall if $A^* > 0$, and 0 otherwise. The vector $\mathbf{z}$ represents variables that affect the likelihood to adapt such as the characteristics of the operating farm (e.g., soil fertility and erosion); farm head and farm household's characteristics (e.g., farmer head's age, gender, education, marital status, off-farm job, and farm household size); the presence of assets (e.g., machinery and animals); past climatic

---

[10] A more comprehensive model of climate change adaptation is provided by Mendelsohn (2000).

factors (e.g., rainfall and temperature); the experience of previous extreme events (e.g., droughts and floods); whether farmers received information on climate; government and farmer-to-farmer extensions, which can be used as measures of access to information about adaptation strategies. It is also important to address the role of access to credit. Households that have limited access to credit can have less capital available to be invested in the implementation of more costly adaptation strategies (e.g., soil conservation measures). We approximate experience by age and education.

*Stage II – Endogenous Switching Regression Model of Downside Risk Exposure* How do we measure risk exposure and its interplay with adaptation? In the second stage, we estimate the effect of adaptation on the skewness of the distribution of yields. This provides information of the role of adaptation on downside risk exposure. We rely on a moment-based specification of the stochastic production function (Antle 1983; Antle and Goodger 1984; Chavas 2004). This is a very flexible device that has been largely used in agricultural economics to model the implication of weather risk and risk management (Just and Pope 1979; Kim and Chavas 2003; Koundouri et al. 2006; Di Falco and Chavas 2009). Consider a risk averse farm household that produces output $y$ using inputs $\mathbf{x}$ under risk through a production technology represented by a well-behaved (i.e., continuous and twice differentiable) stochastic production function $y = g(\mathbf{x}, \upsilon)$, where $\upsilon$ is a vector of random variables representing risk, that is uncontrollable factors affecting output such as current changes in temperature and rainfall.

We assess the probability distribution of the stochastic production function $g(\mathbf{x}, \upsilon)$ by applying a moment-based approach (Antle 1983), that is risk exposure is represented by the moments of the production function $g(\mathbf{x}, \upsilon)$. We consider the following econometric specification for $g(\mathbf{x}, \upsilon)$:

$$g(\mathbf{x},\upsilon) = f_1(\mathbf{x},\beta_1) + u \qquad (2)$$

where $f_1(\mathbf{x}, \boldsymbol{\beta}_1) \equiv E[g,(\mathbf{x}, \upsilon)]$ is the mean of $g,(\mathbf{x}, \upsilon)$, that is the first central moment, and $u = g,(\mathbf{x}, \upsilon) - f_1(\mathbf{x}, \boldsymbol{\beta}_1)$ is a random variable with mean zero whose distribution is exogenous to farmers' actions.[11] The higher moments of $g(\mathbf{x}, \upsilon)$ are given by

$$E\{[g,,|(\mathbf{x},\upsilon) - f_1(\mathbf{x},,|\beta_1)]^k,,|\mathbf{x}\} = f_k(\mathbf{x},\beta_k) \qquad (3)$$

for $k = 2, 3$. This implies that $f_2(\mathbf{x}, \boldsymbol{\beta}_2)$ is the second central moment, that is the variance, and $f_3(\mathbf{x}, \boldsymbol{\beta}_3)$ is the third central moment, that is the skewness. This approach provides a flexible representation of the impacts of past climatic factors (e.g.,

---

[11] Note that the production function can be estimated by OLS without making any normality assumptions regarding the error distribution. Indeed, if the errors were normally distributed, by construction the distribution would be symmetric, and the third central moment would be zero.

temperature and rainfall averages 1970–2000),[12] inputs, (e.g., seeds, fertilizers, manure, and labour), assets (e.g., machinery and animals), farm household characteristics, and soil characteristics (e.g., soil fertility and erosion level) on the distribution of output under production uncertainty. As mentioned in the introduction we capture the extent of risk exposure by the third moment of the distribution of yields: the skewness. An increase in skewness implies a reduction in downside risk exposure, which implies, a reduction in the probability of crop failure. Reducing downside risk means decreasing the asymmetry (or skewness) of the risk distribution toward high outcome, holding both means and variance constant[13] (Menezes et al. 1980; Di Falco and Chavas 2009).

To account for selection biases we adopt an endogenous switching regression model of downside risk exposure where farmers face two regimes (1) to adapt, and (2) not to adapt defined as follows:

$$\text{Regime 1}: y_{1i} = \mathbf{x}_{1i}\beta_1 + \varepsilon_{1i} \text{ if } A_i = 1 \quad (4a)$$

$$\text{Regime 2}: y_{2i} = \mathbf{x}_{2i}\beta_2 + \varepsilon_{2i} \text{ if } A_i = 0 \quad (4b)$$

where $y_i$ is the third central moment $f_3(\mathbf{x}, \beta_3)$ of production function (2) in regimes 1 and 2, i.e., the skewness; and $\mathbf{x_i}$ represents a vector of the past climatic factors, inputs, assets, farm head's, farm household's and soil's characteristics included in $\mathbf{z}$. In addition, the error terms in Eqs. (1, 4a, and 4b) are assumed to have a trivariate normal distribution, with zero mean and covariance matrix $\mathbf{\Sigma}$, i.e., $(\eta, \varepsilon_1, \varepsilon_2)' \sim N(\mathbf{0},$

$\mathbf{\Sigma})$ with $\Sigma = \begin{bmatrix} \sigma_\eta^2 & \sigma_{\eta 1} & \sigma_{\eta 2} \\ \sigma_{1\eta} & \sigma_1^2 & \cdot \\ \sigma_{2\eta} & \cdot & \sigma_2^2 \end{bmatrix}$, where $\sigma_\eta^2$ is the variance of the error term in the

selection Eq. (1), which can be assumed to be equal to 1 since the coefficients are estimable only up to a scale factor (Maddala 1983, p. 223), $\sigma_1^2$ and $\sigma_2^2$ are the variances of the error terms in the skewness functions (4a and 4b), and $\sigma_{1\eta}$ and $\sigma_{2\eta}$ represent the covariance of $\eta_i$ and $\varepsilon_{1i}$ and $\varepsilon_{2i}$.[14] Since $y_{1i}$ and $y_{2i}$ are not observed simultaneously the covariance between $\varepsilon_{1i}$ and $\varepsilon_{2i}$ is not defined (reported as dots in the covariance matrix $\mathbf{\Sigma}$, Maddala 1983, p. 224). An important implication of the error structure is that because the error term of the selection Eq. (1) $\eta_i$ is correlated

---

[12] It should be noted that the use of averages is conventional in this strand of literature (e.g., Mendelsohn et al. 1994; Deressa and Hassan 2009). Recently, however, a more precise agronomic measure of heat stress has been suggested: degree days. This is a piecewise-linear function of temperature captured by two variable degree days 10–30 °C (Schlenker and Lobell 2010). The appropriate calculation of these requires a large amount of daily weather observations. Unfortunately, we do not have access to such detailed information.

[13] This does not provide information on the role of adaptation on farmer's welfare under uncertainty.

[14] For notational simplicity, the covariance matrix $\mathbf{\Sigma}$ does not reflect the clustering implemented in the empirical analysis.

with the error terms of the skewness functions (4a and 4b) ($\varepsilon_{1i}$ and $\varepsilon_{2i}$), the expected values of $\varepsilon_{1i}$ and $\varepsilon_{2i}$ conditional on the sample selection are nonzero:

$$E[\varepsilon_{1i}|A_i = 1] = \sigma_{1\eta}\frac{\phi(\mathbf{z}_i\alpha)}{\Phi(\mathbf{z}_i\alpha)} = \sigma_{1\eta}\lambda_{1i}, \text{ and } E[\varepsilon_{2i}|A_i = 0] = -\sigma_{2\eta}\frac{\phi(\mathbf{z}_i\alpha)}{\Phi(\mathbf{z}_i\alpha)} = \sigma_{2\eta}\lambda_{2i},$$

where $\phi(.)$ is the standard normal probability density function, $\Phi(.)$ the standard normal cumulative density function, and $\lambda_{1i} = \frac{\phi(\mathbf{z}_i\alpha)}{\Phi(\mathbf{z}_i\alpha)}$, and $\lambda_{2i} = \frac{\phi(\mathbf{z}_i\alpha)}{1-\Phi(\mathbf{z}_i\alpha)}$. If the estimated covariances $\sigma_{1\eta}$ and $\sigma_{2\eta}$ are statistically significant, then the decision to adapt and downside risk exposure are correlated, that is we find evidence of endogenous switching and reject the null hypothesis of the absence of sample selectivity bias. This model is defined as a "switching regression model with endogenous switching" (Maddala and Nelson 1975).

For the model to be identified it is important to use as exclusion restrictions, thus as selection instruments, not only those automatically generated by the nonlinearity of the selection model of adaptation (1) but also other variables that directly affect the selection variable but not the outcome variable. Following Di Falco et al. (2011), we use as selection instruments the variables related to the information sources (e.g., government extension, farmer-to-farmer extension, information from the radio or the neighbourhood and, if received information in particular on climate), which enter in $\mathbf{z}$ but not in $\mathbf{x}$. We establish the admissibility of these instruments by performing the simple falsification test by Di Falco et al. (2011): if a variable is a valid selection instrument, it will affect the adaptation decision but it will not affect the risk exposure among farm households that did not adapt. The information sources can be considered as valid selection instruments: they are statistically significant determinants of the decision on whether to adapt or not to climate change ($\chi^2 = 108.27$) but not of downside risk exposure among farm households that did not adapt (F-stat. = 2.10).

Finally, we estimate Stage I and II simultaneously by full information maximum likelihood estimation (FIML) since this is a more efficient method to estimate endogenous switching regression models than a two-step procedure (Lee and Trost 1978).[15] The logarithmic likelihood function given the previous assumptions regarding the distribution of the error terms is.

$$\ln L_i = \sum_{i=1}^{N} A_i \left[\ln\phi\left(\frac{\varepsilon_{1i}}{\sigma_1}\right) - \ln\sigma_1 + \ln\Phi(\theta_{1i})\right] \\ + (1-A_1)\left[\ln\phi\left(\frac{\varepsilon_{2i}}{\sigma_2}\right) - \ln\sigma_2 + \ln(1-\Phi(\theta_{2i}))\right], \text{ where} \quad (5)$$

---

[15] The two-step procedure (see Maddala 1983, p. 224 for details) not only it is less efficient than FIML but it also requires some adjustments to derive consistent standard errors (Maddala 1983, p. 225), and it poorly performs in case of high multicollinearity between the covariates of the selection equation (1) and the covariates of the skewness equations (4a) and (4b) (Hartman 1991; Nelson 1984; Nawata 1994).

$$\theta_{ji} = \frac{\left(\mathbf{z}_i\alpha + \rho_j\varepsilon_{ji}/\sigma_j\right)}{\sqrt{1-\rho_j^2}}, \ j=1,2,$$ with $\rho_j$ denoting the correlation coefficient between the error term $\eta_i$ of the selection Eq. 1 and the error term $\varepsilon_{ji}$ of Eq. 4a and 4b, respectively.

In addition, we exploit plot level information to deal with the issue of farmers' unobservable characteristics such as their skills. Plot level information can be used to construct a panel data and control for farm specific effects (Udry 1996). We follow Mundlak (1978) and Wooldridge (2002) to control for unobservable characteristics. We exploit the plot level information, and insert in the adaptation Eq. (1), in the production Eq. (2), and in the risk equations Eq. (4a and 4b) the average of plot–variant variables $\overline{\mathbf{S}}_i$ such as the inputs used (seeds, manure, fertilizer, and labor). This approach relies on the assumption that the unobservable characteristics $v_i$ are a linear function of the averages of the plot-variant explanatory variables $\overline{\mathbf{S}}_i$, that is $v_i = \overline{\mathbf{S}}_i\pi + \psi_i$ with $\psi_i \sim IIN(0,\sigma_\psi^2)$ and $E(\psi_i/\overline{\mathbf{S}}_i)=0$, where $\pi$ is the corresponding vector of coefficients, and $\psi_i$ is a normal error term uncorrelated with $\overline{\mathbf{S}}_i$.

## 4.1 Counterfactual Analysis

The main objective of our study is to investigate the effect of having adapted to climate change on downside risk exposure, that is to estimate the treatment effect (Heckman et al. 2001). In absence of a self-selection problem, it would be appropriate to assign to the adapters a counterfactual skewness had they not adapted equal to the average skewness among non-adapters with the same observable characteristics. However, as already mentioned, unobserved heterogeneity in the propensity to adapt affecting also risk exposure creates a selection bias that cannot be ignored. The endogenous switching regression model just described can be applied to produce selection-corrected predictions of counterfactual downside risk exposure (i.e., skewness). It can be used (a) to compare the expected downside risk exposure of farm households that adapted relative to the non-adapters, (b) to investigate the expected downside risk exposure in the counterfactual hypothetical cases that the adapted farm households (i) did not adapt and (ii) that the non adapters adapted. The conditional expectations for downside risk exposure in the four cases are defined as follows:

$$E(y_{1i}|A_i=1) = \mathbf{x}_{1i}\beta_1 + \sigma_{1\eta}\lambda_{1i} \qquad (6a)$$

$$E(y_{2i}|A_i=0) = \mathbf{x}_{2i}\beta_2 + \sigma_{2\eta}\lambda_{2i} \qquad (6b)$$

$$E(y_{2i}|A_i = 1) = \mathbf{x}_{1i}\beta_2 + \sigma_{2\eta}\lambda_{1i} \tag{6c}$$

$$E(y_{1i}|A_i = 0) = \mathbf{x}_{2i}\beta_1 + \sigma_{1\eta}\lambda_{2i} \tag{6d}$$

Equation 6a and 6b represent the actual expectations observed in the sample. Equation 6c and 6d represent the counterfactual expected outcomes. In addition, following Heckman et al. (2001), we calculate the effect of the treatment "to adapt" on the treated (TT) as the difference between (6a and 6c),

$$TT = E(y_{1i}|A_i = 1) - E(y_{2i}|A_i = 1) = \mathbf{x}_{1i}(\beta_1 - \beta_2) + (\sigma_{1\eta} - \sigma_{2\eta})\lambda_{1i} \tag{7}$$

which represents the effect of climate change adaptation on downside risk exposure of the farm households that actually adapted to climate change. Similarly, we calculate the effect of the treatment on the untreated (TU) for the farm households that actually did not adapt to climate change as the difference between (6d and 6b),

$$TU = E(y_{1i}|A_i = 0) - E(y_{2i}|A_i = 0) = \mathbf{x}_{2i}(\beta_1 - \beta_2) + (\sigma_{1\eta} - \sigma_{2\eta})\lambda_{2i} \tag{8}$$

We can use the expected outcomes described in Eq. 6a, 6b, 6c, and 6d to calculate also the heterogeneity effects. For example, farm households that did not adapt may have been exposed to lower downside risk than farm households that adapted regardless of the fact that they decided not to adapt but because of unobservable characteristics such as their abilities. We follow Carter and Milon (2005) and define as "the effect of base heterogeneity" for the group of farm households that decided to adapt as the difference between (6a and 6d),

$$BH_1 = E(y_{1i}|A_i = 1) - E(y_{1i}|A_i = 0) = (\mathbf{x}_{1i} - \mathbf{x}_{2i})\beta_{1i} + \sigma_{1\eta}(\lambda_{1i} - \lambda_{2i}) \tag{9}$$

Similarly for the group of farm households that decided not to adapt, "the effect of base heterogeneity" is the difference between (6c and 6b),

$$BH_2 = E(y_{2i}|A_i = 1) - E(y_{2i}|A_i = 0) = (\mathbf{x}_{1i} - \mathbf{x}_{2i})\beta_{2i} + \sigma_{2\eta}(\lambda_{1i} - \lambda_{2i}) \tag{10}$$

Finally, we investigate the "transitional heterogeneity" (TH), that is whether the effect of adapting to climate change is larger or smaller for the adapters or for the non-adapters in the counterfactual case that they did adapt, that is the difference between Eqs. (7 and 8), i.e., (TT) and (TU).

# 5 Results

Table 2 reports the estimates of the endogenous switching regression model estimated by full information maximum likelihood with clustered standard errors at the *woreda* level.[16] The first column presents the estimation of downside risk exposure by ordinary least squares (OLS) with no switching and with a dummy variable equal to 1 if the farm household adapted to climate change, 0 otherwise. The second, third and fourth columns present, respectively, the estimated coefficients of selection Eq. (1) on climate change adaptation, and of downside risk exposure, which is represented by skewness functions (4a and 4b) (i.e., the third central moments of production function (2) in regimes (1) and (2)), for adapters and non-adapters.[17] Table A3 of the appendix shows the estimation of production function (2) in regimes (1) and (2) from which we derived the third central moments.[18]

The estimation of Eq. (1) suggests that key drivers of farm households' decision to adopt some strategies in response to long-term changes in mean temperature and rainfall are represented by the information sources farm households have access to and the environmental characteristics of the farm. More specifically access to government extension, media, and climate information increase the likelihood to adapt. These findings are very consistent with what has been found elsewhere (e.g., Maddison 2006; Deressa et al. 2009; Hassan and Nhemachena 2008; Gbetibouo et al. 2010; Deressa et al. 2011; Di Falco et al. 2011). Farm households with highly fertile soils are less likely to adapt. This highlights that most adaptation intervention is implemented in medium fertility soils. Rainfall in both rainy seasons displays *U*-shaped behaviour.[19] In addition, we find that literacy has a positive significant effect on adaptation as well as having experienced a flood in the past. This is also consistent with what has been found by Deressa et al. (2009) and Deressa et al. (2011). It may be argued that pooling different crops can induce some bias. There

---

[16] We recognise that it is possible that the error terms of the switching regression model are correlated among the nearby geographical areas. As rightly pointed out by one of the reviewers, this may arise for several reasons. First, interpolation methods were applied to create spatial climate data sets. This procedure may introduce correlation in the errors. Unobserved soil characteristics are also spatially correlated. Therefore, standard errors should be adjusted for the spatial dependence in the residuals. However, we do not have the information on the distance between plots to adjust the standard errors for spatial dependence and we account for the correlation among plots within the same *woreda* by clustering the standard errors. Future research should account also for spatial dependence.

[17] We use the "movestay" command of STATA to estimate the endogenous switching regression model by FIML (Lokshin and Sajaia 2004). We rescaled and divided the skewness by 10 milliards to address convergence issues in the FIML estimation. Dividing a number by a constant does not affect the results.

[18] We refer the reader to Di Falco et al. (2011) for a discussion of the factors affecting the production functions of the adapters and non-adapters.

[19] Di Falco et al. (2011) use current weather as a proxy for climate (while we use climatic variables such as past rainfall and mean temperature), and they do not find an effect of weather on adaptation.

may be some underlying differences in their risk functions, for instance. To control for this possible source of heterogeneity, we included a set of dummy variables to capture the specificity of the different crops.[20]

The question now is whether farm households that implemented climate change adaptation strategies experienced a reduction in downside risk exposure (e.g., a decrease in the probability of crop failure). As described in the previous section, we assess the probability distribution of the stochastic production function by applying a moment-based approach. A simple approach to answer the aforementioned question consists in estimating an OLS model of downside risk exposure that includes a dummy variable equal to 1 if the farm household adapted, 0 otherwise (Table 2, column (1)). An increase in skewness implies a reduction in downside risk exposure. This approach would lead us to conclude that the adaption significantly reduces farm households' downside risk exposure (the coefficient of the dummy variable *adaptation* is positive), although the effect is weak (significant at the 10 percent statistical level). This approach, however, assumes that adaptation to climate change is exogenously determined, while, in fact, it is a potentially endogenous variable. As such, the estimation via OLS would yield biased and inconsistent estimates. In addition, OLS estimates do not explicitly account for potential structural differences between the skewness functions of the adapters and non-adapters. The estimates presented in the last two columns of Table 2 account for the endogenous switching in the skewness function. Both the estimated coefficients of the correlation terms $\rho_j$ are not significantly different from zero (Table 2, bottom row). This implies that the hypothesis of absence of sample selectivity bias may not be rejected.

However, the differences in the coefficients of the skewness functions between the farm households that adapted and those that did not adapt illustrate the presence of heterogeneity in the sample (Table 2, columns (3) and (4)). The skewness function of the adapters is significantly different from the skewness function of the non-adapters (Chow test p-value = 0.000). Among farm households that in the past adapted to climate change, assets such as animals are significantly associated with an increase in the skewness, and so in a decrease in downside risk exposure. Inputs such as seeds display an inverted $U$–shape relationship. The total marginal impact (estimated at the sample mean) is positive. This implies that seeds have a positive effect in reducing downside risk exposure for the group of the adapters. While it is difficult to understand the reasons behind such results, one may speculate that the adapters may have better access to markets for inputs and this allows them to better manage risk of crop failure. Infertile soils are instead associated with an increase in downside risk exposure. However, these factors do not significantly affect the downside risk exposure of farm households that did not adapt.[21] We find instead that climatic factors play a very important role in explaining risk exposure of the group of non-adapters. These non-adapters are, indeed, significantly affected by the rainfall

---

[20] We also have estimated models without the crop dummies. Results are robust, and available upon request.

[21] The exception is seeds which displays some weak statistical significance of the positive portion of $U$-shape behaviour. The marginal impact is, however, negligible.

**Table 2** Parameters estimates of climate change adaptation and downside risk exposure (skewness)

| | (1) | (2) | (3) | (4) |
|---|---|---|---|---|
| Model | OLS | Endogenous switching regression[a] | | |
| | | | Regime 1 (adaptation = 1) | Regime 2 (adaptation = 0) |
| Dependent variable | Skewness pooled sample | Adaptation 1/0 | Skewness adapters | Skewness non-adapters |
| Adaptation 1/0 | 4.402* | | | |
| | (2.539) | | | |
| *Climatic factors* | | | | |
| Average temperature | 11.139 | 0.744 | 0.604 | −0.102 |
| | (8.270) | (0.588) | (1.726) | (0.161) |
| squared average temperature | −0.276 | −0.027* | −0.009 | 0.004 |
| | (0.228) | (0.015) | (0.050) | (0.005) |
| Belg rainfall | −0.044 | −0.013*** | −0.001 | 0.002** |
| | (0.070) | (0.003) | (0.005) | (0.001) |
| Squared Belg rainfall/1000 | 0.046 | 0.017*** | 0.003 | −0.002* |
| | (0.119) | (0.005) | (0.009) | (0.001) |
| Meher rainfall | 0.081 | −0.010*** | 0.013 | 0.001** |
| | (0.053) | (0.002) | (0.009) | (0.001) |
| squared Meher rainfall/1000 | −0.381 | 0.049*** | −0.063 | −0.007*** |
| | (0.276) | (0.011) | (0.051) | (0.003) |
| *Crop varieties* | | | | |
| Barley | 20.588 | −0.237*** | 2.725 | −0.004 |
| | (13.500) | (0.079) | (1.788) | (0.017) |
| Maize | 5.983 | 0.044 | 0.606 | 0.012 |
| | (4.596) | (0.109) | (0.516) | (0.036) |
| Teff | −0.161 | −0.062 | −0.143 | −0.001 |
| | (2.978) | (0.088) | (0.407) | (0.016) |
| Wheat | −0.335 | −0.164 | 0.058 | 0.044 |
| | (4.067) | (0.083) | (0.617) | (0.031) |
| *Soil characteristics* | | | | |
| Highly fertile | −4.913 | −0.190** | −0.724 | 0.004 |
| | (4.583) | (0.076) | (0.716) | (0.016) |
| Infertile | −5.910** | −0.076 | −0.808** | 0.021 |
| | (2.308) | (0.104) | (0.352) | (0.016) |
| No erosion | −1.843 | 0.068 | −0.201 | 0.017 |
| | (6.068) | (0.103) | (0.857) | (0.023) |
| Severe erosion | −3.912 | −0.028 | −0.411 | 0.022 |
| | (8.794) | (0.093) | (1.157) | (0.046) |

(continued)

**Table 2** (continued)

|  | (1) | (2) | (3) | (4) |
|---|---|---|---|---|
| Model | OLS | Endogenous switching regression[a] | | |
| *Assets* | | | | |
| Machinery | −9.344* | 0.877 | −0.974 | −0.029 |
|  | (4.778) | (0.574) | (0.702) | (0.088) |
| Animals | 3.885 | 0.205 | 0.523* | −0.011 |
|  | (2.389) | (0.202) | (0.282) | (0.028) |
| *Inputs* | | | | |
| Labor | −0.047 | | −0.006 | 0.000 |
|  | (0.042) | | (0.005) | (0.000) |
| Squared labor/100 | 0.003 | | 0.0003* | −0.000 |
|  | (0.002) | | (0.0002) | (0.000) |
| Seeds | 0.062*** | | 0.007*** | −0.000 |
|  | (0.011) | | (0.001) | (0.000) |
| Squared seeds/100 | −0.003*** | | −0.0003*** | 0.000* |
|  | (0.001) | | (0.000) | (0.000) |
| Fertilizers | −0.021 | | −0.004 | −0.000 |
|  | (0.018) | | (0.003) | (0.000) |
| Squared fertilizers/100 | 0.0005 | | 0.0001 | 0.000 |
|  | (0.0005) | | (0.000) | (0.000) |
| Manure | 0.006 | | 0.001 | −0.000 |
|  | (0.004) | | (0.000) | (0.000) |
| Squared manure/100 | −0.0001** | | −0.000* | 0.000 |
|  | (0.00003) | | (0.000) | (0.000) |
| *Farm head and farm household characteristics* | | | | |
| Literacy | 11.712 | 0.188* | 1.540 | −0.068* |
|  | (8.323) | (0.101) | (0.992) | (0.033) |
| Male | 0.752 | 0.118 | 0.028 | 0.066 |
|  | (2.361) | (0.271) | (0.310) | (0.068) |
| Married | 4.741 | −0.273 | 0.657 | −0.090 |
|  | (3.014) | (0.371) | (0.405) | (0.097) |
| Age | 0.538 | 0.006 | 0.082 | −0.002* |
|  | (0.386) | (0.005) | (0.053) | (0.001) |
| Household size | −1.355 | 0.042* | −0.187 | 0.000 |
|  | (1.039) | (0.023) | (0.126) | (0.005) |
| Off-farm job | 6.078 | 0.099 | 0.811 | −0.010 |
|  | (6.161) | (0.138) | (0.778) | (0.028) |
| Relatives | −0.009 | 0.0003 | −0.001 | 0.001** |
|  | (0.019) | (0.001) | (0.002) | (0.000) |
| Access to credit | 11.855 | 0.207 | 1.509 | −0.060** |
|  | (10.175) | (0.146) | (1.240) | (0.027) |
| Flood | −12.952 | 0.196* | −1.611 | −0.052 |

(continued)

Table 2 (continued)

|  | (1) | (2) | (3) | (4) |
|---|---|---|---|---|
| Model | OLS | Endogenous switching regression[a] | | |
|  | (9.797) | (0.112) | (1.210) | (0.044) |
| Drought | 0.172 | −0.033 | −0.113 | 0.054 |
|  | (4.750) | (0.234) | (0.496) | (0.101) |
| *Mundlak's fixed effects* | | | | |
| Mean fertilizers | 0.011 | −0.000 | 0.002 | 0.0001 |
|  | (0.012) | (0.001) | (0.001) | (0.0002) |
| Mean seeds | 0.007 | −0.0003 | 0.002 | 0.0001 |
|  | (0.021) | (0.001) | (0.003) | (0.0001) |
| Mean manure | −0.004 | −0.0001 | −0.0003 | 0.0001 |
|  | (0.003) | (0.0002) | (0.0003) | (0.0001) |
| Mean labor | 0.015 | 0.0002 | −0.0002 | −0.0002 |
|  | (0.037) | (0.001) | (0.004) | (0.0003) |
| *Information sources* | | | | |
| Government extension |  | 0.352*** |  |  |
|  |  | (0.128) |  |  |
| Farmer-to-farmer extension |  | 0.098 |  |  |
|  |  | (0.130) |  |  |
| Radio information |  | 0.358*** |  |  |
|  |  | (0.134) |  |  |
| Neighborhood information |  | 0.050 |  |  |
|  |  | (0.120) |  |  |
| Climate information |  | 0.477*** |  |  |
|  |  | (0.178) |  |  |
| Constant | −176.139* | 1.679 | −17.985 | −0.413 |
|  | (88.112) | (5.573) | (13.965) | (1.242) |
| $\sigma_i$ |  |  | 17.943*** | 0.313*** |
|  |  |  | (6.712) | (0.090) |
| $\rho_j$ |  |  | −0.035 | −0.731 |
|  |  |  | (0.029) | (0.333) |

[a]Estimation by full information maximum likelihood at the plot-level. Sample size: 2801 plots. Robust standard errors clustered at the *woreda* level in parentheses. The dependent variable "skewness" refers to the third central moment $f_3(\mathbf{x}, \boldsymbol{\gamma_3})$ (i.e., downside risk exposure) of production function (2), and it has been rescaled by 10 milliards; $\sigma_i$ denotes the square-root of the variance of the error terms $\varepsilon_{ji}$ in the outcome Eq. (4a and 4b), respectively; $\rho_j$ denotes the correlation coefficient between the error term $\eta_i$ of the selection Eq. (1) and the error term $\varepsilon_{ji}$ of the outcome Eq. (4a and 4b), respectively.
*Significant at the 10% level; **Significant at the 5% level; ***Significant at the 1% level.

in both the short and long rainy seasons. The relationship between downside risk exposure and rainfall is *inverted U*-shaped. There is therefore a threshold level after which rainfall does increase the risk of crop failure. This can be due, for instance, to flooding. The adapters, instead, are not (statistically) affected by the climatic factors. This may underscore the fact that the adapters are more successful in managing the risk implications of climate. Besides the climatic variables the number of relatives and access to credit are significantly (at the 5 percent statistical level) correlated with the skewness function of the group of non-adapters. The clear determination of the mechanisms behind these results is not possible in this study as we lack the necessary information. We can, however, offer some interpretations. The estimated coefficient for the variable 'relatives' is positive. Farmers with a larger number of relatives in the village seem to better manage their risk exposure. We can, however, highlight that this may be due to the positive spillovers originated by social networks. Farmers may thus implement agricultural technologies because of social learning or imitation of their relatives (e.g., Bandiera and Rasul 2006; Conley and Udry 2010). The estimated coefficient for access to credit displays, instead, a negative correlation for the group of non-adapters. This is consistent with what has been found in another paper using the same dataset[22] and may indicate that farm households that have accessed credit are those with a lower skewness compared to those that did not access credit.

Table 3 presents the expected downside risk exposure under actual (cells (a) and (b)) and counterfactual conditions (cells (c) and (d)). Cells (a) and (b) represent the expected downside risk exposure observed in the sample of the adapters and non-adapters. The last column presents the treatment effects of adaptation on downside risk exposure. Our results show that adaptation to climate change significantly increases the skewness, that is decreases downside risk exposure, and so the probability of crop failure. In addition, we find that the transitional heterogeneity effect is negative, that is, farm households that did not adapt would have benefited the most in terms of reduction in risk exposure from adaptation. This finding can be explained by analyzing the last row of Table 3, which accounts for the potential heterogeneity in the sample. It shows first, that there is negative selection into choosing to adapt for the adapters, i.e., if the non-adapters had chosen to adapt their risk exposure would have been below that of the adapters; and second, that there is positive selection into not choosing to adapt for the non-adapters, i.e., if the adapters had chosen not to adapt their risk exposure would have been higher than that of the non-adapters.[23] In short, non-adapters are less exposed to downside risk than the adapters both with adaptation and without adaptation.

---

[22] See Di Falco et al. (2011). The same paper investigated the potential endogeneity of access to credit. Testing procedure rejected this hypothesis at the 1 percent statistical level.

[23] Note that $BH_2$ is negative in Table 3 because it is calculated as the difference between (c) minus (d). However, it is positive if interpreted as (d) minus (c).

**Table 3** Average expected downside risk exposure (skewness); treatment and heterogeneity effects

|  | Decision stage | | |
|---|---|---|---|
| Sub-samples | To adapt | Not to adapt | Treatment effects |
| Adapters | (a) 0.814<br>(0.050) | (c) -0.333<br>(0.004) | TT = 1.146***<br>(0.048) |
| Non-adapters | (d) 1.510<br>(0.065) | (b) 0.043<br>(0.002) | TU = 1.466***<br>(0.064) |
| Heterogeneity effects | $BH_1$ = -0.696***<br>(0.083) | $BH_2$ = -0.376***<br>(0.006) | TH = -0.320***<br>(0.084) |

(a) and (b) represent observed skewness (downside risk exposure), that is the third central moment $f_3(\mathbf{x}, \boldsymbol{\beta}_3)$ of production function (2); (c) and (d) represent the counterfactual expected downside risk exposure. (a) $E(y_{1i}|A_i=1)$; (b) $E(y_{2i}|A_i=0)$; (c) $E(y_{2i}|A_i=1)$; (d) $E(y_{1i}|A_i=0)$ where $A_i = 1$ if farm households adapted to climate change; $A_i = 0$ if farm households did not adapt; $y_{1i}$: third central moment if farm households adapted; $y_{2i}$: third central moment if farm households did not adapt; TT: the effect of the treatment (i.e., adaptation) on the treated (i.e., farm households that adapted); TU: the effect of the treatment (i.e., adaptation) on the untreated (i.e., farm households that did not adapt); $BH_i$: the effect of base heterogeneity for farm households that adapted (i = 1), and did not adapt (i = 2); TH = (TT – TU), i.e., transitional heterogeneity
Standard errors in parentheses. ***Significant at the 1% level

## 6 Conclusions

This paper investigated the implications of farm households' past decision to adapt to climate change on current downside risk exposure. We used a moment-based approach that captures the third moment of a stochastic production function as a measure of downside yield uncertainty. Then, we estimated a simultaneous equations model with endogenous switching to account for unobservable factors that influence downside risk exposure and the decision to adapt.

The first step of the analysis highlighted that the risk associated with the environmental characteristics of the farm such as soil fertility and access to information are key determinants of adaptation. These findings are consistent with Di Falco et al. (2011) on climate change adaption and food productivity, and Koundouri et al. (2006) on irrigation technology adoption under production uncertainty. Koundouri et al. (2006) emphasize that farm households that are better informed may value less the option to wait, and so are more likely to adopt new technologies than other farmers. This implies that waiting for gathering more and better information might have a positive value, and the provision of information on climate change might reduce the quasi-option value associated with adaptation. In addition, in this study we find that also education and past climatic factors significantly affect the adaptation decision. In particular, rainfall in both rainy seasons displays an $U$-shape behaviour, being literate or having experienced a flood in the past has a positive effect on the likelihood to adapt. Development policies that aim to increase education level can have positive spillovers in terms of adaptation and technology adoption in general.

We can draw four main conclusions from the results of this study on the effects of climate change adaptation on downside risk exposure. First, past climate change adaptation reduces current downside risk exposure. Farm households that implemented climate change adaptation strategies obtained benefits in terms of a decrease in the risk of crop failure. Second, adaptation would have been more beneficial to farm households that previously did not adapt if they adapted. This group would have had a larger reduction on downside risk exposure compared to the group of adapters. This leads us to the third finding, namely, there are some important sources of heterogeneity and differences between adapters and non-adapters that make the non-adapters less exposed to downside risk than the adapters irrespective to the issue of climate change. These differences represent sources of variation between the two groups that the estimation of an OLS model including a dummy variable for adapting or not to climate change cannot take into account. Last but not least, climate change adaptation is a successful risk management strategy that makes the adapters more resilient to climatic conditions. The non-adapters are significantly affected by the rainfall in both the short and long rainy seasons while the adapters are much less affected by climatic factors.

It should be stressed, however, that there are very important caveats to our findings. First, our results derive from cross-sectional and plot level analysis. This does not allow an analysis of the dynamic aspects of risk management decisions. This is an important limitation of our study. Panel data would be required to explore such issues. To our knowledge, there is no climate change survey where the same household has been interviewed in different point in time. Future research should therefore be allocated to the construction of such panel data. This will allow to adequately addressing the dynamic dimension of the problem. A second important limitation of our study is that we do not distinguish among different types of adaptation. Di Falco and Veronesi (2013) find that, in Ethiopia, adaptation based upon a portfolio of strategies is significantly more effective than the adoption of strategies in isolation. Arguably some strategies may be more successful than others in dealing with risk exposure (e.g., changing crop varieties, implementing water harvesting technologies). Future research should thus also distinguish how different strategies may affect risk exposure.

# Appendix

**Table A1** Variables definition

| Variable name | Definition |
|---|---|
| *Dependent variables* | |
| Adaptation | Dummy =1 if the farm household adapted to climate change, 0 otherwise |
| Skewness | Downside risk exposure: third central moment $f_3(\mathbf{x}, \boldsymbol{\beta}_3)$ of production function (2)/10 milliards |
| *Explanatory variables* | |
| *Climatic factors* | |
| Average temperature | Average temperature (°C) 1970–2000 |
| Belg rainfall | Rainfall rate in Belg, short rainy season (mm) 1970–2000 |
| Meher rainfall | Rainfall rate in Meher, long rainy season (mm) 1970–2000 |
| *Crop varieties* | |
| Barley | Dummy = 1 if the farm household grows barley, 0 otherwise |
| Maize | Dummy = 1 if the farm household grows maize, 0 otherwise |
| Teff | Dummy = 1 if the farm household grows teff, 0 otherwise |
| Wheat | Dummy = 1 if the farm household grows wheat, 0 otherwise |
| *Soil characteristics* | |
| High fertility | Dummy =1 if the soil has a high level of fertility, 0 otherwise |
| Infertile | Dummy =1 if the soil is infertile, 0 otherwise |
| No erosion | Dummy = 1 if the soil has no erosion, 0 otherwise |
| Severe erosion | Dummy = 1 if the soil has severe erosion, 0 otherwise |
| *Assets* | |
| Machinery | Dummy =1 if machineries are used, 0 otherwise |
| Animals | Dummy = 1 if farm animal power is used, 0 otherwise |
| *Inputs* | |
| Labor | Labor use per hectare (adult days) |
| Seeds | Seeds use per hectare (kg) |
| Fertilizers | Fertilizer use per hectare (kg) |
| Manure | Manure use per hectare (kg) |
| *Farm head and farm household characteristics* | |
| Literacy | Dummy =1 if the household head is literate, 0 otherwise |
| Male | Dummy =1 if the household head is male, 0 otherwise |
| Married | Dummy =1 if the household head is married, 0 otherwise |
| Age | Age of the household head |
| Household size | Household size |
| Off-farm job | Dummy =1 if the household head took an off-farm job, 0 otherwise |

(continued)

**Table A1** (continued)

| Variable name | Definition |
|---|---|
| Relatives | Number of relatives in the *woreda* |
| Access to credit | Dummy =1 if the farm household has access to formal credit, 0 otherwise |
| Flood | Dummy =1 if the farm household experienced a flood during the last 5 years |
| Drought | Dummy =1 if the farm household experienced a drought during the last 5 years |
| *Information sources* | |
| Government extension | Dummy =1 if the household head received information/advice from government extension workers, 0 otherwise |
| Farmer-to-farmer extension | Dummy =1 if the household head received information/advice from farmer-to-farmer extension, 0 otherwise |
| Radio information | Dummy =1 if the household head received information from the radio, 0 otherwise |
| Neighborhood information | Dummy =1 if the household head received information from the neighborhood, 0 otherwise |
| Climate information | Dummy =1 if extension officers provided information on expected rainfall and temperature, 0 otherwise |

**Table A2** Parameter estimates – Test on the validity of the selection instruments

|  | Model 1 | Model 2 |
|---|---|---|
|  | Adaptation 1/0 | Skewness non-adapters |
| *Information sources* | | |
| Government extension | 0.526*** | −0.044 |
|  | (0.112) | (0.072) |
| Farmer-to-farmer extension | 0.492*** | 0.050 |
|  | (0.143) | (0.085) |
| Radio information | 0.464*** | −0.050 |
|  | (0.173) | (0.043) |
| Neighborhood information | 0.002 | −0.070* |
|  | (0.178) | (0.032) |
| Climate information | 0.488** | 0.147 |
|  | (0.201) | (0.103) |
| Constant | −1.173*** | 0.056 |
|  | (0.398) | (0.055) |
| Wald test on information sources | $\chi^2$ = 108.27*** | F-stat. = 2.10 |
| Sample size | 2801 | 868 |

Model 1: Probit model (Pseudo $R^2$ = 0.323); Model 2: ordinary least squares ($R^2$ = 0.070). Other covariates include climatic factors, crop varieties, soil characteristics, assets, inputs, farm head and farm household characteristics as specified in Eqs. (1), (4a) and (4b). Estimation at the plot-level. Standard errors clustered at the *woreda* level in parentheses
*Significant at the 10% level; **Significant at the 5% level; ***Significant at the 1% level

**Table A3** Parameters estimates of production function (2)

| Dependent variable<br>Quantity produced per hectare | Adapters | Non-adapters |
|---|---|---|
| *Climatic factors* | | |
| Average temperature | −202.129 | 268.006 |
| | (300.619) | (291.700) |
| Squared average temperature | 4.868 | −7.600 |
| | (7.573) | (7.231) |
| Belg rainfall | 4.952* | 0.686 |
| | (2.433) | (1.278) |
| Squared Belg rainfall/1000 | −8.367** | −3.602 |
| | (3.514) | (2.131) |
| Meher rainfall | 1.070 | 1.744** |
| | (1.062) | (0.678) |
| Squared Meher rainfall/1000 | −6.665 | −7.935** |
| | (6.016) | (3.369) |
| *Crop varieties* | | |
| Barley | 288.879** | 10.664 |
| | (109.089) | (60.176) |
| Maize | 461.443** | 222.103** |
| | (171.111) | (83.758) |
| Teff | −22.076 | −47.638 |
| | (109.614) | (66.694) |
| Wheat | 98.186 | 53.065 |
| | (88.497) | (54.796) |
| *Soil characteristics* | | |
| Highly fertile | 126.428 | 37.858 |
| | (73.932) | (63.311) |
| Infertile | −150.982*** | −40.251 |
| | (44.538) | (64.474) |
| No erosion | −21.523 | −12.402 |
| | (73.277) | (33.784) |
| Severe erosion | 52.975 | −46.906 |
| | (134.095) | (87.457) |
| *Assets* | | |
| Machinery | −218.916* | −37.570 |
| | (155.387) | (92.297) |
| Animals | 203.901** | 146.169** |
| | (94.438) | (63.333) |
| *Inputs* | | |
| Labor | 3.888*** | 3.739*** |
| | (1.129) | (1.005) |
| Squared labor/100 | −0.139*** | −0.327*** |
| | (0.072) | (0.087) |

(continued)

**Table A3** (continued)

| Dependent variable<br>Quantity produced per hectare | Adapters | Non-adapters |
|---|---|---|
| Seeds | 1.805** | 0.588 |
| | (0.843) | (0.798) |
| Squared seeds/100 | 0.064* | 0.245 |
| | (0.036) | (0.161) |
| Fertilizers | 1.298*** | 1.088** |
| | (0.330) | (0.441) |
| Squared fertilizers/100 | −0.020*** | −0.026** |
| | (0.006) | (0.010) |
| Manure | 0.186*** | −0.021 |
| | (0.046) | (0.136) |
| Squared manure/100 | −0.002** | 0.004** |
| | (0.001) | (0.002) |
| *Farm head and farm household characteristics* | | |
| Literacy | −22.475 | −118.383** |
| | (53.907) | (51.700) |
| Male | 224.332 | 334.423*** |
| | (166.045) | (90.036) |
| Married | −28.748 | −224.175 |
| | (126.850) | (143.732) |
| Age | −3.076 | −3.323* |
| | (2.157) | (1.763) |
| Household size | 4.958 | 7.465 |
| | (15.826) | (10.927) |
| Off-farm job | 168.830* | −8.177 |
| | (85.889) | (62.114) |
| Relatives | 0.162 | −1.087 |
| | (0.185) | (2.020) |
| Access to credit | −50.871 | −264.125*** |
| | (88.492) | (47.731) |
| Flood | −64.011 | −107.933 |
| | (80.790) | (114.596) |
| Drought | −102.393 | 61.738 |
| | (82.838) | (189.641) |
| *Mundlak's fixed effects* | | |
| Mean fertilizers | −0.534* | −0.103 |
| | (0.262) | (0.388) |
| Mean seeds | 0.915 | 0.423 |
| | (0.654) | (0.416) |
| Mean manure | −0.021 | −0.015 |
| | (0.054) | (0.172) |
| Mean labor | −1.419** | −0.992 |

(continued)

Table A3 (continued)

| Dependent variable Quantity produced per hectare | Adapters | Non-adapters |
|---|---|---|
| | (0.606) | (0.581) |
| Constant | 1269.038 | −2547.265 |
| | (3097.326) | (2894.935) |
| Sample size | 1933 | 868 |
| Adj. $R^2$ | 0.304 | 0.328 |

Estimation by Ordinary Least Squares at the plot-level. Sample size: 2801 plots. Robust standard errors clustered at the *woreda* level in parentheses
*Significant at the 10% level; **Significant at the 5% level; ***Significant at the 1% level

# References

Antle JM (1983) Testing the stochastic structure of production: A flexible moment-based approach. J Business and Economic Statistics 1:192–201

Antle JM, Goodger WM (1984) Measuring stochastic technology: the case of tulare milk production American J Agricultural Economics 66:342–350

Bandiera O, Rasul I (2006) Social networks and technology adoption in northern Mozambique. Economic J 116(514):869–902

Carter DW, Milon JW (2005) Price knowledge in household demand for utility services. Land Economics 81(2):265–283

Chavas J-P (2004) Risk analysis in theory and practice. Elsevier London

Cline WR (2007) Global warming and agriculture impact estimates by country. Washington DC: Center for Global Development and Peter G Peterson Institute For International Economics

Conley T, Udry C (2010) Learning about a new technology: Pineapple in Ghana. American Economic Review 100(1):35–69

Daly C (2006) Guidelines for assessing the suitability of spatial climate datasets. International J Climatology 26:707–721

Dercon S (2004) Growth and shocks: evidence from rural Ethiopia. J Development Economics 74(2):309–329

Dercon S (2005) Risk, poverty and vulnerability in Africa. J African Economies 14(4):483–488

Deressa TT, Hassan RM, Ringler C, Alemu T, Yesuf M (2009) Determinants of farmers' choice of adaptation methods to climate change in the Nile Basin of Ethiopia. Global Environmental Change 19(2):248–255

Deressa TT, Hassan RH (2009) Economic impact of climate change on crop production in Ethiopia: Evidence from cross-section measures. J African Economies 18(4):529–554

Deressa TT, Hassan RM, Ringler C (2011) Perception of and adaptation to climate change by farmers in the Nile Basin of Ethiopia. J Agricultural Science 149(1):23–31

Di Falco S, Chavas J-P (2009) On crop biodiversity, risk exposure and food security in the Highlands of Ethiopia. American J Agricultural Economics 91(3):599–611

Di Falco S, Veronesi M, Yesuf M (2011) Does adaptation to climate change provide food security? A micro-perspective from Ethiopia. American J Agricultural Economics 93(3):829–846

Di Falco S, Veronesi M (2013). How African agriculture can adapt to climate change? A counterfactual analysis from Ethiopia. Land Economics

Dinar A, Hassan R, Mendelsohn R, Benhin J et al (2008) Climate change and agriculture in Africa: Impact assessment and adaptation strategies. London: EarthScan

Gadisso BE (2007) Drought assessment for the Nile Basin using Meteosat second generation data with special emphasis on the upper Blue Nile Region. PhD Thesis International Institute for Geo-Information Science and Earth Observation. Eschede: The Netherlands

Gbetibouo G, Hassan R (2005) Economic impact of climate change on major South African field crops: A Ricardian approach. Global and Planetary Change 47:143–152

Gbetibouo G, Hassan R, Ringler C (2010) Modelling farmers' adaptations strategies to climate change and variability: The case of the Limpopo Basin, South Africa. Agrekon 49(2):217–234

Hagos F, Pender J, Gebreselassie N (1999) Land degradation in the highlands of Tigray and strategies for sustainable land management. (First edition) Socio-economics and Policy Research Working Paper 25 ILRI (International Livestock Research Institute). Addis Ababa Ethiopia 80 pp

Hartman RS (1991) A Monte Carlo analysis of alternative estimators in models involving selectivity. J Business and Economic Statistics 9:41–49

Hassan R, Nhemachena C (2008) Determinants of African farmers' strategies for adaptation to climate change: Multinomial choice analysis. African J Agricultural and Resource Economics 2(1):83–104

Heckman JJ, Tobias JL, Vytlacil EJ (2001) Four parameters of interest in the evaluation of social programs. Southern Economic J 68(2): 210–233

International Food Policy Research Institute (IFPRI) (2010) Ethiopia Nile Basin climate change adaptation dataset. Food and water security under global change: Developing adaptive capacity with a focus on rural Africa, Washington DC

Just RE, Pope RD (1979) Production function estimation and related risk considerations. American J Agricultural Economics 61:276–284

Kim K, Chavas J-P (2003) Technological change and risk management: An application to the economics of corn production. Agricultural Economics 29:125–142

Koundouri P, Nauges C, Tzouvelekas V (2006) Technology adoption under production uncertainty: Theory and application to irrigation technology. American J Agricultural Economics 88(3):657–670

Kurukulasuriya P, Mendelsohn R (2008) Crop switching as an adaptation strategy to climate change. African J Agriculture and Resource Economics 2:105–125

Kurukulasuriya P, Kala N, Mendelsohn R (2011) Adaptation and climate change impacts: A structural Ricardian model of irrigation and farm income in Africa. Climate Change Economics 2(2):149–174

Lautze S, Aklilu Y, Raven-Roberts A, Young H, Kebede G, Learning J (2003) Risk and vulnerability in Ethiopia: Learning from the past, responding to the present, preparing for the future. Report for the US Agency for International Development Addis Ababa, Ethiopia

Lee LF, Trost RP (1978) Estimation of some limited dependent variable models with application to housing demand. J Econometrics 8:357–382

Lobell DB, Burke MB, Tebaldi C, Mastrandrea MM, Falcon WP, Naylor RL (2008) Prioritizing climate change adaptation needs for food security in 2030. Science 319:607–610

Lokshin M, Sajaia Z (2004) Maximum likelihood estimation of endogenous switching regression models. Stata J 4(3):282–289

Maddala GS (1983) Limited dependent and qualitative variables in econometrics. Cambridge, UK: Cambridge University Press

Maddala GS, Nelson FD (1975) Switching regression models with exogenous and endogenous switching. Proceeding of the American Statistical Association (Business and Economics Section) 423–426

Maddison D (2006) The perception of and adaptation to climate change in Africa. CEEPA Discussion Paper No 10 Centre for Environmental Economics and Policy in Africa Pretoria, South Africa: University of Pretoria

Mendelsohn R (2000) Efficient adaptation to climate change. Climatic Change 45:583–600

Mendelsohn R, Dinar A (2003) Climate, water, and agriculture. Land Economics 79:328–341

Mendelsohn R, Nordhaus W, Shaw D (1994) The impact of global warming on agriculture: A Ricardian analysis. American Economic Review 84(4):753–771

Menezes C, Geiss C, Tressler J (1980) Increasing downside risk. American Economic Review 70:921–932

MoFED (Ministry of Finance and Economic Development) (2007) Ethiopia: Building on progress: A plan for accelerated and sustained development to end poverty (PASDEP) . Annual Progress Report: Addis Ababa, Ethiopia

Mundlak Y (1978) On the pooling of time series and cross section data. Econometrica 46(1):69–85

Nawata K (1994) Estimation of sample selection bias models by the maximum likelihood estimator and Heckman's two-step estimator. Economics Letters 45:33–40

Nelson FD (1984) Efficiency of the two-step estimator for models with endogenous sample selection. J Econometrics 24:181–196

Orindi V, Ochieng A, Otiende B, Bhadwal S, Anantram K, Nair S, Kumar V, Kelkar U (2006) Mapping climate vulnerability and poverty in Africa. In PK Thornton, PG Jones, T Owiyo, RL Kruska, M Herrero, P Kristjanson, A Notenbaert, N Bekele, A Omolo Mapping climate vulnerability and poverty in Africa. Report to the Department for International Development. International Livestock Research Institute (ILRI), Nairobi

Parry M, Rosenzweig C, Livermore M (2005) Climate change, global food supply and risk of hunger. Phil Trans Royal Soc B 360:2125–2138

Relief Society of Tigray (REST) and NORAGRIC at the Agricultural University of Norway (1995) Farming systems, resource management and household coping strategies in Northern Ethiopia: Report of a social and agro-ecological baseline study in central Tigray Aas. Norway

Rosenzweig C, Parry ML (1994) Potential impact of climate change on world food supply. Nature 367:133–138

Schlenker W, Lobell DB (2010) Robust negative impacts of climate change on African agriculture. Environmental Research Letters 5:1–8

Seo SN, Mendelsohn R (2008a) An analysis of crop choice: Adapting to climate change in Latin American farms. Ecological Economics 67:109–116

Seo SN, Mendelsohn R (2008b) Measuring impacts and adaptations to climate change: A structural Ricardian model of African livestock management. Agricultural Economics 38(2):151–165

Seo N, Mendelsohn R, Dinar A, Hassan R, Kurukulasuriya P (2009) A ricardian analysis of the distribution of climate change impacts on agriculture across agro-ecological zones in Africa. Environmental and Resource Economics 43(3):313–332

Stige LC, Stave J, Chan K, Ciannelli L, Pattorelli N, Glantz M, Herren H, Stenseth N (2006) The effect of climate variation on agro-pastoral production in Africa. PNAS 103;3049 3053

Udry C (1996) Gender, agricultural production, and the theory of the household. J Political Economy 104(5):1010–1046

Wahba G (1990) Spline models for observational data. Philadelphia: Society for Industrial and Applied Mathematics

Wooldridge JM (2002) Econometric analysis of cross section and panel data. Cambridge, MA: MIT Press

World Bank (2006) Ethiopia: Managing water resources to maximize sustainable growth. A World Bank Water Resources Assistance Strategy for Ethiopia. BNPP Report TF050714. Washington DC

World Bank (2010) World development report. Development and climate change. The International Bank for Reconstruction and Development/The World Bank 1818 H Street NW Washington DC 20433

**Open Access** This chapter is distributed under the terms of the Creative Commons Attribution-NonCommercial-ShareAlike 3.0 IGO license (https://creativecommons.org/licenses/by-nc-sa/3.0/igo/), which permits any noncommercial use, duplication, adaptation, distribution, and reproduction in any medium or format, as long as you give appropriate credit to the Food and Agriculture Organization of the United Nations (FAO), provide a link to the Creative Commons license and indicate if changes were made. If you remix, transform, or build upon this book or a part thereof, you must distribute your contributions under the same license as the original. Any dispute related to the use of the works of the FAO that cannot be settled amicably shall be submitted to arbitration pursuant to the UNCITRAL rules. The use of the FAO's name for any purpose other than for attribution, and the use of the FAO's logo, shall be subject to a separate written license agreement between the FAO and the user and is not authorized as part of this CC-IGO license. Note that the link provided above includes additional terms and conditions of the license.

The images or other third party material in this chapter are included in the chapter's Creative Commons license, unless indicated otherwise in a credit line to the material. If material is not included in the chapter's Creative Commons license and your intended use is not permitted by statutory regulation or exceeds the permitted use, you will need to obtain permission directly from the copyright holder.

# Diversification as Part of a CSA Strategy: The Cases of Zambia and Malawi

Aslihan Arslan, Solomon Asfaw, Romina Cavatassi, Leslie Lipper, Nancy McCarthy, Misael Kokwe, and George Phiri

**Abstract** Climate variability, associated with farm-income variability, is recognized as one of the main drivers of livelihood diversification strategies in developing countries. In this chapter, we present a synthesis of two comprehensive studies from Zambia and Malawi on the drivers of diversification and its impacts on selected welfare outcomes with a specific attention to climatic variables and institutions. We use geo-referenced farm-household-level data merged with data on historical rainfall and temperature as well as with administrative data on relevant institutions. The two case studies demonstrate that diversification is clearly an adaptation response, as long term trends in climatic shocks have a significant effect on livelihood diversification, albeit with different implications. Whereas the long term variation in growing period rainfall is associated with increased crop, labour and income diversification in Malawi, it is only associated with increased livestock diversification in Zambia. With regard to institutions, we find that access to extension agents positively and significantly correlates with crop diversification in both countries, underlining the role of extension in promoting more resilient farming systems in rural Zambia and Malawi. Fertilizer subsidies are among the most important agricultural policies in both countries, where they significantly affect incentives for income diversification – though in opposing ways – providing important policy implications. The two case studies document distinct ways in which incentives for livelihood diversification (measured along different dimensions) are shaped by increased

---

A. Arslan (✉) • R. Cavatassi
International Fund for Agricultural Development (IFAD), Rome, Italy
e-mail: a.arslan@ifad.org

S. Asfaw
FAO of the UN, Rome, Italy

L. Lipper
ISPC-CGIAR, Rome, Italy

N. McCarthy
Lead Analytics Inc., Washington, DC, USA

M. Kokwe
FAO of the UN, Lusaka, Zambia

G. Phiri
FAO of the UN, Lilongwe, Malawi

variability in rainfall and rural institutions. The results also demonstrate that diversification can be an effective adaptation response and the risk-return trade-offs are not as pronounced as might be expected.

## 1 Introduction

Livelihood diversification strategies are implemented by households in rural environments as a response to threats and opportunities to manage risk and increase or stabilize income and consumption. Most households in rural areas of developing countries rely on rain-fed agriculture for their livelihoods and, as such, are highly dependent on climatic conditions. Recent evidence and projections indicate that global climate change is likely to increase the incidence of natural hazards, including the variability of rainfall, temperature and occurrences of climatic shocks (IPCC 2014). As a consequence, all aspects of food security may be potentially threatened by the effects of changes in climate, including food availability, access, utilization, and stability (e.g., Challinor et al. 2010; IPCC 2014). In this context, diversification strategies play a crucial role in ensuring food security under climate change, as they have the potential to address two of the CSA pillars by contributing to food security and adaptation to climate change.

Economic theory, however, suggests that there may be potential tradeoffs between food security and adaptation (i.e. between risk and return), specifically related to diversification behaviour. The potential for tradeoffs and synergies depends on the type of diversification in question and the factors that drive it including climatic and institutional factors. We present a synthesis of two comprehensive studies on the drivers of diversification as well as selected welfare outcomes with a specific attention to climatic variables and institutions in this chapter.

We first provide an overview of the literatures on livelihood diversification, vulnerability and climate change to situate diversification in the CSA agenda. We then present empirical evidence from Zambia and Malawi to better understand the linkages between climate shocks, diversification and welfare outcomes with a goal to highlight potential policy entry points to incentivize the types of diversification that help households to improve food security and resilience to climate shocks. We close with a synthesis of results and policy implications.

## 2 Concepts in the Literature

### 2.1 Livelihood Diversification and Vulnerability

Diversification strategies in the presence of imperfect information and risk are acknowledged among the most fundamental theoretical insights in economics. The economic theory of expected utility maximization leads to diversification under risk aversion even when credit and insurance markets function (Alderman and Paxson 1992). Whereas this result applies in many different sectors (e.g. finance, industrial production), the particularities of agricultural production (seasonality in demands for inputs, heterogeneity in land quality or spatial constraints on allocation of resources, dependence on weather patterns) set it apart from other sectors. Specifically, diversification in agricultural production can arise even without invoking risk or under conditions where specialization would be expected (Just and Pope 2001; Alderman and Sahn 1989; Pope and Prescott 1980). The conditions that lead to diversification are further amplified in rural economies, where credit and insurance markets are missing/imperfect, as diversification takes on a role to fill in the risk-management needs left unmet by these markets (Binswanger 1983; Reardon 1997).

Agricultural households in rural economies can adopt diversification leading to better risk-management and smoother income streams *ex-ante* (Smit and Wandel 2006) but also as an involuntary *ex-post* short-term adjustment to smooth consumption in the wake of shocks or crisis, when *ex-ante* risk mitigation strategies are insufficient (Davies and Hossain 1997; Murdoch 1995). The ability of a livelihood system to respond to shocks through coping strategies is thus a key determinant of livelihood resilience and vulnerability, together with *ex-ante risk* mitigation (Adger 1999; Bryceson 1996, 1999; Delgado and Siamwalla 1999; Toulmin et al. 2000; Barrett et al. 2001a; Adger et al. 2005; Folke 2006).

These two types of diversification can be on-farm (e.g. planting a crop or variety mix, or combining crop and livestock operations) or off-farm (e.g. differentiating income sources through wage employment on others' farms or in other sectors, starting own business or migration of a household member). The classifications along on-farm vs. off-farm sectors are still used in the literature despite Barret's (2001) call for a unified diversification classification along sectoral and spatial lines. Regardless of the terminology, what matters is that the returns to the chosen bundle of assets, activities and incomes should ideally be perfectly negatively correlated or just not perfectly correlated with each other to be able to act as a smoothing strategy.

The extensive literature on the drivers of diversification tends to classify the drivers into push and pull factors (Reardon 1997; Barret 2001). Push factors include imperfect credit and insurance markets, stagnation in the agricultural sector, high transaction costs, as well as adverse shocks, hence the diversification that is driven by them need not necessarily improve average incomes (Barrett et al. 2001a; Reardon et al. 2007; Lay et al. 2009). Pull factors, on the other hand, include a booming non-farm sector or new/improved technologies in the farm sector, which lead to diversification that is more likely correlated with improved average outcomes, as well as reduced variability of those outcomes (Reardon et al. 2007; Bandyopadhyay and Skoufias 2013).

When pull factors dominate, livelihood diversification can be a phase in the transition from subsistence to commercial agriculture or non-farm activities, and implicitly a transition out of poverty (Pingali and Rosengrant 1995). Pull factors, however, tend to dominate for wealthier and more educated households, or in areas where access to markets, infrastructure and urban centers are better (Lanjouw et al. 2001;

Fafchamps and Shilpi 2003, 2005; Deichmann et al. 2008; Babatunde and Qaim 2009; Davis et al. 2010; Losch et al. 2011). The majority of empirical evidence on rural households in Sub-Saharan Africa suggest that pull factors dominate for income and labour diversification, so that wealth, education and access to densely populated areas are correlated with higher labour and income diversification, whereas poverty is correlated with higher crop diversification and lower income and labour diversification (Barrett et al. 2001a; Lanjouw et al. 2001; Babatunde and Qaim 2009; Dimova and Sen 2010; Asmah 2011). Though more difficult to establish due to endogeneity issues, the empirical evidence also suggests that more diversified households have higher incomes and greater consumption per capita (Ersado 2003; Babatunde and Qaim 2009; Asmah 2011).

A better understanding of the factors driving diversification by rural households would therefore provide insights into the role of diversification in poverty reduction, food security and development. It would also help design policies that explicitly address diversification as possible determinants of future levels of welfare and foster institutions to support welfare-improving diversification (Barrett et al. 2001b).

The relationship between diversification and vulnerability at the household level seems conceptually clear at first: as the motivation to spread risk over multiple activities is at the heart of diversification, vulnerability should decline as diversification increases. However, while this may be true for deliberate *ex-ante* diversification that leads to less variable incomes, the opposite may be true for forced or *ex-post* diversification (Barrett et al. 2001a; Bandyopadhyay and Skoufias 2013). Here we have just defined vulnerability as "variability in incomes;" however, there are multitudes of vulnerability definitions and measures that complicate the issue even further (Moret 2014). Disentangling the cause and effect linkages between diversification and vulnerability is very difficult given the dynamic relationships between them: while the more vulnerable may be more likely to diversify today to prevent negative effects of shocks in the future, the fact that they diversify may allow them to build-up assets/human capital that leads them to be less vulnerable in the future. This difficulty is amplified in the absence of longitudinal data covering an identifiable shock (idiosyncratic or systemic) to track the patterns of household diversification and welfare outcomes over time. Empirical analyses of these complex relationships based on cross-sectional data, therefore, need to be very careful in attributing causality, as in the case studies presented in this chapter.

## 2.2 How Does Climate Change Enter the Picture?

Agriculture is exposed to various forms of risk ranging from weather variability to pests and diseases to price volatility in output, input and factor markets. For agricultural households that rely on rainfall and face imperfect market conditions that characterize rural economies, these risks take greater prominence as they lack the means to manage risk effectively (e.g. by investing in irrigation, buying insurance or using credit to smooth income and consumption). Climate change multiplies these risks by increasing the probability and severity of unfavorable weather conditions that affect the livelihoods of households in various ways. Direct effects include the decreases in agricultural productivity (crops, livestock, fisheries and forestry), and indirect effects include a decrease in demand for labour, increased local prices, decreased access to markets due to negative impacts on infrastructure, among others. Climate change not only decreases incomes today, but also makes them less predictable by changing the probability distributions in ways that are difficult for households to incorporate into their decision-making (Lipper and Thornton 2014).

Climate change is expected to have generally negative effects on developing-country agriculture, hence on food security. Climate shocks such as drought, flooding, and extreme temperatures are expected to increase in frequency and intensity, and these impacts are projected to increase over time (Nelson and van der Mensbrugghe 2013; IPCC 2012). In the absence of measures to reduce the vulnerability to, and impacts of, such extreme events, they can be expected to generate significant negative impacts on food security (FAO 2010; Foresight 2011).

The impacts of climate change can be generally classified as push factors for diversification as risk-averse farmers implement *ex-ante* risk management strategies (by diversifying crops, other agricultural activities or incomes) and trade a part of their expected earnings with a lower variability in income (Alderman and Paxson 1992; Reardon et al. 1998, 2000, 2007; Barrett et al. 2001a). While climate variability associated with farm-income variability is already recognized as one of the main drivers of diversification in developing countries, the above-mentioned impacts of climate change give further incentives for diversification into activities that are less susceptible to disruption from climatic shocks (Newsham and Thomas 2009).

Empirical evidence on the role of diversification as an adaptation strategy is growing. Crop diversification is shown to help farmers deal with droughts in Nigeria (Mortimore and Adams 2001) and other shocks leading to crop failure in Ethiopia (Di Falco and Chavas 2009; Cavatassi et al. 2011), while income and livelihood diversification are shown to help households deal with weather shocks in Zimbabwe and Nicaragua (Ersado 2003; Macours et al. 2012). This chapter contributes to this literature with two case studies based on nationally representative data as well as high resolution historical data on climatic shocks.

## 2.3 Diversification as CSA

The above discussion on diversification, vulnerability and climate change naturally leads to the realm of CSA, as these concepts are directly concerned with the food security and adaptation pillars of CSA. Adaptation is defined by the IPCC fourth assessment report as "the adjustment in natural or human systems in response to actual or expected climatic stimuli or their effects" (IPCC 2007). This implies a permanent change in the livelihood system leading to better risk-management or coping capacity in the long-run (Smit and Wandel 2006). Diversification at household, village, landscape and national levels is one of the ways of adapting to the changes in climatic patterns and thus of building resilience to climate change, hence it is frequently mentioned in the international CSA policy discourse (FAO 2010;

FAO 2013; Campbell et al. 2014). At the national level, thirteen countries that have submitted National Adaptation Programmes of Action (out of 48) to the United Nations Framework Convention on Climate Change (UNFCCC) have projects focused explicitly on diversification (of crops, livestock, fisheries, livelihoods) as an adaptation strategy.[1] Eleven out of these thirteen are in Sub-Saharan Africa (SSA), where about 30–50% of rural households rely on non-farm income for their total income (Ellis 1998; Reardon 1997; Reardon et al. 1998).[2] Many countries in SSA, including Zambia and Malawi, have also made diversification part of their national agricultural investment strategies/plans and aim to build the necessary enabling environment to support the types of diversification that build resilience.

The ideal enabling environment for diversification choices would consist of institutions and markets that turn push factors into pull factors by facilitating higher income levels with lower levels of variability under the expected climatic shocks. For example, while households may diversify their crops by incorporating legumes into maize plots to buffer maize from rainfall and temperature shocks (especially when inorganic fertilizer use is negligible), this strategy may result in lower incomes if there is no established market for legumes. Improving access to markets and value chains for legumes would be part of a CSA strategy in this context as it would both improve incomes and make them more resilient to weather shocks. Such a strategy has also the potential to contribute to the mitigation pillar, as legume intercropping (by fixing nitrogen in the soil) would decrease the need for inorganic fertilizers, the production and inefficient use of which contribute to the emissions produced by agriculture. These types of mitigation potentials, however, should be considered a co-benefit only in rural environments based on small-scale agriculture, where food security and adaptation are the development priorities.

## 3 Empirical Evidence from Malawi and Zambia

In what follows, we synthesize the results of two empirical studies that investigate the factors driving diversification and the relationship with vulnerability in Malawi and Zambia.[3] These case studies form part of the evidence base for a project on CSA that was funded by the European Commission (EC) and implemented by the Economic and Policy Innovations for CSA (EPIC) programme in FAO during 2012–2015.

---

[1] UNFCCC established a work programme for least developed countries (LDC) in 2001 that include national adaptation programmes of action (NAPA), to support LDCs to address the challenge of climate change given their particular vulnerability. NAPAs provide a process for LDCs to identify priority activities that respond to their urgent and immediate needs to adapt to climate change – those for which further delay would increase vulnerability and/or costs at a later stage. For further information: http://unfccc.int/adaptation/workstreams/national_adaptation_programmes_of_action/items/7572.php.

[2] http://unfccc.int/adaptation/workstreams/national_adaptation_programmes_of_action/items/4583.php.

[3] The Malawi analysis synthesized here is based on Asfaw et al. (2015).

This project was the first of its kind focused on evidence based development intended for policy support to CSA to improve the efficiency of policy making and targeting for sustainable improvements in food security under climate change. By combining two case studies in a comparative analysis and linking them closer with CSA, this chapter provides a broader perspective on the role of diversification as part of a CSA approach to agricultural development policy.

Both Malawi and Zambia already face the negative impacts of climate change manifested in increasing frequency of droughts and floods, as well as increased temperatures in certain parts of both countries (Thurlow et al. 2012; Kanyanga et al. 2013). This chapter provides an insight into the role of climatic shocks in driving diversification, vulnerability outcomes and the types of institutions that may help support diversification and adaptation in SSA, inasmuch as the climatic, socio-economic and political conditions in these two countries are characteristic of SSA.

## 3.1 Country Background

***Zambia*** ranks 15th in the list of countries that are most vulnerable to climate change (Wheeler 2011). The agricultural sector accounts for approximately 20% of the GDP, and around 80% of the rural population lives below the poverty line (World Bank 2013; Chapoto et al. 2011). Furthermore, the fact that 64% of the total population lives in rural areas that primarily depend on rain-fed subsistence agriculture provides a glimpse into the rural vulnerability to various shocks, be it weather shocks or other shocks typical of the agricultural sector (input/output price shocks).

Temperatures in Southern Africa are projected to increase by 0.6–1.4 °C by 2030 and by 1.5–3.5 °C during 2040–2069 (Lobell et al. 2008; Kihara et al. 2015). Rainfall predictions are more ambiguous, with models suggesting either reduced or increased precipitation (Lobell et al. 2008). Regional models, however, agree more on the prediction of decreased rainfall for Southern Africa (Kihara et al. 2015).

Zambia has four distinct agro-ecological regions (AER) and the predicted impacts of climate change differ across AERs (Fig. 1). The western and southern parts of the country (AER I) are exposed to low, unpredictable and poorly distributed rainfall in general, whereas the central part of the country (AER IIa & b) has the highest agricultural potential, with well distributed rainfall (Jain 2007). Zambia-specific climate models predict that rainfall will decrease and temperatures will increase in AER I and II, while rainfall will increase in the northern parts of the country (AER III) (Kanyanga et al. 2013). Combined with projections of prolonged drought and dry spells, maize production is expected to be severely affected in these regions that cover the majority of Zambia's maize growing area. Increased rainfall on the already leached soils of AER III that are also acidic is expected to have a negative impact on crop production. It is also predicted that climate variability will increase, which has reduced the country's economic growth by four percentage points over the last 10 years pulling an additional 2% of the population into poverty (Thurlow et al. 2012). Empirical analyses show that agricultural technologies

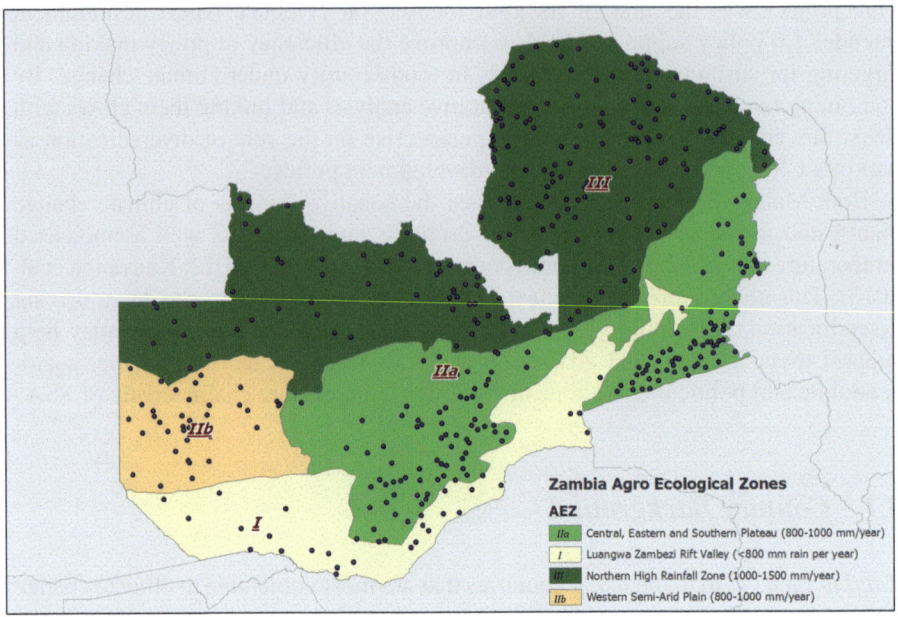

**Fig. 1** Zambia's AER overlaid with the household data points

promoted in rural Zambia, including sustainable agricultural practices as well as the use of modern inputs, are not suited to deal with various shocks expected to get worse under climate change and a more tailored approach is needed to support agricultural growth and food security (Arslan et al. 2015).

The recent Zambia Vulnerability and Needs Assessment Report (VNAR) prepared as a response to prolonged droughts in the 2015 season shows that agriculture is the main income source for 60% of the population and that droughts increased food insecurity in 31 of 48 districts assessed, as approximately 800,000 people were in need of food relief (VAC 2015). It was also observed that costly risk-coping mechanisms were commonly adopted in response, leading to the recommendation that "livelihood diversification programmes be scaled up to reduce dependency on agriculture based activities in view of climate shocks" (VAC 2015). By providing detailed insight into the drivers of diversification under climate change and how institutions may help foster diversification to decrease vulnerability, this chapter provides timely evidence to support policy in Zambia.

*Malawi* is ranked the world's 12th most vulnerable country to the adverse effects of climate change (Wheeler 2011). As in the case of Zambia, projected impacts of climate change combined with the prominence of subsistence farming makes livelihoods vulnerable to climate-related stressors in a number of ways. These include increased exposure to extreme climate events, such as droughts, dry spells, floods, as well as erratic and unreliable rainfall (Chinsinga 2012). Predicted climate change impacts in Malawi are likely to significantly affect smallholders, who depend on rainfall (Denning et al. 2009).

A synthesis of climate data by the World Bank indicates that over the period of 1960 to 2006, mean annual temperature in Malawi increased by 0.9 °C (World Bank 2012). This increase in temperature is concentrated in the rainy summer season (December–February), and is expected to increase further. However, long-term rainfall trends are difficult to characterize due to the highly variable inter-annual rainfall pattern in Malawi. It should be also noted that assessments of climate-change impacts on Malawian agriculture are highly variable across agro-ecological zones (Boko et al. 2007; Seo et al. 2009). Still, given that agricultural production remains the main source of income for most rural communities, the increased risk of crop failure due to projected increases in the frequency of extreme climate events poses a major threat to food security. Adaptation of the agricultural sector to the adverse effects of climate change is thus an important priority for food security (Bradshaw et al. 2004; Wang et al. 2009).

Malawi is one of the least diversified economies in the world, where 84% of the working population is employed in agriculture (the Welfare Monitoring Survey – ILO 2010). In terms of income sources, about 50% of the households derive their income mainly from agriculture and another 25% from a second source (FinScope survey as reported in ILO 2010). Privately owned businesses are common, providing income for over 20% of households, and around 15% have salary or wage income, whereas other sources of income altogether are less than 10% (ILO 2010). Although there is a discrepancy between different surveys, contract labour is reported to be the main source of income for 1–15% of individuals.

The government of Malawi has been trying to address the challenges associated with climate change in various ways. The National Adaptation Programme of Action (NAPA), formulated in 2006, is one of the key climate-change policy documents (GoM 2006; Chinsinga 2012). The Ministry of Agriculture and Food Security operationalizes NAPA priorities through the Agriculture Sector Wide Approach (ASWAp), which identifies several strategies, including diversification, to increase the resilience of rural areas to climate change (GoM 2008; Chinsinga 2012). In-depth studies like the one synthesized here are critical for the efficient design and implementation of such strategies.

## 3.2 Data Sources

The case studies presented in this chapter are based on three main data sources: nationally representative household surveys, historical rainfall and temperature data at high resolution from publicly available data sources, and administrative data on relevant institutions that were collected as part of the project.

For the case of Zambia, the household data come from the 2012 Rural Agricultural Livelihoods Survey (RALS) collected by the Central Statistics Office (CSO) in collaboration with Michigan State University (MSU) and the Indaba Agricultural Policy Research Institute (IAPRI). The data set is nationally representative and includes detailed information on agriculture (crop and livestock) practices, other sources of off-farm rural activities along with household demographic characteristics as well as social capital indicators. The sample consists of more than 8,000 farmers, which are representative at the province level (and at the district level in the Eastern province).

For Malawi, the household data are from the World Bank's Third Malawi Integrated Household Survey (IHS3), which was conducted from March 2010 to March 2011. The IHS3 survey is nationally representative and covers information on various aspects of community and household composition, characteristics and socio-economic status, as well as agriculture-specific production characteristics. The final sample includes a total of 12,271 households that are representative at the district-level IHS (2012).[4]

The RALS and IHS3 data were merged with a set of rainfall and temperature variables that characterise the historical trends as well as current period shocks in these variables, which are closely linked with agricultural production. Rainfall variables are based on data from the Africa Rainfall Climatology version 2 (ARC2) of the National Oceanic and Atmospheric Administration's Climate Prediction Center (NOAA-CPC) for the period of 1983–2012. ARC2 data are based on the latest estimation techniques on a daily basis and have a spatial resolution of 0.1 degrees (~10 km).[5] We also use data from the Harmonized World Soil Database (HWSD) with a resolution of 30 arc-seconds to control for the effects of soil quality on incentives for diversification.[6]

Lastly, administrative data on rural institutions including extension and other sources of agricultural information, credit sources, local community groups, were collected at district level in both countries to better understand the rural institutions that play a role in household livelihood strategies. These data on the availability of rural institutions provide an opportunity to deal with the endogeneity issue in self-reported access variables from household surveys.

## 3.3 Empirical Model

Diversification outcomes at the household level are the result of household optimisation decisions subject to multiple constraints (e.g. imperfect labour, land, credit or insurance markets, and transaction costs) as in standard agricultural household models (Singh et al. 1986; de Janvry et al. 1991). Given the imperfect market conditions pervasive in rural areas of developing countries and the multiple push and pull factors explained above that drive households to diversify their income

---

[4] Malawi IHS3 Basic Information Document. Last accessed 21 October 2014 at: http://siteresources.worldbank.org /INTLSMS/Resources/3358986–1233781970982/5800988–1271185595 871/IHS3.BID.FINAL.pdf.

[5] See http://www.cpc.ncep.noaa.gov/products/fews/AFR_CLIM/AMS_ARC2a.pdf for more information on ARC2.

[6] See http://webarchive.iiasa.ac.at/Research/LUC/External-World-soil-database/HTML/ for more information.

generating activities (both within the farm and off-farm sectors), the observed diversity outcomes can be modelled as functions of endowments and indicators of push and pull factors to test various hypotheses on the drivers of diversification (van Dusen and Taylor 2005).

We use the following estimating equation to understand the drivers of diversification including climatic variables as well as relevant institutions in each country:

$$D_{ij} = \beta_0 + \beta_1 C_k + \beta_2 X_i + \beta_3 G_k + \beta_4 I_d + \varepsilon_i \tag{1}$$

where $D_{ij}$ is the diversification index for household $i$ for the dimension $j$ analysed (e.g., crop, livestock, labour or income), $C_k$ are climatic variables at ward or enumeration area (EA) level (respectively for Zambia and Malawi), $X_i$ are household level variables including socio-demographic characteristics and wealth and social capital indicators, $G_k$ are variables that capture community characteristics at the ward or EA level, and $I_d$ are institutional variables at the district level. In the remainder of this chapter, we first present a descriptive analysis for both countries and then the results of the diversification models described in Eq. (1), before we close with synthesis and policy recommendations.

## 3.4 Descriptive Analysis

### 3.4.1 Zambia

Diversification can be measured along many dimensions using a variety of different indices. Given the high share of agriculture in total incomes of households in our sample (72% on average), the importance placed on diversification into livestock activities as well as diversification of livelihoods in general in the national policy (e.g. NAIP, VNAR, INDC), we measure diversification along three dimension: crops, livestock and income.[7] Given the AER-specific rainfall regimes and predicted climate change impacts, as well as distinct soil structures, one might expect distinct incentives for crop, livestock and income diversification in each AER. We first present descriptive statistics on diversification by AER to provide an understanding of the livelihood structures across the country.

Table 1 summarizes the shares of total agricultural income (from crops and livestock) and livestock income in total income (only for those that have livestock income) by AER to demonstrate the importance of the dimensions along which we analyse diversification in Zambia. Almost three quarters of total income comes from agriculture in our sample, with a variation between 60 percent in AER I and 76 percent in AER IIa. Livestock income is most important in AER I contributing a

---

[7] The income categories used are based on the IAPRI methodology of defining income sources and consist of income from crops, livestock, businesses, remittances, agricultural wages and non-agricultural wages.

quarter of agricultural income (and 14% of total income) as expected given the fact that it covers the provinces where majority of traditional livestock herders live, and least important in AER III with a share of 9% (5%) of agricultural (total) income.

Diversification is measured by different types of indices in the literature, ranging from simple count indices (Jones et al. 2014) or income shares from different sources (Lay et al. 2008; Davis et al. 2010), to more complex indices usually borrowed from biology literature (Smale 2006), which account for evenness, abundance or both. We use the Gini-Simpson index defined as ($1 - \sum_{i}^{i} w_i^2$), where $w_i$ is the number of distinct diversity units in the corresponding index $i$.[8] These are: (a) the area share allocated to different crop species for crop diversification, (b) the shares of different livestock species' contributions to the total livestock holdings measured by Tropical Livestock Units (TLU) for livestock diversification,[9] and (c) the monetary shares of income sources disaggregated into six categories for income diversification (see footnote 7).

The main criteria used to distinguish the AER in Zambia is the average rainfall, which combined with different trends in both rainfall and temperature leads to distinct projections in climate models. Given that climatic shocks are one of the important push factors into livelihood diversification, we first discuss the status of diversification by AER. Table 2 shows both the count and Gini-Simpson indices by AER. AER III is the most diversified in terms of crops with more than

Table 1 Share of agricultural and livestock incomes by AER

| AER | Ag. Inc./ Total Inc. | Lvsk. Inc./ Ag. Inc. | Lvsk. Inc./ Total Inc. |
|---|---|---|---|
| I | 0.60 | 0.25 | 0.14 |
| IIa | 0.76 | 0.14 | 0.09 |
| IIb | 0.72 | 0.15 | 0.09 |
| III | 0.72 | 0.09 | 0.05 |
| Total | 0.73 | 0.13 | 0.08 |

Table 2 Average count and Gini-Simpson indices of diversification by AER

| | Count indices | | | Gini-Simpson indices | | |
|---|---|---|---|---|---|---|
| AER | Crops | Livestock | Income sources | Crops | Livestock | Income sources |
| I | 1.94 | 1.71 | 2.61 | 0.28 | 0.27 | 0.34 |
| IIa | 2.44 | 1.75 | 2.66 | 0.40 | 0.27 | 0.31 |
| IIb | 2.15 | 0.79 | 2.20 | 0.39 | 0.12 | 0.28 |
| III | 2.74 | 1.10 | 2.64 | 0.43 | 0.14 | 0.28 |
| Total | 2.51 | 1.37 | 2.61 | 0.40 | 0.20 | 0.30 |

[8] Count, Simpson and Berger-Parker indices were also constructed and used in analyses for robustness checks. We present results based on the Gini-Simpson index which performed the best.

[9] TLU is created using the following weights for livestock species: horse (0.8), cattle (0.7), donkey (0.5), pig (0.2), sheep and goat (0.1), chicken, duck and fowl (0.01).

2.7 crop species per household, followed by AER IIa and IIb (2.4 and 2.2 species, respectively). AER IIa is the most diversified region in terms of livestock as expected with an average of 1.75 types of livestock per household, followed by AER I and AER III. Households in all AERs have on average at least two income sources. AER IIa has the highest count index of income diversification, followed by AER III. The income diversification is the only dimension that switches the rankings going from count index to Gini-Simpson index, as AER I has the highest Gini-Simpson index for income diversification, indicating that the income shares are more equally distributed in this region contributing more to diversity (measured by proportional abundance) even though it is the third most diverse by the count index.

The observed diversification patterns are the results of both push and pull factors, and the AER classification provides only a broad insight into the climatic push factors into diversification. For example, given the projections of higher temperatures and even lower rainfall in AER II, if the push factors dominate we might expect increased income diversification with lower welfare in this region. AER IIa, however, also includes the urban centers of Lusaka and Eastern provinces, which provide opportunities for pull factors that might be associated with higher diversification at higher welfare levels. Similarly, AER III is projected to have increased rainfall on soils that are already highly leached, but it also includes Copperbelt province with significant mining activity providing potential pull factors. Understanding which factors dominate in driving diversification and what types of welfare outcomes might be expected requires analyses at higher resolution that control for all potential factors as we do below.

We first look at district level climatic variables and diversification outcomes before moving to household level analysis. Figure 2 shows the distribution of long term average of seasonal rainfall and its coefficient of variation (CoV), and Fig. 3 shows the diversification indices by district. Whereas the long run average rainfall in our data conforms to the classification of AER, there is more heterogeneity across districts within AERs in terms of CoV of rainfall indicating climate risk management strategies need to be based on site-specific analyses. It is interesting to note that households seem to diversify their crops more in areas with higher long term average seasonal rainfall, and similarly livestock diversification seems higher in areas where the long term variation in rainfall is higher. Income diversification on the other hand shows no clear pattern correlated with the weather variables plotted in Fig. 3. The heterogeneity within AERs in climatic variables (especially for the variation in rainfall over time) and diversification, provides further evidence that agricultural development planning at the AER level may not be able to capture all factors at play in shaping livelihood decisions. The unconditional averages plotted in these figures provide suggestive evidence only, as it remains to be seen whether and how weather shock variables drive diversification outcomes controlling for other variables that affect livelihood decisions and risk attitudes.

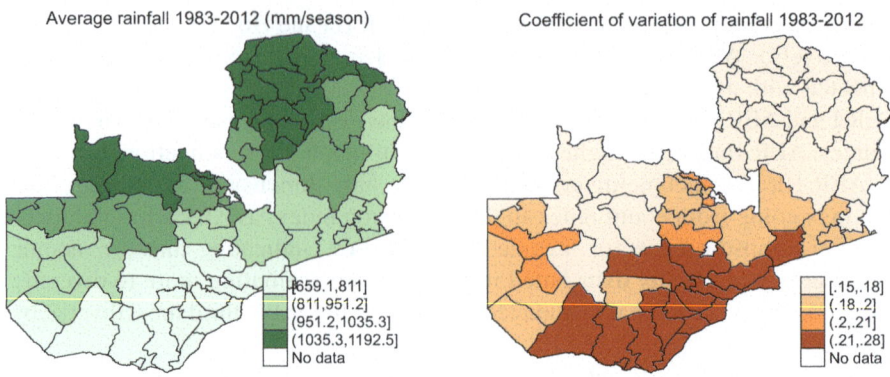

**Fig. 2** Average growing season rainfall and its coefficient of variation over 1983–2012

Table 3 presents the descriptive statistics of all control variables used in the analyses on the determinants of diversification.[10] Our climate variables include the long term (1983–2012) coefficient of variation of rainfall during the cropping season and the current period rainfall anomaly constructed as the deviation of the rainfall in the season covered by the survey from the long term average. While the coefficient of variation captures the effect of long term variation in rainfall on ex-ante incentives, the current period anomaly captures the immediate effect of shocks on diversification (e.g. household being *pushed* into petty jobs to substitute for agricultural income lost due to a shock). Around 24% of household heads are female, and this variable may be expected to have a negative effect on diversification a priori, as female-headed households would find it more difficult to access resources that enable them to take advantage of pull opportunities for diversification (Ellis 1998; Davies and Hossain 1997). However, based on evidence in the literature to suggest that women are more risk averse (Hartog et al. 2002; Borghans et al. 2009), which should "push" them into diversification, the combined effect of gender on diversification is ambiguous and may differ by types of diversification analysed here. Number of household members is a proxy for labour availability and the average household in our sample has 5.4 members. We use operated land size in hectares (2.8) and a household wealth index constructed by principal component analysis based on data on dwelling characteristics as well as the ownership of a large set of assets as wealth indicators.

Social capital and market access can act as pull factors for diversification as households share information and knowledge in groups or in market places that act as information hubs (Cavatassi et al. 2012). We use the share of households in an SEA that participate in farmer cooperatives, women's groups or savings and loan

---

[10] The control variables in both countries are carefully constructed to control for potential endogeneity issues as much as possible in cross-sectional studies. Institutional variables are taken from the district/enumeration area level dataset rather than from household's self-reported values and wealth indices are constructed using the ownership of pre-determined durables. Given the cross-sectional nature of the analyses, this is the best that can be done to control potential endogeneity.

# Diversification as Part of a CSA Strategy: The Cases of Zambia and Malawi 541

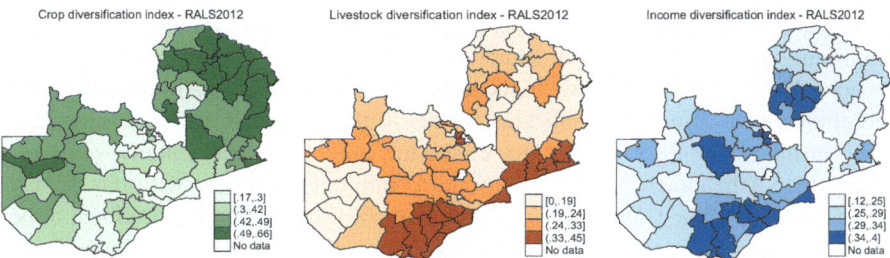

**Fig. 3** Diversification indices in RALS 2012 data by district

**Table 3** Descriptive statistics of control variables (Nr. of observations = 8,219)

| Variable | Mean | Std.Dev | Min | Max |
|---|---|---|---|---|
| **Climate variables** | | | | |
| CoV of Oct-Apr rainfall 1983–2012 | 19.52 | 3.01 | 13.52 | 29.61 |
| Rainfall anomaly during 2010–11 season | 0.08 | 0.09 | 0.00 | 0.38 |
| **Household socio-demographic** | | | | |
| Head is female | 0.24 | 0.43 | 0.00 | 1.00 |
| Age of household head | 44.66 | 15.57 | 17.00 | 111.00 |
| Number of household members | 5.40 | 2.54 | 1.00 | 29.00 |
| Avg adult yrs. of education | 5.59 | 2.84 | 0.00 | 18.00 |
| **Household wealth** | | | | |
| Land size in hectares | 2.77 | 3.82 | 0.00 | 71.56 |
| Wealth index (PCA excluding livestock) | −0.54 | 1.86 | −2.46 | 26.42 |
| **Social capital & market access** | | | | |
| Group membership share in SEA | 0.49 | 0.24 | 0.00 | 1.00 |
| Head/spouse is kin of chief | 0.11 | 0.32 | 0.00 | 1.00 |
| Head/spouse is kin of headman | 0.49 | 0.50 | 0.00 | 1.00 |
| Distance to road (km) | 32.91 | 38.01 | 0.00 | 247.00 |
| Distance to established market place (km) | 27.46 | 23.35 | 0.00 | 153.30 |
| **Ward/district characteristics** | | | | |
| Moderate soil constraint | 0.37 | 0.48 | 0.00 | 1.00 |
| Severe/very severe soil constraint | 0.36 | 0.48 | 0.00 | 1.00 |
| District poverty rate | 0.56 | 0.13 | 0.16 | 0.86 |
| District population density (person/km2) | 0.02 | 0.03 | 0.00 | 0.67 |
| **Institutions** | | | | |
| FISP access (share in SEA) | 0.36 | 0.24 | 0.00 | 0.95 |
| FRA depots in district (nr.) | 10.57 | 11.17 | 0.00 | 48.00 |
| Extension agents from all sources (nr.) | 0.26 | 0.14 | 0.00 | 0.83 |
| Banks in district (nr/100 km$^2$) | 0.03 | 0.07 | 0.00 | 1.44 |
| Tobacco & Cotton Buyers in District (nr.) | 0.82 | 1.02 | 0.00 | 3.00 |

societies, as well as household's kinship ties to the chief and the headman of the community as a proxy for social capital. In an average SEA in our sample 50% of the households participate in any of the groups mentioned above. Almost half of the households have a member with kinship ties to the headman, whereas only 11% have kinship ties to the chief. Village chiefs in Zambia are representatives of their tribe, whereas headmen are elected by the community and deal with day to day activities in the village. We, therefore, expect the kinship ties to the headmen to be stronger drivers of diversification outcomes. Access to urban centers and markets is one of the frequently cited pull factors for diversification as summarized above. We use the distance to a tarmac road and an established marketplace with many buyers and sellers to test this hypothesis.

Given the role that institutions can play in driving diversification outcomes, we use a set of variables to capture the access to relevant institutions. The Farmer Input Support Subsidy Programme (FISP) is one of the most important programmes in Zambia, accounting for around 60% of the poverty reduction programme budget of the ministry of agriculture. It provides fertilisers and seeds to "vulnerable but viable" farmers (i.e. those that have the ability to produce at least 0.5 ha of maize) that are members of cooperatives/farmer groups (Mason et al. 2013). Depending on the specific interventions, such programmes can increase or decrease incentives for diversification. In Zambia, only hybrid maize seed was distributed along with fertilizers until 2009, after which rice, sorghum, cotton and groundnuts were included (Mason et al. 2013). We use the share of households in a given SEA who received FISP support to control for the effect of FISP on diversification.

The Food Reserve Agency (FRA) is another important government programme that takes up the rest of the ministry of agriculture's poverty reduction programme budget (Mason et al. 2013). FRA buys maize from farmers at above market prices, aiming to take some of the price risk away from farmers. By making maize incomes less risky, it increases incentives to grow maize, and hence may be expected to decrease crop diversification. However, it may also increase crop diversification if farmers experiment with other crops given the improved security about their maize income, making the a-priori expectations ambiguous. FRA's effect on other indices of diversification is ambiguous as well, as it depends on other factors at play. We use the number of FRA depots in the district to understand these interactions.

Access to credit is very limited in rural Zambia. Only 15% of households in our sample received a loan from any source during the 2010/11 season. Around 11% of these were from out-grower Schemes (65% of all loans in our sample), while only 0.25% were from commercial banks. Rather than using access to loans as reported by households, which is likely to be endogenous, we use the number of banks per 100 $km^2$ and the number of tobacco and cotton buyers, who are the main suppliers of agricultural credit, to control for the role of credit. Whereas each district has almost one (0.82) cotton or tobacco buyer on average, the average number of banks per 100 $km^2$ is only 0.03. Last but not least, we also use the number of extension agents in each district to understand the impacts of the availability of the information and technical assistance provided by all available extension sources in driving diversification choices.

Finally, we include a number of district and ward level variables, primarily to mitigate potential "placement effects" bias on the coefficients for the institutional variables. Thus, we include measures of soil nutrient availability as defined by the HWSD at the ward level (around 70% of wards have moderate/severe/very severe constraints), and population density and poverty rate (56%) at the district level from the latest census.[11]

### 3.4.2 Malawi

The Malawian case study uses the Margalef index to measure household livelihood diversification. The Margalef index (MI) is computed according to the following formula: $D_i = \dfrac{(S_i - 1)}{\ln(N_i)}$, where $S_i$ is the number of farmer-managed units of diversity (i.e. count) for household $i$ and $N_i$ is the total population count over all farmer-managed units of diversity. The index has a lower limit of zero if only one unit of diversity is observed. We analyse diversification along three dimensions: crop, labour and income.[12]

We use information on the number of crop types planted and the total area planted during the 2009–10 agricultural season for crop diversification and the time (measured in person-hours per year) allocated to three main working activities (i.e. on-farm, off-farm wage labour and self-employment in household enterprises) for labour diversification. We distinguish between nine main sources of aggregate household income for income diversification index: farm agricultural wage, off-farm non-agricultural wage, on-farm livestock income, on-farm temporary and permanent crop income, on-farm fishery income, income from self-employment in household enterprise, public and private transfers, and income from other non-labour sources.

Figure 4 shows the long term average rainfall and its variability measured by the coefficient of variation and Fig. 5 shows the distribution of diversification patterns across Malawian EAs. We observe that the Northern provinces experience relatively higher levels of average rainfall, as compared to the Southern and Central provinces. While rainfall averages are fairly distinct across the three regions (decreasing from north to south), this is not the case for its variability. While the Northern region has more favourable rainfall conditions, farmers are exposed to significant variability within the region. Farmers in the Southern provinces are particularly vulnerable to weather conditions given the lower amount of average rainfall combined with the highest rainfall variability. Though crop diversification does not show a clear pattern across Malawi, labour diversification tends to be higher in the South. Income diversification is particularly low in the southern-most

---

[11] See CSO Census Web Site for details: http://catalog.ihsn.org/index.php/catalog/4124http://catalog.ihsn.org/index.php/catalog/4124.

[12] Count, Gini-Simpson and Berge-Parker indices were also used in analyses. The results are robust to the choice of index and Margalef index provided the best fit for the data.

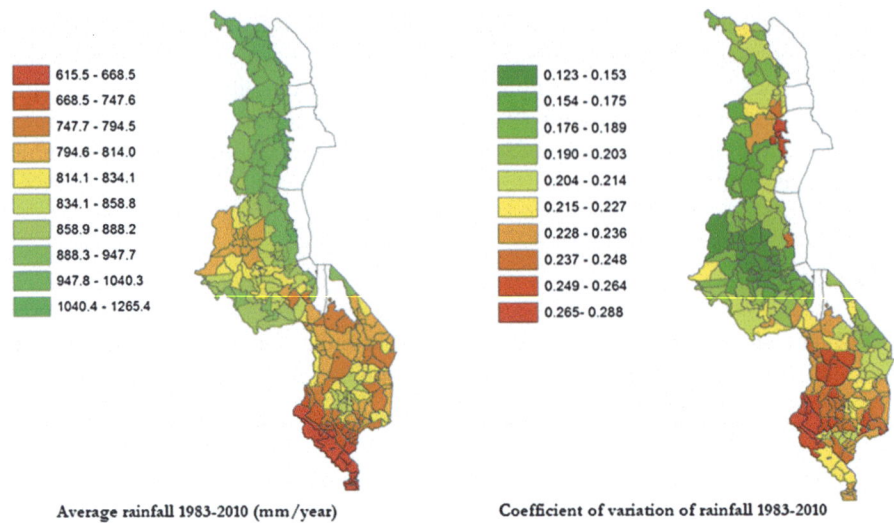

**Fig. 4** Average growing season rainfall and its coefficient of variation over 1983–2010

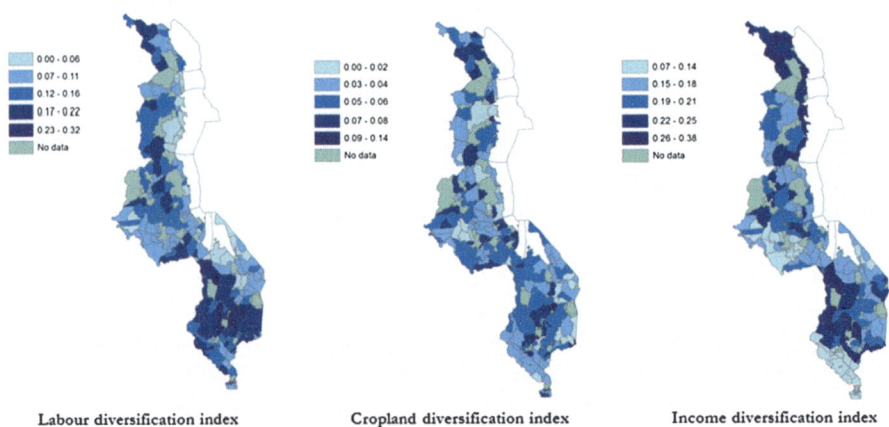

**Fig. 5** Diversification indices by enumeration area (EA)

part of the country and tends to be higher in the central-south as well as in the northern section of the area around Lake Malawi.

Table 4 presents the descriptive statistics of the variables used in the analysis of diversification patterns in Malawi. Similar to the case in Zambia, about 25% of household heads are female, and wealth indices exhibit a right-skewed distribution indicating a high inequality in the distribution of asset ownership.

The institutional variables we use for the Malawi case study capture issues related to access to information and infrastructure (including markets, roads, irrigation schemes and migration flows), as well as primary administrative data on a num-

**Table 4** Descriptive statistics of control variables (Nr. of observations = 7862)

| | Mean | Std. Dev. | Min | Max |
|---|---|---|---|---|
| **Climate variables** | | | | |
| CoV of Nov-may rainfall 1983–2010 | 0.211 | 0.035 | 0.123 | 0.288 |
| Average rainfall 1983–2010 (dm) | 8.5 | 1.065 | 6.096 | 12.654 |
| Rainfall anomaly 2009–10 | −0.086 | 0.092 | −0.369 | 0.2 |
| **Household socio-demographic** | | | | |
| Age of household head | 42.965 | 16.738 | 15 | 110 |
| Head is male | 0.748 | 0.434 | 0 | 1 |
| HH size (Adult Equivalent -AE) | 3.886 | 1.828 | 0.97 | 15.68 |
| Education of the head (yrs.) | 4.848 | 3.94 | 0 | 19 |
| Sex ratio | 1.126 | 1.009 | 0 | 8 |
| Dependency ratio | 1.105 | 0.946 | 0 | 11 |
| Nr of HH members hospitalized in the past 12 months | 0.176 | 0.439 | 0 | 7 |
| **Household wealth** | | | | |
| Wealth index | −0.502 | 1.37 | −1.45 | 12.053 |
| Agricultural implements access index | 0.374 | 1.378 | −3.272 | 8.265 |
| GPS based land size (acre) | 2.479 | 2.571 | 0 | 44.35 |
| **Community characteristics** | | | | |
| In-migration in the community (1 = yes) | 0.54 | 0.498 | 0 | 1 |
| Out-migration in the community (1 = yes) | 0.13 | 0.336 | 0 | 1 |
| Irrigation scheme in the community (1 = yes) | 0.202 | 0.401 | 0 | 1 |
| Road density in 10 km radius ('000 metres) | 9.546 | 2.537 | 0 | 11.274 |
| Number of months main road was passable by a truck | 9.696 | 3.539 | 0 | 12 |
| Ln(price of fertilizer/price of maize) | 1.121 | 0.836 | −2.708 | 5.339 |
| Ln(wage rate for casual labour/price of maize) | 1.63 | 1.161 | −3.401 | 6.032 |
| **Institutions** | | | | |
| Agricultural extension/development officers in district (nr) | 9.546 | 3.9 | 0 | 22 |
| Microfinance institutions in district (nr.) | 2.813 | 1.639 | 0 | 6 |
| Fertilizers distributed per household in district (MT) | 1.269 | 0.518 | 0.305 | 2.249 |
| Ln(MASAF wages paid in 2008/09 season) (mill. MKW/hh) | 0.004 | 0.002 | 0.001 | 0.013 |

ber of government and non-government institutions relevant for understanding incentives for livelihood diversification strategies. These include the number of agricultural extension and development officers, the number of microfinance projects and institutions and the amount of subsidized fertilizer distributed by district. We also control for the total amount of cash paid out in the 2008/09 season in exchange of labour from the Malawi Social Action Fund (MASAF), which is a government social safety net programme, to control for its effects on diversification decisions. By creating a fall-back option, hence a risk-coping mechanism, an active MASAF programme is expected to increase incentives for risk-taking and *ex-ante* diversification.

## 3.5 Econometric Analysis

### 3.5.1 Zambia

We present the coefficients of the models explaining the determinants of crop, livestock and income source diversification in Table 5. All models are estimated using tobit model specification given the fact that the Gini-Simpson index is bounded between 0 and 1 by definition.

The long term variation in season rainfall measured by the coefficient of variation is positively and significantly correlated with livestock diversification, whereas it is negatively and significantly correlated with income diversification. This suggests that households in areas with highly variable seasonal rainfall perceive livestock diversification as an *ex-ante* risk management strategy.[13] Contrary to the expectations, income diversification decreases as rainfall variation increases, suggesting that under highly variable rainfall conditions households revert back to subsistence activities and therefore that pull factor drivers fade away. Current season rainfall deviation from the long term average is not significantly correlated with diversification, suggesting that households are not able respond to immediate shocks to rainfall using the types of diversification analysed here.[14]

In terms of socio-demographic characteristics, female-headed households are less likely to be diversified in terms of crops and livestock but more likely to be diversified in terms of income. These results suggest that female-headed households are not able to take advantage of on-farm diversification opportunities, perhaps due to a gender imbalance in agricultural extension service staff in Zambia (McCarthy, pers. comm.). Greater income diversification in female-headed households may be driven by their higher risk aversion, which leads them to manage risk by engaging in off-farm income opportunities. Education seems to facilitate pull factors into income source diversification by opening up non-farm income opportunities as expected. Of our wealth indicators, land size is positively correlated with crop and livestock diversification but it does not affect income diversification significantly. On the other hand, a higher wealth index – which excludes land – leads to lower crop diversity, but higher livestock and income diversity.

---

[13] Our livestock diversification captures diversification within livestock types. Preliminary analysis of diversification into livestock activities (especially for ruminants) confirms the finding that higher rainfall diversification is significantly and positively correlated with diversification *into* livestock as well as *within* livestock activities.

[14] It should be noted here that rainfall anomalies were, for the most part, not very pronounced during the 2010–2011 growing season. Diversification in response to shocks, primarily of income sources, might still occur with greater anomalies.

**Table 5** Determinants of crop, livestock and income diversification in rural Zambia

| | Simple Models | | | Interaction Models | | |
|---|---|---|---|---|---|---|
| | Crop | Livestock | Income | Crop | Livestock | Income |
| **Climate variables** | | | | | | |
| CoV of rainfall 1983–2012 | −0.004 | 0.027*** | −0.007*** | 0.001 | 0.031*** | −0.000 |
| Rainfall anomaly 2010–11 | 0.100 | −0.091 | 0.038 | 0.086 | −0.101 | 0.047 |
| **Household socio-demographic variables** | | | | | | |
| Head is female | −0.023*** | −0.056*** | 0.029*** | −0.024*** | −0.056*** | 0.029*** |
| Age of household head | 0.001*** | 0.001*** | −0.000 | 0.001*** | 0.001*** | −0.000 |
| HH members | 0.008*** | 0.016*** | 0.001 | 0.008*** | 0.016*** | 0.001 |
| Education (avg) | 0.002 | −0.001 | 0.005*** | 0.002 | −0.000 | 0.004*** |
| **Household wealth** | | | | | | |
| Land size in hectares | 0.006*** | 0.004*** | −0.000 | 0.006*** | 0.004*** | −0.000 |
| Wealth index (PCA excluding livestock) | −0.012*** | 0.018*** | 0.026*** | −0.012*** | 0.018*** | 0.026*** |
| **Social capital & market access** | | | | | | |
| Group membership | 0.134** | 0.151** | 0.017 | 0.135** | 0.150** | 0.030 |
| Kin of chief | −0.001 | 0.005 | 0.024*** | −0.003 | 0.006 | 0.023** |
| Kin of headman | 0.033*** | 0.010 | 0.016** | 0.032*** | 0.011 | 0.016** |
| Distance to road (km) | −0.018 | 0.037 | 0.027** | −0.019 | 0.037 | 0.029*** |
| Distance to market place (km) | 0.127*** | 0.092*** | −0.058*** | 0.122*** | 0.089*** | −0.058*** |
| **Ward/district characteristics** | | | | | | |
| Moderate soil constraint | 0.030 | −0.010 | 0.010 | 0.026 | −0.009 | 0.006 |

(continued)

Table 5 (continued)

|  | Simple Models | | | Interaction Models | | |
| --- | --- | --- | --- | --- | --- | --- |
|  | Crop | Livestock | Income | Crop | Livestock | Income |
| Severe/v.Severe soil constraint | 0.028 | 0.016 | 0.015 | 0.023 | 0.017 | 0.011 |
| District poverty rate | −0.204** | 0.067 | 0.036 | −0.212** | 0.064 | 0.042 |
| District population density (person/km$^2$) | 0.590*** | 0.475*** | 0.370** | 0.900* | 1.043* | −0.132 |
| **Institutions & their interactions** | | | | | | |
| FISP access (share in SEA) | −0.005 | −0.039 | −0.070** | 0.180 | 0.198 | −0.240* |
| FRA depots in district | 0.001 | 0.000 | 0.000 | 0.002 | 0.000 | 0.001 |
| Extension agents in district | 0.014** | 0.008 | 0.001 | −0.028 | 0.037 | 0.023 |
| Banks in district | −0.156* | −0.048 | −0.079 | −0.822 | −1.452 | 1.216 |
| Tobacco & Cotton Buyers in District | −0.033*** | −0.006 | 0.001 | 0.084 | −0.047 | 0.121*** |
| FISP * CoV Rain |  |  |  | −0.010 | −0.012 | 0.008 |
| Extension * CoV Rain |  |  |  | 0.002 | −0.002 | −0.001 |
| Banks * CoV Rain |  |  |  | 0.025 | 0.055 | −0.052* |
| Tobacco/cotton buyers * CoV Rain |  |  |  | −0.006* | 0.002 | −0.006*** |
| N | 8072 | 6779 | 8219 | 8072 | 6779 | 8219 |
| Pseudo R$^2$ | 0.270 | 0.184 | 0.240 | 0.274 | 0.185 | 0.247 |

Note: Standard errors are clustered at the SEA level. *** $p<0.01$, ** $p<0.05$, * $p<0.1$

Membership in cooperatives, farmers', women's or savings and loan groups seems to be effective in facilitating crop and livestock diversification, while it is not significantly correlated with income diversification. These groups would need to be supported to increase their connections with other sectors to facilitate different income generating opportunities if they were to be used as policy entry points to Table 5 (continued) increase income diversification as a risk management strategy. On the other hand, having a kinship tie to the village chief or the headman facilitates income diversification.

The coefficients of the distance to market variable suggest that market constraints/transaction costs act as push factors into crop and livestock diversification as households are significantly more likely to be diversified along these dimensions the farther they are from markets. At the same time, income diversification decreases as the distance to market increases as expected. Distance to an all-weather road, on the other hand, is positively correlated with income diversification, suggesting that while local markets give incentives to diversify income sources, having access to urban centers via all-weather roads gives incentives for specialization.

The institutional variables we use cover the most important institutions that shape households' incentives in rural Zambia, ranging from the most important government programmes to support (particularly maize) farmers, to those that address information and credit constraints. Controlling for all other variables, the higher the proportion of households in the SEA that accessed FISP the less diversified are incomes. This provides suggestive evidence that by giving incentives to cultivate maize (and lately legumes as well) FISP decreases incentives to diversify incomes. FISP and FRA do not have a significant impact with any other diversification outcomes, contrary to the expectations.

The availability of extension agents is positively correlated with crop diversification only, suggesting that they mostly assist farmers on crop production in spite of efforts to improve livestock activities in rural Zambia. Credit constraints seem to act as a push factor into crop diversification as households diversify their crops significantly less in districts with more banks and tobacco and cotton buyers that provide credit. The corollary however is not true, as the number of banks and other credit providers are not positively correlated with livestock and income diversification, suggesting that the credit available is only enough to specialize on farm rather than acting as a pull factor into other activities.

Table 5 also presents the results of the models where we included interaction variables between institutional variables and the coefficient of variation in rainfall to investigate whether and how these institutions perform under highly variable rainfall conditions. This is important if these institutions are to act as policy entry points to decrease vulnerability to climate shocks by facilitating diversification. The coefficient of the FISP variable in income diversification model remains significantly negative and is bigger in magnitude, however its interaction with rainfall variation is not significant (although positive) indicating that FISP does not play a different role under highly variable rainfall conditions.

The role of extension also does not differ by rainfall variation, nor does the role of the availability of banks in the district – except for income diversification. The interaction term between banks and rainfall variation is negative and significant in the income diversification model, indicating that they do not currently act as catalysts for income diversification where agricultural income is highly vulnerable to rainfall shocks. This is similarly true for tobacco and cotton buyers, as the interaction variable with rainfall variation is also negative and significant. The interaction term models point towards a missed opportunity in terms of using these institutions as channels through which household incentives for diversification can be improved especially under highly unpredictable rainfall conditions in order to decrease vulnerability.

### 3.5.2 Malawi

Table 6 presents the results of crop, labor and income diversification models. We present the results that are estimated using OLS in Asfaw et al. (2015), which are robust to functional form specification.[15] We find that the coefficient of variation of rainfall is positively correlated with all three diversification indices indicating that rainfall variability is a push factor into these dimensions of diversification in Malawi. Higher average rainfall is associated with greater diversification in income, but not for crop or labour diversification as expected, indicating that more favourable average rainfall conditions are a pull factor that enables households to secure income from a wider range of sources. A higher rainfall anomaly experienced in the last season reduces income diversification, indicating that households cannot respond quickly to recent shocks. It is important to note that, as in Zambia, the anomaly was mostly moderate during that particular season, suggesting that households are not pursuing income or labour diversification strategies to cope with moderate shocks.

Male-headed households have higher total labour diversification, indicating a potential barrier in labor markets for female headed households. Unlike in Zambia, female headed households are more likely to diversify their crops, but income diversification is higher in female headed households in both countries providing suggestive evidence to support the findings in literature on higher female-risk aversion (Hartog et al. 2002; Borghans et al. 2009). Crop diversification increases with land size and decreases with wealth index just as in Zambia. The existence of an irrigation scheme in the community, which can be associated with less risky crop production, decreases labour and crop diversification, as expected. The number of months during which the main road was passable by a truck is positively and significantly correlated with in labour and income diversification, indicating that lower transaction costs favour these types of diversification.

With respect to institutions, results show that the availability of extension has a positive impact on all diversification measures, indicating that extension information enables pull factors into both on- and off-farm opportunities. Availability of

---

[15] Margalef index has a lower limit of 0 and (unlike the Gini-Simpson index) is not bounded from above. We compared the OLS results with the results of a tobit specification and confirmed that the results are robust. We present the results as in Asfaw et al. (2015) here, as the purpose of this chapter is to synthesize evidence rather than present new results in the case of Malawi.

**Table 6** Determinants of crop, labour and income diversification in rural Malawi

| | Crop | Labour | Income |
|---|---|---|---|
| **Climate variables** | | | |
| CoV of rainfall 1983–2010 | 3.946*** | 1.570*** | 3.438*** |
| Average rainfall 1983–2010 (dm) | 0.005 | 0.003 | 0.230*** |
| Rainfall anomaly 2009–10 | 0.352 | 0.079 | −0.755*** |
| **Household socio-demographic** | | | |
| Age of household head | −0.003*** | −0.003*** | 0 |
| Head is male | −0.115*** | 0.048** | −0.066** |
| Household size | 0.004 | 0.019*** | 0.065*** |
| Household head highest level of education | 0.019*** | 0.017*** | 0.013*** |
| Sex ratio | −0.008 | −0.003 | 0 |
| Dependency ratio | 0.026* | −0.002 | 0.005 |
| HH members hospitalized in the past 12 months | 0.021 | 0.046** | 0.100*** |
| **Household wealth** | | | |
| Wealth index | −0.048*** | 0.088*** | −0.026** |
| Agricultural implements access index | 0.133*** | 0.01 | 0.170*** |
| GPS based land size (acre) | 0.189*** | 0.001 | 0.067*** |
| **Community characteristics** | | | |
| In migration in the community (1 = yes) | −0.084 | 0.029 | 0.012 |
| Out migration in the community (1 = yes) | 0.004 | 0.037 | −0.026 |
| Irrigation scheme in the community (1 = yes) | −0.140*** | −0.052** | −0.072 |
| Road density in 10 km radius ('000 metres) | 0.01 | 0.004 | 0 |
| Number of months road was passable by a lorry | −0.004 | 0.007** | 0.010* |
| Ln(price of fertilizer/price of maize) | 0.134*** | 0.027 | 0.126*** |
| Ln(wage rate for casual labour/price of maize) | −0.070*** | −0.01 | −0.147*** |
| **Institutions** | | | |
| Extension/development officers in district (nr.) | 0.017*** | 0.009*** | 0.022*** |
| Fertilizers distributed/hh in district (MT) | 0.139*** | −0.021 | 0.110** |
| Microfinance institutions in district (nr.) | −0.105*** | 0.019** | −0.046*** |
| Ln(MASAF wages paid in 2008–09 season) | −26.823*** | 1.037 | 12.854 |
| Constant | 0.138 | −0.196 | −1.525*** |
| Observations | 7255 | 7862 | 7768 |
| R-squared | 0.26 | 0.082 | 0.20 |

Note: Standard errors are clustered at the EA level. *** $p < 0.01$, ** $p < 0.05$, * $p < 0.1$

fertilizer subsidies per capita also increases cropland and income diversification. Availability of microfinance institutions and social safety nets, both of which can help farmers to cope with poor weather *ex post*, reduce cropland diversification. Credit availability also reduces income diversification but increases labour diversification, indicating that although it helps farmers secure more diverse set of working activities, this comes at the expense of the diversity in other income sources captured by our index.[16]

As with Zambia, we have done the analysis including interaction terms between our institutional variables and the CoV of rainfall (results not presented here). In this case, none of the interaction terms were statistically significant, indicating that though the institutions lead to greater diversification on average, they are not performing relatively better in enabling diversification in high rainfall risk environments.

## 4 Diversification and Vulnerability

Linking the diversification model results with household vulnerability outcomes empirically is inherently fraught with endogeneity problems (due to both reverse causality and selection/omitted variables bias) as household diversification outcomes are the results of actions taken in response to vulnerability of income/consumption under imperfect market conditions and risk aversion. Therefore an analysis of the dynamic concept of vulnerability – however defined – as a function of diversification indices using cross-sectional data would very likely produce biased results. Here we present only a descriptive analysis of the correlations between vulnerability indicators in our data and diversification measures.

### 4.1 Zambia

We use three variables as indicators of vulnerability in Zambia: the logarithm of income per capita and its variance, and the number of months the household did not have enough food during the survey year. The levels of a welfare outcome (consumption or assets) and its variance are used as the components of vulnerability in the vulnerability to poverty literature (Christiaensen and Subbarao 2005; Chaudhuri et al. 2002). RALS data does not have a detailed consumption module, therefore we use total income and its variance estimated from a regression of income determinants as components of vulnerability to income poverty. We also use the income poverty line from the Zambian Living Conditions Monitoring Report (CSO 2010) to calculate the Foster-Greer-Thorbecke (FGT) poverty measures.

Table 7 reports the simple correlations between diversification measures and vulnerability indicators. Income per capita is positively, and its variance is negatively, correlated with all diversification measures as expected. Number of food deficit months on the other hand is positively correlated with income diversification,

---

[16] The income diversification includes five different on-farm income sources, whereas labour diversification only has one on-farm labour category.

**Table 7** Correlation coefficients between diversification and vulnerability indicators

|  | Crop Div. | Livestock Div. | Income Div. | Income per capita (ln.) | Variance of Income | Food deficit months |
|---|---|---|---|---|---|---|
| Crop div. | 1 | | | | | |
| Livestock div. | 0.06 | 1 | | | | |
| Income div. | −0.10 | 0.05 | 1 | | | |
| Income per capita (ln.) | 0.05 | 0.09 | 0.11 | 1 | | |
| Variance of Income | −0.14 | −0.03 | −0.01 | 0.00 | 1 | |
| Food deficit months | −0.04 | −0.13 | 0.05 | −0.21 | 0.02 | 1 |

**Table 8** Food deficit categories and diversification

| Food deficit | Crop Div. | Livestock Div. | Income Div. |
|---|---|---|---|
| Less than 3 months | 0.41 | 0.15 | 0.29 |
| 3–6 months | 0.37 | 0.09 | 0.33 |
| More than 6 months | 0.35 | 0.09 | 0.33 |
| Total | 0.41 | 0.14 | 0.30 |

suggesting that income diversification may act as a coping strategy to deal with transient shocks.

In order to unpack the relationship between vulnerability to food shortages and diversification, Table 8 reports the average diversification indices by different categories of food deficit months. Households that had less than 3 months of food deficit have the highest crop and livestock diversification and the lowest income diversification. On the other hand, those who had more than 6 months of food deficit have the lowest crop and livestock diversification and the highest income diversification, providing further evidence that income diversification results from push factors in rural Zambia, at least in terms of food availability. That income diversification is a coping strategy rather than voluntary choice in rural Zambia is a finding supported by earlier descriptive literature (Karttunen 2009). Higher incomes per capita, then, do not necessarily translate into the ability to purchase the same amount of food as is available to households with larger landholdings and thus their own production. Given the subjective nature of this result, however, more research is needed to establish this correlation.

Finally, we look at the distribution of diversification and vulnerability measures across AERs, which shape the thinking about climate change and its impacts on agriculture and livelihoods in Zambia (Table 9).

AER I, which is the region with the lowest rainfall that also has the highest variability across years, has the lowest crop diversification and highest income diversification. It also has the second lowest income per capita with the highest variance as well as highest rate and depth of poverty. Given the importance of livestock in the incomes of households in AER I and the fact that rainfall is projected to decrease with increased unpredictability, combined with our finding that increased rainfall

**Table 9** Diversification, vulnerability and poverty by AER

| AER | Crop Div. | Live-stock Div. | Income Div. | Income per capita | Var. of Income | Food deficit months | Poverty Rate | Depth of Poverty |
|---|---|---|---|---|---|---|---|---|
| I | 0.28 | 0.17 | 0.34 | 137,262.83 | 0.76 | 1.75 | 0.79 | 0.62 |
| IIa | 0.40 | 0.18 | 0.31 | 170,519.35 | 0.62 | 1.23 | 0.70 | 0.57 |
| IIb | 0.39 | 0.05 | 0.28 | 135,814.12 | 0.69 | 3.35 | 0.79 | 0.59 |
| III | 0.44 | 0.11 | 0.29 | 168,005.51 | 0.60 | 1.52 | 0.67 | 0.52 |
| **Total** | **0.41** | **0.14** | **0.30** | **163,935.84** | **0.63** | **1.57** | **0.70** | **0.55** |

variation increases livestock diversification indicates that policies that can facilitate diversification under the predicted impacts of climate change are needed to address the compounded issues of poverty and vulnerability in the region. This finding becomes more important taking into account that income diversification is negatively correlated with income and is a coping strategy for the poorest and most food insecure in this region.

AER IIb also stands out with its low incomes with high variance, high average food deficit months and poverty rate, and lowest livestock and income diversification. Projected impacts of climate change in this region (including decreased rainfall and increased temperatures and unpredictability) underline the importance of actions to improve the capacity to diversify income sources and, where possible, integration of livestock into agriculture.

## 4.2 Malawi

As for the Zambia analysis, we conclude our analysis with an exploratory investigation into the correlations between diversification and various consumption/vulnerability indicators. Given the detailed consumption module in IHS data, we calculate the main components of vulnerability (levels and variance) using consumption data (Christiaensen and Subbarao 2005; Chaudhuri et al. 2002). We also use other poverty indicators such as the different types of FGT indices (i.e. poverty rate, depth of poverty and severity of poverty).

All diversification indices are negatively correlated with the variance of consumption. While labor and income diversification are also positively correlated with expected consumption, this correlation is negative for crop diversification, suggesting that diversification of labour and income are driven by pull factors, whereas crop diversification is mainly a result of push factors. The latter indicates that crop diversification is a risk management strategy, leading to lower, but more stable, crop production. All three diversification strategies are negatively correlated with all poverty indicators, providing suggestive evidence that they have potential to contribute to food security and adaptation (Table 10).

To conclude our exploratory analysis we look at the heterogeneity of poverty and diversification strategies across the three regions of the country. The table confirms

Diversification as Part of a CSA Strategy: The Cases of Zambia and Malawi 555

**Table 10** Correlation coefficients between diversification and vulnerability indicators

|  | Labor Div. | Income Div. | Crop Div. | Var. of Consumption | Expected Consumption | Poverty rate | Poverty gap | Poverty severity |
|---|---|---|---|---|---|---|---|---|
| Labor div. | 1.00 | | | | | | | |
| Income div. | 0.30 | 1.00 | | | | | | |
| Crop div. | 0.03 | 0.25 | 1.00 | | | | | |
| Variance of consumption | −0.02 | −0.09 | −0.10 | 1.00 | | | | |
| Expected consumption | 0.17 | 0.01 | −0.03 | −0.04 | 1.00 | | | |
| Poverty rate | −0.15 | −0.08 | −0.06 | 0.02 | −0.51 | 1.00 | | |
| Poverty gap | −0.15 | −0.14 | −0.08 | 0.06 | −0.52 | 0.77 | 1.00 | |
| Poverty severity | −0.12 | −0.15 | −0.09 | 0.07 | −0.46 | 0.59 | 0.95 | 1.00 |

**Table 11** Diversification, vulnerability and poverty by region

| Variable | North | Central | South | Total |
|---|---|---|---|---|
| Labor div. | 0.043 | 0.041 | 0.044 | 0.043 |
| Income div. | 0.228 | 0.186 | 0.198 | 0.199 |
| Crop div. | 0.125 | 0.133 | 0.168 | 0.148 |
| Variance of consumption | 0.250 | 0.252 | 0.227 | 0.240 |
| Expected consumption | 10.696 | 10.804 | 10.646 | 10.710 |
| Poverty rate | 0.525 | 0.420 | 0.543 | 0.495 |
| Poverty gap | 0.184 | 0.139 | 0.202 | 0.176 |
| Poverty severity | 0.086 | 0.063 | 0.098 | 0.083 |

that Malawi is a rather homogenous country (as opposed to Zambia) with similar distributions in these variables across the three regions. The southern region is slightly more diversified in terms of labor and crops, and is slightly lower levels of consumption than the other two regions, although it's more stable. The central region, on the other hand, is less diversified and has lower FGT poverty measures, but presents higher levels as well as variability of consumption (Table 11).

## 5 Synthesis of Cross-Country Evidence and Conclusions

The two case studies presented in this chapter demonstrate that diversification is clearly an adaptation response as long term trends in climatic shocks have a significant effect on livelihood diversification, albeit with different implications. Whereas the long term variation in growing period rainfall acts as a push factor into all three types of diversification in Malawi, it only acts as a push factor into livestock diversification in Zambia. The findings in Malawi are as expected based on both theoretical and empirical literature predicting an increase in diversification with increases in riskiness in agricultural activity (Barrett et al. 2001a; Reardon et al. 2007; Brown 2008). The effect of this variable on income diversification has the opposite sign in Zambia, where households revert back to subsistence crop production activities instead of diversifying incomes. The fact that this effect of rainfall variation disappears when we control for its interactions with institutional variables suggests that a focus on on-farm income generation is facilitated by FISP and credit access from various sources that incentivize agricultural production – potentially at the expense of long term livelihood resilience. Diversification into and within livestock activities has long been promoted in Zambia as a way to address vulnerability, and our results show that rainfall stress increases the incentives to do so. Further research on the implications of these activities for vulnerability based on panel data is needed to devise targeted policies to support livelihoods under climate stress.

Female headed households are found to be more likely to have diversified income sources in both countries, which seems to be driven by women's higher risk aversion observed in the literature. Whereas female headed households seem not to be able to benefit from pull factors into crop diversification in Zambia, those in Malawi are more diversified in terms of crops. Crop diversification in Malawi, however, is potentially driven by push factors as suggested by descriptive analysis, indicating that female headed households are likely to be disadvantaged in terms of benefiting from pull factors there as well.

Higher education acts as a pull factor into income diversification in both countries consistent with the literature (Reardon et al. 2007 and the references within). The more members a household has, the more likely it is to have higher crop and livestock diversification in Zambia, and higher labour and income diversification in Malawi. These differences suggest structural differences between the rural labour markets and other income generating activities in these countries. Perhaps due to its size, Malawi seems to have more active pull factors into diversification beyond the farm than Zambia. This finding is also supported by the positive and significant correlation between labour and income diversification and the number of months the road was passable by a truck in Malawi, whereas income diversification increases with distance to an all-weather road in Zambia.

Households with larger land size are significantly more likely to diversify their crops suggesting potential barriers to diversification for smallholders. Better targeting for smallholders in crop diversification interventions would be needed, especially in cases where climate variability is expected to negatively affect the subsistence crop production they heavily depend on. Another indicator of wealth measured by the wealth index has the same negative correlation with crop diversification in both countries, whereas it correlates with income diversification in opposite ways in Malawi and Zambia. Households with higher wealth seem to specialize in a couple of income generating activities in Malawi, but they diversify income sources more in Zambia. Whereas this finding in Zambia is consistent with most previous findings in Africa (Reardon 1997; Barrett and Reardon 2000; Burke and Lobell 2010; Martin and Lorenzen 2016), Malawi seems to follow the evidence from Latin America, which is explained by the availability of low-barrier-to-entry labor-intensive jobs, high population density and unequal landholdings in the literature (Reardon et al. 2000).

With regard to institutions, we find that access to extension agents positively and significantly correlates with crop diversification in both countries, underlining the role of extension in promoting more resilient farming technologies in rural Zambia and Malawi. Fertilizer subsidies are among the most important agricultural policies in both countries and we find that they significantly affect incentives for income diversification – though in opposing ways. Whereas income diversification is positively correlated with subsidized fertilizer distribution in Malawi, this effect is negative in Zambia (more so under average rainfall variability). If income diversification is a policy goal to decrease vulnerability to climate change as stated in recent national policies and programmes, research to better understand how these subsidy programmes can be reformed to achieve this goal is necessary. Lastly, access to credit is found to decrease crop diversification, especially under highly variable rainfall conditions in Zambia, which requires special attention in the context of climate change as rural development policies strive to improve the functioning of credit markets.

The two case studies in this chapter document distinct ways in which incentives for livelihood diversification (measured along different dimensions) are shaped by increased variability in rainfall and rural institutions. The results also demonstrate that diversification can be an effective adaptation response and the risk-return tradeoffs are not as pronounced as might be expected. The differences across types of diversification and drivers in shaping the tradeoffs and synergies underline the importance of identifying and promoting the desirable diversification options for specific country circumstances. Given the predicted impacts of climate change on rainfall patterns, the implied changes in livelihood diversification merit special attention as part of a climate smart approach to agricultural development. Diversification has the potential to improve food security as well as contribute to adaptation efforts by decreasing vulnerability; however disentangling these multidimensional and dynamic relationships requires panel data analyses planned for future research. Establishing causality among the multiple diversification strategies, institutions and climatic shocks using cross-sectional data is not feasible, hence the results presented here should be interpreted with this caveat in mind.

# References

Adger, W. N. (1999). Social vulnerability to climate change and extremes in coastal Vietnam. World Development, 27, 249–269. doi:10.1016/S0305-750X(98)00136-3

Adger, W. N., Arnell, N. W., & Tompkins, E. L. (2005). Successful adaptation to climate change across scales. Global Environmental Change, 15, 77–86. doi:10.1016/j.gloenvcha.2004.12.005

Alderman, H. and D.E. Sahn.1989. "Understanding the Seasonality of Employment, Wage, and Income," Ch.6 in D.E. Sahn (ed.), Seasonal Variability in Third World Agriculture: The Consequences for Food Security, Baltimore, MD: John Hopkins Press, pp. 81–106.

Alderman, H. and Paxson, C. 1992. "Do the poor insure? A synthesis of the Literature on Risk and Consumption in Developing Countries," Ch. 3 in Bacha, E. L. (ed.), Economics in a Changing World, eds. Vol. 4: Development, Trade and the Environment, S. Martin's Press, pp. 48–78.

Arslan, A., McCarthy, N., Lipper, L., Asfaw, S. Cattaneo, A. Kokwe, M. 2015. "Climate Smart Agriculture? Assessing the Adaptation Implications in Zambia," Journal of Agricultural Economics, 66, 3, 753–780.

Asfaw, S., McCarthy, N., Arslan, A., Lipper, L. and Cattaneo, A. (2015): Diversification, Climate risk and Vulnerability to Poverty: Evidence from Rural Malawi. FAO-ESA Working Paper 15-02.

Asmah, E.E. 2011. Rural livelihood diversification and agricultural household welfare in Ghana. Journal of Development and Agricultural Economics 3(7): 325–334.

Babatunde, R.O. & Qaim, M. 2009. Patterns of income diversification in rural Nigeria: determinants and impacts. Quarterly Journal of International Agriculture 48: 305–320.

Bandyopadhyay, S. & Skoufias, E. 2013. Rainfall variability, occupational choice, and welfare in rural Bangladesh. Policy Research Working Paper 6134. Washington, DC, World Bank.

Barrett, C.B. & Reardon, T. 2000. Asset, activity, and income diversifications among African agriculturalist: Some practical issues. Project report. USAID BASIS CRSP.

Barrett, C.B., Reardon, T. & Webb, P. 2001a. Nonfarm income diversification and household livelihood strategies in rural Africa: concepts, dynamics and policy implications. Food Policy 26(4): 315–331.

Barrett, C.B., Bezuneh, M. & Aboud, A. 2001b. Income diversification, poverty traps and policy shocks in Côte d'Ivoire and Kenya. Food Policy 26(4): 367–384.

Binswanger, H.P. 1983. "Agricultural Growth and Rural Non-farm Activities," Finance & Development, pp. 38–40.

Boko, M., I. Niang, A. Nyong, C. Vogel, A. Githeko, M. Medany, B. Osman-Elasha, R. Tabo & P. Yanda 2007. Africa. In M.L. Parry, O.F. Canziani, J.P. Palutikof, P.J. van der Linden & C.E. Hanson, eds. Climate change 2007: Impacts, adaptation and vulnerability. Contribution of Working Group II to the Fourth Assessment Report of the Intergovernmental Panel on Climate Change. Cambridge University Press, Cambridge, UK, 433–467.

Borghans, L., Golsteyn, B. H. H., Heckman, J. J. and Meijers, H. 2009. "Gender Differences in Risk Aversion and Ambiguity Aversion," Journal of the European Economic Association, MIT Press, 7(2–3): 649–658.

Bradshaw, B., Dolan, A. & Smit, B. 2004. Farm-level adaptation to climatic variability and change: crop diversification in the Canadian prairies. Climatic Change 67(1): 119–141.

Brown, M. E. 2008. "The Impact of Climate Change on Income Diversification and Food Security in Senegal," Chapter 3 in Land Change Science in the Tropics: Changing Agricultural Landscapes, pp. 33–52, Springer US, doi: 10.1007/978-0-387-78864-7_3

Bryceson, D. 1996. Deagrarianization and rural development in Sub-Saharan Africa: A sectoral perspective. World Development 24(1): 97–111.

Bryceson, D. 1999. African rural labour, income diversification and livelihood approaches: A long-term development perspective. Review of African Political Economy 26(80): 171–189.

Burke M. and Lobell, D. 2010. "Food Security and Adaptation to Climate Change: What Do We Know?" Chapter 8 in Climate Change and Food Security, D. Lobell and M. Burke (eds.), Advances in Global Change Research 37, pp. 133–153, doi: 10.1007/978-90-481-2953-9_8

Campbell, B. M., Thornton, P., Zougmoré, R., van Asten, P. and Lipper, L. 2014. "Sustainable intensification: What is its role in climate smart agriculture?" Current Opinion in Environmental Sustainability, 8:39–43. doi:10.1016/j.cosust.2014.07.002

Cavatassi, R. Lipper, L. and Narloch, U. 2011. "Modern variety adoption and risk management in drought prone areas: Insights from the sorghum farmers of eastern Ethiopia." Agricultural Economics; 42(3):279–292. doi:10.1111/j.1574-0862.2010.00514.x

Cavatassi, R., Lipper, L. and Winters, P. 2012. "Sowing the seeds of social relations: social capital and agricultural diversity in Hararghe, Ethiopia," Environment and Development Economics 17(5): 547–578. doi:10.1017/ S1355770X12000356

Challinor, A., Simelton, W., Fraser, E., Hemming, D., Collins, C. (2010) Increased crop failure due to climate change: assessing adaptation options using models and socio-economic data for wheat in China. Environmental Research Letters 5(3): 034012.

Chapoto, A., Banda, D., Haggblade, S. and Hamukwala, P. 2011. "Factors Affecting Poverty Dynamics in Rural Zambia." Food Security Research Project, Working Paper No. 55, Lusaka.

Chaudhuri, S., Jalan, J. & Suryahadi, A. 2002. Assessing household vulnerability to poverty from cross-sectional data: A methodology and estimates from Indonesia. Discussion Paper No. 010252, New York, USA, Columbia University.

Chinsinga, B. 2012. The political economy of agricultural policy processes in Malawi: A case study of the fertilizer subsidy programme. Future Agricultures Consortium Working Paper 39. Brighton, UK.

Christiaensen, L. & Subbarao, K. 2005. Towards an understanding of household vulnerability in rural Kenya. Journal of African Economies 14(4): 520–558.

Davis, B., Winters, P., Carletto, G., Covarrubias, K., Quiñones, E.J., Zezza, A. & DiGiuseppe, S. 2010. A cross-country comparison of rural income generating activities. World Development 38(1): 48–63.

Davies, S. and Hossain, N. 1997. "Livelihood Adaptation, Public Action and Civil Society: A Review of the Literature," IDS Working Paper No.57, Brighton: Institute of Development Studies.

Deichmann, U., Shilpi, F. & Vakis, R. 2008. Spatial specialization and farm-nonfarm linkages. World Bank Policy Research Working Paper No. 4611. Washington, DC, World Bank.

De Janvry, A., M. Fafchamps and E. Sadoulet. 1991. Peasant household behaviour with missing markets – some paradoxes explained, Economic Journal 101: 1400–1417.

Delgado, C. & Siamwalla, L. 1999. Rural economy and farm diversification developing countries. In G. H. Peters, & J. von Braun, eds. Food security, diversification and resource management, refocusing the role of agriculture. International Associations of Agricultural Economists. Brookfield, USA, Ashgate.

Denning, G., Kabambe, P., Sánchez, P., Malik, A., Flor, R., Harawa, R., Nkhoma, P., Zamba, C., Banda, C., Magombo, C., Keating, M., Wangila, J. & Sachs, J. 2009. Input subsidies to improve smallholder maize productivity in Malawi: Toward an African green revolution. PLoS Biology 7(1): 2–10.

Di Falco, S. & Chavas, J.P. 2009. On crop biodiversity, risk exposure, and food security in the highlands of Ethiopia. American Journal of Agricultural Economics 91(3): 599–611.

Dimova, R. & Sen, K. 2010. Is household income diversification a means of survival or a means of accumulation? Panel data evidence from Tanzania. Brooks World Poverty Institute Working Paper No. 12210. BWPI, University of Manchester.

Ellis, F. 1998. Household strategies and rural livelihood diversification. Journal of Development Studies 35(1): 1–38.

Ersado, L. 2003. Income diversification in Zimbabwe: Welfare implications from Urban and Rural Areas., FCND Discussion Paper No. 152. International Food Policy Research Institute, Food Consumption and Nutrition Division.

Fafchamps, M. & Shilpi, F. 2003. The spatial division of labour in Nepal. The Journal of Development Studies 39(6): 23–66.

Fafchamps, M. & Shilpi, F. 2005. Cities and specialisation: evidence from South Asia. The Economic Journal 115(503): 477–504.

FAO. 2010. "Climate-Smart" Agriculture. Policies, Practices and Financing for Food Security, Adaptation and Mitigation. Food and Agriculture Organization of the United Nations, Rome, Italy.

FAO 2013. Climate-Smart Agriculture Sourcebook. Food and Agriculture Organization of the United Nations, Rome, Italy.FAO, IFAD and WFP (Food and Agriculture Organization of the United Nations, International Fund for Agricultural Development and World Food Programme). 2014. State of Food Insecurity in the World 2014: Strengthening the enabling environment for food security and nutrition. FAO, Rome.

Folke, C. (2006). Resilience: The emergence of a perspective for social–ecological systems analyses. Global Environmental Change, 16, 253–267. doi:10.1016/j.gloenvcha.2006.04.002

Foresight International Dimensions of Climate Change (2011). Final Project Report. The Government Office for Science, London.

GoM (Government of Malawi) 2006. Malawi growth and development strategy 2006–2011, Ministry of Economic Planning and Development, Lilongwe.

GoM. 2008. Agricultural Development Programme (ADP), Lilongwe, Ministry of Agriculture and Food Security.

Hartog, J., Ferrer-i-Carbonell, A. and Jonker, N. 2002. "Linking Measured Risk Aversion to Individual Characteristics." Kyklos, 55, 3–26.

IHS (Integrated Household Survey, Malawi). 2012. Household socio-economic characteristics report. National Statistical Office, Lilongwe, Malawi.

IPCC. (2007). Climate Change 2007: Impacts, Adaptation and Vulnerability. In M.L. Parry, O.F. Canziani, J.P. Palutikof, P.J. van der Linden & C.E. Hanson, Eds., Contribution of Working Group II to the Fourth Assessment Report of the Intergovernmental Panel on Climate Change. Cambridge: Cambridge University Press.

IPCC (Intergovernmental Panel on Climate Change). 2012. Managing the risks of extreme events and disasters to advance climate change adaptation. A special report of Working Groups I and II of the Intergovernmental Panel on Climate Change. C.B. Field, V. Barros, T.F. Stocker, D. Qin, D.J. Dokken, K.L. Ebi, M.D. Mastrandrea, K.J. Mach, G-K. Plattner, S.K. Allen, M. Tignor & P.M. Midgley, eds., Cambridge, UK, and New York, USA, Cambridge University Press.

IPCC. 2014. Working Group II contribution to the IPCC Fifth Assessment Report Climate Change 2014: Impacts, Adaptation, and Vulnerability.

ILO (International Labour Organisation). 2010. Employment diagnostic analysis on Malawi. Prepared for the Government of Malawi by Professor Dick Durevall and Dr. Richard Mussa, with assistance from the International Labour Organisation. Geneva, Switzerland.

Jain, S. 2007. "An empirical economic assessment of impacts of climate change on agriculture in Zambia", Policy Research Working Paper No. 4291, The World Bank Development Research Group, Washington D.C.

Jones, A. D., Shrinvas, A., & Bezner-Kerr, R. (2014). Farm production diversity is associated with greater household dietary diversity in Malawi: Findings from nationally representative data. Food Policy, 46, 1–12. doi:10.1016/j.foodpol.2014.02.001

Kanyanga, J., Thomas, T. S., Hachigonta, S. and Sibanda, L.M. 2013. "Zambia" in Southern African Agriculture and Climate Change, eds. Hachigonta, S., Nelson, G.C., Thomas, T.S. and Sibanda, L.M. International Food Policy Research Institute, Washington, DC.

Karttunen, K. 2009. "Rural income generation and diversification: A case study in Eastern Zambia." PhD Dissertation, University of Helsinki Department of Economics and Management, Publication No 47, Agricultural Policy, Helsinki.

Kihara, J., MacCarthy, D.S., Bationo, A., Koala, S., Hickman, J., Koo, J., Vanya, C., Adiku, S., Beletse, Y., Masikate, P., Rao, K.P.C., Mutter, C.Z., Rosenzweig, C. and Jones, J. W. 2015. "Perspectives on climate effects on agriculture: The international efforts of AgMIP in Sub-Saharan Africa," in Handbook of Climate Change and Agroecosystems: The Agricultural Model Intercomparison and Improvement Project (AgMIP), Part 2. C. Rosenzweig, and D. Hillel, Eds., ICP Series on Climate Change Impacts, Adaptation, and Mitigation Vol. 3. Imperial College Press, 3–24, doi: 10.1142/9781783265640_0013

Lanjouw, P., Quizon, J. & Sparrow, R. 2001. Non-agricultural earnings in peri-urban areas of Tanzania: evidence from household survey data. Food Policy 26(4): 385–403.

Lay, J., Mahmood, T.O. & M'mukaria, G. M. 2008. Few opportunities, much desperation: The dichotomy of non-agricultural activities and inequality in Western Kenya. World Development 36(12): 2713–2732.

Lay, J., Nahrloch, U. & Omar Mahmoud, T. 2009. Shocks, structural change, and the patterns of income diversification in Burkina Faso. African Development Review 21(1): 36–58.

Lobell, D., Burke, M., Tebaldi, C., Mastrandrea, M., Falcon, W. and Naylor, R. 2008. "Prioritizing climate change adaptation needs for food security in 2030," Science, 319: 607–610. doi: 10.1126/science.1152339

Losch, B., Freguingresh, S. & White, E. 2011. Rural transformation and late developing countries in a globalizing world: a comparative analysis of rural change. Final Report of the RuralStruc Program, Revised Version, Washington, DC, World Bank.

Lipper, L. & Thornton, P. 2014. How Does Climate Change Alter Agricultural Strategies to Support Food Security? IFPRI Discussion Paper 01340.

Macours, K., Premand, P. & Vakis, R. 2012. Transfers, diversification and household risk strategies: experimental evidence with lessons for climate change adaptation. Policy Research Working Paper 6053, Washington, DC, World Bank.

Martin, S. M. and Lorenzen, K. 2016. "Livelihood Diversification in Rural Laos." World Development (in press). Doi: 10.1016/j.worlddev.2016.01.018

Mason, N. M., Jayne, T.S. and Mofya-Mukuka, R. 2013. "Zambia's input subsidy programs," Agricultural Economics, 44: 613–628. doi: 10.1111/agec.12077

Moret, Whitney. 2014. "Vulnerability Assessment Methodologies: A Review of the Literature." Report commissioned by the USAID, ASPIRES.: http://www.fhi360.org/resource/vulnerability-assessment-methodologies-review-literature

Mortimore, M.J. & Adams, W.M. 2001. Farmer adaptation, change and crisis in the Sahel. Global Environmental Change 11(1): 49–57.

Murdoch, J. (1995). Income smoothing and consumption smoothing. The Journal of Economic Perspectives, 9(3), 103–114. doi:10.1257/jep.9.3.103

Nelson, G.C., van der Mensbrugghe, D. 2013. "Public Sector Agricultural Research Priorities for Sustainable Food Security: Perspectives from Plausible Scenarios." Background paper for the conference "Food Security Futures: Research Priorities for the 21st Century," April 11–12 2013, Dublin.

Newsham, A. & Thomas, D. 2009. Agricultural adaptation, local knowledge and livelihoods diversification in north-central Namibia. Tyndall Working Paper 140.

Pingali, P. & Rosegrant, M. 1995. Agricultural commercialization and diversification: processes and polices. Food Policy 20(3): 171–185.

Pope, R. D., & Prescott, R. (1980). Diversification in relation to farm size and other socio-economic characteristics. American Journal of Agricultural Economics, 62, 554–559. doi:10.2307/1240214

Reardon, T. 1997. Using Evidence of Household Income Diversification to Inform Study of the Rural Nonfarm Labour Market in Africa. World Development 25(5): 735–747. doi:10.1016/S0305-750X(96)00137-4

Reardon, T., Stamoulis, K., Balisacan, A., Cruz, M.E., Berdegue, J. & Banks, B. 1998. Rural Nonfarm Income in Developing Countries. Special Chapter in The State of Food and Agriculture 1998. Rome, FAO.

Reardon, T., Taylor, J. E., Stamoulis, K., Lanjouw, P., & Balisacan, A. 2000. "Effects of non-farm employment on rural income inequality in developing countries: An investment perspective." Journal of Agricultural Economics, 51, pp 266–288.

Reardon, T., Berdegué, J., Barrett, C.B. & Stamoulis, K. 2007. Household Income Diversification into Rural Nonfarm Activities. In S. Haggblade, P. Hazell & T. Reardon, eds. Transforming the Rural Nonfarm Economy. Baltimore, MA, USA, Johns Hopkins University Press.

Richard E. and Pope, Rulon D. 2001. The Agricultural Producer: Theory and statistical Measurement. Chapter 12 in Handbook of Agricultural Economics, Volume 1, Edited by B. Gardner and G. Rausser, Elsevier Science B.V.

Seo, S., Mendelsohn, R., Dinar, A., Hassan, R. & Kurukulasuriya, P. 2009. A Ricardian analysis of the distribution of climate change impacts on agriculture across agro-ecological zones in Africa. Environmental and Resource Economics 43(3): 313–332.

Singh, I., Squire, L. & Strauss, J. (eds.) 1986. Agricultural Houseold Models. Baltimore: The Johns Hopkins University Press.

Smale, M., ed. 2006. Valuing crop diversity: on-farm genetic resources and economic change. Ch. 1, Wallingford, UK, CABI Publishing.

Smit, B., & Wandel, J. (2006). Adaptation, adaptive capacity and vulnerability. Global Environmental Change, 16(3), 282–292. doi:10.1016/j.gloenvcha.2006.03.008

Toulmin, C., Leonard, R., Brock, K., Coulibaly, N., Carswell, G. & Dea, D. 2000. Diversification of livelihoods: evidence from Mali and Ethiopia. Research Report 47, Brighton, UK, Institute of Development Studies.

Thurlow, J., Zhu, T. and Diao, X. 2012. "Current Climate Variability and Future Climate Change: Estimated Growth and Poverty Impacts for Zambia." Review of Development Economics 16(3), 394–411. doi:10.1111/j.1467-9361.2012.00670.x

Van Dusen, M.E. & Taylor, J.E. 2005. Missing markets and crop diversity: evidence from Mexico. Environment and Development Economics 10(04): 513–531.

Wang, J., Mendelsohn, R., Dinar, A. & Huang, J. 2009. How do China's farmers adapt to climate change? Paper presented at the International Association of Agricultural Economics Conference, Beijing.

Wheeler, David. 2011. "Quantifying Vulnerability to Climate Change: Implications for Adaptation Assistance." CGD Working Paper 240. Washington, D.C.: Center for Global Development. http://www.cgdev.org/content/publications/detail/1424759

World Bank 2012. Mainstreaming adaptation to climate change and natural resource management. Washington, DC.

World Bank 2013. "Poverty headcount ratio at $1.25 a day (PPP)" and "Agriculture, value added (% of GDP)" in World Development Indicators, Zambia. Accessed on October 15 2013 at: http://data.worldbank.org/country/zambia

**Open Access** This chapter is distributed under the terms of the Creative Commons Attribution-NonCommercial-ShareAlike 3.0 IGO license (https://creativecommons.org/licenses/by-nc-sa/3.0/igo/), which permits any noncommercial use, duplication, adaptation, distribution, and reproduction in any medium or format, as long as you give appropriate credit to the Food and Agriculture Organization of the United Nations (FAO), provide a link to the Creative Commons license and indicate if changes were made. If you remix, transform, or build upon this book or a part thereof, you must distribute your contributions under the same license as the original. Any dispute related to the use of the works of the FAO that cannot be settled amicably shall be submitted to arbitration pursuant to the UNCITRAL rules. The use of the FAO's name for any purpose other than for attribution, and the use of the FAO's logo, shall be subject to a separate written license agreement between the FAO and the user and is not authorized as part of this CC-IGO license. Note that the link provided above includes additional terms and conditions of the license.

The images or other third party material in this chapter are included in the chapter's Creative Commons license, unless indicated otherwise in a credit line to the material. If material is not included in the chapter's Creative Commons license and your intended use is not permitted by statutory regulation or exceeds the permitted use, you will need to obtain permission directly from the copyright holder.

# Economic Analysis of Improved Smallholder Paddy and Maize Production in Northern Viet Nam and Implications for Climate-Smart Agriculture

Giacomo Branca, Aslihan Arslan, Adriana Paolantonio, Romina Cavatassi, Nancy McCarthy, N. VanLinh, and Leslie Lipper

**Abstract** Adoption of improved agricultural practices is shown to vary based on rainfall variability and long-term average maximum temperature, and although such practices increase productivity and profitability on average, their impacts also vary based on climatic conditions. This paper presents a case study on impacts and implications for adoption of Climate Smart Agriculture (CSA) solutions in the Northern Mountainous Region (NMR) of Viet Nam. We use primary data collected through *ad hoc* household and community surveys to conduct profitability estimates of comparative technologies using crop financial models based on partial budget analysis and a study of the determinants of adoption and of yields. In particular, we find that the majority of farmers in NMR rely on 'conventional' farming despite indications that sustainable land management practices such as Minimum Tillage (MT) applied to upland maize production, and Fertilizer Deep Placement (FDP) and Sustainable Intensification for Paddy (SIP) production are more profitable. Adoption of MT is greater where long-term variation in rainfall during critical growing periods for maize is higher; FDP and SIP adoption is greater in places where the long-term

---

G. Branca (✉)
Department of Economics, University of Tuscia, Viterbo, Italy
e-mail: branca@unitus.it

A. Arslan • A. Paolantonio • R. Cavatassi
International Fund for Agriculture Development (IFAD), Rome, Italy
e-mail: a.arslan@ifad.org; a.paolantonio@ifad.org; r.cavatassi@ifad.org

N. McCarthy
Lead Analytics Inc., Washington, DC, USA
e-mail: nmccarthy@leadanalyticsinc.com

N. VanLinh
Food and Agriculture Organization of the United Nations (FAO), Hanoi, Viet Nam

L. Lipper
ISPC-CGIAR, Rome, Italy
e-mail: leslie.lipper@fao.org

average of maximum temperatures is higher during critical periods for rice growth. Finally, these improved practices have higher labour and input costs compared to conventional practices, which may prevent or slow adoption.

**JEL Classification** Q12 • Q16 • Q54 • Q55 • O33

# 1 Introduction

Viet Nam is forecasted to be among the countries hardest hit by climate change (CC) with expected negative effects on agricultural production, caused mainly by changes in rainfall and temperatures and rising sea levels (Yu et al. 2010). The Northern Mountainous Region (NMR) is a particularly challenging region (FAO 2011) and has poverty rates among the highest and most widespread in the country.[1] CC is expected to exacerbate the instability of food production in the region, where agriculture is the main employer of rural labour force. Unfortunately, region-specific evidence of vulnerability to CC and its impacts is scarce.

An important question for the NMR is thus the extent to which improving agricultural practices may mitigate the negative impacts of CC and further improve resilience indicators. The literature on sustainable agricultural practices indicates that improved farming practices could increase food production without degrading soil and water resources – important elements towards adaptation to CC (World Bank 2006; Pretty 2008; Woodfine 2009). In reviewing 160 studies with field data on yield effects, Branca et al. (2013) found that adoption of Sustainable Land Management (SLM) generally leads to increased yields, although the magnitude and variability of results varies by specific practice and agro-climatic conditions. Many of these practices can also deliver co-benefits in the form of reduced greenhouse gas emissions, enhanced carbon storage in soils and biomass, increased soil fertility and water storage capacity, and strengthen the mechanisms of elemental cycling. Thus, sustainable farming technologies may be Climate-Smart Agriculture (CSA) options for smallholders in fragile environments like NMR.

To assess the possible role of adoption of sustainable farming technologies in the NMR, detailed analyses on their production costs and profitability as well as on the determinants and impacts of adoption are needed (see FAO 2010). This chapter presents a case study conducted in the provinces of Yen Bai, Son La and

---

[1] Poverty rates change, depending on the methodology employed, but in every event, suggesting poverty is the highest and more widespread in the North West, the area of our interest. The headcount ratio suggests that the poor residing in the North West Mountains of Viet Nam ranges from 60.1% from the General Statistics Office of Viet Nam and the World Bank to 39.1% from official estimates (World Bank 2012).

Dien Bien located in the NMR of Viet Nam. It uses primary data coupled with historical climate information using partial budget and econometric analyses. Special attention is paid to the impact of long- and short-term climate variations during critical periods for key food crops in the area, namely maize and rice, during their growing period. The study:

1. documents the type of practices and technology systems used by farmers in NMR for different crops and agro-ecologies;
2. estimates productivity and profitability of improved versus 'conventional' agriculture systems;
3. analyses the determinants of practices' adoption; and
4. assesses the potential of sustainable farming technologies as adaptive response to changes in climate.

## 2 Background

The NMR region of Viet Nam (see Fig. 1) is 103,000 km$^2$, about one third of the country area, and hosts about 12 million people, corresponding to 15% of the national population, living in more than 2000 communes (administrative villages), with a large share consisting of ethnic minority groups (Tran 2003). The region is almost exclusively highland, ranging between very steep (slopes of greater than 25°) and steep (slopes ranging between 15 and 20°), where the former covers 62% and the latter 16% of cultivable land (Le Ba Thao 1997). Due to the varied and fractured topography, there is a wide range of ecosystems (Tran 2003) with a series of mountain ranges and several large intermountain basins. The NMR is affected by the tropical monsoon climate, characterized by hot rainy summers and dry cold winters.

The NMR has poor infrastructure and is less urbanized and more dependent on agriculture than any other region of the country. Almost all farmers are smallholders, which diversify production to some degree. Mechanization is not yet broadly developed and is currently mainly practiced for rice threshing, land preparation in big plain areas, and occasionally for tea and coffee harvesting and/or processing.

Smallholder cropping systems in the study provinces include both rainfed and irrigated annual crop production. The upland environment provides a range of agro-ecological conditions that allow farmers to grow rice, maize, millet, peanuts, vegetables, beans and cassava. Beans, peanuts and vegetables are mainly produced for self-consumption. Cassava and maize are generally produced as cash crops. Rice is the primary staple crop in NMR as in the rest of the country, which is produced both for self-consumption and cash income. Lowland irrigated rice (paddy) plays a major role in most households' food security (Castella and Erout 2002).

Farmers grow rice in the intermountain basins, river valleys, and bunded terraces as wetland/lowland paddy, as well as on the sloping uplands as direct seeded upland rice.

**Fig. 1** The provinces in the NMR

Paddy rice is intensively cultivated in plains, where two cropping seasons per year can be grown. After harvesting the second crop of paddy, upland food crops (potato, sweet potato, legumes, and vegetables) can also be produced in some areas of these plains. Upland rice system still persists in areas under slash-and-burn practices (shifting cultivation). The substantial increase in the productivity of irrigated rice, combined with the ban on slash-and burn cultivation, have brought about a major decrease in upland rice cultivation. In spite of progressively declining upland yields due to shortening fallow periods (Husson et al. 2000) upland rice remains the primary food production strategy for a number of households.

With increasing scarcity of good quality land, farmers are turning upland rice to other food crops (maize, soybean, cassava). Maize is one of the most important cash crops, especially in Son La province, and is now the dominant upland crop (Castella et al. 2002). This is mainly attributable to an increase in the demand for maize from the feed industry, increase in yields and profitability of maize due to the use of improved varieties, and decrease in upland rice yields (Wezel et al. 2002; Doanh and Tiem 2001). Tea and coffee are the most widely produced perennial crops in the area. Tea is grown in all three provinces, but mostly in Yen Bai. Arabica coffee is produced only in Son La and Dien Bien. Regenerated forests of acacia and eucalyptus are common on steep slopes at high altitudes, mostly at places where soil fertility is low, for their value in generating timber.

Climatic patterns are characterized by (i) cold winters, with diurnal temperatures between 12 and 14 °C and hoarfrost on high belts, and (ii) early summers relative to other regions, with night temperatures increasing to 30 °C in March and reaching their maximum in June (41.1–42.5 °C). The region has two monsoons during the wet season from April to October. Total annual rainfall is about 2000 mm (over 85% falls during the rainy season), and its temporal and spatial distribution is highly unevenly (Nguyen 2006). Thus, the role of climate on adoption decision and cropping patterns focuses on rainfall regime and temperature variability.

We analyse the differences of climate depending on crop type and its "critical" growing periods. A critical growing period for maize is the 10-day period after sowing when too little rain would prevent seed germination. This corresponds to late March or early April in our case.[2] Climate data show that Son La historically receives much higher rainfall during the 10-day period after maize sowing than Yen Bai and Dien Bien. In 2013, while Son La experienced higher than average rainfall during this period, Dien Bien and Yen Bai received much less rainfall than their historical average and were more vulnerable to unpredictable rain during this period than Son La.

A second "critical period" is the heading stage of paddy rice when too high temperatures can damage production (Zhu and Trinh 2010). In our case, this corresponds to late May or early June. While Dien Bien has historically lower temperatures during this period, it experienced much higher maximum temperatures in 2013 compared to its long-term average. The other two provinces experienced lower temperatures during this critical period in 2013. The long-run variation in this variable is much higher in Dien Bien, in spite of the fact that it has more favourable temperatures on average, underlining the importance of monitoring the differences in both levels and long-term trends between and among different locations to assist farmers in dealing with various shocks.

## 3 Data Sources

### *3.1 Survey Design and Primary Data Description*

A survey at the household and community level in the study area was conducted in 2014, using Stratified Random Sampling (SRS) with purposively designed strata on an *ad hoc* universe of households and communities to ensure all relevant data could be collected. A qualitative analysis was conducted through literature review, key informants interviews and stocktaking of data and information related to projects and interventions that included adoption and dissemination of potential CSA. Communes where such interventions had been conducted were included in the sampling frame in parallel with comparable communities where no interventions or projects of such types had been conducted. In each commune a full list of households was obtained, including farmers practicing both improved and 'conventional'

---

[2] "Critical periods" for the two crops of concern in the present study have been identified through deep analysis from literature but above all from discussion with experts in the study area.

agriculture.[3] In the process of generating the list of households to be interviewed, an effort was made to stratify respondents according to specific farming practices (or a combination of practices and crops) in order to have a balanced number of observations for each target practice. Disproportionate stratified sampling procedure was used.[4] Actual respondents were randomly selected within each strata to be interviewed.

Questionnaires were designed to collect detailed primary data on benefits and costs of agricultural practices at household and community levels in addition to other relevant socio-economic and agriculture related data. Agricultural data refers to the 2013–2014 production year. Data was geo-referenced to enable merging with climatic information at commune level, as well as institutional data collected at provincial level (see Branca et al. 2015).

The sample covers 900 farmers in 25 communes distributed across the three provinces as follows: 235 in Dien Bien, 314 in Son La, and 351 in Yen Bai. Data collected include key crop production variables[5] related to smallholders (average land size in the sample is between 1 and 2.65 ha) practicing SLM and 'conventional' farming practices. The main crops considered include paddy, upland rice, maize, cassava, coffee and tea.

A list of improved farming practices with CSA potential (see Pham et al. 2014) was developed after literature review and through consultations and validation with the Viet Nam Ministry of Agriculture and Rural Development (MARD) and scientists from the local partner institute Northern Mountain Agriculture and Forestry Science Institute (NOMAFSI). These include:

1. sustainable intensification for paddy (SIP), i.e. transplanting young seedlings according to specific distance or space between plants using straight-row method and irrigation management to increase production efficiency[6];

---

[3] This includes: the Viet Nam Household Living Standards Surveys (VHLSS), conducted by The World Bank and the General Statistics Office of Viet Nam, constituting a panel dataset for the years 2002, 2004 and 2006; and the Viet Nam Access to Resources Household Surveys (VARHS), conducted by the Central Institute for Economic Management (CIEM).

[4] Disproportionate stratified sampling is a stratified sampling procedure in which the number of elements sampled from each stratum is not proportional to their representation in the total population. Given the sometimes low rate of improved farming adoption, using proportionate stratification could have caused the sample size of a stratum to be very small. Proportionate allocation may have not yielded sufficient number of observations for a specific farming technology applied to different crops making it difficult to meet the objectives of the study. The solution was to oversample the small or rare strata; oversampling creates a disproportional distribution of the strata in the sample when compared to the population.

[5] Data contain information on: farmland use, inputs (hybrid and open-pollinated variety seeds, chemicals, organic fertilizer, water for irrigation) quantities and unit costs, labour use in different management activities, labour costs estimated at the prevailing wage rate, inputs acquisition sources and subsidized prices, investment and establishment costs, crop yields, and output prices. The questionnaire includes specific sections on cropland management to capture key information about the agriculture management practices adopted (including sustainable land management practices).

[6] Farmers may apply different subsets of other more well-known and promoted systems such as

2. Fertilizer Deep Placement (FDP), i.e. use of potassium and nitrogen fertilizers mixed and compressed into larger fertilizer granules that are physically placed under the soil surface[7];
3. minimum tillage (MT), i.e. direct sowing without mechanical seedbed preparation and with minimal soil disturbance after harvest of the previous crop;
4. intercropping, i.e. cropping of different legumes (black beans, mung beans, rice beans, soybeans, groundnuts) or other crops (e.g. pumpkins) together with coffee or tea;
5. mini-terracing, i.e. vegetative strips created in sloping fields in order to allow growing a crop on a single row on each terrace to reduce soil erosion.

Based on the qualitative analysis, we define "conventional agriculture" as: fields are ploughed (tillage system), plant residues are piled and burnt or cleared out of the field, and no specific control method for input use is adopted. These practices are a source of land degradation exacerbated by soil erosion and sediment loss due to surface runoff in response to rainfall patterns especially in steep slope areas, such as the NMR (Tran 2003). Further, these practices reduce both productivity and resilience of the system.

The household level survey captured the socio-economic structure of the household as well as the agricultural production including costs, benefits, inputs and technology used by crop and plot. The community questionnaire collected relevant information at village and/or commune level including: (i) average costs of labour, (ii) average time required to perform field tasks, (iii) input sources and prices, (iv) seed types, sources and prices, (v) input subsidies provided to farmers, (vi) output prices at local markets, (vii) access to infrastructures, to extension and to information services, and (viii) perceptions on rainfall and temperature patterns.

The surveys were conducted immediately after harvest in order to minimize recall errors. Annex 1 provides detailed information on the structure of the household and community questionnaires.

---

System of Rice Intensification (SRI), Integrated Crop Management (ICM) and Integrated Pest Management (IPM). These agro-ecological methodologies are supposed to increase the yield of the rice produced in irrigated farming by changing the management of plants, soil, water and nutrients. They are based on a combination of practices aimed at increasing the efficiency of paddy productivity and reducing the use of resources and inputs (choose appropriate varieties and use quality seeds, improve transplanting modalities, balance chemicals and fertilizer application, control water irrigation use). However, since almost no farmer in the three regions applies the whole set of practices that form these systems, for the purpose of the study a new category has been identified under the name of 'SIP' (Sustainable Intensification for Paddy) in order to represent these sets of practices and prevent confusion with other systems.

[7] This is an innovative technique aimed at reducing fertilizer losses and increasing efficiency of fertilizer use.

**Table 1** Critical periods for rainfall and temperature shocks for maize and paddy in Dien Bien, Son La and Yen Bai

| Variable name | Critical periods for maize |
|---|---|
| maize_first10d_rain | First 10 days after sowing: too little rain prevents seed germination (the most critical period for maize) |
| maize_flower_rain | Flowering stage: too much rain is damaging (spring and autumn) |
| maize_midseason_rain | Between 60 and 80 days after sowing: 20 days of good rainfall is necessary |
| | **Critical periods for paddy** |
| rice_midseason_tmin | 50–60 days after planting too low temperatures are damaging (only in the spring season) |
| rice_heading_tmax | Heading stage: 70–80 (50–70) days after planting in spring (autumn) too high temperatures are damaging |
| rice_harvest_rain | Ripening stage: 30 days before harvest heavy rains are damaging |

*Source*: Own elaboration based on expert consultations, May 2015

## 3.2 Climate Data

Household data have been complemented with commune-level data on historical rainfall and temperature patterns from the European Centre for Medium Range Weather Forecast (ECMWF) in 10-daily intervals for the period of 1989–2013.[8] Using the ECMWF ERA-Interim data, a comprehensive set of variables to control for impacts and role of key climatic variables were created, including "critical growing periods" of agriculture and food security in the season of interest. These key climatic variables reflect crop- and province-specific within season shocks and were created during an interactive workshop with experts from the MARD and DARD from all study provinces. These variables are considered to provide a detailed representation of location- and phase-specific shocks for the provinces and crops of interest compared to general findings based on intensively managed experimental stations in the literature (Welch et al. 2010). Table 1 summarizes the variables used to measure long-term trends as well as within-season shocks specifically created.[9]

Long-term coefficients of variation in these variables shape farmer incentives to adopt practices that may help them dealing with climate shocks, and hence are

---

[8] ERA-Interim is the latest global atmospheric reanalysis produced by the ECMWF with a resolution of 0.25° (~28 km) in 10-day intervals. Re-analysis is a process by which model information and observations of many different sorts are combined in an optimal way to produce a consistent, global best estimate of the various atmospheric, wave and oceanographic parameters.

[9] Growing seasons for rice and maize may vary, but they mostly are as follows: (i) Maize. Spring-summer season: sowing from late February to March, and harvesting in July-August. Summer-autumn season: sowing from late July to early August, and harvesting in October- November; (ii) Rice. Spring-summer season: cropping period goes from March-April to June-July. Summer-autumn season: cropping period goes from June-July to September.

used as determinants in adoption analysis. The shocks specific to the crop seasons covered by the primary data are used in yield analyses, since they affect yields directly as well as indirectly (through interactions with the effects of various practices).

## 4 Empirical Analyses

### 4.1 Gross-Margin Analysis

The comparative profitability of the different technologies is estimated using crop financial models based on partial budget analysis (Brown 1980, Swinton and Lowenberg-DeBoer 2013).[10] The following assumptions have been made: (i) cost of the land is not taken into account since it is a fixed production cost and it does not vary depending on the different practices; (ii) farm-gate prices of inputs and outputs are those prevailing during the production season covered by the study and are assumed to be equal for all farmers; (iii) all quantitative information (input and output quantities and prices) are computed on-farm; (iv) economic results are obtained at the farm-gate level.

Profitability outcomes used in the comparison include: gross margin (GM), net income, production costs per unit of output, returns to capital, returns to labour and incremental value-cost ratios. These indicators have been estimated for each combination of crop and technology over the time frame of a 1-year production cycle per 1 hectare of land, using the following equations:

$$TR_{jT} = P_j Q_{jT} \tag{1}$$

$$TVC_{jT} = \sum_{n}^{i=1} P_{X_i} X_{ijT} \tag{2}$$

$$GM_{jT} = TR_{jT} - TVC_{jT} \tag{3}$$

$$TC_{jT} = TVC_{jT} + LC_{jT} \tag{4}$$

$$NI_{jT} = TR_{jT} - TC_{jT} \tag{5}$$

$$UC_{jT} = TC_{jT} / Q_j \tag{6}$$

---

[10] This is a short-term analysis. Resources and technologies are assumed to be fixed, and management decisions are made among existing alternatives which may be limited in the selected timeframe. Long-term changes in the technologies, policies, availability and productivity of the natural resource-base are not taken into account and are out of the scope of this analysis.

$$RC_{jT} = TR_{jT} / TVC_{jT} \tag{7}$$

$$RL_{jT} = TR_{jT} / Total\, labor_{jT} \tag{8}$$

$$L_{jT} = Q_j / Total\, labor_{jT} \tag{9}$$

$$BCR_{jT} = (TR/TVC)_{jT} \tag{10}$$

Where:

- $TR_{jT}$ = total revenue ($/ha) for crop $j$ under technology $T$
- $P_j$ = farm-gate price of crop $j$ ($/kg)[11]
- $Q_{jT}$ = yield of crop $j$ under technology $T$ (kg/ha)
- $TVC_{jT}$ = total variable costs ($/ha) for crop $j$ under technology $T$
- $P_{X_i}$ = farm-gate price of input $i$ ($/unit)
- $X_{ijT}$ = quantity of input $i$ (per ha) used in production of crop $j$ under technology $T$
- $GM_{jT}$ = gross margin ($/ha) for crop $j$ under technology $T$
- $TC_{jT}$ = total costs ($/ha) for crop $j$ under technology $T$
- $LC_{jT}$ = cost of family labour ($/ha) for crop $j$ under technology $T$
- $NI_{jT}$ = net income ($/ha) for crop $j$ under technology $T$
- $UC_{jT}$ = production costs per unit of output ($/kg) for crop $j$ under technology $T$
- $RC_{jT}$ = returns to cash capital ($/$) for crop $j$ under technology $T$
- $RL_{jT}$ = returns to labour ($/person day) for crop $j$ under technology $T$
- $L_{jT}$ = labour productivity (kg/person day) for crop $j$ under technology $T$
- $BCR_{jT}$ = benefit-cost ratio for crop $j$ under technology $T$.

Total variable costs are those directly applicable to the crop on each field and include all cash inputs (e.g. seeds and seedlings, fertilizers, manure, herbicides, insecticides, fungicides). Costs of depreciation of fixed assets, land, labor, and capital costs (e.g. interest) are excluded from GM calculations, because they are either negligible or no inputs, other than family, are used.[12] However, labour costs are taken into account in computing total costs at an imputed agricultural wage rate (unit cost of hired labour) estimated on the basis of field data and kept equal for all

---

[11] Allowance should be made for the time of selling, as price fluctuates throughout the year. However, since it has been verified that among smallholders interviewed almost all sales happen immediately after harvest time, a stable 'average' price is used in the analysis.

[12] Land is seen as a household resource, with different productive activities competing for its use. Including the cost of land in the analysis would make all GMs lower, but would not affect the relative attractiveness of the different crops and technologies. Also, it should be noted that in Viet Nam smallholders in rural areas do not pay a rent for the land. Although it is true that the cost of land will become increasingly important for smallholders in densely populated areas and in areas close to urban centres, this element falls out of the boundaries of the analysis.

crops grown. Since the study concerns small family farms, fixed costs in our analysis only include family labour.[13]

In principle, net income represents the return to the farmer for management and interest on land and capital (i.e. what accrues to management, capital and land). Since we are considering smallholders who have very limited capital invested (the only capital available is the cash used for input purchase), net income is what accrues to land and farm management. However, since farmers do not pay for land, net income is mostly remuneration of management activities.

Production cost per unit of output is one of the most important components of short-term economic results of agricultural activity. Comparing per unit production costs for a given crop and practice is a good indicator of the inherent suitability of a certain practice in a given area.

Return to capital is constructed from the ratio of total revenues to cash inputs. For example, a return to capital ratio of 3.5 means that for each Vietnamese Dong (VND) invested, 3.5 VND are obtained. Return to labour is constructed by the ratio of GM (excluding all costs of labour) or net income to total labour input. The parameter indicates how much is earned for each day of work attributed to the farm, irrespective of who provided labour. When the return to labour is lower than the prevailing wage rate of daily labour, hiring labour implies that the costs outweigh the returns. Labour productivity is calculated by the ratio of crop yields over the total amount of labour needed for that crop under the specific technology used.

## 4.2 Determinants of Adoption and Yield Impacts

We employ econometric analysis of the determinants of adoption and yield implications of the sustainable agricultural practices to address the following questions:

1. What are the determinants of/barriers to adoption of practices deemed to be profitable by the above analysis?
2. What are the marginal effects of practices on yields controlling for all other factors that affect yields?
3. Do the yield implications differ under different climatic shock conditions?

The following estimating equations are used to understand the determinants of adoption, and the effects of practices on yields, with specific focus on the climatic shock variables (see Sect. 3.2):

$$A_{ij} = \alpha_1 + \beta_1 X_i + \gamma_1 W_c + u_i \tag{11}$$

---

[13] This approach will apportion only family labour costs related to field operations in crop production, overcoming, to some extent, the limitations of gross margins which fail to take into account fixed cost changes when comparing different farming practices. Other fixed costs that have to be borne regardless of production (e.g. depreciation, interest payments, administration) are not considered.

$$Y_i = \alpha_2 + \beta_2 X_i + \gamma_2 W_c + \delta A_{ij} + \epsilon_i \tag{12}$$

$A_{ij}$ is the indicator variable for the adoption of SLM practices: it equals one if the household $i$ adopted practice $j$ (i.e. MT, FDP or SIP) on at least one plot for the crop in question (maize and paddy) during the 2014 growing season. $X_i$ is a vector of variables that affect households' incentives to adopt a specific SLM practice including demographic characteristics, wealth indicators, access to credit, extension and other types of government support. $Y_i$ is the productivity per hectare of maize or rice for household $i$. $u_i$ and $\varepsilon_i$ are normally distributed error terms of the adoption and yield models, respectively. $W_c$ is a vector of variables defined based on the climatic shock variables in Table 1, which vary between adoption and yield analyses.

In estimating the adoption probabilities (i.e. Eq. 11), $W_c$ includes long-term coefficients of variation of the variables in Table 1 in order to capture the effects of long-term trends in shocks on incentives to adopt sustainable agricultural practices. We expect, in general, that higher long-term variation of shock variables increase incentives to adopt practices that are perceived/promoted to help deal with these shocks. Adoption of MT, for example, would be positively correlated with increased variability of average rainfall during critical periods. This is because MT has the potential to buffer crops from water stress. In case of SIP/FDP however, the expectations are ambiguous as these practices are not necessarily promoted to deal with shocks but rather to increase yields as captured in the yield equations used in this analysis.

In estimating the productivity model (i.e. Eq. 12), $W_c$ includes the values of the specific shocks during the cropping seasons covered by our data in order to capture the direct yield effects of these shocks. We estimate Eq. (12) for maize and rice using two specifications: one simple specification including the climatic shock variables, and one with interaction variables between adoption indicators and climatic shock variables relevant for the crop. The interaction model helps us investigating whether and how the adoption of SLM practices changes the effects of shocks on yields (Arslan et al. 2015). We expect the direct effects of the shocks on yields to be positive (negative) if the specific shock definition indicates lower (higher) values to be detrimental to crop growth. The signs of interaction variables vary depending on the shock and practice combination, but overall the detrimental effects of shocks are expected to be mitigated by those practices that provide adaptation benefits.

## 5  Results and Discussion

### 5.1  Gross Margin Analysis

Diffusion of farming practices by type among farmers in the sample is reported in Table 2. Of note, the vast majority of surveyed farmers mainly rely on 'conventional' farming systems, especially for upland rice production.[14] Some households,

---

[14] It should be noted that these figures do not reflect the overall adoption shares in these provinces, as the sample selection was such to ensure enough numbers of adopters and a corresponding number of non-adopters in each commune to be able to conduct some analysis (both from "intervention" communes and 'comparable' communes).

**Table 2** Diffusion of sustainable farming and 'conventional' practices among farmers in the sample

| Practice surveyed | Details of the practice | % of Households adopting | | | | Avg. nr. of years of adoption |
|---|---|---|---|---|---|---|
| | | Dien Bien | Son La | Yen Bai | Total | |
| Sustainable paddy production intensification | FDP | 0 | 0 | 40 | 16 | 2.12 |
| | SIP | 9 | 20 | 5 | 11 | 2.34 |
| MT | (with or without any residue management) | 0 | 47 | 39 | 32 | 4.63 |
| Agronomy | Intercropping | 15 | 24 | 17 | 19 | 3.35 |
| | Crop rotation | 3 | 4 | 1 | 3 | 4.25 |
| Soil and water conservation structures | Mini-terracing | 17 | 18 | 8 | 14 | 6.11 |
| Agroforestry | Agroforestry | 4 | 9 | 23 | 13 | 3.73 |
| Conventional | None of the above | 94 | 91 | 79 | 87 | |

*Source*: Branca et al. (2015)

however, also apply a combination of sustainable farming practices to various crops. More specifically, MT applied to upland maize production is the most common among the sustainable farming practices surveyed (32% of adopters located in Son La and Yen Bai provinces). FDP and SIP methods are used in irrigated rice production. FDP is adopted only in Yen Bai province where 40% of the sampled farmers reported its use, whereas SIP is found in all three provinces though with a much higher incidence in Son La (20% of adopters compared to 9% in Dien Bien and only 5% in Yen Bai).

In terms of agronomic practices, crop rotation shows very limited diffusion (only 3% of adopters) whereas intercropping is a more common principle with 19% of households associating different crops. Soil and water conservation (namely mini-terracing) and agroforestry show similar adoption rates in our sample (14 and 13% of adopters, respectively). The first one is applied to perennial crops such as coffee and tea on sloping lands and it is found in all three provinces (with a lower share of adoption in Yen Bai). On the other hand, agroforestry diffusion is much higher in Yen Bai (23%) compared to Son La and Dien Bien (9 and 4%, respectively).

GM analysis finds that FDP and SIP on irrigated rice and MT on rainfed maize are the most profitable practices (see Fig. ?)

Gross margins and profitability indicators described in Eqs. (1 to 10) for improved and 'conventional' practices for paddy in both growing seasons (spring-summer and summer-autumn seasons, denoted as season 1 and 2, respectively) are reported in Tables 3a and 3b. FDP, which is practiced mostly in Yen Bai, is more profitable than 'conventional' paddy production in both seasons (see columns A and B). SIP (column C), which is found in all three provinces albeit with a much more limited diffusion compared to 'conventional' systems, generates higher yields than both 'low' and 'high' intensity 'conventional' practices in both seasons (columns D and E). SIP and 'conventional' high intensive systems (columns C and E) are also more profitable than low intensity ones (column D). However, cash input costs are higher

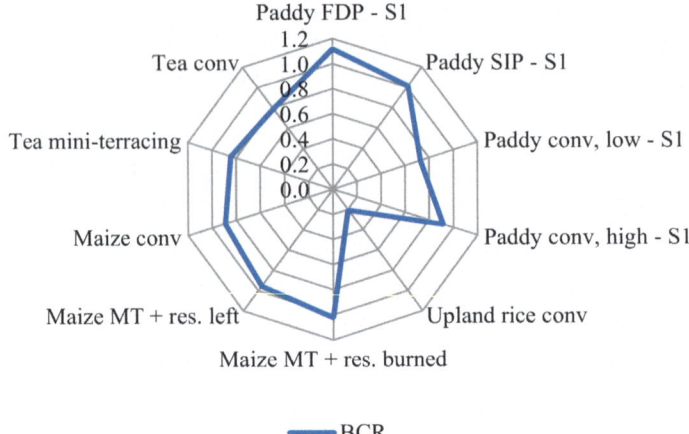

**Fig. 2** Benefit-Cost Ratio (BCR) comparison between different crops and management options (Note: S1 denotes season 1 (*i.e.* spring-summer season); *Source:* Branca et al. (2015))

under SIP (columns C and E) and FDP compared to 'conventional' high intensity systems. In particular, SIP requires more fertilizer, which is partially offset by fewer seeds, and FDP requires more labor and fertilizer in the preparation of FDP briquettes. Combined though, the increased yields of SIP and FDP still guarantee higher gross margins.

Table 4 presents the results of gross margins and profitability indicators for maize grown on uplands, also providing a comparison with rice. Results show rather clearly that upland rice is not a profitable crop (see column A); whereby maize provides much better outcomes in terms of profitability, returns, BCR, and overall production costs, especially under MT systems (columns B and C). MT on upland maize requires less cash inputs and family labour than 'conventional' systems (column D). MT on maize is a labour-saving technology suitable in areas where labour availability is a binding constraint for rural households like the area under study (Castella et al. 2002). It is important to consider that MT would be more profitable and sustainable if combined with residue management as results from Column C indicate. However given the higher labour required in managing residues, MT is mostly (57% of households in our sample) combined with crop residue burning (column B).

The evidence from our study suggests that, on average, households that mechanize both land preparation and post-harvest processing activities gain higher yields compared to those performing these activities manually.[15] Specifically, mechanization allows an average savings of about 20 person-days of family labour per hectare for

---

[15] Different hypotheses on the mechanization scenarios, however, revealed the results on conventional systems to be much more robust whereas in the case of FDP, for instance, mechanization did not always prove to be an effective choice in terms of net income from paddy gains compared to manual production.

Table 3a Paddy, gross margins and profitability indicators: comparison between 'conventional' and improved practices, spring-summer season

| Paddy, manual, flood irrigated | Season 1, Yen Bai | | | Season 1, Yen Bai, Dien Bien and Son La | | |
|---|---|---|---|---|---|---|
| | FDP – Transplanting, high intensity | Conventional – Transplanting, high intensity | | SIP – Transplanting, high intensity | Conventional – Broadcasting, low intensity | Conventional – Transplanting, high intensity |
| | (A) | (B) | | (C) | (D) | (E) |
| Yield (kg/ha) | 5713 | 4899 | | 5404 | 3519 | 4929 |
| (1) Total revenue ($/ha) | 2177 | 1867 | | 2059 | 1341 | 1879 |
| (2) Cash inputs ($/ha) | 479 | 416 | | 492 | 312 | 396 |
| (3) Gross margin ($/ha) | 1698 | 1451 | | 1567 | 1029 | 1482 |
| Cost of family labour ($/ha) | 1466 | 1369 | | 1536 | 1526 | 1654 |
| (4) Total costs ($/ha) | 1945 | 1785 | | 2028 | 1838 | 2050 |
| (5) Net income ($/ha) | 232 | 82 | | 31 | −497 | −172 |
| (6) Production costs per unit of output ($/kg) | 0.34 | 0.36 | | 0.38 | 0.52 | 0.42 |
| (7) Returns to cash capital ($/$) | 4.55 | 4.49 | | 4.18 | 4.29 | 4.74 |
| Total family labour (person days/ha) | 256 | 239 | | 269 | 267 | 289 |
| (8) Return to family labour ($/person day) | 6.62 | 6.06 | | 5.83 | 3.85 | 5.12 |
| (9) Labour productivity (kg/person day) | 22.28 | 20.46 | | 20.11 | 13.18 | 17.04 |
| (10) BCR | 1.12 | 1.05 | | 1.02 | 0.73 | 0.92 |

Source: Branca et al. (2015)

**Table 3b** Paddy, gross margins and profitability indicators: comparison between 'conventional' and improved practices, summer-autumn season

| Paddy, manual, flood irrigated | Season 2, Yen Bai | | | Season 2, Yen Bai, Dien Bien and Son La | |
|---|---|---|---|---|---|
| | FDP – Transplanting, high intensity | Conventional – Transplanting, high intensity | SIP – Transplanting, high intensity | Conventional – Broadcasting, low intensity | Conventional – Transplanting, high intensity |
| | (A) | (B) | (C) | (D) | (E) |
| Yield (kg/ha) | 4783 | 4326 | 4280 | 3268 | 3542 |
| (1) Total revenue ($/ha) | 1823 | 1649 | 1631 | 1246 | 1350 |
| (2) Cash inputs ($/ha) | 462 | 428 | 442 | 318 | 363 |
| (3) Gross margin ($/ha) | 1361 | 1221 | 1189 | 927 | 986 |
| Cost of family labour ($/ha) | 1454 | 1380 | 1569 | 1452 | 1573 |
| (4) Total costs ($/ha) | 1916 | 1808 | 2011 | 1770 | 1936 |
| (5) Net income ($/ha) | –93 | –159 | –380 | –524 | –586 |
| (6) Production costs per unit of output ($/kg) | 0.40 | 0.42 | 0.47 | 0.54 | 0.55 |
| (7) Returns to cash capital ($/$) | 3.95 | 3.86 | 3.69 | 3.91 | 3.72 |
| Total family labour (person days/ha) | 254 | 241 | 274 | 254 | 275 |
| (8) Return to family labour ($/person day) | 5.35 | 5.06 | 4.33 | 3.65 | 3.59 |
| (9) Labour productivity (kg/person day) | 18.80 | 17.92 | 15.60 | 12.87 | 12.88 |
| (10) BCR | 0.95 | 0.91 | 0.81 | 0.70 | 0.70 |

*Source*: Branca et al. (2015)

**Table 4** Upland rice and maize, gross margins and profitability indicators: comparison between 'conventional' and MT practices

| | Upland rice, local | Maize, hybrid | | |
|---|---|---|---|---|
| Upland rice, all three provinces<br>Maize, Son La and Yen Bai provinces | Conventional<br>(A) | MT, residues burned<br>(B) | MT, residues left on field<br>(C) | Conventional, residues burned<br>(D) |
| Yield (kg/ha) | 1246 | 4475 | 4710 | 4768 |
| (1) Total revenue ($/ha) | 475 | 1173 | 1234 | 1249 |
| (2) Cash inputs ($/ha) | 207 | 342 | 408 | 378 |
| (3) Gross margin ($/ha) | 268 | 831 | 826 | 871 |
| Cost of family labour ($/ha) | 2136 | 807 | 886 | 1025 |
| (4) Total costs ($/ha) | 2343 | 1149 | 1294 | 1403 |
| (5) Net income ($/ha) | −1868 | 23 | −60 | −153 |
| (6) Production costs per unit of output ($/kg) | 1.88 | 0.26 | 0.27 | 0.29 |
| (7) Returns to cash capital ($/$) | 2.29 | 3.43 | 3.02 | 3.31 |
| Total family labour (person days/ha) | 374 | 141 | 155 | 179 |
| (8) Return to family labour ($/person day) | 0.72 | 5.88 | 5.33 | 4.86 |
| (9) Labour productivity (kg/person day) | 3.33 | 31.69 | 30.39 | 26.6 |
| (10) BCR | 0.2 | 1.02 | 0.95 | 0.89 |

*Source:* Branca et al. (2015)

land preparation and about 15 person-days for post-harvesting. However, the costs of mechanization can be very high (about 6 million VND) and are not affordable for poor smallholders. Figure 3 shows the returns to family labour per person day, corresponding to each crop and technology.

Innovative farming technologies such as FDP and SIP for paddy in both seasons as well as MT with rainfed maize can improve labour productivity (addressing food security) and increase returns to labour. Under these systems, hiring external labour is feasible (e.g. labour productivity is higher than average wage rate to hire external labour) addressing the labour availability constraint, and allows resource-constrained smallholders to expand farm activity and improve their overall productive potential.

Results from partial budget estimates suggest different results for the crops of interest. With regard to paddy rice most of the SLM practices seem to perform better in terms of yields in each of the provinces and seasons. Nonetheless, there is not widespread adoption possibly due to lack of knowledge diffusion and to access to

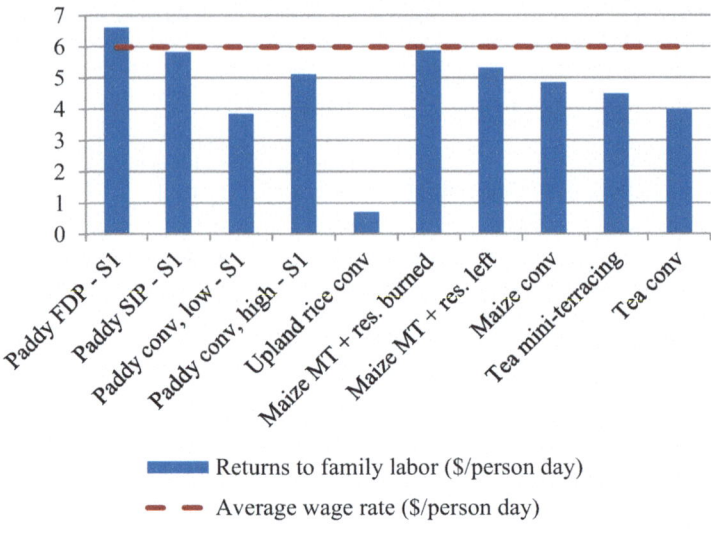

**Fig. 3** Returns to family labour, comparison between different crop and management options (Note: S1 denotes season 1 (*i.e.* spring-summer season); *Source:* Branca et al. (2015))

inputs. On the other hand, results for maize show limited difference across adoption of technologies possibly due to the fact that MT is not combined as it should be with proper residue management, likely due to labour constraints and lack of knowledge. Whereas knowledge could be increased through a more effective and widespread extension service, labour constraints remain an issue not easy to address given mechanization and labour costs.

## 5.2 Econometric Analyses

The next analytical step aims at examining the effect of weather patterns during "critical periods" on the productivity and adoption of the various practices, which is key to assessing the climate smart characteristics of the practices. This econometric analysis complements the GM analysis, which could not control for detailed consideration of climatic shocks in the region.[16] In fact, one of the novel features of this analysis is the specific attention paid to the creation of context specific rainfall and

---

[16] Regression results presented here should be interpreted as representative of the provinces where the data come from. Regression results on Yen Bai restricted sample are not reported for reason of space, and are available upon request.

Economic Analysis of Improved Smallholder Paddy and Maize Production in Northern... 581

**Table 5** Rainfall and temperature during critical periods for maize and rice by province

|  | Dien Bien | Son La | Yen Bai | Total |
|---|---|---|---|---|
| **Rainfall** | | | | |
| maize_first10d_rain | 23.82 | 79.96 | 22.69 | 42.27 |
| maize_flower_rain | 64.65 | 65.93 | 27.87 | 50.17 |
| maize_midseason_rain | 205.05 | 206.49 | 207.30 | 206.44 |
| LT CV of maize_first10d_rain | 0.94 | 0.32 | 0.81 | 0.68 |
| LT CV of maize_flower_rain | 0.52 | 0.51 | 0.58 | 0.54 |
| LT CV of maize_midseason_rain | 0.30 | 0.35 | 0.31 | 0.32 |
| rice_harvest_rain, season 1 | 62.63 | 212.29 | 295.88 | 207.61 |
| rice_harvest_rain, season 2 | 86.54 | 116.22 | 347.79 | 202.46 |
| LT CV of rice_harvest_rain, season 1 | 0.32 | 0.23 | 0.32 | 0.29 |
| LT CV of rice_harvest_rain, season 2 | 0.38 | 0.54 | 0.24 | 0.38 |
| **Temperature** | | | | |
| rice_midseason_tmin, season 1 | 14.55 | 16.79 | 19.50 | 17.31 |
| rice_heading_tmax, season 1 | 29.63 | 25.93 | 25.56 | 26.74 |
| rice_heading_tmax, season 2 | 26.87 | 25.65 | 29.72 | 27.61 |
| LT CV of rice_midseason_tmin, season 1 | 14.27 | 5.40 | 9.10 | 9.19 |
| LT CV of rice_heading_tmax, season 1 | 12.28 | 5.86 | 6.62 | 7.82 |
| LT CV of rice_heading_tmax, season 2 | 2.59 | 3.09 | 3.03 | 2.94 |

Note: LT CV denotes the long-term (1989–2013) coefficient of variation
*Source: own elaboration*

temperature shocks during critical crop growth periods. Table 5 summarizes the "critical period" climatic variables both in levels during the 2013 season and their long-term coefficients of variation (LT CV) by province.

During the 2013 season, Son La had the highest rainfall amount (79.96 mm) and lowest variability of rainfall over years (LT CV of 0.32) during the critical period for maize. On the other hand, Son La also reported very high rainfall amount during flowering season (65.93) when it can damage crop growth.

Yen Bai experienced very high rainfall during the 30-day period before harvest in both rice seasons (295.88 and 347.79 mm in season 1 and 2, respectively), which imply high probability of damaging rice. During the heading stage of rice in the spring-summer season (season 1), Dien Bien recorded the highest average temperatures (29.63 °C), whereas in summer-autumn season (season 2) Yen Bai had the highest temperatures (29.72 °C). Dien Bien shows the highest across-year variability in terms of low and high temperature shocks that matter for rice during the spring-summer season (LT CV of 14.27 and 12.28, respectively). Long-term measures of variability of these variables are used in adoption models, and their 2013 values are used in yield models.

Table 6 presents average sample values of the dependent and independent variables used in our analyses by province. Forty per cent of maize plots in our sample is under MT (only in Son La and Yen Bai provinces). Paddy rice plots on which

**Table 6** Averages of dependent and independent variables by province

| | Dien Bien | Son La | Yen Bai | Total |
|---|---|---|---|---|
| **Dependent variables** | | | | |
| % of maize plots under MT | 0 | 47 | 51 | 40 |
| % of paddy plots under FDP | 0 | 0 | 51 | 21 |
| % of paddy plots under SIP | 10 | 31 | 5 | 14 |
| **Crop/Land characteristics** | | | | |
| Total land operated throughout year (ha) | 0.45 | 0.8 | 0.48 | 0.58 |
| Nr. of seasons | 1.28 | 1.12 | 1.53 | 1.33 |
| Plot slope (weighted) | 2.61 | 2.51 | 2.15 | 2.39 |
| Dummy household has certificate of land | 0.47 | 0.98 | 0.75 | 0.76 |
| Altitude (m asl) | 780.28 | 555.92 | 359.25 | 534.01 |
| Nr. of crop units (plots/seasons) | 1.87 | 1.79 | 2.07 | 1.93 |
| **Socio-economic characteristics** | | | | |
| Age of household head | 41.23 | 43.25 | 45.01 | 43.45 |
| Education of household head | 3.22 | 2.57 | 2.76 | 2.83 |
| Dummy female headed household | 0.03 | 0.07 | 0.13 | 0.08 |
| Nr. adults working on farm | 2.28 | 3.09 | 2.56 | 2.66 |
| Nr. children working on farm | 0.02 | 0.01 | 0.04 | 0.03 |
| Dummy Kinh ethnicity | 0.07 | 0.00 | 0.38 | 0.17 |
| Dummy Thai ethnicity | 0.82 | 0.72 | 0.22 | 0.54 |
| Dummy H'mong ethnicity | 0.10 | 0.06 | 0.20 | 0.13 |
| **Institutions** | | | | |
| Dummy household received ext. advice on MT | 0.22 | 0.45 | 0.54 | 0.43 |
| Dummy household received ext. advice on FDP/SIP | 0.89 | 0.38 | 0.82 | 0.72 |
| Dummy participation to farmer union | 0.83 | 0.67 | 0.75 | 0.74 |
| Dummy support for fertilizer received in 2013 | 0.08 | 0.01 | 0.02 | 0.03 |
| Dummy support for seeds received in 2013 | 0.33 | 0.04 | 0.11 | 0.14 |
| Dummy access to formal credit | 0.73 | 0.44 | 0.22 | 0.43 |
| **Wealth/Income** | | | | |
| Dummy household has income from other sources | 0.28 | 0.19 | 0.20 | 0.22 |
| Household asset index | 0.04 | 0.44 | −0.32 | 0.03 |
| Tropical Livestock Units (TLU) owned | 2.08 | 1.55 | 1.40 | 1.63 |

*Source: own elaboration*

farmers use FDP account for 21% (only in Yen Bai), whereas SIP rice is adopted in all three provinces on 14% of paddy plots.

With respect to independent variables, average sample values show that households have operated 0.6 hectares of land throughout the year during 1.3 seasons, three-fourths of the households have a land-use certificate (almost 100% in Son La), and Dien Bien has the lowest share of those with a certificate and the highest weighted plot slope.

We control for ethnic group due to higher rates of poverty that are expected to affect the adoption of new technologies. Kinh households (the dominant ethnic group in Nam) represent only 17% of our sample, Thai minority is 54% (mostly located in Dien Bien), and H'mong is 13% (mostly located in Yen Bai).

In terms of institutions, 43% of households in the sample have received advice on MT and 72% on FDP and/or SIP. Seventy-four per cent of households have a member that belongs to a farmer union. Only 3% of the households received any support for fertilizers; 14% received seed support (more than 1/3 in Dien Bien); and 43% had access to formal credit, with the highest concentration in Dien Bien (73%). Also, distribution of wealth indicators differ across provinces. Dien Bien has the highest percentage of households with income sources other than agriculture and livestock measured by Tropical Livestock Units (TLU), whereas Son La has the highest asset index.[17]

Table 7 reports the results of the analysis on the determinants of adoption of MT in maize systems (columns A and B), and FDP and SIP in rice systems (columns C to F), using probit specifications as per Eq. (11). We estimate two different specifications for each model: one includes the long-term coefficients of variation (LT CV) of climatic variables (columns A, C and D), and the other includes also long-term averages (LT AVG) (columns B, E and F). These variables capture the potential impact of long-term average values of climatic variables that cannot be obtained from the standardized value of variation using CV.

Results from columns A and B suggest: (i) households that operate plots on higher slopes are significantly more likely to adopt MT; (ii) none of the household socio-economic characteristics significantly affects adoption, suggesting adoption is very much driven by agronomic indicators; (iii) extension advice is significantly and positively correlated with higher probability of adoption as expected; (iv) a positive relation between the share of households adopting MT and the relative diffusion of MT in the same communes is a sign of positive spillovers of effective adoption; (v) access to formal credit significantly and positively affects adoption, which is especially important for ethnic minorities with limited access to credit (and extension) compared to the Kinh majority (Do and Nguyen 2015); and (vi) having received support for improved seeds is negatively associated with the probability of adoption of MT.

Controlling for long-term averages in rainfall shocks that matter for maize (column B), we find that the probability of adoption is significantly lower in places where the variation in rainfall during the first 10 days of maize season is higher. On the other hand, the probability of adoption is significantly higher where the long-term variation in rainfall during the flowering season is higher, indicating that farmers' incentives to adopt MT are more sensitive to long-term variation in rainfall when excessive rain can damage the crop and could be particularly problematic in high slopes.

---

[17] The household asset index is constructed using principal component analysis. It includes key agricultural assets owned by the household.

**Table 7** Determinants of adoption of sustainable crop management practices

| | Maize | | | Paddy rice | | | |
|---|---|---|---|---|---|---|---|
| | MT(A) | MT w/LT AVG(B) | FDP(C) | SIP(D) | FDP w/LT AVG(E) | SIP w/LT AVG(F) |
| ln(Total area operated throughout year) | −0.013 | −0.011 | 0.039 | 0.002 | 0.031 | 0.002 |
| | (0.038) | (0.038) | (0.040) | (0.005) | (0.038) | (0.003) |
| Plot slope (weighted) | 0.187*** | 0.204*** | −0.058* | −0.010* | −0.053* | −0.006* |
| | (0.054) | (0.054) | (0.030) | (0.005) | (0.029) | (0.003) |
| Dummy household has certificate of land | −0.074 | −0.122 | 0.070 | 0.026 | 0.063 | 0.013 |
| | (0.097) | (0.101) | (0.067) | (0.019) | (0.069) | (0.011) |
| ln(Age of household head) | 0.070 | 0.067 | −0.271 | 0.066** | −0.298 | 0.033* |
| | (0.114) | (0.115) | (0.200) | (0.031) | (0.200) | (0.018) |
| Years of Education of household head (median) | 0.014 | 0.013 | 0.153** | −0.003 | 0.159** | −0.003 |
| | (0.037) | (0.037) | (0.067) | (0.011) | (0.064) | (0.007) |
| Dummy female headed household | −0.098 | −0.108 | 0.102 | −0.010 | 0.080 | −0.003 |
| | (0.102) | (0.099) | (0.140) | (0.015) | (0.141) | (0.010) |
| Nr. adults working on farm | −0.000 | 0.007 | −0.040 | −0.008 | −0.033 | −0.005 |
| | (0.027) | (0.027) | (0.049) | (0.008) | (0.049) | (0.005) |
| Nr. children working on farm | 0.117 | 0.078 | 0.210 | 0.123*** | 0.182 | 0.072*** |
| | (0.116) | (0.115) | (0.206) | (0.027) | (0.196) | (0.016) |
| Dummy Kinh ethnicity | −0.030 | −0.106 | −0.274** | 0.222*** | −0.245** | 0.165*** |
| | (0.125) | (0.114) | (0.122) | (0.060) | (0.105) | (0.047) |
| Dummy Thai ethnicity | 0.116 | 0.191** | 0.229** | −0.008 | 0.343*** | −0.001 |
| | (0.093) | (0.091) | (0.111) | (0.015) | (0.133) | (0.008) |

| | | | | | |
|---|---|---|---|---|---|
| Dummy H'mong ethnicity | −0.066 | 0.072 | −0.532*** | | −0.717*** |
| | (0.178) | (0.181) | (0.161) | | (0.236) |
| Dummy household received ext. advice on MT/SIP | 0.561*** | 0.566*** | 0.454*** | 0.083*** | 0.432*** | 0.049*** |
| | (0.056) | (0.060) | (0.108) | (0.014) | (0.107) | (0.009) |
| Dummy participation to farmer union | −0.059 | −0.069 | 0.057 | −0.044* | 0.067 | −0.027* |
| | (0.065) | (0.062) | (0.099) | (0.022) | (0.095) | (0.014) |
| Dummy support for fertilizer received in 2013 | 0.060 | 0.140 | −0.357** | −0.018 | −0.336* | −0.013 |
| | (0.097) | (0.095) | (0.175) | (0.041) | (0.174) | (0.020) |
| Dummy support for seeds received in 2013 | −0.189*** | −0.216*** | 0.473*** | −0.002 | 0.485*** | −0.002 |
| | (0.067) | (0.068) | (0.153) | (0.037) | (0.150) | (0.021) |
| Dummy access to formal credit | 0.158** | 0.160** | 0.072 | 0.011 | 0.088 | 0.005 |
| | (0.070) | (0.067) | (0.093) | (0.021) | (0.091) | (0.012) |
| Dummy household has income from other sources | 0.035 | 0.027 | −0.051 | 0.006 | −0.056 | 0.005 |
| | (0.072) | (0.072) | (0.084) | (0.023) | (0.085) | (0.014) |
| Household asset index | −0.043 | −0.050 | 0.092 | 0.000 | 0.101 | −0.001 |
| | (0.034) | (0.035) | (0.068) | (0.010) | (0.066) | (0.006) |
| Tropical Livestock Units (TLU) owned | 0.022 | 0.023 | −0.002 | 0.001 | 0.003 | 0.000 |
| | (0.019) | (0.019) | (0.016) | (0.003) | (0.018) | (0.002) |
| Share of households adopting MT in the community | 1.120*** | 1.133*** | | | | |

(continued)

Table 7 (continued)

| | Maize | | Paddy rice | | | |
|---|---|---|---|---|---|---|
| | MT(A) | MT w/LT AVG(B) | FDP(C) | SIP(D) | FDP w/LT AVG(E) | SIP w/LT AVG(F) |
| LT CV of maize_first10d_rain | 0.022 | −0.118** | | | | |
| | (0.231) | (0.242) | | | | |
| LT CV of maize_flower_rain | −0.010 | 0.704* | | | | |
| | (0.020) | (0.054) | | | | |
| LT CV of maize_midseason_rain | 0.014 | −0.173 | | | | |
| | (0.042) | (0.400) | | | | |
| LT AVG of maize_first10d_rain | | −0.445 | | | | |
| | (0.060) | (0.381) | | | | |
| LT AVG of maize_flower_rain | | 0.196 | | | | |
| | | (0.581) | | | | |
| LT AVG of maize_midseason_rain | | 0.334 | | | | |
| | | (0.306) | | | | |
| LT CV of rice_heading_tmax, season 1 and/or season 2 | | (0.363) | 0.003 | 0.003 | −0.054 | 0.004 |
| LT CV of rice_harvest_rain, season 1 and/or season 2 | | | (0.142) | (0.003) | (0.121) | (0.003) |
| | | | −0.008 | −0.000 | −0.117 | −0.000 |
| LT CV of rice_midseason_tmin, season 1 and/or season 2 | | | (0.061) | (0.001) | (0.107) | (0.001) |
| | | | | −0.040* | | −0.019 |
| | | | | (0.023) | | (0.013) |

| | | | | | 0.253** | 0.010** |
|---|---|---|---|---|---|---|
| LT AVG of rice_heading_tmax, season 1 and/or season 2 | | | | | (0.113) | (0.005) |
| LT AVG of rice_harvest_rain, season 1 and/or season 2 | | | | | -4.677* | -0.032 |
| | | | | | (2.798) | (0.038) |
| LT AVG of rice_midseason_tmin season 1 and/or season 2 | | | | | | 0.032 |
| | | | | | | (0.023) |
| Number of observations | 504 | 504 | 697 | 697 | 1458 | 1458 |
| Pseudo R² | 0.43 | 0.44 | 0.41 | 0.42 | 0.43 | 0.43 |
| Log-Likelihood | -198.47 | -195.47 | -283.19 | -331.45 | -274.19 | -324.88 |

Notes: Standard errors clustered at commune level in parentheses. Paddy rice analysis is done at plot-season level
Source: *own elaboration*
Significance levels: .01 – ***; .05 – **; .1 – *

In terms of adoption incentives for technologies in rice cropping (columns C to F), we find that FDP adoption is positively correlated with education and Thai ethnicity, while it is negatively correlated with Kinh and H'mong ethnicities. Kinh ethnic group, on the other hand, is significantly more likely to adopt SIP. Having received extension advice on improved rice management technologies is positively and significantly associated with both FDP and SIP adoption. FDP adoption is sensitive to support on fertilizers and seeds: fertilizer support significantly decreases it (this was expected as FDP is also a fertilizer saving technology); and seed support increases probability of adoption significantly.

Long-term coefficients of variation in rain and temperature shocks are not significantly correlated with the adoption of rice technologies (columns C and D); however, the higher the long-term average of maximum temperatures during the heading season, the higher the probability of adoption of both technologies (columns E and F). This suggests that farmers may perceive them as potential adaptation measures for high temperatures. We also find that the higher the long run average of rainfall during the rice harvest season, the lower the adoption of FDP.

Table 8 shows the results of the yield models specified in Eq. (12) used to investigate the effects of climatic shocks, sustainable practices and their interactions on maize and rice yields. Column A suggests that the effect of MT adoption on yields depends on the length of implementation period. Contrary to expectations, we find that the square of the duration variable is negatively correlated with yields. Upon closer inspection, we find that the average duration in our sample is more than 10 years. Discussions with experts suggested that after a very long time of MT, application yields may decline as the soils lose fertility in the absence of mulching (which is common in our sample). With respect to paddy rice, column C shows that SIP is positively correlated with a yield increase of 8%; and the use of high yielding varieties is associated with an increase of 10%. FDP seems to have no effect on rice yields when the regression model is run on the three provinces sample. However, when restricting our sample to Yen Bai (the only province where sampled farmers adopt FDP for paddy), we find that the use of FDP is significantly associated with an increase of about 6% in paddy yields.

We also find that yields are significantly affected by excess rainfall and high temperatures: 10% more rain in the first 10-day period after sowing is associated with a more than 30% increase in maize yields; 10% more rain during flowering is negatively correlated with maize productivity leading to a decrease of about 30% in yields (column A); higher maximum temperature during heading stage of paddy is associated with slightly lower yields (about 1% decrease) (column C).

The effects of some of these shocks interact significantly with the effects of adoption, which is analysed using interaction variables (in columns B and D). While the positive effect of rainfall during the first 10 days of maize is amplified for MT practitioners, the negative effects of rainfall during maize flowering are worsened under MT (column B). Looking at paddy rice, we find that the negative effects on paddy rice yields of excessively high temperatures are ameliorated by the practice of SIP (column D). On the other hand, the interaction variable between rice tech-

Table 8 Maize and rice yield models with adoption and interaction variables

| | Maize | | Paddy rice | |
|---|---|---|---|---|
| | MT(A) | MT w/ interactions(B) | FDP/ SIP(C) | FDP/SIP w/ interactions(D) |
| Dummy MT in at least one plot/season | −0.070 | 0.398 | | |
| | (0.060) | (0.328) | | |
| Years of MT use for those who used in 2013 | 0.047*** | 0.043** | | |
| | (0.015) | (0.017) | | |
| Years of MT use for those who used in 2013 | −0.005*** | −0.005*** | | |
| | (0.001) | (0.001) | | |
| Dummy FDP | | | 0.049 | 2.106*** |
| | | | (0.032) | (0.796) |
| Dummy SIP | | | 0.087* | −0.080 |
| | | | (0.047) | (0.428) |
| Years of FDP use for those who used in 2013 | | | 0.005 | −0.008 |
| | | | (0.012) | (0.012) |
| Years of SIP use for those who used in 2013 | | | −0.002 | 0.004 |
| | | | (0.012) | (0.010) |
| maize_first10d_rain | 0.398* | 0.348* | | |
| | (0.206) | (0.202) | | |
| maize_flower_rain | −0.295* | −0.202 | | |
| | (0.155) | (0.161) | | |
| Dummy MT*maize_first10d_rain | | 0.148* | | |
| | | (0.085) | | |
| Dummy MT*maize_flower_rain | | −0.270** | | |
| | | (0.130) | | |
| rice_midseason_tmin, season 1 and/or season 2 | | | 0.003 | |
| | | | (0.002) | |
| rice_heading_tmax, season 1 and/or season 2 | | | −0.011* | −0.010 |
| | | | (0.006) | (0.008) |
| rice_harvest_rain, season 1 and/or season 2 | | | 0.099 | 0.142* |
| | | | (0.079) | (0.079) |
| SIP*rice_heading_tmax, season 1 and/or season 2 | | | | 0.032** |
| | | | | (0.013) |

(continued)

**Table 8** (continued)

| | Maize | | Paddy rice | |
|---|---|---|---|---|
| | MT(A) | MT w/ interactions(B) | FDP/ SIP(C) | FDP/SIP w/ interactions(D) |
| FDP*rice_heading_tmax, season 1 and/or season 2 | | | | 0.002 |
| | | | | (0.009) |
| SIP*ln(rice_harvest_rain, season 1 and/or season 2) | | | | −0.142*** |
| | | | | (0.044) |
| FDP*ln(rice_harvest_rain, season 1 and/or season 2) | | | | −0.366** |
| | | | | (0.145) |
| Inputs use per ha (seeds, fertilizer, labour) | Yes | Yes | Yes | Yes |
| Controls for crop/land characteristics | Yes | Yes | Yes | Yes |
| Controls for socio-economic characteristics | Yes | Yes | Yes | Yes |
| Controls for institutions | Yes | Yes | Yes | Yes |
| Controls for wealth/income | Yes | Yes | Yes | Yes |
| Constant | 6.547*** | 6.415*** | 7.086*** | 6.902*** |
| | (0.872) | (0.888) | (0.534) | (0.536) |
| Number of observations | 465 | 465 | 1604 | 1604 |
| Adjusted R2 | 0.28 | 0.29 | 0.38 | 0.39 |
| Log-Likelihood | −137.33 | −135.42 | 69 | 84.25 |

Notes: Standard errors clustered at commune level in parenthesis. Paddy rice analysis is done at plot-season level
*Source: own elaboration*
Significance levels: .01 – ***; .05 – **; .1 – *

nologies and the rainfall during the harvest time (when it is damaging) is significant and negative, suggesting that these practices do not generate adaptation benefits. This finding suggests a potential trade-off between higher yields and stability of yields under this type of climatic shock, underlining the importance of integrating ways to address climatic patterns and risk in extension programmes in areas where these practices are promoted.

There are some caveats in interpreting the econometric analysis results. Given the non-random and cross-sectional nature of the sample, results have to be cautiously interpreted as correlations rather than causations, since potential endogeneity in data can only be controlled using instrumental variables, quasi-experimental or panel data techniques. Another caveat is related to our climate data source. Even though re-analysis data from ECMWF offer advantages over collected data in regions with sparse stations with long-term coverage, it relies on the assumptions of climate models, which can be restrictive. Future research should conduct similar analyses using various validation methodologies to improve the robustness of evidence. In spite of these caveats, the strong correlations between adoption of sustainable practices and expected increased yields, as well as potential adaptation benefits documented here underline the importance of such studies for agricultural policy to improve food security accounting for climate.[18]

## 6 Conclusions

Our analyses show that while sustainable farming practices improve productivity and profitability on average, the timing and variations of climatic conditions significantly impact results, and are even shown in some cases to have a negative impact. This means that achieving adaptation benefits for individual households requires sufficient understanding of specific climate patterns, particularly during "critical growing periods" of crops. Our results indicate the high returns to including climate change effects directly into agricultural development planning and investments. The findings of this study imply that NMR agricultural policies should prioritize MT for upland maize, especially where the rainfall at the beginning of the season is a constraint, and SIP on paddy in more productive irrigated flat lands especially where high temperatures during heading stage are a limiting factor. However, sustainable practices often have higher upfront capital and labour requirements, which may prevent or impede adoption.

Our findings suggest the importance of local climate and socio-economic contexts in determining which practices will actually be climate-smart. In some cases we find that sustainable land management practices will be the best CSA option – however in others this is not the case. For example, SIP generates benefits under high temperatures, but is not a good option in places where the long-term average of maximum temperatures during critical periods for rice growth is high. MT is effective under low rainfall conditions and thus could reduce the negative impact of changes in rainfall variation at critical stages of maize cropping. These results indicate the importance of using climate information for targeting the promotion of improved practices, and building adaptive capacity amongst the farming population.

Another important finding of this work is the role of extension. Access to extension information is among the major enablers of adoption identified in the analysis. The results suggest that extension is found to have important spillover effects as adoption is higher where the proportion of adopters in the commune is higher. Returns to extension investments could be quite high in terms of increasing adoption and adaptive capacity of farmers.

---

[18] Further analysis using climate modeling and taking into account the expected changes in weather shocks would significantly strengthen the results of our analysis. This may be taken into consideration for future research work.

Some caveats about the current study are warranted. Our results confirm the importance of credit and labour constraints in impeding adoption in the NMR, implying the need for a regional approach. Nevertheless, related economic and institutional issues are omitted here. Also, sustainable practices are expected to generate environmental benefits (mitigation, water savings, reduced erosion). These benefits are in the form of positive externalities generated by (upstream) farmers toward (downstream) farmers and all of society. Some of these practices show synergies with food security goals. For example, in paddy production, SIP could help reduce overuse of irrigation, which is lowering groundwater levels, and FDP may hold further environmental benefits. It is also worth noting that paddy production is highly dependent on secure water flow availability, which is not a limiting factor. However, foreseen climatic changes may alter this equilibrium and make water-saving techniques (e.g. SIP) more convenient. While we have not explicitly considered these environmental issues and externalities in the analysis, they are clearly important aspects to be considered at the policy level.

**Acknowledgments** This work has been conducted within the FAO Project "Climate Smart Agriculture (CSA): capturing the synergies between adaptation, mitigation and food security", funded by the European Commission over the 2012–15 period and conducted by FAO in partnership with the Viet Nam Ministry of Agriculture and Rural Development (MARD) and the Northern Mountainous Agriculture and Forestry Science Institute (NOMAFSI). The authors wish to thank Pham Thi Sen and her team of enumerators from NOMAFSI for their support with data collection and preparation. The study has also benefited of various comments received during technical meetings and of continuous support provided by all members of the FAO-EPIC Programme.

# Annex 1: Structure of the Household and Community Questionnaires

| Questionnaire sections | Key data collected |
|---|---|
| *Household questionnaire* | |
| Household identification | Location (Province, District, Commune, Village) and contacts |
| Socio economic status of household | Demographic characteristics, assets, access to resources and food security status, access to input support and extension |
| Inventory of fields cropped and collection of data on cropland use and management, by household/field/cropping season | Field and farm size, crops cultivated (annual and perennial), management practices, irrigation, land characteristics, quantity of inputs used, crop yields, input and output prices, family and hired labour use for different practices, |
| Input acquisition | Sources of access to seeds and other inputs |
| Agroforestry, soil and water conservation | Typology of interventions, tree species, labour and input costs, revenues from sales |

| Questionnaire sections | Key data collected |
|---|---|
| Livestock (cattle, buffaloes, poultry, pigs) and forage production | Stock inventory and dynamics (acquisition, sales), feeding and health, labour use, grass production (feed) and grazing management |
| Other income sources and access to credit | Incomes from self-employment and wages, other income sources (pension, rental, external support), credit and loans |
| Institutions and extension | Membership of associations, access to extension services |
| *Community questionnaire* | |
| Community identification | Location (Province, District, Commune, Village) and contacts |
| Village labour costs | Unit costs of hired manual labour, animal draft power, mechanical power, land (rental) |
| Average crop management inputs | Average time required to perform field activities for different management types |
| Access to input and output markets | Input sources and prices; seed types used, purchase source and price; input subsidies provided to farmers; output prices at local market level (village or commune) |
| Access to services and infrastructures | Access to extension and information services, service providers, dissemination methods, access to other services and infrastructures |
| Climate-related information | Perception about rainfall and temperature patterns |

*Source:* Branca et al. 2015

# References

Arslan, A., McCarthy, N., Lipper, L., Asfaw, S., Cattaneo, A., and Kokwe, M. (2015). Climate Smart Agriculture? Assessing the adaptation implications in Zambia. Journal of Agricultural Economics, 66(3): 753–780.

Branca, G. et al. (2015). Benefit-cost analysis of sustainable farming practices for CSA systems in Northern Mountainous Region of Viet Nam. Final report. FAO-CSA Project. June.

Branca, G., McCarthy N, Lipper L, Jolejole MC. (2013). Food security, climate change and sustainable land management. A review. Agronomy for sustainable development, 33:635–650, doi: 10.1007/s13593-013-0133-1

Brown, M.L. (1980). Farm budgets: from farm income analysis to agricultural project analysis. World Bank Staff Occasional Papers, International Bank for Reconstruction and Development 1980. ISBN 0-8018-2387-0

Castella, J.C., Erout A. (2002). Montane paddy rice: the cornerstone of agricultural production systems in Bac Kan Province, Vietnam. In: (J.C. Castella and Dang Dinh Quang eds.) Doi Moi in the Mountains. Land use changes and farmers' livelihood strategies in Bac Kan Province, Vietnam. The Agricultural Publishing House, Ha Noi, Vietnam. 175–195.

Castella, J.C., Boissau S., Nguyen H.T., Novosad P. (2002) Impact of forestland allocation on agriculture and natural resources management in Bac Kan Province, Vietnam. In: (J.C. Castella and Dang Dinh Quang eds.) Doi Moi in the Mountains. Land Use Changes and Farmers' Livelihood Strategies in Bac Kan Province, Vietnam. The Agricultural Publishing House, Ha Noi, Vietnam. 197–220.

Do, X.L., Nguyen T.L. (2015) Credit Access in the Northern Mountainous Region of Vietnam: Do Ethnic Minorities Matter? International Journal of Economics and Finance; Vol. 7, No. 6; 2015

Doanh, L.Q., Tiem L.V. (2001). Feasible solutions for sustainable land use in sloping areas. Project Review and Planning Meeting, 10–14 December 2001, Hanoi, Vietnam.

FAO (2010) "Climate-smart" agriculture. Policies, practices and financing for food security, adaptation and mitigation. Food and Agriculture Organization of the United Nations, Rome

FAO (2011). Strengthening Capacities to Enhance Coordinated and Integrated Disaster Risk Reduction Actions and Adaptation to Climate Change in Agriculture in the Northern Mountain Regions of Vietnam. Project UNJP/VIE/037/UNJ. Hanoi, 2011.

Husson, O., Tuan H.D., Lienhard P., Tham D.H. (2000). Development of "direct sowing" techniques as alternatives to slash-and-burn practices in the mountainous areas of North Vietnam. Preliminary results of SAM – Cropping Systems project. EC workshop on sustainable rural development in the Southeast Asian mountainous region. Hanoi, 28–30 November 2000

Le Ba Thao (1997). Vietnam, the country and its geographical regions. The Gioi Publishers, 1997. Hanoi.

Nguyen, T.M. (2006) Country Pasture/Forage Resource Profiles: Vietnam. FAO: Rome. http://www.fao.org/ag/agp/AGPC/doc/Counprof/PDF per cent20files/Vietnam.pdf

Pham, T.S., Nguyen Q.T., Branca G. (2014). A review of sustainable farming practices in Yen Bai, Son La and Dien Bien provinces of Vietnam. Final report. FAO-EPIC. Rome

Pretty, J.N. (2008) Agricultural sustainability: concepts, principles and evidence. Phil Trans R Soc London B 363(1491):447–466. doi: 10.1098/rstb.2007.2163

Swinton, S.M., Lowenberg-DeBoer J. (2013). Evaluating the Profitability of Site-Specific Farming. Journal of Production Agriculture Vol. 11 No. 4, p. 439–446. doi: 10.2134/jpa1998.0439

Tran, D.V. (2003). Culture, Environment, and Farming Systems in Viet Nam's Northern Mountain Region, Southeast Asian Studies, Vol. 41, No. 2: 180–205, September 2003.

Welch, J. R.; Vincent, J. R., Auffhammer, M., Moya, P. F., Dobermann, A. and Dawe, D. (2010). Rice yields in tropical/subtropical Asia exhibit large but opposing sensitivities to minimum and maximum temperatures, PNAS: 107 (33), pp. 14562–14567; doi:10.1073/pnas.1001222107

Wezel, A., et al. (2002). Temporal Changes of Resources Use, Soil Fertility and Economic Situation in Upland Northwest Vietnam. Land Degradation & Development 13 (2002): 33–44.

Woodfine, A. (2009) The potential of sustainable land management practices for climate change mitigation and adaptation in sub-Saharan Africa. Food and Agriculture Organization of the United Nations, Rome.

World Bank (2006) Sustainable land management: challenges, opportunities, and trade-offs. The World Bank, Washington.

World Bank (2012) Poverty assessment Report Vietnam : Well Begun, Not Yet Done: Vietnam's Remarkable Progress on Poverty Reduction and the Emerging Challenges, The World Bank, Washington.

Yu, B., Zhu T., Breisinger C., Manh Hai N. (2010). Impacts of Climate Change on Agriculture and Policy Options for Adaptation. The Case of Vietnam. IFPRI Discussion Paper 01015. August 2010.

Zhu, T., Trinh M.V. (2010). Climate Change Impacts on Agriculture in Vietnam. In: Proceedings of the International Conference on Agricultural Risk and Food Security, June 11–12, 2010, Beijing.

**Open Access** This chapter is distributed under the terms of the Creative Commons Attribution-NonCommercial-ShareAlike 3.0 IGO license (https://creativecommons.org/licenses/by-nc-sa/3.0/igo/), which permits any noncommercial use, duplication, adaptation, distribution, and reproduction in any medium or format, as long as you give appropriate credit to the Food and Agriculture Organization of the United Nations (FAO), provide a link to the Creative Commons license and indicate if changes were made. If you remix, transform, or build upon this book or a part thereof, you must distribute your contributions under the same license as the original. Any dispute related to the use of the works of the FAO that cannot be settled amicably shall be submitted to arbitration pursuant to the UNCITRAL rules. The use of the FAO's name for any purpose other than for attribution, and the use of the FAO's logo, shall be subject to a separate written license agreement between the FAO and the user and is not authorized as part of this CC-IGO license. Note that the link provided above includes additional terms and conditions of the license.

The images or other third party material in this chapter are included in the chapter's Creative Commons license, unless indicated otherwise in a credit line to the material. If material is not included in the chapter's Creative Commons license and your intended use is not permitted by statutory regulation or exceeds the permitted use, you will need to obtain permission directly from the copyright holder.

# Part VI
# Policy Synthesis and Conclusion

# Devising Effective Strategies and Policies for CSA: Insights from a Panel of Global Policy Experts

**Patrick Caron, Mahendra Dev, Willis Oluoch-Kosura, Cao Duc Phat, Uma Lele, Pedro Sanchez, and Lindiwe Majele Sibanda**

**Abstract** In this section, we present the results of a consultation with a panel of leading thinkers on agricultural and climate change policy. We interviewed these experts using a set of questions based on the main findings, conclusions, insights and questions that emerged from our set of case studies and conceptual papers. This section is divided into four parts, each focusing on a set of questions relating to the conclusions that emerged from the case study and conceptual chapters. (i) Focus on changes in production systems as adaptation: priorities and policy actions; (ii) Incorporating climate change into agricultural research and extension; (iii) Taking a close look at national policies affecting risk management: index insurance, safety nets and input subsidies and (iv) Priorities for the future and summary of main points. Overall, there is a fairly high level of agreement amongst the panel members in responding to most of the interview questions, although with some difference in emphasis or applications. However there are also some differences of opinion that emerge from their responses. In this chapter, we discuss the main points made on each of the issues addressed, highlighting the areas of agreement, as well as differences.

---

P. Caron (✉)
CIRAD, Montpellier, France
e-mail: patrick.caron@cirad.fr

M. Dev
Centre for Economic and Social Studies, Hyderabad, India

W. Oluoch-Kosura
University of Nairobi, Nairobi, Kenya

C.D. Phat
Agriculture and Rural Development, Hanoi, Vietnam

U. Lele
World Bank, Washington, DC, USA

P. Sanchez
Earth Institute, Columbia University, New York, NY, USA

L.M. Sibanda
Food, Agriculture and Natural Resource Policy Network, Pretoria, South Africa

**About panel of leading thinkers on agricultural and climate change policy**

**Patrick Caron,** CIRAD Chair of the High Level Panel of Experts/HLPE of the committee on world food security (CFS).

**Mahendra Dev,** Director, Centre for Economic and Social Studies, Hyderabad India

**Willis Oluoch-Kosura,** Professor of Agricultural Economics, University of Nairobi

**Cao Duc Phat,** Minister of Agriculture and Rural Development Socialist Republic of Vietnam

**Uma Lele, Ramesh Deshpande and Inder Abrol,** Uma Lele is an independent researcher, and former senior advisor in the World Bank. Ramesh Deshpande is former Principal Financial Operations Specialist of the World Bank, presently CEO at IAG International. Inder Abrol, is former Deputy Director General of ICAR and former Facilitator of the Rice-Wheat Consortium.

**Pedro Sanchez,** Director of the Agriculture and Food Security Center and Senior Research Scholar at Columbia University's Earth Institute

**Lindiwe Majele Sibanda,** Chief Executive Officer and Head of Mission Food, Agriculture and Natural Resource Policy Network

# 1 Focus on Changes in Production Systems as Adaptation: Priorities and Policy Actions

Several of the case studies presented in the book give indications of changes in agricultural practice management that are effective adaptation actions. These include a wide range of practices that fall under the general categories of sustainable land and water management (SLWM), as well as diversification of farming systems and livelihoods. These practices are already known and available, and yet adoption rates are generally not very high. As shown in the case studies there are considerable barriers to their adoption, such as increased labor/capital inputs as compared to 'conventional' technologies, or up-front costs of investing in soil health and farm structures, which may take several years to bear fruit. The case studies also indicate that farmers located in areas facing greater climate risks are more likely to diversify agricultural production, labor and incomes, which decrease their vulnerability to extreme weather events. However, as with adoption of SLWM practices, evidence suggests that it is often the wealthiest and more educated farmers who are able to take advantage of opportunities to diversify. The case studies presented, as well as more general literature on adaptation, indicate that water management is a key issue for climate change adaptation and increasing resilience in agriculture. It can be a successful – and essential – adaptation strategy but it requires substantial public investments, which can be problematic when resources are scarce. Managing irrigation schemes after the initial investment can also lead to smaller gains than originally anticipated.

We asked our panel to respond to three questions related to these findings:

(i) *How important do you think it is for policy-makers to promote SLWM practices and what role does policy play in promoting it?*

Cao Duc Phat considers SLWM important to address climate change and improve the sustainability of natural resource use. This is particularly important in densely populated rural areas as in Vietnam, where land pressures are rising due to urbanization and industrialization, further exacerbated by sea level rise. SLWM allows for sustainable intensification of production systems, and thus is essential to ensure livelihoods and stable living conditions for rural residents.

Public-private-partnerships (PPP) are an important means of promoting adoption of such techniques. The public sector can invest in infrastructure and enhance private sector investments with improved access to credit and insurance. It also has an important role to play in developing flexible land use policies that are needed to enable widespread adoption. Sanchez sees the development of an enabling value chain as essential for promoting SLWM across the entire value chain, from services to production, to value added, transport, market, consumption and consideration of environmental effects. He states:

> I think what is really needed is to bundle many of these services in a way that provides good tools to farmers so they don't have to worry about things like credit or where to sell their crops.

He also raises the role of private sector in this effort, citing the example of farmers in Kenya that are contracted to private sector companies where they obtain inputs, fertilizers, improved varieties, credit, crop insurance, and market. This leaves farmers to focus on farming. Sanchez notes that better leadership and outreach activities that establish sustainable social norms are important. The Millennium Villages provide examples of how leadership has helped spread SLWM practices among all strata of farmers.

Lele and colleagues give considerable weight to the need for better soil management – particularly improved nutrient management. They consider case of conservation agriculture (CA) as an important part of the solution in India. However there are major constraints to its adoption amongst smallholder farmers: (i) competing use of crop residues in rain fed areas, (ii) weed management strategies, particularly for perennial species, (iii) localized insect and disease infestation, and (iv) likelihood of lower crop productivity if site-specific complementary technologies are not adopted.

They argue that:

> For wider adoption of CA, there is an urgent need for policy makers, researchers and farmers to change their mindset and explore these opportunities in a site- and situation-specific manner for local adaptation.

A policy framework that recognizes the value (or costs) associated with externalities is important according to Sanchez and Caron. Sanchez argues that positive externalities like soil carbon, improved ecosystems for wildlife, and increase food security (by enhancing resilience) and therefore agriculture should be compensated. Caron also emphasized the need for policies to provide incentives to engage in activities that provide social goods and reduce negative externalities which are particularly important in the SLWM context.

Sibanda states that investment in SLWM is a "must-do" – noting that it is the first pillar for the Comprehensive African Agriculture Development Program (CAADP) because its importance was well recognized. However it hasn't been fully implemented – due to limited funding, but also institutional issues such as land tenure systems.

> Because we're focusing on smallholder farmers, you are dealing with land that is communally owned: they are not titled to land, you are dealing with farmers that are sitting on 1 hectare but relying on public irrigation facilities; you are dealing with farmers who are relying on lands which is not clearly demarcated as owned by the individuals. Now, how likely is it a farmer will pour money into such a situation?

She argues that the solution is to revisit the issue of land tenure to build incentives from the bottom-up so farmers who are willing to invest will be guaranteed that they, their children and grandchildren will have use of the land. They will also be able to use the title to borrow money – an important aspect Cao Duc Phat raises in the Vietnamese context as well.

Kosura sums it up as follows:

> Promoting secure land tenure regimes especially by governments is a key prerequisite for investment in SLWM.

(ii) *What types of public investments or policy options do you believe would improve poor farmers' ability to diversify?*

Caron stressed that it is important to realize that even the wealthier and educated farmers in the areas considered by the case studies are relatively poor, and poor farming households are generally fairly diversified. Dev concurs adding that in India small farmers allocate a larger proportion of their cultivated land to high-value crops like fruits and vegetables. The issue is not so much to promote diversification amongst the poorest, but to build mechanisms to help them take advantage of opportunities. Dev notes that a number of innovative institutional models are emerging to help support opportunities for small and marginal farmers in India. These include institutions relating to (a) land and water management, (b) group or cooperative approach for inputs and marketing and, (c) value chains and supermarkets that can enhance productivity, sustainability and incomes of small holding agriculture.

According to Kosura, public investments in infrastructure (rural roads, market places, storage facilities) and related services are needed to reduce transactions costs faced by the poor – which private and public sector partnerships can foster. Sibanda focuses on the public sector role in building market infrastructure, as well as public investments in land and water management that reduce impacts of extreme events to compliment farmers' own actions. She also thinks that it is important to consider the results of climate models and to the foreseen impacts of climate change in deciding policy priorities for diversification. For example, areas traditionally known for being bread baskets may become food hunger spots, and may therefore need to rely on food imports. There needs to be an update of the mapping of who-produces-what with important implications in terms of crop diversification. Trade therefore plays an important role in facilitating this process.

Dev lays out four key areas of institutional support needed to support diversification for smallholders: (1) enabling farmers' groups and cooperatives to help smallholders access high-value markets through, for instance, improved rural – urban linkages; (2) a wider range of viable and attractive financial and risk management tools; (3) increasing information dissemination needed for smallholders to increase knowledge and technical skills to take advantage of diversification strategies, and (4) ensuring livelihoods are protected in the aftermath of severe weather events through social safety net programs.

Cao Duc Phat broadened the discussion on diversification to consider the importance of creating more employment opportunities in rural areas, including non-agriculture based opportunities, as an essential component of diversification. Specific technical guidance, production support programs, and state-targeted support in terms of capital and extension services are needed to enable farmers, particularly the less wealthy, to take advantage of a wider range of economic opportunities. He also stresses the need for both the public and private sector involvement in expanding access to non-farm income opportunities in rural areas.

Dev and Kosura also take up the issue of the role of trade in adaptation, and it implications for diversification. They point out that climate change effects are varied across locations, and thus opens a possibility of exploiting new comparative advantages in trading. Changes in both domestic and foreign trade may be appropriate, with regions or provinces shifting production patterns in response to the types of risks they face. However Dev also points out that trade can impose risks, through market volatility, and this is an important aspect for policy to address, in order for trade to support increased food security under climate change.

(iii) *What priority should be given to irrigation expansion as an adaptation strategy?*

Dev points out that irrigation expansion is the most important priority as it protects farmers and other people from climate risks. Sibanda says that Africa cannot continue depending on rainfed agriculture, as it is not a climate-smart strategy, it does not build resilience, and does not contribute to lower risk. Kosura argues that irrigation is critical considering the erratic nature of rainfall in recent years. Irrigation not only alleviates water stress but also has the potential to expand the opportunities for switching planting dates and crop varieties, as well as increasing returns on investments in fertilizer and other inputs. Caron thinks that irrigation is key for production but with different roles in different areas. Where irrigation has already expanded, there are issues of sustainability and competition with alternative uses (industrial and urban). On the contrary, in other areas – for example in low density areas where agriculture has been expanded through an extensive process rather than through intensification – irrigation may not have been used as a way to increase production because of investment barriers. Dev says that in many countries it may be necessary to develop big irrigation projects as micro-irrigation may not be enough, but replacement and rehabilitation mechanisms have to be in place when large scale irrigation is developed. In other contexts, there may be the need to develop micro-irrigation projects which better suit local conditions and capabilities. For example, watershed development (e.g. small check dams) can particularly help small farmers.

Kosura would support low-cost small irrigation systems through financing and technical assistance. He thinks that there is much scope for expansion through micro projects, which require local management capabilities, and improved management of existing irrigation schemes. Research by agencies like International Water Management Institute (IWMI) has revealed a growing trend for individual and community-owned agricultural water management systems. Encouraging a cooperative farming approach (collective action) to irrigation would reduce costs and allow greater benefits to the farmers. Sanchez agrees with this line of thinking, arguing that for developing countries it is better to have small scale irrigation where people owning the systems (e.g. Farmers' Associations) are more likely to efficiently manage the resource. Sibanda mentions that Africa is still bearing the cost of establishing big irrigation schemes which show below-average rate of investment returns and high maintenance costs. It is not surprising therefore that the extension of irrigated land in Africa is very limited. Policies should promote technologies that will lower the cost of establishing irrigation infrastructure and its maintenance. In order to have an irrigation scheme commercially operational and to improve returns for the farmers, three issues should be considered: governance arrangements (water access, infrastructure management, property rights); institutions (extension services, water management committees); markets (farmers accessing irrigation should grow high value crops and have access to market economy).

The priority for Sanchez is the so called "green water" associated with soil moisture that represents 2/3 of the water used by agriculture. The priority is to use this more efficiently – which means getting improved production practices in place. He provides the example of rainfed maize production in Malawi to illustrate the concept. At the current levels of about 1 metric ton per hectare, about 80% of the water is lost to evaporation, while the other 20% or so is transpired through the plant making biomass that is harvested. If you tripled that yield, you could get about 80% of that soil moisture going to transpiration and losing only 20% to evaporation. Why? Because a crop cover holds down evaporation; and then the roots of these plants that are fertilized, can reach water at greater depths so that soil moisture is the source of green water, which is to Sanchez the most important.

Cao Duc Phat lays out the current thinking on irrigation investment and adaptation in Vietnam:

> After a period focusing on building large scheme irrigation infrastructure serving for production development, there is now increased interest in building irrigation works that serve for adaptation to climate change with new priorities: enhance local needs, adjust irrigation management appropriately to transform the production practices, protect landscape, conserve water resources and take the most effective use of designed facilities' capacity; reform the operational mechanism, better utilize existing infrastructure systems and improve water use and management efficiency; consider multiple water uses and promote water-saving practices; implement PES schemes to share part of the water users' profits with forest planters and protectors to regenerate water resource in the watershed upstream; invest in critical disaster prevention infrastructures (e.g. flood control and drainage); strengthen management of small and medium irrigation infrastructures throughout capacity building for the local officials and people.

Institutions are key in proper water management and for improving efficiency in adaptation. Dev says that mere increases in water pricing may not result in financial sustainability unless institutions are in place to recover water charges. Reforming institutional structures in favor of Participatory Irrigation Management (PIM) and Water User Associations (WUA) have to be strengthened, together with the promotion of participatory monitoring and evaluation. Dev also thinks that the development of groundwater markets would take care of the equity problems to a large extent. But the evolution of water markets is possible only in those regions where groundwater is available in sufficient quantities. Also, profitability of groundwater exploitation should be raised and users should be involved in the management of irrigation systems.

Most experts agree that a big role in promoting the sustainable use of water for adaptation can be played by technology innovation. Sanchez thinks that, regardless of how that water gets to the field, shifting from furrow irrigation and gravity flows, into sprinklers or drip irrigation, irrigation water can be used a lot more efficiently and in a climate-friendly way. Lele and Dev agree that drip irrigation, which is often described as a water "saving" technology, can be of great help in saving resources and increasing water productivity. However, Lele also warns that technologies that seem water conserving, like drip irrigation, can actually increase overall demand for water:

> Investments in irrigation expansion for years have not increased the amount of irrigated areas nor improved timely reliable supply of water for agriculture over decades (Lele 2013; Lele et al 2013). The result is overexploitation of groundwater and it has been hard to tame the groundwater anarchy.[1] Recent technological development has been the growth of drip irrigation which is often described as water "saving". However recent evidence suggests that drip irrigation is likely to increase rather than save water for at least two reasons. First, increase in crop productivity has an almost a one to one relationship with transpiration (Steduto et al), and increased profitability of investment in drip irrigation is already leading to the rapid spread of drip irrigation through the so-called "Jevon's effect".

---

[1] Shah, T. (2009). Taming the anarchy groundwater governance in South Asia. Washington, DC: Resources for the Future. http://site.ebrary.com/id/10570436.

## 2 Incorporating Climate Change into Agricultural Research and Extension

A second major finding that emerged from the case studies is the need for explicitly incorporating climate change effects into agricultural research and extension activities. The case study findings suggest that (i) managing more integrated and accessible climate and agriculture data at different levels, (ii) expanding research to identify farming practices adapted to the specific climate and farming characteristics (e.g. crop variety breeding programs; farm practices adapted to labor constraints; soil and water management investments adapted to local agro-ecological conditions) and (iii) supporting informed, and continually updated, training and extension programs can increase resilience and food security of agricultural households.

We asked our policy experts their views on these activities.

### 2.1 Climate Data

Starting with the discussion on climate data and its accessibility, there was unanimous agreement that enhancing access to climate data for agricultural producers (including farmers, fishers, foresters and livestock keepers) is quite important and should be given higher priority. However there were differences in which aspects of access should be emphasized, from improving production of the data, to better integration with agricultural data and better delivery of the data, and finally to improving the capacity of the farmers to actually use the data.

Dev points out that currently there is a disconnect between climate and agriculture data, with little integration between the two – including from local to global scale. Technologies such as satellite and remote sensing can play an important role in generating integrated data – but cooperation at global and regional levels will be needed to achieve this.

Delivering and transferring the information is as important as generating it. Caron notes that enhancing the current means that farmers get information, for example through the provision of additional information delivered through cell phones, is an important means of increasing access. Cao Duc Phat raises the importance of reaching remote and isolated farming communities that are highly vulnerable to climate risks, and which currently do not have good access to climate information. In stressing the importance of climate information delivery systems, Lele and colleagues provide the example of India's Meteorological Department (IMD) which has developed a framework for reaching climate information to farmers and fisher people through the use of cell phones. A key element of the IMD program is the Agro-meteorological Advisory Service (AAS), which customizes climate information relevant to the district in which it operates. Despite the benefits this program has generated in terms of reducing losses (including of lives), it is severely hampered by low percentage of cell phone ownership amongst farmers as well as limited awareness of the availability of climate information and inadequate technical capacity at the AAS district level to generate local level forecasts. Lack of human capacity in institutions at local level is also highlighted as a key issue by Sibanda and Kosura.

The lack of capacity of farmers (or any type of agricultural producer) to utilize climate information and thus the need for education at the farm level was raised by several of the policy experts. Sanchez notes that the more smallholders know about the dangers of climate change the better, and education as well as joint actions between climate and agricultural technical agencies is needed. Lele and colleagues point out the need for enhancing capacity amongst women in particular, as they play decision-making roles in ensuring household food security.

Sibanda reports on the results of a study done by FANRPAN in 2015 in 15 countries that indicated lack of human capacity as a key limiting factor. Not only the capacity of farmers to access knowledge and absorb it, but also the capacity of institutions that lack facilities as well as knowledge to cope with these new and complex issues. Pedro Sanchez argues that we need education not only for farmers, but for the general public which influences policies.

## 2.2 Farming Systems Research

All the policy experts felt that farming systems research with integration of adaptation merits greater priority. Cao Duc Phat, Sibanda and Dev emphasized the importance (and difficulty) of shifting away from research and technical assistance focused on single commodities, to a more integrated and systems based approach to analyzing farming systems.

Lele et al. argue that rapidly changing conditions facing agriculture require system thinking including both farm and non-farm aspects. They write:

> There is a growing recognition among developing countries' public sector research institutions that given the changed environment in which agricultural sector now works, coping with challenges such as reduced availability of quality water, nutrient deficiency in soils, climate change, farm energy availability, loss of biodiversity, emergence of new pest and diseases, fragmentation of farms, rural-urban migration, coupled with new IPRs and trade regulations, agricultural research programs must undergo a paradigm shift fully harnessing the potential of modern science, encourage innovations in technology generation, and provide an enabling policy and investment support. And in this research, priority must be given to some of the critical areas such as genomics, molecular breeding, diagnostics and vaccines, nanotechnology, secondary agriculture, farm mechanization, energy and technology dissemination.[2]

Caron makes the point that we already have a tremendous amount of information to support better farming systems research but we need better coordination to

---

[2] Vision 2050 document of ICAR's Indian Institute of Farming Systems Research (IIFSR), Meerut, UP, India. http://www.icar.org.in/Vision%202050%20IIFSR,%20Meerut.pdf.

effectively access it. Kosura cites the difficulties of building good and representative typologies of farming systems, considering the huge variability in biodiversity, scales, management systems, cultural diversity and resource base – although these may be overcome by more targeted investment and training. Lele also raises the problem of several different, and in some cases contradictory, approaches that involve or invoke farming systems research, including Climate Smart Agriculture, sustainable agricultural intensification, Conservation Agriculture and others. She calls for a common understanding and definition of sustainable intensification as an important means of improving the effectiveness of farming systems approaches. Caron argues that FSR should not only be a means of looking at what is out in the field today – but also a means of reflecting on possible options for moving forward.

## 2.3 Extension

On the discussion on extension, Lele et al. point out that the term "Extension" which signified a top-down, uni-directional approach to technology transfer has long been replaced by "Agricultural knowledge information systems" (AKIS) and later by Agricultural Advisory Services. All of the experts agreed that it is absolutely essential to move away from top-down systems to ones where knowledge flows in multiple directions.

Caron argues for new institutional approaches to extension based on the challenges it is now facing. He says:

> In the past, the agriculture revolution has been based on major disruptive innovations, such as genetics, mechanization or chemical inputs. We know that the future transition or the future revolution of agriculture will have to take stock of many, many different types of innovation and that it will be knowledge and information intensive. It will be important to bring disruptive innovation and technologies together with farmers' know-how to be in a position of making the best choice out of that. Of course extension is the way through which all this information can be used and can be put into practice. It means that extension needs institutional arrangements that allow for information exchanges amongst stakeholders.

Lele et al. put the role of extension in historical perspective. Since the Green Revolution there has been tension between commodity-oriented extension and extension oriented towards farming systems. In part that is also related to the changing roles of the public and the private sectors. Studies in India show that only 6.5% of the information farmers get is from public extension, 20% is from farmer to farmer contacts and 20–29% from newspapers, radios and TV. As research and input delivery has moved into private hands and inputs and market access have become important in a diversified agricultural production system, private dealers have become an important source of information for farmers on niche commodities, livestock, poultry, fruits, vegetables and edible oils, and new private sector extension systems have emerged as part of the growing value chains. Through experience and contract farming the emerging input providers are learning to provide integrated services to farmers, albeit with many hitches on the way. They conclude by calling for a redefinition of the public extension system:

> The role of public extension system, and of governments in technology transfer, now needs a clear redefinition, which many extension systems currently lack. With the growing emphasis on sustainable agriculture, that emphasis should be on natural resource management in the overall farming systems, including in water, soils, agro-forestry and the mother of them all, climate change. Traditional extension systems, by and large, used technical staff that were specialized in a particular branch of agricultural science such as agronomy, plant pathology, soil science, plant breeding, animal husbandry, fishery, without necessarily having a comprehensive understanding of agriculture using a farming systems approach. Being supply-driven, the public sector extension services have proved to be ineffective in terms of disseminating information to a whole farm management in a timely manner, going beyond farmer needs or expectations to manage externalities that spill over small fields and farms.

Sibanda also calls for a redefinition of extension in the African context:

> Yes, our extension services need greater strengthening but let's revisit the drawing board in terms of what type of extension service is required to deal with the multi-sectorial, multi-causal problems'. You are no longer dealing with an agent who knows everything, you need an extension agent who learns from the farmers, who promotes learning; you are dealing with an agent who will be able to bring information outside agriculture in a way that can be absorbed and understood by farmers; you are also dealing with an advisory system whereby we promote farmer-to-farmer learning; all this is different from the way the old policy for extension services was designed, i.e. top-down. We now need a bottom-up, cross-learning and inter-sectoral learning.

Both Kosura and Dev emphasize the need for building proper incentives into extension systems to promote higher quality services and better interactions and exchanges with farmers. Kosura gives some concrete examples of how this could be done, including making funding conditional upon the development of effective links between researchers and farmers through adaptive research and extension programs, the use of innovative approaches such as vouchers for advisory services, which could be given to farmer groups to source extension services from private sector providers, and the use of ICT for information and advisory services.

The lack of political will is perhaps the most important constraint to achieving more effective research and extension system, an issue that raised by almost all the experts. Pedro Sanchez provides a different and more optimistic view of the possibility of garnering political will at this time. He notes:

> Right now we are at a very, very positive point in this whole struggle, because (i) the MDGs have finished last year and they have been eminently successful, (ii) there is a new set of sustainable development goals (SDGs) which are better, more sophisticated to keep the world together, and (iii) the Paris Agreement on climate change. All came about at the same time, on the same year, and it gives a tremendous opportunity to really link agriculture and climate change.

Even if the political will to take action on agriculture and climate change in an integrated fashion is indeed increasing, tackling the problem requires policy coordination with multiple sectors beyond these two. Cao Duc Phat comments:

There is a lack of consistency between sustainable agricultural development activities and general development orientation and with other sectors (infrastructure, science and technology, urban development, development of non-agricultural economic industries).

Sibanda also raised the issue of coordination with sectors outside of climate change and agriculture, because the problem is multi-causal and the solutions multi-sectoral.

Even within the more circumscribed context of coordination across climate change and agriculture, there are significant barriers. In the Indian context, Lele et al. note that the lack of convergence among different agencies – local, regional or national – dealing with climate change and agriculture is a major problem. The absence of effective convergence involves huge administrative overheads, reduced outlays on real sector development, and absence of a cohesive approach to climate change mitigation and adaptation.

## 3 Taking a Close Look at National Policies Affecting Risk Management: Index Insurance, Safety Nets and Input Subsidies

Index insurance, safety net programs and input subsidy policies are all development policies that have effects on risk management, which is an important facet of adaptation, although they are not designed with adaptation explicitly in mind. The case studies in the book indicate these policies can have both positive and negative effects on adaptation. They may also not be very effective under changing climate as well as broader development conditions.

Index insurance has been hailed as an important tool for increasing resilience in smallholder agriculture livelihoods – but the case study findings indicate that subsidies are essential for the program to be operational (in absence of subsidies the program is too expensive for the farmer). Extending any type of insurance to individuals in remote locations will likely be of extreme difficulty, even subsidized products.

In the last decade, there has been an expansion of safety-net programs in African countries with the aim of reducing poverty and increasing food security: in most cases targeting focused on economic vulnerability rather than climate vulnerability. However, the case study findings indication that a cash transfer program is effective in managing climate risk and potentially mitigating the effects of climate change.

Input subsidy programs have been promoted against the background of bad weather affecting production and with an aim of increasing resource-poor smallholder farmers' access to improved agricultural inputs. However, programs have not been fully exploited to address constraints associated with climate risk. It is also often criticized for poor targeting at the farmer level.

We asked our panel to comment on each of these policies in terms of their potential role in adaptation, and the types of adjustments that may be needed to realize their effectiveness.

## 3.1 Index Insurance

Index insurance is an important tool for managing climate risk according to Sanchez, but certainly at the very beginning, for all these interventions, they need for some type of subsidy to be successful. Sibanda takes this further citing the cases of subsidized weather-based index insurance in Uganda, Zambia and Swaziland conducted by FANRPAN last year. The results of that study indicate the potential for weather based insurance, but also some key factors to ensure its success, including the importance of organizing farmers into groups. Subsidizing the insurance is an important way of getting people into a new way of doing things. It is key to build the human capacity needed for effective management of such schemes: by training local insurers on the businesses of insurance in agriculture and at the same time helping people to understand what it means to keep records, subscribe as a group, and work through group ownership. She says:

> What is exciting is that through insurance you're now creating a business of a bankable industry whereby you're introducing services that would actually escalate beyond primary entry point which is agriculture.

Dev notes that although crop insurance schemes have not worked in many parts of the world, in recent years these schemes are becoming more effective. In the past, measurement of losses was costlier and he argues that weather index-based insurance can make it cost effective for farmers. Recently, India introduced a new crop insurance called Pradhan Mantri Fasal Bhima Yojana (PMFBY) (Prime Minister's Crop Insurance Scheme). In the previous schemes, premiums were high and coverage in terms of sum insured (SI) was inadequate. The new scheme corrects these two problems. It also broadens the definition of risk to include yield losses, preventive sowing, and post-harvest losses. Farmers now have to pay a uniform premium depending on the types of crops. The gap between the actual premiums and the rates payable by farmers would be fully met by the government. He points out:

> The new crop insurance can be a game changer if the conditions of low premiums and the SI covering the gross value of output are met along with quick claim settlements with mobile and satellite technology.

For Kosura building capacity in the insurance sector as well as amongst farmers is important and thus he advises:

> Insurance programs face barriers since providers are still reluctant to deal with agriculture. Working with insurers to understand the risks and mitigation strategies in agriculture is important. Demonstrating successful farm ventures under different risk scenarios would help reduce the fear of insurers.

Cao Duc Phat also stressed the importance of building effective management capacity for insurance programs and how it needs to be integrated with government policy. He calls for:

> To improve the sustainability of public finances, insurance should be combined as a risk management product invested by the private financing agencies with the poverty reduction policies of the Government.

Lele and colleagues question the benefits of index insurance to manage climate risk. It could end up increasing the cost of credit to smallholders, and moral hazard problems exist with respect to the lending institution's incentive for strong loan management practices. The bottom line is:

> Whether governments should support index insurance schemes for small holders will be a fiscal policy issue as to whether the subsidy is well targeted and that it is the most efficient use of government resources or the aid provided by a foreign agency or a NGO. There may be other ways to facilitate small holders' risk management and coping strategies more efficiently. Answers will vary from country to country.

## 3.2 Cash Transfer Programs

All of the policy experts note that while the overall concept of using safety nets as one tool to manage climate risk is sound, the effectiveness will be determined by the program design – and here there were several different aspects considered. One important one is that the design of safety net programs to support the management of climate risk needs to be tailored to local conditions. Caron cites results from recent reports of the High Level Panel of Experts (HLPE) of the Committee on Food Security (CFS) that indicated considerable variation in the types of risks, tools and programs and institutional arrangement amongst programs and the importance of considering these conditions in designing effective programs.

Sanchez raises the potential benefits from conditional transfers, where cash payments are based on using the right type and amount of fertilizer or the right variety, or sending your children to school. Cao Duc Phat also noted the importance of accompanying cash transfer programs with technical advice on how best to spend funds, as well as establishing a technical service system to provide agricultural services to meet the needs of farmers (such as seed, chemicals, maintenance, consumer guides and more) to help ensure wise use of the transfers.

Dev cited the asset creation benefits of some of the present social protection programs which are beneficial to the development of climate resilient agriculture. He gives the example of India's public works program MGNREGA. A study by Indian Institute of Science, Bangalore in India quantifies the environmental and socio-economic benefits generated by the works implemented under MGNREGA and assesses the potential of these benefits to reduce vulnerability of agricultural production and livelihoods of the beneficiaries, post-implementation (2011–12) as compared to pre-MGNREGA (2006–07), to current climate variability. Agricultural and livelihood vulnerability indices developed showed reduction in vulnerability due to implementation of works under the Act and resulting environmental benefits.

Sibanda points out the importance of understanding which households should receive transfers. She argues that we need to understand the current endowments of the household, including human capacity, education, and health. Do they have a support system that will allow them to utilize cash transfers to engage in Climate Smart Agriculture? What are their natural capital assets such as land and water? Lele et al. agree that there is a strong possibility of using cash transfer programs as a climate management policy but the approach to targeting of this subsidy to eligible beneficiaries could vary from a landscape to landscape.

However Kosura questions the capacity of safety nets to actually mitigate risks, depending on the amount of cash transfer they actually involve. He gives an example from Kenya, where both the Hunger Safety Net Programme (HSNP) and Cash for Asset/Work programs advance households about USD 25 per month. Considering a very poor and vulnerable household with about six family members, the cash advanced will likely not be enough to even meet household food needs.

## 3.3 Input Subsidies

Dev states that subsidies are not sustainable and therefore need to be designed as temporary measures. They also may encourage waste of resources, as is the case in India with water and land. Subsidies lead to inefficient resource allocation by supporting inefficient input sector (e.g. India's domestic fertilizer industry). Furthermore fertilizer subsidies may lead to unsustainable use of land. Some examples of effective subsidies are when they use transfer payments to poor farmers (e.g. a minimum amount of fertilizer for small plots), subsidize valuable technologies when credit markets don't work and the technology generates positive externalities (e.g. drip irrigation).

FISP type programs can have positive impacts by increasing yields and incomes resulting in farmers expanding their financial capital and knowledge base according to Kosura. They can enhance the uptake of valuable technologies, but in the long-run they result in moral hazards, and even corruption, because subsidies become transfer policies and serve to benefit more influential and politically connected farmers. Caron has the same concerns highlighting the role of subsidies in increasing resilience through exposure and learning, but worries about their long-term effect. Sibanda thinks that FISP solves a short-term constraint, but improved productivity in the longer-run requires complementary inputs, like seeds. Given limited budgets, subsidies need to be targeted based on household level vulnerability, need, and productivity gains. It may be worthwhile to consider several types of subsidies, not only fertilizer, but also seeds and the need for establishing effective extension. The big challenge of subsidy program design is overcoming corruption. Therefore adhering to targeting criteria both improves efficiency and improves corruption.

Sibanda believes that FANRPAN targeting criteria provides a step in the right direction.

Lele et al. state:

> Developing countries such as India provide subsidies to farm households indirectly, either through free supply of or reduced prices for inputs such as water, power, seeds, fertilizers and interest- free bank loans. These subsidies tend to benefit wealthier farmers more than poorer farmers who do not necessarily get sufficient access to these inputs either because of the lack of purchasing power or supply constraints. By and large, existing indirect subsidy programs would need to be modified or replaced by new programs that target subsidies mainly to small and marginal farmers adopting new conservation agriculture technologies. It is also important to ensure that the prevailing leakages in subsidy programs are checked by providing cash subsidies against actual purchase of subsidized inputs directly in their bank accounts.

They go on to give examples from India on how this is being accomplished:

> In India, the government has advanced considerably in eliminating middlemen in the provision of subsidies directly to farm households, including particularly cash subsidies to small and marginal farmers, by way of direct deposit to eligible farm household's bank accounts. Similar reforms are needed in all types of existing subsidy schemes and redirect the resources so released to support those farmers which shift from conventional to climate-smart conventional agriculture for a limited number of years, i.e., until farmers adopting new technologies are able to restore any productivity or income losses and begin to benefit from using new technologies.

Sanchez acknowledges the problem of corruption, but thinks the benefits of subsidies may outweigh it. He argues that farmers are subsidized in the developed countries, and there should be no reason why they shouldn't be in developing countries. He points out that Malawi subsidy program effectively addressed the food security problem. He argues:

> Of course, it didn't alleviate poverty and there is some corruption, but overall it provided more resources and improved the health and capability of the poor. Now that the program reached a certain threshold of performance, it can be modified to address other objectives (for example diversifying diets, increasing resilience, etc).

Phat recognizes the immediate benefit of fertilizers but warn against the tendency of subsidies to lead to distorted market prices and overuse of fertilizer. Indeed in Vietnam farmers have over applied fertilizers and pesticides and the government now informs farmers of recommended dosage and tries to avoid subsidization.

## 4 Priorities for the Future and Summary

In this final section, we asked the policy experts to comment on the case study findings indicating the need for better coherence between climate change, agriculture and development policies and suggest means for achieving this. We also asked them to give us their opinion on the priority actions for near term and provide their direct quotes from their replies. This section concludes with a summary of the main points of agreement and divergence amongst our panel in responding to all of the interview questions.

## 4.1  Policy Coherence

Most experts acknowledge the importance of integration and harmonization of climate change consideration into agricultural and non-agricultural sectors to achieve better outcomes. They note that often government ministries work in "silos" and this often works against not only inter-sectoral convergence, but also against intra-sectoral convergence. Incentive mechanisms should be put in place to encourage coordination and harmonization among government ministries and also for many actors to adjust behaviors. The need for convergence in climate change activities has to be recognized in policymaking both at center and province levels and in implementation at different levels and building a supportive evidence base as well as explicit recognition of trade-offs and the need for compromises is important to achieving effective coordination.

Dev notes that the silo mentality works against not only inter-sectoral convergence, but also against intra-sectoral convergence. To fully support the agriculture sector requires coordination among the ministries of agriculture, rural development, and commerce, as well as among the various Ministries and Departments relating to food, irrigation, fertilizer and power. He also raises the possibility of inefficiency and disruption arising in trying to build policy coherence, if is it not well done. For example, multiple departments and multiple schemes can cause confusion among staff. The incentive question is important. Officials think they will lose some of their power, if convergence is pursued with other departments and this issue needs to be addressed directly.

Lele and colleagues make many of the same points as Dev, pointing out that for successful implementation of climate change initiatives, it is important to rationalize/harmonize various government regulations, credit policies, subsidy programs and land tenure laws, and get these initiatives effectively integrated into sector planning, budgeting and development. It is also necessary to bring about convergence among different government departments dealing with climate change and their local offices at the landscape level, to be able to effectively implement climate change adaptation planning and implementation using community/participatory methods at least cost. Reducing duplication and redundancy is an important facet here. For example, in India, the existing multi-agency institutional framework involves huge administrative overheads, reduced outlays on real sector development, and much less impact in terms of outputs and outcomes.

The use of evidence based approaches to policy planning and programming and promotion of multi-stakeholder and multi-agency participation in these processes is key for Kosura. The need for institutional capacity to take part in the climate change adaption planning process varies from country to country but generally, there is a need for (i) human capital development through relevant training and skills enhancement; (ii) financial capital through targeted resource mobilization for priority projects meant to promote Climate Smart Agriculture for Development; (ii) formulating a clear policy and regulatory framework as well as shaping political will and (iv) regular public-private sector meetings and round table discussions must also be sustained in order to assure political will that is critically essential for success of the policies that require reforms in institutions especially in legislation and resource mobilization strategies.

Sanchez stresses the need for more communications between the climate and agricultural scientists. There are many institutions involved in production and dissemination of information and thus it's imperative to have a policy framework that encourages interactions between the different sectors, Ministries, private companies and farmer associations. There are going to be trade-offs and synergies between promoting productivity and environmental issues and an enabling government environment is needed to handle these in a reasonable way. He notes the importance of education and information to promote this process especially in the developing countries.

Caron starts out by noting that agriculture is at the heart of social transformation and thus a key part of the solution – and not just the problem. He also raises the issue of trade-offs and the need for compromises and thinks these have to be acknowledged to build the conceptual, intellectual and operational framework that puts agriculture as a lever for change in other sectors. He gives the example of the Paris Agreement on Climate, where the word 'agriculture' was not in the final agreement even though the sector plays an essential role in the intended nationally determined contributions (INDCs) to the agreement. He notes that Climate Smart Agriculture is built to address trade-offs between food security, mitigation and adaptation to climate change. Building on this strong conceptual basis looking at the trade-offs, and at the gaps, is a strong avenue towards thinking about agriculture in the future in addressing climate issues.

Cao Duc Phat stress the importance of integration of climate change considerations into sector planning and development. Vietnam is currently conducting agricultural restructuring, in which the long-term plan, strategy, policy, organizational innovation, and improvement of public investment are adjusted and implemented synchronously both inside and outside the sector, at all levels of management, not just some policy changes. He also points out the need to improve and enhance communication and advocating for changing a way of thinking of management people from central to local levels. Forming an evidence-based mechanism and public support should also be integral part of decisions for managing natural resources efficiently. Both require good scientific information and research activities. Lastly, forming the unified coordination system under long-term action plans and effective cross-sectoral and regional coordination mechanism is key for promoting effective integration.

The need to reduce duplication and consider the incentives (and disincentives) for cooperating amongst government agencies is emphasized by Sibanda. She stresses the need for (i) strong leadership that points to the directions that people need to go, and (ii) an analysis that looks at what is needed to be added, and what we need to get rid of. She acknowledges that harmonization is not easy and it is important to focus on institutional change that is going to be relevant. Wedding of co-function analysis and co-institutional analysis requires resolute leadership that will pull the trigger where things need to be dropped, and be bold enough to say: 'this we don't know, we need help'. The area of harmonization of policies is a new area and calls for a new way of doing business, which we will need both leadership and mapping to achieve.

## 4.2 Policy Priorities for the Next 20 Years

This section is composed of direct quotes from each of the panel members.

Cao Duc Phat: The priority is to undertake joint scientific research programs to support countries to improve animal and plant breeds, farming systems, technical systems that have better resistance to extreme and unusual climate conditions. This will require support to increasing the effectiveness of South-South cooperation under the 3-sided triangle, in order to transfer experiences, lesson learnt, best technologies and policies among countries with similar conditions or with common problems to be solved. In addition, building operational mechanisms to perform payments for environmental services (for example carbon emissions trading, forest cover, biodiversity levels, etc.) is needed. Strengthening international cooperation in sustainable resource management – especially in the Mekong Delta (e.g. transboundary and multiple country partnership management) supported by transparent information exchange, discussions and cooperation. An important priority for Vietnam is the development of a GHG inventory systems, applying tier 2 and 3 level analysis, for agriculture in order to develop appropriate baselines and carbon footprints – as well as GHG reduction scenarios and development programs that ensure the achievement of development goals, increase productivity, efficient and sustainable uses of natural resources.

Caron: There is an incredible challenge to build intelligence and understanding of the context of where we are. That's even more complicated because we do not know necessarily where we are going. How can we build the capacity, the knowledge, the understanding capacity, the knowledge and the technology that we will be needing in 20 or 30 years' time? There is a need for very strong investment in research that addresses three challenges: better liaison between policy-making and science, secondly to get strong research communities in all parts of the world to address both local and global challenges and third a more global need for investment in research that puts us in a position of preparing what we will need in the future.

Dev: Policy makers, researchers and the international community should recognize that climate change is real and Climate Smart Agriculture should be the present and future priority and work towards achieving climate related adaptation and mitigation measures. Announcement of Sustainable Development Goals (SDGs) provides an opportunity for global level cooperation. The Paris CoP21 agreement has to be enforced. There are many promises but not firm commitments.

Kosura: Given the dynamic nature of climate change and diversity of cultural practices and environments, innovative and responsive research to seek for timely solutions should be a priority agenda. Marshaling investment resources for research, infrastructure and information dissemination to avoid possible disasters brought about by climate change is critical. Institutional innovations to minimize institutional failure, moral hazards and corruption should be prioritized. In this way, farmers and stakeholders in general will have the incentives to adopt available technologies to respond to adverse climate change effects.

Lele, Deshpande and Abrol: Our effort should be to work directly with the farmers over a long (10–15 year) time horizon to convince them about the benefits of CA. For this, involvement of social scientists from the very beginning is critical. The Rice Wheat Consortium in the Indo-Gangetic plains, the 'bread bowl' of India and India's neighboring countries was such a program. It was the most successful eco-regional program receiving the King Baudouin Award on behalf of regional NARS. It was closed and the reasons behind its closing are unclear. It reflects the tragedy of international cooperation.

Sanchez: My main focus is on Sub-Sahara Africa. The goal would be in the next 20 years that Africa is producing at a 3 tons per hectare level on maize or equivalent and all this sort of thing. I think very strongly that tackling climate change has to be made into a positive business, where people will make money out of it, either smallholder farmers or big farms. I've been advocating fertilizers a lot: there is a climate price tag to that because manufacturing fertilizers produces methane and negative things on climate. I think it could be lovely if we could do this in a more natural way which is biological nitrogen fixation by legumes. The science is there and it is very positive. However, the adoption has been miserable. Partly, I think, it is because there was no subsidy of any kind. This is the issue that has been mentioned above, i.e. how to enable farmers to get through this two to four-year period in which you're not going to get anything out of it but you're spending money? This has to be arranged, or subsidized or (supported) with long-term credit or whatever. But if we could have more of these nitrogen-fixing trees, they can be used to partially replace nitrogen fertilizers it would be great.

Sibanda: To me the key is the leadership. I think the next 15–30 years require bold leadership and leadership that doesn't lead for today but leads for tomorrow. What that will take is: leadership that has a plan informed by where we are now, where we want to go and how we're going to get there and who is going to get us there. And when I talk about 'who is going to get us there' is the partnership for finance, knowledge and bottom-up policies, i.e. the policy that serves the home ground.

## 4.3 Summary Conclusions

Overall, there is a fairly high level of agreement amongst the panel members in responding to most of the interview questions, although with some difference in emphasis or applications. However there are also some differences of opinion that emerge from their responses. In the following section, we summarize the main points made on each of the issues addressed, highlighting the areas of agreement, as well as differences.

1. There is a high level of agreement that promoting sustainable land and water management in agriculture, including diversification is a high policy priority, not only for the adaptation benefits they can provide, but also as a key response to improving rural livelihoods under rapidly changing conditions. It was also widely agreed amongst the panel that policy has a fundamental role to play in building the enabling conditions for a major transformation to more sustainable land and water management.
2. The panel indicated that one of the most important policy measures for promoting sustainable and Climate Smart Agriculture is through value chain development – on both the input and output side. Value chains need to be extended and strengthened, but perhaps most importantly repositioned in order to better incorporate both environmental and social externalities. Coordinating collective action through cooperatives, and providing better incentives for sustainable management through improved land and water tenure systems were also considered priority policy actions.
3. Irrigation and improved water management were considered a very high priority for adaptation by the panel, but with much greater emphasis on small scale systems where the users have a high degree of control that can be managed for more than one purpose.
4. There is overall agreement amongst panel members that adaptation to climate change needs to be explicitly integrated into agricultural data and research system, with priorities ranging from building capacity of agricultural technical staff to use climate data to improving systems of communicating and disseminating climate information.
5. Agricultural extension is considered an essential element for Climate Smart Agriculture by the panel – but it needs major rethinking and reform. Building systems that allow for bottom up as well as top down interactions and well as getting correct incentives for extension workers – and building their capacity to use climate data are important.
6. The potential for index insurance as a tool for managing climate risk was generally regarded as positive by the panel but with some skepticism about whether or not it can be scaled up and if it will always need subsidization.
7. The panel considered cash transfer programs as a potentially important tool for managing climate risk for farmers, but its effectiveness depends on good targeting.

8. Probably the most divergence of views amongst panel members was related to the potential role of input subsidies in Climate Smart Agriculture. On the negative side, they are associated with corruption and inefficiency. On the positive side they have been effective in raising productivity as well as other benefits. Actions to reduce corruption, such as direct deposit payments and improve targeting and eligibility rules can make them more climate smart.
9. There is very strong agreement amongst panel members that greater coherence and integration is needed between agriculture and climate change policies that can lead to reduction in duplication, bureaucracy and costs.
10. Strengthening multi-disciplinary and long term systems research was considered a high priority for several panel members, as was better bridging of the policy-research divide. Developing the political will to actually enforce agreements and fostering institutional innovations to see their effective implementation in the field also emerged as priority actions.

**Open Access** This chapter is distributed under the terms of the Creative Commons Attribution-NonCommercial-ShareAlike 3.0 IGO license (https://creativecommons.org/licenses/by-nc-sa/3.0/igo/), which permits any noncommercial use, duplication, adaptation, distribution, and reproduction in any medium or format, as long as you give appropriate credit to the Food and Agriculture Organization of the United Nations (FAO), provide a link to the Creative Commons license and indicate if changes were made. If you remix, transform, or build upon this book or a part thereof, you must distribute your contributions under the same license as the original. Any dispute related to the use of the works of the FAO that cannot be settled amicably shall be submitted to arbitration pursuant to the UNCITRAL rules. The use of the FAO's name for any purpose other than for attribution, and the use of the FAO's logo, shall be subject to a separate written license agreement between the FAO and the user and is not authorized as part of this CC-IGO license. Note that the link provided above includes additional terms and conditions of the license.

The images or other third party material in this chapter are included in the chapter's Creative Commons license, unless indicated otherwise in a credit line to the material. If material is not included in the chapter's Creative Commons license and your intended use is not permitted by statutory regulation or exceeds the permitted use, you will need to obtain permission directly from the copyright holder.

# Conclusion and Policy Implications to "Climate Smart Agriculture: Building Resilience to Climate Change"

**David Zilberman**

**Abstract** The efforts to adapt to climate change in developing countries are in their infancy, and hopefully CSA will be a major contributor to these efforts. But CSA itself is evolving, and there is a growing need to refine and adapt it to the changing realities. This section of the book focus on the implications of the empirical findings for devising effective strategies and policies to support resilience and the implications for agriculture and climate change policy at national, regional and international levels. This section is built upon the analysis provided in the case studies as well as short "think" pieces on specific aspects of the policy relevance issues from policy makers as well as leading experts in agricultural development and climate change. The case study provided concrete illustrations of the conceptual and theoretical framework, taking into account the high level of diversity in agro-ecological and socioeconomic situations faced by agricultural planners and policy-makers today. While the case studies demonstrate the diversity of challenges facing farmers around the world, they also indicate unifying characteristics imposed by climate change on agricultural decision making and the potential for the CSA approach to address them.

Smallholder farms and rural communities in developing countries are especially vulnerable to the impacts of climate change. Climate change will exacerbate existing challenges of resource scarcity, credit constraints, infrastructure limitations, and incomplete information and markets. There is already evidence of the perception and reality of climate shocks, and a growing need for effective adaptation strategies. Climate Smart Agriculture (CSA) is a framework for developing decision support systems at the farm and policy level. It aims to provide principles to identify technologies, management tools, and policies that will enable farmers to adapt to challenges of climate change while maintaining and improving societal wellbeing.

---

D. Zilberman (✉)
Department of Agriculture and Resource Economics, University of California Berkeley, Berkeley, CA, USA
e-mail: zilber11@berkeley.edu

CSA is based on the recognition of *heterogeneity* among farmers and regions in terms of socio-economic and agro-ecologic conditions, which emphasizes the need to understand the *distribution* of impacts arising from severe weather events and shifting climate. In general, heterogeneity and the randomness of impacts increase the value of having access to a wide range of differentiated strategies, rather than to uniform prescriptions. It also recognizes the high degree of *uncertainty* and the *dynamic* nature of climate change, and thus emphasizes the importance of continuous learning and strategic adaptation to changing conditions and new information. Because we are at the early stages of climate change, we emphasize the capacity to adapt to increased likelihood of extreme events, while recognizing that climate change may require more transformative changes in technologies and relocation of production practices in response to permanent and significant changes in weather patterns.

This book starts with an overview of major themes including the evolution of CSA, mechanisms of innovation and institutional change that will influence CSA, and the aspects of climate change that may addressed by CSA. The main part of the book consists of case studies from many regions around the world that analyze adaptation decisions, as well as document barriers to adoption of effective adaptation actions. The emphasis is on developing countries, although we also bring examples from the U.S. to demonstrate that even in highly commercialized systems using advanced technologies, gains can be achieved from access to better information and enhanced ability to adapt to changes proactively. While the case studies demonstrate the diversity of challenges facing farmers around the world, they also indicate unifying characteristics imposed by climate change on agricultural decision making and the potential for the CSA approach to address them.

*Targeted Solutions to Specific Problems* Heterogeneity suggests that we cannot expect universally applicable solutions, but rather encourage a process to develop solutions that are most appropriate for a given location. More frequent weather extremes and uncertainty regarding longer-term changes in weather mean that a range of targeted solutions – both on and off farm -must be developed that enable farmers to flexibly respond to current conditions and adapt to shifts in climate patterns.

*Quantitative Evidence-Based Solutions* To identify differentiated solutions best fit to specific situations requires quantitative analysis based on empirical data and appropriate analytical tools. In particular, more emphasis must be given to understanding the distribution of impacts, instead of relying on average impacts on a "representative" farm household. CSA aims to capitalize on growing sources of data and analytical tools to utilize them, including integrating ever more sophisticated GIS information into more traditional econometric analyses and simulation modelling. Solutions are derived both by quantifying technological feasibilities, consumer demand, and biophysical and behavioral constraints.

*Adaptive Learning* Because of ongoing processes of climate change and technological progress, information is accumulated and new opportunities arise. Thus optimal solutions are changing over time and across locations. The case studies indicate several means of enhancing adaptive learning amongst producers as well as policy-makers including improved analytical tools, improving information channels

Conclusion and Policy Implications... 623

between producers, policy-makers and analysts, and building flexibility into agricultural support services such as agricultural knowledge and extension services as well as input and output markets.

*Opportunity and Social Costs* The analysis in the case studies indicates that climate change already has some impacts on the opportunity costs associated with alternative agricultural development pathways – and is likely to have even more in the future. Approaches to evaluating alternative solutions and new opportunities that explicitly consider changes inopportunity costs imposed by climate change at different locations can be achieved through better utilization of modelling tools and innovative datasets.

*Risk and Risk Aversion* We have long been aware that the economic well-being of farmers is affected significantly by risky outcomes and their actions are impeded by risk and loss aversion. Climate change augments the importance of building institutional capacity for dealing with risks and uncertainty. CSA emphasizes introduction of institutions that provide enhanced information to reduce risk as well as institutions, such as insurance markets, that will allow farmers to reduce the cost of risk and loss aversion.

*Input Use Efficiency and Precision* Not all applied inputs are utilized productivity. The residual is frequently a source of pollution – as well as a cost to the producers. Improving input use efficiency under increasing uncertainty climate change imposes is clearly an area where considerable social and private gains can be realized. Technologies that enhance precision of farming enable farmers to adapt input use to variability in climatic conditions could offer significant improvements in terms of both higher net revenues and lower yield variability. Policies that lead to develop and enhance adoption of affordable technologies that increase precision and input use efficiency may enable farmers in developing countries to "leap-frog" past conventional, often wasteful and costly, input application.

*No Regrets Policies* Given the uncertainty of climate predictions and risk aversion, it is a priority in CSA to identify activities that will address climate change risks but will enhance wellbeing and improve livelihood regardless of changes in climatic conditions.

*Flexibility* Given changes in climatic, technological and socioeconomic conditions and a high degree of uncertainty, CSA strategies aim to avoid costly irreversible choices in favor of making decisions that allow modification in response to changing conditions.

*Resilience* Because climate change may expose farms to severe climatic and economic shocks, CSA encourages developing the capacity to withstand, or rebound afterwards, to these shocks. Resilience can be enhanced through better technologies, improved infrastructure, and reliance on institutional mechanisms (e.g. access to financial products).

*Innovative Capacity* A key for CSA is having innovative capacity that can produce new solutions taking into account new scientific knowledge and understanding of climate change. Innovations may be both technological and institutional. Implementation of innovations requires enabling policies, including investment in infrastructure and extension, and reducing transaction costs that will enable establishment of supply chains and organizations to implement innovations.

*Market-Based Solutions* Effectively governed markets enhance trade opportunities that can increase efficiency in resource allocation, which in turn is important for the diffusion of low cost solutions and reductions in variability of supply. The CSA approach encourages evaluating the role of trade and trade regulations in capturing these benefits, allowing for innovative market solutions to address risks and environmental side effects of environmental activities.

*Supply Chains* Farmers and other actors in agriculture are linked across space and time by supply chains. Adaptation to climate change occurs throughout supply chains, and effective farm level adaptation is dependent on effective adaptation throughout the supply chain. Thus greater integration of supply chain governance is needed in the design of farm level adaptation strategies.

The analyses throughout the book emphasize the importance of designing effective policies. Climate smart policies will develop mechanisms to monitor climate and other conditions, assess situations, and be able to respond to changing realities. Furthermore, policies need to enhance resilience and capacity to adapt to changing agro-climatic conditions. These policies will be part of an overall policy environment that aims at sustainable development, namely assuring that the current generation will continue experiencing increased food security while the next generation will not be worse off than the current one.

Improving knowledge systems to meet climate challenges will require investment in infrastructure that allows for collection of spatial data on climatic conditions, agricultural performance and economic conditions at various scales. There is a need to invest in analytic capacity to utilize the data to develop better quantitative understanding of weather patterns, and related behavioral and agro-ecological responses. Furthermore, utilizing this knowledge will require, first an investment in outreach capacity will disseminate new knowledge and update information at different levels of decision making, and second, an investment in response capacity. This capacity will combine both short term capabilities that enable early warning and response systems as well as long term capabilities that will enhance resilience, adaptation, and contribute to sustainable development.

Adaptation capacity begins with investments in and incentives for innovation. This implies both access and utilization of new technologies and management practices developed throughout the world. Access to new technologies means establishing policies and institutions to reduce intellectual property rights and regulatory barriers. In particular, regulations need to balance gains from emerging technologies with risk considerations. Further, local research and outreach capacity is needed to fit technologies and management practices to local conditions.

Rapid response to crisis and long-term adaptation are hindered by lack of roads, electricity, water, and storage capacity. While generally investments in these forms of infrastructure are 'no regret' policies, it is important to use sound analysis integrating effects of climate change to take into account net social benefit and cost.

Namely, the consideration of viability of certain locations in the long-run, and the environmental and social implications of investments.

Development and resilience in many regions is constrained by lack of access to markets (inputs and outputs), as well as financial constraints. Investment in physical infrastructure can reduce some of these constraints by reducing the cost of doing business, but there is a need for improved institutional capacity. There is a need to expand and improve the supply chains of credit and farm-level inputs and outputs. Developing such supply chains requires strong involvement of the private sector, sometimes in partnership with the public sector, within an improved policy environment. For example, private investment in storage and product processing capacity can be augmented and coordinated with public investment in improved physical infrastructure and training. Public-private partnerships can be established to share risk and obtain finance for joint projects.

Climate smart policies will emphasize incentives and capabilities to encourage improved decision-making at the farm-level. This includes the adoption of best feasible technologies, improved input use, and post-harvest practices. Establishment of extension and improved supply chains may go a long way to meet this objective. Governments may also consider introducing insurance schemes with low transaction costs and moral hazard potential to reduce the cost of risk and risk aversion. Further, governments may provide input subsidies in short-term situations in which learning-by-doing is needed, as well as insured and subsidized credit. These activities should be designed to induce transition to sustainable and economically viable practices.

Climate change is a dynamic process marked with random shocks that may result in significant short-term losses and may make some regions economically unviable. Furthermore, policy design will combine both efficiency and distributional considerations. Climate smart policies may consist of cash transfers that sustain individuals at a minimum level of income and promote transition to more sustainable livelihood, which may include migration.

These policies may be costly and one of the major challenges is to optimize the use of funds given budget and credit constraints. Developing evaluation procedures to assess outcomes on efficiency and equity measurements will allow for creating targeting criteria. Thus policies will vary across location and over time to reflect differences in expected net benefit. Furthermore, one of the challenges of climate smart policies is to develop financial mechanisms and political initiatives that will expand the range of resources available for investment.

This book aims to present the state of the art of CSA, both conceptually and by bringing together case studies and perspectives that will improve the management of agriculture in the era of climate change. The efforts to adapt to climate change in developing countries are in their infancy, and hopefully CSA will be a major contributor to these efforts. But CSA itself is evolving, and there is a growing need to refine and adapt it to the changing realities. We look forward to further efforts in this area as part of the increasing commitment and effort to address the challenges of climate change and sustainable development.

**Open Access** This chapter is distributed under the terms of the Creative Commons Attribution-NonCommercial-ShareAlike 3.0 IGO license (https://creativecommons.org/licenses/by-nc-sa/3.0/igo/), which permits any noncommercial use, duplication, adaptation, distribution, and reproduction in any medium or format, as long as you give appropriate credit to the Food and Agriculture Organization of the United Nations (FAO), provide a link to the Creative Commons license and indicate if changes were made. If you remix, transform, or build upon this book or a part thereof, you must distribute your contributions under the same license as the original. Any dispute related to the use of the works of the FAO that cannot be settled amicably shall be submitted to arbitration pursuant to the UNCITRAL rules. The use of the FAO's name for any purpose other than for attribution, and the use of the FAO's logo, shall be subject to a separate written license agreement between the FAO and the user and is not authorized as part of this CC-IGO license. Note that the link provided above includes additional terms and conditions of the license.

The images or other third party material in this chapter are included in the chapter's Creative Commons license, unless indicated otherwise in a credit line to the material. If material is not included in the chapter's Creative Commons license and your intended use is not permitted by statutory regulation or exceeds the permitted use, you will need to obtain permission directly from the copyright holder.

# Index

**A**

Adaptation, 4, 6–11, 17, 18, 21–24, 27, 33, 36, 38–42, 44, 50, 51, 53, 54, 56, 58–65, 67–70, 78, 138–145, 147–158, 183, 191, 228, 253, 254, 267, 268, 280–282, 285–287, 293–300, 302–304, 308, 310–330, 337, 347–350, 354, 355, 362, 372, 376, 380, 381, 387–389, 391, 395, 396, 400, 401, 405, 428, 439, 446, 449, 450, 455, 457, 459, 461, 467–469, 471, 472, 478, 480, 481, 485–487, 498–520, 528, 531–533, 535, 554, 555, 557, 564, 574, 588, 590–592, 600–607, 610, 611, 615–617, 619, 621–624

Adaptive capacity, 4, 5, 7–11, 21, 32, 33, 41–43, 241, 309, 312, 330, 389, 404, 426, 427, 429, 439, 486, 591, 592

Adaptive learning, 43, 622

Adoption, 5, 8, 10, 11, 15, 21, 25–27, 35, 37, 38, 40–42, 44, 53, 55–58, 61, 65–68, 111, 139, 140, 142, 149, 164, 176, 186, 188, 190, 228, 252, 254, 255, 258, 260, 262, 264–266, 296, 302, 308, 314, 319, 323, 327, 329, 337, 354, 363, 372, 375, 376, 379, 380, 387, 389, 391–393, 395, 396, 398–403, 407–410, 412–414, 427–430, 432, 439, 446, 450–452, 457–467, 469, 471, 472, 478, 481–493, 501, 517, 518, 564, 565, 567, 568, 571, 573–575, 580, 582–592, 600–602, 618, 622–624

Agricultural development, 4, 5, 11, 15, 18, 22, 24, 25, 27, 28, 32, 34, 43, 227, 253, 404, 407, 412, 471, 472, 533, 539, 557, 591, 610, 622

Agricultural productivity, 36, 78, 138, 161, 162, 252, 258, 278, 346, 389, 446, 480, 493, 501, 531

Agricultural systems, 9, 20, 21, 32, 37, 39, 40, 42, 50, 148, 254, 261, 268, 280, 308, 309, 313, 317, 329, 330, 347, 401, 480

**B**

Behavioral constraints, 38, 622
Bio-physical constraints, 63

**C**

Cash transfer programmes, 8, 228–248

Climate change, 3–7, 9–11, 14–25, 27, 28, 31, 32, 34–40, 42–44, 49–70, 78, 79, 105, 106, 114–130, 138–143, 145–158, 161–167, 169, 170, 174–178, 181, 183, 184, 188, 191, 201, 202, 204, 208, 209, 211, 214, 215, 223, 237, 247, 248, 252, 254, 262, 267, 268, 279–287, 293, 294, 300, 302–304, 308, 310–328, 330, 335–350, 354, 355, 357, 359, 361–364, 366, 367, 369–372, 375, 376, 379–381, 387–389, 391, 393, 395, 404, 405, 407, 412, 426, 430, 439, 440, 445–447, 449–451, 453–457, 459–461, 467–472, 478–481, 485–487, 492, 493, 498–503, 505, 508–519, 528, 530–535, 537, 553, 554, 557, 564, 591, 599–601, 603, 605–610, 614–625

Climate impact assessment, 283, 309, 320, 323

Climate information, 11, 381, 405, 505, 511, 515, 520, 565, 591, 606, 607, 619
Climate smart agriculture (CSA), 4–6, 8–11, 14, 18–28, 36–40, 43, 44, 50, 56, 58, 59, 63, 65–67, 70, 78–80, 98, 101, 102, 174–176, 252–268, 308, 309, 311–313, 316, 321–330, 335, 336, 354, 355, 361, 364, 367, 368, 370–372, 374, 375, 386–412, 414, 426–428, 528–533, 535–537, 539, 540, 542–544, 546, 547, 549–557, 564, 567, 568, 591, 592, 621–623, 625
Climate-smart policies, 10, 18, 591, 603
Climate variability, 95, 174, 228, 240, 247, 252, 256, 261–263, 267, 278, 330, 388, 439, 486, 531, 533, 556, 612
Conservation agriculture, 260, 280, 286, 303, 357, 394, 601, 614
Context specific, 6, 44, 311, 581
Cost-benefit analysis, 7, 100, 171
Crop-livestock system, 9, 10, 309–311, 321–330, 386–404, 406–410
Crop yields, 78–81, 83–92, 94–98, 100–102, 116, 138, 161, 162, 164–166, 168, 169, 171, 257, 263, 279, 282, 284, 285, 289–292, 302, 320, 325, 326, 391, 393, 394, 397, 406, 439, 446, 452, 454–457, 471, 499, 568, 573, 593

## D

Developing countries, 3, 14–17, 22–24, 28, 31, 34, 36, 42, 55–58, 63–66, 68, 78, 80, 266, 336–339, 341–349, 356, 386–410, 412, 446, 449, 502, 528, 531, 536, 604, 607, 614, 616, 621–623, 625
Distributional constraints, 5, 44
Diversification, 8, 9, 11, 53, 129, 144, 147, 153, 155, 158, 231, 256, 261, 309, 311, 319, 326, 339, 341, 357–361, 367–370, 372, 375, 379–381, 387, 404, 428, 454, 528–540, 542–557, 600, 602, 603, 619
Downside risk exposure, 10, 11, 499, 500, 503, 505–519

## E

Early warning system, 6, 101, 102, 210, 228, 248
Economic models, 9, 38, 308, 314, 320, 412
Efficiency, 20–22, 24–26, 34, 55–58, 175, 217, 256, 257, 260, 266, 278, 386, 388–390, 394, 396, 402, 403, 406, 409, 449, 450, 462, 533, 568, 569, 605, 613, 623, 625
Evidence-based solutions, 622
Ex-ante risk management, 252, 253, 255–263

Ex-post risk management, 8, 237–244, 531, 546
Extreme events, 6, 32, 42, 43, 50, 52, 53, 60, 61, 82, 84, 98, 101, 102, 106, 107, 145, 326, 405, 506, 531, 603, 621

## F

Food and agricultural organization (FAO), 3, 4, 8, 14, 18–21, 23, 24, 27, 36, 94, 229, 252–254, 268, 278, 279, 335, 379, 380, 388, 390, 397, 398, 400–403, 426, 428, 449, 450, 454, 458, 459, 531, 532, 564, 592
Food security, 3–6, 8, 10, 14, 18, 20–22, 24, 27, 28, 31, 32, 35–37, 39, 40, 42, 44, 78, 79, 83, 101, 102, 109, 129, 138, 162, 212, 228, 231, 232, 237, 240, 241, 244–247, 253, 254, 278, 302, 308, 311, 336, 339, 348, 379, 386, 388, 396, 402, 403, 406, 410, 412, 413, 426, 439, 446, 447, 449–451, 454, 460, 470–472, 478, 480, 528, 530–535, 554, 557, 565, 570, 579, 591, 592, 602, 603, 606, 607, 610, 614, 616, 624

## G

Greenhouse gas emissions, 4, 9, 20, 22, 38, 55, 124, 187, 267–269, 308, 312, 330, 427, 446, 480, 564

## H

Heterogeneity, 5, 6, 38, 43, 44, 51, 53–55, 66, 70, 139, 140, 142, 143, 229, 260, 311, 314, 316, 327, 412, 430, 499, 500, 509, 510, 512, 516–518, 529, 539, 554, 621
Household welfare, 238

## I

Index-based Livestock Insurance (IBLI), 7, 210–215, 219–223
Innovation in agriculture, 6
Innovative capacity, 623
Input subsidy programs (ISP), 8, 252–267
Institutional constraints, 39
Institutional framework, 62, 68, 615
International Panel on Climate Change (IPCC), 3, 14, 22, 31, 37, 41, 52, 124, 252, 262, 268, 280, 426, 446, 453, 478, 479, 528, 531

## M

Market-based solutions, 623

Index

Mitigation, 4–6, 10, 14, 15, 17, 18, 20–25, 27, 28, 34–36, 50, 53, 56, 61, 65, 67–70, 78, 83, 89, 94, 189, 228, 253, 254, 262, 268, 269, 308, 311, 312, 330, 342–344, 346, 388–390, 396, 397, 400–402, 405, 406, 410, 446, 447, 449, 451, 472, 478, 529, 532, 592, 610, 611, 616, 617

**N**
No regrets policies, 623, 624

**O**
Opportunity cost, 215, 266, 299, 324, 432
Optimization, 5, 37, 39, 42–44, 98, 288, 300, 343, 354

**P**
Paris Agreement, 17, 609, 616
Policies, 4–7, 10, 11, 14–24, 27, 28, 32–40, 42, 43, 54, 59–61, 64, 66–70, 83, 90, 95, 100, 109, 110, 112, 115, 125, 128, 129, 139, 158, 162, 170, 176, 187, 190, 202–204, 208, 209, 229, 236, 253, 258, 264, 267, 278–281, 288, 302–304, 308, 310, 320, 330, 336, 337, 339, 345, 348–350, 354, 355, 359–361, 363, 364, 369, 370, 379–381, 389, 403, 407, 412, 414, 426, 428, 430, 439, 440, 445–447, 449, 450, 452, 453, 459, 470–472, 480, 482, 487, 492, 493, 499, 517, 528, 530, 531, 533–535, 537, 549, 554, 556, 557, 571, 591, 592, 599–607, 609–625
Political constraints, 37–40
Public-private-partnerships, 148, 150, 152, 154, 156, 448, 601

**R**
Rainfall, 11, 32, 33, 52, 92, 97, 98, 101, 106, 107, 110, 112, 113, 116, 119, 124, 126, 144–147, 153, 154, 156, 158, 161, 175, 210, 216, 219, 240, 241, 278, 289, 292, 302, 309, 315, 322, 327, 354, 364, 366, 369, 370, 372, 380, 392, 395, 400–402, 405–407, 429, 445, 452, 453, 467, 498, 500–507, 511–513, 516–521, 528, 530, 532–541, 543–547, 549–557, 564, 567, 569, 570, 574, 581, 583, 588, 590, 591, 593, 603
Resilience, 4–6, 8, 9, 11, 20, 21, 24–26, 33, 37–41, 52, 70, 90, 95, 98, 101, 102, 228, 229, 231–233, 237, 241, 244–247, 252, 253, 257, 261, 262, 268, 269, 280, 303, 304, 308–311, 317–319, 324, 328–330, 337, 343, 345, 386, 389–391, 397–399, 401, 405, 406, 410, 413, 426–432, 434, 437, 439, 440, 445, 446, 454, 471, 478, 480, 481, 487, 489, 493, 528, 529, 531, 532, 535, 556, 564, 569, 601–603, 606, 610, 613, 614, 624
Risk aversion, 53, 143, 216, 220, 356, 368, 369, 431, 528, 529, 546, 550, 552, 556, 622, 623, 625
Risk management, 6, 8, 11, 41, 57, 78, 101, 139, 208, 213, 218, 228, 237, 240, 241, 247, 252, 253, 255–264, 319, 337, 354, 355, 359, 360, 362–364, 372, 379, 439, 446–450, 452–455, 468, 498–500, 506, 518, 539, 549, 554, 600, 603, 610–614

**S**
Safety-net programs, 41, 610
Satellite information, 78–81, 83, 84, 86, 87, 89–92, 94–98, 100–102, 218, 222
Smallholder agriculture, 24, 354, 355, 610
Social costs, 622
Social protection, 202–204, 207, 208, 228–231, 241, 248, 612
Sub-Saharan Africa, 8, 35, 41, 227, 266, 267, 386, 445
Supply chain, 9, 39, 40, 58, 63, 261, 262, 335–350, 623–625
Sustainable agriculture, 5, 14–27, 609
Sustainable development goals (SDGs), 3, 609, 617
Sustainable land and water management (SLWM), 10, 446–472, 600–602
Synergies, 10, 18, 20, 22, 25, 27, 36, 42, 101, 248, 388, 394, 397, 398, 412, 427, 528, 557, 592, 616
System-level response, 40, 42

**T**
Temperature, 3, 6, 7, 11, 17, 33, 50–52, 54, 55, 78–81, 83, 84, 86–92, 94–98, 100–102, 106, 110, 111, 113–117, 119, 121–126, 128–130, 138, 144–147, 149, 153, 154, 156, 158, 161–171, 179, 281, 288, 290, 292, 302, 309, 315, 322, 323, 326, 348, 364, 366, 369, 386, 391, 404, 453, 478, 479, 484–486, 489–491, 498, 500–507, 511, 513, 519–521, 528, 531–533, 535, 536, 538, 539, 554, 564, 566, 567, 569, 570, 581, 588, 590, 591, 593

Trade-off, 7, 10, 11, 20, 25, 32, 36, 37, 42, 43, 174, 181, 186–188, 191, 223, 255, 261, 264, 303, 341, 388, 389, 394, 397, 398, 408, 410, 412–414, 590, 615, 616

**U**
Uncertainty, 5, 6, 11, 14, 35, 37, 38, 40, 41, 43, 44, 51–53, 56, 60, 64, 66, 79, 80, 124–127, 129, 183, 215, 216, 222, 261, 280–282, 286–289, 291, 292, 294, 302, 303, 321, 353–355, 359, 360, 369, 405, 498, 507, 517, 621–623

**V**
Value chains, 39, 261, 319, 338, 532, 602, 608
Vulnerability, 4–7, 9, 10, 18, 21, 32, 36, 37, 39, 41, 42, 78, 138, 169, 204, 208, 228, 248, 263, 279, 303, 304, 308–330, 338–344, 348–350, 381, 388, 405, 426, 427, 429, 430, 439, 445, 480, 481, 500, 501, 528–534, 549, 550, 552–557, 564, 600, 610, 612, 613

**W**
Weather index insurance, 354, 355, 363, 376, 380, 381

The manufacturer's authorised representative in the EU is Springer Nature Customer Service Centre GmbH, Europaplatz 3, 69115 Heidelberg, Germany. If you have any concerns regarding our products, please contact ProductSafety@springernature.com

Printed and bound by CPI Group (UK) Ltd, Croydon, CR0 4YY
23/03/2026
02076669-0003

Printed in the USA
CPSIA information can be obtained
at www.ICGtesting.com
JSHW060723150923
48529JS00001B/1